Design and Applications of Hydroxyapatite-Based Catalysts

Design and Applications of Hydroxyapatite-Based Catalysts

Edited by Doan Pham Minh

WILEY-VCH

Editor

Dr. Doan Pham Minh
Université de Toulouse
IMT Mines Albi, UMR CNRS 5302
Campus Jarland
81013 Albi cedex 09
France

Cover Image: Doan Pham Minh

All books published by **WILEY-VCH** are carefully produced. Nevertheless, authors, editors, and publisher do not warrant the information contained in these books, including this book, to be free of errors. Readers are advised to keep in mind that statements, data, illustrations, procedural details, or other items may inadvertently be inaccurate.

Library of Congress Card No.: applied for

British Library Cataloguing-in-Publication Data
A catalogue record for this book is available from the British Library.

Bibliographic information published by the Deutsche Nationalbibliothek
The Deutsche Nationalbibliothek lists this publication in the Deutsche Nationalbibliografie; detailed bibliographic data are available on the Internet at <http://dnb.d-nb.de>.

© 2022 WILEY-VCH GmbH, Boschstr. 12, 69469 Weinheim, Germany

All rights reserved (including those of translation into other languages). No part of this book may be reproduced in any form – by photoprinting, microfilm, or any other means – nor transmitted or translated into a machine language without written permission from the publishers. Registered names, trademarks, etc. used in this book, even when not specifically marked as such, are not to be considered unprotected by law.

Print ISBN: 978-3-527-34849-7
ePDF ISBN: 978-3-527-83018-3
ePub ISBN: 978-3-527-83020-6
oBook ISBN: 978-3-527-83019-0

Cover Design ADAM DESIGN, Weinheim, Germany
Typesetting Straive, Chennai, India
Printing and Binding CPI Group (UK) Ltd, Croydon, CR0 4YY

Printed on acid-free paper

Contents

Preface *xiii*

1 Introduction to Hydroxyapatite-based Materials in Heterogeneous Catalysis *1*
Doan Pham Minh
1.1 Generality *1*
1.2 Hydroxyapatite: A New Family of Catalytic Materials in the Heterogeneous Catalysis *4*
1.2.1 Possible High Porous Volume and High Specific Surface Area *4*
1.2.2 High Thermal Stability *4*
1.2.3 Exceptional Ion Exchange Capacity *6*
1.2.4 Tunable Acid-base Properties *7*
1.2.5 High Affinity with Organic Compounds *8*
1.2.6 Formulation of HA-based Materials *8*
1.3 Opportunities and Challenges *11*
References *14*

2 Synthesis and Characterization of Hydroxyapatite and Hydroxyapatite-Based Catalysts *19*
Yousra EL Jemli, Karima Abdelouahdi, Doan Pham Minh, Abdellatif Barakat, and Abderrahim Solhy
2.1 Introduction *19*
2.2 HA Synthesis and Characterization *20*
2.2.1 HA Synthesis Routes *20*
2.2.1.1 Coprecipitation Method *20*
2.2.1.2 Sol–Gel Method *22*
2.2.1.3 Emulsion Methods *24*
2.2.1.4 Hydrolysis Methods *28*
2.2.1.5 Hydrothermal Methods *30*
2.2.1.6 Microwave (MW)-Assisted Methods *31*
2.2.1.7 Ball-Milling Method *32*
2.2.1.8 Sonochemical Method *34*
2.2.1.9 Dry Methods *35*

2.2.1.10	Other Methods 36	
2.2.2	HA Structure 42	
2.2.3	Physicochemical and Thermal Properties of HA 44	
2.2.3.1	Thermal Stability 44	
2.2.3.2	Solubility of HA 44	
2.2.3.3	HA's Surface Functional Groups 45	
2.2.3.4	Non-stoichiometric Calcium-Deficient or Calcium-Rich Hydroxyapatites 46	
2.2.4	Substitutions in the Structure of HA 47	
2.2.5	Modification and Functionalization of the HA's Surface 49	
2.3	A Concise Overview on Synthesis and Characterization of HA-Based Catalysts 51	
2.4	Summary and Conclusions 55	
	References 55	
3	**Structure and Surface Study of Hydroxyapatite-Based Materials** 73	
	Guylène Costentin, Christophe Drouet, Fabrice Salles, and Stéphanie Sarda	
3.1	Introduction 73	
3.2	Structure and Surface Properties of Hydroxyapatite: Overview 75	
3.2.1	Apatite Structure and Model Studies 76	
3.2.2	Specificities of Nonstoichiometric and/or Biomimetic Apatites 81	
3.2.3	Relevance of Apatites in Catalysis 84	
3.3	Advances in the Characterization of Structural and Surface Properties of Hydroxyapatite: Experimental and Computational Approaches 87	
3.3.1	Structural and Compositional Characterization 87	
3.3.2	Thermodynamic Properties and Thermal Stability 95	
3.3.2.1	Overview of Apatites Thermodynamics 95	
3.3.2.2	Thermal Behavior 98	
3.3.3	Physicochemical and Interfacial Properties 100	
3.3.3.1	Solubility and Evolution in Solution 100	
3.3.3.2	Surface Charge 102	
3.3.3.3	Interfacial Tension 103	
3.3.4	Surface Reactivity 106	
3.3.4.1	Nature of Acid and Base Sites 106	
3.3.4.2	Influence of Substitution on Surface Reactivity 109	
3.3.4.3	Low Temperature Ion Immobilization and Adsorption Properties in Aqueous Media or Wet Conditions 110	
3.4	Conclusions 117	
	References 117	
4	**Hydroxyapatite-Based Catalysts: Influence of the Molar Ratio of Ca to P** 141	
	Zhen Ma	
4.1	Introduction 141	
4.2	Influence of Ca/P Ratio on the Performance of HA 143	

4.2.1	Relatively Simple Reactions	*143*
4.2.2	More Complex Reactions	*147*
4.3	Influence of Ca/P Ratio on the Performance of HA-Supported Catalysts	*152*
4.3.1	Relatively Simple Reactions	*153*
4.3.2	More Complex Reactions	*155*
4.4	Concluding Remarks	*156*
	References	*158*

5 Kinetics and Mechanisms of Selected Reactions over Hydroxyapatite-Based Catalysts *163*

U.P.M. Ashik, Nurulhuda Halim, Shusaku Asano, Shinji Kudo, and Jun-ichiro Hayashi

5.1	Introduction	*163*
5.2	Oxidative Coupling of Methane	*164*
5.3	Partial Oxidation of Methane	*169*
5.4	Acetone to Methyl Isobutyl Ketone	*173*
5.5	Ethanol Coupling Reaction	*177*
5.6	Ethanol to Gasoline	*180*
5.7	Glycerol to Lactic Acid	*182*
5.8	Benzene to Phenol	*183*
5.9	Transesterification	*186*
5.10	Conclusion	*188*
	References	*189*

6 Aerobic Selective Oxidation of Alcohols and Alkanes over Hydroxyapatite-Based Catalysts *201*

Guylène Costentin and Franck Launay

6.1	Introduction	*201*
6.2	Liquid Phase Reactions: Selective Aerobic Oxidation of Alcohols	*202*
6.2.1	Apatite-Based Catalysts Efficient in the Aerobic Oxidation of Alcohols	*203*
6.2.2	Apatite/Ru(III) Catalysts	*204*
6.2.3	Apatite/Pd(0) Catalysts	*211*
6.3	Gas Phase Reactions	*214*
6.3.1	Partial Oxidation of Methane	*214*
6.3.2	Alkane Oxidative Dehydrogenation Reactions	*217*
6.3.2.1	Catalytic Performance of the Metal-Modified Hydroxyapatite in the ODH Reactions	*218*
6.3.2.2	Metal Ion Modifications	*219*
6.3.2.3	Activation Site and Mechanism	*227*
6.4	Conclusions and Perspectives	*229*
	References	*232*

7 Selective Hydrogenation and Dehydrogenation Using Hydroxyapatite-Based Catalysts 241
Vijay K. Velisoju, Hari Padmasri Aytam, and Venugopal Akula

7.1 Introduction 241
7.2 HA as Catalyst Support in Hydrogenation Reactions 242
7.2.1 Hydrogenation of Biomass-Derived Compounds to Fuels and Fine Chemicals 242
7.2.2 Hydrogenation of Olefins and Nitro Compounds 245
7.2.3 Hydrogenation of Benzene, Phenol, and Diols 247
7.2.4 Selective Catalytic Reduction of Nitric Oxide 249
7.2.5 Higher Alcohol Synthesis by Simultaneous Dehydrogenation and Hydrogenation Reactions 249
7.2.6 Hydrogenation of Carbon Dioxide 252
7.2.6.1 CO_2 Methanation 253
7.2.6.2 CO_2 Fisher–Tropsch (FT) Synthesis 253
7.2.6.3 Alcohol Synthesis 255
7.2.6.4 Water–Gas Shift and Reverse Water–Gas Shift 255
7.2.7 Partial Conclusions 257
7.3 HA as Support in Dehydrogenation Reactions 257
7.4 Summary and Conclusions 262
Acknowledgments 262
References 263

8 Reforming Processes Using Hydroxyapatite-Based Catalysts 269
Zouhair Boukha, Rubén López-Fonseca, and Juan R. González-Velasco

8.1 Introduction 269
8.2 Overview on the Nature of the Interactions of HA with Transition Metal Catalysts 271
8.3 HA-Supported Non-noble Metal Catalysts for Methane Reforming Reactions 274
8.3.1 Suitability of the HA-Based Catalysts 274
8.3.2 Effect of the Composition on the Performance of HA in the Reforming Reactions 281
8.3.3 Bimetallic Catalysts 284
8.4 Noble Metal Catalysts 286
8.5 Reforming of Other Hydrocarbons 287
8.6 Summary and Remarks 290
Acknowledgments 291
References 291

9	**Hydroxyapatite-Based Catalysts for the Production of Energetic Carriers** *299*	

Othmane Amadine, Karim Dânoun, Younes Essamlali, Said Sair, and Mohamed Zahouily

9.1	Introduction *299*	
9.2	Biodiesel Production *300*	
9.2.1	Transesterification Reactions *301*	
9.2.2	Esterification Reaction for Biodiesel Production *307*	
9.2.3	Other Esterification Reactions *308*	
9.3	Hydrogen Production *312*	
9.3.1	Water–Gas Shift Reactions *312*	
9.3.2	Borohydride Hydrolysis Reaction *316*	
9.3.3	Ammonia Borane Hydrolysis Reaction *318*	
9.4	Catalytic Production of High Value-Added Energy Additives *319*	
9.4.1	n-Butanol and Its Derivative Chemicals *320*	
9.4.2	Fuel Additives from Furfural *324*	
9.4.3	Organic Carbonates Agents *326*	
9.4.4	Energy Additives from Alcohols via Guerbet Reaction *327*	
9.4.5	Other Value-Added Chemicals *329*	
9.5	Conclusion *329*	
	References *330*	

10	**Hydroxyapatite-Based Catalysts in Organic Synthesis** *345*	

Michel Gruselle, Kaia Tõnsuaadu, Patrick Gredin, and Christophe Len

10.1	Introduction *345*	
10.2	Synthesis and Characterization of HA and HA-Based Catalysts *346*	
10.2.1	Synthesis *346*	
10.2.1.1	Stoichiometric and Nonstoichiometric Apatites *346*	
10.2.1.2	Apatites as Catalyst Supports *346*	
10.2.1.3	HA as Macro-ligands for Catalytic Moieties *347*	
10.3	Apatites as Catalysts in C—C Bond Formation *347*	
10.3.1	Cross-coupling Reactions *347*	
10.3.2	Nucleophilic Carbon–Carbon Bond Forming Reactions *350*	
10.3.3	Multicomponent Reaction *352*	
10.4	Conclusions *361*	
	References *364*	

11	**Electrocatalysis and Photocatalysis Using Hydroxyapatite-Based Materials** *373*	

Eric Puzenat and Mathieu Prévot

11.1	Photocatalysis with Hydroxyapatite-Based Materials *373*	
11.1.1	Basic Photocatalysis Principles *373*	

11.1.2	Hydroxyapatite Structure and Properties Implication in Photocatalysis *376*
11.1.3	Single-Phase HA for Photocatalysis *377*
11.1.4	Doped Photocatalytic HA *378*
11.1.5	Multiphasic HA-Containing Photocatalyst *379*
11.1.5.1	HA–TiO_2 Biphasic Composites *380*
11.1.5.2	HA–TiO_2 Multiphasic Composites *381*
11.1.5.3	Other Photocatalytic HA-Containing Composites *381*
11.1.6	Summary and Outlook *382*
11.2	Electrocatalysis with Hydroxyapatite-Based Materials *383*
11.2.1	Charge Transport Mechanism in Hydroxyapatites *384*
11.2.2	Electrocatalytic Sensors *386*
11.2.3	Fuel Cell Application *396*
11.2.4	Electrocatalytic Water Oxidation *401*
11.2.5	Summary and Outlook *403*
	References *404*

12 Magnetic Structured Hydroxyapatites and Their Catalytic Applications *413*
Tasnim Munshi, Smriti Rawat, Ian J. Scowen, and Sarwat Iqbal

12.1	Introduction *413*
12.2	Magnetic HA *414*
12.2.1	Synthesis Route of Magnetic HA *415*
12.3	Catalysis *420*
12.3.1	Magnetic HA Nanoparticles as Active Catalysts for Organic Reactions *420*
12.3.2	HA Analogs and Their Catalytic Applications *425*
12.3.3	HA Catalysts and Green Chemistry *426*
12.4	Summary and Conclusions *430*
	References *431*

13 Materials from Eggshells and Animal Bones and Their Catalytic Applications *437*
Abarasi Hart and Elias Aliu

13.1	Introduction *437*
13.2	Chemical Composition and Properties of Eggshell and Animal Bones *441*
13.3	Eggshell and Animal Bones Materials *444*
13.3.1	Calcium Carbonate/Oxide/Phosphate *445*
13.3.2	Calcium Supplement *445*
13.3.3	Biofilter (Adsorbent) Biomaterial *446*
13.3.4	Hydroxyapatite Material *448*
13.4	Catalytic Applications of Eggshell and Animal Bones *451*
13.4.1	Catalytic Material Preparation from Eggshells and Animal Bones *451*
13.4.2	Catalytic Applications of Catalyst Derived from Eggshell and Animal Bones *454*

13.4.2.1	Selective Catalytic Oxidation	*455*
13.4.2.2	Gasification of Biomass for Hydrogen Production	*458*
13.4.2.3	Reactive Carbon Dioxide Capture (Calcium Looping)	*459*
13.4.2.4	Water–Gas Shift (WGS) Reaction	*460*
13.4.2.5	Transesterification Reaction for Biodiesel Production	*461*
13.4.2.6	Eggshell Membranes (ESM) in Fuel Cell Applications	*464*
13.4.2.7	Catalytic Materials from Eggshells and Animal Bones in Organic Synthesis	*465*
13.4.2.8	Other Catalytic Applications	*466*
13.5	Conclusions	*467*
	References	*468*

14 Natural Phosphates and Their Catalytic Applications *481*

Karima Abdelouahdi, Abderrahim Bouaid, Abdellatif Barakat, and Abderrahim Solhy

14.1	Introduction	*481*
14.2	Preparation and Characterization of Catalysts or Catalyst Supports from NP	*482*
14.3	Organic Synthesis Using NP and NP-Supported Catalysts	*487*
14.3.1	Condensation Reactions	*487*
14.3.1.1	Knoevenagel Reaction	*487*
14.3.1.2	Claisen–Schmidt Condensation	*491*
14.3.1.3	Michael Addition	*494*
14.3.2	Transesterification Reaction	*496*
14.3.3	Friedel–Crafts Alkylation	*497*
14.3.4	Suzuki–Miyaura Coupling Reaction	*499*
14.3.5	Hydration of Nitriles	*501*
14.3.6	Synthesis of α-Hydroxyphosphonates	*504*
14.3.7	Multicomponent Reactions (MCRs)	*505*
14.3.7.1	Biginelli Reaction	*505*
14.3.7.2	Synthesis of α-Aminophosphonates	*506*
14.3.8	Oxidation Reactions	*506*
14.3.8.1	Oxidative Cleavage of Cycloalkanones	*506*
14.3.8.2	Oxidation of Benzyl Alcohol	*509*
14.3.8.3	Epoxidation of Electron-Deficient Alkenes	*510*
14.3.9	Hydrogenation Reactions	*511*
14.3.9.1	Selective Hydrogenation of Crotonaldehyde	*511*
14.3.9.2	Reduction of Aromatic Nitro Compounds	*512*
14.3.10	Reforming of Methane	*514*
14.3.11	Photooxidation of VOC Model Compounds	*515*
14.4	Conclusions	*516*
	References	*516*

Index *533*

Preface

Well-known by its biocompatibility and its important applications in the field of biomaterials, hydroxyapatite ($Ca_{10}(PO4)_6(OH)_2$), called thereafter HA, has recently gained a lot of attention of researchers, thanks to its interesting physicochemical properties required for a catalytic material. The number of publications on HA-based catalysts per year exponentially increases during the last two decades and reached around 220 papers in 2020. HA can be used as a catalyst or a catalyst support in various chemical processes such as selective hydrogenation, dehydrogenation, selective oxidation, reforming of hydrocarbons, organic synthesis, electrocatalysis, and photocatalysis, among others. However, HA and HA-based materials are still considered as a new family of catalytic materials, despite their high potential of application in the heterogeneous catalysis. Only some reviews have been done on HA-based catalysts, which are not enough to spread knowledge and accelerate the exploration of these materials.

This is the first book dedicated to the design and applications of HA-based catalysts. It covers different aspects of the catalysis using HA-based materials. Through its 14 chapters, readers can find the basis of synthesis, characterization, surface study, kinetic and mechanism aspects, and the latest advancements on HA-based catalytic systems, which were contributed by different research groups around the world. Thus, it is expected that this book will be the last puzzle piece in the literature to boost research on HA-based catalysts.

I gratefully thank the authors of the chapters and the reviewers who dedicated their precious time and expertise to this book. I also thank all the staff members at Wiley for their assistance in editing and publishing this book.

October 2021
Albi, France

Doan Pham Minh

1

Introduction to Hydroxyapatite-based Materials in Heterogeneous Catalysis

Doan Pham Minh

Université de Toulouse, IMT Mines Albi, UMR CNRS 5302, Centre RAPSODEE, Campus Jarlard, F-81013 Albi, Cedex 09, France

1.1 Generality

As recently reviewed by Wisniak [1] and summarized in the book of Zecchina and Califano [2] on the history of the catalysis, the first known work seemed to be carried out in 1552 by Valerius Cordus, for the synthesis of ether from alcohol catalyzed by sulfuric acid. During the next centuries, several examples were recorded [1], such as the hydrolysis of potato starch in water catalyzed by potassium hydrogen tartrate or by acetic acid (1781); hydrogen production from alcohol and concentrated sulfuric acid catalyzed by alumina, silica, or clay without external heating (1796); starch conversion into gum, dextrin, and raisin sugar catalyzed by inorganic acids (1811); and starch fermentation into alcohols via sugar formation (1816). However, the term *catalysis* was only introduced for the first time by Berzelius in 1835 [1, 3], which means "the property of exerting on other bodies an action which is very different from chemical affinity. By means of this action, they produce decomposition in bodies, and form new compounds into the composition of which they do not enter." In 1836, Berzelius stated "the catalytic power seems actually to consist in the fact that substances are able to awake affinities, which are asleep at a particular temperature, by their mere presence and not by their own affinity" [2, 4].

The first large-scale industrial application of the catalysis could be assigned to the production of sulfuric acid over platinum catalyst at the end of the nineteenth century [2]. This process was later improved (around 1920) by replacing platinum catalyst by vanadium pentoxide, which is still used today for sulfuric acid production. The beginning of the twentieth century was also marked by the industrialization of the ammonia oxidation over platinum catalyst to produce nitric acid (Ostwald–Brauer process), ammonia production from direct hydrogenation of molecular nitrogen over iron-based catalysts (Haber–Bosch process), and liquid fuel synthesis from syngas over iron, cobalt, or ruthenium catalysts (Fisher–Tropsch process) [2]. Then, the catalysis passed through its golden period with the petroleum and polymer eras, with the discovery and industrialization of many important

Design and Applications of Hydroxyapatite-Based Catalysts, First Edition. Edited by Doan Pham Minh.
© 2022 WILEY-VCH GmbH. Published 2022 by WILEY-VCH GmbH.

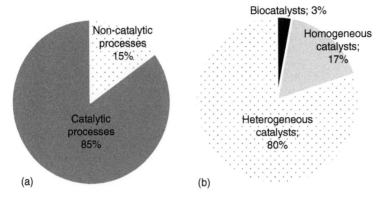

Figure 1.1 Pie charts showing: (a) the percentages of catalytic processes *versus* non-catalytic processes; (b) and the percentages of all industrial processes that entail the use of heterogeneous, homogeneous and bio-catalysis. Source: Adapted from Thomas and Harris [5] with permission of the Royal Society of Chemistry.

catalytic processes such as catalytic cracking, isomerization, alkylation, reforming, hydrodesulfurization, and polymerization [2]. Nowadays, the catalysis has a pivotal role in the modern society, and it is present in ca. 85% of manufacturing processes (Figure 1.1a), as recently reviewed by Thomas and Harris [5].

Generally, three categories of catalysts are distinguished: heterogeneous catalysts that are usually solid materials, homogeneous catalysts that are usually soluble salts or complexes, and enzymatic catalysts. Among them, heterogeneous catalysts occupy an important place in comparison with the counterparts, mostly because of their easy separation from products and their applicability in large ranges of temperature and pressure. Thus, among catalytic processes, heterogeneous catalysts represent up to 80%, much higher than the parts of homogeneous and enzymatic catalysts (Figure 1.1b) [5].

More specifically, other catalyst classifications have been also proposed in the literature. For example, by key properties responsible for their catalytic behavior, the following catalyst families can be classified: redox catalysts, acid-base catalysts, or polyfunctional catalysts [6]. Misono [7] classified catalysts by their main components. Thus, different families of catalysts are classified such as metals, metal oxides, metal salts, metal coordination compounds, organic molecules, organic polymers, and biocatalysts.

Considering the periodical table of elements, the most well-known catalysts are found within the elements of the columns 1–14, which are principally metals and metal-based compounds. Among the most available non-metallic elements (e.g. C, Si, N, P, O, S, F, Cl, Br, I), O is usually present in catalytic materials; C constitutes a large family of catalyst supports such as activated carbon, carbon nanotube, carbon nanofiber, and graphene; Si also builds different types of catalysts (zeolites) or catalyst supports (SiO_2-based materials); F, Cl, Br, I, N, and S have been applied as acid catalysts (in the form of their inorganic acids) or catalyst additives (in the form of their inorganic acids or salts) to modify acid-base properties of another material. On the other hand, P is more particular in comparison with the other non-metallic elements, since it provides a large number of water soluble and insoluble compounds, of simple or complex oligomer or polymer structures, e.g. orthophosphates (PO_4^{3-}), pyrophosphates ($P_2O_7^{4-}$), cyclophosphates, and polyphosphates [8]. Among them, phosphate is the most popular form of phosphorus compounds and also constitutes a large number of materials that have potential applications in the catalysis of various chemical processes such as dehydration and dehydrogenation of alcohols, hydrolysis reactions, oxidation and aromatization reactions, isomerization, alkylation, Knoevenagel reaction, Claisen–Schmidt condensation, epoxidation, nitrile hydration, cycloaddition, Michael addition, water splitting, and conversion of biomass-derived monosaccharides [9–13]. For example, Vieira et al. [13] studied direct conversion of glucose and xylose to HMF (5-hydroxymethylfurfural) and furfural over the niobium phosphates as efficient bifunctional catalysts. The acidity of the catalysts, expressed as the ratio of Lewis to Brønsted acid sites (L/B) could be tuned by varying the molar ratio of P to Nb, which allows controlling the catalyst performance. Thus, both the conversion of monosaccharides (glucose and xylose) and the formation of furans linearly increases with the L/B ratio; the later increases by decreasing the P to Nb molar ratio. In another example, Kim et al. [14] successfully synthesized cobalt pyrophosphate ($CoP_2O_7^{2-}$) and cobalt phosphate ($CoPO_4^-$) based materials as new electrocatalysts for water splitting reaction. High electrocatalytic activity and stability during 100 cycles were observed, explained by the structural stabilities of the investigated materials.

Stoichiometric calcium phosphate hydroxyapatite (chemical formula: $Ca_{10}(PO_4)_6(OH)_2$), denoted thereafter HA, belongs to the family of apatites, which is a category of phosphate compounds. A given apatite has the general chemical formula of $M_{10}(XO_4)_6Y_2$, where M is generally a bivalent metal cation (Ca^{2+}, Mg^{2+}, etc.), XO_4 is generally a trivalent anion (PO_4^{3-}, VO_4^{3-}, AsO_4^{3-}, etc.), and Y is typically a monovalent anion (OH^-, F^-, Cl^-). However, other cations and anions with different valences can also be present in an apatitic compound. For example, carbonate anion (CO_3^{2-}) can partially replace both Y and XO_4 anions, while Na^+, K^+, Al^{3+}, etc., can partially replace M cations. More details on the apatite composition and structure will be discussed in Chapter 3 of this book.

In heterogeneous catalysis, HA-based materials are still considered as a new potential family of catalysts for various applications, as recently reviewed by Gruselle [15], and Fihri et al. [16]. Gruselle [15] focused his review on the catalytic performance of HA-based materials in the organic synthesis, while Fihri et al. [16] enlarges their review to other processes including the photocatalysis, hydration, hydrogenation, hydrogenolysis, transesterification, and multicomponent reactions. On the one hand, HA can be used itself as a solid catalyst. For example, in the Knoevenagel condensation reaction conducted by Sebti et al. [17], undoped HA allowed obtaining high yields (e.g. 80–98%) at room temperature and atmospheric pressure. In many cases, HA-based materials were usually found to be competitive in terms of catalytic performance in comparison with conventional catalysts. On the other hand, HA can be used as a catalyst support to disperse a catalytically active phase, such as metal nanoparticles, or metal cations. For instance, Phan et al. [18] demonstrated the superior catalytic activity of HA supported bimetallic

Co-Ni catalysts in the dry reforming of methane for syngas production. As will be discussed in Chapter 3 of this book, the first publication on HA as catalyst appeared around 1940s, but, research on HA catalysts has only accelerated during the last two decades, with an exponential increase of the annual number of publications. The Section 1.2 will focus on the main reasons making HA-based materials as new promising candidates in the heterogeneous catalysis.

1.2 Hydroxyapatite: A New Family of Catalytic Materials in the Heterogeneous Catalysis

Well-known for their application in the field of biomaterial [19–21], HA and apatite-based materials have gained increasing attention during the last two decades owing to their structural, physicochemical, and thermal properties [15, 16, 22, 23]. The most important properties of HA and HA-based materials are summarized next, which are detailed and analyzed in different chapters of this book.

1.2.1 Possible High Porous Volume and High Specific Surface Area

In the heterogeneous catalysis, high specific surface area is generally required since it favors the contact of solid catalyst surface with chemical compounds in both the liquid and gas phases. *It is possible to synthesize HA with high specific surface area and porosity* by different methods, which will be detailed in Chapter 2 of this book. For example, Verwilghen et al. [24] obtained HA powder with a specific surface area of 156 m^2 g^{-1} by coprecipitation of monoammonium dihydrogen phosphate ($NH_4H_2PO_4$) and calcium nitrate ($Ca(NO_3)_2$) at 75 °C. Starting from calcium acetate (($CH_3COOH)_2Ca$) and diammonium hydrogen phosphate (($NH_4)_2HPO_4$) as calcium and phosphate precursors, Nagasaki et al. [25] successfully synthesized HA exhibiting specific surface area up to 90 m^2 g^{-1} and containing mostly mesopores with mean diameter of around 2.5 nm. It is usually observed that the specific surface area and the porosity of HA strongly depends on a set of multiple factors including the nature of calcium and phosphate precursors, the nature of solvent, the maturation time, the agitation speed, the pH, and the post-treatment by drying.

1.2.2 High Thermal Stability

HA can be heated to high temperature (ca. > 850 °C, this threshold temperature being a function of different parameters such as the stoichiometry of HA, the atmosphere of heat treatment, and the pressure of the heat treatment) without significant modification of its chemical composition and structure. *Thus, HA can be used as catalyst or catalyst support in most of the chemical reactions, which are usually performed below this threshold temperature.* Above this threshold, partial dehydroxylation occurs, leading to the formation of oxyhydroxyapatite ($Ca_{10}(PO_4)_6(OH)_{2-2x}O_x$). Then, oxyhydroxyapatite can decompose into tricalcium phosphate (TCP) and tetracalcium phosphate (TTCP). The impact of this dehydroxylation on catalytic

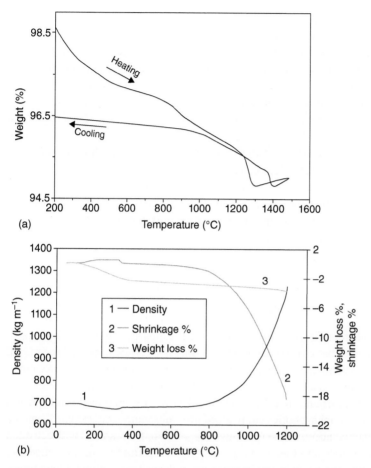

Figure 1.2 (a) Thermal behavior of a commercial stoichiometric HA under the air atmosphere. Source: Liao et al. [26]. Reproduced with permission of Elsevier. (b) Evolution of bulk density, weight loss, and shrinkage during HA heating. Source: Bailliez and Nzihou [27]. Reproduced with permission of Elsevier.

properties also merits to be further studied in high temperature processes. Liao et al. [26] studied the decomposition of a commercial stoichiometric HA under the air (Figure 1.2a). The weight losses below 200 °C and between 200 and 400 °C are assigned to adsorbed water and lattice water, respectively. The latter causes a contraction in the a-lattice dimension. This dehydroxylation seemed to take place progressively up to around 1000–1360 °C. Then, oxyhydroxyapatite decomposed into TCP and TTCP between 1400 and 1500 °C. TCP and TTCP can be partially recombined forming oxyhydroxyapatite and HA by cooling down the material. The flexibility of HA and HA-based materials toward dehydroxylation/hydration makes them more resistant in catalytic systems involving water vapor, which is the case for various chemical processes.

On the other hand, heating HA to high temperature (e.g. 1000 °C) generally leads to a decrease of its porosity and specific surface area by thermal sintering.

For example, heating an HA to 1200 °C decreased its specific surface area from ca. 105 to ca. 5 m² g⁻¹ and led to the formation of non-porous and dense HA (Figure 1.2b) [27]. For high temperature catalytic processes, this evolution of textural properties of HA can provoke a catalytic deactivation. Alternative solution can be a stabilization of HA by thermal treatment, as demonstrated by Rego de Vasconcelos et al. [28]. In this work, HA support was sintered at 1000 °C for 5 hours, before the deposition of Ni nanoparticles by a conventional impregnation method. This allowed limiting the catalyst deactivation due to the thermal sintering during the chemical reaction at high temperature (dry reforming of methane). Nevertheless, this also caused a decrease of the initial catalyst activity due to the reduction of the specific surface area and the suppression of the porosity of HA support. *A compromise between the activity and the thermal stability must be found for catalytic applications of HA and HA-based materials at high temperatures.*

1.2.3 Exceptional Ion Exchange Capacity

Calcium cations, phosphate, and hydroxyl anions of a given HA can be partially replaced by other cations and anions [23]. Cation exchange capacity of HA with various metallic cations such as alkalis, alkaline earth metals, or transition metals in aqueous media differentiates it from other catalyst supports, at the exception of zeolites. The cation exchange usually takes place quickly on HA surface (few seconds to few minutes), with higher sorption capacities (Q_e) in comparison to reference materials such as activated carbon, ion exchange resins [29, 30], or even zeolites [31]. For instance, it was found that the Q_e of an HA powder synthesized from calcium carbonate and orthophosphoric acid reached up to 625 $mg_{Pb^{2+}}$ g_{HA}^{-1} for Pb^{2+} removal from an aqueous solution, which was much higher than that obtained with activated carbon, resin, or biochar (15–250 $mg_{Pb^{2+}}$ $g_{adsorbent}^{-1}$) [32]. High cation exchange rate and high Q_e offer the flexibility to improve metal deposition and metal loading for the preparation of HA supported catalysts. *This is an important advantage of HA, in comparison with the conventional catalyst supports, particularly when highly dispersed metal supported catalysts are targeted.* For example, bimetallic Ru–Zn/HA could be easily obtained by cation exchange method by setting in contact an HA powder with an aqueous solution of $RuCl_3$ and $ZnSO_4$ at room temperature under stirring for two hours [33]. After drying and activation under H_2 at 300 °C for 1 hour, nanoparticles of Ru smaller than 2.4 nm were formed on HA surface, which were found to be active and particularly stable in the selective gas-phase hydrogenation of benzene at 150 °C and 50 bar. Kaneda's group [34–36] performed the deposition of monomeric Zn, Ru, or Pd species on HA-based supports, for selective alcohol oxidation, using a simple ion exchange approach, in which HA-based supports were set in contact with an aqueous solution of $RuCl_3$ or $Zn(NO_3)_2$, or an acetone solution of $[PdCl_2(PhCN)_2]$. No further activation was applied to these catalysts, leading to the dispersion of isolated metal cations on HA surface. In fact, extended X-ray absorption fine structure (EXAFS) analysis proved that the active phase existed as monomeric species, and no metal–metal bond was detected. These catalysts were found particularly active in alcohol oxidation. More recently, Akri et al. [37] have reported the preparation of

Ni$_{SA}$/HA catalysts by strong electrostatic adsorption method. This method consists in the cation exchange of Ca^{2+} with Ni^{2+} at a controlled pH in an aqueous solution. A mixture of isolated Ni species, small Ni clusters, and Ni nanoparticles could be obtained, and the contribution of each Ni species depended on the Ni loading and the thermal treatment (e.g. drying or H_2 reduction). At low Ni loading (0.5 wt%), only isolated Ni atoms were deposited on HA surface after the drying step, while the reduction under H_2 at 500 °C mainly led to a deposition of Ni isolated atoms (ca. 94%) and small amounts of Ni clusters (ca. 6%). The synthesized catalyst exhibited high specific activity in dry reforming of methane to produce syngas at 750 °C and 1 bar.

It is worth noticing that ion exchange is usually carried out in an aqueous solution. HA generally has a very low solubility product in water. However, in the literature, different values were reported, because the solubility depends on various parameters such as the pH, the synthesis method, and the stoichiometry of HA [38]. As examples, at a pH of 8.25, Bell et al. [39] found a mean value of the solubility product of $1\,830\,000 \times 10^{-58}$ for a stoichiometric HA, while at a pH of 4.90, the solubility product strongly decreases to 3.15×10^{-58}. Ito et al. [40] reported much lower values of solubility product for a carbonated apatite, which varied from 10^{-119} at pH of 4.9 to 10^{-130} at pH of 4.1. Despite this inconsistency, it is important to take into consideration the high sensitivity of HA and HA-based materials to pH values in order to design catalytic formulations meant for aqueous-phase processes. In fact, a solution presenting a very low pH such as that of 1 M HNO_3 acid can completely dissolve an HA powder.

1.2.4 Tunable Acid-base Properties

The possibility of tuning the acid-base properties of HA and HA-based materials is another interesting advantage that justifies their use in various catalytic processes. HA bears both acidic and basic sites on its surface. Acidic sites are principally attributed to PO–H from HPO_4^{2-} species (Brønsted sites) and Ca^{2+} cations (Lewis sites), while basic sites are due to surface PO_4^{3-} and OH^- groups (and possibly other phases such as CaO, $Ca(OH)_2$, and $CaCO_3$) [41, 42]. Particularly, the acido-basicity of HA can be tuned by controlling the Ca/P molar ratio [41–43]. Globally, calcium-deficient HA ($Ca_{10-Z}(HPO_4)_Z(PO_4)_{6-Z}(OH)_{2-Z}$, with $0<Z<1$), with a Ca/P molar ratio smaller than 1.67, behave as acid materials. Stoichiometric HA (Ca/P = 1.67) and calcium-surplus HA (global Ca/P > 1.67, with the presence of other phases such as CaO, $Ca(OH)_2$, and $CaCO_3$), which depend on the synthesis conditions, behave as basic materials. Moreover, the acido-basicity of HA-based materials can also be controlled by promoter addition (Mg^{2+}, Ba^{2+}, Sr^{2+}, etc.) while keeping their apatitic structure [43–45]. *The tunability of the surface chemistry of HA and HA-based materials is an important property, suitable for various processes,* since the acid-base catalysis covers a large number of chemical reactions such as isomerization, alkylation, esterification, hydration, dehydration, oligomerization, cracking, and acylation. For example, Silvester et al. [41] synthesized different HA-based materials having different Ca/P molar ratios by using the precipitation method under controlled conditions of temperature (Table 1.1). Expectedly, increasing the Ca/P molar ratio leads to an increase in basicity and a decrease in acidity. The insertion of Na^+ and CO_3^{2-} also modified the acidity and basicity of the materials. In another work, Tsuchida et al. [46] also demonstrated the dependency of the acidity and basicity of different HAs on the Ca/P molar ratio. These catalysts were evaluated in the transformation of ethanol into other valuable chemicals. At the iso-conversion of ethanol of 10 and 20% (by varying the reaction temperature and keeping other conditions unchanged such as gas hourly space velocity [GHVS]), the acidity and basicity strongly impacted the selectivity of the reaction, where different mechanisms pathways were found to be involved. Ethylene was the main product over strongly calcium-deficient HA, exhibiting high acidity that favors dehydration reaction, while 1-butanol was predominant over stoichiometric HA possessing high basicity that favors the Guerbet condensation.

1.2.5 High Affinity with Organic Compounds

HA and HA-based materials are efficient not only for the fixation of metallic cations but also for the immobilization of organic compounds on their surface [47]. Adeogun et al. [48] synthesized HA from poultry eggshells (calcium source) and ammonium dihydrogen phosphate (phosphate source). In the removal of reactive yellow 4 dye from the liquid phase, the resulting materials showed a high sorption capacity of 127.9 mg g^{-1}. Sebei et al. [49] investigated the sorption of catechol on an HA synthesized from $CaCO_3$ and KH_2PO_4. Interesting sorption capacity of 24.6 mg g^{-1} could be obtained. HA synthesized from $Ca(NO_3)_2$ and $NH_4H_2PO_4$ was also found to be efficient for the removal of fulvic acid in the liquid phase, and its sorption capacity (90.2 mg g^{-1}) was even much higher than that of other materials such as zeolite, activated sludge, chitosan hydrogen beads, carbon nanotubes, and vermiculite [50]. In the gas phase, HA deposited on a polyamide film also presented interesting reactivity in the removal of formaldehyde [51].

High affinity of HA and HA-based materials to organic compounds will be beneficial for the design of efficient catalytic systems employing these molecules. Moreover, the reactivity of the surface of HA and HA-based materials can be improved by different routes (e.g. mechanical treatment, chemical modification). As examples, HA doped with Zn^{2+} was more efficient than a bare HA in the sorption of catechol in the liquid phase [49], while HA activated by mechanical treatment, which led to the creation of surface defects and oxygen vacancies, was more efficient than untreated HA in the oxidation of volatile organic compounds [47].

1.2.6 Formulation of HA-based Materials

HA can be synthesized by different routes as will be detailed in the Chapter 2 of this book. Depending on the synthesis conditions, fine micro- and/or nanoparticles of HA can be formed [52, 53]. Laboratory investigations of HA and HA-based materials in the catalysis can be performed using fine HA powders. For an eventual scale-up, HA powders must be processed to obtain materials of controlled form

Table 1.1 Quantification of the density and nature of acid and basic sites of various HA-based materials as functions of the Ca/P molar ratio.

		Basic sites						Acid sites			
		TPD-CO$_2$		Benzoic acid adsorption						Number and distribution of acid sites	
Materials	Ca/P atomic ratio (ICP)	Total basic sites (μmol g^{-1})	Specific basicity (μmol m^{-2})	Total basic sites (μmol g^{-1})	Specific basicity (μmol m^{-2})	Strong basic sites (μmol m^{-2})	Weak basic sites (μmol m^{-2})	Number of acid sites (μmol g^{-1})	Specific acidity (μmol m^{-2})	Lewis (μmol m^{-2})	Brønsted (μmol m^{-2})
HapD	1.62	41.33	0.33	259.9	2.09	1.88	0.21	956	7.73	5.16	2.57
Hap	1.69	100.67	0.87	194.3	1.70	1.45	0.25	763.3	6.73	5.43	1.30
Hap-CO$_3$	1.70	136	1.27	499.5	4.66	4.66	0	658	6.13	5.58	0.54
HapNa-CO$_3$	1.72	155.33	1.4	134.7	1.24	0.83	0.41	604	5.53	5.17	0.36
HapE-CO$_3$	1.90	112.67	1.47	310.4	4.08	3.60	0.48	324	4.27	4.01	0.26
HapE-Na-CO$_2$	2.39	104	1.47	351.6	4.88	4.29	0.59	141.3	1.87	1.78	0.09

HapD: calcium-deficient hydroxyapatite; Hap: stoichiometric hydroxyapatite; Hap-CO$_3$: carbonated apatite; HapNa-CO$_3$: sodium-containing carbonate apatite; HapE-CO$_3$: sodium-free carbonate-rich apatite; HapE-Na-CO$_3$: sodium-containing carbonate-rich apatite; TPD: temperature programed desorption.
Source: Silvester et al. [41]. Reproduced with permission of the Royal Society of Chemistry.

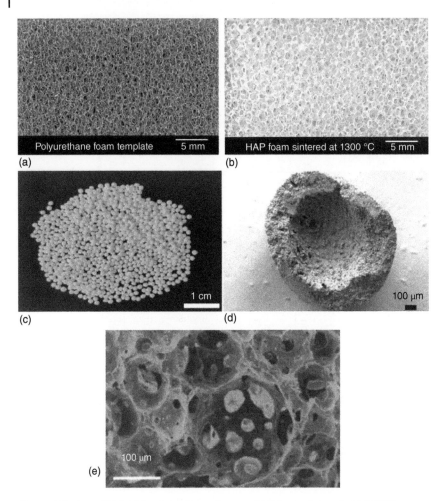

Figure 1.3 (a and b) Initial polyurethane foam and HA foam sintered at 1300 °C, respectively. This HA foam was obtained by impregnating polyurethane foam in an HA slurry followed by drying and sintering. Source: Munar et al. [58]. Reproduced with permission of The Japanese Society for Dental Materials and Devices; (c and d) optical image of beads, and SEM image of hollow structure of a Cu/HA catalyst prepared by a rapid gelling process, using guar gum as gelatinizing agent, and sintered at 1100 °C. Source: Hui et al. ([59]. Reproduced with permission of Elsevier; (e) HA foam obtained from HA powder, polyacrylate derivatives (Dispex A40) as a dispersing agent, acrylic monomers as gel agent, and a nonionic surfactant (Tergitol TMN10) as foam stabilizer. Source: Sepulveda et al. [54]. Reproduced with permission of John Wiley and Sons.

and size, such as beads and pellets. To date, macro-porous HA ceramics have been largely reported in the literature for application in the field of biomaterials [54–57]. The global approach is to use fine HA powder and at least a foaming agent and a stabilizer to perform the gelation step, usually in the aqueous medium. Then, the gel is dried and the resulting solid is sintered at high temperature (> 1000 °C) to burn out organic components, leading to the formation of macro-porous materials. Figure 1.3a, b shows examples of macro-porous HA ceramics reported in the

Figure 1.4 Examples of HA blocks after drying (a) and after sieving to 125–315 μm fraction (b). Source: Phan [43]. Reproduced with permission of Phan Thanh Son.

literature. Using this approach, beads of a Cu/HA catalyst could also be obtained as illustrated in Figure 1.3c, d [59].

However, HA foams prepared by this approach generally lose specific surface area, which negatively impacts their catalytic activity. An alternative solution can be the control of HA synthesis conditions to directly obtain HA large particles [43]. In a recent PhD work, conducted by Phan [43], HA was synthesized by the conventional precipitation method using $Ca(NO_3)_2$ and $NH_4H_2PO_4$ at 40 °C for 48 hours under vigorous stirring. Fine particles of HA were formed, which were carefully filtered on filter paper to form large blocks of HA. The latter were dried at 105 °C overnight, crushed, and sieved into particles of desired size (125–315 μm in this case), as illustrated in Figure 1.4. By this way, the resulting HA materials maintain their initial textural properties including meso-porosity (mean pore diameter: 19.3–20.4 nm) and specific surface area (66–90 $m^2\,g^{-1}$), which were suitable for the deposition of nickel nanoparticles for the reforming of methane at high temperatures (700–800 °C) [43]. Another possibility is to simply pelletize HA powders into pellets [60–62]. However, this pelletization step can provoke a partial loss of catalytic performance in comparison with the corresponding powders. For example, Thrane et al. [60] compared the catalytic activity of the powder and pellets of 10 wt% MoO_3/HA in the oxidation of methanol to formaldehyde. A loss of catalytic activity of 10–23% was observed when the HA support was pressed into industrial-sized pellets with different densities (1.18–1.76 $g\,cm^{-3}$).

1.3 Opportunities and Challenges

The global catalyst market was valued at $33.9 billion in 2019 and can reach $48 billion in 2027 (with CAGR – compounded annual growth rate – estimated at 4.5%) [63]. The prime factor driving the catalyst market growth is still the rising demand for petrochemicals from various end-use industries. However, demand for clean and green fuels has shifted the trend of the energy field with focus on clean resources and processes. Moreover, emission regulation is expected as a strong driver for a wide range of application sectors, and the use of catalysts for environmental and chemical applications is anticipated to have the highest market growth in medium and long terms [63]. Thus, new breakthrough scientific findings in the development of efficient catalysts for the production of intermediates, chemicals, and energy carriers (e.g. syngas, biomethane, renewable hydrogen, and biofuel) from renewable resources (e.g. biomass and biowastes), or even from harmful molecules for environment such as CO_2 [64, 65] are currently of crucial importance. Chemical conversion of CO_2 into useful products is strategically important worldwide for limiting global warming, but to date, the number of chemicals produced industrially using CO_2 is still very limited [65]. Moreover, nowadays, catalytic processes also have their place in pharmaceutical and fine chemical industries, since catalytic coupling reactions (C–C and C–N coupling), C–H bond activation, hydrogenation of C=C, C=O, and C=N groups, etc., are efficient for the synthesis of new substances [66, 67]. *Multicatalysis*, term employed by Martínez et al. [68], including cooperative, domino, and relay catalysis, is also an emerging research, in which multiple catalysts enable chemical transformations that cannot be performed through classical approaches. Downsizing metallic particle sizes up to the isolated metal atoms also constitutes a very hot topic (single atom catalysis – SAC) during the past 10 years, which opens new prospects in the catalysis [69–72]. Strong impact on the catalyst market is expected in the upcoming decades by the development and deployment of new catalysts and catalytic processes of high efficiency and low environmental impact. Consequently, HA and HA-based materials, having suitable physicochemical and thermal properties, as presented earlier, and still being considered as a new family of catalytic materials, have great opportunities to be developed in this field, either as catalysts or as catalyst supports. The capacity and flexibility of HA structure to incorporate one or several metals is a major advantage in the design of new catalytic materials. Recent works on HA-based nanocatalysts (metal nanoparticles supported on HA) reported their superior catalytic performance in comparison with reference materials in various reactions. For instance, Rego de Vasconcelos et al. [28] compared the catalytic performance Ni/HA with Ni/MgAl$_2$O$_4$ in dry reforming of methane to produce syngas. The catalysts prepared on non-sintered support showed comparable catalytic performance, while Ni supported on sintered HA was more active than that supported on sintered MgAl$_2$O$_4$. Similarly, Munirathinam et al. [73] evidenced the large superiority of a Co/HA catalyst versus a Co/Al$_2$O$_3$ catalyst in the Fisher–Tropsch synthesis under the same reaction conditions (Figure 1.5). More generally, Fihri et al. [16] reviewed recent works related to HA-based catalysts and pointed out the versatility and efficiency of these materials in many chemical reactions.

Despite the numerous advantages of HA-based catalysts as mentioned earlier, research conducted on HA and HA-based materials still faces several challenges [16]:

- High sensibility to the acid medium: HA can be partially or completely dissolved by an aqueous solution at low pH (see Section 1.2.3). This limits the utilization of HA-based catalysts in strong acid media.

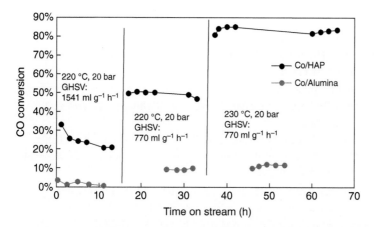

Figure 1.5 CO conversions in Fisher–Tropsch (FT) synthesis using Co/HAP and Co/Alumina catalysts reduced at 350 °C under reaction conditions of 20 bar pressure with different gas hourly space velocities (GHSV: 1540 or 770 ml g^{-1} h^{-1}) and temperatures (220 or 230 °C). Source: Munirathinam et al. [73]. Reproduced with permission of Elsevier.

- Low mechanical resistance: The formulation of HA powders into HA monoliths is possible as discussed in the Section 1.2.6. Pure HA powders usually result in ceramics of low mechanical resistance. The latter can be improved by using additives, usually metal oxides such as bio-glass (a mixture of SiO_2, Na_2O, K_2O, MgO, CaO, P_2O_5, B_2O_3) [74]. However, the presence of additives can negatively impact the catalytic properties of the modified materials, particularly for selectivity control.
- Strong dependence of the textural and structural properties on the synthesis conditions: many synthesis routes can be applied for the synthesis of HA. However, textural and structural properties of the final HA strongly depend on synthesis conditions: nature of calcium and phosphate precursors, solvent, precursor concentration, method of the precursor addition (calcium precursor to phosphate precursor, or phosphate precursor to calcium, or simultaneously both precursors), pH, maturation time, temperature, pressure, washing protocol, drying protocol, presence or not of dissolved oxygen and carbon dioxide in the synthesis medium, etc. This issue will be discussed in Chapter 2 of this book.
- High reactivity of HA surface: though the apatitic bulk structure of HA is generally well known, the structure of the first surface layers of HA is difficult to be determined because of high reactivity and sensibility of HA surface with the environment (such as the presence of humidity and CO_2). Especially, for nanocrystalline nonstoichiometric apatites, highly reactive amorphous surface layers generally exist, of which the composition is usually different from the bulk composition [75]. This feature will be discussed in Chapter 3 of this book. To date, few reports have been devoted to the impact of the surface composition of HA and apatite compounds on their catalytic properties or on the preparation of supported catalysts.

References

1. Wisniak, J. (2010). The history of catalysis. From the beginning to Nobel Prizes. *Educación Química* 21 (1): 60–69.
2. Zecchina, A. and Califano, S. (2017). *The Development of Catalysis: A History of Key Processes and Personas in Catalytic Science and Technology*. Hoboken, NJ: John Wiley & Sons, Inc.
3. Berzelius, J.J. (1835). Sur un Force Jusqu'ici Peu Remarquée qui est Probablement Active Dans la Formation des Composés Organiques, Section on Vegetable Chemistry. *Jahres-Bericht* 14: 237.
4. Berzelius, J.J. (1836). Einige Ideen über eine bei der Bildung organischer Verbindungen in der lebenden Natur wirksame, aber bisher nicht bemerkte Kraft. *Jahres-Bericht uber die Fortschritte der Chemie* 15: 237–245.
5. Thomas, J.M. and Harris, K.D.M. (2009). Some of tomorrow's catalysts for processing renewable and non-renewable feedstocks, diminishing anthropogenic carbon dioxide and increasing the production of energy. *Energy & Environmental Science* 9: 687–708.
6. Cavani, F. and Trifiro, F. (1997). Classification of industrial catalysts and catalysis for the petrochemical industry. *Catalysis Today* 34: 269–279.
7. Misono, M. (2013). *Heterogeneous Catalysis of Mixed Oxides Perovskite and Heteropoly Catalysts*, 8. Elsevier.
8. Van Wazer, J.R. (1958). *Phosphorus and Its Compounds: Chemistry.*, Volume 1. John Wiley & Sons Inc.
9. Clearfield, A. and Thakur, D.S. (1986). Zirconium and titanium phosphates as catalysts: a review. *Applied Catalysis* 26: 1–26.
10. Moffat, J.B. (1978). Phosphates as catalysts. *Catalysis Reviews. Science and Engineering* 18 (2): 199–258.
11. Sebti, S., Zahouily, M., and Lazrek, H.B. (2008). Phosphates: new generation of liquid-phase heterogeneous catalysis in organic chemistry. *Current Organic Chemistry* 12 (3): 203–232.
12. Rekha, P., Yadav, S., and Singh, L. (2021). A review on cobalt phosphate-based materials as emerging catalysts for water splitting. *Ceramics International* 47 (12): 16385–16401.
13. Vieira, J.L., Paul, G., and Iga, G.D. (2021). Niobium phosphates as bifunctional catalysts for the conversion of biomass-derived monosaccharides. *Applied Catalysis A: General* 617: 118099. https://doi.org/10.1016/j.apcata.2021.118099.
14. Kim, H., Park, J., and Park, I. (2015). Coordination tuning of cobalt phosphates towards efficient water oxidation catalyst. *Nature Communication* 6: 8253. https://doi.org/10.1038/ncomms9253.
15. Gruselle, M. (2015). Apatites: a new family of catalysts in organic synthesis. *Journal of Organometallic Chemistry* 793: 93–101.
16. Fihri, A., Len, C., and Varma, R.S. (2017). Hydroxyapatite: a review of syntheses, structure and applications in heterogeneous catalysis. *Coordination Chemistry Reviews* 347: 48–76.

17 Sebti, S., Tahir, R., Nazih, R. et al. (2002). Hydroxyapatite as a new solid support for the Knoevenagelreaction in heterogeneous media without solvent. *Applied Catalysis A: General* 228: 155–159.

18 Phan, T.S., Sane, A.R., Rêgo de Vasconcelos, B. et al. (2018). Hydroxyapatite supported bimetallic cobalt and nickel catalysts for syngasproduction from dry reforming of methane. *Applied Catalysis B: Environmental* 224: 310–321.

19 Palmer, L.C., Newcom, C.J., Kaltz, S.R. et al. (2008). Biomimetic systems for hydroxyapatite mineralization inspired by bone and enamel. *Chemical Reviews* 108: 4754–4783.

20 Naderi, A., Zhang, B., Belgodere, J.A. et al. (2021). Improved biocompatible, flexible mesh composites for implant applications via hydroxyapatite coating with potential for 3-dimensional extracellular matrix network and bone regeneration. *ACS Applied Materials & Interfaces* 13: 26824–26840.

21 Haider, A., Haider, S., Han, S.S., and Kang, I.K. (2017). Recent advances in the synthesis, functionalization and biomedical applications of hydroxyapatite: a review. *RSC Advances* 7: 7442–7458.

22 Nzihou, A. and Sharrock, P. (2010). Role of phosphate in the remediation and reuse of heavy metal polluted wastes and sites. *Waste and Biomass Valorization* 1: 163–174.

23 Ibrahim, M., Labaki, M., Marc Giraudon, J.M. et al. (2020). Hydroxyapatite, a multifunctional material for air, water and soil pollution control: a review. *Journal of Hazardous Materials* 383: 121139. https://doi.org/10.1016/j.jhazmat.2019.121139.

24 Verwilghen, C., Rio, S., Nzihou, A. et al. (2007). Preparation of high specific surface area hydroxyapatite for environmental applications. *Journal of Materials Science* 42: 6062–6066.

25 Nagasaki, T., Nagata, F., and Sakurai, M. (2017). Effects of pore distribution of hydroxyapatite particles on their protein adsorption behavior. *Journal of Asian Ceramic Societies* 5: 88–93.

26 Liao, C.J., Lin, F.H., Chen, K.S. et al. (1999). Thermal decomposition and reconstitution of hydroxyapatite in air atmosphere. *Biomaterials* 20: 1807–1813.

27 Bailliez, S. and Nzihou, A. (2004). The kinetics of surface area reduction during isothermal sintering of hydroxyapatite adsorbent. *Chemical Engineering Journal* 98: 141–152.

28 Rego de Vasconcelos, B., Pham Minh, D., Martins, E. et al. (2020). A comparative study of hydroxyapatite- and alumina-based catalysts in dry reforming of methane. *Chemical Engineering Technology* 43 (4): 698–704.

29 Vahdat, A., Ghasemi, B., and Yousefpour, M. (2019). Synthesis of hydroxyapatite and hydroxyapatite/Fe3O4 nanocomposite for removal of heavy metals. *Environmental Nanotechnology, Monitoring & Management* 12: 100233. https://doi.org/10.1016/j.enmm.2019.100233.

30 Zhou, C., Wang, X., Wang, Y. et al. (2021). The sorption of single- and multi-heavy metals in aqueous solution using enhanced nano-hydroxyapatite assisted with ultrasonic. *Journal of Environmental Chemical Engineering* 9 (3): 105240. https://doi.org/10.1016/j.jece.2021.105240.

31 Khaleque, A., Alam, M.M., Hoque, M. et al. (2020). Zeolite synthesis from low-cost materials and environmental applications: a review. *Environmental Advances* 2: 100019. https://doi.org/10.1016/j.envadv.2020.100019.

32 Pham Minh, D., Sebei, H., Nzihou, N. et al. (2012). Apatitic calcium phosphates: synthesis, characterization and reactivity in the removal of lead(II) from aqueous solution. *Chemical Engineering Journal* 198–199: 180–190.

33 Zhang, P., Wu, T., and Jiang, T. (2013). Ru–Zn supported on hydroxyapatite as an effective catalyst for partial hydrogenation of benzene. *Green Chemistry* 15: 152–159.

34 Yamaguchi, K., Mori, K., Mizugaki, T. et al. (2000). Creation of a monomeric Ru species on the surface of hydroxyapatite as an efficient heterogeneous catalyst for aerobic alcohol oxidation. *Journal of the American Chemical Society* 122: 7144–7145.

35 Mori, K., Yamaguchi, K., Hara, T. et al. (2002). Controlled synthesis of hydroxyapatite-supported palladium complexes as highly efficient heterogeneous catalysts. *Journal of the American Chemical Society* 124: 11572–11573.

36 Mori, K., Mitani, Y., Hara, T. et al. (2005). A single-site hydroxyapatite-bound zinc catalyst for highly efficient chemical fixation of carbon dioxide with epoxides. *Chemical Communications* 3331–3333.

37 Akri, M., Zhao, S., Li, X. et al. (2020). Atomically dispersed nickel as coke-resistant active sites for methane dry reforming. *Nature Communications* 10: 5181.

38 Larsen, S. (1966). Solubility of hydroxyapatite. *Nature* 212: 605.

39 Bell, L.C., Mika, H., and Kruger, B.J. (1978). Synthetic hydroxyapatite-solubility product and stoichiometry of dissolution. *Archives of Oral Biology* 23 (5): 329–336.

40 Ito, A., Maekawa, K., Tsutsumi, S. et al. (1997). Solubility product of OH-carbonated hydroxyapatite. *Journal of Biomedical Materials Research* 36 (4): 522–528.

41 Silvester, L., Lamonier, J.F., Vannier, R.N. et al. (2014). Structural, textural and acid–base properties of carbonate-containing hydroxyapatites. *Journal of Materials Chemistry A* 2: 11073–11090.

42 Boukha, Z., Pilar Yeste, M., Ángel Cauqui, M. et al. (2019). Influence of Ca/P ratio on the catalytic performance of Ni/hydroxyapatite samples in dry reforming of methane. *Applied Catalysis A: General* 580: 34–45.

43 Phan, T.S. (2020). Élaboration, caractérisation et mise en œuvre d'un catalyseur dans le reformage du biogaz en vue de la production d'hydrogène vert. PhD thesis. IMT Mines Albi. http://theses.fr/2020EMAC0007 (accessed 30 August 2021).

44 Moussa, S.B., Lachheb, J., Gruselle, M. et al. (2017). Calcium, barium and strontium apatites: a new generation of catalysts in the Biginelli reaction. *Tetrahedron* 73: 6542–6549.

45 Ben Moussa, S., Mehri, A., and Badraoui, B. (2020). Magnesium modified calcium hydroxyapatite: an efficient and recyclable catalyst for the one-pot Biginelli condensation. *Journal of Molecular Structure* 1200: 127111. https://doi.org/10.1016/j.molstruc.2019.127111.

46 Tsuchida, T., Kuboa, J., Yoshioka, T. et al. (2008). Reaction of ethanol over hydroxyapatite affected by Ca/P ratio of catalyst. *Journal of Catalysis* 259: 183–189.

47 Xin, Y. and Shirai, T. (2021). Noble-metal-free hydroxyapatite activated by facile mechanochemical treatment towards highly-efficient catalytic oxidation of volatile organic compound. *Scientific Reports* 11: 7512.

48 Adeogun, A.I., Ofudje, E.A., Idowu, M.A. et al. (2018). Biowaste-derived hydroxyapatite for effective removal of reactive yellow 4 dye: equilibrium, kinetic, and thermodynamic studies. *ACS Omega* 3: 1991–2000.

49 Sebei, S.I., Pham Minh, D., Lyczko, N. et al. (2017). Hydroxyapatite-based sorbents: elaboration, characterization and application for the removal of catechol from the aqueous phase. *Environmental Technology* 38 (20): 2611–2620.

50 Wei, W., Yang, L., Zhong, W. et al. (2015). Poorly crystalline hydroxyapatite: a novel adsorbent for enhanced fulvic acid removal from aqueous solution. *Applied Surface Science* 332: 328–339.

51 Kawai, T., Ohtsuki, C., Kamitakahara, M. et al. (2006). Removal of formaldehyde by hydroxyapatite layer biomimetically deposited on polyamide film. *Environmental Science & Technology.* 40 (13): 4281–4285.

52 Ferraz, M.P., Monteiro, F.J., and Manuel, C.M. (2004). Hydroxyapatite nanoparticles: a review of preparation methodologies. *Journal of Applied Biomaterials & Biomechanics* 2: 74–80.

53 NMohd Puad, A.S., Abdul Haq, R.H., Mohd Noh, H. et al. (2020). Synthesis method of hydroxyapatite: a review. *Materials Today: Proceedings* 29: 233–239.

54 Sepulveda, P., Ortega, F.S., Innocentini, M.D.M. et al. (2000). Properties of highly porous hydroxyapatite obtained by the gelcasting of foams. *Journal of the American Ceramic Society* 83 (12): 3021–3024.

55 Sarda, S., Nilsson, M., Balcells, M. et al. (2003). Influence of surfactant molecules as air-entraining agent for bone cement macroporosity. *Journal of Biomedical Materials Research* 65A (2): 215–221.

56 Chevalier, E., Chulia, D., Pouget, C. et al. (2008). Fabrication of porous substrates: a review of processes using pore forming agents in the biomaterial field. *Journal of Pharmaceutical Sciences* 97: 1135–1154.

57 Khallok, H., Elouahli, A., Ojala, S. et al. (2020). Preparation of biphasic hydroxyapatite/β-tricalcium phosphate foam using the replication technique. *Ceramics International* 46: 22581–22591.

58 Munar, M.L., Udoh, K.I., Ishikawa, K. et al. (2006). Effects of sintering temperature over 1,300°C on the physical and compositional properties of porous hydroxyapatite foam. *Dental Materials Journal* 25 (1): 51–58.

59 Hui, Y., Dong, Z., Wenkun, P. et al. (2020). Facile synthesis of copper doping hierarchical hollow porous hydroxyapatite beads by rapid gelling strategy. *Materials Science & Engineering C* 109: 110531. https://doi.org/10.1016/j.msec.2019.110531.

60 Thrane, J., Mentzel, U.V., Thorhauge, M. et al. (2021). Hydroxyapatite supported molybdenum oxide catalyst for selective oxidation of methanol to formaldehyde: studies of industrial sized catalyst pellets. *Catalysis Science & Technology* 2021 (11): 970–983.

61 Dasireddy, V.D.B.C., Friedrich, H.B., and Singh, S. (2015). A kinetic insight into the activation of n-octane with alkaline-earth metal Hhdroxyapatites. *South African Journal of Chemistry* 68: 195–200.

62 Ogo, S., Maeda, S., and Sekine, Y. (2017). Coke resistance of Sr-hydroxyapatite supported Co catalyst for ethanol steam reforming. *Chemistry Letters* 46: 729–732.

63 Grandreviewresearch (2020). Catalyst Market Size, Share & Trends Analysis Report By Raw Material (Chemical Compounds, Zeolites, Metals), By Product (Heterogeneous, Homogeneous), By Application, By Region, And Segment Forecasts, 2020–2027. https://www.grandviewresearch.com/industry-analysis/catalyst-market (accessed 16 August 2021).

64 Arakawa, H., Aresta, M., Armor, J.N. et al. (2001). Catalysis research of relevance to carbon management: progress, challenges, and opportunities. *Chemical Reviews* 101 (4): 953–996.

65 Burkart, M.D., Hazari, N., Tway, C.L. et al. (2019). Opportunities and challenges for catalysis in carbon dioxide utilization. *ACS Catalysis* 9 (9): 7937–7956.

66 Blakemore, D.C., Castro, L., Churcher, I. et al. (2018). Organic synthesis provides opportunities to transform drug discovery. *Nature Chemistry* 10: 383–394.

67 Grayson, I. (2016). Challenges and opportunities in catalysis. *Catalysis & Biocatalysis Chimica Oggi – Chemistry Today* 34 (5): 12–13.

68 Martínez, S., Veth, L., Lainer, B. et al. (2021). Challenges and opportunities in multicatalysis. *ACS Catalysis* 11: 3891–3915.

69 Wang, A., Li, J., and Zhang, T. (2018). Heterogeneous single-atom catalysis. *Nature Reviews Chemistry* 2 (6): 65–81.

70 Liu, L. and Corma, A. (2018). Metal catalysts for heterogeneous catalysis: from single atoms to nanoclusters and nanoparticles. *Chemical Reviews* 118 (10): 4981–5079.

71 Mitchell, S., Thomas, J.M., and Pérez-Ramírez, J. (2017). Single atom catalysis. *Catalysis Science & Technology* 7 (19): 4248–4249.

72 Kaiser, S.K., Chen, Z., Faust, A.D. et al. (2020). Single-atom catalysts across the periodic table. *Chemical Reviews* 120 (21): 11703–11809.

73 Munirathinam, R., Pham Minh, D., and Nzihou, A. (2020). Hydroxyapatite as a new support material for cobalt-based catalysts in Fischer-Tropsch synthesis. *International Journal of Hydrogen Energy* 45 (36): 18440–18451.

74 Kemiha, M., Pham Minh, D., Lyczko, N. et al. (2014). Highly porous calcium hydroxyapatite–based composites for air pollution control. *Procedia Engineering* 83: 394–402.

75 Euw, S.V., Wang, Y., Laurent, G. et al. (2019). Bone mineral: new insights into its chemical composition. *Scientific Reports* 9: 8456.

2

Synthesis and Characterization of Hydroxyapatite and Hydroxyapatite-Based Catalysts

Yousra EL Jemli[1], Karima Abdelouahdi[1], Doan Pham Minh[2], Abdellatif Barakat[3,4], and Abderrahim Solhy[3]

[1] University Cadi Ayyad, IMED-Lab, FST-Marrakech, Av. A. Khattabi, BP 549, 40000. Marrakech, Morocco
[2] Université de Toulouse, IMT Mines Albi, UMR CNRS 5302, Centre RAPSODEE, Campus Jarlard, F-81013 Albi, Cedex 09, France
[3] IATE, Montpellier University, INRAE, Agro Institut, 34060 Montpellier, France
[4] Mohamed VI Polytechnic University, Lot 660 - Hay Moulay Rachid, 43150 Ben Guerir, Morocco

2.1 Introduction

Apatites constitute a family of minerals defined by the chemical formula $M_{10}(XO_4)_6(Y)_2$ in which M generally represents a bivalent cation (Ca^{2+}, Pb^{2+}, Cd^{2+}, etc.), XO_4 a trivalent anionic group (PO_4^{3-}, VO_4^{3-}, AsO_4^{3-}, etc.), and Y a monovalent anion (OH^-, Cl^-, F^-, etc.) [1]. Apatites are therefore a large family of isomorphic inorganic materials. Most of the studies conducted on apatites have focused on their biocompatibility and their specific crystalline structure allowing various cation and/or anion substitutions. Indeed, apatites often have a strong capacity to substitute ions when brought in contact with aqueous solutions, where Ca^{2+} ions can be substituted by cations such as transition metal ions and PO_4^{3-} ions can be replaced by anions such as AsO_4^{3-}, CO_3^{2-}, etc. [2]. Hydroxyapatite (HA, $Ca_{10}(PO_4)_6(OH)_2$), belongs to this family of inorganic minerals. It is currently the subject of intense interest in many areas of research [3, 4]. Moreover, due to its similarity in chemical composition to the mineral phase of bone tissues, it has found applications in medicine as synthetic bone substitutes [5]. Many experimental studies were performed with the aim of systematically evaluating potential improvements in the fixation of implants to bone, using an HA coating, under pathological and mechanical conditions similar to the clinical conditions [6–8]. In addition to its important application as biomaterials, HA is studied for various applications such as fluorescent lamps [9], materials for fuel cell [10], adsorbents for the removal of metals in aqueous solution [11], and catalytic materials either as catalysts or catalytic supports [12–14].

It is well known that this phosphate can actively catalyze many reactions [15–17]. The catalytic properties are closely linked to its structure, stoichiometry, porosity, and morphology. However, these main characteristics are influenced by a number

Design and Applications of Hydroxyapatite-Based Catalysts, First Edition. Edited by Doan Pham Minh.
© 2022 WILEY-VCH GmbH. Published 2022 by WILEY-VCH GmbH.

of factors such as the synthesis methods, the post-synthesis treatments, and the substitution of its building elements (Ca^{2+}, PO_4^{3-}, OH^-) by others to modulate its catalytic properties. In the past, several methods of synthesizing HA have already been described [14, 18]. The application of HA in heterogeneous catalysis is reinforced by the ease of modifying the material by the introduction of other elements such as transition metals instead of calcium, resulting in a profound change in catalytic properties [12, 14]. Thus, this chapter focuses on the synthesis and the characterization of HA and its derived materials for catalytic applications. This chapter is structured and organized as follows: first, its physicochemical properties and synthesis methods are highlighted. Consequently, the characterization of HA powders doped with other elements is discussed, with a focus on the interactions between HA and substituted elements. Conclusively, the direction for future outlook and perspectives are presented to inform and provide insight for further investigation.

2.2 HA Synthesis and Characterization

2.2.1 HA Synthesis Routes

The structures, morphologies, and textures of HA materials differ depending on the adopted synthesis process. The different HA synthesis routes developed and reported in the literature can be classified in four categories: (i) wet methods (coprecipitation, sol–gel, emulsion, hydrolysis, hydrothermal methods, etc.), (ii) dry methods, (iii) assisted methods (microwave [MW], ball milling, or ultrasound), and (iv) other nonconventional methods. In each category, there are several variants depending on the synthesis conditions and the reagents deployed. In this part of the chapter, the most important HA synthesis methods are explored.

2.2.1.1 Coprecipitation Method

The main principle of this method is to add drop by drop one reagent to the other reagent in aqueous phase under stirring at a temperature between 25 and 100 °C for 5–24 hours and under a fixed pH (from 3 to 12). There are two types of reagents commonly used: (i) those for the direct neutralization such as $Ca(OH)_2$ with H_3PO_4 [19] or $CaCO_3$ with H_3PO_4 [20], which do not generate any residual counterions but the resulting HA is often carbonated; and (ii) those for the double decomposition by continuous precipitation that include $Ca(NO_3)_2$ with $(NH_4)_2HPO_4$ [21], traces of nitrate and ammonium ions being generally present in the final HA that can be removed by heating.

The addition order plays an important role in the formation of HA. Experimentally, it is possible to perform HA precipitation by: (i) adding the calcium source to the phosphate source, where calcium phosphate compounds progressively evolve in acidic medium; (ii) adding the phosphate source to the calcium source, where calcium phosphate compounds progressively evolve in basic medium; (iii) and adding together calcium and phosphate sources at the stoichiometric ratio at a controlled pH, temperature, and atmosphere, where the formation of stoichiometric HA can

2.2 HA Synthesis and Characterization

theoretically be achieved along the precipitation. This last way is generally preferred and is used to produced stoichiometric HA at the industrial scale.[1]

The synthesis of HA by this method is generally influenced by different parameters including the order and the rate of reactant addition, the nature of starting reagents, the pH, the temperature, the stirring speed, and the control of the atmosphere during the synthesis to avoid the carbonation reactions. The simultaneous control of all these parameters is usually delicate and sometimes leads to a difficult reproduction of the synthesis. On the other hand, this method allows varying all these parameters to synthesize HA materials having different physicochemical properties such as porosity, morphology, stoichiometry, and acid-basic properties.

As example, Santos et al. [19] synthesized HA from $Ca(OH)_2$ and H_3PO_4 by precipitation at 40 °C for 1 hour under vigorous stirring, according to the following theoretical equation (Eq. 2.1). Two solutions of 0.3 and 0.5 M H_3PO_4 were used, which were slowly added to a suspension of $Ca(OH)_2$. The pH of the final solution was adjusted by using a 1 M NH_4OH solution. By this way, the authors highlighted the presence of CaO in the final HA powder, since $Ca(OH)_2$ was surplus at the beginning of the precipitation. A chemical post-treatment with H_3PO_4 should be done to transform this residual CaO into HA.

$$10Ca(OH)_2 + 6H_3PO_4 \rightarrow Ca_{10}(PO_4)_6(OH)_2 + 18H_2O \tag{2.1}$$

Another example on the HA precipitation is the work performed by Pham Minh et al. [20]. HA was synthesized by progressively adding $CaCO_3$ powder to an aqueous solution of H_3PO_4 under vigorous stirring at 25 °C for 48 hours. Despite the long synthesis time, the reaction could not be finished because $CaCO_3$ still remained in the final product. Thus, "core-shell" product was obtained with the residual $CaCO_3$ in the core and HA in the shell of the particles. A mechanism for the formation of these core-shell was also proposed as reported in Figure 2.1. The nature of the reactants was also studied in this work by using various orthophosphates and calcium carbonate [20]. The authors concluded that the most interesting source of phosphate was orthophosphoric acid since it favors calcium carbonate dissolution.

The synthesis of HA was also carried out by the double decomposition method [21] (Eq. 2.2). The pH of the solution of diammonium hydrogen phosphate, $(NH_4)_2HPO_4$,

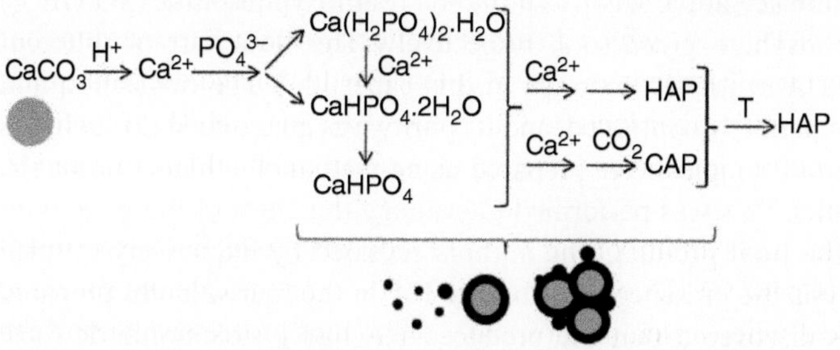

Figure 2.1 Reaction pathway of HA synthesis by addition of $CaCO_3$ powder to H_3PO_4 solution at 25 °C. Source: Pham Minh et al. [20]. Reproduced with permission of Elsevier.

---

1 https://www.prayon.com/en/index.php.

was maintained at around 12, by adding ammonium hydroxide (NH_4OH). Then, a solution of calcium nitrate, $Ca(NO_3)_2$, was dropped under constant stirring. After the double decomposition process, the suspension was refluxed for four hours for maturation and the obtained powder was washed, dried, and calcined to obtain HA.

$$6(NH_4)_2HPO_4 + 10Ca(NO_3)_2 + 8NH_4OH \rightarrow Ca_{10}(PO_4)_6(OH)_2 \\ + 20NH_4(NO_3) + 6H_2O \quad (2.2)$$

The important role of pH in the coprecipitation method for the production of HA was also revealed [22], resulting in the formation of different morphologies of nanostructured carbonated HA at pH 8 and pH 12 by using urea. At pH 8, the formation of β-tricalcium phosphate (β-TCP) and HA after heat treatment at 900 °C were achieved, while at pH 12, the HA phase was more stable and no secondary phase was detected by X-ray diffraction (XRD). The as-prepared powders at pH 8 possessed a needle-like shape (average size ca. 20 nm) with substitution of PO_4^{3-} by CO_3^{2-} at low content. This CO_3^{2-} can be easily removed by heat treatment above 800 °C. Additionally, the as-prepared powders at pH 12 exhibited rod-like structures of 50 nm average diameter and 300 nm average length.

2.2.1.2 Sol–Gel Method

Another technique used in synthesizing the HA powder is the sol–gel method. The sol–gel mineralization method has the advantage of producing porous and nanostructured HA under "soft chemistry" conditions. This is significantly lower than those of conventional synthesis routes in terms of temperature, making it an energy-efficient method [23]. In light of this, Liu et al. synthesized HA ceramics via this process by using triethoxyphosphine ($PO(OEt)_3$) and calcium nitrate ($Ca(NO_3)_2$) as phosphorus and calcium precursors, respectively [23]. This synthesis procedure has been developed in the field of biomaterials because the resulting HA product has porous structure with pore size ranging from 0.3 to 1 μm, having controlled sinter ability at low temperatures (e.g. 200–300 °C). Thus, this method offers an alternative to thermal spraying for the manufacture of coatings made of thin layers, which is a process difficult to control and requires high temperatures (e.g. 6 000–10 000 °C). Jillavenkatesa and Condrate [24] synthesized HA by the sol–gel method using calcium acetate ($Ca(CH_3COO)_2$) and triethoxyphosphine ($PO(OEt)_3$) as calcium and phosphate precursors, respectively. The flowchart of different steps followed for HA synthesis is shown in this paper [24]. The nucleation and growth of HA were perfectly controlled, and its purity was guaranteed through this process. Four different samples were prepared using methanol, ethanol, propanol, and water as solvents. This was performed to evaluate the effect of the solvent on the properties of the final product. The authors revealed by inductively coupled plasma (ICP) analysis the presence of carbonate HA in the four samples prepared in alcohols. It was discovered that the produced HA had a stoichiometric Ca/P weight ratio of 2.157. The substitution of PO_4^{3-} ligand by carbonate ions causes an increase in the Ca/P ratio in the sample. The Ca/P weight ratio of the samples is found to vary from 2.167 to 2.497 for the sample prepared in propanol. The authors concluded that the formation of carbonate apatite occurs during the treatment of

HA powder with a 0.01 M HCl solution, and that the incorporation of carbonate in the HA lattice is related to the incomplete removal of these carbonates during the heat treatment step.

On the other hand, fibrous HA was prepared by spinning sols prepared from a 2-butanol solution of phosphorous pentoxide (P_2O_5) and calcium acetate ($Ca(CH_3COO)_2$) solution in distill water [25]. The phosphorous solution was prepared by stirring P_2O_5 in 2-butanol for 45 minutes, while the calcium source solution was prepared by dissolving $Ca(CH_3COO)_2$ in distilled water. Both solutions were then mixed under continuous stirring maintaining the Ca/P ratio at 1.67 and room temperature. In order to avoid the Ca and P precipitation during mixing, the authors have added acetic acid. Also, added is lactic acid as a spinning aid, and then, the mixture was refluxed for eight hours. The required sol must undergo and possess polymerization and spinning viscosity properties necessary for the generation of HA fibers. In other words, the sol should exhibit a Newtonian behavior and the viscosity must be optimal to avoid droplet formation and the appearance of gelling. The as-prepared HA fibers were dried and calcined at three different temperatures – 500, 700, and 1000 °C. However, optimal crystallization was obtained when the material was calcined at 1000 °C.

Chen et al. [26] reported the use of trimethyl phosphite (($CH_3O)_3P$) and calcium nitrate as precursors to synthesize different calcium phosphates (HA, β-TCP, and biphasic calcium phosphate [BCP]) via sol–gel route and by changing the molar ratio of Ca/P (1.4, 1.5, and 1.67). The XRD analysis showed that pure phase of β-TCP was produced with the initial Ca/P ratio of 1.4, whereas the one with the initial Ca/P ratio of 1.5 was a mixture of phases such as BCP, β-TCP, and HA (HA was the major component), and subsequently the last sample with the initial Ca/P ratio of 1.67 was only an HA phase. Furthermore, Velu and coworkers synthesized HA via the same process by using $CaCO_3$ and ammonium dihydrogen phosphate ($NH_4H_2PO_4$) as calcium and phosphorus sources and alginic acid (biopolymer) as a chelating agent [27]. It was found that both pure and crystalline HA was obtained by this method at moderate temperature of 300 °C. The completely formed crystallites are hexagonal in shape with average particles size in the range of 50–100 nm. This method illustrates the effect of alginic acid as a chelating agent in sol–gel synthesis, since this natural polysaccharide has hydrophilic properties capable of chelating cations.

Basically, the rate of gel formation depends strongly on: (i) the nature of the solvent, (ii) the temperature and pH used during the process, and (iii) the chemical nature of the reagents used. However, lack of control of certain parameters during the growth of HA may cause the formation of secondary phases such as CaO, $Ca_2P_2O_7$, $Ca_3(PO_4)_2$, and $CaCO_3$ [26, 28, 29]. It should be noted that the non-alkoxide sol–gel process for synthesis of HA, without need to adjust the pH, has also been developed, using precursors such as $Ca(NO_3)_2$, $(NH_4)_2HPO_4$, H_3PO_4, $PO(OEt)_3$, $PO(Me_3O)_3$, or KH_2PO_4 [30–32]. The synthesis of nanocrystalline HA powder via these processes can be easily achieved. Moreover, the degree of crystallinity, particle size, and agglomeration increased with increasing calcination temperature.

2.2.1.3 Emulsion Methods

Emulsion templating is a versatile method for the preparation of highly porous nanostructured HA powder [33–35], by allowing precise control of the morphology and distribution of grain size. In general, the technique involves forming a high internal phase emulsion, and locking in the structure of the continuous phase, usually by reaction-induced phase separation (e.g. sol–gel chemistry). Subsequent removal of the internal phase gives rise to a porous replica of the emulsion. Shum et al. have developed an innovative method for the preparation of HA based on the creation of microreactors and formulating double emulsion droplets: water-in-oil-in-water (W/O/W) double emulsion drops [33]. They exploited the capillary microfluidic techniques, having an easy control of the size and geometry of these double emulsion droplets, which gives the possibility to form highly versatile microreactors (Figure 2.2). The inner drops of the double emulsion consist of an aqueous solution of calcium nitrate and phosphoric acid, which are the calcium and the phosphorus precursors. The oil shells surrounding the inner drops are made up of an inert oil phase consisting of a surfactant (Dow Corning 749 fluid, Midland, MI) in a silicone oil (Dow Corning 200 Fluid, 5 cSt, Midland, MI), and the outer phase was a 5 wt% polyvinyl alcohol (PVA) aqueous solution. It was concluded that the HA is formed in the inner drops of the emulsion depending on the increase in pH by the addition of ammonium hydroxide. The porosity of the as-synthesized HA can be tuned by adjusting the reactant concentration in the inner drop, with the surface area and the total pore volume of the powders prepared are as high as 162.8 m^2 g^{-1} and 0.499 ml g^{-1}, respectively. The as-formed HA is made up of smaller spherical particles of about 1 μm in size, as shown by scanning electron microscopy (SEM). Each spherical particle has an open, feathery morphology shown in Figure 2.2. Transmission electron microscopy (TEM) clearly shows the fibrous structure in the form of hairs of the HA synthesized by this method (Figure 2.2).

Zhou and coworkers investigated the nanoemulsion technique without surfactant for the synthesis of spherical morphology HA [34]. The synthesis process is illustrated as follows: a solution of Ca(NO$_3$)$_2$·4H$_2$O dissolved in acetone was mixed with an aqueous solution of (NH$_4$)$_2$HPO$_4$ and NH$_4$HCO$_3$ at a molar ratio of Ca^{2+}/PO$_4^{3-}$/CO$_3^{2-}$ = 1.67/1/0.5 using a magnetic stirrer. The pH of the mixture was adjusted to 11 by adding 1 M NaOH solution. This process was conducted at 4, 25, 37, or 55 °C. This led to the formation of slightly milky nanoemulsions [34]. The resultant nanoprecipitates were filtered, washed, and then freeze-dried to obtain dry powder; the as-synthesized nanoparticles before and after freeze-drying are shown in this work [34]. Then, the powders were heat treated at 900 °C. It was discovered that all samples synthesized at different temperatures were amorphous without heat treatment. The sample prepared at 25 °C and calcined at 900 °C revealed a very high degree of crystallinity corresponding to an HA's structure. The as-synthesized nanoparticles were spherical in shape and their sizes were in the range of 10–30 nm.

Ethirajan et al. reported a novel biomimetic approach, which is conceptually mimicking the biomineralization process for the synthesis of hybrid HA/gelatin nanoparticles, which shows a high potential for application in tissue engineering [35]. This method consists of synthesizing an HA inside cross-linked gelatin

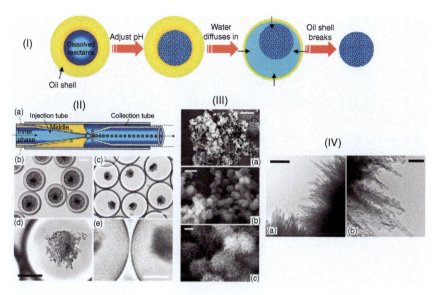

Figure 2.2 (I) Schematic representation of the double-emulsion-templated synthesis of hydroxyapatite adopted by the authors. (II) Glass microcapillary device used for generating double emulsion droplets, and (b–e) optical microscope images of double emulsion drops. (III) SEM images of dried hydroxyapatite powder at different magnifications. (IV) TEM image of HA powder. Source: Shum et al. [33]. Reproduced with permission of American Chemical Society.

nanoparticles, which serve as nanoreactors for HA crystallite growth in the aqueous phase. Gelatin nanoparticles prepared in inverse mini-emulsions and transferred to the water phase serve as an ideal molecular organic template providing the spatial confinement whereby promoting a different growth environment for the formation of HA nanoparticles. However, freeze-dried gelatin nanoparticles already prepared by inverse mini-emulsion technique were used to precipitate HA nanoparticles from $CaCl_2$ and Na_2HPO_4. The pH of the mixture was adjusted to 10.3 using NaOH. To form those mini-emulsions, p-xylene was used as the organic phase and poly[(butylene-co-ethylene)-b-(ethyleneoxide)] [P(B/E-b-EO)] as a stabilizer. The TEM photographs of the hybrid gelatin particles and the particle size distribution obtained from TEM analysis were presented in this work [35], showing the influence of calcium phosphate loading (strong contrast is clearly visible). The corresponding average particle size is 280 ± 53 nm. The authors reported that these nanoreactors promote the growth of spherical HA nanoparticles because of the confinement inside the nanoreactors based on Ostwald's ripening process.

In another work, nanostructured HA was prepared by reverse microemulsion technique using calcium nitrate $((Ca(NO_3)_2 \cdot 4H_2O)$ and orthophosphoric acid (H_3PO_4) as starting materials using cyclohexane, hexane, and isooctane as organic solvents [36]. To generate emulsions, different surfactants were tested in this study, namely, dioctyl sulfosuccinate sodium salt $(C_{20}H_{37}NaO_7S)$, dodecyl phosphate $(C_{12}H_{27}O_4P)$, NP5 (poly(oxyethylene)5 nonylphenol ether, and NP12 (poly(oxyethylene)12 nonylphenol ether. The effect of several synthesis parameters

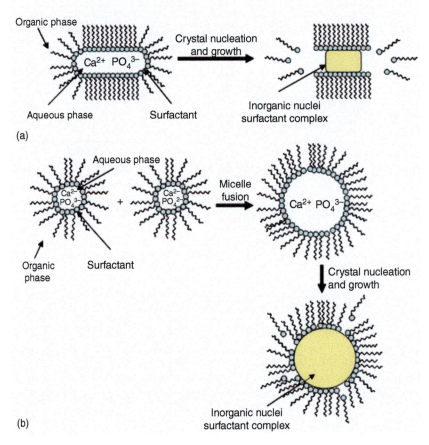

Figure 2.3 (a) Formation of HA nanoparticle in the core of the cylindrical reverse micelles in system with A/O = 1:5 and (b) formation of HA nanoparticle in spherical reverse micelles of non-uniform size produced by the random fusion of smaller micelle in the system with A/O = 1:15. Source: Saha et al. [36] Reproduced with permission of Elsevier.

on HA properties such as surfactant nature, water/organic (A/O) ratio, pH, and temperature were studied. Consequently, both aqueous and organic phases were mixed in a volume ratio of 1:5, 1:10, and 1:15 to obtain the reverse micelle. The morphology of the as-synthesized HA was found dependent on the size and the shape of the core of reversed micelles (Figure 2.3). Different morphologies such as spherical, needle shaped, or rod-like were obtained by adjusting the conditions of the emulsion system. Generally, the polymerization of concentrated oil-in-water (O/W) emulsions provides a direct synthetic route to a variety of porous HA. However, a significant disadvantage is that concentrated O/W emulsion techniques are very solvent intensive [36].

In 2005, Guo et al. investigated the synthesis of HA nanoparticles via reverse microemulsion route by using TX-100 and Tween 80 as surfactants, n-butanol

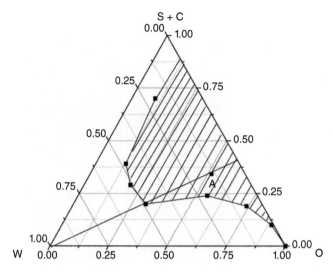

Figure 2.4 Partial phase diagram established for the system of cyclohexane-mixed surfactants and mixed cosurfactants – 0.5 M Ca(NO$_3$)$_2$ aqueous solution at room temperature; the microemulsion region is marked as shaded area. Source: Guo et al. [37]. Reproduced with permission of Elsevier.

and n-hexanol as cosurfactants, and cyclohexane as oil phase [37]. In the first step, a phase diagram was established for the system of oil-mixed surfactants and mixed cosurfactant-aqueous phase containing 0.5 M Ca(NO$_3$)$_2$ solution (Figure 2.4). Thus, various reverse microemulsions were prepared beneath different hydrophile–lipophile balance (HLB) values through the variation of the weight ratio of TX-100 and Tween 80. The effect of different HLB values on the particle size was also reported. The authors concluded that the reverse microemulsion route led to a significant improvement in the HA particle size in comparison with HA prepared by precipitation method.

The same research group has synthesized HA nanoparticles with controlled morphology through reverse microemulsion systems (aqueous solution/TX-100/n-butanol/cyclohexane), followed by hydrothermal treatment [38]. The strategy adopted in this study is illustrated in Figure 2.5. The authors reported that the concentration of cetyltrimethylammonium bromide (CTAB) in the aqueous solution and pH value exerted significant effect on the morphology and crystal phases of the produced HA powders. When the pH is higher than 8, only HA phase is obtained, and when the pH is lower than 8, other phases were obtained in addition to HA. It was also reported that the morphology of HA nanoparticles was sensitive to the mixture pH, as highlighted in Figure 2.5(II). Hence, different shapes and sizes of HA particles were obtained at different pH. In addition, the concentration of CTAB surfactant also influences the morphologies of HA particles. Though CTAB promoted the growth of HA along the a-axis, the growth along the c-axis under those conditions was inhibited, leading to various shapes and sizes of the product.

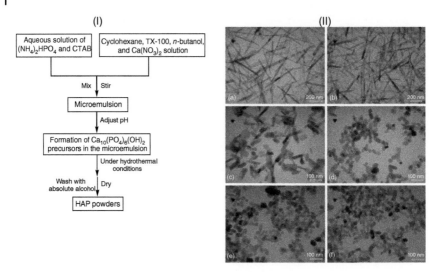

Figure 2.5 (I) Flowchart for the synthesis of HA by the emulsion technique under hydrothermal conditions. (II) TEM images of the products obtained at different pH values corresponding to sample (a) 10, pH = 7.0, (b) 11, pH = 7.5, (c) 12, pH = 8.0, (d) 1, pH = 9.0, (e) 13, pH = 10.0, and (f) 14, pH = 11.0. Source: Sun et al. [38]. Reproduced with permission of Elsevier.

2.2.1.4 Hydrolysis Methods

Calcium phosphate hydrolysis is one of the typical methods for the synthesis of HA in aqueous solutions [39, 40]. The principle of this method involves the dissolution of a precursor, which is often a calcium phosphate, followed by precipitation [41–44]. The hydrolysis reactions allow the production of HA with better control of composition and morphology than precipitation processes, because of a smaller number of synthesis parameters such as temperature, reaction duration, ratio and composition of the starting reagents, and pH. However, the key parameter in the hydrolysis of calcium phosphates is the stoichiometry of the initial compound. The additions of other calcium and phosphate sources are sometimes required in order to adjust HA stoichiometry. Different calcium phosphates can be used as precursors to prepare HA by this method especially $CaHPO_4 \cdot 2H_2O$, $CaHPO_4$, $\alpha\text{-}Ca_3(PO_4)_2$, $\beta\text{-}Ca_3(PO_4)_2$, $Ca_8H_2(PO_4)_6 \cdot 5H_2O$, and $Ca_4(PO_4)_2O$ [45, 46].

Lin et al. investigated the dissolution evolution of the various types of TCP and an HA ceramic soaked in water, buffered at pH 7.4, and kept at 37 °C for 1–28 days [47]. The ceramics exhibited different dissolution rates and did not convert to HA after being immersed in distilled water. But, petal-like apatite crystals formed on the surface of HαβTCP ceramic (α-TCP/β-TCP/HA) after two weeks of immersion, which was due to the precipitation of Ca^{2+} and PO_4^{3-} ions released from HαβTCP dissolution. This suggests that the HA acted as a nucleus for apatite precipitation during immersion. Likewise, Yubao et al. studied the behavior of alpha-tricalcium phosphate (α-TCP) in water [48]. The authors revealed that α-TCP was transformed into a special nonstoichiometric apatite called apatite-TCP, which has an apatite crystal

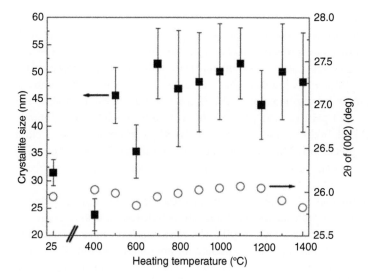

Figure 2.6 Crystallite size and (002) reflection of the HA powders of Ca/P = 1.5 synthesized at 75 °C for 1 hour by hydrolysis and heated from 25 to 400–1500 °C. (■ average crystallite size, ♦ (002) peak position). Source: Shih et al. [49]. Reproduced with permission of Elsevier.

structure but possesses the Ca/P ratio of TCP (1.50). The hydrolysis and phase conversion of α-TCP into apatite-TCP can be described by the following equation:

$$3Ca_3(PO_4)_2 + H_2O \rightarrow Ca_9(OH)(HPO_4)(PO_4)_5 \tag{2.3}$$

Shih et al. synthesized HA from $CaHPO_4 \cdot 2H_2O$ dicalcium phosphate dihydrate (DCPD) and $CaCO_3$ in basic medium (NaOH) at 75 °C [49]. The evolution of crystallinity was determined by in situ XRD between 25 and 1500 °C. The authors highlighted the evolution of the crystallite size and the (002) peak position, which is the HA's fingerprint (Figure 2.6). The HA was crystallized at 600 °C and maintained as the major phase until 1400 °C.

Sakamoto and coworkers have adopted the hydrolysis route to convert α-TCP into HA [43]. The reaction was achieved in a binary system of hydrophobic organic solvent and water. The following organic solvents such as ethanol, 1-butanol, 1-hexanol, 1-octanol, n-octane, cyclooctanol, cyclooctane, benzyl alcohol, and toluene were tested. The system's initial pH was adjusted in the range 11.0–12.3 by the addition of ammonia, 2-aminoethanol, diethanolamine, or triethanolamine. In the study, the as-prepared HA was deficient in calcium (Ca/P = 1.56–1.61, a-axis = 0.9440 nm, c-axis = 0.6880 nm) depending on the organic solvent used, with formula: $Ca_{10-z}(HPO_4)_z(PO_4)_{6-z}(OH)_{2-z}(H_2O)_y$ (where z = 0.34–0.64). The authors reported that the polarity of the organic solvent also had an effect on the morphology of the HA synthesized by this method. Thus, the HA particles produced were in the form of needles (length 1–4 μm, width 0.1 μm). In presence of a hydrocarbon solvent such as n-octane, the HA particles were less than 1.0 μm, contrary to those produced in the presence of polar solvent as 1-octanol.

2.2.1.5 Hydrothermal Methods

Hydrothermal (or solvothermal) syntheses occur in a confined environment with a higher temperature and pressure greater than autogenously ambient pressure, inside an autoclave under subcritical or supercritical conditions. It is widely used for the growth of crystalline HA [50–52]. Zhang et al. used this method to synthesize HA nanocrystallites having a nanorod-like shape (ca. 80 nm length and 15 nm width) at 100 °C and pH 10 in the presence of alanine ($C_3H_7NO_2$) and glutamic acid ($C_5H_9NO_4$) [50]. In the synthesis, $CaCl_2$ and H_3PO_4 were employed as calcium and phosphorus precursors. The authors suggested that the addition of the amino acid could help to control the growth of HA nanocrystallites with a perfectly homogeneous morphology.

Cao et al. used the hydrothermal process in the presence of a biomolecule such as $ATPNa_2$ (adenosine 5′-triphosphate disodium salt hydrate) as phosphate source in order to synthesize nanostructured HA [51]. The calcium source used in this study was $CaCl_2$ and the pH was adjusted by adding an ammonia solution. The maturation process is ensured by a hydrothermal treatment that consisted of putting the mixture in a sealed Teflon-lined autoclave, heated to 110 °C, and kept at this temperature for 1–120 hours. The mechanism of nucleation and growth of HA is illustrated in Figure 2.7. The authors mentioned that the first stage is the nucleation process, that is, the initial reaction between Ca^{2+} and ATP^{2-} ions, which is too fast to generate the HA nuclei, and subsequently followed by the growth stage. It was found that HA synthesized by this method has obvious anticancer properties. As a result, this HA was tested in human cervical cancer cells (HeLa) as a cell model in vitro. The inhibition of HeLa cell proliferation by as-synthesized HA is measured by tetrazolium dye MTT 3-(4,5-dimethylthiazol-2-yl)-2,5-diphenyltetrazolium bromide (MTT) colorimetric analysis and a decrease in cell viability as a function of time was observed. It should be noted, however, that the MTT colorimetric is an established method of determining viable cell number in proliferation and cytotoxicity studies, based on the cleavage of the yellow tetrazolium salt (MTT).

In another study, Lee and coworkers showed that the growth of well-defined hollow hexagonal prisms of HA could be achieved via the hydrothermal reaction of $CaCO_3$ and H_3PO_4 in the presence of hydrazine ($N_2H_4 \cdot H_2O$) in a Teflon-lined autoclave, which was sealed into a stainless steel tank at 170 °C for 6 hours [53]. The size and morphology of HA crystallites depend largely on temperature and reaction time. It was observed that hollow hexagonal prisms HA [53] were easily formed at 170 °C after 6 hours under the hydrothermal conditions used.

On the other hand, Lin et al. proposed an innovative strategy to control HA growth into different morphologies (from simple zero-dimensional morphology to complex three-dimensional architecture) using similarly structured hard-precursors of $CaCO_3$ nanoparticles, xonotlite nanowires, and hollow $CaCO_3$ microspheres in Na_3PO_4 solution, respectively [54]. The mixtures undergo a hydrothermal treatment without any templates or organic solvents. The strategy adopted in this work for controlling HA growth is shown in this paper [54]. Briefly, under hydrothermal treatment, a dissolution of the precursors occurred, accompanied by the release of the following ions into the solutions: Ca^{2+}, CO_3^{2-} (from the precursor $CaCO_3$) and

2.2 HA Synthesis and Characterization | 31

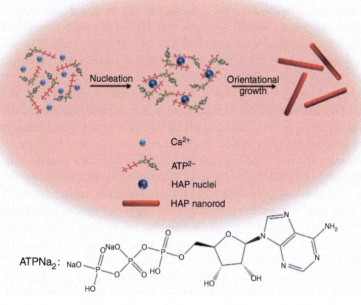

Figure 2.7 Reaction pathway of HA nanocrystallite synthesis by hydrothermal method. Source: Cao et al. [51]. Reproduced with permission of American Chemical Society.

Ca^{2+}, SiO_3^{2-} (from the xonotlite precursor). When the medium is saturated with Ca^{2+} and PO_4^{3-}, the nucleation takes place to form HA on the altered precursor surfaces. During this process, the precursors themselves played the roles of Ca^{2+} ion source and the HA crystal nucleation sites. The continuing release of these ions promote the continuous HA crystals growth, according to the extension of the maturation period until the complete disappearance of the precursors. It was concluded that with this method, the morphologies of the precursors can be well preserved.

Consequently, the high temperature and pressure in hydrothermal treatment facilitate the formation of the chemical bonds and the creation of nuclei, which ensure the formation of closely stoichiometric and highly crystalline HA [55, 56]. The control of nucleation and growth steps determine the formation of micro- or nano-sized crystallites, with calibrated morphology and porosity [57, 58]. The variety of precursors used in this method is quite large. For calcium precursors, conventional $Ca(NO_3)_2$ [59, 60] and $Ca(OH)_2$ [22, 56] but less common $CaCO_3$ also [54, 61, 62] could be used. In some cases, non-apatitic calcium phosphates such as $Ca(H_2PO_4)_2$ and $CaHPO_4$ can also be employed. For phosphorus precursors, H_3PO_4, Na_3PO_4, Na_2HPO_4, and $(NH_4)_2HPO_4$ are generally used. However, this method usually requires long synthesis times to obtain pure HA (up to 120 hours).

2.2.1.6 Microwave (MW)-Assisted Methods

MW-assisted synthesis route can provide rapid, affordable, and easy-to-upscale approaches to particularly overcome the problems associated with the control of

HA nucleation and growth [63]. The review published by Hassan and coworkers [63] showcased the state of the art in MW-assisted HA design and the potential implementation of this technology. The authors reported that this technique is often combined with other classical synthesis methods such as MW-wet precipitation, MW-hydrothermal and solvothermal, MW-combustion synthesis, MW-solid state, ultrasonic-assisted MW, and reflux-assisted MW, using several kinds of reagents [63].

During the irradiation, dielectric heating creates a heating effect on reagents, and its efficiency depends on the ability of these reagents to interact with MW radiation and produce heat without creating any major thermal slope and annihilation in the structure. However, the exact interaction between MW and reagents is still not well understood and clarified. It should be noted that MW frequencies, which are between radiowaves and infrared frequencies in the electromagnetic spectrum, lie in the range of 300 MHz to 300 GHz with wavelengths from 1 m to 1 mm [63]. Moreover, the MW electric field polarizes the molecules, which may couple with the rapid reversal of the electric field [63]. The main benefits of using MW-assisted strategy for the synthesis of HA are as follows: high yield, enhanced maturation, high reproducibility, and expanded reaction conditions, since reaction conditions can be easily optimized in comparison with the conventional methods.

Thus, this technique offers a versatile control to have a perfectly crystalline powder, homogeneous size, porosity and morphology [64–67], and enhanced surface area and also calibrated pore size, especially with the use of a template [68, 69]. Solhy's [68] research group reported a simple method of HA synthesis that consisted of a cooperative self-assembly of the calcium precursor with the zwitterionic surfactant molecules (lauryl dimethylaminoacetic acid). These latter reacted with PO_4^{3-} ligand, followed by MW-assisted maturation, which led to the formation of mesoporous nano-hydroxyapatite (mn-HA) with narrow pore sizes (36 nm mean diameter) (Figure 2.8) [68].

The same research group has developed a process for mn-HA synthesis by using a pseudo-sol–gel MW-assisted protocol in the presence of two novel templates, namely, sodium lauryl ether sulfate and linear alkylbenzenesulfonate [69]. This route enabled the precise control of pore size in a narrow-size distribution range (35 nm mean diameter) [69].

2.2.1.7 Ball-Milling Method

In this process, a mixture of calcium and phosphorus precursors is placed in a ball mill, which is subjected to a high-energy collision from the balls that can be made of different natures such as ceramic, flint pebbles, or stainless steel. This process can be carried out using different apparatus, namely, attritor, planetary mill, or a horizontal ball mill. The factors that affect the quality of HA include the milling time, rotational speed, size of balls, balls/reagents amount ratio, and post-synthesis processing such as thermal treatment [70]. During the ball-milling process, the collision between the tiny rigid balls in a concealed container will generate localized high pressure and temperature, which drive the chemical reaction between reactants. This process has been used to synthesize HA with some advantages such as its simplicity, reproducibility, and large-scale production [71]. However, high energy

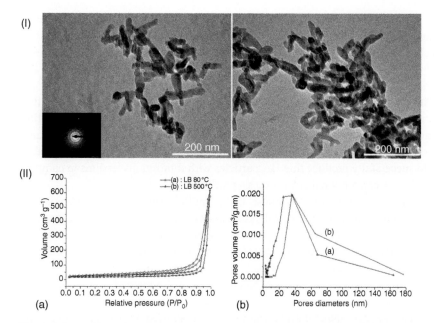

Figure 2.8 (I) TEM micrographs and selected area electron diffraction (SAED) pattern of calcined mn-HA at 500 °C for 10 hours. (II) (A) Nitrogen adsorption isotherms on mn-HA powders: dried at 80 °C (a), and calcined at 500 °C (b); (B) BJH (Barrett–Joyner–Halenda [pore size and volume analysis]) pore size distribution for mn-HA by action of 0.01 M of Lauryl betaine or laury dimethylaminoacetic acid betaine (LB) surfactant: (a) dried material at 80 °C, and (b) calcined material at 500 °C. Source: Amer et al. [68]. Reproduced with permission of Elsevier.

consumption and long reaction time have been identified as disadvantages of this synthesis route.

Honarmandi et al. applied a dry mechanochemical process to produce HA nanoparticles [72]. The authors used the following commercial reagents $Ca(OH)_2$, $CaHPO_4$, and $CaCO_3$, for the preparation of HA powders through the following two distinct reactions:

$$6CaHPO_4 + 4Ca(OH)_2 \rightarrow Ca_{10}(PO_4)_6(OH)_2 + 6H_2O \tag{2.4}$$

$$6CaHPO_4 + 4CaCO_3 \rightarrow Ca_{10}(PO_4)_6(OH)_2 + 4CO_2 + 2H_2O \tag{2.5}$$

These reactions were carried out on a planetary mill, with a mass ratio of reagents to ball about 1/20, for different reaction times. The synthesis was performed in sealed tempered chrome steel and polyamide-6 vials using zirconia balls (20 mm diameter) under ambient air atmosphere with a rotational speed of 600 rpm. HA nanoparticles were successfully formed via the reactions (2.4) and (2.5) after 40 hours of milling, with excellent purity, but with lower crystallinity due to reaction (2.5) than reaction (2.4). The presence of CO_2 and H_2O as by-products in Eq. (2.5) led to the formation of the carbonate ions, and thus, the formation of carbonated HA. The milling time and the calcium precursor influenced the morphology of the crystallites.

El Briak-Ben Abdeslam et al. [73] studied HA synthesis by milling dicalcium phosphate dihydrate and calcium oxide in the presence of water or without water (dry mechanosynthesis) (Eq. 2.6):

$$6CaHPO_4 \cdot 2H_2O + 4CaO \rightarrow Ca_{10}(PO_4)_6(OH)_2 + 14H_2O \qquad (2.6)$$

The authors reported that the dry mechanosynthesis was easier and swifter than the wet milling. The addition of water slowed down the reaction rate. In addition, grinding in wet conditions increased contamination of the final powder (HA) due to erosion of mill components. It was found that this method (dry mechanosynthesis) could successfully produce fine HA particles with uniform dispersions in comparison with the conventional HA synthesis processes.

Silv et al. investigated HA synthesis by ball-milling method using different reactants according to the reactions (2.4), (2.5), and (2.7) [74]. In this study, the synthesis was performed in a Fritsch Pulverisette 6 planetary mill, equipped with air-sealed stainless steel vials and balls and 370 rpm rotary speed. The reactant mixtures with the starting stochiometric ratio of Ca to P were used. The mass ratio of reagents to balls of ca. 1/6 was adapted. HA powders prepared by this method exhibited different degrees of crystallinity, and grain size, which depend on each reaction and grinding time. The powder obtained according to Eq. (2.4) showed a Ca/P ratio close to the theoretical ratio 1.67.

$$6(NH_4)H_2PO_4 + 10CaCO_3 \rightarrow Ca_{10}(PO_4)_6(OH)_2 + 8H_2O + 10CO_2 + 6NH_3 \qquad (2.7)$$

In accordance to the Eq. (2.6), Yeong et al. prepared HA powders by milling process using a mixture of calcium oxide and brushite powders in the molar ratio of calcium to phosphate equal to 3:2, in a high-energy shaker mill for various reaction times [75]. A milling of about 20 hours led to the formation of β-TCP after calcining the material at 800°C. On the other hand, increasing the grinding time to 30 hours followed by calcination at 800°C resulted in a perfectly crystallized HA.

2.2.1.8 Sonochemical Method

Sonochemistry is experiencing a revival of interest due to its various applications in several fields [76], among others, the synthesis of HA without bulk high temperatures, high pressures, or long reaction time [77, 78]. Several phenomena are responsible for sonochemistry. It must be emphasized that the chemical and physical effects of ultrasound arise not from a direct interaction between chemical species and sound waves, but rather from the physical phenomenon of acoustic cavitation: the formation, growth, and implosive collapse of bubbles [79, 80]. The extreme transient conditions induced by ultrasound produces unique hot spots that can achieve temperatures above 5000 K, pressures exceeding 1000 atm, and heating and cooling rates in excess of $1010\,K\,s^{-1}$ [81]. These conditions are distinct from other conventional synthetic routes. So, this method allowed the control of

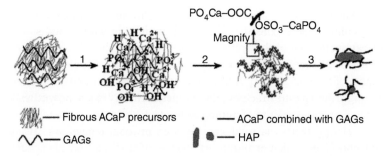

Figure 2.9 Illustrative scheme of the growth of HA nanoparticles stabilized by GAGs under ultrasonic irradiation. Source: Han et al. [91]. Reproduced with permission of Elsevier.

HA nucleation and growth with homogeneous morphology and increased porosity [82–86]. The hot spots produced by ultrasound can further accelerate, depending on the wave intensity, the reaction of calcium and phosphorus precursors to obtain HA for a shorter reaction time compared to other HA synthesis techniques [87–89].

As examples, Cao et al. examined the synthesis of HA by the ultrasound-assisted precipitation method by using $Ca(NO_3)_2$ and $NH_4H_2PO_4$ as calcium and phosphorus precursors and carbamide (NH_2CONH_2) as precipitating agent [90]. In this work, the power and amplitude of sonication were adjusted to provide the output power of 200 W and the irradiation time was varied from 1 to 4 hours. This process resulted in the formation of needle-like HA crystallites. The authors mentioned that HA yield increases with the temperature and ultrasound irradiation time and the addition of carbamide had an important role in the growth of HA nanoparticles, by adjusting the pH of the solution to around 7.4.

At the same time, Han et al. investigated the efficacy of this process in preparing nanostructured HA from $Ca(H_2PO_4)_2$ and $Ca(OH)_2$ as starting materials and glycosaminoglycans (GAGs) as a structuring agent whose role is of crucial importance in the stabilization of HA nanoparticles (Figure 2.9) [91]. Thus, the authors tested various GAGs concentration for different reaction times in an ultrasonic cleaner at 40 kHz and 250 W. They observed that the HA nanoparticles with perfectly homogeneous size were obtained with a concentration of GAGs between 0.35 and 0.45 g l^{-1}, and the increase of the ultrasound irradiation time in the presence of GAGs enhanced the crystallinity of the HA nanoparticles.

2.2.1.9 Dry Methods

Dry process or solid-state reaction for preparing HA involves grinding mixtures of precursors using a mortar, followed by calcination at high temperatures (900 and 1100 °C). High purity solid reagents in stoichiometric quantities are used [92]. In the work of Rao and coworkers [92], heat treatment from 600 to 1250 °C of a mixture of commercial TCP and calcium hydroxide in various proportions ranging from 3 : 0 to 3 : 4 M ratios, resulted in the production of pure stoichiometric β-TCP or HA as well as their two-phase composite powder mixtures (HA and β-TCP). Notably, the difference between solid-state reaction and dry mechanosynthesis lies in the milling method itself. The solid-state reaction method, which is the oldest technique adopted by crystallographers, consists of manual grinding in a mortar and subsequent heat treatments. One of the bottlenecks of the method is sensitivity to the variation in the stoichiometry and homogeneity of the reagent mixture, which influences the quality of the powder [93]. Additionally, it is an energy-intensive process and requires long grinding time to achieve the synthesis. The dry mechanosynthesis, on the other hand, involves the use of milling technology.

For instance, HA synthesis via solid-state reaction at room temperature using diammonium phosphate, calcium nitrate as sources of calcium and phosphate, and sodium bicarbonate was performed according to the procedure illustrated in Figure 2.10 [94]. The influence of calcination temperature on the crystallinity and composition of HA phase was investigated. The authors revealed that an increase in the calcination temperature had a major impact on the HA crystallization, since the uncalcined sample is an amorphous solid, and the crystallinity was increased progressively as the calcination temperature was raised up to 600 °C. However, between 650 and 900 °C, the HA decomposed into β-TCP and CaO, respectively. This method of synthesis was further studied by Fazan et al., in which only mix-grinding dry powders of β-TCP, $Ca(OH)_2$, and $CaCO_3$ from pure commercial chemicals or derived from natural limestone or from seashells were used with a total calcium/phosphorus molar ratio between 1.5 and 2.0 and a particle size of less than 10 μm [95]. The process requires a heat treatment between 600 and 1250 °C under air or in controlled atmospheric condition in order to produce HA having very high degree of crystallinity.

Similarly, Fowler [96] synthesized a stoichiometric and well-crystallized HA via the solid-state method induced by thermal treatments according to the reaction schemes described by Eqs. (2.8)–(2.10). These reactions were carried out in a controlled atmosphere in a tube inserted in a furnace and required relatively high temperatures (above 900 °C) and long treatment times. By the similar method, the authors could also elaborate Ba-HA and Sr-HA, by using the corresponding Ba and Sr precursors instead of Ca precursor.

$$2CaHPO_4 \xrightarrow[\text{Air}]{1000\,°C} Ca_2P_2O_7 + H_2O \tag{2.8}$$

$$3Ca_2P_2O_7 + 4CaCO_3 \xrightarrow[\text{Vacuum/20 h}]{1100\,°C} 10CaO\cdot3P_2O_5 + 4CO_2 \tag{2.9}$$

$$10CaO\cdot3P_2O_5 \xrightarrow[\text{Vacuum/24 h}]{1100\,°C} Ca_{10}(PO_4)_6(OH)_2 \tag{2.10}$$

2.2.1.10 Other Methods

Recently, abundant literature has focused on the development of new HA synthesis processes using alternative reagents (fish and scale bone, bone from bovine and ovine animals, etc.) [97]. Mohd Pu'ad et al. summarized recent works on the extraction of HA from natural sources, including mammalian, aquatic or marine sources, shellfish, plants and algae, and also from mineral sources (Figure 2.11) [98]. The extraction methods used to obtain HA are also described in the review [98].

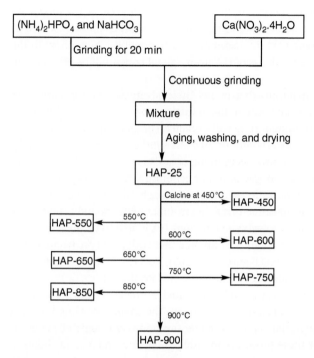

Figure 2.10 Scheme of HA synthesis by solid-state method. Source: Guo et al. [94]. Reproduced with permission of Elsevier.

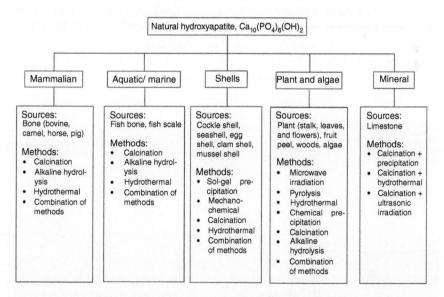

Figure 2.11 Summary of processes for synthesizing natural HA. Source: Mohd Pu'ad et al. [98]. https://www.sciencedirect.com/science/article/pii/S2405844018368944 (accessed 11 May 2021). CC BY-SA 4.0.

The effect of the extraction process and the natural waste source on HA properties such as the ratio: Ca/P, crystallinity, and the existence of other phases, as well as particle size and morphology are discussed and commented by the authors in the review. More details on the synthesis of HA from animal bones will be presented in Chapter 14 of this book.

The properties of HA materials that propel their advances and applications in various fields are of interest and will be taken into account. It is believed that HA synthesized from natural biomaterials will enhance compatibility particularly in tissue engineering applications. In light of this, HA synthesis methods inspired by the biomimetic mineralization process in living organisms are developed [99]. Yin et al. reported the elaboration of HA/bacterial cellulose (BC) nanocomposites via an optimal biomimetic mineralization synthesis approach for bone tissue engineering application. The BC has ultrafine three-dimensional network and is negatively charged with the adsorption of polyvinylpyrrolidone (PVP) to initiate the nucleation of HA, which was grown in vitro along the nanofiber network of BC via dynamic simulated body fluid (SBF) treatment (Figure 2.12) [100]. This resulted into rod-like HA nanoparticles (100–200 nm) homogeneously deposited on the surface of PVP-BC. It was also discovered that HA synthesized by biomimetic mineralization method was carbonated, with Ca/P ratio ranged from 1.37 to 1.59. The challenge of biomimetism in the development of materials is to overcome the traditional logic of materials design and to reconsider these processes in order to operate under mild chemical conditions (ambient pressure and temperature, aqueous solvent), using a limited number of elementary bricks composed of abundant chemical elements [101].

Xu et al. [102] used zwitterionic structuring agent such as zwitterionic poly(3-carboxy-N,N-dimethyl-N-(3'-acrylamidopropyl) propanaminium inner salt) (PCBAA) for synthesizing HA nanoparticles, denoted as Z-nHA. The sources of calcium and phosphate used in this work were $Ca(NO_3)_2 \cdot 4H_2O$ and $(NH_4)_2HPO_4$, and the pH was adjusted by using an aqueous solution of NH_4OH. The results showed that Z-nHA could be formed at all pH ranges in the presence of PCBAA.

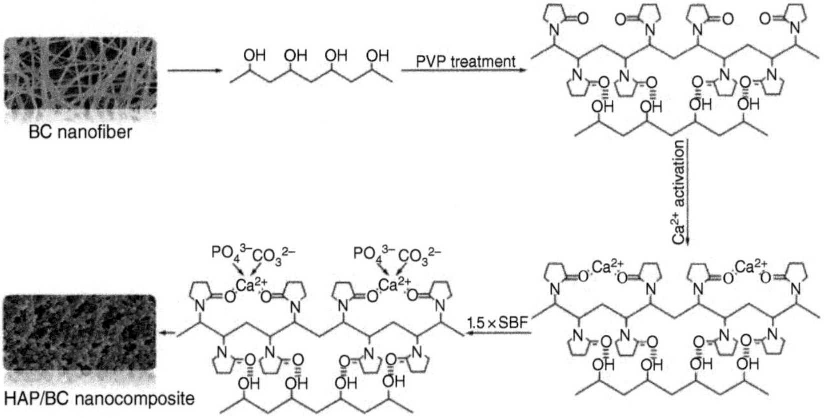

Figure 2.12 Scheme illustrating the mechanism of formation of HA/bacterial cellulose nanocomposites. Source: Yin et al. [100]/with permission of Elsevier.

In addition, a high pH value was beneficial for the formation of Z-nHA with a small particle size. The Z-nHA exhibited a fiber-like structure after the first two hours, and became rod-like after five hours, whereas pure HA prepared in water displayed an irregular flake-like structure. A possible mechanism for the formation of Z-nHA in the presence of PCBAA has been proposed [102]. The behavior of the zwitterionic PCBAA, which has anionic and cationic groups exhibit electrically neutral properties but extremely high polarity. Thus, the interaction of the positive quaternary ammonium group ($-R_3N^+$) with PO_4^{3-} ligand and negative carboxyl group ($-COO^-$) with Ca^{2+} ions is decisive for the growth of Z-nHAs in a radial direction and then the transformation of their morphology [102].

Moreover, self-assembly process allows the production of hierarchical structured HA [103–106]. Deng et al. studied the synthesis of bundle-like carbonated apatite using agar-gelatin hybrid hydrogels [107]. To understand the cooperative roles played by the gelatin and polysaccharide in the biomineralization, agar hydrogel, gelatin, and agar-gelatin hybrid hydrogel were, respectively, introduced as mineralization matrix for the in vitro growth of apatite in the study. The authors used $Ca(NO_3)_2 \cdot 4H_2O$, $Na_2HPO_4 \cdot 12H_2O$, NaOH, EtOH, agar powder (($C_{12}H_{18}O_9)_n$), and gelatin (from porcine) as chemicals in order to synthesize hybrid HA. They investigated also the in vitro biocompatibility of the as-prepared nanostructured HA using cell counting kit (CCK)-8 assay and alkaline phosphatase activity of osteoblast-like MC3T3-E1. The authors concluded that the obtained HA in agar-gelatin hybrid hydrogel provided significantly higher cell viability and alkaline phosphatase activity, in comparison with HA synthesized by precipitation route. A possible self-assembling mechanism of nanostructured carbonated apatite formation in hydrogels is proposed by the authors [107]. This mechanism is based on the chelating ability of the hybrid agar-gelatin hydrogel containing Ca^{2+} and PO_4^{3-} ions, which aided and controlled the nucleation and growth process.

HA synthesis has also been carried out by combustion, pyrolysis, and spray pyrolysis methods [108–110]. For instance, Ghosh et al. adopted the combustion process using urea ($CO(NH_2)_2$) and glycine (NH_2-CH_2-COOH) as fuels for the synthesis of HA from calcium nitrate and diammonium hydrogen phosphate [108]. The mixture of these reagents was introduced into a muffle furnace preheated to the desired temperature (300–700 °C). The procedure is illustrated in Figure 2.13. In the study, the authors investigated the effects of fuel nature (urea and glycine), fuel to oxidizer ratio, and heat treatment temperature on the combustion behavior and HA properties such as specific surface area, particle size, and powder morphology. Both systems of combustion produced poorly crystalline and carbon-contaminated powders (black powders). Additional heat treatment is needed to obtain a pure crystalline HA, especially in the case of hyper-stoichiometric fuel use.

The same research group studied the effect of addition of a small quantity of glucose to fuel (urea or glycine) on the combustion behavior and the characteristics of resulting powders [109]. The use of stoichiometric amounts of urea and glycine produced flames at temperatures of 817 and 888 °C, respectively. The reason being that the heat of combustion of glycine is almost three times higher than that of urea. Moreover, the addition of a small amount of glucose significantly reduces the

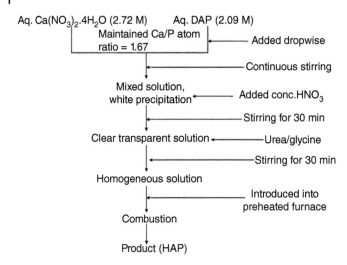

Figure 2.13 Illustration of solution combustion process for the synthesis of HA. Source: Ghosh et al. [108]. Reproduced with permission of Elsevier.

temperature of the flame and a vigorous reaction becomes almost burning in nature. However, it should be noted that glucose decomposes at a much lower temperature (<200 °C) and generates a large volume of gas. This rapidly dissipates the heat from the system and causes a decrease in temperature and the separation of the entire reaction mass into porous agglomerates. The slow combustion mode was characterized by a relatively slow and flameless reaction with the formation of a large and voluminous powder compared to self-propagating reactions with the flame. It was found that all combinations of urea-glycine resulted in a well-crystalline HA powder. The addition of a small amount of glucose (<1 g) also produced a well-crystallized HA, but higher amount of glucose (~1 g) resulted in the formation of an amorphous carbon-contaminated powder, which requires additional heat treatment to crystallize the HA. It has also been revealed that the crystallite size of the as-formed HA decreases significantly by increasing glucose addition. Consequently, the addition of glucose induced a positive effect on HA surface area, which increases according to the amount of glucose added. This can be attributed to the reduction of the flame temperature and as a consequence the system temperature during the combustion reaction. Thus, the sintering of the particles and subsequently the grain growth could be prevented. On the other hand, the release of gas due to the combustion of glucose also led to the formation of a highly porous product structure.

Sasikumar et al. tested both citric acid and succinic acid separately and in mixture as fuels, while nitrate and nitric acid were used as oxidants to synthesize HA following the flowchart shown in Figure 2.14 [110]. The calcium and phosphate sources were calcium nitrate and diammonium hydrogen phosphate. Both citric acid and succinic acid act as a fuel for combustion and also form a polymer matrix, which inhibits the precipitation of ions. However, self-ignition in the case of the use of citric acid was at 185 °C resulting in a black precursor, whereas in the case of succinic acid, it was at 425 °C forming a pale yellow precursor, and when a mixture of the two

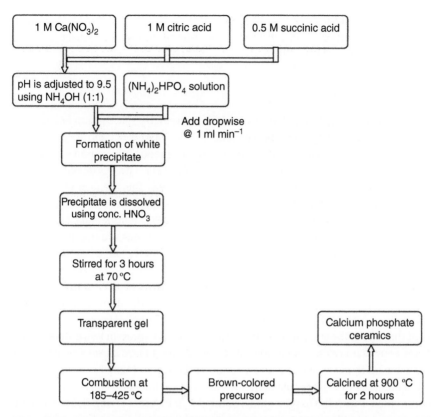

Figure 2.14 Flowchart of the synthesis of hydroxyapatite by self-propagating combustion method. Source: Sasikumar et al. [110]. Reproduced with permission of Elsevier.

acids was used, the self-ignition temperature was 290 °C independent of the ratio between succinic acid and citric acid. In all cases, the transformation of the solution into a gel took place within 2 hours, and thereafter reached the autoignition temperature in about 20 minutes. As a result, the combustion reaction took about one to two minutes depending on the fuel used. However, the observed swelling of the gel before ignition is due to the release of CO_2 produced by the decomposition of organic matter. The emission of brown fumes before ignition indicates the formation of NO_2, attributed to the oxidation of the gel by nitric acid. Finally, the powder obtained is calcined at 900 °C for 2 hours. It was discovered that HA phase was formed when either citric acid or succinic acid was separately used as fuel, but β-TCP was formed when a mixture of citric acid and succinic acid was used. The average crystallite size of the synthesized powder determined by Debye–Scherrer formula was found in the range of 55–65 nm.

The fundamental principle of the combustion process is based on the thermochemical concepts used in the field of propellant and explosive chemistry. A self-sustained exothermic redox reaction occurs in an aqueous phase between precursors and a suitable organic fuel (e.g. glycerin, urea, sucrose, citric acid, and succinic acid) [111–114]. The pyrolysis process consists of evaporated liquid reactants (precursors of calcium and phosphate) [115–117], and involves spraying the precursor solutions into a flame or hot zone of an electric furnace using an ultrasound generator [118, 119].

2.2.2 HA Structure

HA crystallizes in the hexagonal system with a symmetry space group P6$_3$/m, with the following crystallographic parameters: $a = 9.418$ Å, $c = 6.881$ Å, $\beta = 120\,°C$ (JCPDS No. 9-432) [120], and more rarely in monoclinic symmetry (P2$_1$/b) ($a = 9.4214$ Å, $b = 2a$, $c = 6.8814$ Å, and $\gamma = 120°$) [121]. The hexagonal symmetry is often obtained by wet synthesis methods, whereas the monoclinic crystal is generally obtained by heating the hexagonal crystal system under air at 850 °C. The crystal structure of the HA is a compact assembly of PO$_4$ tetrahedral group, whose P^{5+} ions are in the center of the tetrahedrons and the tops are occupied by four oxygen atoms. Each PO$_4$ tetrahedron is shared by a column and delimits two types of unconnected canals (Figure 2.15a).

The stability of the apatite's structural lattice is ensured by the assembly of metal ions and PO$_4^{3-}$ ions independent of the ions located in the tunnels. This compact arrangement of the PO$_4^{3-}$ groups reveals two types of tunnels. However, it should be noted that one of the structural particularities of HA is to present two different types of non-equivalent calcium sites noted as Ca1 and Ca2 (Figure 2.15a). The properties of HA can be tuned by specific modification of these sites [122]. So, the first tunnel with a diameter of 2.5 Å is ringed by Ca1 sites (4 per unit cell), each of them under coordination 9 with the oxygen atoms of the PO$_4$ tetrahedra giving rise to a polyhedron as shown in Figure 2.15b. The second tunnel that has a larger diameter than the previous one (3–4.5 Å) holds six Ca2 sites. The latter are positioned on the periphery of the second tunnel. These Ca2 sites are localized in two dimensions 1/4 and 3/4 of the unit cell along c-axis and form alternate equilateral triangles around the helical senary axis of the apatite lattice. Their coordination is 7, and they are surrounded by six atoms of oxygen belonging to PO$_4$ tetrahedron and one OH$^-$ anion in position 2a (Figure 2.15c) [120]. These tunnels host OH$^-$ groups along the c-axis to balance the positive charge of the matrix. The OH$^-$ ions are present in columns perpendicular to the unit cell face at the center of the second tunnel. The oxygen present in the hydroxyl group is located at 0.4 Å out of the plane formed by the calcium ion, and the hydrogen of the hydroxyl group is located at 1 Å, which is almost on the triangle plane of calcium. The dimension of the tunnels endows certain mobility to these ions and consequently allows their circulation along of these tunnels in the direction of O_z axis (Figure 2.15c). However, for a monoclinic HA, all the OH$^-$ of a single column parallel to the c-axis are oriented in the same direction but the OH$^-$ of two adjacent columns are oriented in opposite directions, which will lead to a doubling of the b parameter and then create a monoclinic crystal system [123]. More details on the structure of apatitic compounds will be presented in Chapter 3 of this book.

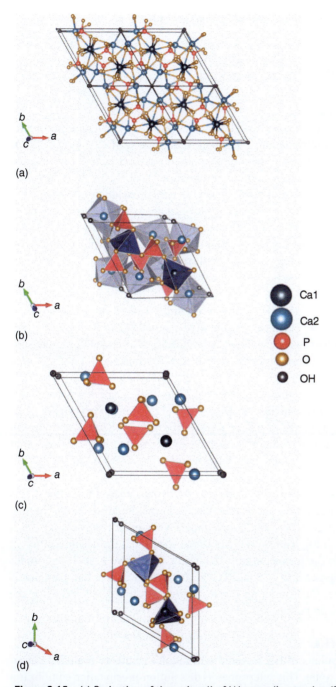

Figure 2.15 (a) Projection of the unit cell of HA according to plan (001) by using Vesta software; (b) projection showing the sequence of octahedral [Ca1O$_6$] and [Ca2O$_6$], and also tetrahedral [PO$_4$] in the HA structure, by using Vesta software; (c) projection showing the arrangement of octahedrons [Ca(1)O$_6$] in the HA structure; (d) projection showing the sequence of octahedral [Ca(1)O$_6$] and tetrahedral [PO$_4$] in the HA structure by using Vesta software. Source: Abderrahim SOLHY.

2.2.3 Physicochemical and Thermal Properties of HA

In this section, the physicochemical and thermal properties of bulk HA that support and promote its catalytic applications will be summarized. A detailed structure and surface properties of HA-based materials have been presented in Chapter 3.

2.2.3.1 Thermal Stability

HA is thermodynamically stable and decomposes into tricalcium phosphate $Ca_3(PO_4)_2$ (TCP) and tetracalcium phosphate monoxide $Ca_4(PO_4)_2O$ (TCPM) at high temperatures. Its thermal stability depends on several factors that include particle size, Ca/P ratio, and water vapor pressure [124]. It has been reported that the decomposition of HA starts first with its dehydroxylation into oxyapatite $(Ca_{10}(PO_4)_6O)$ at 850 °C [124]. However, at 1400 °C it decomposes into tetracalcium phosphate monoxide (TCPM) and α-TCP that is an allotropic variety of TCP stable above 1180 °C. On the other hand, it has been reported that oxyapatite can undergo a small rehydration to give rise to an oxy-hydroxy apatite stable in air at ordinary temperatures, and above 1050 °C, and could decompose into a mixture of TCP and TCPM [125]. Indrani et al. investigated the effects of heat treatment at various temperatures (100, 900, or 1300 °C) on the crystallinity and phase composition of HA synthesized by the chemical precipitation method [126]. It was found that samples treated at 100 or 900 °C exhibited a crystallinity of 48 and 68%, respectively, without an additional phase observed, while samples treated at 1300 °C had a crystallinity of 72% with the presence of dicalcium phosphate $(CaHPO_4)$ and TCP. Conversely, Goldberg et al. studied the effect of Al content on specific surface area, morphology, and thermal stability of HA powder [127]. The incorporation of Al in HA structure in the range of 0.5–1.0 mol% improved its thermal stability. On the other hand, the incorporation of 5–10 mol% of Al resulted in the formation of additional phases such as whitlockite-like structure $(Ca_9Al(PO_4)_7)$ at 900 °C and apatite and α-TCP phases at 1200–1400 °C. At 1400 °C, with the introduction of 20 mol% of Al, an isomorphic Al-substituted whitlockite phase $Ca_9Al(PO_4)_7$ was formed. However, there have been several debates in the literature on the thermal stability of HA materials [128, 129]. For example, Weng et al. reported that HA is stable up to 1250 °C [130], while Wang and Chaki stated that the decomposition of HA at 1100 °C led to the formation of β-TCP [131], and above 1150 °C and under vacuum, HA decomposes into TCP and TCPM. It is worth to note that parameters such as particle size, Ca/P ratio, and water vapor pressure could influence the stability and thermal properties of the HA.

2.2.3.2 Solubility of HA

The catalytic applications of HA material will require solubility evaluation especially its behavior in water. This determines its dissolution-precipitation equilibrium constants. This property is influence by the conditions of studies, HA synthesis method, its stoichiometry, and also its degree of crystallinity [122, 132–135]. This equilibrium is controlled not only by the thermodynamic dissolution-precipitation equilibrium but also by the acidic/basic equilibrium of the phosphate and calcium ions, in addition to the interfacial equilibrium of adsorption–desorption [136–138]. It has been found that HA in aqueous solution is capable of releasing and fixing non-negligible amounts of calcium and phosphate ions before reaching the dissolution-precipitation equilibrium [139–141]. In reality, these ions are responsible to some extent for the dissolution equilibrium, and as a consequence, has become the origin of the controversial constants observed by some authors [142, 143]. This abnormal behavior of HA in water can be explained by the formation of complexes on the surface of the solid [144–146].

2.2.3.3 HA's Surface Functional Groups

The catalytic reactivity of solid materials is related to the properties and their surface structure [147, 148]. Those surfaces are heterogeneous in nature at the atomic scale. This is illustrated by the anarchic arrangement of the surface particles according to the terrace-step-notch model for a rough surface. The atoms are present in plane terraces, steps, and notches. However, punctual defects, vacancies, and adsorbed atoms can be also present. These varied surface sites reach their equilibrium surface concentration through the process of atom transport along the surface. The crystal lattice of the surface of catalytic materials is broken, leaving the atoms on the surface unsaturated from the point of view of their coordination [149, 150]. These structural changes on the surface are unique and associated with the two-dimensional and anisotropic environment to which surface groups must adapt. During a catalytic reaction, the tendency to satisfy the coordination of these ions is the key factor of chemisorption (one of the steps of the catalytic process), which leads to the dissociation of adsorbates, and then their combination, leaving the surface functional groups [151]. It has been reported that HA presents on its surface different sites related to the appearance of phosphate and hydroxyl functional groups [152]. Wu et al. concluded that the surface of HA has amphoteric properties and possesses the following sites: $\equiv$Ca–OH and $\equiv$PO$_2$H, which are susceptible to being protonated or deprotonated to give rise to $\equiv$Ca–OH$_2^+$ and $\equiv$PO$^-$, respectively [153]. In a recent work, single-pulse ^{1}H, ^{31}P, and ^{31}P MAS NMR were used to characterize HA surface sites, which were defined as $\equiv$PO$_x$, $\equiv$PO$_x$H, and $\equiv$PO$_x$H$_2$ (x = 1, 2, or 3). This defines the unprotonated sites $\equiv$O$_3$PO, $\equiv$O$_2$PO^{2-}, and $\equiv$OPO$_3^{2-}$ and a number of possible protonated ($\equiv$O$_2$PO$_2$H, $\equiv$OPO$_3$H$^-$, and $\equiv$PO$_3$H$_2$) sites [154]. Potentiometric acid/base titrations of the HA were also performed to confirm the surface reactions shown in Scheme 2.1 [155]. These results further revealed that the predominant processes on the HA surface are protonation/deprotonation reactions.

$\equiv$CaOH + H$^+$ ⇌ $\equiv$CaOH$_2^+$

$\equiv$OPO$_3$H$_2$ ⇌ $\equiv$OPO$_3$H$^-$ + H$^+$

$\equiv$OPO$_3$H$_2$ + Na$^+$ ⇌ $\equiv$OPO$_3$Na$^-$ + 2H$^+$

Scheme 2.1 HA's surface reactions during acid–base titration. Source: Bengtsson et al. [155]. Reproduced with permission of Elsevier.

In another study, other types of sites were determined using simultaneous measurements of ^{45}Ca and ^{32}P radioactivity [156], with the protonation and deprotonation equilibria presented in Scheme 2.2.

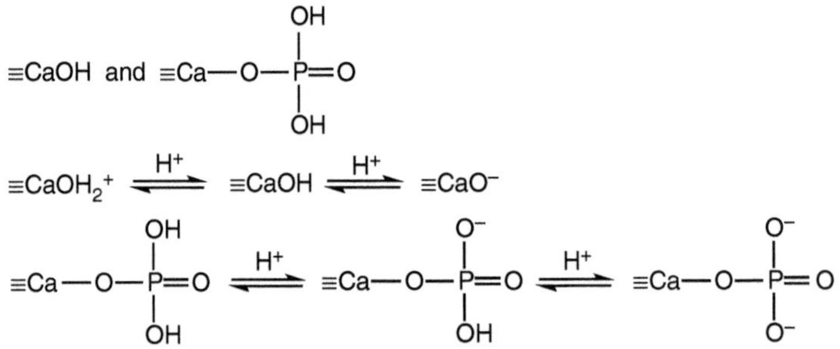

$$\equiv CaOH \text{ and } \equiv Ca-O-\overset{\overset{\displaystyle OH}{|}}{\underset{\underset{\displaystyle OH}{|}}{P}}=O$$

$$\equiv CaOH_2^+ \underset{}{\overset{H^+}{\rightleftharpoons}} \equiv CaOH \underset{}{\overset{H^+}{\rightleftharpoons}} \equiv CaO^-$$

$$\equiv Ca-O-\overset{\overset{\displaystyle OH}{|}}{\underset{\underset{\displaystyle OH}{|}}{P}}=O \underset{}{\overset{H^+}{\rightleftharpoons}} \equiv Ca-O-\overset{\overset{\displaystyle O^-}{|}}{\underset{\underset{\displaystyle OH}{|}}{P}}=O \underset{}{\overset{H^+}{\rightleftharpoons}} \equiv Ca-O-\overset{\overset{\displaystyle O^-}{|}}{\underset{\underset{\displaystyle O^-}{|}}{P}}=O$$

Scheme 2.2 HA's surface reactions. Source: Kukura et al. [156]. Reproduced with permission of American Chemical Society.

The authors of these studies reported that the phosphate groups are attached to a single surface calcium atom to ensure the electroneutrality of the interface [156]. However, most of the models developed to quantify the calcium and phosphorus sites on the HA surface consider only the contribution of the three planes (100), (011), and (101) [156, 157]. The predominance plan (001) that is atypical for most hexagonal close-packed symmetries, given that the growth along the central axis of (001) plane is usually the most favorable, that is, primarily along the c-axis. As an illustration, the model developed by Kukura et al. proposes 3.33 calcium sites and 2 phosphorus sites available per plane (100) and per unit cell [156]. On the other hand, the model developed by Misra et al. mentions three easily accessible calcium sites per plane (100) and per unit cell [157].

2.2.3.4 Non-stoichiometric Calcium-Deficient or Calcium-Rich Hydroxyapatites

HA stoichiometry is defined by the Ca/P molar ratio, which varies from 1.50 to 1.67 [158]. This ratio can be expressed as Ca/P < 1.67 for a calcium-deficient HA (sub-stoichiometry) or Ca/P = 1.67 for stoichiometric HA. HA-based materials can have the Ca/P molar ratio higher than 1.67 when other calcium-containing compounds such as CaO, Ca(OH)$_2$, and CaCO$_3$ coexist. The stoichiometry is extremely dependent on the synthesis conditions and also on the starting materials [159–161]. As an illustration, pH, synthesis temperature, and maturation time were found as key parameters to control the HA's stoichiometry [162, 163]. Moreover, Tsuchida et al. reported that the final molar ratio of Ca/P closely increases with pH [164]. They have argued this observation by the reason that PO$_4^{3-}$ ligands are more stable than HPO$_4^{2-}$ and H$_2$PO^{4-} at very high pH. In another study, it was found that an increase in the synthesis temperature causes an increase in the Ca/P molar ratio, even when the pH of the medium and the initial Ca/P ratio were both fixed [165]. This suggests that careful control of the synthesis temperature and pH is necessary to control HA's stoichiometry [166]. Nevertheless, it is difficult to propose unique and

Table 2.1 Proposed formula for calcium-deficient HA's structures.

Formula		Reference
$Ca_{10-x}(HPO_4)_{2x}(PO_4)_{6-2x}(OH)_2$	$0 \leq x \leq \sim 2$	[170, 171]
$Ca_{10-x}(HPO_4)_x(PO_4)_{6-x}(OH)_{2-x}$	$0 \leq x \leq 2$	[172, 173]
	$0 \leq x \leq 1$	[167, 174]
$Ca_{10-x-y}(HPO_4, CO_3)_x(PO_4)_{6-x}(OH)_{2-x-2y}$	$0 \leq x \leq 2, y \leq 1 - x/2$	[175]
$Ca_{10-x-y}(HPO_4)_x(PO_4)_{6-x}(OH)_{2-x-2y}$	$0 \leq x \leq 2, y \leq 1 - x/2$	
$Ca_{9-x}(HPO_4)_{1+2x}(PO_4)_{5-2x}(OH)_2$	$0 \leq x \leq 1$	[176, 177]

simple justification for the origin of the nonstoichiometry of HA. Studies have shown that this phenomenon can be due to: (i) calcium deficiency [167, 168], (ii) the significant protonation of the phosphate groups on the surface [169], (iii) the substitution of PO_4^{3-} groups by other ions, and (iv) the coexistence of other secondary crystalline or amorphous phases besides HA. For example, octacalcium phosphate phase [$Ca_8H_2(PO_4)_6 \cdot 5H_2O$], having a lower Ca/P molar ratio (1.33) than that of the stoichiometric HA, can coexist and can be found in the crystalline structure of the HA under certain synthesis conditions, which would decrease the overall Ca/P molar ratio. Nonstoichiometry, however, remains most often associated with calcium deficiency, related to a problem of solubility of the calcium precursor as a function of the pH of synthesis. It should be noted that the calcium-deficient HA electroneutrality is maintained by the presence of two protons in the apatite structure for each missing calcium. The chemical composition of the Ca-deficient HA was expressed by a general formula $Ca_{10-z}(HPO_4)_z(PO_4)_{6-z}(OH)_{2-z} \cdot nH_2O$ [166]. However, several formulas for the composition of calcium-deficient HA have been proposed to interpret this phenomenon while keeping the neutrality of the structure (Table 2.1).

The decrease in the Ca/P ratio associated with the introduction of cationic vacancies will be compensated by the replacement of PO_4^{3-} ions by HPO_4^{2-} ions and the decrease of the OH^- content in the core of the material. In addition, as stated earlier, the origin of calcium-rich HA (over-stoichiometry: $1.67 < Ca/P < 1.83$) is particularly due to an excess of calcium in the powder in the form of $Ca(OH)_2$, CaO, and $CaCO_3$. [122, 178]. It has been shown that a ratio Ca/P = 1.71 would formally correspond to a CaO mass content of 1.5 wt% [165]. Obtaining a calcium-rich HA can also be explained by the decrease in the number of PO_4^{3-} relative to calcium due to its substitution by carbonates (CO_3^{2-}) or eventually other anions, which would lead to an increase in the Ca/P ratio [179, 180]. Milev et al. showed that an HA with a Ca/P = 1.82 ratio had a carbonate level almost twice as high as that of a stoichiometric HA, even after treatment at 500 °C [180].

2.2.4 Substitutions in the Structure of HA

One of the characteristics of HA is the ability of its specific structure to accept a large number of substituents by forming solid solutions [181, 182]. Thus, the bivalent Ca^{2+}

cations in HA structure can be substituted by other bivalent cations: Cd^{2+}, Pb^{2+}, Sr^{2+}, Ba^{2+}, Fe^{2+}, Zn^{2+}, etc., but also by monovalent cations: Na^+, K^+, Li^+; trivalent: La^{3+}, Eu^{3+}, Ga^{3+}, Al^{3+}, etc.; or vacancies (□) [181–187]. The anionic groups PO_4^{3-} can also be substituted by trivalent anions: VO_4^{3-} AsO_4^{3-}, MnO_4^{3-} [188–191]; bivalent anions: CO_3^{2-}, SO_4^{2-}, HPO_4^{2-} [192]; or tetravalent anions: SiO_4^{4-}, GeO_4^{4-} [193]. OH^- anions can also be replaced by monovalent anionic ions: F^- [122, 194], Cl^-, I^-, Br^- [122]; bivalent anions: CO_3^{2-}, O^{2-}, S^{2-}; or vacancies (□) [195]. The substitution of Ca^{2+}, PO_4^{3-}, and OH^- ions by ions of identical valences can be total, with the exception of Mn^{2+} and Mg^{2+} ions [183], while the incorporation of ions of different valences is limited and requires charge compensation in order to preserve the electroneutrality of the structure. This is obtained by coupled substitutions or by the creation of cationic and/or anionic vacancies. It should be mentioned that the XO_4^{3-} sites are always saturated at six ions per mesh regardless of the differences in stoichiometry caused by the multiple substitutions. The substitution of phosphate anions by SiO_4^{4-} anions is accompanied by a decrease in the hydroxyl group content in order to obtain an HA of the general formula: $Ca_{10}(PO_4)_{6-x}(SiO_4)_x(OH)_{2-x}$, while the substitution of phosphate anions by carbonate or sulfate anions leads to a decrease in the amount of Ca^{2+} cations [193]. This type of substitution enables the chemical properties of HA to be modified by bringing new properties (e.g. redox properties, density and strength of acid-basic sites) and also to modify its crystallinity. The substitution of OH^- or PO_4^{3-} ions by CO_3^{2-} ions leads to carbonated HA of A-type or B-type, respectively. A-type carbonated HA are non-stoichiometric and lacunar structures. The structure can be described by the following formula [122]: $Ca_{10-x+y}\square_{x-y}(PO_4)_{6-x}(CO_3)_x(OH)_{2-x+2y}\square_{x-2y}$ (with $0 \leq x \leq 2$, $0 \leq 2y \leq x$, and □: vacancy). The characterization of carbonated HA by Fourier transform infrared (FTIR) spectroscopy is largely detailed in the literature [196, 197]. The characteristic FTIR bands of A-type carbonated HA are similar to those of the dental tissue structure that is found at 1548, 879 cm^{-1} [122]. Nevertheless, A-type substitution is relatively limited in HA synthesized by coprecipitation route [122]. B-type carbonated HA, having the general formula: $Ca_{10-x}(PO_4)_{6-x}(CO_3)_x(OH)_{2-x}$ with $0 \leq x \leq 2$, is characterized by FTIR bands at 873, 1412, and 1465 cm^{-1} [122]. Barralet et al. showed that HA synthesis by coprecipitation with a low carbonate content (<4 wt%) gives rise to A-type carbonated HA, and that a higher carbonatation (>4 wt%) would mainly lead to the formation of B-type carbonated HA [197]. Cheng et al. reported that the pH of HA synthesis influences the nature of the carbonatation sites. For a pH of 5.94, the substitution degree to form B-type carbonated HA was maximal. Then, the formation of A-type carbonated HA closely increases with the increase of pH [198]. On the other hand, Driessens et al. confirmed by FTIR the existence of both A-type and B-type carbonated HA, synthesized via dry mechanosynthesis route [199]. The FTIR bands were found at 1547, 1468, 1415, 880, and 874 cm^{-1}, which translates into simultaneous substitution of the PO_4^{3-} and OH^- anions by CO_3^{2-} anions. It should be noted that the A-type substitution causes an elongation along the a-axis and a contraction along the c-axis, whereas B-type substitution induces the opposite behavior. In addition, the B-type substitution leads to a reduction in the crystallinity and HA particle size [200].

2.2.5 Modification and Functionalization of the HA's Surface

The functionalization of the surface of HA to obtain an organofunctionalized HA has received great interest in the development of bio-composites with exceptional properties [201, 202]. In this context, the precipitation of a functionalized HA was achieved by using 2-amino-2-ethyl phosphate ($H_2PO_4(CH_2)_2NH_2$) and calcium nitrate, leading to a poorly crystallized HA with a molar ratio of Ca/P = 1.33 [203]. Delpech et al. were able to graft hydroxyethylmethacrylate (HEMA) and methylmethacrylate (MMA) onto HA surface in order to obtain a functionalized HA after the radical copolymerization of the HA-HEMA or HA-MMA compound [204]. This functionalization led to hydrophilic or hydrophobic materials according to the nature of the monomer with the general formula: $Ca_{8.93}(HPO_4)_{1.56}(PO_4)_{4.44}(PO_4R)_{0.63}(OH)_{0.14}$ and subsequently improved the mechanical properties and biocompatibility of the surgical cement. In addition, Kandori et al. were able to graft oleyphosphonate (OP) onto the HA surface during its synthesis by the coprecipitation of the mixture of $Ca(OH)_2$ and H_3PO_4 in the presence of dissodium oleyphosphate (DSOP: $C_{18}H_{35}Na_2O_4P$) [205]. These authors concluded that the particles were monocrystallized with a rod-like morphology and variable sizes as a function of the DSOP concentration. The hydrophobicity of HA-OP particles was improved with increasing DSOP concentration. This high surface hydrophobicity was confirmed by water adsorption experiments. HA-OP adsorbed much less water than HA particles produced without DSOP. The Ca/P molar ratios of the HA-OP ranged from 1.55 to 1.61 and that of HA produced without DSOP reached 1.57. Furthermore, Tanaka et al. [206] have shown that pyridine, n-butylamine, and acetic acid can irreversibly adsorb onto HA via the formation of hydrogen bonds with P–OH groups of the HA's surface. In another work, the same authors studied the surface modification of HA with alkyl-phosphates such as monohexyl- and monodecyl-phosphates in acetone at 25 °C [207] and in water-acetone solutions [208, 209]. They concluded that the number and type of surface P-OH species were altered by surface modification.

HA used as a catalyst or catalyst support is usually treated at high temperature; this treatment can influence the surface structure of HA, in particular the surface P–OH groups and their interaction with reagent molecules. Tanaka et al. [210] were able to modify the surface of a nonstoichiometric HA (Ca/P = 1.62) by the action of hexamethyldisilazane [$(CH_3)_3Si]_2NH$ (HMDS). $Si(CH)_3$ groups were observed on the surface of the HA resulting from the reaction of the P–OH groups on the surface with HMDS, so that the HA becomes hydrophobic. Hence, heat treatment of the modified HA at 500 °C caused the $-CH_3$ groups to disappear and Si–OH groups to be formed on their surface. The authors also reported a second functionalization of this HA after the heat treatment at 500 °C, carried out by grafting HMDS on the P–OH groups but also on the Si–OH groups formed on the surface after the first functionalization. Thus, the specific surface area of the HA decreases according to the amount of HMDS grafted. A stoichiometric HA (Ca/P = 1.67) was also functionalized via the silanization reaction at the surface using three trimethoxylesilanes: $(CH_3O)_3SiR$ with R = $(CH_2)_3NH_2$, $(CH_2)_3NH(CH_2)_2NH_2$, and

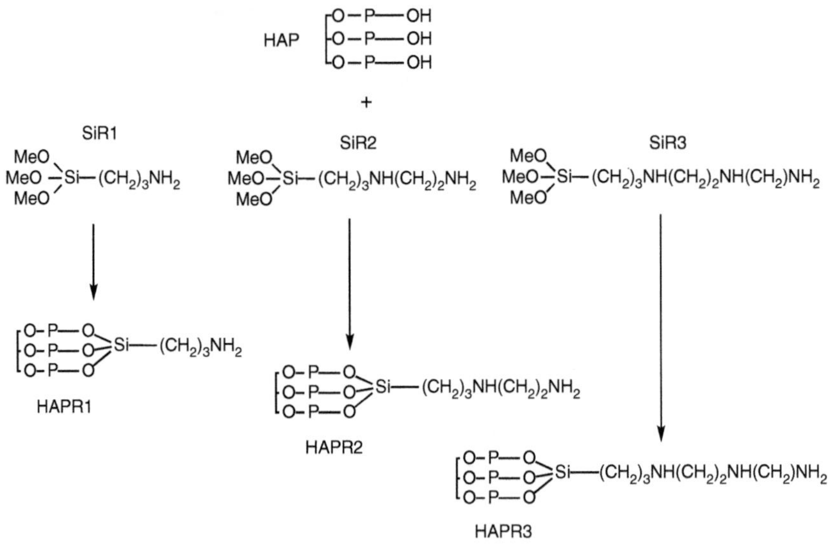

Figure 2.16 Illustration of the elaboration of organofunctionalized HA. Source: Da Silva et al. [211]. Reproduced with permission of Elsevier.

$(CH_2)_3NH(CH_2)_2NH(CH_2)_2NH_2$ (Figure 2.16) [211]. These authors confirmed the silanization reaction on the surface of the HA with three organosilanes by using elemental analysis, thermogravimetric analysis (TGA), FTIR, and nuclear magnetic resonance (NMR) (^{13}C and ^{31}P CP/MAS). These results showed that an increase in organosilane chain length induced an increase in the amount of molecule on the surface (0.75 mmol g^{-1} for HA-R1, 4.71 mmol g^{-1} for HA-R2, and 7.45 mmol g^{-1} for HA-R3). These organofunctionalized phosphates showed a significant ability to adsorb Co^{2+} and Cu^{2+} cations. It was confirmed that an increase in amine functions on the chain displayed a very noticeable effect on this process.

In 2014, Russo et al. studied the silanization of the surface of carbonated HA by mixtures of organosilanes in order to optimize the potential for immobilization of bioactive molecules onto the surface of HA [212]. In this study, it was concluded that an organosilane mixture such as 3-aminopropyltriethoxylesilane (APTES) and tetraethoxysilane (TEOS) is more effective than APTES alone. Grafting of APTES should allow homogeneous coupling between the OH$^-$ groups of the substrate surface and the silanol groups of APTES. However, experimental results show that the formation of Si–O-type bonds between the silanol groups of APTES and the OH$^-$ groups of the substrate is accompanied by physisorption, which may also involve molecular inversion of APTES. The silanization of the surface of HA with APTES has also been used for the grafting of poly(N-isopropylacrylamide) (PNIPAM) [213] and (g-benzyl-L-glutamate N-carboxyanhydride) (BLG-NCA) [214]. This grafting was easy owing to the surface amine functions generated by the functionalization with APTES. Jacobsen et al. used (3-glycidyloxypropyl) trimethoxylesilane (GPTMS) to functionalize a calcium phosphate [215]. The epoxy-like terminal function of this organosilane reacted with cyclodextrin to form an ether-like bond with this cage molecule used for continuous release of active substance. Based on the functionalization of silica, several works have been developed to functionalize the surface of HA via the silanization reaction in order to graft other molecules covalently [216–219].

2.3 A Concise Overview on Synthesis and Characterization of HA-Based Catalysts

HA and HA-based materials are arousing great interest for applications in heterogeneous catalysis, either as an active phase or as a support [12, 14, 220]. As previously stated, HA-based materials exhibit tunable acido-basicity, which allows their use as acidic, basic, or bifunctional acido-basic catalysts [12, 14, 220]. Moreover, supported catalysts are constituted of a solid support that serves to disperse an active phase usually in the form of nanoparticles or sub-nanoparticles [221, 222]. The use of the support not only improves the efficiency of the active phase, in particular for noble metals, but also contributes to the performance of the catalyst via its physicochemical and textural properties such as acid-basic sites, specific surface area, porosity, pore size, thermal stability, mechanical stability, and metal-support interaction. [223]. For example, an adequate metal-support interaction inhibits or limits aggregation phenomena of the active phase and improves catalyst stability [224]. Its use as solid catalyst also facilitates separation, recuperation, and recycling [225]. The synthesis of HA-based catalysts consists of several steps, each of them can modify the properties of the final catalyst [12, 14, 220]. As stated in Chapter 1, the main advantages of using HA as a catalyst support are: (i) a possible high specific surface area and tailored porosity, (ii) a high thermal stability, (iii) a modular acidity-basicity via the control of the Ca/P molar ratio, and (iv) particularly a large number of possibilities of substitution of Ca^{2+}, PO_4^{3-}, and OH^- ions by other ions leading to new modifiable properties [226–228]. Many studies have been devoted to the development of HA-supported catalysts [12, 14, 220]. The most widely used method for producing HA-supported catalysts is the impregnation [12, 14, 220]. There are three essential steps in this method: (i) impregnation of HA with a solution of metal precursor (nitrates, carbonates, sulfates, organometallic complexes, etc.), (ii) drying of the impregnated solid, and sometimes (iii) calcination before the activation of the active phase by hydrogen reduction [229–231]. Solhy et al. were able to boost the performance of HA by doping it with sodium. After impregnating HA with sodium nitrate (with Na/HAP ratio = 1/2 w/w) and drying the resulting mixture, the material was calcined at 900 °C under air for 1 hour [230]. However, this modification of HA, by doping, caused the decrease of the specific surface area from 38 m^2 g^{-1} for HA to 2 m^2 g^{-1} for Na-HA. This material was used in this way to catalyze the synthesis of α-hydroxyphosphonates by condensation of carbonyl compounds onto dialkyl phosphites under solvent-free conditions. The control of the growth of semiconductor oxide (e.g. TiO_2@HA) simultaneously during HA synthesis or of semiconductor oxide in the HA pores is also used as a method for the synthesis of HA-based photocatalysts [232–234]. During impregnation, the Ti precursor interacts with the HA (support) by different types of bonds (e.g. van der Waals, covalent, or ionic), favoring adsorption onto the surface of the HA [235–245].

Impregnation sometimes results from anionic [246, 247] or cationic [248–250] exchange operations, after contacting the powder with a solution of a metal salt. After drying, it should be noted that ionic compounds strongly linked to the surface of the HA are present, with a high binding energy. Dasireddy et al. reported the impregnation of HA by V_2O_5 with loadings of 2.5 and 10 wt% [247]. These catalysts were used for oxidative dehydrogenation reactions of n-octane in a continuous flow fixed bed reactor. The authors concluded that selectivity toward desired products was dependent on the phase composition of the catalyst and the *n*-octane to oxygen molar ratios. Another point that was made in this work was that for 2.5 wt% loadings, vanadium is in the form of V_2O_5 with a homogeneous distribution of vanadium species on the HA surface, while an additional pyrovanadate phase exists for 10 wt% loadings. Mori et al. prepared a catalyst such as HA-bound lanthanum complex (LaHA) via cation-exchange method. An equimolar substitution of La^{3+} for Ca^{2+} occurred by the treatment of stoichiometric HA with an aqueous solution of $La(OTf)_3$, yielding a monomeric LaHA [248]. It has been reported in this work that La K-edge X-ray absorption fine structure (XAFS) proved that a monomeric La^{3+} phosphate complex was generated on HA's surface. The authors used this catalyst for the Michael reaction of 1,3-dicarbonyls with enones under aqueous or solvent-free conditions. In another work, the authors exploited the ability of HA to exchange cations or anions to insert zinc into the HA framework using $Zn(NO_3)_2.6H_2O$ (Zn content: 0.3 mmol g^{-1}) and stoichiometric HA [249]. The authors used X-ray photoelectron spectroscopy (XPS) to have the binding energy values of the ZnHA and ZnO, which were 1021.5 and 1021.4 eV for Zn 2p3/2, respectively. It was concluded that the edge position of the Zn K-edge X-ray absorption near edge structure (XANES) spectra of the ZnHA was also comparable with that of ZnO. The authors reported that the Zn species in the ZnHA exist in a tetrahedral geometry having a $+2$ oxidation state. The lack of peaks assignable to the Zn–O–Zn bond in the Fourier transform (FT) of the k3-weighted Zn K-edge extended X-ray absorption fine structure (EXAFS) spectrum, which was detected for ZnO at around 2.8 Å, showed that the Zn species is monomeric [249]. Thus, it revealed the existence of a Zn—O bond with an interatomic distance and coordination number of 1.88 Å and 4.0, respectively. This indicates that the interaction between Zn and O atoms in ZnHA might be stronger due to the structural contraction around zinc. This can be translated into the formation of a monomeric ZnII phosphate complex surrounded by four oxygen atoms in tetrahedral coordination on the HA surface.

The same research team was able to design another type of catalyst through the ability of HA to exchange Ca^{2+} cations [250]. So, $RuCl_3.nH2O$ was used to impregnate HA and gives rise to RuHA. Based on ICP, XPS, energy-dispersive X-ray spectroscopy (EDX), and Ru—K edge XAFS analyses, the authors concluded the appearance of an equimolar substitution of Ru^{3+} for Ca^{2+} to generate a monomeric Ru^{3+} species surrounded by four oxygen atoms and one chlorine atom (Figure 2.17a). The cationic RuHA-(I) and -(II) were prepared by treatment of the RuHA, at room temperature under argon atmosphere, with an aqueous solution of AgX (1.1 equiv.

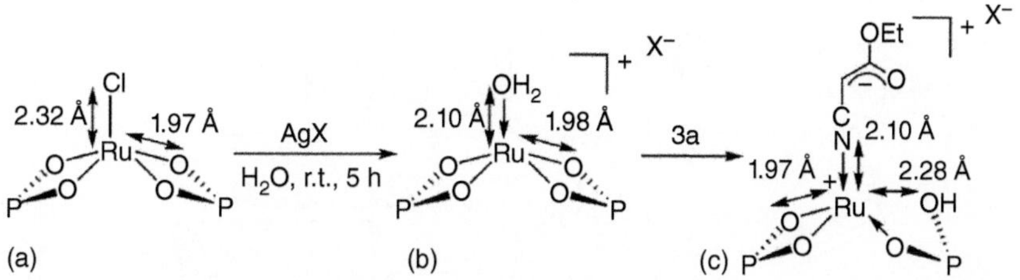

Figure 2.17 Proposed structures of (a) RuHAP, (b) cationic RuHAP-(I) and -(II) (I, X = SbF6; II, X = OTf), and (c) Ru-enolate intermediate (X = OTf). Source: Mori et al. [250]. Reproduced with permission of American Chemical Society.

of Ru; X), SbF^{6-}, and TfO$^-$, respectively. In addition, XPS analysis of the cationic RuHAs confirmed the absence of chlorine. The authors reported that the Ru K-edge XANES spectrum was quite similar to that of the parent RuHA, which means Ru species exist in the 3$^+$ oxidation state. In the FT of k^3-weighted Ru K-edge EXAFS, there were no peaks around 3.5 Å due to the presence of contiguous Ru sites. The inverse FTs of the cationic RuHAs were well fitted by replacing the Ru—Cl bond (2.32 Å) in the RuHA with a Ru—O bond (2.10 Å) assignable to a weakly coordinated aqua ligand. Thereafter, the authors found that a well-defined cationic Ru phosphate complex can be created on the HA's surface, as illustrated in Figure 2.17b [250].

In some cases, impregnation followed by calcination at high temperature (500–800 °C) under air led to changes in catalytic performance (e.g. Na@HA and Li@HA) [229–231]. During the calcination process, various transformations could take place: (i) thermal decomposition of the precursors, resulting in the release of volatile products [251]; and (ii) modification of the structure and texture (specific surface area, pore volume) [229–231]. Figure 2.18 shows in a non-exhaustive way some selected catalytic systems developed from HA support.

For HA-supported catalysts, the maximization of the metal dispersion is strongly sought [12–14]. To date, some first works have been successful in the development of single atom catalysts using HA support [249, 252, 253]. Akri et al. reported the development of a catalyst based on dispersed Ni single atoms stabilized by interaction with Ce-substituted HA [252]. These are the highly active and coke-resistant catalytic sites for dry methane reforming. The HA and Ce-substituted HA were synthesized by coprecipitation method by using the following reagents: calcium nitrate (Ca(NO$_3$)$_2$.4H$_2$O), ammonium dihydrogen phosphate ((NH$_4$)$_2$HPO$_4$), and cerium nitrate (Ce(NO$_3$)$_2$.6H$_2$O). The dried samples were subsequently calcined at 400 °C (ramp rate 10 °C min^{-1}) in muffle furnace for 4 hours, for having HA and Ce-doped HA. The deposition of Ni on HA and Ce-doped HA was achieved by varying the Ni loadings (0.5, 1, 2, and 10 wt%) on the previous supports by strong electrostatic adsorption (conventional impregnation) at room temperature. The authors used nickel nitrate (Ni(NO$_3$)$_2$·6H$_2$O), which was dissolved in 50 ml of water at pH adjusted to 10 with the addition of ammonia. After impregnation and drying, the authors calcined the resulting solids at 500 °C (ramp rate 10 °C min^{-1}) for 4 hours. Through the analyses performed, catalytic tests achieved, and also

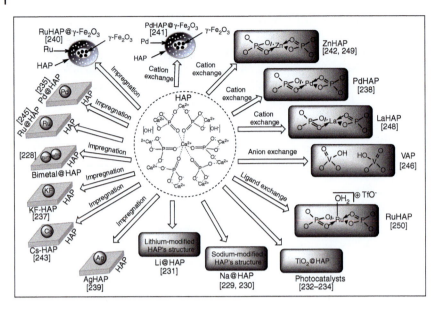

Figure 2.18 Selected HA-based catalysts developed over the past 20 years. Source: Abderrahim SOLHY.

computational studies using density functional theory (DFT) methods, the authors reported that the isolated Ni atoms, and well dispersed, are inherently resistant to coke due to their unique ability to activate only the first C—H bond in CH_4, thus avoiding the deep decomposition of methane into carbon. However, dispersion is not the only constraint in the HA-based catalysts development process. Depending on the nature of the reaction under consideration, the localization of the active sites is also an important parameter. In addition, shaping HA-based catalysts by extrusion or other technologies for eventual scale-up phase can be also a challenge. Many analysis techniques were used to characterize HA-based catalysts to get an insight about its structure (X-ray diffraction [XRD], FTIR, Raman spectroscopy, NMR), its texture (nitrogen adsorption desorption isotherms analysis, mercury porosimetry), its surface properties (temperature programmed desorption [TPD], temperature programmed reduction [TPR], temperature programmed oxidation [TPO], and XPS), and its morphology (SEM, TEM, HR-TEM). The contribution of XANES and EXAFS spectroscopies in the analysis of HA-based catalysts is undeniable [12]. These two spectroscopic techniques allow the determination of the coordination and length of cation-ligand (Ca–PO_4^{3-}) bonds [12]. Moreover, these techniques demonstrate that, in some HA-based catalysts, the metals are not randomly arranged on the HA surface [12]. However, obtaining high-performance HA-based catalysts is a matter for the know-how of each laboratory. It has always been difficult to establish the influence of the steps in the process of elaboration of this type of catalysts and to understand the importance of the choice: of the method of HA synthesis, of the metal precursors, of the solvent, and of the post-synthetic stage, on the activity and selectivity of HA-based catalysts.

2.4 Summary and Conclusions

This chapter provides an original contribution to the synthesis and characterization of HA-based materials for application in the catalysis. The main synthesis methods and their impact on the intrinsic properties of the HA materials have been discussed. The characteristics of HA such as thermal stability, solubility, surface functional groups, stoichiometry, structure substitution, and surface functionalization have been reviewed. Particular attention should be paid on its exceptional capacity of substitution by many elements, either during its synthesis or during catalyst preparation from HA materials. Tunable acidity/basicity, high thermal stability, and cation exchange capacity are the main strengths of HA. Particularly, ion exchange capacity of HA offers the possibility to disperse and stabilize isolated metal atoms, which can lead to unprecedented catalytic properties and so open new perspectives in the heterogeneous catalysis. Also, the functionalization of the surface of the HA by grafting functional molecules is controllable, which gives rise to hybrid organic/inorganic composites (organo-HA). Unfortunately, no catalytic applications of these hybrid composites have been reported yet, but it can be the subject of future work, taking into account the fact that HA is still considered a new family of catalytic materials. This conclusion would encourage us to prepare, in the near future, new hybrid materials (organo-HA) for potential applications in catalysis (e.g. heterogeneous asymmetric catalysis).

References

1 Owen, C.L., Nash, G.R., Hadler, K. et al. (2019). Apatite enrichment by rare earth elements: a review of the effects of surface properties. *Advances in Colloid and Interface Science* 265: 14–28.
2 Cacciotti, I. (2016). Cationic and anionic substitutions in hydroxyapatite. *Handbook of Bioceramics and Biocomposites* 145–211.
3 Pawłowski, L. (2018). Chapter 7 – synthesis, properties and applications of hydroxyapatite. In: *Industrial Chemistry of Oxides for Emerging Applications* (ed. L. Pawłowski and P. Blanchart). John Wiley & Sons Ltd.
4 Sridhar, T.M., Mudali, U.K., and Subbaiyan, M. (2003). Preparation and characterisation of electrophoretically deposited hydroxyapatite coatings on type 316L stainless steel. *Corrosion Science* 45: 237–252.
5 Zhao, B., Hu, H., Mandal, S.K., and Haddon, R.C. (2005). A bone mimic based on the self-assembly of hydroxyapatite on chemically functionalized single-walled carbon nanotubes. *Chemistry of Materials* 17: 3235–3241.
6 Lu, Y.-P., Li, S.-T., Zhu, R.-F. et al. (2003). Formation of ultrafine particles in heat treated plasma-sprayed hydroxyapatite coatings. *Surface and Coatings Technology* 165: 65–70.
7 Karachalios, T. (2014). *Bone-Implant Interface in Orthopedic Surgery: Basic Science to Clinical*. Springer Science & Business Media.

8 Epinette, J.-A. and Manley, M.T. (2013). *Fifteen Years of Clinical Experience with Hydroxyapatite Coatings in Joint Arthroplasty*. Springer Science & Business Media.

9 Wagner, D.E., Eisenmann, K.M., Nestor-Kalinoski, A.L., and Bhaduri, S.B. (2013). A microwave-assisted solution combustion synthesis to produce europium-doped calcium phosphate nanowhiskers for bioimaging applications. *Acta Biomaterialia* 9: 8422–8432.

10 Wei, X. and Yates, M.Z. (2012). Yttrium-doped hydroxyapatite membranes with high proton conductivity. *Chemistry of Materials* 24: 1738–1743.

11 Watanabe, Y., Ikoma, T., Suetsugu, Y. et al. (2006). The densification of zeolite/apatite composites using a pulse electric current sintering method: a long-term assurance material for the disposal of radioactive waste. *Journal of the European Ceramic Society* 26: 481–486.

12 Kaneda, K. and Mizugaki, T. (2009). Development of concerto metal catalysts using apatite compounds for green organic syntheses. *Energy & Environmental Science* 2: 655–673.

13 Tsuchida, T., Sakuma, S. (2011). U.S. Patent 8080695 B2, 2011.

14 Fihri, A., Len, C., Varma, R.S., and Solhy, A. (2017). Hydroxyapatite: a review of syntheses, structure and applications in heterogeneous catalysis. *Coordination Chemistry Reviews* 347: 48–76.

15 Ibrahim, M., Labaki, M., Giraudon, J.-M. et al. (2019). Hydroxyapatite, a multifunctional material for air, water and soil pollution control: a review. *Journal of Hazardous Materials* 383: 121139.

16 Munirathinam, R., Minh, D.P., and Nzihou, A. (2018). Effect of the support and its surface modifications in cobalt-based Fischer-Tropsch synthesis. *Industrial & Engineering Chemistry Research* 57 (48): 16137–16161.

17 Xu, Z., Huang, G., Yan, Z. et al. (2019). Hydroxyapatite-supported low-content Pt catalysts for efficient removal of formaldehyde at room temperature. *ACS Omega* 4 (26): 21998–22007.

18 Mohd Pu'ad, N.A.S., Abdul Haq, R.H., Mohd Noh, H. et al. (2020). Synthesis method of hydroxyapatite: a review. *Materials Today* 29: 233–239.

19 Santos, M.H., de Oliveira, M., de Freitas Souza, L.P. et al. (2004). Synthesis control and characterization of hydroxyapatite prepared by wet precipitation process. *Materials Research* 7: 625–630.

20 Pham Minh, D., Lyczko, N., Sebei, H. et al. (2012). Synthesis of calcium hydroxyapatite from calcium carbonate and different orthophosphate sources: a comparative study. *Materials Science and Engineering: B* 177: 1080–1089.

21 Solhy, A., Tahir, R., Sebti, S. et al. (2010). Efficient synthesis of chalcone derivatives catalyzed by re-usable hydroxyapatite. *Applied Catalysis A: General* 374: 189–193.

22 Wu, Y.-S., Lee, Y.-H., and Chang, H.-C. (2009). Preparation and characteristics of nanosized carbonated apatite by urea addition with coprecipitation method. *Materials Science and Engineering: C* 29: 237–241.

23 Liu, D.-M., Troczynski, T., and Tseng, W.J. (2001). Water-based sol-gel synthesis of hydroxyapatite: process development. *Biomaterials* 22: 1721–1730.

24 Jillavenkatesa, A. and Condrate, R.A. Sr., (1998). Sol-gel processing of hydroxyapatite. *Journal of Materials Science* 33: 4111–4119.

25 Ramanan, S.R. and Venkatesh, R. (2004). A study of hydroxyapatite fibers prepared via sol-gel route. *Materials Letters* 58: 3320–3323.

26 Chen, J., Wang, Y., Chen, X. et al. (2011). A simple sol-gel technique for synthesis of nanostructured hydroxyapatite, tricalcium phosphate and biphasic powders. *Materials Letters* 65 (12): 1923–1926.

27 Velu, G. and Gopal, B. (2009). Preparation of nanohydroxyapatite by a sol-gel method using alginic acid as a complexing agent. *Journal of the American Ceramic Society* 92 (10): 2207–2211.

28 Eshtiagh-Hosseini, H., Housaindokht, M.R., and Chahkandi, M. (2007). Effects of parameters of sol-gel process on the phase evolution of sol-gel-derived hydroxyapatite. *Materials Chemistry and Physics* 106 (2–3): 310–316.

29 Hsieh, M.-F., Perng, L.-H., Chin, T.-S. et al. (2001). Phase purity of sol-gel-derived hydroxyapatite ceramic. *Biomaterials* 22 (19): 2601–2607.

30 Rajabi-Zamani, A.H., Behnamghader, A., and Kazemzadeh, A. (2008). Synthesis of nanocrystalline carbonated hydroxyapatite powder via nonalkoxide sol-gel method. *Materials Science and Engineering: C* 28 (8): 1326–1329.

31 Feng, W., Mu-Sen, L., Yu-Peng, L. et al. (2005). A simple sol-gel technique for preparing hydroxyapatite nanopowders. *Materials Letters* 59 (8–9): 916–919.

32 Kim, I.-S. and Kumta, P.N. (2004). Sol-gel synthesis and characterization of nanostructured hydroxyapatite powder. *Materials Science and Engineering: B* 111 (2–3): 232–236.

33 Shum, H.C., Bandyopadhyay, A., Bose, S. et al. (2009). Double emulsion droplets as microreactors for synthesis of mesoporous hydroxyapatite. *Chemistry of Materials* 21 (22): 5548–5555.

34 Zhou, W.Y., Wang, M., Cheung, W.L. et al. (2008). Synthesis of carbonated hydroxyapatite nanospheres through nanoemulsion. *Journal of Materials Science: Materials in Medicine* 19 (1): 103–110.

35 Ethirajan, A., Ziener, U., Chuvilin, A. et al. (2008). Biomimetic hydroxyapatite crystallization in gelatin nanoparticles synthesized using a miniemulsion process. *Advanced Functional Materials* 18 (15): 2221–2227.

36 Saha, S.K., Banerjee, A., Banerjee, S. et al. (2009). Synthesis of nanocrystalline hydroxyapatite using surfactant template systems: role of templates in controlling morphology. *Materials Science and Engineering: C* 29 (7): 2294–2301.

37 Guo, G., Sun, Y., Wang, Z. et al. (2005). Preparation of hydroxyapatite nanoparticles by reverse microemulsion. *Ceramics International* 31 (6): 869–872.

38 Sun, Y., Guo, G., Tao, D. et al. (2007). Reverse microemulsion-directed synthesis of hydroxyapatite nanoparticles under hydrothermal conditions. *Journal of Physics and Chemistry of Solids* 68: 373–377.

39 Fernandez, E., Gil, F.J., Ginebra, M.P. et al. (1999). Calcium phosphate bone cements for clinical applications. Part I: solution chemistry. *Journal of Materials Science: Materials in Medicine* 10: 169–176.

40 Ginebra, M.P., Fernandez, E., Driessens, F. et al. (1999). Modeling of the hydrolysis of α-tricalcium phosphate. *Journal of the American Ceramic Society* 82: 2808–2812.

41 Sturgeon, J.L. and Brown, P.W. (2009). Effects of carbonate on hydroxyapatite formed from CaHPO4 and $Ca_4(PO_4)_2O$. *Journal of Materials Science: Materials in Medicine* 20 (9): 1787–1794.

42 Park, H.C., Baek, D.J., Park, Y.M. et al. (2004). Thermal stability of hydroxyapatite whiskers derived from the hydrolysis of α-TCP. *Journal of Materials Science* 39 (7): 2531–2534.

43 Sakamoto, K., Yamaguchi, S., Nakahira, A. et al. (2002). Shape-controlled synthesis of hydroxyapatite from α-tricalcium bis(orthophosphate) in organic-aqueous binary systems. *Journal of Materials Science* 37 (5): 1033–1041.

44 Durucan, C. and Brown, P.W. (2000). α-Tricalcium phosphate hydrolysis to hydroxyapatite at and near physiological temperature. *Journal of Materials Science: Materials in Medicine* 11 (6): 365–371.

45 Graham, S. and Brown, P.W. (1996). Reactions of octacalcium phosphate to form hydroxyapatite. *Journal of Crystal Growth* 165 (1–2): 106–115.

46 De Maeyer, E.A.P., Verbeeck, R.M.H., and Pieters, I.Y. (1996). Effect of K^+ on the stoichiometry of carbonated hydroxyapatite obtained by the hydrolysis of Monetite. *Inorganic Chemistry* 35 (4): 857–863.

47 Lin, F.H., Liao, C.J., Chen, K.S. et al. (2001). Petal-like apatite formed on the surface of tricalcium phosphate ceramic after soaking in distilled water. *Biomaterials* 22: 2981–2992.

48 Yubao, L., Xingdong, Z., and De Groot, K. (1997). Hydrolysis and phase transition of alpha-tricalcium phosphate. *Biomaterials* 18: 737–741.

49 Shih, W.-J., Wang, J.-W., Wang, M.-C. et al. (2006). A study on the phase transformation of the nanosized hydroxyapatite synthesized by hydrolysis using in situ high temperature X-ray diffraction. *Materials Science and Engineering: C* 26: 1434–1438.

50 Zhang, G., Chen, J., Yang, S. et al. (2011). Preparation of amino-acid-regulated hydroxyapatite particles by hydrothermal method. *Materials Letters* 65 (3): 572–574.

51 Cao, H., Zhang, L., Zheng, H. et al. (2010). Hydroxyapatite nanocrystals for biomedical applications. *Journal of Physical Chemistry C* 114 (43): 18352–18357.

52 Zhu, K., Yanagisawa, K., Onda, A. et al. (2009). Morphology variation of cadmium hydroxyapatite synthesized by high temperature mixing method under hydrothermal conditions. *Materials Chemistry and Physics* 113 (1): 239–243.

53 Lee, D.K., Park, J.Y., Kim, M.R. et al. (2011). Facile hydrothermal fabrication of hollow hexagonal hydroxyapatite prisms. *CrystEngComm* 13 (17): 5455–5459.

54 Lin, K., Liu, X., Chang, J. et al. (2011). Facile synthesis of hydroxyapatite nanoparticles, nanowires and hollow nano-structured microspheres using similar structured hard-precursors. *Nanoscale* 3 (8): 3052–3055.

55 Zhang, H. and Darvell, B.W. (2011). Morphology and structural characteristics of hydroxyapatite whiskers: effect of the initial Ca concentration, Ca/P ratio and pH. *Acta Biomaterialia* 7 (7): 2960–2968.
56 Guo, X., Xiao, P., Liu, J. et al. (2005). Fabrication of nanostructured hydroxyapatite via hydrothermal synthesis and spark plasma sintering. *Journal of the American Ceramic Society* 88 (4): 1026–1029.
57 Tsiourvas, D., Tsetsekou, A., Kammenou, M.-I. et al. (2011). Controlling the formation of hydroxyapatite nanorods with dendrimers. *Journal of the American Ceramic Society* 94 (7): 2023–2029.
58 Zhang, H. and Zhang, M. (2011). Phase and thermal stability of hydroxyapatite whiskers precipitated using amine additives. *Ceramics International* 37 (1): 279–286.
59 Earl, J.S., Wood, D.J., and Milne, S.J. (2006). Hydrothermal synthesis of hydroxyapatite. *Journal of Physics: Conference Series* 26: 268–271.
60 Andrés-Vergés, M., Fernández-González, C., and Martínez-Gallego, M. (1998). Hydrothermal synthesis of calcium deficient hydroxyapatites with controlled size and homogeneous morphology. *Journal of the European Ceramic Society* 18: 1245–1250.
61 Rodríguez-Lugo, V., Angeles-Chavez, C., Mondragon, G. et al. (2003). Synthesis And Structural Characterization Of Hydroxyapatite Obtained From CaO And CaHP04 By A Hydrothermal Method. *Materials Research Innovations* 9 (1): 20–22.
62 Zhang, X. and Vecchio, K.S. (2007). Hydrothermal synthesis of hydroxyapatite rods. *Journal of Crystal Growth* 308: 133–140.
63 Hassan, M.N., Mahmoud, M.M., Abd El-Fattah, A. et al. (2016). Microwave-assisted preparation of Nano-hydroxyapatite for bone substitutes. *Ceramics International* 42 (3): 3725–3744.
64 Farzadi, A., Solati-Hashjin, M., Bakhshi, F. et al. (2011). Synthesis and characterization of hydroxyapatite/β-tricalcium phosphate nanocomposites using microwave irradiation. *Ceramics International* 37 (1): 65–71.
65 Kumar, A.R., Kalainathan, S., and Saral, A.M. (2010). Microwave assisted synthesis of hydroxyapatite nano strips. *Crystal Research and Technology* 45 (7): 776–778.
66 Kalita, S.J. and Verma, S. (2010). Nanocrystalline hydroxyapatite bioceramic using microwave radiation: synthesis and characterization. *Materials Science and Engineering: C* 30 (2): 295–303.
67 Tang, Q.-L., Wang, K.-W., Zhu, Y.-J. et al. (2009). Single-step rapid microwave-assisted synthesis of polyacrylamide-calcium phosphate nanocomposites in aqueous solution. *Materials Letters* 63 (15): 1332–1334.
68 Amer, W., Abdelouahdi, K., Ramananarivo, H.R. et al. (2013). Synthesis of mesoporous nano-hydroxyapatite by using zwitterions surfactant. *Materials Letters* 107: 189–193.
69 Amer, W., Abdelouahdi, K., Ramananarivo, H.R. et al. (2014). Microwave-assisted synthesis of mesoporous nano-hydroxyapatite using surfactant templates. *CrystEngComm* 16: 543–549.

70 Kim, W., Zhang, Q., and Saito, F. (2000). Mechanochemical synthesis of hydroxyapatite from $Ca(OH)_2$-P_2O_5 and CaO-$Ca(OH)_2$-P_2O_5 mixtures. *Journal of Materials Science* 35: 5401–5405.

71 Fathi, M.H. and Mohammadi Zahrani, E. (2009). Mechanical alloying synthesis and bioactivity evaluation of nanocrystalline fluoridated hydroxyapatite. *Journal of Crystal Growth* 311 (5): 1392–1403.

72 Honarmandi, P., Honarmandi, P., Shokuhfar, A. et al. (2010). Milling media effects on synthesis, morphology and structural characteristics of single crystal hydroxyapatite nanoparticles. *Advances in Applied Ceramics* 109 (2): 117–122.

73 El Briak-BenAbdeslam, H., Ginebra, M.P., Vert, M. et al. (2008). Wet or dry mechanochemical synthesis of calcium phosphates? Influence of the water content on DCPD-CaO reaction kinetics. *Acta Biomaterialia* 4 (2): 378–386.

74 Silva, C.C., Pinheiro, A.G., de Oliveira, R.S. et al. (2004). Properties and in vivo investigation of nanocrystalline hydroxyapatite obtained by mechanical alloying. *Materials Science and Engineering: C* 24 (4): 549–554.

75 Yeong, B., Junmin, X., and Wang, J. (2001). Mechanochemical synthesis of hydroxyapatite from calcium oxide and brushite. *Journal of the American Ceramic Society* 84: 465–467.

76 Bang, J.H. and Suslick, K.S. (2010). Applications of ultrasound to the synthesis of nanostructured materials. *Advanced Materials* 22: 1039–1059.

77 Han, Y.C., Wang, X.Y., and Li, S.P. (2009). Change of phase composition and morphology of sonochemically synthesised hydroxyapatite nanoparticles with glycosaminoglycans during thermal treatment. *Advances in Applied Ceramics* 108 (7): 400–405.

78 Stanislavov, A.S., Sukhodub, L.F., Sukhodub, L.B. et al. (2018). Structural features of hydroxyapatite and carbonated apatite formed under the influence of ultrasound and microwave radiation and their effect on the bioactivity of the nanomaterials. *Ultrasonics Sonochemistry* 42: 84–96.

79 Brenner, C.R. (1995). *Cavitation and Bubble Dynamics*. Oxford: Oxford University Press.

80 Leighton, T.G. (1994). *The Acoustic Bubble*. London: Academic Press.

81 Suslick, K.S. (1988). *Ultrasound: Its Chemical, Physical, and Biological Effects*. New York: Wiley-VCH.

82 Giardina, M.A. and Fanovich, M.A. (2010). Synthesis of nanocrystalline hydroxyapatite from $Ca(OH)_2$ and H_3PO_4 assisted by ultrasonic irradiation. *Ceramics International* 36 (6): 1961–1969.

83 Nikolaev, A.L., Gopin, A.V., Severin, A.V. et al. (2018). Ultrasonic synthesis of hydroxyapatite in non-cavitation and cavitation modes. *Ultrasonics Sonochemistry* 44: 390–397.

84 Kim, W. and Saito, F. (2001). Sonochemical synthesis of hydroxyapatite from H_3PO_4 solution with $ca(OH)_2$. *Ultrasonics Sonochemistry* 8 (2): 85–88.

85 Qi, C., Zhu, Y.-J., Wu, C.-T. et al. (2016). Sonochemical synthesis of hydroxyapatite nanoflowers using creatine phosphate disodium salt as an organic phosphorus source and their application in protein adsorption. *RSC Advances* 6: 9686–9692.

86 Jevtić, M., Mitrić, M., Škapin, S. et al. (2008). Crystal structure of hydroxyapatite nanorods synthesized by sonochemical homogeneous precipitation. *Crystal Growth & Design* 8 (7): 2217–2222.

87 Rouhani, P., Taghavinia, N., and Rouhani, S. (2010). Rapid growth of hydroxyapatite nanoparticles using ultrasonic irradiation. *Ultrasonics Sonochemistry* 17 (5): 853–856.

88 Itatani, K., Iwafune, K., Howell, F.S. et al. (2000). Preparation of various calcium-phosphate powders by ultrasonic spray freeze-drying technique. *Materials Research Bulletin* 35: 575–585.

89 Varadarajan, N., Balu, R., Rana, D. et al. (2014). Accelerated Sonochemical synthesis of calcium deficient hydroxyapatite nanoparticles: structural and morphological evolution. *Journal of Biomaterials and Tissue Engineering* 4: 295–299.

90 Cao, L.-Y., Zhang, C.-B., and Huang, J.-F. (2005). Synthesis of hydroxyapatite nanoparticles in ultrasonic precipitation. *Ceramics International* 31 (8): 1041–1044.

91 Han, Y., Li, S., Wang, X. et al. (2007). Sonochemical preparation of hydroxyapatite nanoparticles stabilized by glycosaminoglycans. *Ultrasonics Sonochemistry* 14 (3): 286–290.

92 Rao, R.R., Roopa, H.N., and Kannan, T.S. (1997). Solid state synthesis and thermal stability of HAP and HAP – β-TCP composite ceramic powders. *Journal of Materials Science: Materials in Medicine* 8: 511–518.

93 Korber, F. and Trömel, G.Z. (1932). Investigation on lime-phosphoric acid and lime-phosphoric acid-silicic acid compounds. *Electrochemistry* 38: 578–580.

94 Guo, X., Yan, H., Zhao, S. et al. (2013). Effect of calcining temperature on the particle size of hydroxyapatite synthesized by solid-state reaction at room temperature. *Advanced Powder Technology* 24: 1034–1038.

95 Fazan, F. and Shahida, K.B.N. (2004). Fabrication of synthetic apatites by solid-state reactions. *Medical Journal of Malaysia* 59: 69–70.

96 Fowler, B.O. (1974). Infrared studies of aatites. II. Preparation of normal and isotopically substituted calcium, strontium, and barium hydroxyapatites and SpectraStructure-composition correlations. *Inorganic Chemistry* 13 (1): 207–214.

97 Boskey, A.L. (2013). *Natural and Synthetic Hydroxyapatites. Biomaterials Science: An Introduction to Materials*, 3ee. Elsevier.

98 Mohd Pu'ad, N.A.S., Koshy, P., Abdullah, H.Z. et al. (2019). Syntheses of hydroxyapatite from natural sources. *Heliyon* 5 (5): e01588.

99 Palmer, L.C., Newcomb, C.J., Kaltz, S.R. et al. (2008). Biomimetic systems for hydroxyapatite mineralization inspired by bone and enamel. *Chemical Reviews* 108 (11): 4754–4783.

100 Yin, N., Chen, S.-y., Ouyang, Y. et al. (2011). Biomimetic mineralization synthesis of hydroxyapatite bacterial cellulose nanocomposites. *Progress in Natural Science: Materials International* 21 (6): 472–477.

101 Sanchez, C., Arribart, H., and Giraud Guille, M.M. (2005). Biomimetism and bioinspiration as tools for the design of innovative materials and systems. *Nature Materials* 4: 277–288.

102 Xu, M., Ji, F., Qin, Z. et al. (2018). Biomimetic mineralization of a hydroxyapatite crystal in the presence of a zwitterionic polymer. *CrystEngComm* 20: 2374–2383.

103 Xu, D., Wan, Y., Li, Z. et al. (2020). Tailorable hierarchical structures of biomimetic hydroxyapatite micro/nano particles promoting endocytosis and osteogenic differentiation of stem cells. *Biomaterials Science* 8: 3286–3300.

104 Myung, S.W., Ko, Y.M., and Kim, B.H. (2013). Effect of plasma surface functionalization on preosteoblast cells spreading and adhesion on a biomimetic hydroxyapatite layer formed on a titanium surface. *Applied Surface Science* 287: 62–68.

105 Du, K., Shi, X., and Gan, Z. (2013). Rapid biomimetic mineralization of hydroxyapatite-g-PDLLA hybrid microspheres. *Langmuir* 29 (49): 15293–15301.

106 Zhou, C., Ye, X., Fan, Y. et al. (2014). Biomimetic fabrication of a three-level hierarchical calcium phosphate/collagen/hydroxyapatite scaffold for bone tissue engineering. *Biofabrication* 6 (3): 035013.

107 Deng, Y., Zhao, X., Zhou, Y. et al. (2013). In vitro growth of bioactive nanostructured apatites via agar-gelatin hybrid hydrogel. *Journal of Biomedical Nanotechnology* 9: 1972–1983.

108 Ghosh, S.K., Roy, S.K., Kundu, B. et al. (2011). Synthesis of nano-sized hydroxyapatite powders through solution combustion route under different reaction conditions. *Materials Science and Engineering: B* 176: 14–21.

109 Ghosh, S.K., Prakash, A., Datta, S. et al. (2010). Effect of fuel characteristics on synthesis of calcium hydroxyapatite by solution combustion route. *Bulletin of Materials Science* 33: 7–16.

110 Sasikumar, S. and Vijayaraghavan, R. (2008). Solution combustion synthesis of bioceramic calcium phosphates by single and mixed fuels – a comparative study. *Ceramics International* 34: 1373–1379.

111 Sasikumar, S. and Vijayaraghavan, R. (2010). Synthesis and characterization of bioceramic calcium phosphates by rapid combustion synthesis. *Journal of Materials Science & Technology* 26: 1114–1118.

112 Ghosh, S.K., Pal, S., Roy, S.K. et al. (2010). Modelling of flame temperature of solution combustion synthesis of nanocrystalline calcium hydroxyapatite material and its parametric optimization. *Bulletin of Materials Science* 33: 339–350.

113 Trommer, R.M., Santos, L.A., and Bergmann, C.P. (2009). Nanostructured hydroxyapatite powders produced by a flame-based technique. *Materials Science and Engineering: C* 29: 1770–1775.

114 Tas, A.C. (2000). Combustion synthesis of calcium phosphate bioceramic powders. *Journal of the European Ceramic Society* 20: 2389–2394.

115 Vallet-Regí, M., Gutiérrez-Rios, M.T., Alonso, M.P. et al. (1994). Hydroxyapatite particles synthesized by pyrolysis of an aerosol. *Journal of Solid State Chemistry* 112: 58–64.

116 An, G.-H., Wang, H.-J., Kim, B.-H. et al. (2007). Fabrication and characterization of a hydroxyapatite nanopowder by ultrasonic spray pyrolysis with salt-assisted decomposition. *Materials Science and Engineering: A* 449: 821–824.

References

117 Hwang, K.-S., Jeon, K.-O., Jeon, Y.-S. et al. (2007). Hydroxyapatite forming ability of electrostatic spray pyrolysis derived calcium phosphate nano powder. *Journal of Materials Science: Materials in Medicine* 18: 619–622.

118 Aizawa, M., Hanazawa, T., Itatani, K. et al. (1999). Characterization of hydroxyapatite powders prepared by ultrasonic spray-pyrolysis technique. *Journal of Materials Science* 34: 2865–2873.

119 Wakiya, N., Yamasaki, M., Adachi, T. et al. (2010). Preparation of hydroxyapatite-ferrite composite particles by ultrasonic spray pyrolysis. *Materials Science and Engineering: B* 173: 195–198.

120 Kay, M.I., Young, R.A., and Posner, A.S. (1964). Crystal structure of hydroxyapatite. *Nature* 204: 1050–1052.

121 Raynaud, S., Champion, E., Bernache-Assollant, D. et al. (2002). Calcium phosphate apatites with variable Ca/P atomic ratio I. Synthesis, characterization and thermal stability of powders. *Biomaterials* 2: 1065–1072.

122 Elliott, J.C. (1994). *Structure and Chemistry of the Apatites and Other Calcium Orthophosphates*, 1ee, vol. 18, 389. Amsterdam: Elsevier Science.

123 Nakamura, S., Takeda, H., and Yamashita, K. (2001). Proton transport polarization and depolarization of hydroxyapatite ceramics. *Journal of Applied Physics* 89: 5386.

124 Riboud, P.V. (1973). Composition et stabilité des phases à structure d'apatite dans le système $CaO-P_2O_5$-Oxyde de fer-H_2O à haute température. *Annales de Chimie* 8: 381–390.

125 Trombe, J.C. and Montel, G. (1978). Some features of the incorporation of oxygen in different oxidation states in the apatitic lattice-I on the existence of calcium and strontium oxyapatites. *Journal of Inorganic and Nuclear Chemistry* 40: 15–21.

126 Indrani, D.J., Soegijono, B., Adi, W.A. et al. (2017). Phase composition and crystallinity of hydroxyapatite with various heat treatment temperatures. *International Journal of Applied Pharmaceutics* 9 (2): 87–91.

127 Goldberg, M.A., Protsenko, P.V., Smirnov, V.V. et al. (2020). The enhancement of hydroxyapatite thermal stability by Al doping. *Journal of Materials Research and Technology* 9 (1): 76–88.

128 Kannan, S. and Ferreira, J.M.F. (2006). Synthesis and thermal stability of hydroxyapatite–β-tricalcium phosphate composites with cosubstituted sodium, magnesium, and fluorine. *Chemistry of Materials* 18 (1): 198–203.

129 Tõnsuaadu, K., Gross, K.A., Plūduma, L. et al. (2012). A review on the thermal stability of calcium apatites. *Journal of Thermal Analysis and Calorimetry* 110: 647–659.

130 Weng, J., Liu, X., Zhang, X. et al. (1994). Thermal decomposition of hydroxyapatite structure induced by titanium and its dioxide. *Journal of Materials Science Letters* 13: 159–161.

131 Wang, P.E. and Chaki, T.K. (1993). Sintering behavior and mechanical-properties of hydroxyapatite and dicalcium phosphate. *Journal of Materials Science: Materials in Medicine* 4: 150–158.

132 McDowell, H., Gregory, T.M., and Brown, W.E. (1977). Solubility of $Ca_5(PO_4)_3OH$ in the system $Ca(OH)_2$-H_3PO_4-H_2O at 5, 15, 25, and 37 °C. *Journal of Research of the National Bureau of Standards* 81A: 273–281.

133 Mahapatra, P.P., Mishra, H., and Chickerur, N.S. (1982). Solubility and thermodynamic data of cadmium hydroxyapatite in aqueous media. *Thermochimica Acta* 54: 1–8.

134 Moreno, E.C., Gregory, T.M., and Brown, W.E. (1968). Preparation and solubility of hydroxyapatite. *Journal of Research of the National Bureau of standards Section A. Physics and Chemistry* 72A: 773–782.

135 Avnimelech, Y., Moreno, E.C., and Brown, W.E. (1972). Solubility and surface properties of finely divided hydroxyapatite. *Journal of Research of the National Bureau of Standards Section A. Physics and chemistry* 77A: 149–153.

136 Clark, J.S. (1955). Solubility criteria for the existence of hydroxyapatite. *Canadian Journal of Chemistry* 33: 1696–1700.

137 Jacques, J.K. (1963). The heats of formation of fluorapatite and hydroxyapatite. *Journal of the Chemical Society* 1963: 3820–3822.

138 Egan, E.P. Jr., Wakefield, Z.T., and Elmore, K.L. (1951). Low-temperature heat capacity and entropy of hydroxyapatite. *Journal of the American Chemical Society* 73: 5579–5580.

139 Dorozhkin, S.V. (2002). A review on the dissolution models of calcium apatites. *Progress in Crystal Growth and Characterization of Materials* 44: 45–61.

140 Mafe, S., Manzanares, J.A., Reiss, H. et al. (1992). Model for the dissolution of calcium hydroxyapatite powder. *Journal of Physical Chemistry* 96: 861–866.

141 Shimabayashi, S. and Nakagaki, M. (1983). Concurrent adsorption of calcium ion and hydroxyl ion on hydroxyapatite. *Chemical and Pharmaceutical Bulletin* 31: 2976–2985.

142 Dorozhkin, S.V. (1997). Surface reactions of apatite dissolution. *Journal of Colloid and Interface Science* 191: 489–497.

143 Smith, A.N., Posner, A.M., and Quirk, J.P. (1974). Incongruent dissolution and surface complexes of hydroxyapatite. *Journal of Colloid and Interface Science* 48: 442–449.

144 Rootare, H.M., Deitz, V.R., and Carpenter, F.G. (1962). Solubility product phenomena in hydroxyapatite-water systems. *Journal of Colloid Science* 17: 179–206.

145 Levinskas, G.J. and Neuman, W.F. (1955). The solubility of bone mineral. I. Solubility studies of synthetic hydroxylapatite. *Journal of Physical Chemistry* 59: 164–168.

146 LaMer, V.K. (1962). Solubility behavior of hydroxyapatite. *Journal of Physical Chemistry* 66: 973–978.

147 Busca, G. (2014). *Heterogeneous Catalytic Materials. Solid State Chemistry, Surface Chemistry and Catalytic Behaviour*. Elsevier.

148 Fu, Q. and Baoa, X. (2017). Surface chemistry and catalysis confined under two-dimensional materials. *Chemical Society Reviews* 46: 1842–1874.

149 Harold, H.K. (1989). Chapter 4: surface coordinative unsaturation. In: *Transition Metal Oxides. Surface Chemistry and Catalysis*, vol. 45, 1–282. Elsevier.

150 Liu, P., Qin, R., Fu, G., and Zheng, N. (2017). Surface coordination chemistry of metal nanomaterials. *Journal of the American Chemical Society* 139 (6): 2122–2131.

151 Margelefsky, E.L., Zeidan, R.K., Dufaud, V. et al. (2007). Organized surface functional groups: cooperative catalysis via thiol/sulfonic acid pairing. *Journal of the American Chemical Society* 129 (44): 13691–13697.

152 Bengtsson, Å. and Sjöberg, S. (2009). Surface complexation and proton-promoted dissolution in aqueous apatite systems. *Pure and Applied Chemistry* 81 (9): 1569–1584.

153 Wu, L., Forsling, W., and Schindler, P.W. (1991). Surface complexation of calcium minerals in aqueous solution. *Journal of Colloid and Interface Science* 147: 178–185.

154 Jarlbring, M., Sandström, D.E., Antzutkin, O.N. et al. (2006). Characterization of active phosphorus surface sites at synthetic carbonate-free fluorapatite using single-pulse ^{1}H, ^{31}P, and ^{31}P CP MAS NMR. *Langmuir* 22: 4787–4792.

155 Bengtsson, Å., Shchukarev, A., Persson, P. et al. (2009). A solubility and surface complexation study of a non-stoichiometric hydroxyapatite. *Geochimica et Cosmochimica Acta* 73: 257–267.

156 Kukura, M., Bell, L.C., Posner, A.M. et al. (1972). Radioisotope determination of the surface concentrations of calcium and phosphorous on hydroxyapatite in aqueous solution. *Journal of Physical Chemistry* 76: 900–904.

157 Misra, D.N., Bowen, R.L., and Wallace, B.M. (1975). Adhesive bonding of various materials to hard tooth tissues, VIII. Nickel and copper ions on hydroxyapatite - Role of ion exchange and surface nucleation. *Journal of Colloid and Interface Science* 51: 36–43.

158 Al-Qasas, N.S. and Rohani, S. (2005). Synthesis of pure hydroxyapatite and the effect of synthesis conditions on its yield, crystallinity, morphology and mean particle. *Separation Science and Technology* 40: 3187–3224.

159 López-Macipe, A., Rodríguez-Clemente, R., Hidalgo-López, A. et al. (1998). Wet chemical synthesis of hydroxyapatite particles from nonstoichiometric solutions. *Journal of Materials Synthesis and Processing* 6: 21–26.

160 Rodríguez-Lugo, V., Salinas-Rodríguez, E., Vázquez, R.A. et al. (2017). Hydroxyapatite synthesis from a starfish and β-tricalcium phosphate using a hydrothermal method. *RSC Advances* 7: 7631–7639.

161 Gibson, I.R. and Bonfield, W. (2002). Novel synthesis and characterization of an AB-type carbonate-substituted hydroxyapatite. *Journal of Biomedical Materials Research* 59 (4): 697–708.

162 Jingbing, L., Xiaoyue, Y., Hao, W. et al. (2003). The influence of pH and temperature on the morphology of hydroxyapatite synthesized by hydrothermal method. *Ceramics International* 29 (6): 629–633.

163 Mekmene, O., Quillard, S., Rouillon, T. et al. (2009). Effects of pH and Ca/P molar ratio on the quantity and crystalline structure of calcium phosphates obtained from aqueous solutions. *Dairy Science and Technology* 89: 301–316.

164 Tsuchida, T., Kubo, J., Yoshioka, T. et al. (2009). Influence of preparation factors on Ca/P ratio and surface basicity of hydroxyapatite catalyst. *Journal of the Japan Petroleum Institute* 52: 51–59.

165 Raynaud, S., Champion, E., Bernache-Assollant, D. et al. (2002). Calcium phosphate apatites with variable Ca/P atomic ratio I. Synthesis, characterisation and thermal stability of powders. *Biomaterials* 23: 1065–1072.

166 Monma, H., Ueno, S., and Kanazawa, T. (1981). Properties of hydroxyapatite prepared by the hydrolysis of tricalcium phosphate. *Journal of Chemical Technology & Biotechnology* 31: 15–24.

167 Joris, S.J. and Amberg, C.H. (1971). Nature of deficiency in nonstoichiometric hydroxyapatites. I. Catalytic activity of calcium and strontium hydroxyapatites. *Journal of Physical Chemistry* 75: 3167–3171.

168 Meyer, J.L. and Fowler, B.O. (1982). Lattice defects in nonstoichiometric calcium hydroxyapatite. A chemical approach. *Inorganic Chemistry* 21: 3029–3035.

169 Arends, J., Christoffersen, J., Christoffersen, M.R. et al. (1987). A calcium hydroxyapatite precipitated from an aqueous solution: an international multimethod analysis. *Journal of Crystal Growth* 84: 515–532.

170 Posner, A.S. and Perloff, A. (1957). Apatites deficient in divalent cations. *Journal of Research of the National Bureau of Standards* 58: 279–286.

171 Stutman, J.M., Posner, A.S., and Lippincott, E.R. (1962). Hydrogen bonding in the calcium phosphates. *Nature* 193: 368–369.

172 Winand, L., Dallemagne, M.J., and Duyckaerts, G. (1961). Hydrogen bonding in apatitic calcium phosphates. *Nature* 1961 (190): 164–165.

173 Winand, L. and Dallemagne, M.J. (1962). Hydrogen bonding in the calcium phosphates. *Nature* 1962 (193): 369–370.

174 Berry, E.E. (1967). The structure and composition of some calcium-deficient apatites. *Journal of Inorganic and Nuclear Chemistry* 29: 317–327.

175 Kuhl, G. and Nebergall, W.H. (1963). Hydrogenphosphat- und Carbonatapatite. *Zeitschrift für Anorganische und Allgemeine Chemie* 324: 313–320.

176 Berry, E.E. (1967). The structure and composition of some calcium-deficient apatites-II. *Journal of Inorganic and Nuclear Chemistry* 29: 1587–1590.

177 Joris, S.J. and Amberg, C.H. (1971). Nature of deficiency in nonstoichiometric hydroxyapatites. II. Spectroscopic studies of calcium and strontium hydroxyapatites. *Journal of Physical Chemistry* 75: 3172–3178.

178 Tourbin, M., Brouillet, F., Galey, B. et al. (2020). Agglomeration of stoichiometric hydroxyapatite: impact on particle size distribution and purity in the precipitation and maturation steps. *Powder Technology* 360: 977–988.

179 Elfeki, H., Rey, C., and Calcif, M.V. (1991). Carbonate ions in apatites: infrared investigations in the $\nu 4$ CO_3 domain. *Calcified Tissue International* 49: 269–274.

180 Milev, A.S., Kamali Kannangara, G.S., and Wilson, M.A. (2004). Strain and microcrystallite size in synthetic lamellar apatite. *Journal of Physical Chemistry B* 108: 13015–13021.

181 Šupová, M. (2015). Substituted hydroxyapatites for biomedical applications: a review. *Ceramics International* 41: 9203–9231.

182 Cacciotti, I. (2019). Multisubstituted hydroxyapatite powders and coatings: the influence of the codoping on the hydroxyapatite performances. *International Journal of Applied Ceramic Technology* 16: 1864–1884.

183 Legeros, R.Z., Taheri, M.H., Quirolgico, G.B., et al. (1980). Formation and stability of apatite: effects of some cationic substituents. *Proceedings, 2nd International Congress on Phosphorus Compounds*, Boston, 89–103.

184 Panda, A. and Sahu, B. (1991). Preparation and lattice constant measurement of (Ca + Cd + Pb) hydroxylapatites. *Journal of Materials Science Letters* 10: 638–639.

185 Panda, A., Sahu, B., Patel, P.N. et al. (1991). Calcium-lead-copper and calcium-lead-cadmium hydroxylapatite solid solutions: preparation, infrared and lattice constant measurements. *Transition Metal Chemistry* 16: 476–477.

186 Zuo, K.H., Zeng, Y.P., and Jiang, D. (2012). Synthesis and magnetic property of iron ions-doped hydroxyapatite. *Journal of Nanoscience and Nanotechnology* 12: 7096–7100.

187 Oviedo, M.J., Contreras, O., Vazquez-Duhalt, R. et al. (2012). Photoluminescence of europium-activated hydroxyapatite nanoparticles in body fluids. *Science of Advanced Materials* 4: 558–562.

188 Boechat, C.B., Eon, J.-G., Rossi, A.M. et al. (2000). Structure of vanadate in calcium phosphate and vanadate apatite solid solutions. *Physical Chemistry Chemical Physics* 2: 4225–4230.

189 Sugiyama, S., Osaka, T., Hirata, Y. et al. (2006). Enhancement of the activity for oxidative dehydrogenation of propane on calcium hydroxyapatite substituted with vanadate. *Applied Catalysis A: General* 312: 52–58.

190 Cazalbou, S., Combes, C., Eichert, D. et al. (2004). Adaptative physico-chemistry of bio-related calcium phosphates. *Journal of Materials Chemistry* 14: 2148–2153.

191 Audubert, F., Carpena, J., Lacout, J.L. et al. (1997). Elaboration of an iodine-bearing apatite iodine diffusion into a $Pb_3(VO_4)_2$ matrix. *Solid State Ionics* 95: 113–119.

192 Veiderma, M., Tõnsuaadu, K., Knubovets, R. et al. (2005). Impact of anionic substitutions on apatite structure and properties. *Journal of Organometallic Chemistry* 690: 2638–2643.

193 Gibson, I.R., Best, S.M., and Bonfield, W. (1999). Chemical characterization of silicon-substituted hydroxyapatite. *Journal of Biomedical Materials Research* 44: 422–428.

194 Hidouri, M., Bouzouita, K., Kooli, F. et al. (2003). Thermal behaviour of magnesium-containing fluorapatite. *Materials Chemistry and Physics* 80: 496–505.

195 Trombe, J.C. and Montel, G. (1978). Some features of the incorporation of oxygen in different oxidation states in the apatitic lattice-II on the synthesis and properties of calcium and strontium peroxiapatites. *Journal of Inorganic and Nuclear Chemistry* 40: 23–26.

196 Lafon, J.P., Champion, E., and Bernache-Assollant, D. (2008). Processing of AB-type carbonated hydroxyapatite $Ca_{10-x}(PO_4)_{6-x}(CO_3)_x(OH)_{2-x-2y}(CO_3)_y$ ceramics with controlled composition. *Journal of the European Ceramic Society* 28: 139–147.

197 Barralet, J., Best, S., and Bonfield, W. (1998). Carbonate substitution in precipitated hydroxyapatite: an investigation into the effects of reaction temperature and bicarbonate ion concentration. *Journal of Biomedical Materials Research* 41: 79–86.

198 Cheng, Z.H., Yasukawa, A., Kandori, K. et al. (1998). FTIR study of adsorption of CO_2 on nonstoichiometric calcium hydroxyapatite. *Langmuir* 14: 6681–6686.

199 Driessens, F.C.M., Verbeeck, R.M.H., and Kiekens, P. (1983). Mechanism of substitution in carbonated apatites. *Zeitschrift für Anorganische und Allgemeine Chemie* 504: 195–200.

200 Madupalli, H., Pavan, B., and Tecklenburg, M.M.J. (2017). Carbonate substitution in the mineral component of bone: discriminating the structural changes, simultaneously imposed by carbonate in A and B sites of apatite. *Journal of Solid State Chemistry* 255: 27–35.

201 Hammami, K., Elloumi, J., Aifa, S. et al. (2013). Synthesis and characterization of hydroxyapatite ceramics organofunctionalized with ATP (adenosine triphosphate). *Journal of Advances in Chemistry* 9 (1): 1787–1797.

202 Pereira, M.B.B., Honório, L.M.C., Lima-Júnior, C.G. et al. (2020). Modulating the structure of organofunctionalized hydroxyapatite/tripolyphosphate/chitosan spheres for dye removal. *Journal of Environmental Chemical Engineering* 8 (4): 103980.

203 Zahidi, E., Lebugle, A., and Bonel, G. (1985). On a new class of biomaterials for osseous or dental prothesis. *Bulletin de la Société chimique de France* 4: 523–527.

204 Delpech, V. and Lebugle, A. (1990). Calcium phosphate and polymer interfaces a orthopaedic cement. *Clinical Materials* 5: 209–216.

205 Kandori, K., Fujiwara, A., Yasukawa, A. et al. (1999). Preparation and characterization of hydrophobic calcium hydroxyapatite particles grafting oleylphosphate groups. *Colloids and Surfaces A: Physicochemical and Engineering Aspects* 150: 161–170.

206 Tanaka, H., Watanabe, T., and Chikazawa, M. (1997). FTIR and TPD studies on the adsorption of pyridine, n-butylamineand acetic acid on calcium hydroxyapatite. *Journal of the Chemical Society, Faraday Transactions* 93: 4377–4381.

207 Tanaka, H., Yasukawa, A., Kandori, K. et al. (1997). Surface modification of calcium hydroxyapatite with hexyl and decyl phosphates. *Colloids and Surfaces A: Physicochemical and Engineering Aspects* 125: 53–62.

208 Ishikawa, T., Tanaka, H., Yasukawa, A. et al. (1995). Modification of calcium hydroxyapatite using ethyl phosphates. *Journal of Materials Chemistry* 5: 1963–1967.

209 Tanaka, H., Yasukawa, A., Kandori, K. et al. (1997). Modification of calcium hydroxyapatite using alkyl phosphates. *Langmuir* 13: 821–826.

210 Tanaka, H., Watanabe, T., Chikazawa, M. et al. (1998). Surface structure and properties of calcium hydroxyapatite modified by Hexamethyldisilazane. *Journal of Colloid and Interface Science* 206: 205–211.

211 Da Silva, O.G., Filho, E.C.D.S., Fonseca, M.G.D. et al. (2006). Hydroxyapatite organofunctionnalized with silylating agents to heavy cation removal. *Journal of Colloid and Interface Science* 302: 485–491.

212 Russo, L., Taraballi, F., Lupo, C. et al. (2014). Carbonate hydroxyapatite functionalization: a comparative study towards (bio)molecules fixation. *Interface Focus* 4 (1): 20130040.

213 Wei, J., He, P., Liu, A. et al. (2009). Surface modification of hydroxyapatite nanoparticles with thermal-responsive PNIPAM by ATRP. *Macromolecular Bioscience* 9: 1237–1246.

214 Wei, J., Liu, A., Chen, L. et al. (2009). The surface modification of hydroxyapatite nanoparticles by the ring opening polymerization of g-benzyl-L-glutamate N-carboxyanhydride. *Macromolecular Bioscience* 9: 631–638.

215 Jacobsen, P.A.L., Nielsen, J.L., Juhl, M.V. et al. (2012). Grafting cyclodextrins to calcium phosphate ceramics for biomedical applications. *Journal of Inclusion Phenomena and Macrocyclic Chemistry* 72: 173–181.

216 Durrieu, M.C., Pallu, S., Guillemot, F. et al. (2004). Grafting RGD containing peptides onto hydroxyapatite to promote osteoblastic cells adhesion. *Journal of Materials Science: Materials in Medicine* 15: 779–786.

217 Nelson, M., Balasundaram, G., and Webster, T.J. (2006). Increased osteoblast adhesion on nanoparticulate crystalline hydroxyapatite functionalized with KRSR. *International Journal of Nanomedicine* 1: 339–349.

218 Wang, V., Misra, G., and Amsden, B. (2008). Immobilization of a bone and cartilage stimulating peptide to a synthetic bone graft. *Journal of Materials Science: Materials in Medicine* 19: 2145–2155.

219 Zurlinden, K., Laub, M., and Jennissen, H.P. (2005). Chemical functionalization of a hydroxyapatite based bone replacement material for the immobilization of proteins. *Materialwissenschaft und Werkstofftechnik* 36: 820–827.

220 Sebti, S., Zahouily, M., Lazrek, H.B. et al. (2008). Phosphates: new generation of liquid-phase heterogeneous catalysts in organic chemistry. *Current Organic Chemistry* 12: 203–232.

221 Munnik, P., de Jongh, P.E., and de Jong, K.P. (2015). Recent developments in the synthesis of supported catalysts. *Chemical Reviews* 115 (14): 6687–6718.

222 van Deelen, T.W., Mejía, C.H., and de Jong, K.P. (2019). Control of metal-support interactions in heterogeneous catalysts to enhance activity and selectivity. *Nature Catalysis* 2: 955–970.

223 Sherrington, D.C. and Kybett, A.P. (2001). *Supported Catalysts and their Applications*. Royal Society of Chemistry.

224 Li, S. and Gong, J. (2014). Strategies for improving the performance and stability of Ni-based catalysts for reforming reactions. *Chemical Society Reviews* 43: 7245–7256.

225 Benaglia, M. (2009). *Recoverable and Recyclable Catalysts*. John Wiley & Sons.

226 de Vasconcelos, B.R., Minh, D.P., Martins, E. et al. (2020). A comparative study of hydroxyapatite- and alumina-based catalysts in dry reforming of methane. *Chemical Engineering & Technology* 43 (4): 698–704.

227 Martin, R.I. and Brown, P.W. (1995). Mechanical properties of hydroxyapatite formed at physiological temperature. *Journal of Materials Science: Materials in Medicine* 6: 138–143.

228 Guo, J., Yu, H., Dong, F. et al. (2017). High efficiency and stability of au-cu/hydroxyapatite catalyst for the oxidation of carbon monoxide. *RSC Advances* 7: 45420–45431.

229 Sebti, S., Solhy, A., Tahir, R. et al. (2002). Modified hydroxyapatite with sodium nitrate: an efficient new solid catalyst for the Claisen-Schmidt condensation. *Applied Catalysis A: General* 235 (1–2): 273–281.

230 Solhy, A., Sebti, S., Tahir, R. et al. (2010). Remarkable catalytic activity of sodium-modified-hydroxyapatite in the synthesis of α-hydroxyphosphonates. *Current Organic Chemistry* 14 (14): 1517–1522.

231 Sebti, S., Solhy, A., Smahi, A. et al. (2002). Dramatic activity enhancement of natural phosphate catalyst by lithium nitrate. An efficient synthesis of chalcones. *Catalysis Communications* 3 (8): 335–339.

232 Maocong, H., Zhenhua, Y., Xuguang, L. et al. (2018). Enhancement mechanism of hydroxyapatite for photocatalytic degradation of gaseous formaldehyde over TiO_2/hydroxyapatite. *Journal of the Taiwan Institute of Chemical Engineers* 85: 91–97.

233 Yao, J., Zhang, Y., Wang, Y. et al. (2017). Enhanced photocatalytic removal of NO over titania/hydroxyapatite (TiO_2/HAp) composites with improved adsorption and charge mobility ability. *RSC Advances* 7: 24683–24689.

234 Zhang, X. and Yates, M.Z. (2018). Enhanced photocatalytic activity of TiO_2 nanoparticles supported on electrically polarized hydroxyapatite. *ACS Applied Materials & Interfaces* 10 (20): 17232–17239.

235 Takeda, A., Kamijo, S., and Yamamoto, Y. (2000). Indole synthesis via palladium-catalyzed intramolecular cyclization of alkynes and imines. *Journal of the American Chemical Society* 122 (23): 5662–5663.

236 Murata, M., Hara, T., Mori, K. et al. (2003). Efficient deprotection of N-benzyloxycarbonyl group from amino acids by hydroxyapatite-bound pd catalyst in the presence of molecular hydrogen. *Tetrahedron Letters* 44: 4981–4984.

237 Smahi, A., Solhy, A., El Badaoui, H. et al. (2003). Potassium fluoride doped fluorapatite and hydroxyapatite as new catalysts in organic synthesis. *Applied Catalysis A: General* 250 (1): 151–159.

238 Bulut, A., Aydemir, M., Durap, F. et al. (2018). Palladium nanoparticles supported on hydroxyapatite nanospheres: highly active, reusable and green catalyst for Suzuki-Miyaura cross coupling reactions under aerobic conditions. *ChemistrySelect* 3: 1569–1576.

239 Mitsudome, T., Arita, S., Mori, H. et al. (2008). Supported silver-nanoparticle-catalyzed highly efficient aqueous oxidation of phenylsilanes to silanols. *Angewandte Chemie International Edition* 47 (41): 7938–7940.

240 Mori, K., Kanai, S., Hara, T. et al. (2007). Development of ruthenium-hydroxyapatite-encapsulated superparamagnetic γ-Fe_2O_3 nanocrystallites as an efficient oxidation catalyst by molecular oxygen. *Chemistry of Materials* 19 (6): 1249–1256.

241 Hara, T., Kaneta, T., Mori, K. et al. (2007). Magnetically recoverable heterogeneous catalyst: palladium nanocluster supported on hydroxyapatite-encapsulated γ-Fe_2O_3 nanocrystallites for highly efficient dehalogenation with molecular hydrogen. *Green Chemistry* 9: 1246–1251.

242 Solhy, A., Clark, J.H., Tahir, R. et al. (2006). Transesterifications catalysed by solid, reusable apatite-zinc chloride catalysts. *Green Chemistry* 8: 871–874.

243 Gupta, M., Gupta, R., and Anand, M. (2009). Hydroxyapatite supported caesium carbonate as a new recyclable solid base catalyst for the Knoevenagel condensation in water. *Beilstein Journal of Organic Chemistry* 5 (68): 1–7.

244 Yamaguchi, K., Mori, K., Mizugaki, T. et al. (2000). Creation of a monomeric Ru species on the surface of hydroxyapatite as an efficient heterogeneous catalyst for aerobic alcohol oxidation. *Journal of the American Chemical Society* 122 (29): 7144–7145.

245 Akbayrak, S., Erdek, P., and Özkâr, S. (2013). Hydroxyapatite supported ruthenium(0) nanoparticles catalyst in hydrolytic dehydrogenation of ammonia borane: insight to the nanoparticles formation and hydrogen evolution kinetics. *Applied Catalysis B: Environmental* 142–143: 187–195.

246 Hara, T., Kanai, S., Mori, K. et al. (2006). Highly efficient C–C bond-forming reactions in aqueous media catalyzed by monomeric vanadate species in an apatite framework. *The Journal of Organic Chemistry* 71 (19): 7455–7462.

247 Dasireddy, V.D.B.C., Singh, S., and Friedrich, H.B. (2013). Activation of n-octane using vanadium oxide supported on alkaline earth hydroxyapatites. *Applied Catalysis A: General* 456: 105–117.

248 Mori, K., Oshiba, M., Hara, T. et al. (2006). Creation of monomeric La complexes on apatite surfaces and their application as heterogeneous catalysts for Michael reactions. *New Journal of Chemistry* 30: 44–52.

249 Mori, K., Mitani, Y., Hara, T. et al. (2005). A single-site hydroxyapatite-bound zinc catalyst for highly efficient chemical fixation of carbon dioxide with epoxides. *Chemical Communications* 3331–3333.

250 Mori, K., Hara, T., Mizugaki, T. et al. (2003). Hydroxyapatite-bound cationic ruthenium complexes as novel heterogeneous Lewis acid catalysts for Diels-Alder and Aldol reactions. *Journal of the American Chemical Society* 125 (38): 11460–11461.

251 Sebti, S., Solhy, A., Tahir, R. et al. (2003). Application of natural phosphate modified with sodium nitrate in the synthesis of chalcones: a soft and clean method. *Journal of Catalysis* 213 (1): 1–6.

252 Akri, M., Zhao, S., Li, X. et al. (2019). Atomically dispersed nickel as coke-resistant active sites for methane dry reforming. *Nature Communications* 10: 5181.

253 Akri, M., El Kasmi, A., Batiot-Dupeyrat, C., et al. (2020). Highly active and carbon-resistant nickel single-atom catalysts for methane dry reforming. *Catalysts* 10: 630. doi: https://doi.org/10.3390/catal10060630.

3

Structure and Surface Study of Hydroxyapatite-Based Materials

Experimental and Computational Approaches

Guylène Costentin[1], Christophe Drouet[2], Fabrice Salles[3], and Stéphanie Sarda[2]

[1]Sorbonne Université, CNRS, Laboratoire de Réactivité de Surface (LRS), 4 place Jussieu, 75005 Paris, France
[2]CIRIMAT, Université de Toulouse, CNRS, Université Toulouse 3 Paul Sabatier, ENSIACET, 31030 Toulouse, France
[3]ICGM, University of Montpellier, CNRS, ENSCM, 8 Rue de l'École Normale, 34090 Montpellier, France

3.1 Introduction

Heterogeneous catalysis plays a key role in the development of green and sustainable chemistry. It appears as an alternative method in organic synthesis as it presents several advantages over homogeneous processes, including simple product isolation while allowing catalyst separation and recycling. In addition to designing catalytic materials ever more active and selective, the emergence of new classes of greener catalysts remains very challenging. Due to their eco-compatibility (low production costs using green chemistry in aqueous medium and excellent storage abilities) and their sorption ability toward organic molecules and ionic species selected to tune surface properties, the apatite family system was investigated as catalysts from the 1960s [1–3], but their increasing attractiveness appeared essentially from the 2000s, as illustrated by the growing number of publications dedicated to the use of this type of calcium phosphate as heterogeneous catalyst [4–6] (Figure 3.1).

Such inexpensive and environment-friendly apatitic calcium phosphate as hydroxyapatite (HA, $Ca_{10}(PO_4)_6(OH)_2$) and their nonstoichiometric and/or substituted counterparts can develop large surface areas and various morphologies, depending on their mode of preparation. HA (in its stoichiometric form) is stable not only in aqueous medium (typically above pH ~4.5) but also at high temperature [7–9] and can thus operate in large condition ranges, being of interest for both liquid medium (aqueous [10–14] or solvent free [15–17]) and gas phase reactions [18–20]. Moreover, it was reported to be stable over time on stream, resistant to sintering [18, 21–23] and to leaching [10, 12, 24–29], and recyclable [11, 17, 23, 25–27, 30–40]. Also, as mentioned earlier, several subtypes of apatitic compounds (stoichiometric or nonstoichiometric, nanocrystalline or not, substituted or not, surface grafted or not) may be obtained depending on the mode of preparation, thus providing samples with different features of relevance to catalysis.

Design and Applications of Hydroxyapatite-Based Catalysts, First Edition. Edited by Doan Pham Minh.
© 2022 WILEY-VCH GmbH. Published 2022 by WILEY-VCH GmbH.

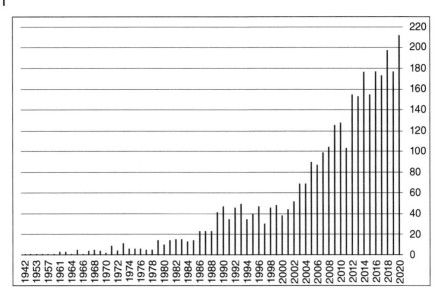

Figure 3.1 Referenced publications per year for "hydroxyapatite and catalysis." Source: Data from SciFinder Source, 2020.

One of the most interesting characteristics of apatitic calcium phosphates is their surface reactivity. The great modulation of composition by substitution on both its cationic and anionic sites offers endless opportunities for the design of calcium phosphate compounds with the apatite structure, eventually modified by heteroatoms, to be used by themselves as bulk tunable active phases [8], or as suitable supports for the controlled surface deposition of highly dispersed catalytic metallic centers [10, 15, 18, 30, 41–47]. Such an amazing tunable composition ability, including its nonstoichiometry properties, confers to the apatite system a very unique property: it is a rare example of metallic oxide surface exposing not only several types of acid sites, both Brønsted PO–H (proton donor groups) and Lewis (coordinatively unsaturated cationic metallic centers: Ca^{2+} or other exchanged or anchored cation), but also several types of efficient basic sites, whose nature will be discussed in this chapter. Their relative density and strength are tunable by synthesis and activation conditions. Moreover, the substitution ability participates to the adjustment of the acid–base balance properties [48, 49]. Also, when associated to transition metal modification, it also provides additional redox properties [19, 48, 50–58], making this catalytic material a multifunctional very powerful and tunable system efficient for many classes of catalytic reactions, including C—C formation [14, 59] (Knoevenagel [11, 15, 17, 60], Claisen–Schmidt [61, 62], Michael [47, 63, 64], Guerbet [8, 20, 59, 65–76] reactions), cycloaddition [77], N-arylation [78], epoxidation [79, 80], selective [19, 26, 50, 53–55, 81–83] or total [51, 84–88] oxidation, hydrogenation [56, 89], dehydrogenation [90–93], dehydration [2, 56, 91, 94–99], water gas shift [23, 43, 100], reduction of NO_x [52, 57, 58, 101], steam reforming [102], dry reforming [18, 30, 103–105], etc. Despite being thus of prime interest to answer the challenges or many societal issues, going from organic syntheses [14, 32, 60, 106] to environmental reactions (for depollution [6, 10, 12, 52, 57, 58, 84–86, 101, 107] or valorization of bio-sourced molecules [35, 56, 93–95, 108–111]) and reactions for energy [27, 34, 36, 65, 89, 107, 112], most of the studies essentially focused so far on the catalytic performances, while structure–reactivity relationship approaches needed for rationalized optimization still remain quite limited, especially in the case of liquid phase reactions.

In this view, this chapter aims at providing tools to help rationalize the behavior of apatite-based catalysts. We will give details about the HA structure, surface properties, and reactivity of apatitic compounds, whether stoichiometric or nonstoichiometric and potentially substituted or biomimetic. A special attention will be dedicated to the latest progress in the characterization of their structural properties (bulk and surface/interphase), which have a strong impact on their reactivity, through both experimental and computational approaches. By combining molecular simulations and experiments, the elucidation of the mechanisms occurring in the structure and at the surface as well as the thermodynamic and dynamical properties (in order to take into account the chemical versatility) is also considered to provide quantitative structure–properties relationships [113, 114]. Remaining characterization challenges for apatite-based catalysts will also be discussed.

3.2 Structure and Surface Properties of Hydroxyapatite: Overview

Apatites are crystalline ionic compounds that share common structural features to which the term "apatite" refers [115] – although some local modifications may arise depending on their exact chemical composition [116]. We will focus in this chapter on phosphate apatites, but it may be mentioned that the apatite family also encompasses vanadates, arsenates, silicates, and sulfates among other minerals [117]. The generic formula often encountered nowadays for apatites is of the form:

$$M_{10}(XO_4)_6 Y_2 \tag{3.1}$$

where M represents a metal cation, X stands here for phosphorous, and Y refers typically to a monovalent anion such as OH^-, F^-, or Cl^-. In these conditions, there is one formula unit per unit cell ($Z = 1$). In some domains as in mineralogy, however, the notation $M_5(XO_4)_3Y$ is otherwise frequently used. In this case, the atomic arrangement remains of course unchanged, but $Z = 2$. Although, strictly speaking, the notation in M_5 is more straightforward from a crystallographic point of view, the M_{10} notation is easier to handle when tackling possible nonstoichiometry aspects (Section 3.2.2), HPO_4^{2-}-for-PO_4^{3-} substitutions or else hydrolysis-bound transformations into apatite from possible precursor phases like amorphous calcium phosphate (ACP) or octacalcium phosphate (OCP). Therefore, this convention will be used in the following.

Depending on the conditions of preparation and/or of post-treatments, a large variety of apatitic compounds can be obtained, thus exhibiting different features. The term "apatite" thus refers in fact to a large family of compounds with some specificities. Stoichiometric apatites should in particular be distinguished from non-stoichiometric apatites with surface properties that could present interesting features for catalysis applications. Apatites may also be classified on the basis of the size of their constituting crystals, nanocrystalline apatites exposing a larger contact surface area than micron/millimeter-sized crystals. A subcategory of apatites that cumulate both the nonstoichiometry and the nanocrystalline character is found in biomineralizations as in bone or dentin but may also be obtained synthetically at low temperature (e.g. near room temperature) and can be referred to as "biomimetic" apatites – although applications of use may extend way beyond the biomedical field. Also, apatite can be substituted with various cations and anions that may confer additional properties, e.g. in terms of catalytic activity. Finally, by way of adsorption of molecular species, it is possible to expose on the surface of the nanocrystals some functional groups, e.g. exploiting the exchange capacity of surface ions.

In the following, we will first address stoichiometric apatites and present the main characteristics of the apatite structure (Section 3.2.1). We will then discuss the cases of nonstoichiometric and/or biomimetic apatites (Section 3.2.2).

3.2.1 Apatite Structure and Model Studies

Most apatites crystallize in the hexagonal system (typically in the $P6_3/m$ space group) [115, 116], although a few apatite compounds such as chlorapatite and HA [118] exhibit a monoclinic structure. Figure 3.2 represents the typical crystallographic/symmetry features of hexagonal calcium HA $Ca_{10}(PO_4)_6(OH)_2$. For HA, typical unit cell parameters are $a = 9.418$ Å and $c = 6.881$ Å, leading to a theoretical density of ca. $3.16\,\text{g cm}^{-3}$ [119, 120]. Using molecular simulations, both the hexagonal and monoclinic structures have been determined as well as the theoretical localization of the OH^- ions localized along the c-axis [113, 121]. The hexagonal and monoclinic symmetries are in fact strongly linked: starting from the hexagonal symmetry $P6_3/m$, the requirement to access the monoclinic structure is first to fix the position of OH^- groups in a channel along the c-axis, limiting the symmetry of the solid to $P6_3$ by losing the mirror between the OH^- ions, and then by doubling the unit cell and imposing the alternation between two adjacent channels for the orientation of OH^- ions, leading to the $P2_1$ type of symmetry [122].

Besides Ca^{2+} and OH^-, a wealth of other cations (such as Mg^{2+}, Sr^{2+}, Cu^{2+}, Zn^{2+}, Na^+, Pb^{2+}, Cd^{2+}, etc.) and anions (F^-, Cl^-, CO_3^{2-}, etc.) can also be accommodated, at least to some extent, in the apatite structure that is known to be very "open" to many substituents, thus exhibiting a high chemical versatility [117, 123]. Such substitutions may occur either at the time of synthesis (in dry or humid conditions) or during subsequent steps. For example, upon heating HA at high temperature (typically in the range 900–1200 °C depending on external conditions [124]), HA dehydroxylation leads to oxyapatite $Ca_{10}(PO_4)_6O$ where one O^{2-} ion "replaces" two OH^-. It may also be mentioned that some end-members can be called using another terminology than "apatite"; one typical example being pyromorphite for designating chlorinated lead apatite.

The three-dimensional cohesion of the structure is ensured by strong electrostatic interactions between the constitutive anions (here essentially the large

Figure 3.2 Representation of the hydroxyapatite structure (projection on the [*a*, *b*] plane), evidencing the crystallographic sites Ca1 and Ca2. Source: Ptáček [117]/IntechOpen/ Licensed under CC BY 3.0.

phosphate ions) and cations – therefore lower stability is to be expected in the case of nonstoichiometric apatites (Section 3.2.2). The determination of partial charges by quantum calculations also confirmed that the interaction between phosphate and calcium ions is mainly ionic [125] and that the monoclinic structure seems only slightly more stable than the hexagonal one [126], even if the resulting energy difference is not enough for some authors to affirm a particular preference. Concerning the structure, it has been shown that the coordination number of the different elements plays an important role in the structure. For instance, the different calcium sites have an influence on the charges carried by the cation and on the related strength of the Ca—O interaction [127], which results in different adsorption properties with guest molecules. To go further, quantum calculations coupled with nuclear magnetic resonance (NMR) results showed the impact of the presence of OH$^-$ ions on the migration of Ca^{2+} as well as the rotation of PO$_4^{3-}$ groups [128]. In the HA structure, phosphate ions PO$_4^{3-}$ are the largest components of the structure. As such, they provide a primary backbone via 3D compact-like piling, generating interstitial sites for the other ions of the structure. Metal cations such as Ca^{2+} occupy two types of crystallographic sites, denoted Ca1 (4 sites per unit formula) and Ca2 (6 sites per unit formula), see Figure 3.2. Ca1 sites are localized along the trigonal axis with the coordinates (1/4, 3/4, 1/2) and (3/4, 1/4, 1/2), forming linear columns along the c-axis. Ca2 sites in contrast form equilateral triangles at $z = 1/4$ and $z = 3/4$ on the sixfold screw (senary) 6$_3$ axes. Adjacent Ca1 and Ca2 sites are linked through shared oxygen atoms from PO$_4$ tetrahedra. Ca1 sites generate (distorted) polyhedra where the metal ion is in ninefold coordination, while in Ca2 sites it is in sevenfold coordination.

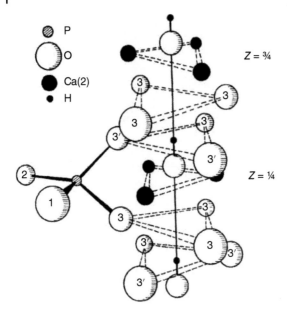

Figure 3.3 Detail of the hydroxyapatite structure: apatitic channels along the c-axis, hosting the OH⁻ ions and partially delimited by Ca2 sites. Source: Elliott [115]. Reproduced with permission of Elsevier.

The Ca2 triangles contribute to delimit so-called apatitic channels where ions such as OH⁻ are axially located (Figure 3.3) [116, 120]. In calcium HA, the Ca2 triangular sites correspond to the narrow portion of the channels with a diameter of c. 2.73 Å. At $z = 1/2$, the channels are delimited by a distorted hexagon of oxygen atoms from PO_4^{3-} ions, thus slightly enlarging the channel diameter (2.85 Å) [76]. The approximate "volume" of the channel in calcium HA may be evaluated to roughly 28 Å^3 [129], but the channels should better be seen as a linear series of ovoid cavities rather than straight cylinders [130]. A modulation of unit cell parameters e.g. by substitution of calcium/phosphates with other larger/smaller ions may modify the dimensions of these channels by either expanding or compressing the lattice (e.g. [131]). For instance, A-type carbonation increases the "a" unit cell parameter and slightly decreases the "c" parameter compared to raw HA [132]. The existence of these channels may allow the exchange of channel ions such as OH⁻ if adequate conditions are in place (e.g. substitution during the synthesis or if enough energy is provided to the system for example by heating). This explains why these compounds are sometimes compared to zeolites [133], although in apatites the channels remain rather small and monodimensional. Besides ions like Cl⁻, F⁻, OH⁻, and A-type CO_3^{2-} (i.e. substituting OH⁻ ions), small molecular species can also be incorporated in these channels such as peroxide, glycine, and acetate. There are essentially two processes by which molecular species may be entrapped into apatitic channels: via the intracrystalline decomposition of unstable lattice ions (e.g. molecular oxygenated species) and by synthesis/treatment in the presence of molecular entities to be incorporated [133].

Chemical exchange capabilities involving channel entities may prove helpful for catalysis purposes for providing active species (if conditions allowing the exchange and related ionic mobility are verified). Also, chemical substitution in the apatitic network itself can be another strategy for conferring/improving catalytic properties (e.g. vanadate substitution for the oxidative dehydrogenation of propane [50]). It may also be noted that proton diffusion along the channels may come into play in some conditions, allowing the transport of protons across the OH$^-$ channels [130, 134].

Due to the substitution capabilities of apatites, molecular simulations are a relevant complementary tool to determine expected structural changes and the evolution of the unit cell parameters for a large range of compositions, and then compare with experimental data such as X-ray diffraction (XRD). This is the main validation for force fields specifically developed for stoichiometric apatites, and when simulation conditions are validated, thermodynamic, dynamic, or mechanical properties induced by substitutions can be calculated [135, 136]. In the case of nonstoichiometric samples, it is required to localize the position of the charge deficit (vacancies) and try to compensate the charges in order to obtain a neutral structure. This last situation is more difficult to address due to the models issued from the numerous substitution possibilities [137]. Some theoretical results have shown that, for nonstoichiometric apatites, the vacancies are most probably situated in Ca2 sites with a charge compensation resulting from the incorporation of H in the structure (H_2O instead of HO^- and HPO_4^{2-} instead of PO_4^{3-}). Such calculations thus concluded that hydration has a strong impact on the apatite structure [138].

After understanding the structure of HA as a reference apatitic compound, the description of the substitution positions (in simple cases where only ions with similar charges are considered) or several substitution-charge compensations concerted mechanisms have been elucidated using molecular simulations. This is the case of carbonate ions (CO_3^{2-}), one of the main substitutions occurring in apatites, either on purpose (precipitation in the presence of a source of carbonates) or by contact with atmospheric CO_2 in some preparation conditions. Carbonate ions can be found in substitution of hydroxyl groups (type-A defaults) or phosphate groups (type-B defaults). Quantum calculations have determined that substitutions could occur in both default sites, strongly modifying the electrostatic potential of the solid and therefore the adsorption performances [139, 140]. In addition, the corresponding charge compensation mechanisms were investigated by invoking the presence of supplementary OH^- groups or the substitution of Ca^{2+} by alkaline ions [141, 142] (Figure 3.4). As a typical illustration, the interaction between Na^+ and CO_3^{2-} seems to play an important role in the limitation of structural distortion. From the comparison between calculations with substitutions in A, B, or simultaneously in A and B sites (see Section 3.2.2), it appeared that occupancy of CO_3^{2-} in both sites could result as a configuration with a minimum energy [141], confirming that this mechanism is plausible while other authors have shown that other configurations (involving only sites A or B) are preferential. This discrepancy is mainly based on the chosen charge compensation mechanisms [140, 143]. The substitution of OH^- groups by halogens (F^-, Cl^-, Br^-) has also been the subject of many theoretical studies. In particular, fluorapatite was widely investigated, demonstrating that the structure is essentially ionic in this case [144]. In addition, the transport of F^- ions in the structure seems to impose a concerted mechanism along the c-axis, implying both framework and interstitial sites: one interstitial site replaces an ion in a structural site, which can replace an interstitial [145], and so on. Regarding mixtures of halogens (F^- and Cl^-), the

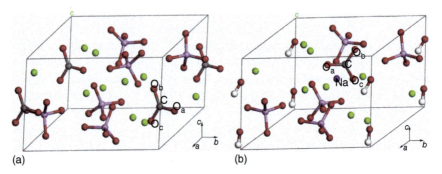

Figure 3.4 Illustration of the most energetically favored configuration for type-A carbonated apatite (a) and the most stable structure of type-B carbonated apatite with charge compensation by cosubstituting a Na^+ ion for a Ca^{2+} ion (b), after geometry optimization. Source: Ren et al. [142]. Reproduced with permission of Elsevier.

most stable structure was defined by distinct channels containing either F^- or Cl^-, with an equimolar composition. In the case of combined F^-/CO_3^{2-} substitution, the fluoride ions preferentially adopt interstitial positions [127].

The substitution of Ca^{2+} by other divalent cations has also been characterized by molecular simulations in link with the possibility to enhance the catalytic properties [146]. From molecular simulations, it has been concluded that the position of the substituents is strongly dependent on the nature of the cation (mainly its charge, radius, and electronegativity) and on the spatial constraints: for instance Zn^{2+} preferentially occupies the Ca2 sites (see Figure 3.5 as a typical illustration), which is less obvious for Mg^{2+} [122, 127, 148, 149]. In contrast, substituting Sr^{2+} cations is preferentially situated in Ca2 sites for an amount above 10 at% but in Ca1 sites below 5 at% [150], even if some authors proposed a different threshold value between the two preferential sites (e.g. 20 at%) [151]. A similar behavior was observed for Ba^{2+} [152]. In the case of Pb^{2+}, a reorganization of the oxygen atoms around the cations was

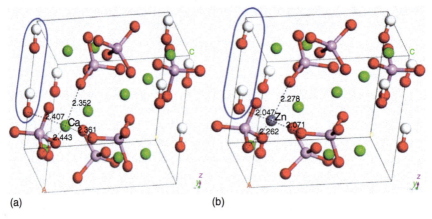

Figure 3.5 DFT-optimized structures for (a) HA and (b) Zn-doped HA. Source: Shang et al. [147]. Reproduced with permission of Elsevier.

observed (in both Ca1 and Ca2 sites) and the substitution energy appears strongly impacted by the formation of covalent-like bond between divalent cations and oxygen [152]. Finally, few studies have focused on the codoping for cations [148] or cations and anions [153]. The doping in Ag^+ and Sr^{2+} has for example been investigated from a modeling point of view and, while the two cations individually replace Ca^{2+} in sites 2, the mixture has a different behavior with Sr^{2+} placed in Ca1 sites and Ag^+ in Ca2 sites [154].

Although progresses have been made in the last decades on the modeling of apatite compounds, additional work is needed in this field. Due to the numerous possibilities of substitutions and therefore compensation mechanisms (vacancies for OH^- and Ca^{2+}, substitution of Ca^{2+} cations, substitution of PO_4^{3-} by other anions, etc.) combined with different database sets and approaches, theoretical calculations have sometimes led to very different results considering the same apatitic compounds [140, 141, 143]. It is therefore still necessary to clarify the mechanisms involved during these substitutions by coupling experiments and theoretical calculations in order to fix the conditions leading to the calculated results and to better ascertain their accuracy.

Knowing the bulk structure, the determination of the surface structure(s) is of primary relevance in the catalysis field, so as to be able to analyze the surface activity. Then, the elucidation or prediction of the ionic conductivity and polarization mechanisms can be of great importance to better apprehend the catalysis activity of doped apatites. For that purpose, molecular simulations should bring appropriate solutions to clarify the different components present in the solid phase and involved in the catalytic mechanisms. However, to date, few simulations have been performed in this way [155]. Indeed, while the simulation of apatitic structures doped with trace elements has already allowed adjusting the mechanical or antibacterial properties of apatites for biomedical applications, similar calculations to extract the electronic properties and potential interactions should be proposed for catalysis systems [156, 157].

3.2.2 Specificities of Nonstoichiometric and/or Biomimetic Apatites

As mentioned previously, apatite compounds can exhibit different characteristics that strongly depend on their synthesis conditions. While low temperature synthesis processes (lower than 100°C, typically close to room temperature) usually generate nanocrystalline nonstoichiometric apatites, also referred to as "biomimetic" [158–160], apatite samples prepared at higher temperatures are often close to stoichiometry (providing that precipitation is achieved in basic medium) as the thermal activation of ions allows for faster diffusion and greater crystallinity. When short heating periods are used for sample preparation at high temperature, apatites composed of nanosized crystals may sometimes be obtained; however, in most other cases, significantly larger crystals are formed.

The physicochemical and morphological features of biomimetic apatites have been explored in detail in the recent years due to technological evolutions of materials characterization techniques. They exhibit specific properties mainly related to their surface structure that exhibits non-apatitic features in the form of an ionic hydrated layer [161, 162]. From a chemical point of view, the composition of biomimetic apatites thus strongly differs from that of HA [163]. In comparison with stoichiometric HA (molar ratio of Ca/P equal to 1.67), which is the most stable and least soluble calcium phosphate at ambient conditions, biomimetic apatites present a nonstoichiometric composition (Ca/P lower than 1.67).

One important factor in the comprehension of apatites behavior is the possible departure from stoichiometry. Nonstoichiometry is related to vacancies in M sites and Y sites (see Eq. (3.1) in Section 3.1). Several types of nonstoichiometric apatites can be distinguished depending on the substituents and vacancies present in these structures. The nonstoichiometry starting from the HA composition may be explained on the basis of vacancies in both Ca and OH sites, concomitantly to the substitution of PO_4^{3-} ions by bivalent ions, like HPO_4^{2-} or CO_3^{2-} [158]. This incorporation of bivalent ions are mainly compensated by a complex defect associating of anionic and cationic vacancies, partially confirmed by Rietveld analysis [164]. Although some variations can be observed, such nonstoichiometric HA generally respond to the global chemical composition [163]:

$$Ca_{10-x}(PO_4)_{6-x}(HPO_4 \text{ or } CO_3)_x(OH \text{ or } \tfrac{1}{2} CO_3)_{2-x} \tag{3.2}$$

with $0 \leq x \leq 2$. It can be noticed that carbonate ions may occupy different types of sites historically named as A and B, where sites A correspond to carbonate substitution in the monovalent anionic sites of the apatite structure (substitution of some OH^- ions in the case of HA) and sites B to trivalent anionic sites (substitution of some PO_4^{3-} ions).

The nanocrystalline character is another important feature to be taken into account when dealing with apatite compounds. HA generally exhibits typical rod-like or platelet-like morphology (depending on the synthesis method), elongated along the c-axis of the hexagonal apatitic structure, and nanosize dimensions of the crystals allow increasing the extent of surface area being potentially activated and therefore impact surface interaction [165]. For example, the mean crystallite dimensions of biomimetic apatites reach about 15–30 nm length and 6–9 nm width, as opposed to micron-sized crystals for sintered HA [159]. The crystallinity degree of these apatites is then rather low, and they are thus generally referred to as "poorly crystalline" apatites in the literature [159, 162]. It may, however, be remarked that, in some cases, unsintered HA precipitated close to the boiling point of water may also be composed of nanosized crystals. Therefore, several approaches can be developed to modulate the surface area of apatitic samples.

In addition to the nanosized dimensions and nanorods platelet morphology, biomimetic apatites exhibit on their surface a non-apatitic ionic and hydrated layer that provides very specific surface reactivity. The existence of this structured but non-apatitic surface layer was evidenced especially by spectroscopic techniques as Fourier transform infrared (FTIR) [166] and solid-state NMR spectroscopy [167] (see Section 3.3.1). Such biomimetic apatite nanocrystals may then be described as an apatitic core (often nonstoichiometric) covered by an ionic and hydrated layer containing rather labile ions (Ca^{2+}, HPO_4^{2-}, CO_3^{2-}, etc.) [159], as illustrated in

Figure 3.6 Schematic model of the peculiar structure of biomimetic apatite nanocrystals. Source: Stéphanie Sarda, Christophe Drouet.

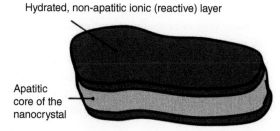

Figure 3.6, that can take part in exchange or adsorption reactions. This hydrated layer should not be confused with a Stern double layer. Instead, this is the result of the precipitation process of biomimetic apatites. Moreover, the hydrated layer, particularly important in freshly precipitated apatites, progressively transforms into the more stable apatitic lattice upon aging in aqueous solution and progressively disappears as the stable apatite domains (in the core of the crystals) develop with time, incorporating ions from the hydrated layer in the apatitic lattice (maturation) [161, 162, 168]. This "maturation" phenomenon can be slowed down when inhibitors of apatite crystal growth like Mg^{2+} or CO_3^{2-} are present in the hydrated layer but cannot be stopped [161]. The presence of this hydrated surface layer is responsible for most of the properties of biomimetic apatites and in particular their high surface reactivity. The ionic substitution and molecular adsorption capabilities of the nanocrystals are greater than those of stoichiometric HA due to the high mobility of the mineral ions from the hydrated layer [159, 161], dependently on the composition of the hydrated layer and on the maturation stage (see Sections 3.3.4.1 and 3.3.4.2).

In computational studies, the activities and description of apatite surfaces have been generally characterized using a slab or nanopore model. However, these models are not perfectly adapted for studying the processes occurring on apatite surfaces. More recently, the understanding of the structure of biomimetic nanocrystalline apatites has focused researchers' attention, revealing the coexistence on the nanocrystals of an ordered core and a disordered surface [169], in good agreement with experimental evidence. The interest of the molecular dynamics (MD) simulations is also to probe the impact of the temperature on the apatite structure and to quantify the disorder of the nanocrystals [169, 170]. To better understand the chemical composition of surface, density functional theory (DFT) calculations have been combined with NMR results and the estimation of chemical displacements was compared to experimental data, evidencing the impact of the atoms localization, as surface atoms exhibit a larger displacement compared to atoms from the bulk. The disorder observed for OH groups modifies the signal of P and allowed proposing a description of HA interfaces [128].

The distribution of acidic sites and basic sites on the apatitic surface depends on the apatite conditions of formation and characteristics such as Ca/P ratio and the presence of substituents. Ionic substitutions in the apatitic structure may indeed allow regulating the acidity and alkalinity of the crystal surface [171, 172] (see Section 3.3.2.2). Thus, the tunable composition and surface properties of substituted (often nonstoichiometric) apatites have been widely used as acid–base catalyze for many classes of catalytic reactions (hydrolysis, alcoholysis, esterification, etc.) [76, 95, 173]. However, the excellent physicochemical and surface properties of biomimetic (i.e. both nonstoichiometric, nanocrystalline and presenting a non-apatitic surface layer) have essentially been exploited to date for biomaterials applications [163] but have only recently developed in heterogeneous catalysis, which is thus still a domain to examine.

3.2.3 Relevance of Apatites in Catalysis

Bulk and/or surface composition to account for acid base properties?
The acid–base balance of the surfaces of HA and HA-modified systems are of primary interest for catalytic reaction mechanisms. The originality of the HA system is related to the fact that it can be used as acid catalysts (for example, Friedel–Crafts [41], Diels–Alder [25], and dehydration [2, 56, 91, 94–99, 174]), as basic catalysts (for example, dehydrogenation of alcohol [90, 174], hydrogen transfer [7, 90], Knoevenhagel, [11, 15, 17, 60], Claisen–Schmidt [61, 62], and Michael [47, 63, 64, 175]), as well as bifunctional acid–base catalysts (for example, Aldol reactions [176] and cascade reaction such as Guerbet reaction [8, 20, 59, 65–76]). Interestingly, the surface reactivity over HA is structure sensitive: both the crystallinity and the stoichiometry were found to influence the catalytic behavior. Despite being detrimental to the specific surface area, high crystallinity was reported to promote the performances of the system (even if only the surface is directly involved in the reaction) [177], indicating that the surface arrangement of the different active sites is also important, probably due to the involvement of cooperating effects between different sites working as acid–base pairs in concerted mechanism or in successive elementary steps. Moreover, many catalytic results are discussed in line with the stoichiometry of the bulk material [66, 68, 94, 95, 99, 100, 105, 178–181] using the Ca/P bulk ratio that can be varied by the preparation conditions to account for the modulation of the acid–base properties: globally, the lower the Ca/P molar ratio, the higher is the acidity, and the higher the Ca/P molar ratio, the higher is the basicity [20, 66, 70, 94, 95, 99, 105, 172, 180, 181]. This is illustrated in Figure 3.7 on the basis of temperature programmed desorption (TPD) experiments of basic NH_3 and acid CO_2 molecules that are commonly used to quantify the total amount of surface acid and base sites, respectively [99]. Similar tendencies were reported from the model conversion of isopropanol accounting for acid and base sites: a higher selectivity to acetone (basic dehydrogenation route) than in propylene (acidic dehydration route) is obtained as the Ca/P molar ratio increases [172]. However, from another model reaction, i.e. the basic conversion of 2-methylbut-3-yn-2-ol (MBOH) into acetone and acetylene reaction that is a very sensitive tool to probe basic reactivity [182], the situation appears more complex: the use of the Ca/P descriptor is valid at the first order when comparing an homogeneous series of compounds prepared in very similar conditions [183], but some inconsistencies with the Ca/P macroscopic ratio were encountered when considering samples prepared in different pH conditions, which was ascribed to very different carbonation contents [184].

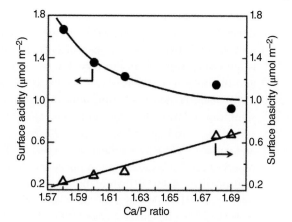

Figure 3.7 Influence of the Ca/P molar ratio on the surface density of acidic and basic sites probed by NH_3 and CO_2 TPD (temperature programmed desorption). Source: Yan et al. [99]. Reproduced with permission of American Chemical Society.

In fact, the origin of the empirical relationships between bulk composition and surface reactivity is very questionable. Based on the simplified $Ca_{10-z}(HPO_4)_z(PO_4)_{6-z}(OH)_{2-z} \cdot nH_2O$ ($0 < z \leq 1$), which is the formula of nonstoichiometric apatites, it may roughly be explained on the one hand by the larger amount of HPO_4^{2-} (likely to account for the PO–H Brønsted acidity on the surface) that is associated with Ca^{2+} deficiency in under-stoichiometric compounds [66] and, on the other hand, to the higher concentration in basic OH^- in stoichiometric compounds [184]. Note that, by analogy with the case of CaO, the basicity was sometimes simply claimed to be associated to the presence of Ca^{2+} [66], which is a misinterpretation since the basic sites in CaO are O^{2-} and surface OH^- [185]. In addition, such a rapid explanation suffers from several limitations:

i) It completely skips the possible influence of cation content (Ca^{2+} or other ions in the case of substitution [coprecipitation or ion exchange] or grafted complex) likely to provide by themselves Lewis acidity (quite weak of Ca^{2+}, but possibly stronger depending on the nature of hetero-cation involved) [49].

ii) There is a lack of data for so-called over-stoichiometric compounds, since the influence of bulk carbonation responsible for this over-stoichiometry is rarely discussed [68]. Moreover, high molar ratio of Cation/P values can result from synthesis carried out in the presence of a very large excess of cation (Ca(+Na)/P) precursor ratio up to 5.5 [172]: this particular case may correspond to highly Ca^{2+} doped materials with formation of CaO-like domains [68, 172]. In this later case, the acid–base properties are no more related to the apatite active phase only.

iii) Due to the complex general formula $Ca_{10-x-B}\square_{Ca, x+B}(PO_4)_{6-x-B}(HPO_4)_x(CO_3)_{A+B}(OH)_{2-x-2A-B}\square_{OH, x+2A+B}$ where A and B account for A- and B-type carbonate contents and $\square$ stands for Ca and OH vacancies ($\square_{Ca}$ and $\square_{OH}$), some antagonistic compensating effects (Ca deficiency and B-type carbonation) are hidden by the global $Ca/P = (10 - x - B)/(6 - B)$ value [184]. In addition, the Ca/P ratio does not account for the A-type carbonation. However, it greatly impacts the concentration in OH^-. This indicator was found to be a better bulk descriptor than the Ca/P molar ratio to account for basic properties, which directs to a key role played by OH^- in the basic properties [184].

iv) Correlation between the surface reactivity and the bulk Ca/P molar ratio value is quite surprising given the very different compositions between the surface and the bulk. The surface layer probed by X-ray photoelectron spectroscopy (XPS) analysis (depth ~ 10 nm) indicates that the surface Ca/P ratio is always significantly lower than the bulk one [66, 172, 186] even when considering crystalline terminations [187, 188]. Such a discrepancy between the bulk and surface Ca/P values seems to be specific of the HA system since it is not observed for the beta-tricalcium phosphate (β-TCP) [66]. Consistently, Ospina et al. predicted from DFT simulations and high-resolution transmission electron microscopy (HRTEM)) that surface terminations of crystalline HA particles can be enriched in phosphate species [189]. This is also confirmed by ion scattering spectroscopy (ISS) that enables to analyze the evolution of the Ca/P ratio of the first atomic surface layers due to a progressive scraping of the surface via He^+ bombardment (few Å probed) concluding to a gradient of concentration of calcium, with a progressive relative enrichment in calcium in the deepest layers: hence part of calcium tends to be relaxed in the sub-surface, rather than at the top surface [9, 187]. Interestingly, in the case of vanadium-substituted $Ca_{10}(VO_4)_x(PO_4)_{6-x}(OH)_2$ stoichiometric apatite solid solution, a reverse tendency was observed, where vanadate groups are relaxed in the sub-surface leading to overexposure of calcium cations (surface $Ca/(P+V) > 2$ for $x \geq 4$ whereas bulk $Ca/(P+V) = 1.67$) [9].

As mentioned earlier, surface characterization approach based on NH_3 and CO_2 TPD measurements is widely used. It leads to quite rough correlation tendencies between the total amount of acidic and basic sites probed with catalytic performances. For instance, in the case of the production of acrylic acid from dehydration of lactic acid [99] or of ethanol to n-butanol Guerbet conversion reaction [68], a volcano-type dependence between the reaction rate and ratio between the amount of acid and basic sites was reported. This underlines the cooperation between acid and base sites in a concerted mechanism or in complex cascade network. The use of CO_2 to titrate the basic sites has, however, several drawbacks and caution should be taken: pretreatment has to be carefully done to desorb the surface carbonate ions prior to adsorption (the infrared (IR) fingerprints of surface carbonate and hydrogencarbonate were discriminated from those of the bulk [190]) and blank experiment or limited temperature should be applied to avoid misleading interpretation due to high temperature bulk carbonate decomposition that would result in overestimating the amount of basic sites. In addition, CO_2 is known to react with the surfaces [94] and, especially, being also a soft Lewis base, it might interact with strong Lewis acid [191] (especially in the case of HAs modified with strong Lewis acid cations). Recently, SO_2, which is a stronger acid molecule than CO_2, was reported to be more convenient to titrate all basic sites over HA [94]. Irrespective of the accuracy of TPD characterization to quantify the acid and basic sites of HA, only the total amount of sites is probed (eventually discussed in terms of weak, medium, or strong strengths), but it does not provide information on the chemical nature of the sites probed. When there are several types of basic or acid sites present on the surface (typical case of HA with Brønsted and Lewis acid sites), it is moreover questionable if the total amount or only a fraction is really involved for a given catalytic reaction.

All these considerations emphasize the need for refined surface characterizations, including *operando* approaches [75] to identify the nature of active species more precisely and to rationalize the surface reactivity.

3.3 Advances in the Characterization of Structural and Surface Properties of Hydroxyapatite: Experimental and Computational Approaches

3.3.1 Structural and Compositional Characterization

The atomic Ca/P molar ratio, equal to 1.67 for stoichiometric (non-carbonated) apatite, is an important feature of the composition of apatitic compounds. For nonstoichiometric non-carbonated apatites (Ca/P lower than 1.67) where some HPO_4^{2-} ions substitute to PO_4^{3-}, the Ca/P molar ratio is directly related to the amount of cationic (and OH) vacancies. For nonstoichiometric carbonated apatites where CO_3^{2-} ions substitute PO_4^{3-} ions (B-type carbonate apatites), cationic vacancies are related to the Ca/(P+C) ratio. In more complex compositions for example with Sr^{2+} cationic substitution, the (Ca+Sr)/(P+C) molar ratio has in turn to be taken into account to relate to the amount of vacancies [120]. In practice, depending on the preparation conditions, except careful attention devoted to the decarbonation of the precursor and base solutions, A- and B-type carbonate, as well as bulk HPO_4^{2-} coexist in the apatitic compounds resulting in complex general formula, $Ca_{10-x-B}\square_{Ca,x+B}(PO_4)_{6-x-B}(HPO_4)_x(CO_3)_{A+B}(OH)_{2-x-2A-B}\square_{OH,x+2A+B}$ with antagonistic effects on the Ca/P ratio and huge impact of the amount of OH vacancies [184]. To determine these ratios, various chemical analyses can be used. The amount of calcium (and other substituting cations) can be determined using several techniques including atomic adsorption spectroscopy (AAS), inductively coupled plasma (ICP) emission spectroscopy, ionic chromatography, ionometry using specific electrodes, complexometry with a complexing agent as ethylene diamine tetracetic acid (EDTA), etc. [192]. The total phosphorus content (included in PO_4^{3-}, HPO_4^{2-}) can be titrated by ICP or ionic chromatography. However, to get the orthophosphate ions amount, titration by UV spectrophotometry of the phospho-vanado-molybdic acid complex is among the most used and most suitable. The amount of HPO_4^{2-}, related to nonstoichiometry, can be subsequently measured by difference using a second run of the same method after thermal conversion of HPO_4^{2-} into pyrophosphate ions [193]. For carbonate content titration, an accurate method is coulometry, exploiting an electrochemical cycle in acidic conditions. It can be noticed that FTIR spectroscopy can be used to determine total carbonate content relatively to phosphate bands [194], but also apatitic and non-apatitic

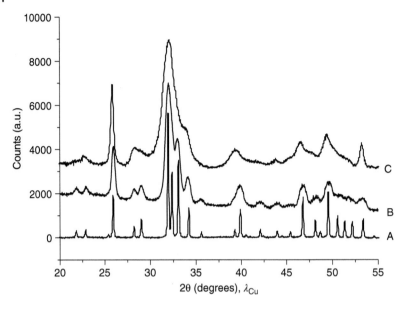

Figure 3.8 Typical XRD patterns for sintered stoichiometric HA (A), freeze-dried stoichiometric HA (B), and for biomimetic apatite (C) evidencing different degrees of crystallinity. Source: Stéphanie Sarda, Christophe Drouet.

PO_4^{3-} and HPO_4^{2-} content (in the case of biomimetic apatite), from mathematical decomposition of the ν_4 band related to the P—O bond of FTIR spectrum [195, 196]. Moreover, a very accurate technique for the determination of Ca/P molar ratio by XRD has been developed (ISO standard 13779) for non-carbonated samples.

As previously detailed in Section 3.2.1, apatites generally crystallize in the hexagonal system ($P6_3/m$ space group) [115, 116] and XRD is an essential method for the characterization of HA and related compounds [197]. The peaks position in XRD patterns allows for the identification of apatitic calcium phosphates as recorded in the databases of diffraction peaks of HA (JCPDS file number 9-0432) (Figure 3.8). In addition to phase recognition, XRD can be used to estimate crystallographic parameters such as the unit-cell dimensions, mean crystallize dimensions, and crystal strain parameters, and thereby evaluate the effects of ion substitutions on the structure. Substitution of calcium or phosphate ions by larger entities induces an increase in the unit-cell parameters; however, the replacement of a monovalent anion (e.g. OH$^-$) by a larger one induces an increase in the "a" parameter and a decrease in the "c" parameter [120, 198]. Moreover, more elaborate treatments using Rietveld analysis of XRD data can be performed for more accurate determination of crystallite dimension, microstrains, unit-cell parameters, and phase composition, but also structural information such as the atomic position of substituted ions or vacancies. Such data treatments have been performed for example to study the orientation of carbonate species and the location of vacancies in nonstoichiometric carbonate apatites [164] or the location of strontium ions in Sr-containing apatites [150].

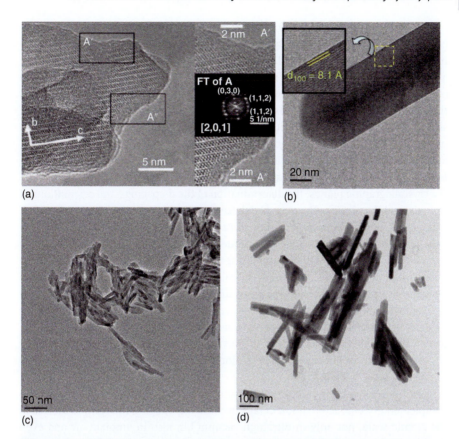

Figure 3.9 HRTEM pictures of (a) an apatite synthesized at 40 °C (Source: Sakhno et al. [199]/with permission of American Chemical Society); (b) a stoichiometric hydroxyapatite obtained by coprecipitation at 80 °C and high pH (after drying at 100 °C and thermal treatment under Ar flow at 350 °C) indicating it is crystalline up to the surface (Source: Ben Osman et al. [188]/with permission of American Chemical Society); TEM observations of (c) a biomimetic nanocrystalline apatite synthesized at ambient temperature (aged in solution for one month) and freeze-dried; (d) typical rod-like morphology of crystalline stoichiometric hydroxyapatite obtained by coprecipitation at 80 °C and high pH and dried at 100 °C (Source: Diallo-Garcia et al. [190]/with permission of American Chemical Society).

Besides XRD, electron microscopy techniques are widely used for the determination of apatitic crystal morphology, size, and organization [120]. Indeed, apatitic calcium phosphates can show a variable morphology depending on their synthesis conditions. While precipitated ACPs appear most generally as spherical units (diameter between 20 and 300 nm), well-crystallized apatites generally exhibit a rod/needle-like morphology with crystals elongated along the c-axis of the hexagonal structure (Figure 3.9). Some approaches such as hydrothermal synthesis may lead to apatite crystals with a very high length/width ratio that can be considered as fibers [200].

In addition to morphological and dimensional information, electron microscopy coupled with energy dispersive spectrometry (EDS) or wavelength dispersive spectrometry (WDS) can also be used for composition determination. These techniques are used for the determination of main elemental components of calcium phosphate compounds (as Ca, Mg, P, F). However, their accuracy depends on the stability and the characteristics of the apatitic sample (surface roughness, conditions of analyses, size), and the analysis of nonstoichiometric apatites remains difficult even with the use of adequate standards. Identification of phases at the microscopic level using electron diffraction, as selected area electron diffraction (SAED), can also be performed on calcium phosphate samples. HRTEM can also allow the identification of the crystallographic planes, the determination of the d-spacings, the orientations of crystals, and the observation of crystal defects of apatitic compounds.

It should, however, be noticed that all of these beam techniques under vacuum can generate sample damages, as frequently observed with hydrated apatitic compounds, especially with biomimetic apatites and may ultimately lead to phase decomposition [201, 202]. Moreover, calcium phosphates are nonconductive compounds, and it is often necessary to limit the charge effect via for example a coating with an electrically conductive layer. Working at low accelerating voltage at short working distance or under low vacuum should then be preferred or the use of environmental microscopes to prevent the alteration of hydrated apatitic compounds. The use of cryogenic conditions during electron microscopy analyses can also be recommended.

In addition to electron microscopy, atomic force microscopy (AFM) may lead to complementary characterization details by allowing imaging the sample surface at the atomic scale, not only in ultrahigh vacuum but also in ambient air and even liquid environments [162]. AFM is based on the detection of a tip/sample surface interaction force, giving a map of surface topography [203]. Moreover, in situ AFM can highlight the interaction of ions or molecules with HA surfaces in aqueous environments at the nanometer scale [204], as for example the interaction between HA in contact with citrate-containing solutions [205] or with proteins [206].

For surface assessments complementary to microscopy techniques, the specific surface area of the samples can be measured according to the BET (Brunauer–Emmett–Teller) method via nitrogen adsorption. However, since an outgassing initial stage is necessary to evacuate pre-adsorbed H_2O/CO_2 and other physisorbed species, care should be taken also when analyzing hydrated (biomimetic) apatites so as to limit crystal degradation.

Spectroscopic techniques (FTIR, Raman, solid-state NMR, etc.) are particularly useful for characterizing apatite compounds as they are in particular sensitive to the local environment of phosphate, carbonate, and hydroxide ions. Vibrational spectroscopies (FTIR, Raman) appear in particular as interesting tools for the identification of such compounds. As explained previously, apatites may exhibit complex structures due to ion substitutions and vacancies and they can be poorly crystallized, which drastically limits the potential of XRD. Also, the principal advantage of vibrational spectroscopies over diffraction techniques is the identification of fine structural details, e.g. concerning the presence and location of PO_4^{3-}, HPO_4^{2-}, CO_3^{2-}, or

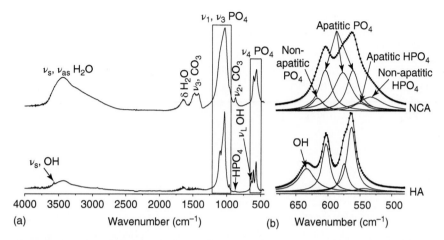

Figure 3.10 (a) FTIR spectra of well-crystallized stoichiometric hydroxyapatite (HA) and biomimetic nanocrystalline carbonated apatite (NCA) (with six days of maturation) and (b) the corresponding decomposition in the ν_4 PO$_4$ domain. Both were prepared by double decomposition, filtered, washed, and freeze-dried. Source: Errassifi et al. [207]. Reproduced with permission of Elsevier.

OH$^-$ groups within the (nano)crystals or on the surface (see Section 3.3.4). Moreover, these techniques are non-invasive and can adapt to a wide range of experimental conditions (temperature, wet conditions, etc.).

Involving transitions between vibrational levels, these methods can particularly detect phosphate, carbonate, hydroxide ions, and water molecules in apatite spectra (Figure 3.10). The PO$_4^{3-}$ groups present four main vibration domains: ν_1 (around 950 cm^{-1}), ν_2 (400–470 cm^{-1}), ν_3 (1000–1150 cm^{-1}), and ν_4 (500–620 cm^{-1}), and the main FTIR and Raman line positions of HA are summarized in Table 3.1. In addition to phosphate groups vibrations, several OH$^-$ ion vibration bands can be distinguished: one domain corresponding to the stretching of the O—H bond (3400–3720 cm^{-1}) and a libration band, corresponding to a rotational energy level of the OH$^-$ ion in the lattice (630–750 cm^{-1}) [120, 196]. The OH$^-$ vibration and libration modes are very sensitive to the nature of the surrounding cations and to hydrogen bonding; it has for example been shown in fluoride- and chloride-containing apatites that the OH bands can be shifted compared to their position in HA [210]. As for water molecules, they exhibit three vibrational modes: the symmetric and antisymmetric stretching vibration, very close in energy and difficult to distinguish, giving a broad absorption band on FTIR spectra (3000–3700 cm^{-1}); and the H-O-H bending vibration displaying a narrower band at about 1640 cm^{-1}. Like OH$^-$, water molecules are very sensitive to hydrogen bonding. In addition to the characteristic fingerprints of OH running along the c channels (3570 cm^{-1}), or water molecules, lower intensive bands in the 3720–3640 cm^{-1} range, detected from self-supported pellet, are ascribed to ν_{PO-H} from bulk defective hydrogenphosphate (3657 cm^{-1}) and surface terminated phosphate groups [211] (see Section 3.3.4).

Table 3.1 FTIR and Raman line positions of phosphate ions in hydroxyapatite (cm^{-1}; m: medium, s: strong, sh: shoulder, v: very, w: weak).

	FTIR (cm^{-1})	Raman (cm^{-1})
ν_3 PO$_4$	1092 s	1077 w
		1064 w
		1057 w
		1048 w
	1040 vs	1041 w
		1034 w
		1029 w
ν_1 PO$_4$		964 vs
	962 w	
ν_4 PO$_4$		614 w
	601 m	607 w
		591 w
	575 m,sh	580 w
	561 m	
ν_2 PO$_4$	472 vw	448 w
	462 sh	433 w

Sources: Fowler et al. [208] and Penel et al. [209]. Adapted with permission of Elsevier.

Ionic substitution in nonstoichiometric HAs can also be evidenced by FTIR and Raman spectroscopies. For example, carbonate ions, present in biological apatites and in carbonate-substituted synthetic apatites as explained in Section 3.2, are active in four main vibrational domains: ν_1 (around 1050 cm^{-1}), ν_2 (820–900 cm^{-1}), ν_3 (1400–1550 cm^{-1}), and ν_4 (650–750 cm^{-1}) [120] (Figure 3.10). But A- and B-type carbonates give vibrations that essentially differ from each other, allowing for more in-depth carbonate analysis. Also, in biomimetic apatites, labile carbonates (LC type) can also be present, although their exact band location is still a matter of controversy. Several slight shifts can also be observed among carbonated apatites due to vacancies or water molecules or combination of the two main carbonate substitutions (A and B sites) [196, 212]. Moreover, besides bands ascribed to bulk A- and B-type carbonates, given the basic properties of HA, there are surface carbonates on crystalline HA surfaces. In the absence of in situ thermal treatment prior to IR spectrum recording, typically general case of KBr diluted samples and

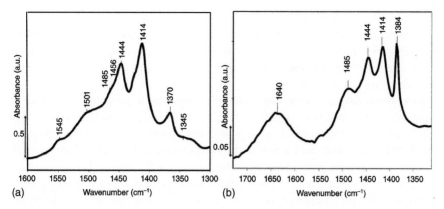

Figure 3.11 Comparison of FTIR spectra of a same crystalline hydroxyapatite sample recorded a) from self-supported pellet pretreated at 523 K and b) from dilution in KBr. Source: Diallo-Garcia et al. [190]. Reproduced with permission of American Chemical Society.

their involvement in the $\square_{CO}$ components have to be also considered to assign the various bulk and surface contributions (Figure 3.11) [190].

As explained previously (Section 3.2.2), one of the main properties of highly reactive biomimetic apatites (whether biological or their synthetic analogues) is the presence of "non-apatitic" environments of mineral ions (Ca^{2+}, HPO_4^{2-}, CO_3^{2-}, etc.) in a structured metastable surface-hydrated ionic layer surrounding an apatitic core (often itself nonstoichiometric) [166, 168]. For example, detailed analyses of the phosphate groups by FTIR reveal the presence of additional bands that cannot be attributed to phosphate groups in a regular apatitic environment and that are not present in stoichiometric HA; it is best observed in the ν_4 PO_4 vibration domain of FTIR spectra of biomimetic apatites where two shoulders can be found beside apatitic bands, respectively assigned to non-apatitic PO_4^{3-} ions (around 617 cm^{-1}) and non-apatitic HPO_4^{2-} ions (around 535 cm^{-1}), and FTIR spectra can be used to provide a sufficiently accurate evaluation of the amounts of such environments (Figure 3.10) [159, 168]. This method may thus prove helpful in the characterization of upcoming catalysts based on biomimetic apatites, in order to evaluate the location of the acidic HPO_4^{2-} sites, keeping in mind that the non-apatitic HPO_4^{2-} ions are exposed on the nanocrystals surface.

Similar observations have been made for carbonated biomimetic apatites (and they have been confirmed by solid-state NMR measurements [213, 214]): a characteristic IR band around 866 cm^{-1} is clearly visible by FTIR in the ν_2 CO_3 vibration domain attributed to the previously mentioned labile (non-apatitic) surface environments of carbonate ions, distinct from apatite domains, and incorporated in the surface-hydrated layer, which can provide quantitative assessment of the amount of non-apatitic carbonate ions in the sample. Moreover, ^{13}C carbonate-containing solution allowed following carbonate species to throw surface exchange reactions, and

FTIR analysis confirmed the existence of different domains for carbonate ions inside the apatitic domains and on the surface in the hydrated layer [196]. In addition, X-ray absorption spectroscopy experiments have shown the existence of non-apatitic environments of calcium ions in the nanocrystals, not observed in well-crystallized stoichiometric apatites [215].

It can be noticed that these non-apatitic features tend to progressively disappear during the aging in solution (named "maturation process" as explained in Section 3.2.2) due to the metastability of such poorly crystallized nonstoichiometric apatites in solution, evolving toward stoichiometry in a thermodynamically driven manner, leading to enhanced crystallinity and the progressive growth of the nanocrystals [216]. Moreover, vibrational spectroscopies used on wet samples (i.e. freshly precipitated) revealed the fine structuration of non-apatitic band that is lost upon drying [217], accompanied by a loss of reactivity of the nanocrystals on aging related to the decrease of proportion of the unstable surface hydrated layer and its labile constituting surface ions [196]. It can finally be noticed that the alteration of the ionic environments in the nanocrystals upon aging can be evidenced by FTIR analysis upon experiments using isotopic substitution experiments [196].

Complementary to vibrational spectroscopies, solid-state NMR constitutes another very powerful technique for the characterization of apatitic compounds. This method allows for the exploration of the short-range environment of different nuclei possessing a magnetic moment, such as ^{1}H, ^{13}C, ^{31}P, and ^{19}F, which can be present in (substituted) apatites [218, 219]. Despite exhibiting much more narrow signals than biomimetic apatites, crystalline stoichiometric or nonstoichiometric HA samples exhibit multicomponents ^{31}P and ^{1}H signals (Figure 3.12). They were identified by combining H—D exchanges procedures, specific NMR sequences including inversion recovery measurements, and 2D Hetcor and DQSQ spectra [188]. Besides main signals associated with adsorbed water, structural phosphates, and OH groups (including two resolved contributions related to up and down orientations of their related protons), additional NMR signals could be resolved. They were ascribed to the bulk defective hydrogenphosphate and to surface components (non-protonated and protonated surface terminating phosphate groups) and to surface OH emerging from the tunnel [188]. As for carbonated substituted apatites, NMR study of carbonate species confirmed the IR conclusions by evidencing the two main types of carbonate sites in the apatitic domain (A and B sites, see Section 3.2.2) and a third one observed in biomimetic apatites and previously denoted as labile surface carbonates. Two-dimensional solid-state NMR analyses of carbonated apatites have also established the location of these labile carbonate species in the hydrated (surface) domain [214]. NMR spin diffusion experiments have then confirmed the concept of the hydrated layer rich in HPO_4^{2-} ions, proposed earlier on the basis of FTIR results [167]. In fluoride-containing carbonated apatites, ^{19}F solid-state NMR studies clearly showed the existence of two crystallographic sites corresponding to fluoride ions in monovalent sites and close to B-type carbonate ions [214]. Moreover, as FTIR and Raman spectroscopies, solid-state NMR appears particularly suitable concerning nonstoichiometric and low crystallinity apatites study, especially for biomimetic apatites characterization. As in vibrational spectroscopies, the presence of non-apatitic environments has

3.3 Advances in the Characterization of Structural and Surface Properties of Hydroxyapatite

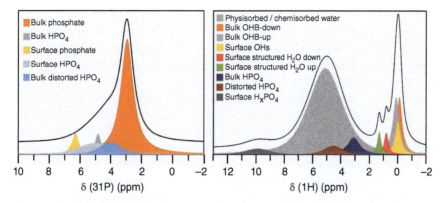

Figure 3.12 Typical multicomponent ^{31}P NMR (left) and ^{1}H spectra (right) of well-crystallized hydroxyapatite sample. The different contributions are schematically reported. For the sake of clarity, unlike the positions of the different contributions, the relative intensities are not represented within a realistic manner. Source: Ben Osman et al. [188]. Reproduced with permission of American Chemical Society.

been evidenced by solid-state NMR: in stoichiometric HA, ^{31}P solid-state NMR shows a single peak at 2.3–2.9 ppm with respect to H_3PO_4 [220–222]; in contrast, in biomimetic apatite an asymmetrical broad ^{31}P peak is often observed [223, 224] involving HPO_4^{2-} ions in addition to PO_4^{3-}, suggesting phosphate environments different from those of stoichiometric HA, which was pointed out by 2D $^1H\leftrightarrow{}^{31}P$ cross-polarization solid-state NMR analyses [223].

In addition to all the aforementioned experimental techniques, in order to investigate the properties of such a large variety of solids, molecular simulations appear as another powerful tool and have allowed to elucidate/confirm the crystallographic structures of apatitic solids and their interfaces (see Section 3.2), as well as adsorption performances and related energy values, electrical and mechanical properties. For that purpose, different levels of molecular simulations can be considered [113]: first-principles or DFT calculations investigating the electronic distribution on the solids with a high precision but on a small number of atoms, and force field based or classical simulations (such as energy minimization or MD simulations) dealing with a large number of atoms but requiring the development of adapted force fields [225] in order to elucidate the thermodynamic and dynamical properties of the solids [136, 226, 227]. It becomes, however, evident that coupling experiments and simulations appears as a key approach to capture the global understanding of the various properties of the solids, due to the large variability of their chemical composition and related structural modifications [148].

3.3.2 Thermodynamic Properties and Thermal Stability

3.3.2.1 Overview of Apatites Thermodynamics

Apatite thermodynamics have been the object of several studies, although addressing essentially end-member compositions. Standard thermodynamic properties of phosphate-bearing apatites with different substituents and degrees of stoichiometry have been recently reviewed (essentially based on experimental data) and analyzed. This allowed the development of the *Therm'AP* predictive thermodynamic model [228] as well as the related calculation freeware. This latter permitted one to estimate the standard formation enthalpy $\Delta H°_f$, entropy $S°$, and Gibbs free energy $\Delta G_f°$ of phosphate apatites based on their chemical composition [229] (Table 3.2), with a relative uncertainty of the order of 0.5–1%. This approach was cross-validated on the basis of the simple salt approximation (SSA) [230]. Lately, the *Therm'AP* model has also been extended to additional metal cations [231] (namely, Ni^{2+}, Co^{2+}, Mn^{2+}, and Fe^{2+}) that may also be relevant to the field of catalysis. Classical and quantum molecular simulations have also been used in a parallel way, to extract thermodynamic properties of apatites such as adsorption or formation enthalpy as a function of their composition. For instance, it has therefore been calculated that fluorapatite is more stable than HA [148, 232] and that the substitution of OH^- ions by F^- is exothermic while the substitution of F^- by Cl^- is strongly endothermic [144]. It has also been possible to calculate the impact of various substitutions (Sr^{2+} for Ca^{2+} or halogens for OH^-) and to compare the stability evolution [153, 226, 233, 234]. Only few data are available to date on the thermodynamics of the surface of substituted apatites. For instance, ionic surface substitution was punctually investigated using MD simulations to understand the mechanisms occurring during the incorporation of Zn^{2+} and Sn^{2+} ions [235]. The comparison shows that adsorption of Zn^{2+} is preferred for the surfaces (010) while the adsorption of Sn^{2+} is weaker. Furthermore, the adsorption of cations appears favored by Ca2 sites.

An increasing departure from stoichiometry – via vacancies in cationic and monovalent anionic sites – has a direct decreasing effect on the overall thermodynamic stability (as followed by the $\Delta G°_f$ value), which may be quantified from the chemical composition of the sample [228, 236]. In turn, this explains the increase in solubility of apatitic compounds departing from stoichiometry (see Section 3.3.3.1.). As far as solubility goes, however, the ionic contents of the solution have to be considered along with the nature of the apatitic phase itself (and pH), as the presence of some ionic species may modify the apatite solubility by affecting the ion activity product.

It may be noted that alkaline earth apatites (Ca^{2+}, Sr^{2+}, Ba^{2+}, etc.) are globally more stable than transition (e.g. Mn, Fe, Co) or post-transition (e.g. Cd, Pb) metal apatites for a similar cation/P molar ratio and monovalent ion type (e.g. OH^-): Ca ~ Sr ~ Ba > Mg > Mn(II) > Zn > Cd ~ Fe(II) > Pb > Co(II) > Ni(II) > Cu. The effect of the incorporation of heavier cations like lanthanides was also scrutinized [237]. Overall, the thermodynamic properties of complex oxides like apatites depend not only on the radius and charge of the constituting ions but also importantly on the affinity of the cation(s) for oxygen (shared by phosphate ions) and the elements own electronegativity [228]. However, it can be reminded that, in the presence of an aqueous medium (e.g. containing metal ions), the relative stability of the aqueous ions themselves is also to be considered when attempting to draw conclusions on the thermodynamic natural evolution path of the system. For example, while the Pb-HA solid phase ($\Delta G_{f,298}° \cong -7515$ kJ mol^{-1}) is noticeably less stable than Ca-HA ($\Delta G_{f,298}° \cong -12\,634$ kJ mol^{-1}), the incorporation of Pb^{2+} is still thermodynamically

Table 3.2 Gibbs free energy, enthalpy and entropy contributions of some constitutive ions in the apatite structure, as determined from the *Therm'AP* predictive approach.

Contributing components in the apatite formula (298 K, 1 bar)	g_i (kJ mol^{-1})	h_i (kJ mol^{-1})	s_i (J mol^{-1} K^{-1})	Contribution to the oxygen stoichiometry
Energetic contributions per divalent cation				
Ca^{2+}	−740	−790	38.8	+1
Sr^{2+}	−740.9	−796.1	53	+1
Mg^{2+}	−634.3	−666.4	23.1	+1
Ba^{2+}	−739.4	−791.9	71.1	+1
Cu^{2+}	−134.6	−171	34.6	+1
Cd^{2+}	−262.4	−317	53.6	+1
Pb^{2+}	−236.2	−282.2	81	+1
Zn^{2+}	−344.5	−394.8	52.4	+1
Ni^{2+}	−221.8	−262.9	38.6	+1
Co^{2+}	−227.7	−266.9	53.3	+1
Mn^{2+}	−403.6	−446.8	60.7	+1
Fe^{2+}	−273.2	−308.5	61.3	+1
Energetic contributions per anion				
PO$_4^{3-}$	−816.15	−861.6	41.05	+2.5
OH$^-$	−140.8	−121.5	80.65	−0.5
F$^-$	−269.5	−237.2	68	−0.5
Cl$^-$	−103.5	−70	95.7	−0.5
Br$^-$	−90.5	−58	118.3	−0.5
Energetic contributions by other species				
H$^+$	−147.75	−187.85	66.2	
H$_2$O (hydration)	−234	−290	50.7	
P$_2$O$_5$	−1632.3	−1723.2	82.1	

Source: Drouet [228] Reproduced with permission of Elsevier.

favored by the very low stability of aqueous lead ions ($\Delta G_{f,298}° \cong -24.43$ kJ mol^{-1}) compared to the highly stable Ca$^{2+}_{(aq)}$ species ($\Delta G_{f,298}° \cong -553.58$ kJ mol^{-1}) [228].

Biomimetic apatites described in Section 3.2.2 are an important subfamily of apatite compounds that exhibit a high (surface) reactivity that will be further developed in Section 3.3.4, allowing in particular rapid surface ion exchange and a large adsorption capability. This reactivity is fairly directly related to the lower stability of these compounds – obtained at "low temperature" typically lower than 100 °C – compared to high-temperature apatites. The hydrated ionic layer on the surface of the nanocrystals is in particular one key feature of such low-temperature apatites, as the presence of water plays a major role in ion mobility and apatite processing [238, 239]. Therefore, the relative metastability of biomimetic apatites can be exploited advantageously for applicative purposes for boosting apatite-based processes, as in catalysis. This increased reactivity can be tailored by controlling scrupulously the conditions of formation and post-treatment of such apatite nanocrystals, leading to more or less reactive crystals [236, 240]. Indeed, upon immersion in aqueous medium, nanocrystalline apatites undergo a process known as "maturation" (aging in solution) that has to be differentiated to regular Ostwald ripening and corresponds in fact to the evolution of the non-apatitic ionic environments from the hydrated surface domains to more stable apatitic environments composing the apatite nanocrystals core (see Section 3.2.2). A mechanistic scheme has been proposed to identify the main stages of this process, which itself involves ion exchanges/equilibration with the surrounding solution [238]. If maturation is undergone in the presence of foreign ions, the latter may be more or less incorporated into the nanocrystals (e.g. [241]) depending on their affinity for the apatite structure and the stability of their aqueous species as mentioned earlier.

3.3.2.2 Thermal Behavior

Apatite compounds are often subjected to post-synthesis heat treatments whether during synthesis/production or processing in view of a wide range of applications. Also, the application itself may provide a heated environment as in several catalytic processes. Therefore, appreciating their thermal behavior is often necessary. Among phosphate compounds, stoichiometric apatites exhibit a rather high thermal stability. HA can for example be heated/sintered at 1200–1250 °C without major alteration of the structural features [120], albeit dehydroxylation into oxyapatite occurs toward high temperatures (>850 °C in vacuum/dry atmosphere) [242] in a way that depends on experimental conditions such as water partial pressure (Figure 3.13, right side chart). Then, oxyapatite decomposes at higher temperatures into TCP and tetracalcium phosphate (TTCP), itself decomposing further to CaO. In the process of hydroxy→oxyapatite transformation, vacancies are progressively created in the apatitic channels generating intermediate oxy/hydroxyapatites $Ca_{10}(PO_4)_6(OH)_{2-2x}O_x\square_x$, and the value of x can then influence the temperature of subsequent decomposition [243–245]. In stoichiometric HA, it may also be noted that surface phosphorus is often considered to be reactive and may form P–OH surface bonds that are dehydroxylated above 400 °C to form P–O–P surface groups [244].

In contrast, nonstoichiometric apatites follow a different thermal decomposition path. In link with the thermodynamic metastability features mentioned earlier, these compounds will evolve differently depending on their chemical composition (Figure 3.13, left chart for non-carbonated apatites and middle chart for carbonated apatites). In all cases, the hydrated layer progressively loses its water by c. 200 °C, then HPO_4^{2-} and CO_3^{2-} decompositions occur upon further heating, ultimately giving rise to high-temperature phases like HA, β-TCP, calcium pyrophosphate (typically β-$Ca_2P_2O_7$), or CaO following the indications given on the figure. The end products and their relative amounts are dictated by the initial composition of the

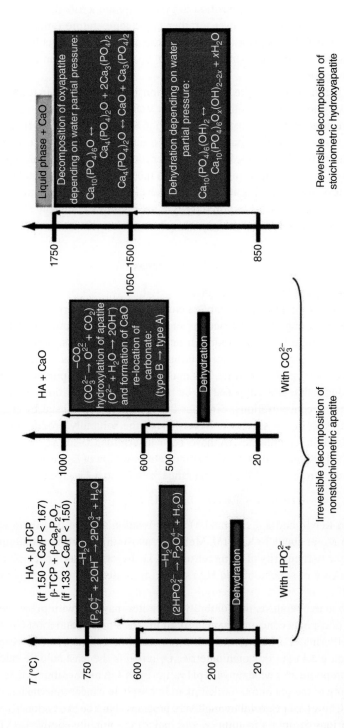

Figure 3.13 Thermal decomposition scheme of (non)stoichiometric apatites. Source: Rey et al. [120]. Reproduced with permission of Elsevier.

nonstoichiometric apatite sample considered. It may, however, be noticed that in the first stages of HPO_4^{2-} decomposition to pyrophosphate ions $P_2O_7^{4-}$, the latter remains in the apatite structure over a rather larger temperature range (typically up to at least 550 °C [246]) prior to witnessing the apatite structure collapse, thus potentially widening the temperature domain of use of these compounds for applicative purposes.

As already mentioned in the preceding subsection, ion substitutions in the apatitic lattice are likely to modify the thermodynamic stability of the phase. It is thus bound to have direct effects on its thermal behavior. Substitution of the OH^- anions by F^- or Cl^- ions within the channels has for example the tendency to increase the apatite thermal stability typically in the order (fluorapatite) FA > (chlorapatite) ClA > HA [243]. Cationic substitutions may also modulate the thermal stability, as in the example of Mg-containing apatites for which decomposition temperatures are clearly downshifted by several hundred degrees taking into account the stabilizing effect of phases such as β-TCP $(Ca,Mg)_3(PO_4)_2$ by Mg^{2+} ions [247].

3.3.3 Physicochemical and Interfacial Properties

3.3.3.1 Solubility and Evolution in Solution

Solubility is one of the most important calcium phosphate properties since it will determine the meaning of some reactions such as their dissolution, precipitation, hydrolysis, and phase transformations in aqueous environments. Solubility and interfacial properties of HA have been studied by several authors [115, 248, 249] and controversies remain numerous. The main reason is that the published values refer to products that are supposed to be pure, but contamination by various ions, even in low concentrations, can alter the stability of the crystal by creating significant stresses in the structure. Nevertheless, the solubility product (K_{sp}) of well-crystallized stoichiometric HA have been determined ($-\log K_{sp} = 116.8$ at 25 °C [120, 250]). Some studies show that the determination of K_{sp} as a function of the temperature (from 5 to 37 °C) in a pH range from 3.7 to 6.7 is given by [250]:

$$\log K_{sp} = -8219.41/T - 1.6657 - 0.098215\,T \tag{3.3}$$

where T is in Kelvin and K_{sp} in $(mol\,l^{-1})^9$. This relation showed that there was a maximum in K_{sp} at about 16 °C [115]. Moreover, above this point, the solubility of apatites has the particularity of being retrograde, i.e. beyond a specific temperature (16 °C in the case of stoichiometric HA), and it decreases when the temperature increases [243].

In addition to temperature, the solubility of apatites may vary with pH as well as calcium and phosphate concentrations in solution. The simplest and most common solubility isotherms as a function of pH encountered in the literature correspond to that of Figure 3.14 [115]; stoichiometric HA remains the least soluble calcium phosphate compound in a wide range of pH values up to 4. A more elaborated 3D diagram according to the pH and the calcium and phosphate concentrations has been proposed by Chow [251]. Several free software programs have been developed to calculate the various equilibria in solution and including solubility equilibria (Visual

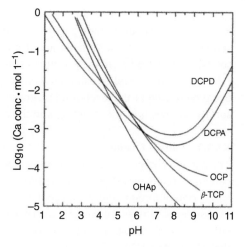

Figure 3.14 Solubility isotherms of calcium phosphate phases in the system $Ca(OH)_2 - H_3PO_4 - H_2O$ at 37 °C. DCPA: dicalcium phosphate anhydrous; DCPD: dicalcium phosphate dihydrate; OCP: octacalcium phosphate; β-TCP: β-tricalcium phosphate; OHAp: stoichiometric hydroxyapatite. Source: Elliott [115]. Reproduced with permission of Elsevier.

MINTEQ, MINEQL, PHREEQC, etc.); these tools are very useful but not suitable for solutions with high solute concentrations. It must be mentioned that the presence of foreign ions in the composition of a phosphocalcic phase or in the surrounding solution may significantly affect the dissolution behavior; for example, the presence of fluoride ions in the apatitic structure decreases the solubility, while the presence of carbonate or magnesium ions increases the solubility.

For the same composition, the phases with a lower density have the highest solubility in solution. These observations are related to the degree of cohesion of the phase structure by assuming that the electrostatic attraction between anions and cations within the structure is the main parameter responsible for the cohesion within the crystals of calcium phosphates. Thus, phosphocalcic apatites presenting smaller unit-cell dimensions have the highest cohesion forces and the lowest solubilities; for example, fluorapatite is less soluble than HA, and calcium phosphate apatites less soluble than strontium phosphate apatites. Similarly, apatitic calcium phosphates with a high vacancies content should be more soluble than those with a low vacancies content [120].

In addition to the considerations concerning stoichiometric HA, in the case of nonstoichiometric apatites and biomimetic apatites, some variability in the dissolution properties has been reported. Nonstoichiometric apatites show a higher solubility than stoichiometric apatites and the dissolution characteristic of the crystals depends on the composition of the samples, itself related to their conditions of formation [159, 252]. Moreover, the solubility of biomimetic apatites depends on the presence of vacancies: nanocrystals with low Ca/(P + C) ratios (where P represents phosphate ions and C the carbonate ions) present a high amount of cationic vacancies and a less cohesive solid with a higher solubility [159]. The amount of labile (non-apatitic) surface species may also influence the solubility of the nanocrystals [248, 253]. The presence of crystal maturation inhibitors, such as carbonate, magnesium, or pyrophosphate, involves small apatitic crystals with high specific surface, rich in labile non-apatitic environments, which promote their dissolution [159]. It should be noted that nonstoichiometric apatites have the ability to mature like bone mineral, as previously explained (Section 3.2.2), and may involve more stable and less soluble compounds upon aging [159]. Moreover, unlike stoichiometric HA, the solubility product of biomimetic apatites could vary according to the amount of solid dissolved, leading to the concept of "Metastable Equilibrium Solubility" (MES) [161, 253–255]. Several explanations can be given for this observation; this behavior could be related to the level of microstrains in the crystals [248] or to the variable surface composition and structure and the existence of a hydrated layer [161].

3.3.3.2 Surface Charge

The surface charge of ionic solids in contact with solutions, at the origin of "zeta potential" (ζ-potential), is considered to be the result of preferential dissolution or adsorption of foreign ions or molecules from the solution, or else of the formation of complexes between lattice ions and species in the solution (which may be formed either at the solid/solution interface or in the solution and subsequently adsorb on the solid surface) [115, 120, 256].

The point of zero charge (pzc) defines the conditions of the solution (the pH value) for which the surface density of positive charges (contribution of cations) equals that of negative charges (anions) [120]. There exist a wide range of pzc values for stoichiometric HA reported in the literature [257, 258]; indeed, significant variations are observed between experimental and calculated values [256], which have been explained by substituted impurities, lattice defects, or lack of equilibrium conditions in the experiments [256]. However, most of them are found in the range of 6.4 and 8.5 [251]. HA surface involves at least three types of potential-determining ions (calcium, phosphate, and hydroxide or proton) [120, 256] and the pH of the solution can influence the solid surface charge by changing the distribution of proton and hydroxyl groups at the interface. To explain the surface behavior of HA as a function of pH, Boix et al. [259] proposed models for HA surface using a mixture of a surface complexation model and a surface hydroxylation model (Figure 3.15).

For ionic solids such as apatitic calcium phosphates, the surface charge is closely related to the concentration of specific ions in the surrounding fluid that can interact with the surface of the solid, often called as "potential-determining." Under constant pH and ionic strength, a change in the calcium or phosphate concentrations can affect the surface charge of HA [120]. Indeed, an increase in calcium concentration will lead to more positive values of surface charge (adsorption of calcium ions on the surface of the solid), while an increase in phosphate concentration will lead to more negative surface charge [260]. Moreover, ionic substitutions and apatite composition also influence surface charge of apatitic compounds. For example, the pzc values of four apatite samples follow the following order: FA < HA < strontium-HA < barium-HA, which corresponds also to the order of increasing solubility [261]. The surface charge of silicated-HA appears more negative than that of HA, which can be related to the presence of tetravalent silicate ions on the apatitic surface instead of phosphate ions [262].

In addition, nonstoichiometry may also affect the surface charge of apatites; biomimetic apatites, more soluble than stoichiometric ones, have been found to exhibit greater zeta potential than stoichiometric HA [263]. However, in the case of

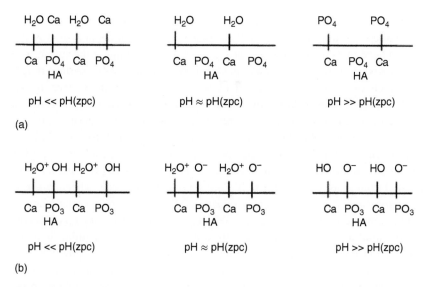

Figure 3.15 Models for HA surface proposed by Boix et al.: (a) model of surface complexation, (b) model of surface hydroxylation. Source: Boix et al. [259]. Reproduced with permission of Elsevier.

nonstoichiometric biomimetic apatites, the existence of a non-apatitic surface layer on the apatite nanocrystals (see Section 3.2.2) leads to a more complex system that makes it difficult to evaluate the amounts of phosphate or calcium ions actually exposed on the surface [163].

Finally, the presence of organic molecules adsorbed on the surface of apatite solids, including nonstoichiometric ones, can significantly modify their surface charge of the particles [261, 264]. For example, the adsorption of acetaminophen on HA leads to a surface noticeably more negatively charged compared to the unfunctionalized one [265], whereas the adsorption of amino acids [266] or an amine-terminated phospholipid [264] on calcium-deficient apatites leads to a more positive surface charge. The adsorption of organic molecules on apatitic surface can often be explained by the replacement of some surface ions by some charged/polarized end-groups of the organic molecule. For instance, molecules exposing negatively charged functional groups may displace phosphate surface ions, while positively charged groups may displace calcium ions [163]; however, the affinity of such functional groups for the surface of the apatite has also a great influence on adsorption properties of organic molecules on the apatitic surface. The mechanism of adsorption and the role of the nature of organic molecules in the binding mechanisms are discussed *in* Section 3.3.4.

3.3.3.3 Interfacial Tension

Surface/interfacial (free) energy or "tension" is another relevant parameter when addressing interactions between a solid surface and the surrounding medium. It may for example affect the adsorption capacities of a solid [267]. Also, in solution, it is linked to the stability of colloidal suspensions and impacts the crystal morphology and growth rate in solution [268]. In turn, interfacial tension plays a role in most theoretical dissolution kinetics models [269]. The surface energy of a solid is also directly related to its wettability (e.g. as a general trend, the higher the surface free energy, the better is the water wettability) and consequently to the surface roughness (an increase of surface roughness leads to an increase in apatite surface energy) [270]. However, data available in the literature on this matter are rather scarce for apatite compounds. Also, variability in reported values may be underlined, in part due to different measuring approaches or interfaces considered [268, 271, 272].

Surface tension can either be determined experimentally (typically from contact angle measurements or during HA dissolution or crystal growth experiments) or be calculated from a simulation approach. Care should thus be taken when extracting or comparing data from literature, as experimental and simulation works generally refer to different thermodynamic states. When considering an apatite phase in contact with liquid water, the initial state is the apatite solid phase in equilibrium with its own vapor (described by $H_S°$ that is related to the surface tension of the solid $\gamma_S°$) while the final state is the solid–liquid interface (described by H_{SL} related to the γ_{SL} surface tension) [273, 274]. In enthalpy terms (a similar equation could be written in terms of free energy), this can be described by way of the enthalpy of immersion ΔH_{imm} as follows:

$$\Delta H_{imm} = H_{SL} - H_S° \tag{3.4}$$

While in experimental studies, the value accessed is ΔH_{imm} or γ_{SL} thus referring to the solid in contact with liquid water; computational approaches rather calculate the $H_S°$ value.

From an experimental viewpoint, surface tension values (γ_{SL} referring to the apatite–liquid water interface) drawn from HA dissolution are generally lower than those determined from HA growth studies [275], which may be related to surface imperfections that may affect both approaches in a different way [276]. In some cases, the adsorption of impurities (e.g. foreign ions present in the medium) onto crystal faces may additionally alter the measured surface free energies of crystal faces, modifying the tendency for ions to be released from the crystal during dissolution experiments [269]. It is thus necessary to use well-established methodologies allowing the characterization of interfacial reactions in well-known reproducible conditions, and comparisons should probably be made on the basis of a common approach so as to limit undesirable bias.

As a general tendency, the surface tension of calcium phosphates seems to correlate rather well with their water solubility, the lowest surface tensions being noticed for the most soluble phases [120, 277]. Interfacial energies with respect to water for hydroxyl- and fluorapatite are generally considered around 9.0–9.3 and 17.1–18.5 mJ m^{-2}, respectively [268, 277, 278]. Since surface tension originates from the physical attractive force between the atoms composing the solid surface [279], ion substitutions in apatite are likely to modify the crystal surface energy. Heavier elements favor larger surface energy, as was evidenced for instance between barium- and strontium-chlorapatite surfaces [279]. Carbonation of HA (3 wt% CO_3) was on the other hand reported to exhibit a surface energy of 9.0 mJ m^{-2}, close to (slightly lower than) that of HA [277]. However, changes were detected on the dispersive and polar components of surface energy, as the polar interaction energy with water was significantly lower on (A-type) carbonated apatite than on HA [280]. In nonstoichiometric apatites, carbonation was found to generate several significant modifications of apatite crystal features, obviously also affecting surface energy [281]. However, the surface energy of nonstoichiometric apatites has not yet been specifically addressed and is still a matter of discussion that will need further consideration. In the case of nonstoichiometric biomimetic apatites, since the nanocrystals exhibit a hydrated non-apatitic layer, their surface features such as surface energy are indeed expected to be deeply modified compared to the regular/stoichiometric HA model: the surface layer is rich in HPO_4^{2-} ions (unless heavily carbonated) and in Ca^{2+} ions but exempt of OH^- ions (solely contained in the crystal core within the apatitic channels); a new model will thus have to be proposed to determine actual surface features.

From a modeling viewpoint, molecular simulations have also been used to calculate the surface energy [282] (referring to the solid in equilibrium with its own vapor) of various apatites and to determine from theoretical calculations the cleavage face or the morphology for crystal growth [283]. Indeed, one of the main goals of recent studies was to propose a description of surfaces (more realistic than bulk model [284] for applications dealing with surfaces). The relevance of involving surface OH groups has been pointed out. On stoichiometric apatite, the surfaces (010) terminated with OH groups have indeed been evidenced by calculations to be the most stable surfaces while the (001) and (010) surfaces terminated with calcium or phosphate groups have shown to be unreactive in presence of water [285]. The evaluation of the surface energy of different apatite compositions has also shown the impact of substitution. For example, in a carbonate-saturated HA, the surface energy was calculated as $900\,mJ\,m^{-2}$ compared to OH-saturated HA ($750\,mJ\,m^{-2}$) using B3LYP [286]. To modify the adsorption properties, incorporation of ions is a possibility: considering different HAs (HA, HA substituted with Ag or Zn), the adhesion performance has been characterized by classical and quantum molecular simulations and the interactions between HA and Ti were found to be enhanced by doping with Zn or Ag [287]. In order to evaluate the activity of the apatite surfaces, an indirect approach considering water molecules as a probe has been envisaged [288, 289]. The adsorption properties for water on HA surfaces have also been evaluated using DFT calculations and showed that, depending on the surface ((001) or (010)), the interaction between water molecules and constituents was based on physisorption ($\sim 80\,kJ\,mol^{-1}$) on the (001) surface but on chemisorption with dissociation of H_2O and formation of Ca–OH and PO–H covalent bonds on the (010) HA surface [290]. Similar results led some authors to consider the (100) surface terminated by Ca^{2+} or phosphate groups as unstable since it strongly interacts with water [114]. It appears therefore that the different surfaces of the apatites present specific behaviors (in terms of hydrophilicity-hydrophobicity, stability, etc.), which can explain the differences observed during crystallization and the resulting morphology of the solids in the presence of external molecules [114, 284]. However, due to the limitation of DFT calculations and to investigate larger systems, adapted force fields have to be developed to reproduce the interactions with water and then analyze the impact of the substitutions on the adsorption properties [291, 292] and characterize, especially in the case of apatitic nanoparticles, the rearrangements of atoms at their surface, which can be accessed during simulations in the presence of water molecules [288].

3.3.4 Surface Reactivity

3.3.4.1 Nature of Acid and Base Sites

Due to the lack of characterization tools for investigating the solid–liquid interface, the surface termination of HA in liquid medium is an open question and depends on the pKa values of the reactant molecules and products versus the pH of the aqueous medium. If recent developments propose methods for quantifying acid (or base) sites in liquid medium [293], the exact nature of sites in real operating conditions is still an unsolved question.

Although challenging, the identification of active sites working at solid–gas interface for reactions implemented at moderate to high temperature conditions benefits of some in situ characterizations at solid–gas interface. Among species likely to be present on the surface of HAs, Ca^{2+} cations (or other coordinatively unsaturated cationic metallic centers depending on the target composition) can act as Lewis acid sites. However, surface relaxation processes may lead to some Ca^{2+} deficiency on the top surface [187]. To ensure the surface charge balance, terminated phosphate groups are protonated [211, 224, 294] and the resulting PO-H are likely to act as Brønsted sites. As far as basic sites are concerned, it is often implicitly assumed that PO_4^{3-} groups should be responsible for the basic properties of HA [97, 100], even if there is a lack of experimental proof for their direct involvement in catalytic reactions. Alternatively, OH^- emerging from the channels should be considered [184].

FTIR characterization may help in the identification of the active sites involved in the gas phase catalytic reactions. However, the main difficulty for investigating the active surface of HA comes from the fact that the same species are present both in the bulk and on the surface, only the latter species being likely to directly play a role in catalytic reactions. In a first step, spectroscopic fingerprints of surface species have to be discriminated from the bulk ones. The spectroscopic features of OH^- and HPO_4^{2-} originating from the bulk and from the surface could be discriminated using isotopic H—D gas phase exchanges successively implemented in mild conditions (180–200 °C), then at higher temperature (300 °C) to first selectively label the lone surface, then to achieve deep bulk deuteration, and finally followed by a last surface reprotonation step (Figure 3.16) [211]. Once IR bands of vibrators located on the surface are identified, their behavior can be monitored via in situ adsorption of probe molecule and even *operando* reaction conditions to complementarily characterize their acid (or basic) properties and discuss their involvement as active sites.

Acid sites

It was found that among the six IR v_{PO-H} vibrators in the 3750–3600 cm^{-1} range, only that at 3657 cm^{-1} corresponds to bulk species, all the others at 3734, 3729, 3680, 3670, and 3636 cm^{-1} correspond to surface species (Figure 3.16). Consistently,

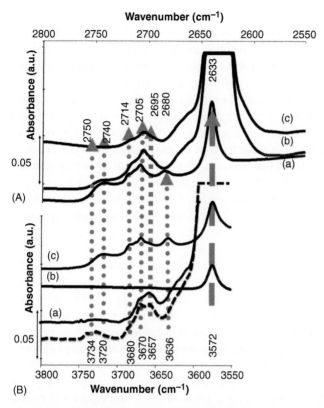

Figure 3.16 FTIR spectra on self-supported wafer of an HA: (A) in the v_{OD} region after surface deuteration (a spectrum), after bulk deuteration (b spectrum), and after surface protonation subsequent to a bulk deuteration (c spectrum); (B) in the v_{OH} region before deuteration (dotted line), after surface deuteration (a spectrum), after bulk deuteration (b spectrum), and after surface protonation subsequent to a bulk deuteration (c spectrum). To evidence the correspondence between the bands in the v_{OD} and v_{OH} regions (vertical arrows), the scale of the two frequencies axis has been adapted, taking into account the v_{OH}/v_{OD} ratio = 1.37. Source: Diallo-Garcia et al. [211]. Reproduced with permission of Elsevier.

their perturbation upon adsorption on the surface of basic gaseous probe molecules, typically CO probe molecule at 77 K (that also generates vibration bands in the 2200–2000 cm^{-1} range), confirmed their surface location as well as their BrØnsted acidic properties [187, 211]. Interestingly, the intensity of the perturbations (negative contribution in different spectra) does not follow that of the absolute bands, which is indicative of a differentiation of the surface PO–H in terms of acidic strengths, accessibility, and/or proximity with the neighboring basic sites [75, 190].

CO can also probe the presence of calcium cations. However, the top surface accessibility of CO adsorption experiments remained quite limited compared to that of PO–H [187]. Consistently, the lack of Lewis sites was also reported from gaseous pyridine adsorption [186]. Adsorption of evaporated glycine concludes to weak interaction through both H bonding between PO–H and the carboxyl group and between

Ca^{2+} and the amino group [295]. The competitive involvement of cations or terminated PO–H sites may thus depend on the one hand on their affinity for interacting with the organic reactant and/or the involved mechanism and on the other hand on the preferential surface exposure of one site over the other [187]. For instance, the higher efficiency of PO–H over Ca^{2+} acid sites was observed for ethanol or MBOH conversion in the gas phase by tuning the relative proportion of exposed PO–H and Ca^{2+} playing on the washing and drying procedures [187]. From *operando* studies, the unique selectivity of the HA system for the ethanol to n-butanol conversion was ascribed to the involvement of the weak PO–H Brønsted acid sites: when poisoned, the reaction stops to acetaldehyde, revealing that the selectivity is greatly dependent on the Brønsted or Lewis nature of the acidic sites involved [75]. Of course, in the case of modified HA, the acidic properties of the heteroatom may be determinant: for instance, the Lewis acidity associated with Ru bound complex promotes the Diels–Alder reaction [25].

Basic sites

According to H—D exchange studies, the IR band at 3572 cm^{-1} was found to gather both the contributions of the main bulk OH located inside the channel and of the minor surface fraction emerging at the surface [211]. The involvement of the surface OH$^-$ as basic partner in acid–base pair with neighboring acidic PO–H is evidenced by the simultaneous perturbation of its respective v_{O-H} vibrator upon adsorption of acetylene [190], and its role as active sites is unambiguously shown from *operando* study [75] in the ethanol conversion to n-butanol. The ethanol conversion is directly dependent on the availability of the surface OH$^-$. At reverse, the involvement of O^{2-} of PO_4^{3-} for gas phase reaction still remains unproved, since the related IR v_{P-O} bands – whose saturation on self-supported wafer could be avoided upon dilution of the sample in diamond sample – were not perturbed upon adsorption of acetylene [190].

The role and identification of basic sites was also recently questioned for the activation of light alkanes, likely to occur via the hydrogen abstraction from the C—H bond. In the absence of any vanadyl group exposed on the surface in vanadate-modified HA, intrinsic basicity provided by the HA was proposed [9]. Obviously, the strength of the OH$^-$ basic sites remains too weak for accounting for the vanadate- or cobalt-modified HAs' efficiency in oxidative dehydrogenation of propane or ethane [296, 297]. Thus, it was assumed that oxygen species derived from the abstraction of a hydrogen atom of the hydroxyl might be involved [50, 297]. This was recently confirmed by diffuse reflectance infrared Fourier transform spectroscopy (DRIFTS) and NMR since upon thermal activation (up to 500 °C), vanadate-substituted HAs [298] progressively dehydrate to form a metastable vanadate-modified oxyhydroxyapatite active phase [9]. This process is assisted by a thermally activated anisotropic proton conduction ability along the c-axis [299] that finally leads to the surface exposure of strongly basic O^{2-} sites that are able to activate the C—H bond of light alkanes [9]. In the case of butan-2-ol oxidative dehydrogenation, it was found that low iron loading leading to Fe^{3+}–O–Ca^{2+} species (instead of Fe^{3+}–O–Fe^{3+} obtained at high iron loading) is favorable for catalytic performance since it avoids a drastic decrease of the basicity [48].

Note also that the fluoroapatite system was reported to exhibit modified basic properties compared to HA: the F$^-$ homologous to OH$^-$ inside the column could impact the acid–base balance, or may even be directly involved as basic sites [60, 63, 64, 78, 91].

3.3.4.2 Influence of Substitution on Surface Reactivity

As mentioned in Section 3.2, due to its peculiar flexibility, the apatite structure accepts both cationic (on Ca1 and/or Ca2 sites) and anionic (both on PO_4^{3-} and OH$^-$ sites) substitutions. These substitution abilities are not limited to ions of similar charges [39] ($Ca^{2+} \rightarrow Mg^{2+}$, Sr^{2+}, Ba^{2+}, Co^{2+}, Ni^{2+}, Cu^{2+}, Zn^{2+}, Pb^{2+}, Cd^{2+}; [67, 70, 86, 183, 300–304] $PO_4^{3-} \rightarrow AsO_4^{3-}$, VO_4^{3-};[298, 305, 306] OH$^- \rightarrow$ Cl$^-$, F$^-$ [307]) but incorporation of ions of different charges is also reported ($Ca^{2+} \rightarrow Ln^{3+}$ [308]; $PO_4^{3-} \rightarrow SiO_4^{4-}$ [309, 310]; CO_3^{2-}; OH$^- \rightarrow CO_3^{2-}$ [190, 299]). Although mostly associated to bulk modifications, such ability of HA structure to accommodate a wealth of cationic or anionic substitutions makes it a very attractive tunable support for a large class of multifunctional catalytic reactions. Various compositions of modified HAs have been prepared incorporating the foreign ions in one pot coprecipitation approach [9, 183, 298] eventually followed by a hydrothermal treatment [49, 76, 305] or by high temperature exchange reactions [311] (see Section 3.2). Depending on the nature and content of the heteroatoms, the crystallinity, morphology [39], textural properties, and surface relaxations [9] may also be greatly impacted. The bulk incorporation can be probed by XRD following the induced distortion of the crystalline cell. Recent literature focuses on DFT calculations and X-ray absorption spectroscopy to more directly address the foreign atom location [312, 313]. As shown earlier, vibrational spectroscopies also allow for the identification of site occupancies in substituted apatites through the position and/or shape of vibration bands; for cationic-substituted phosphate, IR vibration bands are globally shifted toward lower wavenumbers when a large cation replaces a smaller one (as strontium-calcium HA) [196]. Besides ionic radius and electronegativity, the spatial constrain also appears to be a key parameter to assess the preferential occupancy of divalent cations on Ca1 or Ca2 sites. This contributes to explain why the ability for M^{2+} (Mg^{2+}, Zn^{2+}, Cd^{2+}, and Pb^{2+}) bulk doping is favored in the case of carbonated HAs: by modifying the bond length and coordination number of Ca site, CO_3^{2-} enhances the solubility of foreign M^{2+} ions in the lattice [314]. In particular, the relaxation possibilities within the OH channels help the incorporation of many ions. Whereas the preference of Sr^{2+} for Ca1 site prevails for loading lower than 1 at%, for higher Sr^{2+} concentration Ca2 site appears progressively preferred, as visualized by the shift to a higher frequency of the IR v_{OH} stretching mode [151]. At reverse, the related band shifts to lower frequency upon progressive Cd^{2+} incorporation are indicative of the higher covalent character of Cd–OH bonds [304]. Silicate incorporation is associated with OH loss [315], whereas vanadate substitution favors the oxyhydroxyapatite formation, as clearly evidenced not only during in situ thermal treatment (catalytic conditions) but also by the persistent loss in OH once back to room temperature, at least for high vanadium loading (after thermal treatment at 500 °C) [9]. Besides intrinsic properties provided by these new groups (redox in the case of vanadate), because those modifications impact the OH channels, they also influence the strength of basic sites. As an example, the rate of acetone condensation to mesityl oxide that is strongly dependent on the basicity is tuned depending on the nature of the divalent ion close to basic sites: typically, because of too acidic partners within the acid–base pair, Mg^{2+} or Ca^{2+} bind acetone too strongly compared to Sr^{2+} that promotes the reaction [39]. Similarly, methanol dehydrogenation to formaldehyde also depends on the binding strength of methanol on the surface [96]. Moreover, by tuning the acid–base balance, the incorporation of Sr^{2+} and CO_3^{2-} promotes the selectivity in the Guerbet coupling of ethanol [68, 70]. Substituted vanadate anions increase the selectivity of acetaldehyde dimerization into crotonaldehyde [67] and generate strong basic sites able to activate the C—H bond of propane [9].

Despite these proven beneficial impacts on the catalytic performances observed for bulk-modified HA systems, it remains quite difficult to make a fair comparison with other systems since turnover frequencies (referring to activity per surface active site) cannot be accurately determined. Indeed, most introduced heteroatoms being located in the bulk (thus not directly involved as active sites), the quantification of surface active sites remains challenging. In addition, the formation of substituted HA solid solution by coprecipitation method is not always easy to establish, as other phosphate compounds [56, 316] and/or oxide particles [317] may competitively precipitate.

3.3.4.3 Low Temperature Ion Immobilization and Adsorption Properties in Aqueous Media or Wet Conditions

In substituted apatites previously presented, the foreign mineral ions enter the apatite lattice; thus crystal surface is likely to present the same structure and composition than the apatitic core (although relaxation effects/surface energy considerations may lead to some extent to some differentiation in the surface composition [9]). On the contrary, modifications only of the surface of the crystals can be performed in aqueous media without affecting the apatitic structure, as the foreign species do not enter the apatitic domain, and concerns two types of surface phenomena: ion immobilization and molecules functionalization. Surface modifications can involve several mechanisms, as for example dissolution and/or reprecipitation, or ionic exchange, depending on the nature of the foreign species, the characteristics of the apatitic support, and the experimental conditions (pH, concentration, contact time, etc.) [120, 161, 311, 318]. Given the known influence of the apatite composition on the bioactivity of bone tissue [319] (Zn^{2+} enhances antibacterial properties [302, 313, 319], Sr^{2+} contributes to the activation of new bone tissue and reducing fracture risk [320], etc.), such properties are widely investigated for biomaterial applications, as drug delivery [321], for adsorbents for heavy metals in environmental applications [322, 323], or in chromatography column separation systems. The functionalization of HA support by controlled adsorption in aqueous medium is also obviously of particular interest for the development of new catalysts [324] since it is likely to introduce heteroatoms, complexes, or grafted molecules specifically onto the surface. The understanding of the nature of interactions with the apatitic surface can provide fundamental tools to design performant multifunctional supported apatites.

In the catalysis community, the deposition of active ions, molecules, or phase onto a support in aqueous medium is often referred to as wet impregnation (also called impregnation in excess of solution since the support is stirred in the metal containing solution). Both the speciation in solution of these chemical entities (depending on the pH and concentration) and the charge of the surface govern the adsorption, and only species precursor in strong electrostatic interaction with the support remains after washing. It should be noted that although the zeta potential of apatites has been considered to play a major role in the adsorption process and most adsorption behaviors have been related to this surface parameter [120, 161], the correlation between this global physical parameter (describing the total surface charge of apatite particles) and chemical interactions in solution with specific ions or charged groups of molecules at an atomic level is not straightforward. It shall be considered that the global charge of a particle does not give any information on the local distribution of cations and anions on a mineral surface: for example, the adsorption of positively charged molecules on a globally positive apatite surface and inversely negatively charged molecules on a globally negative apatite surface has been described. Surface distribution of a metal on HA surface can be determined by techniques such as SEM-EDX and TEM-EDX.

The dynamical processes occurring at apatite surfaces are also key parameters and have been studied using classical molecular simulations. By this way, water diffusion has been investigated by MD simulations. It results that an anisotropic diffusion inside nanopores ranging from 20 to 240 Å is observed [325, 326], in which the molecules diffuse along the surface noticeably faster than perpendicularly to the surfaces. This observation is justified by the impact of the polarization effect of divalent cations [189, 286] as well as the hydration of the ions present at the (001) surfaces (confirmed by the evaluation of the activation energy for water diffusion close to 20–25 kJ mol^{-1}, which is higher than in liquid water). Self-diffusion coefficients have also been calculated by MD simulations, and the obtained values were found lower than self-diffusion in liquid water [327].

Surface ion immobilization

When applied to the case of HA support, adsorption of metal ions in excess of aqueous solution is often referred to as an "ion exchange procedure." The term "ion exchange" corresponds to a reaction specifically affecting the accessible ions on the surface of the apatitic crystals, and not to describe ion substitutions inside the apatite lattice (in this case, the word "substitution" is used instead of "exchange"). Several studies have investigated the ion exchange capability of HA with metal ions as Cd^{2+} or Zn^{2+} in solution [322, 323, 328]. However, the exact mechanism involved at the solid–liquid interface for such post-synthesis support modification is rarely precisely investigated in the catalysis community. This aspect is better documented in the context of remediation to transition metal cations dissolved in wastewaters by HA used as low-cost and eco-efficient adsorbent [329, 330] or in the biomedical and biomineralization fields [311]. From this literature, depending on the nature of the metal present in solution, its concentration, the pH solution, and the surface charge [331], several immobilization mechanisms are likely to occur over HA: dissolution–precipitation [322], ionic exchange with Ca^{2+} [311], filling of surface cationic vacancies [149], and surface complexation (through surface hydroxyls, carbonates, and/or hydrogenphosphates) [303]. The discrimination between these competitive processes should be made combining structural characterization (XRD, FTIR, Raman, SEM, TEM, etc.) and quantitative analyses of both the composition of the modified surface (ICP vs. XPS) and of the ions present in the solution at the end of the process (concentrations of consumed/remaining vs. released ions).

Dissolution–precipitation mechanism may affect the whole bulk composition. This is typically the case for lead, since the formation of lead-doped HA potentially up to the end-member composition $Pb_{10}(PO_4)_6(OH)_2$ (resembling the pyromorphite phase $Pb_{10}(PO_4)_6Cl_2$) is obtained. Indeed, although the solid Pb-HA phase is significantly less thermodynamically stable than Ca-HA by itself [228, 229], the whole thermodynamic cycle involving also the aqueous species $Pb^{2+}_{(aq)}$ and $Ca^{2+}_{(aq)}$ – the former being significantly less stable than the latter – favors the formation of Pb-HA [303]. If such dissolution–precipitation is limited (typically it should be minimized in neutral to basic medium), the deep surface layers composition may be affected and associated with surface restructuring. On the contrary, if any of the three other immobilization mechanisms is involved, the location of the heteroatoms is restricted to the top monolayer of the HA surface (in agreement between ICP-optical emission spectroscopy (OES) and XPS measurements [39]). According to Gervasini et al., the immobilization of copper in the excess of solution first proceeds via surface complexation (its affinity with the surface depends on the B- or AB-type carbonation, and its interaction strength with the surface decreases as the involved surface anions go from OH^- to CO_3^{2-} and then to HPO_4^{2-}), and then ion exchange with Ca^{2+} cations becomes active at long contact time [303]. From DFT, the substitution of Zn^{2+} on Ca^{2+} site in aqueous medium is less energetically favorable than Zn^{2+} occupation on Ca vacancy site [149]. Zn-bound HA was found very effective as a co-catalyst together with a Lewis base in coupling of CO_2 and epoxides [332]. The occupancy of Ca vacancy site is also invoked in the generation of monomeric La complexes active for Michael reactions [64] or Pd^{II} complexes grafted onto the surface of HA of different stoichiometries active for Heck or Suzuki reactions [42]. These Pd^{II} grafted complexes easily transformed into Pd^0 particles with narrow size distribution active in selective oxidation of alcohols [42].

Interestingly, despite the use of chloride precursor (for the preparation of Ru- or Co-modified HA), most studies conclude the absence of surface Cl^- substitution for OH^- observed at solid–liquid interface, at least for low Ru loading [26, 333, 334]. This has to be related to the lower thermodynamic stability of chlorapatite compared to HA, even in the presence of large Cl^- contents in solution (this is why in bone the apatite phase, yet in the presence of ubiquitous dissolved NaCl, does not get chlorinated). At reverse, in gas phase catalytic reaction, typically partial oxidation of methane [179, 335] or oxidative-coupling reaction [335] operating at high temperature (700 °C) in the presence of tetrachloromethane in the feedstock, surface, and even bulk chlorination occurs, resulting in the formation of chloroapatite. Such modification would promote the selectivity in CO [335].

The possibility to immobilize mineral ions rapidly and reversibly by ion exchange process on apatitic surface was first studied by Neuman et al. for carbonate-HPO_4^{2-} exchanges in biological apatites [336], which was later highlighted by spectroscopic studies and solid-state NMR [196]. Note that concerning well-crystallized stoichiometric apatites, low temperature ion surface immobilization reactions described in aqueous media in most instances involve a dissolution–reprecipitation process [311] (as for example fluoride uptake by HA); the surface ion exchange property is particularly relevant for biomimetic apatites, which display a metastable and non-apatitic hydrated ionic layer, and highly reactive, rich in labile ionic species able to rapidly and reversibly be exchanged by ions in solution. Several types of adsorption sites have been described on apatitic surface, such as carbonate/hydrogenphosphate, Mg/Ca, or Sr/Ca [196, 311]. Analysis of the solid and of the solution ion concentrations confirms that the reactions occur by the replacement of ions of the solid by ions in the solution. By the titration of the remaining solution during adsorption (supernatant), the ion exchange isotherms can be obtained by plotting the quantity of species adsorbed as a function of the remained (equilibrated) concentration in solution. The adsorption isotherms on apatitic surface may often be satisfyingly described by the Langmuir model (for example, for Sr^{2+} on nonstoichiometric apatite in Figure 3.17) [241]. The adsorption parameters K and N, representing the affinity constant to the apatitic surface and the maximum amount adsorbed (at a given temperature), respectively, are deduced from Langmuir equation and depends on the nature of foreign species and on the composition of the apatite and its maturation stage. For example, for identical solution concentrations, the exchange rate of Sr^{2+} ions is always larger than that of Mg^{2+} ions for a given maturation stage of the nanocrystals studied. It can be noticed that some monovalent cations like Na^+ and K^+ could not, in contrast, be incorporated in these conditions [159]. All these exchange reactions are very fast (the equilibrium is reached within few minutes) and irreversible with respect to dilution or simple washing. It has been shown that the quantity of ions exchanged decreased when the maturation time of the nanocrystals used increased, which can be interpreted as a decrease of the amount of potentially exchangeable ions in the hydrated surface layer [160]. However, when a maturation period (aging) occurs after the exchange, several types of ions in the hydrated layer shall be distinguished: ions like Mg^{2+}, for which only a limited number can enter the HA lattice and remain essentially exchangeable in the surface during maturation through ion exchange reactions, and on the contrary ions like Sr^{2+} that can replace calcium ions in the apatitic structure and thus, during the maturation, the growing of apatitic domain progressively incorporates such ions and the amount of this exchangeable ions constantly decreases as the maturation time increases after a primary exchange reaction [311]. Carbonate ions show an intermediate behavior. The incorporation of foreign species from the hydrated layer into the apatitic core of the nanocrystals during maturation can be studied typically by vibrational spectroscopy experiments using isotopic substitution. For example, solution exchange experiments involving ^{13}C-labeled carbonate have been used to follow carbonate species from the hydrated layer during the aging of nanocrystalline apatites using FTIR spectroscopy [196].

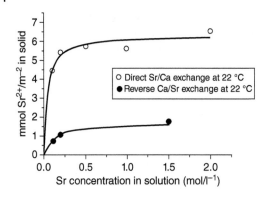

Figure 3.17 Sr^{2+} uptake on nonstoichiometric apatite after direct and inverse exchange with Ca^{2+}. Source: Drouet et al. [241]. Reproduced with permission of Elsevier.

Moreover, such experiments evidenced that only surface species were exchanged and that the apatitic core was not altered during the exchange experiments [160].

Regardless of the mechanisms involved, such preparation of metal-modified HA in excess of solution has many advantages: easy to implement, possible quantification of metal deposited on the surface by elementary analysis, absence of metal leaching [24, 26, 28] due to irreversible immobilization ascribed to strong metal–support interaction, and finally high metal dispersion obtained [42]. Depending on the metal loading, the immobilized ion may remain isolated, resulting in atomically dispersed species, or be a nucleation point for the growth of highly dispersed nanoclusters or nanoparticles. Note that Ca^{2+} exchange process was also suggested by several authors to be the initial stage of metal deposition by incipient wetness impregnation (the volume of the metal-containing solution that is added onto the catalyst support is precisely adapted to its pore volume). In addition, in this later case, larger particle sizes are expected than with the exchange in excess of solution. In particular, the exchangeability properties (or involving other single atom or monomeric complex immobilization mechanisms) in aqueous medium provide a unique opportunity to get the active species atomically dispersed for low metal loading [18, 332, 337]. Such single atom catalysts are of great interest in heterogeneous catalysis as they are stable and durable [18]. The dispersion of cobalt onto ion-exchanged HA surface in the form of isolated Co^{2+} species for 0.2 wt% loading was evidenced by magnetic measurements. However, for higher metal loading, Co_xO_y clustered entities and then Co_3O_4 nanoparticles are formed at higher cobalt loading, resulting in decreased basicity, and hence in the decay of specific activity in the 2-butanol and ethane dehydrogenation reactions [300, 338].

The Ni HA system also illustrates how crucial is the preparation method toward nickel dispersion for the control of selectivity and stability of the catalyst. This parameter is a very critical point for the challenging methane dry reforming reaction to avoid deactivation due to the coke formation on nanoparticles. Large-sized nickel particles (200 nm) supported over HA obtained by incipient wetness impregnation [103] were responsible for irreversible deactivation due to their densification and core-shell carbon formation. Coke deposition was comparatively more limited when nickel and cobalt were simultaneously co-impregnated on the support [104]. Although the coprecipitated Ni–Sr HA system led to promising results for the dry reforming of methane, the simultaneous presence of other nickel phosphate phases makes the origin of fine Ni particles (few nm), formed onto the surface of the spent catalyst, difficult to explain [316]. Recently, Ni deposited by exchange in excess of solution onto a cerium-doped HA previously prepared by coprecipitation led to very well dispersed Ni species, as deduced from aberration-corrected high angle annular dark field scanning transmission electron microscopy [18]. Cerium would prevent Ni sintering, resulting in the formation of particles of only 10 nm for 10 wt% Ni loading. Furthermore, for lower loading, 0.5 wt%, atomically dispersed Ni atoms could be stabilized as revealed by the absence of Ni—Ni bond detected by extended X-ray absorption fine structure (EXAFS). Such isolated Ni atoms were found intrinsically coke resistant due to their unique ability to only activate the first C—H bond in methane. It is proposed that cerium doping induces strong metal-support interaction, which stabilizes Ni single atoms toward sintering resulting in highly stable and performant system for the dry reforming of methane [18].

Adsorption of molecules

As for ions adsorption, the adsorption of molecules on apatite particles affects only the surface of the crystals. In most cases, typically in high-temperature HA compounds or in biomimetic apatites with adsorbing molecule with only a low affinity for apatite surfaces, the adsorption process corresponds to a regular adsorptive event, with the "simple" positioning of the molecule on the crystal surface, establishing essentially electrostatic bonds with the ions constitutive of the surface. However, in some favorable cases involving biomimetic apatites when the adsorbing molecule exposes at least one high-affinity end-group (often anionic end-groups such as phosphate or phosphonate), the adsorption of the molecule was shown to be simultaneously accompanied by an ion exchange process, where the relevant end-group displaces surface ions (e.g. HPO_4^{2-}) for an actual anchoring of the molecule onto the surface hydrated layer. Several publications have indeed pointed out this effect showing, for example, that the adsorption of molecules with negative functional groups is generally related to phosphate (or carbonate) ions released in the solution, evidencing such a concomitant ionic exchange reaction [318, 339, 340]. It may be remarked, however, that this cannot be a priori extended to all molecules containing a phosphate or phosphonate end-group, as other studies have shown the absence of simultaneous ion exchange even on biomimetic apatites [341, 342], and the overall molecular geometry/conformation also probably plays a key role in the molecule/support affinity, which will have to be further investigated by simulation methods.

Adsorption isotherms give essential information on the molecule/apatite interaction. They can be obtained by measuring the concentration of species in solution and in the solids during adsorption process. As for ion exchange, the adsorption of molecules on apatites is generally well represented by a Langmuir-type isotherm (although in certain cases other models such as Sips/Hill lead to a better fit), even in the case of a concomitant ion exchange involving the replacement of mineral ions of the apatite surface by molecular ions from the solution. In the case of such ion exchange simultaneous to adsorption, contrarily to general adsorption cases, no desorption should be expected here by simple dilution of the adsorbent solution (or even upon washing of the apatitic support), except for a low release due to dissolution of

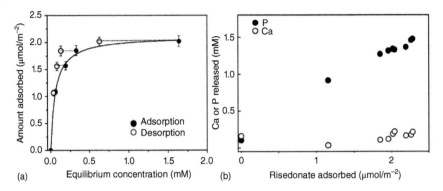

Figure 3.18 (a) Adsorption and desorption isotherms of bisphosphate molecules (risedronate) on stoichiometric HA at pH = 7.4 and $T = 37\,°C$; the dashed lines indicate that no risedronate desorbs upon dilution. (b) Variation of the phosphate and calcium concentrations in the solution versus the amount of risedronate absorbed. Source: Al-Kattan et al. [321]. Reproduced with permission of John Wiley and Sons.

the support. As for ion exchange, the number of ions displaced (e.g. phosphates) per adsorbed molecule depends then on the characteristics of the apatitic surface [340, 343, 344] (Figure 3.18). Whether the adsorption event is associated or not to an ion exchange, the maturation stage and apatite composition play a significant role in the adsorption process, which is particularly relevant for biomimetic apatites. The Langmuir affinity constant also varies according to the characteristics of the apatite substrate, and the amount adsorbed at saturation appears higher for immature nanocrystals with a well-developed hydrated layer and decreases with the maturation time.

Moreover, the nature of the functional end-groups of the molecule exposed is one of the key parameters to be taken into account when considering the interaction with a calcium phosphate surface. In case of adsorption with concomitant ion exchange, the number of (phosphate) ions displaced per adsorbed molecule has been determined by several authors and it appears dependent on the characteristics of the apatitic surface as well as the nature of the molecules adsorbed [340, 343, 344]. Phosphate esters and phosphonate, silicate and sulfate, and similar groups especially interact strongly with apatite surfaces. In general, carboxylate groups present a lower affinity, except when they are associated in polyacids [343, 345]. For example, the very high affinity of phosphoserine for apatite surface compared to that of serine can be related to the existence of the phosphate group that can strongly interact with the apatite surface, whereas the carboxyl group of serine itself presents only a moderate interaction with HA surface. Thus, adsorption of serine can be described by a Freundlich-type isotherm like that of other simple molecules weakly interacting with apatite surfaces such as glycine or acetic acid, for example [344, 346]. Positively charged groups (like $-NH_3^+$) can also adsorb on apatites but generally less strongly than negatively charged ones.

It can be noticed that, as for ion exchanges, several phenomena can be considered as competitive reactions: the possible precipitation of calcium (or phosphate) salts with the adsorbing molecules, the dissolution equilibrium of the apatite and the surface, or hydrolysis reactions of the support. These events do not generally give a Langmuir-type isotherm with a typical saturation domain. The analysis of the solid phase by XRD, spectroscopic analysis, or electron microscopy generally evidences new phase formation [207].

3.4 Conclusions

Apatitic calcium phosphates, such as HA, appear as excellent candidates of new class of greener catalytic materials offering numerous advantages for versatile applications in catalysis: tunable surface properties, recyclability, and high thermal stability. These bio-inspired and eco-compatible materials offer the possibility of generating diverse chemical compositions, degrees of crystallinity, and particle morphologies, depending on their mode of preparation (stoichiometric or nonstoichiometric, nanocrystalline or not, substituted or not, surface-grafted or not), thus providing samples with different features of relevance to catalysis. This chapter provides fundamental tools to help understand the behavior of apatite-based catalysts, based on advances in the characterization of their structural and surface properties, through both experimental and computational approaches, to study their structure as well as their thermodynamic and physicochemical properties and surface reactivity. Based on the characteristics of the apatitic support, it is possible to improve performances by simply adjusting certain parameters to better fit the required application. Heterogeneous catalysis on apatites requires a good understanding of solid catalysts, including of their preparation methods and thorough characterization. Studying structure–properties–reactivity relationships is essential to provide a multi-scale description from microscopic to macroscopic scales, evaluate application performances, and optimize their catalytic perspectives.

References

1 Bett, J.A.S., Christner, L.G., and Hall, W.K. (1967). Studies of the hydrogen held by solids. XII. Hydroxyapatite Catalysts. *Journal of the American Chemical Society* 89 (22): 5536–5541.
2 Bett, J.A.S. and Hall, W.K. (1968). The microcatalytic technique applied to a zero order reaction: the dehydration of 2-butanol over hydroxyapatite catalysts. *Journal of Catalysis* 10: 108–113.
3 Bett, J.A.S., Christner, L.G., and Hall, W.K. (1969). Infrared studies of hydroxyapatite catalysts. Adsorbed carbon dioxide, 2-butanol, and methyl ethyl ketone. *Journal of Catalysis* 13 (3): 332–336.
4 Fihri, A., Len, C., Varma, R.S., and Solhy, A. (2017). Hydroxyapatite: a review of syntheses, structure and applications in heterogeneous catalysis. *Coordination Chemistry Reviews* 347: 48–76.

5 Nayak, B., Samant, A., Misra, P.K., and Saxena, M. (2019). Nanocrystalline hydroxyapatite: a potent material for adsorption, biological and catalytic studies. *Materials Today: Proceedings.* 9: 689–698.

6 Ibrahim, M., Labaki, M., Giraudon, J.-M., and Lamonier, J.-F. (2020). Hydroxyapatite, a multifunctional material for air, water and soil pollution control: a review. *Journal of Hazardous Materials* 383: 121139.

7 Imizu, Y., Kadoya, M., Abe, H. et al. (1982). High-temperature evacuated hydroxyapatite as a base catalyst for the isomerization of 1-butene. *Chemistry Letters* 3: 414–416.

8 Scalbert, J., Thibault-Starzyk, F., Jacquot, R. et al. (2014). Ethanol condensation to butanol at high temperatures over a basic heterogeneous catalyst: how relevant is acetaldehyde self-aldolization? *Journal of Catalysis* 311: 28–32.

9 Petit, S., Thomas, C., Millot, Y. et al. (2020). Activation of C-H bond of propane by strong basic sites generated by bulk proton conduction on V-modified hydroxyapatites for the formation of propene. *ChemCatChem* 12: 2506–2521.

10 Han, Y.F., Phonthammachai, N., Ramesh, K. et al. (2008). Removing organic compounds from aqueous medium via wet peroxidation by gold catalysts. *Environmental Science and Technology* 42: 908–912.

11 Zhang, Y., Zhao, Y., and Xia, C. (2009). Basic ionic liquids supported on hydroxyapatite-encapsulated γ-Fe_2O_3 nanocrystallites: an efficient magnetic and recyclable heterogeneous catalyst for aqueous Knoevenagel condensation. *Journal of Molecular Catalysis A* 306 (1): 107–112.

12 Lv, C., Liang, H., Chen, H., and Wu, L. (2019). Hydroxyapatite supported Co_3O_4 catalyst for enhanced degradation of organic contaminants in aqueous solution: synergistic visible-light photo-catalysis and sulfate radical oxidation process. *Microchemical Journal* 149: 103959.

13 Mitsudome, T., Mikami, Y., Mori, H. et al. (2009). Supported silver nanoparticle catalyst for selective hydration of nitriles to amides in water. *Chemical Communications* 22: 3258–3260.

14 Hara, T., Kanai, S., Mori, K. et al. (2006). Highly efficient C-C bond forming reactions in aqueous media catalyzed by monomeric vanadate species in an apatite framework. *The Journal of Organic Chemistry* 71: 7455–7462.

15 Sebti, S., Tahir, R., Nazih, R. et al. (2002). Hydroxyapatite as a new solid support for the Knoevenagel reaction in heterogeneous media without solvent. *Applied Catalysis A* 228 (1): 155–159.

16 Ichihara, J., Yamaguchi, S., Nomoto, T. et al. (2002). Keggin-type polyacid clusters on apatite: characteristic catalytic activities in solvent-free oxidation. *Tetrahedron Letters* 43 (46): 8231–8234.

17 Pillai, K.M., Singh, S., and Jonnalagadda, S. (2010). Solvent-free Knoevenagel condensation over cobalt hydroxyapatite. *Synthetic Communications* 40: 3710–3715.

18 Akri, M., Shu Zhao, S., Li, X. et al. (2019). Atomically dispersed nickel as coke-resistant active sites for methane dry reforming. *Nature Communications* 10 (1): 1–10.

19 Sugiyama, S., Hashimoto, T., Morishita, Y. et al. (2004). Effects of calcium cations incorporated into magnesium vanadates on the redox behaviors and the catalytic activities for the oxidative dehydrogenation of propane. *Applied Catalysis A.* 270 (1–2): 253–260.

20 Tsuchida, T., Sakuma, S., Takeguchi, T., and Ueda, W. (2006). Direct synthesis of n-butanol from ethanol over nonstoichiometric hydroxyapatite. *Industrial and Engineering Chemistry Research* 45: 8634–8642.

21 Wang, Y., Chen, B.-b., Crocker, M. et al. (2015). Understanding on the origins of hydroxyapatite stabilized gold nanoparticles as high-efficiency catalysts for formaldehyde and benzene oxidation. *Catalysis Communications* 59: 195–200.

22 Tang, H., Wei, J., Liu, F. et al. (2016). Strong metal–support interactions between gold nanoparticles and nonoxides. *Journal of American Chemical Society* 138 (1): 56–59.

23 Boukha, Z., Ayastuy, J.L., González-Velasco, J.R., and Gutiérrez-Ortiz, M.A. (2018). Water-gas shift reaction over a novel Cu-ZnO/HAP formulation: enhanced catalytic performance in mobile fuel cell applications. *Applied Catalysis A* 566: 1–14.

24 Yamaguchi, K., Mori, K., Mizugaki, T. et al. (2000). Creation of a monomeric Ru species on the surface of hydroxyapatite as an efficient heterogeneous catalyst for aerobic alcohol oxidation. *Journal of American Chemical Society* 122 (29): 7144–7145.

25 Mori, K., Hara, T., Mizugaki, T. et al. (2003). Hydroxyapatite-bound cationic ruthenium complexes as novel heterogeneous Lewis acid catalysts for Diels–Alder and Aldol reactions. *Journal of American Chemical Society* 125 (38): 11460–11461.

26 Opre, Z., Ferri, D., Krumeich, F. et al. (2006). Aerobic oxidation of alcohols by organically modified ruthenium hydroxyapatite. *Journal of Catalysis* 241 (2): 287–295.

27 Yan, B., Zhang, Y., Chen, G. et al. (2016). The utilization of hydroxyapatite-supported CaO-CeO2 catalyst for biodiesel production. *Energy Conversion and Management* 130: 156–164.

28 Pang, Y., Kong, L., Chen, D. et al. (2020). Facilely synthesized cobalt doped hydroxyapatite as hydroxyl promoted peroxymonosulfate activator for degradation of rhodamine B. *Journal of Hazardous Materials* 384: 121447.

29 Maaten, B., Moussa, J., Desmarets, C. et al. (2014). Cu-modified hydroxy-apatite as catalyst for Glaser–Hay CC homo-coupling reaction of terminal alkynes. *Journal of Molecular Catalysis A: Chemical* 393: 112–116.

30 Boukha, Z., Kacimi, M., Pereira, M.F.R. et al. (2007). Methane dry reforming on Ni loaded hydroxyapatite and fluoroapatite. *Applied Catalysis A* 317 (2): 299–309.

31 Zahmakıran, M., Tonbul, Y., and Özkar, S. (2010). Ruthenium(0) nanoclusters supported on hydroxyapatite: highly active, reusable and green catalyst in the hydrogenation of aromatics under mild conditions with an unprecedented catalytic lifetime. *Chemical Communications* 46: 4788–4790.

32 Gruselle, M. (2015). Apatites: a new family of catalysts in organic synthesis. *Journal of Organometallic Chemistry* 793: 93–101.

33 Bilakanti, V., Boosa, V., Velisoju, V.K. et al. (2017). Role of surface basic sites in Sonogashira coupling reaction over $Ca_5(PO_4)_3OH$ supported pd catalyst: investigation by diffuse reflectance infrared Fourier transform spectroscopy. *Journal of Physical Chemistry C* 121 (40): 22191–22198.

34 Essamlali, Y., Amadine, O., Larzek, M. et al. (2017). Sodium modified hydroxyapatite: highly efficient and stable solid-base catalyst for biodiesel production. *Energy Conversion and Management* 149: 355–367.

35 Li, C., Xu, G., Liu, X. et al. (2017). Hydrogenation of biomass-derived furfural to tetrahydrofurfuryl alcohol over hydroxyapatite-supported pd catalyst under mild conditions. *Industrial and Engineering Chemistry Research* 56 (31): 8843–8849.

36 Xu, G., Zhang, Y., Fu, Y., and Guo, Q. (2017). Efficient hydrogenation of various renewable oils over Ru-HAP catalyst in water. *ACS Catalysis* 7 (2): 1158–1169.

37 Chaudharya, R. and Dhepe, P.L. (2017). Solid base catalyzed depolymerization of lignin into low molecular weight products†. *Green Chemistry* 19: 778–788.

38 Gao, T., Yin, Y., Fang, W., and Cao, Q. (2018). Highly dispersed ruthenium nanoparticles on hydroxyapatite as selective and reusable catalyst for aerobic oxidation of 5-hydroxymethylfurfural to 2,5-furandicarboxylic acid under base-free conditions. *Journal of Molecular Catalysis* 450: 55–64.

39 Ho, C.R., Zheng, S., Shylesh, S., and Bell, A.T. (2018). The mechanism and kinetics of methyl isobutyl ketone synthesis from acetone over ion-exchanged hydroxyapatite. *Journal of Catalysis* 365: 174–183.

40 Peeters, A., Claes, L., Geukens, I. et al. (2014). Alcohol amination with heterogeneous ruthenium hydroxyapatite catalysts. *Applied Catalysis A* 469: 191–197.

41 Sebti, S., Tahir, R., Nazih, R., and Boulaajaj, S. (2001). Comparison of different Lewis acid supported on hydroxyapatite as new catalysts of Friedel–Crafts alkylation. *Applied Catalysis A* 218: 25–30.

42 Mori, K., Yamaguchi, K., Hara, T. et al. (2002). Controlled synthesis of hydroxyapatite-supported palladium complexes as highly efficient heterogeneous catalysts. *Journal of the American Chemical Society* 124 (39): 11572–11573.

43 Venugopal, A. and Scurrell, M.S. (2003). Hydroxyapatite as a novel support for gold and ruthenium catalysts. Behaviour in the water gas shift reaction. *Applied Catalysis: A* 245 (1): 137–147.

44 Wuyts, S., De Vos, D.E., Verpoort, F. et al. (2003). A heterogeneous Ru–hydroxyapatite catalyst for mild racemization of alcohols. *Journal of Catalysis* 219 (2): 417–424.

45 Mori, K., Hara, T., Mizugaki, T. et al. (2004). Hydroxyapatite-supported palladium nanoclusters: a highly active heterogeneous catalyst for selective oxidation of alcohols by use of molecular oxygen. *Journal of American Chemical Society* 126 (34): 10657–10666.

46 Solhy, A., Clark, J.H., Tahir, R. et al. (2006). Transesterifications catalysed by solid, reusable apatite–zinc chloride catalysts. *Green Chemistry* 8: 871–874.

47 Tahir, R., Banert, K., Solhy, A., and Sebti, S. (2006). Zinc bromide supported on hydroxyapatite as a new and efficient solid catalyst for Michael addition of indoles to electron-deficient olefins. *Journal of Molecular Catalysis* 246 (1–2): 39–42.

48 Khachani, M., Kacimi, M., Ensuque, A. et al. (2010). Iron–calcium–hydroxyapatite catalysts: iron speciation and comparative performances in butan-2-ol conversion and propane oxidative dehydrogenation. *Applied Catalysis A* 388: 113–123.

49 Matsuura, Y., Onda, A., Ogo, S., and Yanagisawa, K. (2014). Acrylic acid synthesis from lactic acid over hydroxyapatite catalysts with various cations and anions. *Catalysis Today* 226: 192–197.

50 Sugiyama, S., Osaka, T., Hirata, Y., and Sotowa, K.I. (2006). Enhancement of the activity for oxidative dehydrogenation of propane on calcium hydroxyapatite substituted with vanadate. *Applied Catalysis A* 312: 52–58.

51 Chlala, D., Giraudon, J.M., Nuns, N. et al. (2016). Active Mn species well dispersed on Ca2+ enriched apatite for total oxidation of toluene. *Applied Catalysis. B, Environmental* 184: 87–95.

52 Tounsi, H., Djemal, S., Petitto, C., and Delahay, G. (2011). Copper loaded hydroxyapatite catalyst for selective catalytic reduction of nitric oxide with ammonia. *Applied Catalysis. B, Environmental* 107 (1): 158–163.

53 Dasireddy, V.D.B.C., Friedrich, H.B., and Singh, S. (2013). Studies towards a mechanistic insight into the activation of n-octane using vanadium supported on alkaline earth metal hydroxyapatites. *Applied Catalysis A.* 467: 142–153.

54 Dasireddy, V.D.B.C., Singh, S., and Friedrich, H.B. (2013). Activation of n-octane using vanadium oxide supported on alkaline earth hydroxyapatites. *Applied Catalysis A* 456: 105–117.

55 Dasireddy, V.D.B.C., Singh, S., and Friedrich, H.B. (2014). Vanadium oxide supported on non-stoichiometric strontium hydroxyapatite catalysts for the oxidative dehydrogenation of n-octane. *Journal of Molecular Catalysis A* 395: 398–408.

56 Carvalho, D.C., Pinheiro, L.G., Campos, A. et al. (2014). Characterization and catalytic performances of copper and cobalt-exchanged hydroxyapatite in glycerol conversion for 1-hydroxyacetone production. *Applied Catalysis A: General* 471: 39–49.

57 Jemal, J., Petitto, C., Delahay, G. et al. (2015). Selective catalytic reduction of NO by NH3 over copper-hydroxyapatite catalysts: effect of the increase of the specific surface area of the support. *Reaction Kinetics, Mechanisms and Catalysis* 114: 185–196.

58 Campisi, S., Galloni, M.G., Bossola, F., and Gervasini, A. (2019). Comparative performance of copper and iron functionalized hydroxyapatite catalysts in NH3-SCR. *Catalysis Communications* 123: 79–85.

59 Moteki, T. and Flaherty, D.W. (2016). Mechanistic insight to C–C bond formation and predictive models for Cascade reactions among alcohols on ca- and Sr-hydroxyapatites. *ACS Catalysis* 6: 4170–4183.

60 Smahi, A., Solhy, A., El Badaoui, H. et al. (2003). Potassium fluoride doped fluorapatite and hydroxyapatite as new catalysts in organic synthesis. *Applied Catalysis A: General* 250: 151–159.

61 Sebti, S., Solhy, A., Tahir, R., and Smahi, A. (2002). Modified hydroxyapatite with sodium nitrate: an efficient new solid catalyst for the Claisen-Schmidt condensation. *Applied Catalysis A.* 235 (1–2): 273–281.

62 Solhy, A., Tahir, R., Sebti, S. et al. (2010). Efficient synthesis of chalcone derivatives catalyzed by re-usable hydroxyapatite. *Applied Catalysis A.* 374 (1): 189–193.

63 Mori, K., Oshiba, M., Hara, T. et al. (2005). Michael reaction of 1,3-dicarbonyls with enones catalyzed by a hydroxyapatite-bound La complex. *Tetrahedron Letters* 46 (25): 4283–4286.

64 Mori, K., Oshiba, M., Hara, T. et al. (2006). Creation of monomeric La complexes on apatite surfaces and their application as heterogeneous catalysts for Michael reactions. *New Journal of Chemistry* 30: 44–52.

65 Tsuchida, T., Yoshioka, T., Sakuma, S. et al. (2008). Synthesis of biogasoline from ethanol over hydroxyapatite catalyst. *Industrial and Engineering Chemistry Research* 47: 1443–1452.

66 Tsuchida, T., Kubo, J., Yoshioka, T. et al. (2008). Reaction of ethanol over hydroxyapatite affected by Ca/P ratio of catalyst. *Journal of Catalysis* 259 (2): 183–189.

67 Ogo, S., Onda, A., and Yanagisawa, K. (2011). Selective synthesis of 1-butanol from ethanol over strontium phosphate hydroxyapatite catalysts. *Applied Catalysis A* 402 (1): 188–195.

68 Silvester, L., Lamonier, J.F., Faye, J. et al. (2015). Reactivity of ethanol over hydroxyapatite-based Ca-enriched catalysts with various carbonate contents. *Catalysis Science & Technology* 5: 2994–3006.

69 Hanspal, S., Young, Z.D., Shou, H., and Davis, R.J. (2015). Multiproduct steady-state isotopic transient kinetic analysis of the ethanol coupling reaction over hydroxyapatite and magneisa. *ACS Catalysis* 5: 1737–1746.

70 Silvester, L., Lamonier, J.F., Lamonier, C. et al. (2017). Guerbet reaction over strontium-substituted hydroxyapatite catalysts prepared at various (Ca+Sr)/P ratios. *ChemCatChem* 9: 2250–2261.

71 Hanspal, S., Young, D., Prillaman, J.T., and Davis, R.J. (2017). Influence of surface acid and base sites on the Guerbet coupling of ethanol to butanol over metal phosphate catalysts. *Journal of Catalysis* 352: 182–190.

72 Young, Z.D. and Davis, R.J. (2018). Hydrogen transfer reactions relevant to Guerbet coupling of alcohols over hydroxyapatite and magnesium oxide catalysts. *Catalysis Science & Technology* 8: 1722–1729.

73 Wu, X., Fang, G., Tong, Y. et al. (2018). Catalytic upgrading of ethanol to n-butanol: progress in catalyst development. *ChemSusChem* 11: 71–85.

74 Cimino, S., Lisi, L., and Romanucci, S. (2018). Catalysts for conversion of ethanol to butanol: effect of acid-base and redox properties. *Catalysis Today* 304: 58–63.

75 Ben Osman, M., Krafft, J.M., Thomas, C. et al. (2019). Importance of the nature of the active acid/base pairs of hydroxyapatite involved in the catalytic

transformation of ethanol to n-butanol revealed by operando DRIFTS. *ChemCatChem* 11 (6): 1765–1778.

76 Ogo, S., Onda, A., Iwasa, Y. et al. (2012). 1-butanol synthesis from ethanol over strontium phosphate hydroxyapatite catalysts with various Sr/P ratios. *Journal of Catalysis* 296: 24–30.

77 Masuyama, Y., Yoshikawa, K., Suzuki, N. et al. (2011). Hydroxyapatite-supported copper(II)-catalyzed azide–alkyne [3+2] cycloaddition with neither reducing agents nor bases in water. *Tetrahedron Letters* 52 (51): 6916–6918.

78 Choudary, B.M., Sridhar, C., Kantam, M.L. et al. (2005). Design and evolution of copper apatite catalysts for N-arylation of heterocycles with Chloro- and Fluoroarenes. *Journal of the American Chemical Society* 127 (28): 9948–9949.

79 Kanai, H., Nakao, M., and Imamura, S. (2003). Selective photoepoxidation of propylene over hydroxyapatite–silica composites. *Catalysis Communications* 4 (8): 405–409.

80 Opre, Z., Mallat, T., and Baiker, A. (2007). Epoxidation of styrene with cobalt-hydroxyapatite and oxygen in dimethylformamide: a green technology? *Journal of Catalysis* 245 (2): 482–486.

81 Singh, S.J. and Sreekanth, B. (2008). Selective oxidation of n-pentane over V_2O_5 supported on hydroxyapatite. *Catalysis Letters* 126 (1–2): 200–206.

82 Boucetta, C., Kacimi, M., Ensique, J.Y. et al. (2009). Oxidative dehydrogenation of propane over chromium-loaded calcium hydroxyapatite. *Applied Catalysis A.* 356: 201–210.

83 Dasireddy, V.D.B.C., Singh, S., and Friedrich, H.B. (2012). Oxidative dehydrogenation of n-octane using vanadium pentoxide-supported hydroxyapatite catalysts. *Applied Catalysis A.* 421-422: 58–69.

84 Chlala, D., Giraudon, J.M., Nuns, N. et al. (2017). Highly active Noble-metal-free copper hydroxyapatite catalysts for the Total oxidation of toluene. *ChemCatChem* 9: 2275–2283.

85 Aellach, B., Ezzamarty, A., Leglise, J. et al. (2010). Calcium-deficient and stoichiometric hydroxypatite promoted by cobalt for the catalytic removal of oxygenated volatile organic compounds. *Catalysis Letters* 135: 197–206.

86 More, R.K., Lavande, N.R., and More, P.M. (2019). Copper supported on Co substituted hydroxyapatite for complete oxidation of diesel engine exhaust and VOC. *Molecular Catalysis* 474: 110414.

87 Dominguez, M.I., Romero-Sarria, F., Centeno, M.A., and Odriozola, J.A. (2009). Gold/hydroxyapatite catalysts. Synthesis, characterization and catalytic activity to CO oxidation. *Applied Catalysis. B, Environmental* 87: 245–251.

88 Chlala, D., Labaki, M., Giraudon, J.-M. et al. (2016). Toluene total oxidation over Pd and Au nanoparticles supported on hydroxyapatite. *CR Chimie* 19 (4): 525–537.

89 Sun, Y.-P., Fu, H.-Y., Zhang, D.-l. et al. (2010). Complete hydrogenation of quinoline over hydroxyapatite supported ruthenium catalyst. *Catalysis Communications* 12 (3): 188–192.

90 Kibby, C.L. and Hall, W.K. (1973). Dehydrogenation of alcohols and hydrogen transfer from alcohols to ketones over hydroxyapatite catalysts. *Journal of Catalysis* 31 (1): 65–73.

91 Boukha, Z., Kacimi, M., Ziyad, M. et al. (2007). Comparative study of catalytic activity of Pd loaded hydroxyapatite and fluoroapatite in butan-2-ol conversion and methane oxidation. *Journal of Molecular Catalysis* 270 (1): 205–213.

92 Zaccheria, F., Scotti, N., and Ravasio, N. (2018). The role of copper in the upgrading of bioalcohols. *ChemCatChem* 10: 1526–1535.

93 Goulas, K.A., Song, Y., Johnson, G.R. et al. (2018). Selectivity tuning over monometallic and bimetallic dehydrogenation catalysts: effects of support and particle size. *Catalysis Science and Technology* 8 (1): 314–327.

94 Stošić, D., Bennici, S., Sirotin, S. et al. (2012). Glycerol dehydration over calcium phosphate catalysts: effect of acidic–basic features on catalytic performance. *Applied Catalysis A.* 447–448: 124–134.

95 Ghantani, V.C., Lomate, S.T., Dongare, M.K., and Umbarkar, S.B. (2013). Catalytic dehydration of lactic acid to acrylic acid using calcium hydroxyapatite catalysts. *Green Chemistry* 15 (5): 1211–1217.

96 Sugiyama, S. and Moffat, J.B. (2002). Cation effects in the conversion of methanol on calcium, strontium, barium and lead hydroxyapatites. *Catalysis Letters* 81 (1–2): 77–81.

97 Cheikhi, N., Kacimi, M., Rouimi, M. et al. (2005). Direct synthesis of methyl isobutyl ketone in gas-phase reaction over palladium-loaded hydroxyapatite. *Journal of Catalysis* 232 (2): 257–267.

98 Rahmanian, A. and Ghaziaskar, H.S. (2013). Continuous dehydration of ethanol to diethyl ether over aluminum phosphate–hydroxyapatite catalyst under sub and supercritical condition. *Journal of Supercritical Fluids* 78: 34–41.

99 Yan, B., Tao, L.Z., Liang, Y., and Xu, B.Q. (2014). Sustainable production of acrylic acid: catalytic performance of hydroxyapatites for gas-phase dehydration of lactic acid. *ACS Catalysis* 4 (6): 1931–1943.

100 Miao, D., Cavusoglu, G., Lichtenberg, H. et al. (2017). Water-gas shift reaction over platinum/strontium apatite catalysts. *Applied Catalysis. B, Environmental* 202: 587–596.

101 Schiavoni, M., Campisi, S., Carniti, P. et al. (2018). Focus on the catalytic performances of cu-functionalized hydroxyapatites in NH3-SCR reaction. *Applied Catalysis A* 563: 43–53.

102 Ogo, S. and S. M, Sekine Y. (2017). Coke resistance of Sr-hydroxyapatite supported Co catalyst for ethanol steam reforming. *Chemistry Letters* 46: 729–732.

103 Rego de Vasconcelos, B., Pham Minh, D., Sharrock, P., and Nzihou, A. (2018). Regeneration study of Ni/hydroxyapatite spent catalyst from dry reforming. *Catalysis Today* 310: 107–115.

104 Phan, T.S., Sane, A.R., Rêgo de Vasconcelos, B. et al. (2018). Hydroxyapatite supported bimetallic cobalt and nickel catalysts for syngas production from dry reforming of methane. *Applied Catalysis. B, Environmental* 224: 310–321.

105 Boukha, Z., Yeste, M.P., Cauqui, M.Á., and González-Velasco, J.R. (2019). Influence of Ca/P ratio on the catalytic performance of Ni/hydroxyapatite samples in dry reforming of methane. *Applied Catalysis A* 580: 34–45.

106 Kaneda, K. and Mizugaki, T. (2009). Development of concerto metal catalysts using apatite compounds for green organic syntheses. *Energy & Environmental Science* 2: 655–673.

107 Sugiyama, S. (2007). Approach using apatite to studies on energy and environment. *Phosphorus Research Bulletin* 21: 1–8.

108 Kuwahara, Y., Ohmichi, T., Kamegawa, T. et al. (2009). A novel synthetic route to hydroxyapatite–zeolite composite material from steel slag: investigation of synthesis mechanism and evaluation of physicochemical properties. *Journal of Materials Chemistry* 19: 7263–7272.

109 Gorbanev, Y.Y., Kegnæs, S., and Riisager, A. (2011). Selective aerobic oxidation of 5-hydroxymethylfurfural in water over solid ruthenium hydroxide catalysts with magnesium-based supports. *Catalysis Letters* 141: 1752–1760.

110 Wang, Q.N., Weng, X.F., Zhou, B.C. et al. (2019). Direct, selective production of aromatic alcohols from ethanol using a tailored bifunctional cobalt–hydroxyapatite catalyst. *ACS Catalysis* 9: 7204–7216.

111 Nishimura, S., Mizuhori, K., and Ebitani, K. (2016). Reductive amination of furfural toward furfurylamine with aqueous ammonia under hydrogen over Ru-supported catalyst. *Research on Chemical Intermediates* 42 (1): 19–30.

112 Lovón-Quintana, J.J., Rodriguez-Guerrero, J.K., and Valença, P.G. (2017). Carbonate hydroxyapatite as a catalyst for ethanol conversion to hydrocarbon fuels. *Applied Catalysis A* 542: 136–145.

113 De Leeuw, N. (2010). Computer simulations of structures and properties of the biomaterial hydroxyapatite. *Journal of Materials Chemistry* 20 (26): 5376–5389.

114 Zhao, W., Xu, Z., Yang, Y., and Sahai, N. (2014). Surface energetics of the hydroxyapatite nanocrystal–water interface: a molecular dynamics study. *Langmuir* 30 (44): 13283–13292.

115 Elliott, J.C. (1994). *Structure and Chemistry of the Apatites and Other Calcium Orthophosphates*. Amsterdam: Elsevier Science BV.

116 White, T.J. and ZhiLi, D. (2003). Structural derivation and crystal chemistry of apatites. *Acta Crystallographica Section B* 59 (1): 1–16.

117 Ptáček, P. (2016). *Introduction to Apatite in Apatites and their Synthetic Analogues-Synthesis, Structure, Properties and Applications*. IntechOpen.

118 Ma, G. and Liu, X.Y. (2009). Hydroxyapatite: hexagonal or monoclinic? *Crystal Growth and Design* 9 (7): 2991–2994.

119 Kay, M.I., Young, R.A., and Posner, A.S. (1964). Crystal structure of hydroxyapatite. *Nature* 204: 1050–1052.

120 Rey, C., Combes, C., Drouet, C., and Grossin, D. (2011). Bioactive ceramics: physical chemistry. In: *Comprehensive Biomaterials*, 187–221. Elsevier.

121 De Leeuw, N. (2001). Local ordering of hydroxy groups in hydroxyapatite. *Chemical Communications* 17: 1646–1647.

122 Almora-Barrios, N., Grau-Crespo, R., and Leeuw, N.H. (2013). A computational study of magnesium incorporation in the bulk and surfaces of hydroxyapatite. *Langmuir* 29 (19): 5851–5856.

123 Balachandran, P.V. and Rajan, K. (2012). Structure maps for AI4AII6 (BO4) 6X2 apatite compounds via data mining. *Acta Crystallographica, Section B: Structural Science* 68 (1): 24–33.

124 Habelitz, S., Pascual, L., and Durán, A. (1999). Nitrogen-containing apatite. *Journal of the European Ceramic Society* 19 (15): 2685–2694.

125 Corno, M., Rimola, A., Bolis, V., and Ugliengo, P. (2010). Hydroxyapatite as a key biomaterial: quantum-mechanical simulation of its surfaces in interaction with biomolecules. *Physical Chemistry Chemical Physics* 12 (24): 6309–6329.

126 Slepko, A. and Demkov, A.A. (2011). First-principles study of the biomineral hydroxyapatite. *Physical Review B* 84 (13): 134108.

127 Šupová, M. (2015). Substituted hydroxyapatites for biomedical applications: a review. *Ceramics International* 41 (8): 9203–9231.

128 Chappell, H., Duer, M., Groom, N. et al. (2008). Probing the surface structure of hydroxyapatite using NMR spectroscopy and first principles calculations. *Physical Chemistry Chemical Physics* 10 (4): 600–606.

129 Goldenberg, J.E., Wilt, Z., Schermerhorn, D.V. et al. (2015). Structural effects on incorporated water in carbonated apatites. *American Mineralogist* 100 (1): 274–280.

130 Uskoković, V. (2015). The role of hydroxyl channel in defining selected physicochemical peculiarities exhibited by hydroxyapatite. *RSC Advances* 5: 36614–36633.

131 Wang, J. (2015). Incorporation of iodine into apatite structure: a crystal chemistry approach using artificial neural network. *Frontiers in Earth Science* 3: 20.

132 Markovic, S., Veselinovic, L., Lukic, M.J. et al. (2011). Synthetical bone-like and biological hydroxyapatites: a comparative study of crystal structure and morphology. *Biomedical Materials* 6 (4): 045005.

133 Rey, C. (2020). Apatite channels and zeolite-like properties. In: *Hydroxyapatite and Related Materials* (ed. W. Paul and B.C. Brown), 257–262. Boca Raton: CRC Press.

134 Yashima, M., Kubo, N., Omoto, K. et al. (2014). Diffusion path and conduction mechanism of protons in hydroxyapatite. *The Journal of Physical Chemistry C* 118 (10): 5180–5187.

135 Camargo, C.L., Resende, N.S., Perez, C.A. et al. (2018). Molecular dynamics simulation and experimental validation by X-ray data of hydroxyapatite crystalline structures. *Fluid Phase Equilibria* 470: 60–67.

136 Mostafa, N.Y. and Brown, P.W. (2007). Computer simulation of stoichiometric hydroxyapatite: structure and substitutions. *Journal of Physics and Chemistry of Solids* 68 (3): 431–437.

137 Zahn, D. and Hochrein, O. (2008). On the composition and atomic arrangement of calcium-deficient hydroxyapatite: an ab-initio analysis. *Journal of Solid State Chemistry* 181 (8): 1712–1716.

138 Sun, J., Song, Y., Wen, G. et al. (2013). Softening of hydroxyapatite by vacancies: a first principles investigation. *Materials Science and Engineering: C* 33 (3): 1109–1115.

139 Ulian, G., Valdrè, G., Corno, M., and Ugliengo, P. (2014). DFT investigation of structural and vibrational properties of type B and mixed AB carbonated hydroxylapatite. *American Mineralogist* 99 (1): 117–127.

140 Astala, R. and Stott, M. (2005). First principles investigation of mineral component of bone: CO_3 substitutions in hydroxyapatite. *Chemistry of Materials* 17 (16): 4125–4133.

141 Fleet, M.E. and Liu, X. (2007). Coupled substitution of type A and B carbonate in sodium-bearing apatite. *Biomaterials* 28 (6): 916–926.

142 Ren, F., Lu, X., and Leng, Y. (2013). Ab initio simulation on the crystal structure and elastic properties of carbonated apatite. *Journal of the Mechanical Behavior of Biomedical Materials* 26: 59–67.

143 Peroos, S., Du, Z., and de Leeuw, N.H. (2006). A computer modelling study of the uptake, structure and distribution of carbonate defects in hydroxy-apatite. *Biomaterials* 27 (9): 2150–2161.

144 De Leeuw, N. (2002). Density functional theory calculations of solid solutions of fluor-and chlorapatites. *Chemistry of Materials* 14 (1): 435–441.

145 Jay, E.E., Rushton, M.J., and Grimes, R.W. (2012). Migration of fluorine in fluorapatite–a concerted mechanism. *Journal of Materials Chemistry* 22 (13): 6097–6103.

146 Sugiyama, S., Sugimoto, N., Hirata, Y. et al. (2008). Oxidative dehydrogenation of propane on vanadate catalysts supported on various metal hydroxyaptites. *Phosphorus Research Bulletin* 22: 13–16.

147 Shang, S., Zhao, Q., Zhang, D. et al. (2019). Molecular dynamics simulation of the adsorption behavior of two different drugs on hydroxyapatite and Zn-doped hydroxyapatite. *Materials Science and Engineering: C* 105: 110017.

148 Wang, M., Wang, Q., Lu, X. et al. (2017). Computer simulation of ions doped hydroxyapatite: a brief review. *Journal of Wuhan University of Technology-Materials Science Edition* 32 (4): 978–987.

149 Ma, X. and Ellis, D.E. (2008). Initial stages of hydration and Zn substitution/occupation on hydroxyapatite (0001) surfaces. *Biomaterials* 29 (3): 257–265.

150 Bigi, A., Boanini, E., Capuccini, C., and Gazzano, M. (2007). Strontium-substituted hydroxyapatite nanocrystals. *Inorganica Chimica Acta* 360 (3): 1009–1016.

151 Terra, J., Dourado, E.R., Eon, J.-G. et al. (2009). The structure of strontium-doped hydroxyapatite: an experimental and theoretical study. *Physical Chemistry Chemical Physics* 11 (3): 568–577.

152 Matsunaga, K., Inamori, H., and Murata, H. (2008). Theoretical trend of ion exchange ability with divalent cations in hydroxyapatite. *Physical Review B* 78 (9): 094101.

153 Wang, Q., Li, P., Tang, P. et al. (2019). Experimental and simulation studies of strontium/fluoride-codoped hydroxyapatite nanoparticles with osteogenic and antibacterial activities. *Colloids and Surfaces. B, Biointerfaces* 182: 110359.

154 Li, P., Jia, Z., Wang, Q. et al. (2018). A resilient and flexible chitosan/silk cryogel incorporated ag/Sr co-doped nanoscale hydroxyapatite for osteoinductivity and antibacterial properties. *Journal of Materials Chemistry B* 6 (45): 7427–7438.

155 Kasamatsu, S. and Sugino, O. (2018). First-principles investigation of polarization and ion conduction mechanisms in hydroxyapatite. *Physical Chemistry Chemical Physics* 20 (13): 8744–8752.

156 Wang, Q., Tang, P., Ge, X. et al. (2018). Experimental and simulation studies of strontium/zinc-codoped hydroxyapatite porous scaffolds with excellent osteoinductivity and antibacterial activity. *Applied Surface Science* 462: 118–126.

157 Ishisone, K., Jiraborvornpongsa, N., Isobe, T. et al. (2020). Experimental and theoretical investigation of WOx modification effects on the photocatalytic activity of titanium-substituted hydroxyapatite. *Applied Catalysis B: Environmental* 264: 118516.

158 Rey, C., Combes, C., Drouet, C., and Glimcher, M. (2009). Bone mineral: update on chemical composition and structure. *Osteoporosis International* 20 (6): 1013–1021.

159 Eichert, D., Drouet, C., Sfihi, H. et al. (2007). Nanocrystalline apatite-based biomaterials: synthesis, processing and characterization. *Biomaterials Research Advances* 93–143.

160 Rey, C., Combes, C., Drouet, C. et al. (2007). Nanocrystalline apatites in biological systems: characterisation, structure and properties. *Materialwissenschaft und Werkstofftechnik* 38 (12): 996–1002.

161 Rey, C., Combes, C., Drouet, C. et al. (2014). Surface properties of biomimetic nanocrystalline apatites; applications in biomaterials. *Progress in Crystal Growth and Characterization of Materials* 60 (6): –3e73.

162 Gómez-Morales, J., Iafisco, M., Delgado-López, J.M. et al. (2013). Progress on the preparation of nanocrystalline apatites and surface characterization: overview of fundamental and applied aspects. *Progress in Crystal Growth and Characterization of Materials* 59 (1): 1–46.

163 Drouet, C., Gomez-Morales, J., Iafisco, M. et al. (2012). Chapter 2: Calcium phosphate surface tailoring technologies for drug delivering and tissue engineering. In: *Surface Tailoring of Inorganic Materials for Biomedical Applications*, 43–111. Bentham Science Publishers.

164 Wilson, R.M., Elliott, J.C., Dowker, S.E., and Rodriguez-Lorenzo, L.M. (2005). Rietveld refinements and spectroscopic studies of the structure of ca-deficient apatite. *Biomaterials* 26 (11): 1317–1327.

165 Kim, H.M., Rey, C., and Glimcher, M.J. (1995). Isolation of calcium-phosphate crystals of bone by non-aqueous methods at low temperature. *Journal of Bone and Mineral Research* 10 (10): 1589–1601.

166 Eichert, D., Sfihi, H., Combes, C., and Rey, C. (2004). Specific characteristics of wet nanocrystalline apatites. Consequences on biomaterials and bone tissue. *Key Engineering Materials* 254: 927–930.

167 Jäger, C., Welzel, T., Meyer Zaika, W., and Epple, M. (2006). A solid state NMR investigation of the structure of nanocrystalline hydroxyapatite. *Magnetic Resonance in Chemistry* 44 (6): 573–580.

168 Rey, C., Combes, C., Drouet, C. et al. (2007). Physico-chemical properties of nanocrystalline apatites: implications for biominerals and biomaterials. *Materials Science and Engineering: C* 27 (2): 198–205.

169 Wu, H., Xu, D., Yang, M., and Zhang, X. (2016). Surface structure of hydroxyapatite from simulated annealing molecular dynamics simulations. *Langmuir* 32 (18): 4643–4652.

170 Xie, Q., Xue, Z., Gu, H. et al. (2018). Molecular dynamics exploration of ordered-to-disordered surface structures of biomimetic hydroxyapatite nanoparticles. *The Journal of Physical Chemistry C* 122 (12): 6691–6703.

171 Lu, Y., Dong, W., Ding, J. et al. (2019). *Hydroxyapatite Nanomaterials: Synthesis, Properties, and Functional Applications*, 485–536. Nanomaterials from Clay Minerals: Elsevier.

172 Silvester, L., Lamonier, J.F., Vannier, R.N. et al. (2014). Structural, textural and acid-base properties of carbonates-containing hydroxyapatites. *Journal of Materials Chemistry A* 2: 11073–11090.

173 Solhy, A., Amer, W., Karkouri, M. et al. (2011). Bi-functional modified-phosphate catalyzed the synthesis of α-α'-(EE)-bis (benzylidene)-cycloalkanones: microwave versus conventional-heating. *Journal of Molecular Catalysis A: Chemical* 336 (1–2): 8–15.

174 Monma, H. (1982). Catalytic behavior of calcium phosphates for decompositions of 2-propanol and ethanol. *Journal of Catalysis* 75: 200–203.

175 Gruselle, M., Kanger, T., Thouvenot, R. et al. (2011). Calcium hydroxyapatites as efficient catalysts for the michael C–C bond formation. *ACS Catalysis* 1 (12): 1729–1733.

176 Rodrigues, E.G., Keller, T.C., Mitchell, S., and Pérez-Ramírez, J. (2014). Hydroxyapatite, an exceptional catalyst for the gas-phase deoxygenation of bio-oil by aldol condensation. *Green Chemistry* 16: 4870–4874.

177 Wang, J.-D., Liu, J.-K., Lu, Y. et al. (2014). Catalytic performance of gold nanoparticles using different crystallinity HAP as carrier materials. *Materials Research Bulletin* 55: 190–197.

178 Kibby, C.L. and Hall, W.K. (1973). Studies of acid catalyzed reactions: XII. Alcohol decomposition over hydroxyapatite catalysts. *Journal of Catalysis* 29 (1): 144–159.

179 Sugiyama, S., Minami, T., Moriga, T. et al. (1996). Surface and bulk properties, catalytic activities and selectivities in methane oxidation on near-stoichiometric calcium hydroxyapatites. *Journal of Materials Chemistry* 6 (3): 459–464.

180 Matsumura, Y. and Moffat, J.B. (1996). Methanol adsorption and dehydrogenation over stoichiometric and non-stoichiomet ric h ydroxyapa tit e cat alysts. *Journal of the Chemical Society, Faraday Transactions* 92: 1981–1984.

181 Tsuchida, T., Kubo, J., Yoshioka, T. et al. (2009). Influence of preparation factors on ca/P ratio and surface basicity of hydroxyapatite catalysts. *Journal of the Japan Petroleum Institute* 52 (2).

182 Lauron-Pernot, H. (2006). Evaluation of surface acido-basic properties of inorganic-based solids by model catalytic alcohol reaction networks. *Catalysis Review* 48: 315–361.

183 Diallo-Garcia, S., Laurencin, D., Krafft, J.M. et al. (2011). Influence of magnesium substitution on the basic properties of hydroxyapatites. *Journal of Physical Chemistry C* 115: 24317–24327.

184 Ben Osman, M., Krafft, J.M., Millot, Y. et al. (2016). Molecular understanding of the bulk composition of crystalline nonstoichiometric hydroxyapatites: application to the rationalization of structure–reactivity relationships. *European Journal of Inorganic Chemistry* 17: 2709–2720.

185 Petitjean, H., Krafft, J.M., Che, M. et al. (2010). Basic reactivity of CaO: investigating active surface sites under realistic conditions. *Physical Chemistry Chemical Physics* 12: 14740–14748.

186 Tanaka, H., Watanabe, T., and Chikazawa, M. (1997). FTIR and TPD studies on the adsorption of pyridine, n-butylamine and acetic acid on calcium hydroxyapatite. *Journal of the Chemical Society, Faraday Transactions* 93 (24): 4377–4381.

187 Ben Osman, M., Diallo Garcia, S., Krafft, J.M. et al. (2016). Control of calcium accessibility over hydroxyapatite by post precipitation steps: influence on the catalytic reactivity toward alcohols. *Physical Chemistry Chemical Physics* 18: 27837–27847.

188 Ben Osman, M., Diallo-Garcia, S., Herledan, V. et al. (2015). Discrimination of surface and bulk structure of crystalline hydroxyapatite nanoparticles by NMR. *Journal of Physical Chemistry C* 119: 23008–23020.

189 Ospina, C., Terra, J., Ramirez, A. et al. (2012). Experimental evidence and structural modeling of nonstoichiometric (0 1 0) surfaces coexisting in hydroxyapatite nano-crystals. *Colloids and Surfaces. B, Biointerfaces* 89: 15–22.

190 Diallo-Garcia, S., Ben Osman, M., Krafft, J.M. et al. (2014). Identification of surface basic sites and acid-base pairs of hydroxyapatite. *Journal of Physical Chemistry C* 118 (24): 12744–12757.

191 Bolis, V., Magnacca, G., Cerrato, G., and Morterra, C. (2001). Microcalorimetric and IR-spectroscopic study of the room temperature adsorption of CO_2 on pure and sulphated t-ZrO_2. *Thermochimica Acta* 379: 147–161.

192 Charlot, G. (1974). *Chimie Analytique Quantitative*. Paris: Masson et Cie.

193 Gee, A. and Deitz, V.R. (1953). Determination of phosphate by differential spectrophotometry. *Analytical Chemistry* 25 (9): 1320–1324.

194 Rey, C., Collins, B., Goehl, T. et al. (1989). The carbonate environment in bone mineral: a resolution-enhanced Fourier transform infrared spectroscopy study. *Calcified Tissue International* 45 (3): 157–164.

195 Combes, C., Rey, C., and Mounic, S. (2000). Identification and evaluation of HPO_4 ions in biomimetic poorly crystalline apatite and bone mineral. *Key Engineering Materials* 192: 143–146.

196 Rey, C., Marsan, O., Combes, C. et al. (2014). Characterization of calcium phosphates using vibrational spectroscopies. In: *Advances in Calcium Phosphate Biomaterials*, 229–266. Springer.

197 Ducheyne, P., Van Raemdonck, W., Heughebaert, J., and Heughebaert, M. (1986). Structural analysis of hydroxyapatite coatings on titanium. *Biomaterials* 7 (2): 97–103.

198 Howie, R. (1974). (D.) McConnell. Apatite: its crystal chemistry, mineralogy, utilization, and geologic and biologic occurrences. In: *Applied Mineralogy*, vol. 5. Vienna and New York: Springer-Verlag.

199 Sakhno, Y., Bertinetti, L., Iafisco, M. et al. (2010). Surface hydration and cationic sites of nanohydroxyapatites with amorphous or crystalline surfaces: a comparative study. *The Journal of Physical Chemistry C* 114 (39): 16640–16648.

200 Ioku, K., Yamauchi, S., Fujimori, H. et al. (2002). Hydrothermal preparation of fibrous apatite and apatite sheet. *Solid State Ionics* 151 (1–4): 147–150.

201 Bres, E., Hutchison, J., Senger, B. et al. (1991). HREM study of irradiation damage in human dental enamel crystals. *Ultramicroscopy* 35 (3–4): 305–322.

202 Suvorova, E. and Buffat, P.-A. (2001). Size effect in X-ray and electron diffraction patterns from hydroxyapatite particles. *Crystallography Reports* 46 (5): 722–729.

203 Wiesendanger, R. (1994). *Scanning Probe Microscopy and Spectroscopy: Methods and Applications*. Cambridge University Press.

204 Siedlecki, C.A. and Marchant, R.E. (1998). Atomic force microscopy for characterization of the biomaterial interface. *Biomaterials* 19 (4): 441–454.

205 Jiang, W., Pan, H., Cai, Y. et al. (2008). Atomic force microscopy reveals hydroxyapatite– citrate interfacial structure at the atomic level. *Langmuir* 24 (21): 12446–12451.

206 Wallwork, M.L., Kirkham, J., Zhang, J. et al. (2001). Binding of matrix proteins to developing enamel crystals: an atomic force microscopy study. *Langmuir* 17 (8): 2508–2513.

207 Errassifi, F., Sarda, S., Barroug, A. et al. (2014). Infrared, Raman and NMR investigations of risedronate adsorption on nanocrystalline apatites. *Journal of Colloid and Interface Science* 420: 101–111.

208 Fowler, B., Moreno, E., and Brown, W. (1966). Infra-red spectra of hydroxyapatite, octacalcium phosphate and pyrolysed octacalcium phosphate. *Archives of Oral Biology* 11 (5): 477–492.

209 Penel, G., Leroy, N., Van Landuyt, P. et al. (1999). Raman microspectrometry studies of brushite cement: in vivo evolution in a sheep model. *Bone* 25 (2): 81S–84S.

210 Engel, G. and Klee, W. (1972). Infrared spectra of the hydroxyl ions in various apatites. *Journal of Solid State Chemistry* 5 (1): 28–34.

211 Diallo-Garcia, S., Ben Osman, M., Krafft, J.M. et al. (2014). Discrimination of infra red fingerprints of bulk and surface POH and OH of hydroxypatites. *Catalysis Today* 226: 81–88.

212 Barroug, A., Rey, C., and Trombe, J.C. (1994). Precipitation and formation mechanism of type AB carbonate apatites analogous to dental enamel. *Advanced Materials Research* (1): 147–154.

213 Beshah, K., Rey, C., Glimcher, M. et al. (1990). Solid state carbon-13 and proton NMR studies of carbonate-containing calcium phosphates and enamel. *Journal of Solid State Chemistry* 84 (1): 71–81.

214 Fraissard, J. and Lapina, O. (2012). *Magnetic Resonance in Colloid and Interface Science*. Springer Science & Business Media.

215 Eichert, D., Salome, M., Banu, M. et al. (2005). Preliminary characterization of calcium chemical environment in apatitic and non-apatitic calcium phosphates

of biological interest by X-ray absorption spectroscopy. *Spectrochimica Acta Partt B-Atomic Spectroscopy* 60 (6): 850–858.

216 Cazalbou, S., Combes, C., Eichert, D. et al. (2004). Poorly crystalline apatites: evolution and maturation in vitro and in vivo. *Journal of Bone and Mineral Metabolism* 22 (4): 310–317.

217 Eichert, D., Combes, C., Drouet, C., and Rey, C. (2005). Formation and evolution of hydrated surface layers of apatites. *Key Engineering Materials* 284: 3–6.

218 Yesinovski, J.P. and Eckert, H. (1987). Hydrogen environments in calcium phosphates: 1H MAS NMR at high spinning speeds. *Journal of the American Chemical Society* 109: 6274–6282.

219 Yesinowski, J.P. (1998). Nuclear magnetic resonance spectroscopy of calcium phosphates. In: *Calcium Phosphates in Biological and Industrial Systems*, 103–143. Springer.

220 Pourpoint, F., Gervais, C., Bonhomme-Coury, L. et al. (2007). Calcium phosphates and hydroxyapatite: solid-state NMR experiments and first-principles calculations. *Applied Magnetic Resonance* 32 (4): 435–457.

221 Aue, W., Roufosse, A., Glimcher, M., and Griffin, R. (1984). Solid-state phosphorus-31 nuclear magnetic resonance studies of synthetic solid phases of calcium phosphate: potential models of bone mineral. *Biochemistry* 23 (25): 6110–6114.

222 Miquel, J., Facchini, L., Legrand, A. et al. (1990). Solid state NMR to study calcium phosphate ceramics. *Colloids and Surfaces* 45: 427–433.

223 Wu, Y., Glimcher, M., and Rey, C. (1994). Unique protonated group in bone mineral not present in synthetic calcium phosphates–identification by phosphorous-31 solid state NMR spectroscopy. *Journal of Molecular Biology* 244: 423–435.

224 Jarlbring, M., Sandström, D.E., Antzutkin, O.N., and Forsling, W. (2006). Characterization of active phosphorus surface sites at synthetic carbonate-free fluorapatite using single-pulse 1H, 31P, and 31P CP MAS NMR. *Langmuir* 22 (10): 4787–4792.

225 Lin, T.-J. and Heinz, H. (2016). Accurate force field parameters and pH resolved surface models for hydroxyapatite to understand structure, mechanics, hydration, and biological interfaces. *The Journal of Physical Chemistry C* 120 (9): 4975–4992.

226 Yuan, Z., Li, S., Liu, J. et al. (2018). Structural, electronic, dynamical and thermodynamic properties of $Ca_{10}(PO_4)_6(OH)_2$ and $Sr_{10}(PO_4)_6(OH)_2$: first-principles study. *International Journal of Hydrogen Energy* 43 (29): 13639–13648.

227 Cheng, X., Wu, H., Zhang, L. et al. (2017). Hydroxyl migration disorders the surface structure of hydroxyapatite nanoparticles. *Applied Surface Science* 416: 901–910.

228 Drouet, C. (2015). A comprehensive guide to experimental and predicted thermodynamic properties of phosphate apatite minerals in view of applicative purposes. *The. Journal of Chemical Thermodynamics* 81: 143–159.

229 Drouet, C. and Alphonse, P. (2015). ThermAP additive model for applied predictive thermodynamics. Therm'AP free calculation program. http://www.christophedrouet.com/thermAP (accessed January 2022).

230 Glasser, L. (2019). Apatite thermochemistry: the simple salt approximation. *Inorganic Chemistry* 58 (19): 13457–13463.

231 Drouet, C. (2019). Applied predictive thermodynamics (ThermAP). Part 2. Apatites containing Ni^{2+}, Co^{2+}, Mn^{2+}, or Fe^{2+} ions. *The Journal of Chemical Thermodynamics* 136: 182–189.

232 Li, C.-X., Duan, Y.-H., and Hu, W.-C. (2015). Electronic structure, elastic anisotropy, thermal conductivity and optical properties of calcium apatite $Ca_5(PO_4)_3X$ (X=F, cl or Br). *Journal of Alloys and Compounds* 619: 66–77.

233 Cruz, F.J.A.L., Canongia Lopes, J.N., Calado, J.C.G., and Minas da Piedade, M.E. (2005). A molecular dynamics study of the thermodynamic properties of calcium Apatites. 1. Hexagonal phases. *The Journal of Physical Chemistry B* 109 (51): 24473–24479.

234 Cruz, F.J.A.L., Canongia Lopes, J.N., and Calado, J.C.G. (2006). Molecular dynamics study of the thermodynamic properties of calcium Apatites. 2. Monoclinic phases. *The Journal of Physical Chemistry B* 110 (9): 4387–4392.

235 Garley, A., Hoff, S.E., Saikia, N. et al. (2019). Adsorption and substitution of metal ions on hydroxyapatite as a function of crystal facet and electrolyte pH. *The Journal of Physical Chemistry C* 123 (27): 16982–16993.

236 Rollin-Martinet, S., Navrotsky, A., Champion, E. et al. (2013). Thermodynamic basis for evolution of apatite in calcified tissues. *American Mineralogist* 98 (11–12): 2037–2045.

237 Hosseini, S.M., Drouet, C., Al-Kattan, A., and Navrotsky, A. (2014). Energetics of lanthanide-doped calcium phosphate apatite. *American Mineralogist* 99 (11–12): 2320–2327.

238 Drouet, C., Aufray, M., Rollin-Martinet, S. et al. (2018). Nanocrystalline apatites: the fundamental role of water. *American Mineralogist: Journal of Earth and Planetary Materials* 103 (4): 550–564.

239 Wopenka, B., Pasteris, J.D., and Yoder, C.H. (2014). Molecular water in nominally unhydrated carbonated hydroxylapatite: the key to a better understanding of bone mineral†. *American Mineralogist* 99 (1): 16–27.

240 Vandecandelaere, N., Rey, C., and Drouet, C. (2012). Biomimetic apatite-based biomaterials: on the critical impact of synthesis and post-synthesis parameters. *Journal of Materials Science. Materials in Medicine* 23 (11): 2593–2606.

241 Drouet, C., Carayon, M.-T., Combes, C., and Rey, C. (2008). Surface enrichment of biomimetic apatites with biologically-active ions Mg2+ and Sr2+: a preamble to the activation of bone repair materials. *Materials Science and Engineering: C* 28 (8): 1544–1550.

242 Trombe, J.C. and Montel, G. (1978). Some features of the incorporation of oxygen in different oxidation state in the apatitic lattice-1 on the existence of calcium and strontium oxyapatites. *Journal of Inorganic and Nuclear Chemistry* 40: 15–21.

243 Tõnsuaadu, K., Gross, K.A., Plūduma, L., and Veiderma, M. (2012). A review on the thermal stability of calcium apatites. *Journal of Thermal Analysis and Calorimetry* 110 (2): 647–659.

244 Tanaka, H., Chikazawa, M., Kandori, K., and Ishikawa, T. (2000). Influence of thermal treatment on the structure of calcium hydroxyapatite. *Physical Chemistry Chemical Physics* 2 (11): 2647–2650.

245 Wang, T., Dorner-Reisel, A., and Müller, E. (2004). Thermogravimetric and thermokinetic investigation of the dehydroxylation of a hydroxyapatite powder. *Journal of the European Ceramic Society* 24 (4): 693–698.

246 Heughebaert, J.C. (1977). *Contribution à l'étude de l'évolution des othophosphates de calcium précipités amorphes en orthophosphates apatitiques*. Toulouse, France: Institut National Polytechnique de Toulouse.

247 Cacciotti, I., Bianco, A., Lombardi, M., and Montanaro, L. (2009). Mg-substituted hydroxyapatite nanopowders: synthesis, thermal stability and sintering behaviour. *Journal of the European Ceramic Society* 29 (14): 2969–2978.

248 Baig, A., Fox, J., Young, R. et al. (1999). Relationships among carbonated apatite solubility, crystallite size, and microstrain parameters. *Calcified Tissue International* 64 (5): 437–449.

249 Brown, W.E., Chow, L.C., and Mathew, M. (1983). Thermodynamics of hydroxyapatite surfaces. *Croatica Chemica Acta* 56 (4): 779–787.

250 McDowell, H., Gregory, T., and Brown, W. (1977). Solubility of $Ca_5(PO_4)_3OH$ in the system $Ca(OH)_2$-H_3PO_4-H_2O at 5, 15, 25, and 37 °C. *Journal of Research of the National Bureau of Standards Section A* 81: 273–281.

251 Chow, L.C. and Eanes, E. (2001). Solubility of calcium phosphates. *Monographs in Oral Science* 18: 94–111.

252 LeGeros, R.Z. (1993). Biodegradation and bioresorption of calcium phosphate ceramics. *Clinical Materials* 14 (1): 65–88.

253 Baig, A.A., Fox, J.L., Hsu, J. et al. (1996). Effect of carbonate content and crystallinity on the metastable equilibrium solubility behavior of carbonated Apatites. *Journal of Colloid and Interface Science* 179 (2): 608–617.

254 Hsu, J., Fox, J.L., Higuchi, W.I. et al. (1994). Metastable equilibrium solubility behavior of carbonated apatites. *Journal of Colloid and Interface Science* 167 (2): 414–423.

255 Chhettry, A., Wang, Z., Hsu, J. et al. (1999). Metastable equilibrium solubility distribution of carbonated apatite as a function of solution composition. *Journal of Colloid and Interface Science* 218 (1): 57–67.

256 Chander, S. and Fuerstenau, D. (1984). Solubility and interfacial properties of hydroxyapatite: a review. In: *Adsorption On and Surface Chemistry of Hydroxyapatite*, 29–49. Springer.

257 El Shafei, G.M. and Moussa, N.A. (2001). Adsorption of some essential amino acids on hydroxyapatite. *Journal of Colloid and Interface Science* 238 (1): 160–166.

258 Wu, L., Forsling, W., and Schindler, P.W. (1991). Surface complexation of calcium minerals in aqueous solution: 1. Surface protonation at fluorapatite—water interfaces. *Journal of Colloid and Interface Science* 147 (1): 178–185.

259 Boix, T., Gomez-Morales, J., Torrent-Burgues, J. et al. (2005). Adsorption of recombinant human bone morphogenetic protein rhBMP-2m onto hydroxyapatite. *Journal of Inorganic Biochemistry* 99 (5): 1043–1050.

260 Misra, D.N. (1984). *Adsorption on and Surface Chemistry of Hydroxyapatite*. Springer.

261 Saleeb, F. and De Bruyn, P. (1972). Surface properties of alkaline earth apatites. *Journal of Electroanalytical Chemistry and Interfacial Electrochemistry* 37 (1): 99–118.

262 Botelho, C., Lopes, M., Gibson, I.R. et al. (2002). Structural analysis of Si-substituted hydroxyapatite: zeta potential and X-ray photoelectron spectroscopy. *Journal of Materials Science. Materials in Medicine* 13 (12): 1123–1127.

263 Ducheyne, P., Pollack, S.R., and Kim, C. (1991). Zeta potential studies in hydroxyapatite. In: *Electromagnetics in Biology and Medicine* (ed. C.T. Brighton and S.R. Pollack), 49–52. San Francisco: San Francisco Press.

264 Bouladjine, A., Al-Kattan, A., Dufour, P., and Drouet, C. (2009). New advances in nanocrystalline apatite colloids intended for cellular drug delivery. *Langmuir* 25 (20): 12256–12265.

265 Mangood, A., Malkaj, P., and Dalas, E. (2006). Hydroxyapatite crystallization in the presence of acetaminophen. *Journal of Crystal Growth* 290 (2): 565–570.

266 Palazzo, B., Walsh, D., Iafisco, M. et al. (2009). Amino acid synergetic effect on structure, morphology and surface properties of biomimetic apatite nanocrystals. *Acta Biomaterialia* 5 (4): 1241–1252.

267 dos Santos, E.A., Farina, M., Soares, G.A., and Anselme, K. (2008). Surface energy of hydroxyapatite and beta-tricalcium phosphate ceramics driving serum protein adsorption and osteoblast adhesion. *Journal of Materials Science. Materials in Medicine* 19 (6): 2307–2316.

268 Wu, W. and Nancollas, G.H. (1999). Determination of interfacial tension from crystallization and dissolution data: a comparison with other methods. *Advances in Colloid and Interface Science* 79: 229–279.

269 Tang, R., Wu, W., Haas, M., and Nancollas, G. (2001). Kinetics of dissolution of beta-tricalcium phosphate. *Langmuir* 17: 3480–3485.

270 Wang, X. and Zhang, Q. (2020). Insight into the influence of surface roughness on the wettability of apatite and dolomite. *Minerals* 10: 114.

271 Rakovan, J. (2020). Growth and surface properties of apatite. *Reviews in Mineralogy and Geochemistry* 48 (1): 51–86.

272 Tung, M.S. and Skrtic, D. (2001). Interfacial properties of hydroxyapatite, fluorapatite and octacalcium phosphate. In: *Octacalcium Phosphate*, Monographs in Oral Science, vol. 18 (ed. L.C. Chow and E.D. Eanes), 112–129. Basel: Karger.

273 Médout-Marère, V., Belarbi, H., Thomas, P. et al. (1998). Thermodynamic analysis of the immersion of a swelling clay. *Journal of Colloid and Interface Science* 202 (1): 139–148.

274 Douillard, J.M. (1996). What can really be deduced from enthalpy of Immersional wetting experiments? *Journal of Colloid and Interface Science* 182 (1): 308–311.

275 Dorozhkin, S.V. (2012). Dissolution mechanism of calcium apatites in acids: a review of literature. *World Journal of Methodology* 2 (1): 1–17.

276 Christoffersen, J., Christoffersen, M.R., and Johansen, T. (1996). Kinetics of growth and dissolution of fluorapatite. *Journal of Crystal Growth* 163 (3): 295–303.

277 Tang, R., Henneman, Z.J., and Nancollas, G.H. (2003). Constant composition kinetics study of carbonated apatite dissolution. *Journal of Crystal Growth* 249: 614–624.

278 Nancollas, G.H., Wu, W., and Tang, R. (1999). The control of mineralization on natural and implant surfaces. *MRS Online Proceedings Library (OPL)* 599: 99–108.

279 Suzuki, T., Hirose, G., and Oishi, S. (2004). Contact angle of water droplet on apatite single crystals. *Materials Research Bulletin* 39 (1): 103–108.

280 Redey, S.A., Nardin, M., Bernache-Assolant, D. et al. (2000). Behavior of human osteoblastic cells on stoichiometric hydroxyapatite and type a carbonate apatite: role of surface energy. *Journal of Biomedical Materials Research* 50: 353–364.

281 Deymier, A.C., Nair, A.K., Depalle, B. et al. (2017). Protein-free formation of bone-like apatite: new insights into the key role of carbonation. *Biomaterials* 127: 75–88.

282 Fleche, J.L. (2002). Thermodynamical functions for crystals with large unit cells such as zircon, coffinite, fluorapatite, and iodoapatite from ab initio calculations. *Physical Review B* 65 (24): 245116.

283 Xie, J., Zhang, Q., Mao, S. et al. (2019). Anisotropic crystal plane nature and wettability of fluorapatite. *Applied Surface Science* 493: 294–307.

284 Wang, X., Zhang, L., Liu, Z. et al. (2018). Probing the surface structure of hydroxyapatite through its interaction with hydroxyl: a first-principles study. *RSC Advances* 8 (7): 3716–3722.

285 Zeglinski, J., Nolan, M., Thompson, D., and Tofail, S.A. (2014). Reassigning the most stable surface of hydroxyapatite to the water resistant hydroxyl terminated (010) surface. *Surface Science* 623: 55–63.

286 Peccati, F., Bernocco, C., Ugliengo, P., and Corno, M. (2018). Properties and reactivity toward water of a type carbonated apatite and hydroxyapatite surfaces. *The Journal of Physical Chemistry C* 122 (7): 3934–3944.

287 Sun, J.P. and Song, Y. (2018). Strengthening adhesion of the hydroxyapatite and titanium Interface by substituting silver and zinc: a first principles investigation. *ACS Applied Nano Materials* 1 (9): 4940–4954.

288 Wang, X., Wu, H., Cheng, X. et al. (2020). Probing the surface activity of hydroxyapatite nanoparticles through their interaction with water molecules. *AIP Advances* 10 (6): 065217.

289 Lou, Z., Zeng, Q., Chu, X. et al. (2012). First-principles study of the adsorption of lysine on hydroxyapatite (100) surface. *Applied Surface Science* 258 (11): 4911–4916.

290 Corno, M., Busco, C., Bolis, V. et al. (2009). Water adsorption on the stoichiometric (001) and (010) surfaces of hydroxyapatite: a periodic B3LYP study. *Langmuir* 25 (4): 2188–2198.

291 Hauptmann, S., Dufner, H., Brickmann, J. et al. (2003). Potential energy function for apatites. *Physical Chemistry Chemical Physics* 5 (3): 635–639.

292 de Leeuw, N.H. (2004). A computer modelling study of the uptake and segregation of fluoride ions at the hydrated hydroxyapatite (0001) surface: introducing a Ca10(PO4)6(OH)2 potential model. *Physical Chemistry Chemical Physics* 6 (8): 1860–1866.

293 Carniti, P. and Gervasini, A. (2013). Liquid-solid adsorption properties: measurement of the effective surface acidity of solid catalysts. In: *Calorimetry and Thermal Methods in Catalysis*, 543–551. Springer.

294 Wilson, R.M., Elliot, J.C., and Dowker, S.E.P. (2003). Formate incorporation in the structure of ca-deficient apatite: rietveld structure refinement. *Journal of Solid State Chemistry* 174: 132–140.

295 Hill, I.M., Hanspal, S., Young, Z.D., and Davis, R.J. (2015). DRIFTS of probe molecules adsorbed on magnesia, zirconia, and hydroxyapatite catalysts. *Journal of Physical Chemistry C* 119: 9186–9197.

296 Elkabouss, K., Kacimi, M., Ziyad, M., and Bozon-Verduraz, F. (2005). Catalytic behaviour of cobalt exchanged hydroxyapatite in the oxidative dehydrogenation of ethane. *Journal de Physique IV* 123: 313–317.

297 Sugiyama, S. and Hayashi, H. (2003). Role of hydroxide groups in hydroxyapatite catalysis for the oxidative dehydrogenation of alkanes. *International Journal of Modern Physics B* 17: 1476–1481.

298 Petit, S., Gode, T., Thomas, C. et al. (2017). Incorporation of vanadium in the framework of hydroxyapatites: importance of the vanadium content and pH conditions during the precipitation step. *Physical Chemistry Chemical Physics* 19: 9630–9640.

299 Tanaka, Y., Kikuchi, M., Tanaka, K. et al. (2010). Fast oxide ion conduction due to carbonate substitution in hydroxyapatite. *Journal of the American Ceramic Society* 93 (11): 3577–3579.

300 Elkabouss, K., Kacimi, M., Ziyad, M. et al. (2004). Cobalt-exchanged hydroxyapatite catalysts: magnetic studied, spectroscopic investigations, performance in 2-butanol and ethane oxidative dehydrogenations. *Journal of Catalysis* 226: 16–24.

301 Sugiyama, S., Matsumoto, H., Hayashi, H., and Moffat, J.B. (2000). Sorption and ion-exchange properties of barium hydroxyapatite with divalent cations. *Colloids and Surfaces, A: Physicochemical and Engineering Aspects* 169 (1): 17–26.

302 Matsunaga, K., Murata, H., Mizoguchi, T., and Nakahira, A. (2010). Mechanism of incorporation of zinc into hydroxyapatite. *Acta Biomaterialia* 6 (6): 2289–2293.

303 Campisi, S., Castellano, C., and Gervasini, A. (2018). Tailoring the structural and morphological properties of hydroxyapatite materials to enhance the capture efficiency towards copper(II) and lead(II) ions. *New Journal of Chemistry* 42: 4520.

304 Terra, J., Gonzalez, G.B., Rossi, A.M. et al. (2010). Theoretical and experimental studies of substitution of cadmium into hydroxyapatite. *Physical Chemistry Chemical Physics* 12 (47): 15490–15500.

305 Ogo, S., Onda, A., and Yanagisawa, K. (2008). Hydrothermal synthesis of vanadate-substituted hydroxyapatites, and catalytic properties for conversion of 2-propanol. *Applied Catalysis A.* 348: 129–134.

306 Zhu, Y., Zhang, X., Long, F. et al. (2009). Synthesis and characterization of arsenate/phosphate hydroxyapatite solid solution. *Materials Letters* 63 (13): 1185–1188.

307 Hughes, J.M., Cameron, M., and Crowley, K.D. (1989). Structural variations in natural F, OH, and Cl apatites. *American Mineralogist* 74: 870–876.

308 Kaur, K., Singh, K.J., Anand, V. et al. (2017). Lanthanide (=Ce, pr, Nd and tb) ions substitution at calcium sites of hydroxyl apatite nanoparticles as fluorescent bio probes: experimental and density functional theory study. *Ceramics International* 43 (13): 10097–10108.

309 Gibson, L.R. and Skakle, J.M.S. (2010). Inventors silicate- substituted hydroxyapatite. Patent WO2010079316A1.

310 Solonenko, A.P. and Golovanova, O.A. (2014). Silicate substituted carbonated hydroxyapatite powders prepared by precipitation from aqueous solutions. *Russian Journal of Inorganic Chemistry* 59 (11): 1228–1236.

311 Cazalbou, S., Eichert, D., Ranz, X. et al. (2005). Ion exchanges in apatites for biomedical application. *Journal of Materials Science. Materials in Medicine* 16 (5): 405–409.

312 Tõnsuaadu, K., Gruselle, M., Villain, F. et al. (2006). A new glance at ruthenium sorption mechanism on hydroxy, carbonate, and fluor apatites: analytical and structural studies. *Journal of Colloid and Interface Science* 304 (2): 283–291.

313 Tang, Y., Chappell, H.F., Dove, M.T. et al. (2009). Zinc incorporation into hydroxylapatite. *Biomaterials* 30 (15): 2864–2872.

314 Saito, T., Yokoi, T., Nakamura, A., and Matsunaga, K. (2020). Formation energies and site preference of substitutional divalent cations in carbonated apatite. *Journal of the American Ceramic Society* 103 (9): 5354–5364.

315 Mostafa, N.Y., Hassan, M.H., and Abd Elkader, O.H. (2011). Preparation and characterization of Na_1, SiO_{44}, and CO_{32}- Co-substituted hydroxyapatite. *Journal of the American Ceramic Society* 94 (5): 1594–1600.

316 Lee, S.J., Jun, J.H., Lee, S.-H. et al. (2002). Partial oxidation of methane over nickel-added strontium phosphate. *Applied Catalysis A.* 230 (1): 61–71.

317 Qu, Z., Sun, Y., Chen, D., and Wang, Y. (2014). Possible sites of copper located on hydroxyapatite structure and the identification of active sites for formaldehyde oxidation. *Journal of Molecular Catalysis A: Chemical* 393: 182–190.

318 Errassifi, F., Menbaoui, A., Autefage, H. et al. (2010). Adsorption on apatitic calcium phosphates: application to drug delivery. In: *8th Pacific Rim Conference on Ceramic and Glass Technology* (ed. A.C. Soc), 159–174. Canada: Vancouver.

319 Bootchanont, A., Sailuam, W., Sutikulsombat, S. et al. (2017). Synchrotron X-ray absorption spectroscopy study of local structure in strontium-doped hydroxyapatite. *Ceramics International* 43 (14): 11023–11027.

320 Bazin, D., Daudon, M., and Chappard c, Rehr JJ, Thiaudière D, Reguer S. (2011). The status of strontium in biological apatites: an XANES investigation. *Journal of Synchrotron Radiation* 18: 912–918.

321 Al-Kattan, A., Errassifi, F., Sautereau, A.M. et al. (2010). Medical potentialities of biomimetic apatites through adsorption, ionic substitution, and mineral/organic associations: three illustrative examples. *Advanced Engineering Materials* 12 (7): B224–B233.

322 Chen, X., Wright, J.V., Conca, J.L., and Peurrung, L.M. (1997). Effects of pH on heavy metal sorption on mineral apatite. *Environmental Science & Technology* 31 (3): 624–631.

323 Xu, Y., Schwartz, F.W., and Traina, S.J. (1994). Sorption of Zn^{2+} and Cd^{2+} on hydroxyapatite surfaces. *Environmental Science & Technology* 28 (8): 1472–1480.

324 Michelot, A., Sarda, S., Audin, C. et al. (2015). Spectroscopic characterisation of hydroxyapatite and nanocrystalline apatite with grafted aminopropyltriethoxysilane: nature of silane–surface interaction. *Journal of Materials Science* 50 (17): 5746–5757.

325 Prakash, M., Lemaire, T., Caruel, M. et al. (2017). Anisotropic diffusion of water molecules in hydroxyapatite nanopores. *Physics and Chemistry of Minerals* 44 (7): 509–519.

326 Honorio, T., Lemaire, T., Di Tommaso, D., and Naili, S. (2019). Anomalous water and ion dynamics in hydroxyapatite mesopores. *Computational Materials Science* 156: 26–34.

327 Lemaire, T., Pham, T., Capiez-Lernout, E. et al. (2015). Water in hydroxyapatite nanopores: possible implications for interstitial bone fluid flow. *Journal of Biomechanics* 48 (12): 3066–3071.

328 Viipsi, K., Sjöberg, S., Tõnsuaadu, K., and Shchukarev, A. (2013). Hydroxy-and fluorapatite as sorbents in cd (II)–Zn (II) multi-component solutions in the absence/presence of EDTA. *Journal of Hazardous Materials* 252: 91–98.

329 Mignardi, S., Corami, A., and Ferrini, V. (2013). Immobilization of Co and Ni in mining-impacted soils using phosphate amendments. *Water, Air, and Soil Pollution* 224: 1447.

330 Chen, X., Wright, J.V., Conca, J.L., and Peurrung, L.M. (1997). Evaluation of heavy metal remediation using mineral apatite. *Water, Air, and Soil Pollution* 98: 57–78.

331 Vila, M., Sánchez-Salcedo, S., and Vallet-Regí, M. (2012). Hydroxyapatite foams for the immobilization of heavy metals: from waters to the human body. *Inorganica Chimica Acta* 393: 24–35.

332 Mori, K., Mitani, Y., Hara, T. et al. (2005). A single-site hydroxyapatite-bound zinc catalyst for highly efficient chemical fixation of carbon dioxide with epoxides. *Chemical Communications* 26: 3331–3333.

333 Mondelli, C., Ferri, D., and Baiker, A. (2008). Ruthenium at work in Ru-hydroxyapatite during the aerobic oxidation of benzyl alcohol: an in situ ATR-IR spectroscopy study. *Journal of Catalysis* 258 (1): 170–176.

334 Opre, Z., Grunwaldt, J.D., Maciejewski, M. et al. (2005). Promoted Ru–hydroxyapatite: designed structure for the fast and highly selective oxidation of alcohols with oxygen. *Journal of Catalysis* 230: 406–419.

335 Sugiyama, S., Minami, T., Hayashi, H. et al. (1996). Enhancement of the selectivity to carbon monoxide with feedstream doping by tetrachloromethane in the oxidation of methane on stoichiometric calcium hydroxyapatite. *Journal of the Chemical Society, Faraday Transactions* 92 (2): 293–299.

336 Neuman, W., Toribara, T., and Mulryan, B. (1956). The surface chemistry of bone. IX. Carbonate: phosphate Exchange1. *Journal of the American Chemical Society* 78 (17): 4263–4266.

337 Akri, M., El Kasmi, A., Batiot-Dupeyrat, C., and Qiao, B. (2020). Highly active and carbon-resistant nickel single-atom catalysts for methane dry reforming. *Catalysts* 10 (6): 630.

338 El Kabouss, K., Kacimi, M., Ziyad, M. et al. (2006). Cobalt speciation in cobalt oxide-apatite materials: structure–properties relationship in catalytic oxidative dehydrogenation of ethane and butan-2-ol conversion. *Journal of Materials Chemistry* 16: 2453–2463.

339 Josse, S., Faucheux, C., Soueidan, A. et al. (2005). Novel biomaterials for bisphosphonate delivery. *Biomaterials* 26 (14): 2073–2080.

340 Mukherjee, S., Huang, C., Guerra, F. et al. (2009). Thermodynamics of bisphosphonates binding to human bone: a two-site model. *Journal of the American Chemical Society* 131 (24): 8374–8375.

341 Choimet, M., Tourrette, A., and Drouet, C. (2015). Adsorption of nucleotides on biomimetic apatite: the case of cytidine 5′ monophosphate (CMP). *Journal of Colloid and Interface Science* 456: 132–137.

342 Hammami, K., El Feki, H., Marsan, O., and Drouet, C. (2015). Adsorption of nucleotides on biomimetic apatite: the case of adenosine 5′ monophosphate (AMP). *Applied Surface Science* 353: 165–172.

343 Pascaud, P., Errassifi, F., Brouillet, F. et al. (2014). Adsorption on apatitic calcium phosphates for drug delivery: interaction with bisphosphonate molecules. *Journal of Materials Science. Materials in Medicine* 25 (10): 2373–2381.

344 Benaziz, L., Barroug, A., Legrouri, A. et al. (2001). Adsorption of O-Phospho-L-serine and L-serine onto poorly crystalline apatite. *Journal of Colloid and Interface Science* 238 (1): 48–53.

345 Barroug, A., Legrouri, A., and Rey, C. (2008). Exchange reactions at calcium phosphates surface and applications to biomaterials. *Key Engineering Materials* 361-363: 79–82.

346 Misra, D.N. (1998). Interaction of some alkali metal citrates with hydroxyapatite: ion-exchange adsorption and role of charge balance. *Colloids and Surfaces A: Physicochemical and Engineering Aspects* 141 (2): 173–179.

4

Hydroxyapatite-Based Catalysts: Influence of the Molar Ratio of Ca to P

Zhen Ma

Fudan University, Department of Environmental Science and Engineering, Shanghai 200438, PR China

4.1 Introduction

Heterogeneous catalysis is very useful for converting fossil fuels and biomass to valuable organic and inorganic chemicals, ablating hazardous environmental pollutants (such as CO, NO_x, VOCs, greenhouse gases), and synthesizing chemicals or materials that are widely used in industry and in daily life. Metal oxides (such as Al_2O_3, Fe_2O_3, TiO_2, ZrO_2), SiO_2, zeolites, and active carbon have been traditionally used in industry and academia to produce different kinds of heterogeneous catalysts. However, metal salts (such as metal chlorides, sulfates, phosphates) have not been used so often in this regard. At least, some metal salts are soluble in water, and they are of low thermal stability, making them unsuitable for application in heterogeneous catalysis.

Some metal phosphates (for instance, $AlPO_4$, $LaPO_4$, $CePO_4$) are not soluble in water, and they have high thermal stability. Thus, they have been used as catalysts or catalyst supports [1–5]. Recently, hydroxyapatite (HA) has emerged as an attractive component in making heterogeneous catalysts. As an inorganic material with adjustable acid/base properties, ion exchange property, and high thermal stability, HA can be used as a heterogeneous catalyst [6, 7], a support for loading other components that are catalytically active [8–11], a precursor for making metal ion-modified HA catalysts [12, 13], or a component for making more complex composite catalysts [14, 15]. Figure 4.1 shows the structure, SEM–EDX mapping images, and SEM/TEM images of a stoichiometric HA used as a catalyst for gas-phase deoxygenation of bio-oil by aldol condensation [16]. Generally, HA-based catalysts can be used in catalyzing a number of organic reactions and inorganic reactions such as cross-coupling reactions, condensation reactions, oxidation reactions, photocatalytic reactions, etc. [17–20], as already mentioned in the previous chapters of this book.

HA refers to a family of calcium HA with the general formula of $Ca_{10-x}(HPO_4)_x$ $(PO_4)_{6-x}(OH)_{2-x}$ where $0 \leq x \leq 2$. When $x = 0$, we get a stoichiometric calcium HA, i.e. $Ca_{10}(PO_4)_6(OH)_2$, and the Ca/P molar ratio reaches 1.67. On the other hand, when $x \neq 0$, we get a nonstoichiometric Ca-deficient HA. For example, for $x = 2$,

Design and Applications of Hydroxyapatite-Based Catalysts, First Edition. Edited by Doan Pham Minh.
© 2022 WILEY-VCH GmbH. Published 2022 by WILEY-VCH GmbH.

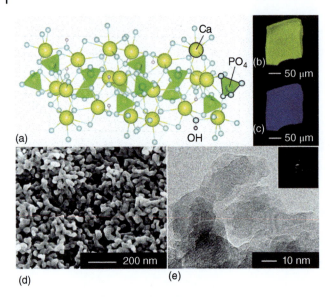

Figure 4.1 Detailed structure (a), SEM–EDX mapping data of Ca (b) and P (c) elements, SEM (d) and TEM (e) images of a stoichiometric HA. Source: Rodrigues et al. [16]. Reproduced with permission of the Royal Society of Chemistry.

we have $Ca_8(HPO_4)_2(PO_4)_4$, and the Ca/P ratio is 1.33. The Ca/P ratio of HA can be tuned by adjusting the composition of Ca and P precursors in the synthesis and the synthesis conditions such as the temperature, the maturation time, the pH, etc. [21, 22], as discussed in Chapter 2 of this book. Of course, the substitution of some Ca^{2+} by metal cations (such as Na^+, Ti^{4+}) can also decrease the Ca/P ratio, and the replacement of some PO_4^{3-} groups by anions such as CO_3^{2-} can increase the Ca/P ratio [23]. Thus, sometimes the Ca/P ratio of HA can be higher than 1.67 or lower than 1.33. This also indicates the complex composition, structure, and surface chemistry of HA-based materials.

In most of the research papers dealing with HA-based catalysts and their catalytic reactions, presumably stoichiometric HA was used to demonstrate the usefulness of the catalytic materials in these studies [24–26]. That way, the research is relatively straightforward, i.e. without having to consider too many parameters that may influence the physicochemical properties and catalytic performance of HA-based materials. However, to further understand the physicochemical properties and active sites required for the targeted reactions and tune the catalytic properties according to wishes, it is necessary to study the difference between stoichiometric and nonstoichiometric HA catalysts in detail.

Since the pioneering work of Hall et al. on the influence of Ca/P ratio of HA on catalytic behavior observed in catalytic conversion of 2-butanol [27], there have been a few valuable publications concerning the effect of Ca/P ratio of HA on catalytic performance. In particular, the catalytic reactions under investigation nowadays are not merely simple probe reactions such as 2-butanol dehydration [27], but may involve complex reaction networks where high selectivity to desired products is emphasized, but not easily achieved. A delicate balance between acid and base properties of HA may be necessary for enhancing the activity and selectivity of these reactions. In this chapter, we summarize some of the key findings in the literature and propose potential research directions in the future.

4.2 Influence of Ca/P Ratio on the Performance of HA

Pure HA without metal ion doping or combination with a catalytic component can be used as a heterogeneous catalyst. Thus, attempts have been made to study the influence of Ca/P ratio on the catalytic performance of HA. That is a research direction that deserves exploration. However, pure HA is not a universal catalyst for any chemical reaction. Thus, most of the research dealing with the influence of Ca/P ratio on the performance of HA is related to the reactions that rely on acid/base properties of HA, and the catalytic reactions under study are almost always organic reactions. On the other hand, pure HA, without being modified by transition metal cations, metal oxides, or noble-metal nanoparticles, is generally not active for inorganic reactions such as CO oxidation. Some organic reactions catalyzed by HA are relatively simple, yielding simple products in a straightforward way, whereas other reactions are relatively complex, often involving complex reaction networks and leading to selectivity issues.

4.2.1 Relatively Simple Reactions

An early work of Hall and coworkers, published in 1967, showed the dramatic effect of Ca/P ratio of HA on catalytic conversion of 2-butanol [27]. Therein, a number of HA catalysts with various Ca/P molar ratios (1.67, 1.63, 1.60, 1.58, 1.57) were prepared via mixing a concentrated H_3PO_4 solution with a saturated $Ca(OH)_2$ solution. Although NH_3-TPD was not used by the authors at that time to characterize the acidities of these catalysts, they probed the acidities by measuring the spectra of adsorbed probes such as triphenylcarbinol, trianisylcarbinol, and tritolycarbinol. These probes, being bases, can react with protons on HA surfaces to yield carbonium ions that show colors. The authors also used catalytic conversion of 2-butanol as a probe reaction to compare the catalytic properties of these catalysts. It is well known that the dehydration of 2-butanol to 1-butene occurs on acid sites, whereas the dehydrogenation route relies on base sites. It was found that on stoichiometric HA (Ca/P = 1.67), both dehydration route (forming 1-butene) and dehydrogenation route (forming methyl ethyl ketone) took place to about an equal extent at 350 °C. As the Ca/P ratio decreased from 1.67 to 1.57, the density of acid sites on catalyst surfaces increased, the rate of gas-phase 2-butanol dehydration at 350 °C increased, and the selectivity to 1-butene increased significantly. The results indicated that the acidity and catalytic properties can be tuned by changing the Ca/P molar ratio of HA.

The authors then reported in more detail about the kinetics of 2-butanol dehydration over stoichiometric HA (Ca/P = 1.67) and Ca-deficient HA (Ca/P = 1.58) in microcatalytic and steady-state flow reactors [28] and used in situ infrared (IR) spectroscopy to monitor the surface chemistry of 2-butanol on these two catalysts upon heating [29]. In these studies, the authors confirmed that both dehydrogenation and dehydration occurred on stoichiometric HA whereas dehydration was the only significant reaction that occurred on Ca-deficient HA. In another work, the physical organic chemistry of alcohol dehydration over stoichiometric HA (Ca/P = 1.67) and Ca-deficient HA (Ca/P = 1.58) was studied by monitoring the reaction rates associated with different alcohols [30]. The authors still found that both dehydrogenation and dehydration occurred on stoichiometric HA, with dehydrogenation being favored at lower temperatures, whereas dehydration was the only reaction observed over Ca-deficient HA.

In another work, Amberg and coworkers found that the catalytic activity of HA in the dehydration of 1-butanol at 350 °C significantly increased when the Ca/P ratio decreased from 1.607 to 1.533, although the activity of 1-butanol, a primary alcohol, was lower than the activity of 2-butanol [31]. Similarly, Dykman found that the activity of HA in the dehydration of trimethylcarbinol, methylphenylcarbinol, and α-hydroxyisobutyric acid and in the decomposition of 4,4-dimethyl-1,3-dioxane at 300 °C increased as the Ca/P ratio decreased [32]. The authors claimed that the Ca-deficient HA catalysts could release phosphoric acid in the form of a film on catalyst surfaces, whereas stoichiometric HA can hardly release phosphoric acid. The catalytic activity of HA samples in the dehydration of trimethylcarbinol, methylphenylcarbinol, and α-hydroxyisobutyric acid and in the decomposition of 4,4-dimethyl-1,3-dioxane was found to increase with the increasing amount of phosphoric acid released by the catalysts.

Dumeignil and coworkers tested the performance of several HA catalysts in conversion of isopropanol [33]. Propylene was formed on acid sites, whereas acetone was formed on basic sites. The authors found that the selectivity to propylene on catalysts, recorded at 20% conversion of isopropanol (typically between 275 and 350 °C), followed the sequence of HapD (Ca/P = 1.62) > Hap (Ca/P = 1.69) > HaP–CO_3 (Ca/P = 1.70) > HapNa–CO_3 (Ca/P = 1.72) > HapE–CO_3 (Ca/P = 1.90) > HapE–Na–CO_3 (Ca/P = 2.39), which is consistent with the acid property of these catalysts (Figure 4.2a) [33]. On the other hand, the selectivity to acetone on catalysts, recorded at 20% conversion of isopropanol, followed a reversed order, which is consistent with the basic property of these catalysts (Figure 4.2b) [33]. The authors proved that HPO_4^{2-} acted as Brønsted acid sites whereas Ca^{2+} or OH^- vacancies (δ^+) acted as Lewis acid sites. This study again demonstrated that the acid/base properties of HA can be tuned via changing the Ca/P ratio, and such a change results in selectivity changes in the catalytic conversion of alcohols.

Tsuchida and coworkers reported catalytic conversion of ethanol using HA with different Ca/P ratios (1.67, 1.65, 1.62, 1.59) [34]. The basic site density was found to increase with the Ca/P ratio [22, 34], and the acid site density was found to decrease with the Ca/P ratio [34]. Figure 4.3 shows the selectivity to ethylene, 1,3-butadiene, and total Guerbet alcohols (total C_4, C_6, C_8, and C_{10} alcohols) at 50% ethanol conversion over HA with different Ca/P molar ratios [34]. The selectivity

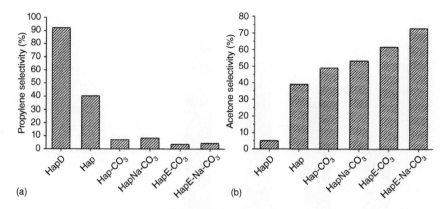

Figure 4.2 Selectivities to propylene and acetone over different HAP catalysts when 20% conversion of isopropanol was achieved [33]. The Ca/P ratios of HapD, Hap, Hap–CO$_3$, HapNa–CO$_3$, and HapE–Na–CO$_3$ were 1.62, 1.69, 1.70, 1.72, 1.90, and 2.39, respectively. Source: Silvester et al. [33]. Reproduced with permission of the Royal Society of Chemistry.

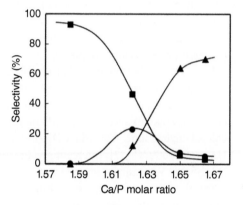

Figure 4.3 Influence of Ca/P ratio of HAP on the catalytic performance of HAP in ethanol conversion. The selectivities to ethylene (■), 1,3-butadiene (●), and total Guerbet alcohols, i.e. total C$_4$, C$_6$, and C$_4$ alcohols (▲) at 50% ethanol conversion. Source: Tsuchida et al. [34]. Reproduced with permission of Elsevier.

to ethylene decreased as the Ca/P ratio increased, due to the decrease of acid site density. The selectivity to 1,3-butadiene first increased with the Ca/P ratio and then decreased. On the other hand, the selectivity to total Guerbet alcohols increased with the increase of Ca/P ratio, correlating with the HA's basic site density. The authors analyzed the Guerbet reaction over a nonstoichiometric HA (Ca/P = 1.64) in more detail in another work, using bioethanol as a reactant [35]. In that work, the authors obtained biogasoline with a research octane number of 99.

Costentin and coworkers reported the conversion of 2-methyl-3-butyn-2-ol (MBOH) over a series of HA catalysts with various Ca/P ratios (1.77, 1.72, 1.67, 1.65) [36]. The reaction products were acetone and acetylene in a 1 : 1 ratio. Because the surface areas of the HA catalysts prepared using different protocols were significantly different, the authors normalized the catalytic conversions by considering different surface areas of HA catalysts. The conversion per surface area of catalysts followed the sequence of HA-1.65 > HA-1.67 > HA-1.77 > HA-1.72, which is consistent with the order of OH$^-$ content of these catalysts (Figure 4.4) [36]. The authors proposed that OH$^-$ concentration of HA should be a better

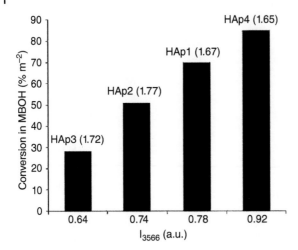

Figure 4.4 Correlation between the MBOH conversion per surface area of HAP catalysts with different Ca/P ratios and the relative OH content (indicated by the intensity of the IR peak at 3566 cm^{-1}). Source: Ben Osman et al. [36]. Reproduced with permission of John Wiley and Sons.

parameter than the Ca/P ratio when accounting for the catalytic activity in that case. Indeed, in this study, the conversion per surface area of catalysts did not simply increase with the decrease of Ca/P ratio. The authors thus argued that the Ca/P ratio is not sufficient to predict the catalytic performance of HA samples prepared under various conditions.

The catalytic conversion of alcohols in the absence of O_2 is a probe reaction to measure the impact of the acid/base properties of catalysts. It is sometimes necessary to convert volatile organic compounds (VOCs) to CO_2, etc. in the presence of O_2. Nishikawa and coworkers initially reported the conversion of gaseous chlorobenzene into CO and CO_2 under oxidative conditions over Ca-deficient HA at 400–500 °C [37]. Interestingly, the authors found that most of the Cl in the chlorinated VOC can be trapped by the Ca-deficient HA without being released.

Shirai and coworkers, after initially demonstrating that stoichiometric HA can catalyze the decomposition of VOCs [38], reported a complete work regarding the decomposition of VOCs (ethyl acetate, isopropanol, acetone) using HA catalysts with different Ca/P ratios (1.70, 1.67, 1.57, 1.37) [39]. The conversion of ethyl acetate or isopropanol at 400 °C over HA catalysts followed the sequence of HA-1.67 > HA-1.70 > HA-1.57 > HA-1.37. The conversion of acetone over different catalysts at 400 °C followed the sequence of HA-1.67 > HA-1.57 > HA-1.70 > HA-1.37. It was concluded by the authors that the observed catalytic activities should be related to several factors such as the generation of oxygen radicals and the abilities of HA to adsorb VOCs, acid sites (HPO_4^{2-}, Ca^{2+}), and base sites (PO_4^{3-}, OH^-). In particular, it was demonstrated by electron paramagnetic resonance (ESR) characterization in another work that much more oxygen radicals were formed on stoichiometric HA upon heating than on a Ca-deficient HA, which explained the catalytic behavior of these materials [38]. However, under the reaction conditions (air/VOC molar ratio = 1 : 1), the selectivity to CO_2 was around 50% for the reactions over HA-1.67 and around 40% for the reactions over HA-1.70 [39]. For the reactions over HA-1.57 and HA-1.37, the selectivity to CO_2 was significantly lower. Although the complete oxidation VOCs was not achieved, which can be possibly due to the low feeding air/VOC ratio and the inferior ability of HA in catalyzing total oxidation reactions, this work demonstrated that stoichiometric HA exhibits the best performance among the HA catalysts studied.

4.2.2 More Complex Reactions

HA not only catalyzes the dehydration, dehydrogenation, and decomposition of molecules via simple steps but also catalyzes the conversion of biomass-derived chemicals (such as ethanol, lactic acid (LA), alkyl lactates) in which multiple steps or pathways are involved and achieving high selectivity to a desired product is difficult.

Xu and coworkers prepared a series of HA_m-T catalysts by adjusting the Ca/P ratio (m) and calcination temperature (T) [40]. Thus, $HA_{1.58}$-360, $HA_{1.60}$-360, $HA_{1.62}$-360, $HA_{1.68}$-360, $HA_{1.69}$-360, $HA_{1.62}$-400, $HA_{1.62}$-500, $HA_{1.62}$-600, and $HA_{1.62}$-700 were prepared. These catalysts were used in the gas-phase dehydration of LA, a biomass-derived chemical, to form acrylic acid (AA) that can be used in the production of acrylate polymers. Acetaldehyde (AD) was a main side product in the conversion of LA. For the HA_m-360 series samples, it was found that the surface acidity, measured by NH_3-TPD, increased as the Ca/P ratio decreased, whereas the surface basicity, measured by CO_2-TPD, decreased concurrently (Figure 4.5a) [40]. For the $HA_{1.62}$-T series samples, the surface acidity first increased with the calcination temperature, reaching a maximum value after the sample was calcined at 500 °C, and then decreased with the calcination temperature (Figure 4.5b) [40]. The surface basicity, being low, was relatively constant as a function of calcination temperature. The authors found that both the reaction rate and AD formation rate increased with the acidity/basicity ratio of HA, whereas a volcano plot was observed for the AA formation rate (Figure 4.6) [40]. Since AA was the desired product and AD was a side product, HA with an intermediate acidity/basicity ratio was desirable in this case. The most efficient $HA_{1.62}$-360 catalyst can result in an AA yield as high as 62% at a reaction temperature of 360 °C. The authors proposed a reaction mechanism involving both acid sites and base sites of HA (Figure 4.7) [40]. However, it should be mentioned that the reaction mechanism proposed in Figure 4.7 only describes the formation of AA starting from LA. The reaction networks starting from LA were found to be complex in reality, and other products such as AD and pyruvic acid (PA) could be obviously formed [40]. This fact emphasizes the importance of selectivity control by choosing the adequate catalyst and reaction conditions.

In another work, Xu and coworkers further explored the application of HA_m-360 catalysts in the gas-phase conversion of methyl lactate (ML) and ethyl lactate (EL) to AA [41]. The reaction mechanism involves the hydrolysis of ML or EL to LA and surface lactate on acid sites of HA (Figure 4.8) [41]. The reaction intermediates then undergo dehydration on acid/base sites to form AA. Of course, side products such as AD, 2,3-pentanedione (2,3-PD), propionic acid (PA), and acrylates can be formed. The authors found that the optimal catalyst for the conversion of ML or EL was

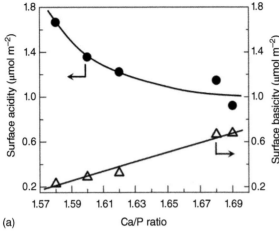

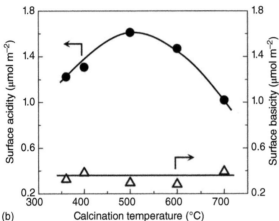

Figure 4.5 (a) Influence of Ca/P ratio (1.58, 1.60, 1.62, 1.68, 1.69) of HAP$_m$-360 catalysts on the surface acidity and basicity of these samples. (b) Influence of calcination temperature (360, 400, 500, 600, 700) of HAP$_{1.62}$-T catalysts on the surface acidity and basicity. Source: Yan et al. [40]. Reproduced with permission of the American Chemical Society.

HA$_{1.60}$-360 with a Ca/P ratio of 1.60 (but not HA$_{1.62}$-360 with a Ca/P ratio of 1.62 for dehydration of LA to AA in their previous work [40]) and the optimal acidity/basicity ratio was ca. 4.5, higher than the optimal value of ca. 3.2 for dehydration of LA to AA [40]. This is because ML or EL has to undergo hydrolysis on acid sites before the formed LA can be converted to AA. This work thus provides another example on the cooperative acid–base catalysis.

Dumeignil and coworkers studied the conversion of ethanol to butanol via the Guerbet reaction over HA catalysts with different Ca/P ratios, i.e. HapD (Ca/P = 1.62), Hap (Ca/P = 1.69), HaP–CO$_3$ (Ca/P = 1.70), HapNaCO$_3$ (Ca/P = 1.72), HapE–CO$_3$ (Ca/P = 1.90), and HapE–Na–CO$_3$ (Ca/P = 2.39) [42]. On the basis of their reaction data and relevant information in the literature, the authors proposed that the whole reaction contains a series of consecutive reactions (Figure 4.9) [42]. First, the dehydrogenation of ethanol on weak/medium basic sites leads to the formation of AD and H$_2$. The formed AD then undergoes aldol condensation on stronger basic sites to yield an intermediate aldol which is not detected during the reaction. The subsequent dehydration of aldol takes place on Brønsted acid sites

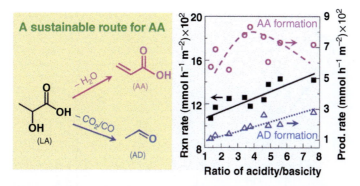

Figure 4.6 A graphic abstract showing the formation of acrylic acid (AA) and side product acetaldehyde (AD) starting from lactic acid (LA), as well as changes of reaction rate, AA formation rate, and AD formation rate as a function of the acidity/basicity ratio of HAP_m-T catalysts. Source: Yan et al. [40]. Reproduced with permission of the American Chemistry Society.

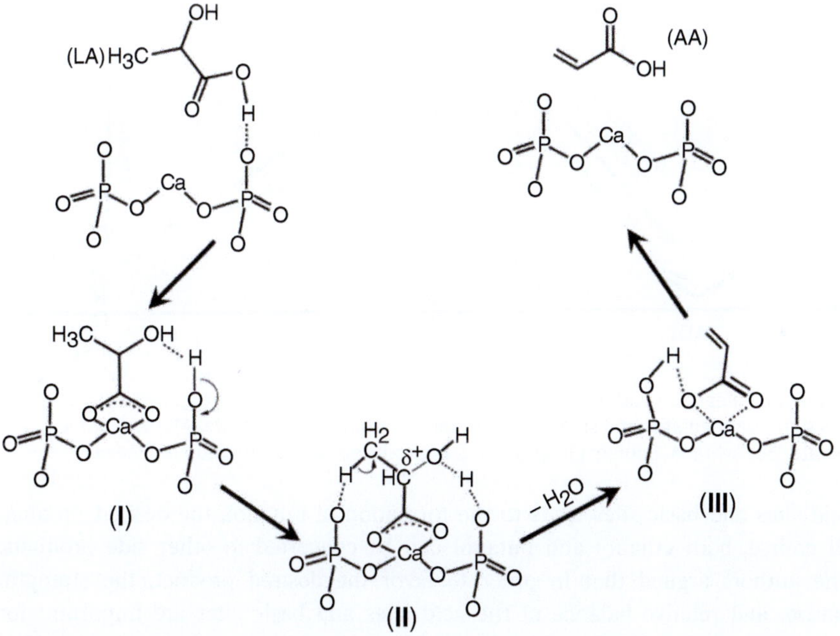

Figure 4.7 A scheme showing the possible reaction mechanism for the formation of acrylic acid (AA) starting from lactic acid (LA) over HAP_m-T catalysts [40]. Note that the reaction mechanism only showed the conversion of AA to LA, whereas the conversion of AA to other products was omitted here. Source: Yan et al. [40]. Reproduced with permission of the American Chemistry Society.

(HPO_4^{2-}), forming crotonaldehyde, which is also not detected during the reaction. Crotonaldehyde then undergoes partial hydrogenation to form 2-buten-1-ol (major product) and butyraldehyde (minor product) over Lewis acid sites (Ca^{2+} and OH^- vacancies) and basic sites. The further hydrogenation of these two chemicals on

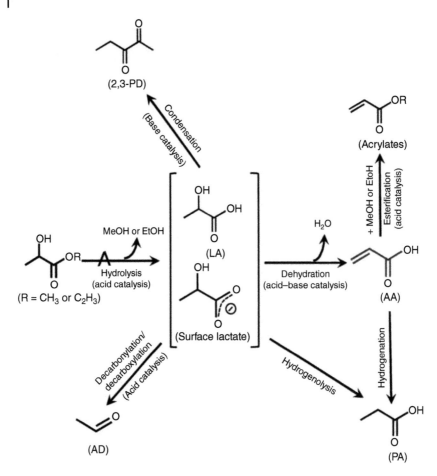

Figure 4.8 Reaction pathways for the formation of acrylic acid (AA) together with several reaction intermediates and side products starting from methyl lactate (ML) and ethyl lactate (EL) over HA. Source: Liu et al. [41]. Reproduced with permission of Elsevier.

acid sites and basic sites leads to the formation of butanol, the desired product. Of course, both ethanol and butanol can be converted to other side products. The authors argued that in order to favor the desired product, the strength, nature, and relative balance of the acid sites and basic sites are important for each step of the Guerbet cycle. The best catalyst in this regard was identified as Hap–CO$_3$ (Ca/P = 1.70) that yielded 30% of heavier alcohols at 40% ethanol conversion. The optimal ratio between the acid site density and basic site density was found to be 5 (Figure 4.10) [42]. When the acidity/basicity ratio was higher than 5, the total yield to alcohols decreased, whereas the ethylene yield increased. Note that the experiments in Figure 4.10 were carried out at the same reaction temperature of 400 °C. In another set of experiments where the reactions were carried out at different temperatures to achieve the same ethanol conversion of 14% over different catalysts, Hap–CO$_3$ still showed the highest selectivity to butanol [42].

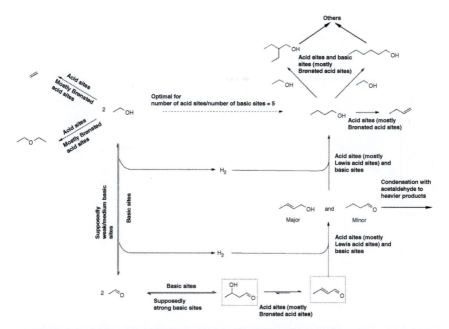

Figure 4.9 Reaction networks for the formation of butanol and other products starting from ethanol over HAP catalysts [42]. The dashed boxes mean the reaction intermediates that were not detected during the reaction but supposed to exist. The basic sites mean PO_4^{3-} and OH^- of HAP, together with CaO when applicable. The Brønsted acid sites mean HPO_4^{2-}, whereas the Lewis acid sites include Ca^{2+} and OH^- vacancies (δ^+). Source: Silvester et al. [42]. Reproduced with permission of the Royal Society of Chemistry.

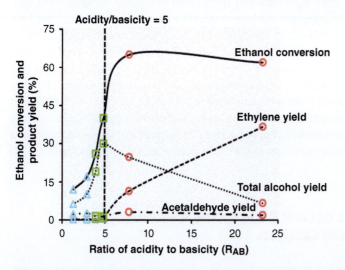

Figure 4.10 Correlation between the acidity/basicity ratio and catalytic performance of HAP catalysts at 400 °C [42]. Starting from left to right, the involved catalysts are HapE–Na–CO_3 (Ca/P = 2.39), Hap–CO_3 (Ca/P = 1.90), HapNa–CO_3 (Ca/P = 1.72), Hap–CO_3 (Ca/P = 1.70), Hap (Ca/P = 1.69), and Hap-D (Ca/P = 1/60). Source: Silvester et al. [42]. Reproduced with permission of the Royal Society of Chemistry.

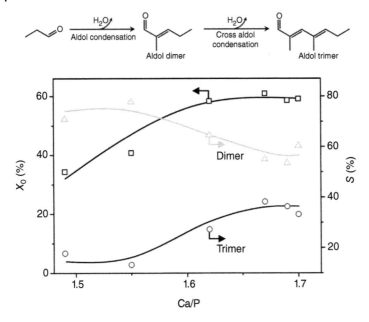

Figure 4.11 Self-condensation of propanol to aldol dimer and aldol trimer over HAP catalysts with different Ca/P ratios (1.49, 1.50, 1.62, 1.67, 1.69, 1.70) at a reaction temperature of 400 °C. Source: Rodrigues et al. [16]. Reproduced with permission of the Royal Society of Chemistry.

Pérez-Ramírez and coworkers used HA to catalyze the gas-phase deoxygenation of propanol (a common light-aldehyde component of bio-oil) via aldol condensation [16]. Several HA catalysts with different Ca/P ratios (1.49, 1.55, 1.62, 1.67, 1.69, 1.70) were studied. As shown in Figure 4.11 [16], the propanol conversion at 400 °C increased with the increase of Ca/P molar ratio and leveled off when the Ca/P ratio was higher than 1.62. The selectivity to aldol dimer decreased with the increase of Ca/P ratio, and the selectivity to aldol trimer increased accordingly. The stoichiometric HA-1.67 catalyst was more active and stable than MgO, Ca(OH)$_2$, CsX zeolite, and Mg–Al hydrotalcite in this reaction. The higher activity of HA-1.67 compared with Ca-deficient HA was ascribed to the presence of a higher basic site density and/or a higher strength of these basic sites as the Ca/P ratio increased, as proved by CO$_2$-TPD experiments.

4.3 Influence of Ca/P Ratio on the Performance of HA-Supported Catalysts

HA can also be used as a support for making supported catalysts. In this regard, support may not be an inert material. It can also play roles in interacting with the active component loaded onto the support and in tuning the catalytic activity and selectivity via its acid/base properties. However, examples on the influence of Ca/P ratio on the performance of HA-supported catalysts are limited in number. Although relevant studies deal with organic reactions and inorganic reactions, the influence of Ca/P of HA-supported catalysts on the selectivity in organic reactions is more interesting.

4.3.1 Relatively Simple Reactions

We previously prepared HA supports by precipitation involving the reaction between aqueous solutions of $(NH_4)_2HPO_4$ and $Ca(NO_3)_2$ under different initial pH values of 10.5, 9.5, and 8.5, respectively [9]. RhO_x was loaded onto these HA supports via wet impregnation. It was found that RhO_x supported on HA prepared under the initial pH of 10.5 showed significantly higher activity in N_2O decomposition ($N_2O = N_2 + 1/2\ O_2$) than the ones prepared under the initial pH of 9.5 or 8.5. For the latter, the support was composed of a mixture of HA and $Ca_2P_2O_7$, meaning that the overall Ca/P ratio was obviously below 1.67. The amounts of basic sites on these three HA supports were determined by CO_2-TPD to be 42.7, 9.5, and 5.2 μmol g^{-1}, respectively, and the sizes of RhO_x nanoparticles formed on these supports were 0.78 ± 0.22, 1.12 ± 0.33, and 1.33 ± 0.50 nm, respectively, highlighting the important impact of basic sites of HA on the average size of RhO_x nanoparticles and on the catalytic activity in N_2O decomposition.

N_2O decomposition mentioned previously is a simple inorganic reaction. Water–gas shift (WGS, $CO + H_2O = CO_2 + H_2$) is also a simple inorganic reaction, which is an important step for H_2 production from coal gasification or natural gas or biogas reforming. Goldbach and coworkers prepared several Pt/HA catalysts for WGS using HA supports prepared under different pH values (10.5, 10, 9.5, 9) [43]. The Ca/P ratios of the catalysts reached 1.76, 1.75, 1.68, and 1.58, respectively. It was found that the activities of these catalysts in WGS followed the sequence of Pt/HA-10.5 ~ Pt/HA-10 > Pt/HA-9.5 > Pt/HA-9, i.e. Pt/HA with a Ca/P ratio of 1.76 or 1.75 showed the highest activity in WGS. The authors proposed that Pt nanoparticles could adsorb and activate CO, whereas HA played an importing role in activating H_2O that needs Ca^{2+} and O atoms of neighboring PO_4^{3-}. Formate was identified by in situ Fourier transform infrared (FTIR) spectroscopy as a reaction intermediate; an associative mechanism involving the formate species was proposed, and a regenerative redox mechanism was ruled out because HA does not have redox properties. Although the activity of the optimal Pt/HA catalysts in WGS was comparable with that of a benchmark Pt/CeO_2 catalyst, these catalysts were generally active at high temperatures (i.e. 350–450 °C), but were not quite active at lower temperatures. Apparently, there are better catalysts that can work for low-temperature WGS.

Dry reforming of methane with CO_2 ($CH_4 + CO_2 = 2\ CO + 2\ H_2$) involves the reaction between an organic molecule and an inorganic molecule to form H_2 and CO. Boukha and coworkers used HA with different Ca/P ratios (1.73, 1.67, 1.62, 1.57) to prepare Ni/HA catalysts for dry reforming of methane with CO_2 [44]. The catalytic activity on different catalysts at a reaction temperature of 750 °C followed the sequence of Ni/HA-D2 (Ca/P = 1.62) > Ni/HA-D1 (Ca/P = 1.57) > Ni/HA-S (Ca/P = 1.67) > Ni/HA-E (Ca/P = 1.73). The authors concluded that the Ca-deficient samples had more abundant strong acid sites, helpful for the dispersion of Ni on HA surfaces and thus leading to higher catalytic activity in dry reforming of methane. On the other hand, the presence of basic sites could allow the adsorption of CO_2 that not only reacts with methane (to form H_2 and CO) but also with carbon deposited on catalyst surface to form CO and consequently to liberate active catalyst surface. The superior performance of Ni/HA-D2 (Ca/P = 1.62) in this reaction was explained by the highest surface area (26 m^2 g^{-1}) among these four catalysts and an abundance of strong acid sites and basic sites.

Pham Minh and coworkers studied dry reforming of methane using Ca-deficient HA-supported Co and Ni catalysts [45]. In that work, one support denoted as HA_N (Ca/P = 1.43) was prepared by wet chemical precipitation involving $Ca(NO_3)_2$ and $NH_4H_2PO_4$, and another support denoted as HA_C (Ca/P = 1.60) was prepared by reacting $CaCO_3$ with H_3PO_4. These two supports were loaded with Co, Ni, or both Co and Ni. It was found that the catalysts prepared using HA_C (Ca/P = 1.60) were slightly less active but more stable than the corresponding catalysts prepared using HA_N (Ca/P = 1.43), at a reaction temperature of 700 °C. HA_N (Ca/P = 1.43) was regarded as a strongly acidic support that may facilitate the formation of carbon and coke deposited on catalyst surface. Thus, HA_C (Ca/P = 1.60) was considered as a more promising support for designing supported Co and Ni catalysts for dry reforming of methane.

The application of HA-supported catalysts in organic catalysis is interesting and useful. Kaneda and coworkers evaluated the catalytic performance of Pd/HA catalysts in liquid-phase organic catalysis [46–49]. HA-0 (Ca/P = 1.67) and HA-1 (Ca/P = 1.50) were used as supports to graft $PdCl_2(PhCN)_2$ dissolved in acetone [46, 47]. EXAFS data showed that on HA-0, the grafted Pd species existed in the form of monomeric $PdCl_2$, whereas on HA-1, the grafted Pd species existed in the form of monomeric Pd^{II} phosphate surrounded by four oxygen atoms (Figure 4.12) [46]. More interestingly, Pd/HA-0 (Ca/P = 1.67) was able to catalyze the aerobic oxidation of a number of benzylic and allylic alcohols such as benzyl alcohols to form carbonyl compounds (the corresponding ketones and aldehydes) with high activity and selectivity due to the in situ formation of Pd nanoparticles during the reaction, whereas Pd/HA-1 (Ca/P = 1.50) was not active for these reactions, and the local structure of the grafted Pd^{II} species was not changed. In contrast, Pd/HA-1 (Ca/P = 1.50) was outstanding for the Heck and Suzuki couplings, whereas Pd/HA-0 (Ca/P = 1.67) was less effective in the Heck reaction. It was suggested that in these cases, the Heck reaction did not proceed via Pd^0/Pd^{II} cycle but proceeded via the Pd^{II}/Pd^{IV} mechanism [46]. However, in another work, the authors identified the formation of Pd nanoparticles after conducting Suzuki–Miyaura coupling of an aryl chloride over Pd/HA-1 (Ca/P = 1.50), indicating that the Pd^0 nanoparticles formed in situ were also active for this reaction.

Chen and coworkers, after demonstrating that MnO_x/HA can be used in catalyzing the reaction between alcohols and amines to form imines [50], explored the catalytic performance of MnO_x/HA catalysts using HA supports prepared under different pH values (7, 8, 8.5, 9, 10, 10.5) [51]. The Ca/P ratio of the resulting catalysts reached 1.06, 1.12, 1.35, 1.59, 1.63, and 1.68, respectively. In catalytic liquid-phase oxidation of benzyl alcohol at 80 °C, the selectivity to benzaldehyde was in the range

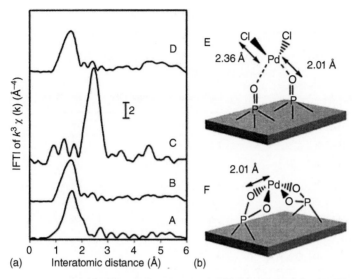

Figure 4.12 (a) Pd K-edge EXAFS data of PdHAP-0 (A), PdHAP-1 (B), recovered PdHAP-0 (C), and recovered PdHAP-1 (D) after catalyzing the oxidation of 1-phenylethanol. (b) Proposed local structures of the grafted Pd^{II} species of PdHAP-0 (E) with a Ca/P ratio of 1.67 and PdHAP-1 (F) with a Ca/P ratio of 1.50. Source: Mori et al. [46]. Reproduced with permission of the American Chemical Society.

of 94–97% using these catalysts, but the conversion greatly differed from each other (Table 4.1) [51]. The most active catalysts were MnO$_x$/HA-10 and MnO$_x$/HA-9.5 with Ca/P ratios of 1.63 and 1.59, respectively. When the Ca/P ratio of MnO$_x$/HA was lower than 1.59 or higher than 1.63, the benzyl alcohol conversion decreased obviously. The authors concluded that the coexistence of acid sites and basic sites should be necessary for this reaction.

4.3.2 More Complex Reactions

Nowadays, the conversion of biomass or biomass-derived chemicals has been a hot topic. Wan and coworkers, after demonstrating that Au/HA was much better than Au/SiO$_2$, Au/MgO, Au/TiO$_2$, Au/hydrotalcite, Au/Al$_2$O$_3$, and Au/ZnO in oxidative esterification of acetol (one of the major light pyrolysis products of biomass) with methanol to form methyl pyruvate (MPA), explored the effect of Ca/P ratio of HA on the performance of Au/HA catalysts [52]. The reaction mechanism involves the conversion of acetol to pyruvaldehyde (PA) in the first step. PA is then converted to different intermediates following different pathways, thus finally leading to the formation of MPA, the desired product (Figure 4.13) [52]. It was found that the acid site densities of Au/HA-1.65, Au/HA-1.62, and Au/HA-1.59 were 2.21, 2.72, and 3.34 µmol g^{-1}, respectively, whereas their base site densities were 4.00, 3.40, and 2.87 µmol g^{-1}, respectively. Au/HA-1.62 with a Ca/P ratio of 1.62 showed the best performance among these Au/HA catalysts, indicating that a suitable acid/base site ratio is necessary for achieving high activity and selectivity. When the Ca/P ratio of Au/HA deviated from 1.62, the selectivity to side products would increase.

Table 4.1 Aerobic oxidation of benzyl alcohol by various catalysts.[a]

Entry	Catalyst	Ca/P ratio	Conv. (%)[b]	Sel. (%)[b]
1	$MnO_x/HAP-7$	1.06	1	97
2	$MnO_x/HAP-8$	1.12	3	94
3	$MnO_x/HAP-8.5$	1.35	10	95
4	$MnO_x/HAP-9$	1.59	47	96
5	$MnO_x/HAP-10$	1.63	48	97
6	$MnO_x/HAP-10.5$	1.68	40	96
7	HAP-10	–	–	–
8	$Mn(OAc)_2 \cdot 4H_2O$ + HAP-10	–	–	–

a) Reaction conditions: benzyl alcohol (0.5 mmol), catalyst (10 mol% based on Mn), toluene (1 ml), 0.4 bar O_2, 2 h, 80 °C.
b) Determined by GC using diphenyl as the internal standard.
Source: Chen et al. [51]. Reproduced with permission of John Wiley and Sons.

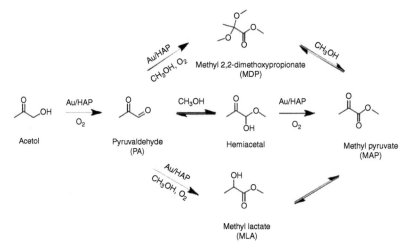

Figure 4.13 Reaction networks for the formation of methyl pyruvate (MPA) starting from acetol over Au/HAP-1.62 [52]. A number of reaction intermediates were formed along different reaction pathways. Source: Reproduced with permission of Elsevier.

4.4 Concluding Remarks

HA belongs to the family of calcium phosphates with the general formula of $Ca_{10-x}(HPO_4)_x(PO_4)_{6-x}(OH)_{2-x}$ where $0 \leq x \leq 2$. Although HA has been used less often in making catalysts than metal oxides, HA-based catalysts have recently attracted growing interest. There are hundreds of publications dealing with heterogeneous catalysis involving HA-based catalysts. However, for convenience, most of the researchers have chosen to use presumably stoichiometric HA (either prepared by these researchers or obtained from commercial companies, even though the Ca/P ratio was not always checked) as a catalyst or catalyst support. The purposes of most of the relevant research were to demonstrate that HA-based catalysts can be used for catalyzing some reactions and to compare the catalytic activities of these catalysts with those of other conventional catalysts. The question arises as whether the subtle change of Ca/P ratio (normally in the range of 1.33–1.67) of HA will make any difference in the catalytic activity of HA-based catalysts.

It is clear from the preceding summary that the answer is certainly yes. The change of Ca/P ratio not only changes the composition of HA but also changes the acid/base properties, thus modifying activity, selectivity, and stability of the catalyst. In addition, varying the Ca/P molar ratio can also influence the dispersion (size) of supported active components such as metal oxides and metal nanoparticles, thus influencing the catalytic performance.

It should be mentioned that there are numerous reactions in the literature. Sometimes nonstoichiometric HA is better than stoichiometric HA, whereas in some other cases, stoichiometric HA is better than nonstoichiometric HA. That means the requirements of different reactions may be different, and a rational choice of catalysts based on reaction evaluating and fundamental understanding is necessary.

There are several potential directions for future research. First, most researchers working in the field of HA-based catalysts and their catalysis have not studied the influence of Ca/P ratio of HA on the catalytic performance in detail. Considering that HA-based catalysts are useful in converting organic chemicals and selectivity may be an issue in this regard, we believe that it would be very interesting to explore the influence of Ca/P ratio of HA on the catalytic performance in a variety of reactions. Perhaps better catalytic performance will be achieved, and "new" catalytic routes or products will be found.

Second, in previous works, the findings on the effect of Ca/P ratio on catalytic performance were usually empirical, i.e. the conclusions were drawn based on simple correlations, but an in-depth and precise understanding has been still missing. In addition, sometimes the correlation between Ca/P ratio and the catalytic performance was not straightforward when HA catalysts with different Ca/P ratios were prepared using different protocols or the sizes and surface areas of these HA catalysts were significantly different [36]. That means the catalytic activity may be influenced by other factors (such as sizes, surface areas, morphologies) of the catalysts. It would be necessary to consider different aspects of the structures, properties, and catalytic mechanisms. IR studies on the acid–base properties of HA catalysts are necessary [53, 54]. Kinetic studies [55–57], *operando* FTIR [58], and DFT calculations [13, 25] are important for understanding the reaction mechanism.

References

1. Ma, Z., Yin, H.F., Overbury, S.H. et al. (2008). Metal phosphates as a new class of supports for gold nanocatalysts. *Catalysis Letters* 126: 20–30.
2. Xie, W.L. and Yang, D. (2012). Transesterification of soybean oil over WO_3 supported on $AlPO_4$ as a solid acid catalyst. *Bioresource Technology* 119: 60–65.
3. Machida, M., Minami, S., Hinokuma, S. et al. (2015). Unusual redox behavior of $Rh/AlPO_4$ and its impact on three-way catalysis. *Journal of Physical Chemistry C* 119: 373–380.
4. Lin, Y., Meng, T., and Ma, Z. (2015). Catalytic decomposition of N_2O over RhO_x supported on metal phosphates. *Journal of Industrial and Engineering Chemistry* 28: 138–146.
5. Liu, H., Lin, Y., and Ma, Z. (2016). $Au/LaPO_4$ nanowires: synthesis, characterization, and catalytic CO oxidation. *Journal of the Taiwan Institute of Chemical Engineers* 62: 275–282.
6. Xu, J., White, T., Li, P. et al. (2010). Hydroxyapatite foam as a catalyst for formaldehyde combustion at room temperature. *Journal of the American Chemical Society* 132: 13172–13173.
7. Mimura, N., Sato, O., Shirai, M. et al. (2017). 5-Hydroxymethylfurfural production from glucose, fructose, cellulose, or cellulose-based waste material by using a calcium phosphate catalyst and water as a green solvent. *ChemistrySelect* 2: 1305–1310.
8. Liu, Y.M., Tsunoyama, H., Akita, T. et al. (2011). Aerobic oxidation of cyclohexane catalyzed by size-controlled au clusters on hydroxyapatite: size effect in the sub-2 nm regime. *ACS Catalysis* 1: 2–6.
9. Huang, C.Y., Ma, Z., Xie, P.F. et al. (2015). Hydroxyapatite-supported rhodium catalysts for N_2O decomposition. *Journal of Molecular Catalysis A-Chemical* 400: 90–94.
10. Huang, C.Y., Jiang, Y.X., Ma, Z. et al. (2016). Correlation among preparation methods/conditions, physicochemical properties, and catalytic performance of Rh/hydroxyapatite catalysts in N_2O decomposition. *Journal of Molecular Catalysis A-Chemical* 420: 73–81.
11. De Vasconcelos, B.R., Zhao, L.L., Sharrock, P. et al. (2016). Catalytic transformation of carbon dioxide and methane into syngas over ruthenium and platinum supported hydroxyapatites. *Applied Surface Science* 390: 141–156.
12. Othmani, M., Bachoua, H., Ghandour, Y. et al. (2018). Synthesis, characterization and catalytic properties of copper-substituted hydroxyapatite nanocrystals. *Materials Research Bulletin* 97: 560–566.
13. Ho, C.R., Zheng, S., Shylesh, S. et al. (2018). The mechanism and kinetics of methyl isobutyl ketone synthesis from acetone over ion-exchanged hydroxyapatite. *Journal of Catalysis* 365: 174–183.
14. Sun, W., Jiang, W., Zhu, G. et al. (2018). Magnetic Cu^0@HAP@g-Fe_2O_3 nanoparticles: an efficient catalyst for one-pot three-component reaction for the synthesis of imidazo[1,2-a] pyridines. *Journal of Organometallic Chemistry* 873: 91–100.

15 Jahanshahi, P., Mamaghani, M., Haghbin, F. et al. (2018). One-pot chemoselective synthesis of novel pyrrole-substituted pyrido [2,3-*d*]pyrimidines using [γ-Fe$_2$O$_3$@HAp-SO$_3$H] as an efficient nanocatalyst. *Journal of Molecular Structure* 1155: 520–529.

16 Rodrigues, E.G., Keller, T.C., Mitchell, S. et al. (2014). Hydroxyapatite, an exceptional catalyst for the gas-phase deoxygenation of bio-oil by aldol condensation. *Green Chemistry* 16: 4870–4874.

17 Gruselle, M. (2015). Apatites: a new family of catalysts in organic synthesis. *Journal of Organometallic Chemistry* 793: 93–101.

18 Fihri, A., Len, C., Varma, R.S. et al. (2017). Hydroxyapatite: a review of syntheses, structure and applications in heterogeneous catalysis. *Coordination Chemistry Reviews* 347: 48–76.

19 Piccirillo, C. and Castro, P.M.L. (2017). Calcium hydroxyapatite-based photocatalysts for environment remediation: characteristics, performances and future perspectives. *Journal of Environmental Management* 193: 79–91.

20 Ibrahim, M., Labaki, M., Giraudon, J.M. et al. (2020). Hydroxyapatite, a multifunctional material for air, water and soil pollution control: a review. *Journal of Hazardous Materials* 383: 121139 (18 pp).

21 Ebrahimpour, A., Johnsson, M., Richardson, C.F. et al. (1993). The characterization of hydroxyapatite preparations. *Journal of Colloid and Interface Science* 159: 158–163.

22 Tsuchida, T., Kubo, J., Yoshioka, T. et al. (2009). Influence of preparation factors on Ca/P ratio and surface basicity of hydroxyapatite catalyst. *Journal of the Japan Petroleum Institute* 52: 51–59.

23 Narasaraju, T.S.B. and Phebe, D.E. (1996). Some physico-chemical aspects of hydroxylapatite. *Journal of Materials Science* 31: 1–21.

24 Sebti, S., Tahir, R., Nazih, R. et al. (2002). Hydroxyapatite as a new solid support for the Knoevenagel reaction in heterogeneous media without solvent. *Applied Catalysis A-General* 228: 155–159.

25 Usami, K. and Okamoto, A. (2017). Hydroxyapatite: catalyst for a one-pot pentose formation. *Organic & Biomolecular Chemistry* 15: 8888–8893.

26 Thanh Son, P., Sane, A.R., de Vasconcelos, B.R. et al. (2018). Hydroxyapatite supported bimetallic cobalt and nickel catalysts for syngas production from dry reforming of methane. *Applied Catalysis B-Environmental* 224: 310–321.

27 Bett, J.A.S., Christner, L.G., and Hall, W.K. (1967). Studies of the hydrogen held by solids. XII. Hydroxyapatite catalysts. *Journal of the American Chemical Society* 89: 5535–5541.

28 Bett, J.A.S. and Hall, W.K. (1968). The microcatalytic technique applied to a zero order reaction: the dehydration of 2-butanol over hydroxyapatite catalysts. *Journal of Catalysis* 10: 105–113.

29 Bett, J.A.S., Christner, L.G., and Hall, W.K. (1969). Infrared studies of hydroxyapatite catalysts – adsorbed CO_2 2-butanol and methyl ethyl ketone. *Journal of Catalysis* 13: 332–336.

30 Kibby, C.L. and Hall, W.K. (1973). Studies of acid catalyzed reactions: XII. Alcohol decomposition over hydroxyapatite catalysts. *Journal of Catalysis* 29: 144–159.

31 Joris, S.J. and Amberg, C.H. (1971). Nature of deficiency in nonstoichiometric hydroxyapatites. 1. Catalytic activity of calcium and strontium hydroxyapatites. *Journal of Physical Chemistry* 75: 3167–3171.

32 Dykman, A.S. (2003). Nature of catalytic activity of defective calcium hydroxyapatites. *Russian Journal of Applied Chemistry* 76: 226–228.

33 Silvester, L., Lamonier, J.F., Vannier, R.N. et al. (2014). Structural, textural and acid-base properties of carbonate-containing hydroxyapatites. *Journal of Materials Chemistry A* 2: 11073–11090.

34 Tsuchida, T., Kubo, J., Yoshioka, T. et al. (2008). Reaction of ethanol over hydroxyapatite affected by Ca/P ratio of catalyst. *Journal of Catalysis* 259: 183–189.

35 Tsuchida, T., Yoshioka, T., Sakuma, S. et al. (2008). Synthesis of biogasoline from ethanol over hydroxyapatite catalyst. *Industrial & Engineering Chemistry Research* 47: 1443–1452.

36 Ben Osman, M., Krafft, J.M., Millot, Y. et al. (2016). Molecular understanding of the bulk composition of crystalline nonstoichiometric hydroxyapatites: application to the rationalization of structure-reactivity relationships. *European Journal of Inorganic Chemistry* 2016: 2709–2720.

37 Nishikawa, H. and Monma, H. (1994). Oxidative decomposition of chlorobenzene over calcium-deficient hydroxyapatite. *Bulletin of the Chemical Society of Japan* 67: 2454–2456.

38 Nishikawa, H., Oka, T., Asai, N. et al. (2012). Oxidative decomposition of volatile organic compounds using thermally-excited activity of hydroxyapatite. *Applied Surface Science* 258: 5370–5374.

39 Xin, Y.Z., Ando, Y., Nakagawa, S. et al. (2020). New possibility of hydroxyapatites as noble-metal-free catalysts towards complete decomposition of volatile organic compounds. *Catalysis Science & Technology* 10: 5453–5459.

40 Yan, B., Tao, L.Z., Liang, Y. et al. (2014). Sustainable production of acrylic acid: catalytic performance of hydroxyapatites for gas-phase dehydration of lactic acid. *ACS Catalysis* 4: 1931–1943.

41 Liu, Z.H., Yan, B., Liang, Y. et al. (2020). Comparative study of gas-phase "dehydration" of alkyl lactates and lactic acid for acrylic acid production over hydroxyapatite catalysts. *Molecular Catalysis* 494: 111098 (10 pp).

42 Silvester, L., Lamonier, J.F., Faye, J. et al. (2015). Reactivity of ethanol over hydroxyapatite-based Ca-enriched catalysts with various carbonate contents. *Catalysis Science & Technology* 5: 2994–3006.

43 Miao, D.Y., Goldbach, A.s., and Xu, H.Y. (2016). Platinum/apatite water-gas shift catalysts. *ACS Catalysis* 6: 775–783.

44 Boukha, Z., Yeste, M.P., Cauqui, M.A. et al. (2019). Influence of Ca/P ratio on the catalytic performance of Ni/hydroxyapatite samples in dry reforming of methane. *Applied Catalysis A-General* 580: 34–45.

45 Thi Quynh, T., Doan Pham, M., Thanh Son, P. et al. (2020). Dry reforming of methane over calcium-deficient hydroxyapatite supported cobalt and nickel catalysts. *Chemical Engineering Science* 228: 115975 (14 pp).

46 Mori, K., Yamaguchi, K., Hara, T. et al. (2002). Controlled synthesis of hydroxyapatite-supported palladium complexes as highly efficient heterogeneous catalysts. *Journal of the American Chemical Society* 124: 11572–11573.

47 Mori, K., Hara, T., Mizugaki, T. et al. (2004). Hydroxyapatite-supported palladium nanoclusters: a highly active heterogeneous catalyst for selective oxidation of alcohols by use of molecular oxygen. *Journal of the American Chemical Society* 126: 10657–10666.

48 Kaneda, K., Mori, K., Hara, T. et al. (2004). Design of hydroxyapatite-bound transition metal catalysts for environmentally-benign organic syntheses. *Catalysis Surveys from Asia* 8: 231–239.

49 Mori, K., Hara, T., Oshiba, M. et al. (2005). Catalytic investigations of carbon-carbon bond-forming reactions by a hydroxyapatite-bound palladium complex. *New Journal of Chemistry* 29: 1174–1181.

50 Chen, B., Li, J., Dai, W. et al. (2014). Direct imine formation by oxidative coupling of alcohols and amines using supported manganese oxides under an air atmosphere. *Green Chemistry* 16: 3328–3334.

51 Chen, B., Zhao, Z.B., Zhang, Y.C. et al. (2020). Hydroxyapatite-supported manganese oxides as efficient non-noble-metal catalysts for selective aerobic oxidation of alcohols. *ChemistrySelect* 5: 4297–4302.

52 Wan, Y., Zheng, C.C., Lei, X.C. et al. (2019). Oxidative esterification of acetol with methanol to methyl pyruvate over hydroxyapatite supported gold catalyst: essential roles of acid-base properties. *Chinese Journal of Catalysis* 40: 1810–1819.

53 Diallo-Garcia, S., Ben Osman, M., Krafft, J.M. et al. (2014). Identification of surface basic sites and acid-base pairs of hydroxyapatite. *Journal of Physical Chemistry C* 118: 12744–12757.

54 Ben Osman, M., Krafft, J.M., Thomas, C. et al. (2019). Importance of the nature of the active acid/base pairs of hydroxyapatite involved in the catalytic transformation of ethanol to *n*-butanol revealed by *operando* DRIFTS. *ChemCatChem* 11: 1765–1778.

55 Ho, C.R., Shylesh, S., and Bell, A.T. (2016). Mechanism and kinetics of ethanol coupling to butanol over hydroxyapatite. *ACS Catalysis* 6: 939–948.

56 Moteki, T. and Flaherty, D.W. (2016). Mechanistic insight to C-C bond formation and predictive models for cascade reactions among alcohols on Ca- and Sr-hydroxyapatites. *ACS Catalysis* 6: 4170–4183.

57 Eagan, N.M., Lanci, M.P., and Huber, G.W. (2020). Kinetic modeling of alcohol oligomerization over calcium hydroxyapatite. *ACS Catalysis* 10: 2978–2989.

58 Wang, S.C., Cendejas, M.C., and Hermans, I. (2020). Insights into ethanol coupling over hydroxyapatite using modulation excitation *operando* infrared spectroscopy. *ChemCatChem* 12: 4167–4175.

5

Kinetics and Mechanisms of Selected Reactions over Hydroxyapatite-Based Catalysts

U.P.M. Ashik[1], Nurulhuda Halim[2], Shusaku Asano[1], Shinji Kudo[1], and Jun-ichiro Hayashi[1,3]

[1] Kyushu University, Institute for Materials Chemistry and Engineering, 6-1 Kasuga Koen, Kasuga 816-8580, Japan
[2] Institut Teknologi Bandung, Department of Metallurgical Engineering, Jl. Ganesa No. 10, Bandung, Jawa Barat 40132, Indonesia
[3] Kyushu University, Transdisciplinary Research and Education Center of Green Technology, Kasuga 816-8580, Japan

5.1 Introduction

Hydroxyapatite (HA) ($Ca_{10}(PO_4)_6(OH)_2$) is a common calcium phosphate compound present in the bone and teeth [1, 2], which is extensively investigated as a biomaterial. Figure 5.1 shows the three-dimensional (3D) visualization of HA structure [3]. As presented in Chapter 3 of this book, the hexagonal HA has a structure of a $P6_3/m$ space group with a thermal stability up to 1000 °C [4–6]. The Ca^{2+} cations and PO_4^{3-} anions are placed with three vertical symmetry axes. Details on the structure of HA-based materials have been presented in Chapter 3 of this book. The structural and catalytic characteristics of HA can be effectively tuned by replacing its Ca^{2+} cation and/or the PO_4^{3-} or OH^- anions according to purpose [7–10]. The complex structure of HA contains two sets of nonequivalent sites in a single unit cell to accommodate 10 Ca^{2+} ions. The 6 Ca^{2+} ions of set 1 coordinate with O-atoms, including that of PO_4 tetrahedra [11, 12].

The unique structure of HA results in its exceptional properties for various application, especially in catalysis [13–18], adsorption [19, 20], and ion transport [21–23]. The presence of diverse pair set of Ca^{2+} ions provides distinctive catalytic features to HA. Hence, it enables substitution of existing metal ions with foreign ions according to the targeted catalytic application [8, 17, 24–26].

For example, Ca^{2+} cations replaced with various metals such as Cu^{2+} [8], Pb^{2+} [27, 28], Sr^{2+} [29], and Ni^{2+} [30] allowed improving catalytic performance in different reactions such as oxidative dehydrogenation of propane, catalytic conversion of 2-propanol to acetone, etc. In addition, HA has two unique facets: positively charged Ca^{2+}-rich basal c-surface and prismatic a-surface enriched by negatively charged phosphate and hydroxide ions. Besides, the large Ca channel can facilitate the

Design and Applications of Hydroxyapatite-Based Catalysts, First Edition. Edited by Doan Pham Minh.
© 2022 WILEY-VCH GmbH. Published 2022 by WILEY-VCH GmbH.

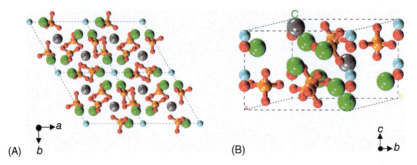

Figure 5.1 Three-dimensional visualization of HA (a) normal to c-axis and (b) along with the c-axis. Ca atoms: gray and green, P atoms: orange; O atoms belonging to PO$_4$ tetrahedra: red; O atoms from hydroxyl groups: cyan. Source: Oh et al. [3]. Reproduced with permission of American Chemical Society.

hydroxyl ion movements and promote proton migration and hence act as a proton conductor at high temperature (500–800 °C) [31]. HA generally contains both acid and basic sites, and their density and strength depend on the synthesis method and the chemical composition, in particular the molar ratio of Ca/P (see more details in Chapters 2 and 4 of this book). HA of high specific surface area and porosity can be also obtained. All these special physicochemical and thermal properties make HA as a special candidate for various catalytic applications, either as a solid catalyst itself or as a catalyst support. To date, the HA-based catalysts are investigated in various catalytic reactions such as oxidative coupling of methane (OCM), partial oxidation of methane (POM), ethane coupling, selective hydrogenation, organic pollutant degradation, photocatalysis and electrophotocatalysis, organic synthesis, etc., and improved performance had been reported. The physicochemical characteristics of HA-based catalysts are extensively investigated, though the active sites responsible for respective reaction are not thoroughly studied yet. Besides, the kinetic and reaction mechanism studies over HA-based catalysts are also limited, since the determination of the exact structure and composition of HA-based catalyst is usually delicate. This chapter discusses the kinetics and mechanisms of selected reactions catalyzed by HA-based catalysts.

5.2 Oxidative Coupling of Methane

C2-hydrocarbons (C_2H_6 and C_2H_4) can be produced through C–C bonds formation in a high temperature oxidative coupling of CH_4 and O_2 [32]. This single step process directly converts the major component of natural gas or biogas into more valuable compounds, and it has attracted extensive research interest in the last couple of decades [33, 34], since the ethylene supply is under threat due to massive hike on oil prices. Hence, it possibly increases the stock value of the least reactive hydrocarbon on the Earth – methane [29]. Optimization of OCM process for an industrially acceptable performance parameter with suitable catalysts can be a breakthrough in methane utilization for development and sustainability. Hence, the development of an active and durable catalyst for OCM reaction is the indispensable need in this research. An enormous number of catalysts based on alkali metals [35, 36], alkaline earth metals [37], rare earth metals [38], and transition metals [39] have been largely investigated for OCM reaction.

The catalytic OCM reaction clearly depicted that two types of active sites are contributing in this catalysis [32]. The surface oxygen species with the capacity to strongly attract hydrogen atoms causes hemolytic cleavage of C–H bonds. This is essential for the methane activation, hence, providing valid clue for catalyst development. However, the type of oxidizing species is also relevant in OCM reaction, making it attractive for the application of HA-based catalysts in methane coupling. This can be attributed to the presence of PO_4^{3-} or OH^- species in the HA, which can be easily substituted with required type of anions in order to achieve the targeted catalytic characteristics [7, 9]. Notably, transition metal-substituted HA [27, 40] is a competitive composition among the investigated OCM catalysts.

As a consequence, Matsumura and Moffat [41] reported a high C2 selectivity at reaction temperatures as low as 700 °C when lead (Pb) incorporated into the HA was used as catalyst. In addition, this catalyst exhibited an enhancement in C2 selectivity about five times magnitude compared to the HA virgin counterpart. Similarly, Lee and coworkers [28, 29] claimed lead incorporated HA as the most effective catalyst for OCM reaction to produce ethane and ethylene. The Pb^{2+} active phase promotes pairwise reaction and subdues the oxidation reaction, which finally lead to the improved C2 selectivity. The Pb-doped catalyst maintains the methane coupling activity for extended period of time at high temperature too [28]. Hence, it can be noticed that the Pb is capable of stabilizing the HA lattices and the high stability of Pb-HA can be attributed to the Pb—O interactions, which induce the covalent bond-like characteristics [42].

It is commonly agreed that the free radical formation triggers OCM reaction, though the reaction mechanism has been a subject of discussion. In this mechanism, the collision of CH_4 with an oxidation site, which has high hydrogen atom affinity [32], releases free methyl radicals ($CH_3\cdot$) and surface hydroxyl groups as shown in Eq. (5.1).

$$O_{(S)} + CH_4 \rightarrow OH_{(S)} + CH_3\cdot \tag{5.1}$$

Alternatively, the C–H bond cleavage occurs at the strong basic sites of the catalyst to produce anions, which will be transformed to electronic state as illustrated in Eqs. (5.2) and (5.3).

$$O_{(s)}^{2-} + CH_4 \rightarrow O^{2-} + H^+ + CH_3^- \tag{5.2}$$

$$CH_3^- \rightarrow CH_3 + e^- \tag{5.3}$$

On the basis of previously mentioned mechanisms, it is concluded that the catalyst with electrophilic oxygen species and electron-accepting promoted structure defects could be suitable for OCM.

Briefly, the methane coupling reaction occurred through a combined homogeneous–heterogeneous reaction network [43, 44]. Initially, methane

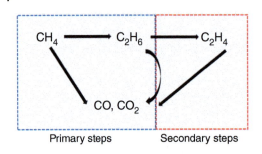

Figure 5.2 Oxidative coupling of methane. Source: Oh et al. [2]. Reproduced with permission of Elsevier.

activation occurs on a catalyst surface to form $CH_3\cdot$ radicals, which primarily combine to form ethane (C_2H_6). In addition, the so-produced C_2H_6 may undergo a dehydrogenation process to form ethylene (C_2H_4) as a secondary product. In parallel, the $CH_3\cdot$ could be deeply oxidized to produce CO_2 and CO, though the final C2 yield is solely dependent on the primary and secondary reactions on the respective catalysts.

The generally accepted reaction pathway for OCM is shown in Figure 5.2, which was drawn from various researches on a series of catalysts [45–47]. Oh et al. [2] investigated the influence of cation and anion substitution in HA on the rate and selectivity of OCM reaction. In addition, they also examined the identity and reaction rate constants of elementary steps in primary reactions of OCM. For this purpose, they prepared Pb–HA (Pb^{2+} cation) and HA–F (HA with partial substitution of OH^- anion by F^- anion) catalysts. They categorized methane activation on HA-based catalysts to form $CH_3\cdot$ and the coupling reaction leading to the formation of C_2H_6 as primary reaction steps, whereas the dehydrogenation of C_2H_6 to C_2H_4 is considered as the secondary reaction step [2].

Consequently, both Eley–Rideal [48–50] and Langmuir–Hinshelwood [51, 52] mechanisms have been accepted to describe the OCM reaction. According to Eley–Rideal mechanism, the rate-determining step (RDS) is either dissociative or molecular type reaction between the gaseous methane and adsorbed oxygen [2, 50, 53]. It is reported that catalysts such as Na_2WO_4–Mn/SiO_2 [54, 55] follows Eley–Rideal mechanism. On the other hand, in Langmuir–Hinshelwood mechanism, the RDS of OCM is the reaction between adsorbed methane and oxygen species. This mechanism is validated for OCM over catalysts such as Sm_2O_3 [56], Na-doped MgO [51], and perovskite [52].

Oh et al. [2] investigated the so-mentioned mechanisms for HA-based catalysts and concluded that HA and HA-F follow the Langmuir–Hinshelwood mechanism, while catalysts substituted with Pb (Pb-HA and Pb-HA-F) follow the Eley–Rideal mechanism. In the case of the Langmuir–Hinshelwood mechanism for HA and HA-F catalysts, the active sites of the catalyst coordinate the interaction between adsorbed methane and oxygen to activate C—H bond, which is the RDS. The detailed steps involved in this mechanism are as follows:

$$\text{Step (1): } O_2 + *^1 \xrightleftharpoons{K_1} O_2^{*1} \tag{5.4}$$

$$\text{Step (2): } CH_4 + *^2 \xrightleftharpoons{K_2} CH_4^{*2} \tag{5.5}$$

Step (3): $CH_4^{*2} + O_2^{*1} \xrightarrow{k_3} CH_3\cdot + HO_2\cdot + *^1 + *^2 (RDS)$ (5.6)

Step (4): $CH_3\cdot + CH_3\cdot \xrightarrow{k_4} C_2H_6$ (5.7)

Step (5): $CH_3\cdot + O_2 \xrightarrow{k_5} CO$ (5.8)

Step (6): $CH_3\cdot + O_2^{*1} \xrightarrow{k_6} CO_2$ (5.9)

Oxygen and methane initially adsorbed on two active HA sites (*1 and *2) to form O_2^* and CH_4^* species in steps 1 and 2, respectively. The activated species reacted to produce $CH_3\cdot$, which is the RDS in this mechanism. Later, the $CH_3\cdot$ detached from the HA surface for coupling, as a consequence, forming C_2H_6. However, the adsorbed $CH_3\cdot$ could interact with the reactive O_2^* to form CO_2, while gaseous O_2 reacts with $CH_3\cdot$, resulting in CO production. The adsorbed $CH_3\cdot$ adsorbed on HA surface reacts with reactive O_2 and results in CO_2 formation (step 6). At the same time, gaseous O_2 reacts with $CH_3\cdot$ to produce CO (step 5). It is seen in the previous section that Eqs. (5.8) and (5.9) are intensively not balanced by the authors of reference [2]. The latter developed the formation rate equations of CO and CO_2. Hence, unknown species are not considered, and the carbon is balanced in each equation.

The rate law for CH_4 consumption has been determined as follows:

$$r_{CH_4} = k_3 \left[K_1 P_{O_2} \left(\frac{n_1}{1 + K_1 P_{O_2}} \right) \right] \left[K_2 P_{CH_4} \left(\frac{n_2}{1 + K_2 P_{CH_4}} \right) \right] \quad (5.10)$$

where n_1 and n_2 are the total number of active sites for type *1 and type *2, respectively. K_1 is the oxygen adsorption equilibrium constant, and K_2 is the methane adsorption equilibrium constant. On the basis of the proposed Langmuir–Hinshelwood mechanism and equations derived, the authors analyzed and fitted the measured kinetic data for HA and HA-F catalysts [2].

Figure 5.3 shows the fitting of the measured rates of CH_4 conversion with respect to P_{CH4} and P_{O2}, respectively, using the Langmuir–Hinshelwood model. As per the plot in Figure 5.3, the CH_4 consumption over HA-based catalysts increases with P_{CH4} and P_{O2} until it reaches its upper limit. The HA exhibited the highest K_1 value among the different investigated catalysts and the K_1 value trend as follows, HA > HA-F > Pb-HA-F ~ Pb-HA. The untreated HA has the highest CH_4 and O_2 adsorption capability. F^- substitution declines O_2 sorption by three times, while CH_4 adsorption was almost intact. At the same time, Pb^{2+} dropped both oxygen and methane adsorption capability.

Irrespective of the HA and HA-F catalysts, the consumption of hydrogen from methane by adsorbed oxygen on the active catalyst sites is the RDS on Pb-HA and Pb-HA-F, which follows the Eley–Rideal mechanism, as illustrated in the following steps:

Step (1): $O_2 + * \xrightleftharpoons{K_1} O_2^*$ (5.11)

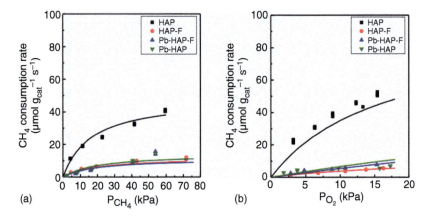

Figure 5.3 Methane consumption rate as a function of CH_4 pressure at $P_{O_2} = 7$ kPa (a) and O_2 pressure at $P_{CH_4} = 25$ kPa (b). The curves in (a) and (b) denote the fitting results using optimized rate constants based on Langmuir–Hinshelwood mechanism at 700 °C. Source: Oh et al. [2]. Reproduced with permission of Elsevier.

$$\text{Step (2): } CH_4 + O_2* \xrightleftharpoons{k_3} CH_3\cdot + HO_2\cdot + * \text{ (RDS)} \tag{5.12}$$

$$\text{Step (3): } CH_3\cdot + CH_3\cdot \xrightarrow{k_4} C_2H_6 \tag{5.13}$$

$$\text{Step (4): } CH_3\cdot + O_2 \xrightarrow{k_5} CO \tag{5.14}$$

$$\text{Step (5): } CH_3\cdot + O_2^* \xrightarrow{k_6} CO_2 \tag{5.15}$$

The rate law for CH_4 consumption is derived as thus:

$$r_{CH_4} = k_3 P_{CH_4} \left[K_1 P_{O_2} \left(\frac{n_1}{(1+K_1 P_{O_2})} \right) \right] \tag{5.16}$$

The K_2 values for HA and HA-F are considerably higher than that of Pb-HA and Pb-HA-F, indicating their high methane adsorption capability. Hence, OCM kinetics over Pb-HA and Pb-HA-F can be validated by Eley–Rideal mechanism, which considers reaction of gaseous CH_4 with adsorbed O_2. Figure 5.4 shows a very good fitting of experimental kinetic data with the Eley–Rideal model, which validates the mechanism considered for both Pb-HA and Pb-HA-F catalysts in OCM reaction.

From this study, it can be concluded that despite the global morphological similarity of HA-based catalysts, the insertion of Pb^{2+} cations and F^- anions on the surface of HA must lead to fundamental change in surface chemistry of the catalytic materials, as previously discussed in Chapter 3 of this book, and thus drives the reaction by another mechanism.

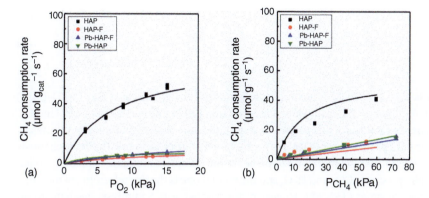

Figure 5.4 Methane consumption rate as a function of O_2, pressure at $P_{CH4} = 25$ kPa (a), and CH_4 pressure at $P_{O2} = 7$ kPa (b), respectively. The curves in (a) and (b) denote the fitting results using optimized rate using the Eley–Rideal model. Source: Oh et al. [2]. Reproduced with permission of Elsevier.

5.3 Partial Oxidation of Methane

Partial oxidation of methane (POM) has received renewed attention in recent times due to its economical and feasible nature to produce syngas over conventional methodologies [57–59]. Irrespective of commercialized steam reforming of methane process, POM does not produce large quantity of CO_2. Moreover, this exothermic process does not need high operating pressure and hence is energy efficient. The high temperature POM process results in the production of syngas (CO and H_2 mixture) as a main product, while coke, CO_2, and H_2O are by-products [60].

Thermodynamically, the reaction is favorable to coke formation at high temperature, thereby reducing the yields of syngas [61, 62].

The overall reactions involved in the POM is given as follows [61]:

$$CH_4 + 2O_2 \rightarrow CO_2 + 2H_2O \tag{5.17}$$

$$CH_4 + \tfrac{1}{2}O_2 \rightarrow CO + 2H_2 \tag{5.18}$$

$$CH_4 + O_2 \rightarrow CO_2 + 2H_2 \tag{5.19}$$

$$CO + H_2O \rightleftharpoons CO_2 + H_2 \tag{5.20}$$

$$CH_4 + H_2O \rightleftharpoons CO + 3H_2 \tag{5.21}$$

$$CH_4 + CO_2 \rightleftharpoons 2CO + 2H_2 \tag{5.22}$$

$$CO + H_2 \rightleftharpoons C + H_2O \tag{5.23}$$

$$CH_4 \rightleftharpoons C + 2H_2 \tag{5.24}$$

$$2CO \rightleftharpoons CO_2 + C \qquad (5.25)$$

$$CO + \tfrac{1}{2}O_2 \rightarrow CO_2 \qquad (5.26)$$

$$H_2 + \tfrac{1}{2}O_2 \rightarrow H_2O \qquad (5.27)$$

It is clear that POM is a complex reaction system as shown in Eqs. (5.17)–(5.27) [61]. Equation (5.18) is the only direct reaction producing syngas from CH_4 and O_2, and hence catalysis of this particular step alone is the challenge. Basically, two mechanisms are proposed for POM: indirect scheme or direct scheme. Indirect scheme is a two-step mechanism, the first one being the complete methane oxidation, which initially produces CO_2 and H_2O, and the second one is the dry reforming of methane to produce syngas [59, 63]. However, direct scheme is a pyrolysis mechanism, in which CH_4 initially dissociates to produce H_2 and carbon. Then the H_2 desorbs from the catalyst surface and the deposited carbon oxidizes to CO upon reacting with surface oxygen species [64, 65].

The mechanism involved depends on both metal and support. As example, Rh/Al_2O_3 and Rh/TiO_2 produce syngas via pyrolysis mechanism, while the reaction route over Rh/SiO_2 and Ir/TiO_2 is a two-step mechanism [66]. Various catalysts have been investigated for POM, including Ni, Rh, Pt, Pd, etc., supported on metal oxides [67]. Rh is known to be the most effective active phase for POM. However, considering the rational performance, economic feasibility, and availability, Ni-based catalysts instead have been explored extensively [68]. The Ni-based catalysts are generally supported on Al_2O_3 or SiO_2, which causes the production of inactive nickel aluminate and silicates. As a result, the performance is inhibited. Hence, different strategies have been developed for POM catalyst synthesis to overcome these limitations, such as pre-coating of alumina or silica supports with MgO, which results in a protective layer by forming magnesium aluminate or silicate [68].

On the other hand, the physicochemical characteristics of HA are favorable for POM; hence, it is being investigated. In light of this, Boukha et al. [69] experimented the Rh(x)/HA catalysts (with $x = 0.5$, 1, and 2 wt%) for POM and compared their performance with that of a commercial 1 wt% Rh/Al_2O_3 catalyst. Nonstoichiometric HA was used to prepare the supported catalysts by impregnation and observed highest coke resistance among the assessed catalysts. The POM performance over Rh(x)/HAP compared with the thermodynamic equilibrium values is shown in Figure 5.5.

In the experimented temperature range between 500 and 727 °C, the Rh(1)/HA presented superior performance consistently. The trend was as follows: Rh(1)/HA (587 °C) > Rh(0.5)/HA (613 °C) > Rh(2)/HA (637 °C). Besides, the best two compositions (Rh(0.5)/HA and Rh(1)/HA) considered for stability, which were sustained during 30 hours experiment at 700 °C. The consistent CH_4 conversion values were around 68% over Rh(0.5)/HA and 76% over the Rh(1)/HA catalyst. The improved catalytic activity of Rh(1)/HA is attributed to its Rh distribution and improved textural properties. This specific characteristic is due to the initial distribution of the Rh species in the freshly calcined sample. The temperature programmed reduction studies showed that Rh distribution at different HA sites is independent of Rh

5.3 Partial Oxidation of Methane

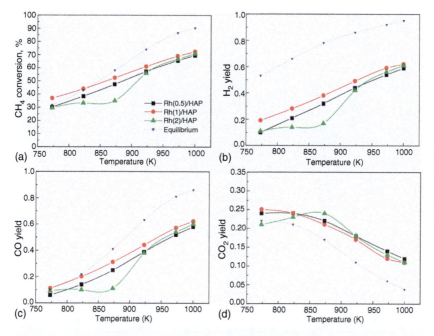

Figure 5.5 POM performance of Rh(x)/HA catalysts with different composition at various reaction temperatures. (a) CH_4 conversion and (b) H_2, (c) CO, and (d) CO_2 yields. Reaction conditions: 19 200 cm³ CH_4 g⁻¹ h⁻¹; W = 0.250 g. Gas feed: 10%CH_4/5%O_2/N_2. Source: Boukha et al. [69]. Reproduced with permission of Elsevier.

content. Rh(1)/HA exhibited the lowest contribution of 17%, which indicates its less prominence of Rh incorporation into the HA structure, compared with other compositions. Hence, Rh(1)/HA is more stable to protect its active surfaces.

In the work of Jun et al. [70], their results demonstrated the excellent performance of HA-based catalysts in POM. They investigated Ni/HA catalysts, containing Ni, NiO, and HA as the major components, and other calcium phosphates as the minor phases, and investigated the POM reaction mechanism by a pulse experiment. The catalyst, having the targeted composition of $Ca_{9.5}Ni_{2.5}(PO_4)_6$, was prepared by the coprecipitation of a mixture of calcium nitrate, nickel nitrate, and dibasic ammonium phosphate at pH 10–11 under rigorous stirring followed by drying at 110 °C and air calcination at 800 °C for 2 hours. It is worth noting that no reduced phase was detected with X-ray diffraction (XRD), but the major components were HA and NiO. The feed gas ($CH_4 + O_2 + Ar$) acted as a reducing agent too and hence converted NiO to metallic form upon POM. It has been observed that during the reaction, some Ni substituted Ca-cations in the HA structure. The size of Ni particles ranges between few nanometers (major quantity) and dozen nanometers (minor quantity). The investigated composition produced only CO_2 and H_2O below 600 °C, but exceptional syngas yield was obtained above 650 °C, indicating the reducibility of Ni^{2+} above 650 °C under the atmosphere of the feed gas.

Then POM reaction mechanism was investigated by measuring temperature profiles in a pulse experiment; the results from $CH_4-O_2-CH_4$ and CH_4/O_2 pulses

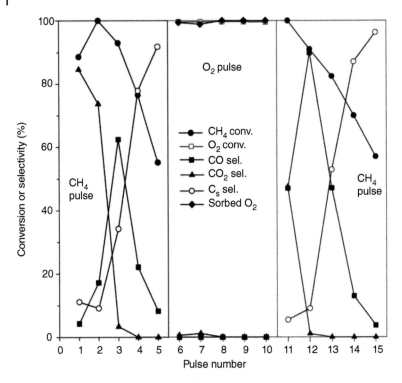

Figure 5.6 Results for sequential pulses of CH_4, O_2, and CH_4 over unreduced $Ca_{9.5}Ni_{2.5}(PO_4)_6$ at 750 °C. Source: Jun et al. [70]. Reproduced with permission of Elsevier.

over reduced and unreduced catalysts were compared. A 30 mg Ni/HA was loaded to the reactor and studied POM with pulses of CH_4, O_2, and/or CH_4/O_2, after cleaning the residual gases by He flow. A 0.25 cm³ (STP) gas pulse was injected at an interval of 10 minutes. The pulse experiment showed a rapid decrease in the CH_4 conversion and CO_2 selectivity from the third pulse. CO_2 selectivity touched 0% by the fourth pulse, while CO selectivity increased sharply to 60% at the third pulse and decreased rapidly afterward (Figure 5.6). It demonstrates the catalyst reduction by reaction between CH_4 and lattice oxygen, resulting in CO_2 formation. Initial high lattice oxygen consumption leaves small quantity of lattice oxygen and hence increases the CO selectivity by preventing deep carbon oxidation. As lattice oxygen consumed for the reduction of Ni^{2+}, carbon deposition occurred causing catalyst deactivation and thus decreasing methane conversion. The following oxygen pulses did not produce significant CO and CO_2, which is irrespective of O_2 consumption. It might be attributed to the stable form of as-produced carbon and to the reoxidation of reduced Ni. Hence, the authors suspect an unknown species other than Ni and as-produced carbon, which is capable to react with injected O_2 [70].

The results presented in Figure 5.6 proved that the POM reaction carried out over HA-based catalyst follows pyrolysis mechanism on the reduced Ni atoms. The first and second pulse gave a high CO_2 selectivity, which sharply decreased afterward. This shows the utilization of lattice oxygen for Ni reduction by CH_4 (Figure 5.6),

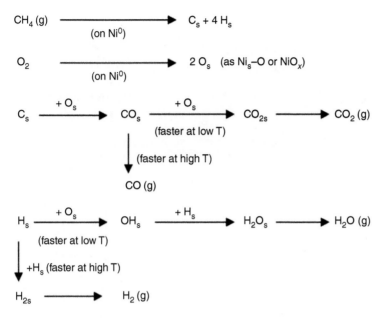

Figure 5.7 POM reaction route over HA catalyst. Source: Jun et al. [70]. Reproduced with permission of Elsevier.

while CO selectivity maintained ~100% from the fourth pulse, ensuring a pyrolysis mechanism operated over reduced Ni-phase.

Figure 5.7 exhibits the proposed mechanism for POM over $Ca_{9.5}Ni_{2.5}(PO_4)_6$ catalyst. The reduced Ni-phase hosts the dissociation of CH_4 to surface carbon and hydrogen. The surface carbon reacts with lattice oxygen to produce surface CO, which will be released as gaseous CO or oxidize to CO_2. It is worth to note that the efficiency of CH_4 conversion depends on the gaseous oxygen supply and high temperature. The completely Ni^0 phase is less active due to the gaseous O_2 consumption, and, thus, both Ni^0 and Ni^{2+} phases are essential for high activity and selectivity for the investigated catalytic system. The POM reaction over Ni catalysts on other supports also follows, basically, similar methane activation and dissociation mechanism. Jin et al. [71] investigated POM mechanism with a pulse experiment over Ni/α-Al_2O_3 catalysts. The O_2 in the feed converts Ni^0 to NiO, and a complete oxidation of CH_4 occurs at the sites as well, which results in a sharp temperature increase. At the critical temperature, CH_4 reduces NiO to Ni^0, producing CO_2 and H_2O. Besides, the dissociative activation of methane forms surface Ni···C species and H_2. The RDS is the formation of CO by reaction between Ni···C species and $Ni^{\delta+}···O^{\delta-}$ species derived from O_2 activation.

5.4 Acetone to Methyl Isobutyl Ketone

Methyl isobutyl ketone (MIBK or 4-methyl-2-pentanone) is a relevant material in various industries such as paint, varnishes, pesticides, etc. It is also extensively

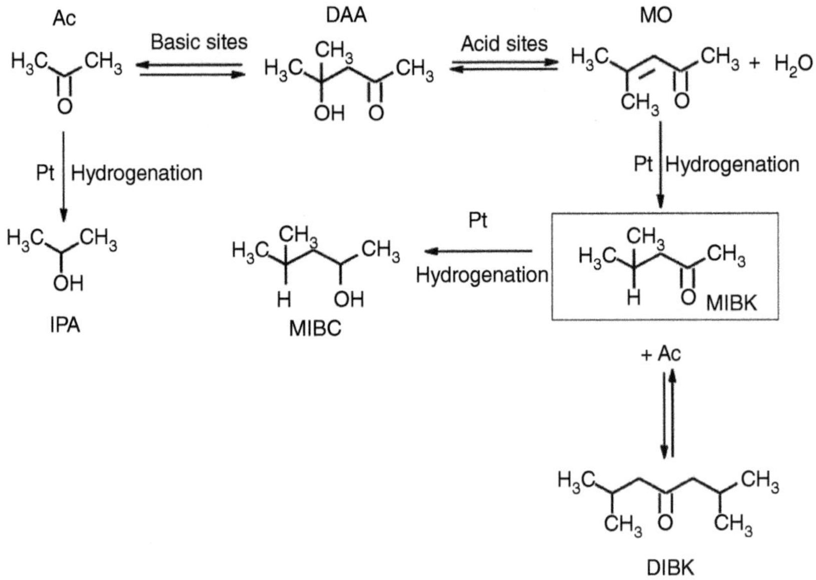

Figure 5.8 Simplified scheme of acetone conversion in the presence of H_2 over Pt/HA catalyst. Source: Takarroumt et al. [72]. Reproduced with permission from Elsevier.

applied in zirconium extraction [72]. Presently, the global MIBK demand has surpassed more than 366 000 t per year. Its production includes a three-stage homogeneous catalysis process. The process is faced with many challenges such as operation pressure, production of unwanted salt wastes, and separation. Hence, a simplified economic route for MIBK synthesis by condensation of acetone under hydrogen stream over suitable catalysts has been developed [73]. It involves a multifunctional heterogeneous catalyst with acid–base characteristic with potential to catalyze three reactions under a suitable condition. In addition, the heterogeneous catalysts are easier to be recovered and regenerated, in comparison with homogeneous catalysts [72]. Irrespective of the series of catalysts investigated for MIBK synthesis, HA-based catalysts exhibited unique features. The presence of tunable basic sites and the exceptional ion exchange properties of HA make them worthy candidate for MIBK synthesis, which is not yet extensively investigated. Takarroumt et al. [72] synthesized HA-supported Pt (0–2.40 wt%) catalysts for MIBK production. It is well known that Pt metal has a good track record of great performance in various hydrogenation reactions [74].

The simplified visualization of MIBK formation starting from acetone is shown in Figure 5.8. In the reaction scheme, Pt/HA catalysts can catalyze either the formation of isopropyl alcohol (IPA) via the direct hydrogenation of acetone or the formation of MIBK via mesityl oxide (MO) as intermediate by providing hydrogenating sites. Among the prepared catalysts with different Pt loadings, the one containing 0.5 wt% Pt showed the highest MIBK selectivity of 74% with a yield of 20% at 150 °C. It was found that the higher Pt loading resulted in the production of less quantity of MO, which is the major intermediate in the MIBK formation. Hence,

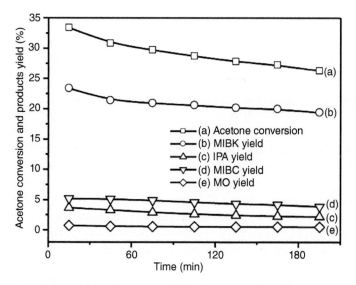

Figure 5.9 Acetone conversion and products yield as a function of reaction versus time over 0.5 wt% Pt/HA. Reaction conditions: 150 °C, molar ratio of H_2/acetone = 2/1, and atmospheric pressure. Source: Takarroumt et al. [72]. Reproduced with permission from Elsevier.

MO hydrogenation at 0.5 wt% Pt loading is kinetically efficient. Thus, the results recommend that tuning catalyst characteristics for MO yield enhancement could lead to improved MIBK yield. The catalytic performance of 0.5 wt% Pt-loaded HA is exhibited in Figure 5.9. A continuous catalyst deactivation occurred, which is attributed to the degradation of catalyst by water and the deposition of carbonaceous by-products. The deactivation mechanism is revealed through characterization of used catalysts.

The acetone condensation to MIBK could follow either enol or enolate mechanism depending on the acidity/basicity of the catalyst support, while the strength of acid/base sites determines the reaction rate [75]. Generally, the correlation study between the catalytic activity and the support features such as acidity or basicity is difficult because of the alteration on the physicochemical features and structures during synthesis. It is believed that the coprecipitation and post-synthetic ion exchange on HA-based catalysts could not alter its structural features. Ho et al. [73] investigated MIBK formation mechanism during acetone coupling over HA-based catalyst; using cation exchange technique, Mg, Ca, Sr, Ba, Cd, and Pb could be deposited on HA surface while keeping intact the bulk HA structure and properties. The MIBK formation rate is 1.5 order at low acetone partial pressure (<0.5 kPa) and high reaction temperature (150 °C) [73]. Hence, they assumed that the RDS may be either the coupling of two acetone-derived surface species or a reaction with an intermediate formed from the bimolecular coupling of two acetone molecules [73]. The proposed reaction pathway is shown in Figure 5.10. This suggests that diacetone alcohol (DAA) is formed via aldol addition, which is then dehydrated into MO, which is thereafter hydrogenated to MIBK. The experimental results reveal that

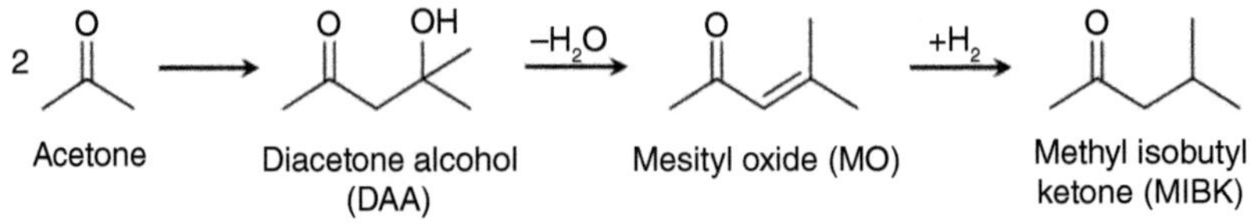

Figure 5.10 Acetone condensation to methyl isobutyl ketone (MIBK). Source: Ho et al. [73]. Reproduced with permission of Elsevier.

aldol addition is a reversible and fast reaction, while DAA dehydration is the RDS. This RDS is controlled by the basic sites of the HA support. The exchanged cations that linked to surface O-atoms regulate the acetone to MO condensation. Hence, the basicity of surface O-atoms is relevant. This reaction pathway was then verified by studying the MIBK formation from different feeding compositions. The reaction of DAA with HA-based catalysts ensured that retroaldol reaction rate is much faster than that of DAA dehydration, indicating a pseudoequilibrium between acetone and DAA. When the reactor was fed with MO, H_2O, and H_2 mixture, MIBK yield reached 99%, showing a rapid hydrogenation and slow DAA dehydration. This observation was validated for various HA-based catalysts studied. Talwalkar et al. [76] studied the single-step production of MIBK in a batch reactor over bifunctional commercial ion exchange resin (Amberlyst CH28) and observed a similar mechanism as discussed previously. The acetone adsorbed on two adjacent active acid sites and desorbed as a dimer. This desorbed dimer dehydrated to give diadsorbed MO, which is then converted to MIBK. They observed that the reaction is a pseudo-zero order with respect to the H_2 concentration in their experimented range.

Among the different catalysts investigated, Sr^{2+}-exchanged HA exhibited the highest performance as shown in Figure 5.11, which was further confirmed by density functional theory (DFT) calculations. This figure showed an interesting correlation of MIBK formation rate with basic sites density (except for Ba), induced by the insertion of bivalent cations studied on the HA surface. In fact, the electronegativity of the cation is considered as the descriptor of basicity. Thus, Sr^{2+}-exchanged HA with the lowest electronegativity leads to the highest basicity, corresponding to the highest MIBK formation rate. However, despite the high basicity of Ba-HA compared with that of Sr-HA, Ba-HA is three times less active. This distinctive activity may be indicating that a suitable balance acid and base strength is necessary for MIBK formation. However, it is not evidently proved, and more investigation is needed to claim this point. Ho et al. [73] deeply researched the physical characteristic variation of HA upon addition of various metals. The surface area of investigated materials ranged between 87 and 92 m^2 g^{-1}, except Pb-HA with 71 m^2 g^{-1}. It is evident that the mass difference between Ca and Pb causes this surface area reduction. It is assumed that Pb^{2+} modifies the HA surface via dissolution/reprecipitation mechanism, which results in the least surface area of 71 m^2 g^{-1}. Authors observed that the XRD patterns provide peaks for nanocrystalline apatite structure only, and no 2θ shift occurred for any doped samples. It evidenced that the bulk HA structure is intact in the presence of metal cations, without structure expansion or contraction. Besides, the metal oxide peaks were also absent, which

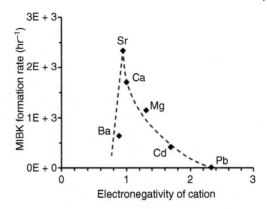

Figure 5.11 Effect of cation substitution on MIBK formation rate normalized by number of acid sites. Conditions: $P_{Acetone} = 15$ kPa, $P_{H2} = 5$ kPa, $T = 100\,°C$, catalyst mass $= 0.01$ g. Source: Ho et al. [73]. Reproduced with permission of Elsevier.

may be attributed to small nanoparticle size or simply to the formation of isolated metal cations species fixed on the HA surface to replace calcium cations. In addition, the degree of exchange in HAP samples is dependent on the exchanged metal and decreases in the order of $Pb^{2+} > Cd^{2+} > Sr^{2+} > Mg^{2+} > Ba^{2+}$ [77].

5.5 Ethanol Coupling Reaction

Upgradation of bioethanol to valuable chemicals (e.g. butanol) is a remarkable technology nowadays. Butanol has attracted a wider interest, thanks to its various applications including paint, chemicals, and polymer industry [78]. In addition, it is a better biofuel than ethanol because of its higher energy density (29.2 MJ l^{-1}), lower miscibility with water, and lower corrosivity in comparison with those of ethanol [79]. Due to the so-mentioned properties, butanol has been utilized in gasoline blending at higher concentrations than ethanol. Besides, its energy content is comparable with petroleum and hence resulted in improved mileage [79, 80]. The oxo process, by hydroformylation, and fermentation are two traditional methods to synthesize *n*-butanol [81], even though the first is more prevalent in petroleum industry. However, the oxo process requires a huge quantity of energy to reform natural gas (methane) with steam to produce syngas, which is required for butanol synthesis via this process.

Irrespective of traditional approaches [82], ethanol coupling is the trending method for butanol preparation over a variety of catalysts [83, 84] such as ruthenium complexes [78], solid base metal oxides (MgO) [85, 86], mixed metal oxides (Mg – Al) [87] Cu/CeO$_2$ [88], zeolites [89], and carbonated hydroxyapatite (HA-CO$_2$) [90]. HA-based catalyst exhibited remarkable performance in ethanol coupling with good activity (10–20%) and high butanol selectivity (>70%) at 300 °C, which is significantly higher than those reported for other catalysts [78, 91, 92]. This high catalytic performance was basically attributed to the unique coexistence of both acid and base sites of HA-based materials.

In reality, as previously detailed in Chapter 3, the Ca/P molar ratio in HA is the key factor to tune the density of basic (and acid) sites, and it can be used to improve and optimize the catalytic performance of HA-based catalysts and butanol selectivity in ethanol coupling reaction [91]. Previous studies [93, 94] have demonstrated that nonstoichiometric HA is an excellent candidate for the reaction due to selective conversion of ethanol to *n*-butanol. The work of Scalbert et al. [95] on the condensation of ethanol to butanol over a commercial HA catalyst in the temperature range of 350–410 °C highlighted the unique characteristics of HA with bifunctional active sites. By analyzing the thermodynamic and kinetic data, including the concentration of water and dihydrogen formed during the conversion, they concluded that the alcohol condensation mechanism(s) taking place over bifunctional solids (containing both metallic and basic sites) such as HA catalyst at lower temperatures would be fundamentally different from that over basic oxides (i.e. MgO, CuO, and MnO) and at high temperatures. Tsuchida et al. [96] used HA catalysts with different Ca/P molar ratios for ethanol conversion. They observed that increasing the Ca/P ratio increased the concentration of basic sites and reduced the density of acid sites. This favored ethanol dehydrogenation, and, consequently, the selectivity to acetaldehyde increased. In contrast, the lowest Ca/P ratio corresponding to the highest density of acid sites was more selective to ethylene due to ethanol dehydration catalyzed by these sites. The findings strongly indicated that there should be a balance between both acidic and basic sites for the HA-based catalyst to produce an optimal ethanol conversion and butanol selectivity.

The reactive phases of HA for ethanol coupling and its mechanism is still a topic of investigation. It is generally agreed that ethanol coupling follows Guerbet mechanism [91, 93]. At the same time, additional mechanism over different catalysts was also proposed. Yang and Meng [89] proposed a direct coupling to form butanol over zeolite catalyst, which was further supported by others [97, 98]. According to Guerbet coupling, initially ethanol dehydrogenates to acetaldehyde, which is then transformed to crotonaldehyde by aldol self-condensation reaction. Finally, crotonaldehyde is hydrogenated to butanol. Meanwhile, Scalbert et al. [95] reported that aldol condensation of acetaldehyde is not kinetically favored above 350 °C. Understanding the structural features of HA is relevant for proposing a mechanism. Hence, the acid–base density on HA can be tuned by synthesis parameters, which strongly contribute toward the catalytic performance. The Ca−OH sites of HA promote ethanol dehydrogenation, while CaO/PO_4^{3-} pairs trigger aldol condensation [78]. Diallo Garcia et al. [99] found that the most active sites that participate in this catalysis are the strong basic OH^- groups. Hence, the acetylene adsorption studies showed that the surface O^{2-} anions of PO_4^{3-} groups are weakly basic and remain intact. Considering the limited data available on kinetic and mechanism studies, there is room for investigation. Hanspal et al. [92] compared the nature, density, and strength of acid–base sites on HA with that of different materials such as MgO and $Sr_3(PO_4)_2$. Among the investigated catalysts, the HA sorbed 2.5 mmol m^{-2} CO_2, which is much higher than that sorbed on other materials, indicating the high base site density of HA. Figure 5.12 shows the total rate of feed conversion and product formation calculated for the ethanol coupling reaction performed at 250 °C. It exhibited a first order at low initial ethanol feed concentration and zero order at higher concentrations of ethanol. The result was further supported by other researchers [78].

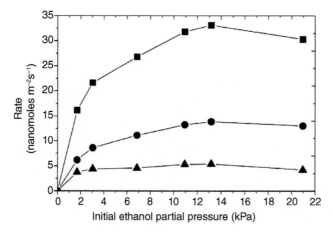

Figure 5.12 Total rate of ethanol conversion (■) and production rate of butanol (●) and acetaldehyde (▲) as a function of the initial ethanol concentration during steady-state conversion of ethanol over HA at 250 °C. Source: Hanspal et al. [92]. Reproduced with permission of Elsevier.

Ho et al. [78] investigated the ethanol condensation mechanism over HA catalyst in a partial pressure range of 3.5–9.4 kPa ethanol and 0.055–0.12 kPa acetaldehyde. They observed an enhancement in the butanol formation rate when acetaldehyde was added into the reaction stream and concluded that the coupling is autocatalytic. In their experimented conditions, condensation reaction followed Guerbet-type mechanism with the following sequence: dehydrogenation, aldol condensation, and hydrogen transfer reactions.

According to Kibby et al. [100] and Tsuchida et al. [91], the preliminary step in ethanol coupling is the acetaldehyde formation by ethanol dehydrogenation. Initially, ethanol is adsorbed on Ca—O sites to form surface ethoxide and hydroxyl species. Later, α-hydrogen atom interacts with surface oxygen and transfers electrons to Ca—O bond to form acetaldehyde. The desorbed acetaldehyde could migrate to nearby CaO/PO_4^{3-} sites, where aldol condensation occurs. In this stage, upon adsorption of acetaldehyde on the respective phase, basic oxygen abstracts α-hydrogen and results in enolate species formation. These enolate species interact with slightly acidic phosphate group and accelerate aldol condensation. The interaction of enolate species with acetaldehyde carbonyl group results in acetaldol, which dehydrates to crotonaldehyde. This latter is then hydrogenated to butanol.

Figure 5.13 simplifies the overall ethanol coupling mechanism. Briefly, dehydrogenation of ethanol occurs on Ca—O sites to produce acetaldehyde, which is then converted to crotonaldehyde on CaO/PO_4^{3-} phase. Finally, the butanol formation occurred via a hydrogen transfer between ethanol and crotonaldehyde. The rate of butanol formation is derived by integrating rate equation for aldol condensation of acetaldehyde to crotonaldehyde. The derived equation is shown as follows [78]:

$$r_{BuOH,app} = \frac{1}{L'} \int_0^{L'} r_{BuOH} dL = \frac{1}{L'} \frac{k_2 K_{AA,2}}{1 + K_{EtOH,2} P_{EtOH}} \int_0^{L'} P_{AA} dL \qquad (5.28)$$

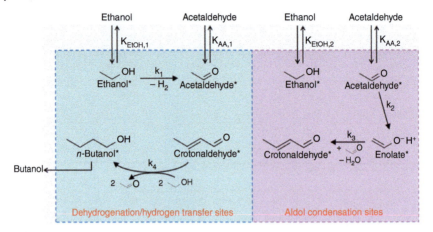

Figure 5.13 Ethanol coupling reaction into butanol: Reaction pathway of Guerbet coupling over HA. Source: Ho et al. [78]. Reproduced with permission of American Chemical Society.

where $r_{BuOH,app}$ is the apparent butanol formation rate, r_{BuOH} is the local butanol formation rate at a distance L from the top of the catalyst bed, and L' is the total length of the bed.

In Guerbet ethanol coupling, the potential role of hydrogen atoms produced during alcohol dehydrogenation on butanol yield remains unclear. The suspected possibilities were either desorption as H_2 after a surface migration and recombination or consumption for hydrogenation in C—C coupling [83]. Hanspal et al. [92] investigated it by co-feeding H_2 with ethanol stream at steady-state reaction at 250 °C. They observed no effect on the feed conversion and butanol selectivity even at high H_2 concentration of >90%. At experimented parameter range, the H_2-molecules do not play any kinetic or thermodynamic role in ethanol coupling over HA catalysts. It may indicate that the H-atoms contributing toward Guerbet coupling reaction are not sourced by H_2-molecules [24]. They also investigated the contribution of by-product water in ethanol coupling to butanol. The presence of co-feed water prevented butanol formation and reduced acetaldehyde formation by 50%. Interestingly, the butanol and acetaldehyde production rate restored to 70% when the water was removed from feed flow. This proves the reversible interaction of water with HA surface. The presence of water could have shifted the equilibrium of butanol formation and/or competitively adsorbing on catalytically active HA sites. This result further confirmed the involvement of PO_4^{3-} for C—C bond formation [92].

5.6 Ethanol to Gasoline

Production of gasoline from an abundance of plant-derived ethanol is an important pathway to reduce global CO_2 emissions from fossil fuel consumption, particularly in the transportation sector. As classified in the Kyoto Protocol, the use of biofuels such as bioethanol as a gasoline substitute does not count as a source of new carbon dioxide emission. The conversion of ethanol into gasoline-family hydrocarbons (C_4–C_{12}) is done by diluting ethanol in an inert gas flow. The gas mixture then passes through a solid catalyst at temperatures between 300 and 500 °C [101]. HA is considered as a right candidate for enhancing the rate of this reaction due to its bifunctional catalytic properties. HA contains both acid and basic sites that promote dehydration and dehydrogenation of ethanol, respectively [96].

Tsuchida et al. [102] seemed to be the first to discover the use of HA as a catalyst in gasoline production. They investigated the mechanism of gasoline synthesis from bioethanol by employing combinations of five different alcohols with ethanol to produce C_5 to C_{10} hydrocarbons in one step over a highly active nonstoichiometric HA catalyst. Their results suggested reactions to propagate carbon numbers occurred by the Guerbet reaction between largely normal alcohols, forming normal and branched alcohols. In contrast, branched aldehydes and olefins were formed by dehydrogenation and dehydration, respectively, of mainly branched alcohols. For example, the reactions between ethanol and 1-butanol are as follows:

$$C_2H_5OH + 1\text{-}C_4H_9OH \rightarrow 1\text{-}C_6H_{13}OH + H_2O \tag{5.29}$$

$$1\text{-}C_4H_9OH + C_2H_5OH \rightarrow C_2H_5CH(C_2H_5)CH_2OH + H_2O \tag{5.30}$$

$$1\text{-}C_4H_9OH + 1\text{-}C_4H_9OH \rightarrow C_4H_9CH(C_2H_5)CH_2OH + H_2O \tag{5.31}$$

$$1\text{-}C_4H_9OH \rightarrow C_3H_7CHO + H_2 \tag{5.32}$$

The main products of the previously mentioned reactions, in order of selectivity (expressed as C wt%), were 1-hexanol (30.9%), 2-ethyl-1-butanol (15.2%), 2-ethyl-1-hexanol (14.7%), and butyraldehyde (6.4%).

In addition, Eagan et al. [103] recently described the kinetics of oligomerization of ethanol into higher alcohols. They studied the reactions of ethanol, 1-butanol, and water–ethanol feeds at 350 °C, 1 bar, and 3–20% conversion over HA catalyst that were obtained. They developed a kinetic model that describes coupling to higher alcohols, condensation to "interrupted coupling products" such as aromatics and diene, and direct unimolecular dehydration to monoenes. The modified step-growth kinetic model consists of nine kinetic parameters describing 647 reactions among 185 alcohols. The model described alcohol distributions at this conversion by using Eq. (5.33):

$$y_n = (1-\alpha)\alpha^{n-1} \tag{5.33}$$

where y_n is the mole fraction of oligomer n and α is the alcohol oligomerization probability, which is a function of ethanol conversion (X). The alcohol distribution (α) was obtained from the slope by plotting $\log(y_n)$ versus n. They also analyzed the kinetic data and found that the formed water shows an inhibition effect that leads to decreases in the rates of reaction and alcohol selectivity as ethanol conversion increases. This result shows a downside of conventional HA catalyst during ethanol conversion.

To improve the performance of HA catalyst, Quintana et al. [101] modified HA by chemical precipitation method in basic media using an alkali-free system to produce carbonate hydroxyapatite (CO_3HA). The yield of ethanol conversion into gasoline (C_4–C_{18+}) achieved 97% under optimum conditions: reaction temperature (500 °C) and modified residence time $W/F_{Ethanol}$ (40 g h mol^{-1}) with the presence of CO_3HA catalyst. They confirmed that above 400 °C, the ethanol conversion is controlled by diffusion. The previously mentioned results highlighted the importance of developing more active HA catalysts to advance this alcohol oligomerization technology.

5.7 Glycerol to Lactic Acid

Lactic acid (LA) is one of the most useful chemicals that can be used as a precursor of green solvents such as ethyl lactate and that of polylactic acid, which is increasingly employed to produce biodegradable packaging. However, LA's conventional production route via fermentative process takes considerable cost and time, limiting its use. The catalytic route to the production of LA from glycerol (GLY), which is a by-product in the biodiesel manufacturing process, is a sustainable alternative to conventional fermentation due to its large availability and productivity.

The conversion of glycerol to LA is commonly performed via hydrothermal methods without or with solid catalyst use. The synthesis of LA by hydrothermal conversion of GLY has been reported by some researchers [104]. Shen et al. [104] examined the effectiveness of alkaline as homogenous catalyst for hydrothermal treatment of glycerin at 300 °C. They employed eight alkali metals, alkaline earth metals, and aluminum hydroxide solutions and found that KOH was superior among other hydroxides since it worked at a lower concentration or within a shorter reaction time to achieve the same LA yield. Hisanori et al. [105] studied the hydrothermal decomposition of GLY with alkali. They found that LA yield of 90% was obtained at 300 °C with glycerol and NaOH concentrations of 0.33 and 1.25 mol l^{-1}, respectively. For the hydrothermal process without solid catalyst addition, low glycerol concentration c. 0.3 mol l^{-1} and high reaction temperature ca. 300 °C become the barriers for industrial application. To overcome the problems, the noble metal catalysts such as $AuPt/TiO_2$, $AuPt/CeO_2$, or Pt/Sn-MFI have been used to decrease the operating temperature to ca. 100 °C and improve the GLY conversion and LA yield [106, 107]. However, considering the price of noble metal catalysts, the investigation of relatively low-cost solid catalysts for glycerol conversion is highly desirable.

Previous researchers [108, 109] reported that metallic copper-based catalysts exhibited catalytic activities for synthesizing LA from GLY in an NaOH aqueous solution. By applying NaOH/GLY mole ratio of 1.1 : 1 at 240 °C for 6 hours, the conversions of GLY (1 mol l^{-1}) over the Cu/SiO_2, CuO/Al_2O_3, and Cu_2O catalysts reached 75.2, 97.8, and 93.6%, respectively [108]. The selectivity into LA reached 79.7, 78.6, and 78.1%, respectively. Although the maximum yield of LA over the metallic copper-based catalyst was less than 60%, the hydrothermal conversion of GLY to LA catalyzed by Cu-based catalysts has room to improve. Yin et al.

[110] investigated HA-supported Cu catalysts prepared by the incipient wetness impregnation method that was used to catalyze the hydrothermal conversion of GLY to LA in NaOH aqueous solution. HA with high basicity and good thermal stability was selected as the support because it could give basic conditions that are needed to convert GLY. The GLY conversion and LA selectivity of 91 and 90% were successfully obtained over Cu(16)/HA (weight ratio of Cu to HA is 16:100) catalyst at 230 °C for 2 hours with the initial concentration of GLY and NaOH of 1.0 and 1.1 mol l^{-1}, respectively. The result demonstrated that HA-supported Cu catalysts have high catalytic activity for the hydrothermal conversion of GLY to LA at a relatively low reaction temperature and high initial glycerol concentration. Yin et al. [110] also performed a kinetic analysis of the conversion of GLY to LA by expressing the reaction rate as a function of the exponential of GLY and NaOH concentration (power law) as shown in Eq. (5.34). They found that the presence of Cu(16)/HA catalyst decreased the intrinsic reaction activation energy, E_a, from 174.0 kJ mol^{-1} (without catalyst) [21] to 117.2 kJ mol^{-1}.

$$-r_A = \frac{dn_A}{m_{cat} dt} = k C_A^a C_B^b \tag{5.34}$$

where m_{cat} is the catalyst loading; n_A is the moles of reacted glycerol; C_A and C_B are the concentrations of glycerol and NaOH, respectively; k is the rate constant; and a and b are the reaction orders for glycerol and NaOH.

A recent study by Shen et al. [111] employed HA and Pd/HA catalyst to convert GLY to LA. The reaction was carried out in 100 ml reaction volume of glycerol (2 mol l^{-1}) and NaOH (2.2 mol l^{-1}) at 230 °C for 90 minutes batch time. In the presence of HA alone, GLY conversion only achieved 16%, indicating that the presence of strong base active sites is not sufficient to activate the hydroxyl groups of GLY. On the other hand, when Pd/HA (3 wt% Pd) was employed as catalyst, GLY conversion and LA selectivity reached 99% and 95%, respectively. These results highlighted that the interactions between the HA (support) and the Pd species (active phase) play a major role to promote GLY transformation reactions. Shen et al. [111] also proposed a reaction pathway for this reaction. In addition to the active phase dispersion, different reaction parameters (reaction temperature, glycerol concentration, NaOH : glycerol molar ratio, and glycerol/metal intake) significantly influence both activity and selectivity of the reaction. Perfect selection of active phase is equally important as the interactions between the support and the active phase for high GLY conversion to LA. Thus, the GLY oxidation to glyceraldehyde or dihydroxyacetone is the RDS of the glycerol conversion. The formation of oxalic-, formic-, and acetic acid proves the formation of the by-products, resulting from undesired C—C splitting reactions. Future studies should explore the application of HA as support for metal catalysts to solve the previously mentioned problems.

5.8 Benzene to Phenol

Phenol (carbolic acid) is an essential precursor for various petrochemicals, agrochemicals, and pharmaceutical products. Phenol is globally produced by the multistage cumene process that consumes much energy and generates a considerable amount of undesirable products. Thus, a simple yet effective method to synthesize phenol has remained highly desirable. The direct hydroxylation of benzene into phenol in both the liquid and gas phases is an attractive approach to the conventional one. This one-step process employs mild oxidants such as oxygen, nitrous oxide, hydrogen peroxide, and a noncorrosive catalyst.

Liptikova et al. [112] studied the direct conversion of benzene to phenol in the gas phase over HA-supported Cu catalysts. HA support was prepared by precipitation method, while Cu deposition was performed by ion exchange or impregnation. The experiments were conducted in a fixed-bed reactor at atmospheric pressure and reaction temperature of 450 °C in the presence of ammonia. The combination of Ca and Cu ions in the cation part of HA resulted in high selectivity, ca. 97%, of phenol formation at about 3.5% conversion of benzene. They also analyzed the outlet gases continuously by online mass spectrometry. They proposed the following reactions to build reaction schemes of the hydroxylation of benzene to phenol:

$$4NH_3 + 5O_2 \rightarrow 4NO + 6H_2O \tag{5.35}$$

$$4NH_3 + 4O_2 \rightarrow 2N_2O + 6H_2O \tag{5.36}$$

$$4NH_3 + 3O_2 \rightarrow 2N_2 + 6H_2O \tag{5.37}$$

$$4NH_3 + 6NO \rightarrow 5N_2 + 6H_2O \tag{5.38}$$

$$N_2O + * \rightarrow O^* + N_2 \tag{5.39}$$

$$C_6H_6 + O^* \rightarrow C_6H_5OH^* \tag{5.40}$$

$$C_6H_5OH \rightarrow C_6H_5OH + * \tag{5.41}$$

The process consists of the sequential reaction of (i) oxidation of ammonia to nitrous oxides and nitrogen (Eqs. (5.35)–(5.38)), (ii) the formation of reactive oxygen species by the decomposition of N_2O (Eq. (5.39)), and (iii) the direct attack of benzene proceeds by reactive oxygen species. The asterisks represent the active centers of the catalyst. As seen in Eqs. (5.39)–(5.41), the decomposition of N_2O on the active sites of HA catalyst leads to the formation of active oxygen species, which directly oxidize benzene molecules to produce phenol. A further study by Liptikova et al. [113] was performed to investigate the effect of various reaction parameters on benzene conversion to phenol. Their results showed that the reaction temperature and the concentration of oxygen and ammonium hydroxide in the feed influenced the selectivity of benzene oxidation to phenol. On the other hand, when the temperature of catalyst calcination was above 580 °C, the phenol yields sharply decreased. Their results also showed that the used HA catalyst experienced no significant deactivation during the first period (one to three hours) of reaction.

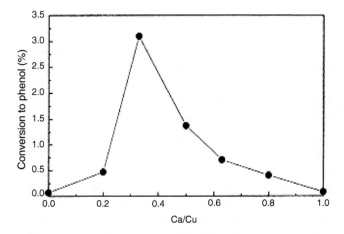

Figure 5.14 Effect of the Ca/Cu molar ratio in the Cu/HA catalysts on the phenol yield from direct benzene hydroxylation. Source: Liptáková et al. [112]. Reproduced with permission of Elsevier.

Table 5.1 Catalytic activities for direct phenol production from benzene by Fe-g-C_3N_4-based catalysts with (+) or without (−) visible light ($\lambda > 420$ nm).

Catalyst	hv	t (h)	Benzene conv. (%)	H_2O_2 conv. (%)	H_2O_2 sel. (%)[a]
Fe-g-C_3N_4	−	4	1.8	99.1	3.2
Fe-g-C_3N_4	+	4	4.8	100	8.3
Fe-g-C_3N_4/SBA-15	−	4	6.7	80.9	14.2
Fe-g-C_3N_4/SBA-15	+	4	11.9	97.6	20.7

a) Moles of produced phenol/mol of reacted $H_2O_2 \times 100$.
Source: Chen et al. [116]. Reproduced with permission of American Chemical Society.

Even though a high selectivity of phenol formation has been obtained via direct hydroxylation, a low conversion of benzene is still a challenge (see Figure 5.14). The low activity is noticed not only on HA-based catalysts but also on other supports. Niwa et al. investigated a one-step conversion of benzene to phenol with a palladium membrane and observed excellent selectivity of 80–97% below 250 °C. However, the benzene conversion observed in this study was low (2–16%) [114]. Semiconductor photocatalytic functions of Fe-based catalysts are promising in organic synthesis. Porous graphitic carbon nitride (g-C_3N_4) is evidently efficient to adsorb benzene and catalyze both Friedel–Crafts reactions and phenol synthesis [115]. Chen et al. [116] experimented Fe-g-C_3N_4/SBA-15 for the conversion of benzene to phenol with H_2O_2 even without the aid of strong acids or alkaline promoters. The results are not promising in terms of activity as shown in Table 5.1. Therefore, a deep understanding of the kinetics of direct conversion of benzene to phenol using HA catalyst, particularly on the determination of the RDS, should be a focus of future studies.

5.9 Transesterification

Due to the increasing environmental concerns arising from excessive exploitation of fossil fuels, it is urgent to discover feasible substitutes for petroleum-derived fuels, especially green and sustainable ones. In such a situation, biodiesel or fatty acid methyl ester (FAME) is getting much interest as an alternative fuel due to its nontoxic nature, renewability, biodegradability, and blending capability with diesel [117]. Biodiesel is conventionally produced by the catalytic transesterification of vegetable oil with C_1–C_2 alcohols by using homogeneous acids/bases and heterogeneous catalysts. The former shows good catalytic activity but harnesses the environment because it requires intensive neutralization and purification steps [118]. On the other hand, the latter offers some advantages, such as easy separation and purification of the reaction products and the ability to be regenerated and reused several times without any significant loss in its performance [117, 119–121]. Solid base catalysts such as alkaline earth and transition metal oxide, alkali-doped materials, clays, HA, anion exchange resins, etc. can be used to produce biodiesel. However, only a few of those solid catalysts are being employed in commercial applications due to their high production cost. Among them, HA has exceptional features. It can be prepared from various wastes, e.g. animal bones and biomass waste, and can be served as a catalyst as well as support. Table 5.2 shows the activity of HA catalysts, which were prepared from various wastes, in transesterification of bio-oil.

As seen in Table 5.2, natural and doped HA catalysts derived from wastes had excellent activity in transesterification of vegetable oil for biodiesel production at relatively low operating temperatures. These results were achieved by controlling the properties of the catalyst, mainly specific surface area and basicity. The increase in specific surface area, generally by varying calcination temperature during HA production from animal bones or biowastes, leads to enhanced catalytic activity for the bio-oil conversion into biodiesel. For basicity, the Hammett indicator method was used to determine the basic strength (H) and basicity of the samples. Naturally, HA has higher basicity compared to the other types of catalyst such as calcium zincate and zinc hydroxy nitrate [118, 120, 124, 125]. The strong basic sites of HA make it suitable for the conversion of bio-oil. However, the reaction still needs a rich-methanol solution to occur due to kinetics limitation, as explained by Sahu [119]. They investigated the reaction order of transesterification of neem oil. The transesterification reaction is simply given by the following equation:

$$T + 3M \leftrightarrow FE + G \quad (5.42)$$

where T, M, FE, and G present triglyceride, methanol, FAME, and glycerol, respectively. The general rate equation can be described by Eq. (5.43):

$$-dC_T/dt = kC_T^y C_M^z \quad (5.43)$$

where C_T and C_M are the concentration of triglyceride and methanol after time t; y and z are the order of triglyceride and methanol, respectively; and k is the rate

Table 5.2 Biodiesel yield using various HA catalysts in the transesterification of vegetable oil.

Feed oil	Source of HA + doped metal	Catalyst amount (wt%)	Reaction temperature (°C)	Reaction time (h)	Methanol-to-oil (molar ratio)	Biodiesel yield (%)	References
Peanut oil	Goat bone	18	60	4	20:1	94	[117]
Rapeseed oil	Goat bone	18	60	4	20:1	96	[117]
Neem oil	Cooked fish bones + Al	3	70	2	30:1	95	[119]
Refined palm oil	Pig bone + K	8	65	1.5	9:1	96.4	[120]
Soybean oil	Marble slurry	6	65	3	9:1	94	[122]
Palm oil	Cow bones	20	65	4	18:1	96.8	[123]

Source: U.P.M. Ashik.

constant. C_T and C_M are functions of conversion (X) and can be written as follows:

$$C_T = C_{T0}(1-X) \tag{5.44}$$

$$C_M = C_{M0}(\theta - X) \tag{5.45}$$

$$\theta = C_{M0}/C_{T0} \tag{5.46}$$

where C_{T0} and C_{M0} are the initial concentration of reactants and θ is the ratio of C_{M0} to C_{T0}. Equation (5.46) can be written as follows:

$$dX/dt = kC_{T0}^{(y+z+1)}(1-X)^y(\theta - 3X)^z \tag{5.47}$$

They calculated the correlation coefficient by comparing the calculated and predicted rate of conversions of eight combinations of the value of y and z. Their results suggested that the reaction order of transesterification was 2 ($y = 0$ and $z = 2$). In other words, the kinetics of bio-oil conversion mainly depends on the concentration of methanol. Thus, an optimum methanol/oil molar ratio should be maintained during the reaction. However, a study by Chen et al. [120] showed that the yield of biodiesel decreased when increasing the methanol/oil ratio beyond its optimum value. This result was mainly due to the following reasons: (i) the excessive methanol makes it challenging to separate the biodiesel from GLY; (ii) the increase in the molar ratio makes the lower relative concentration of bio-oil, causing the lower reaction rate and changing the reaction equilibrium; and (iii) the catalyst could be diluted or deactivated by excessive amount of methanol. Further study on HA-based catalyst should emphasize the understanding of complex interaction among methanol-triglyceride-catalyst to make heterogeneous transesterification feasible to be applied on large scale.

5.10 Conclusion

Considering the unique physicochemical features, HA is actively considered for various catalytic processes including OCM, POM, acetone to MIBK, and ethanol coupling reaction. The researchers evidenced that two types of active sites are contributing to methane coupling and lead to high catalytic stability and activity when HA is used as catalyst. Besides, the hemolytic C—H cleavage is triggered by the oxygen species on the HA surface. It is validated that the Pb-metal incorporation enhances the HA performance by five times with high C2-selectivty. The $CH_3\cdot$ formation triggers OCM reaction. Eley–Rideal mechanism considered dissociative or molecular type reaction between the gaseous methane and adsorbed oxygen as the rate-limiting step, while Langmuir–Hinshelwood mechanism considered the reaction between adsorbed methane and oxygen species as the rate-limiting step. The literatures concluded that HA and HA-F follow the Langmuir–Hinshelwood mechanism, while Pb-loaded HA and HA-F follow the Eley–Rideal mechanism for methane coupling. POM over HA follows distinctive pyrolysis mechanism over HA, in which CH_4 initially dissociates to produce H_2 and carbon. The as-produced carbon oxidizes to CO upon reacting with surface oxygen species. In a pulse experiment, ∼100% CO selectivity was observed from fourth pulse, ensuring the existence of pyrolysis mechanism. Similar or reasonable performance is presented by HA for other catalytic processes too. An industrially relevant MIBK production via simplified economic route by condensation of acetone under hydrogen stream over HA is also discussed in this chapter. Research proved that the strength of acid/base sites on HA determines the MIBK formation rate. Besides, the acetone partial pressure and reaction temperature play vital roles. Hence, the RDS may be either the coupling of two acetone-derived surface species or a reaction with an intermediate formed from the bimolecular coupling of two acetone molecules. Upgradation of bioethanol to butanol is another remarkable reaction catalyzed by HA. The Ca/P ratio on HA can regulate its density of acid–base sites and hence redirect the reaction. The high acid sites at low Ca/P ratio enhance ethylene selectivity due to ethanol dehydration. Hence, the balance between acid–base sites is vital for optimal ethanol conversion and butanol selectivity. As per Guerbet coupling, the ethanol dehydrogenates to acetaldehyde and then converts to crotonaldehyde, which finally hydrogenated to butanol. The ethanol coupling at 250 °C exhibited first order at low initial ethanol feed concentration and zero order at high ethanol concentration. It can be concluded that the HA-based catalysts either superiorly or competitively perform in the investigated catalytic processes compared with conventional catalyst candidates. At the same time, considering the limited data available on kinetic and mechanism studies of catalytic behavior of HA, in depth investigations are necessary for large scale applications.

References

1 Liou, S.-C., Chen, S.-Y., and Liu, D.-M. (2005). Manipulation of nanoneedle and nanosphere apatite/poly(acrylic acid) nanocomposites. *Journal of Biomedical Materials Research Part B: Applied Biomaterials* 73B: 117–122. https://doi.org/10.1002/jbm.b.30193.

2 Oh, S.C., Lei, Y., Chen, H. et al. (2017). Catalytic consequences of cation and anion substitutions on rate and mechanism of oxidative coupling of methane over hydroxyapatite catalysts. *Fuel* 191: 472–485. https://doi.org/10.1016/j.fuel.2016.11.106.

3 Oh, S.C., Xu, J., Tran, D.T. et al. (2018). Effects of controlled crystalline surface of hydroxyapatite on methane oxidation reactions. *ACS Catalysis* 8: 4493–4507. https://doi.org/10.1021/acscatal.7b04011.

4 Aquilano, D., Bruno, M., Rubbo, M. et al. (2014). Low symmetry polymorph of hydroxyapatite. Theoretical equilibrium morphology of the monoclinic $Ca_5(OH)(PO_4)_3$. *Crystal Growth & Design* 14: 2846–2852. https://doi.org/10.1021/cg5001478.

5 Kay, M.I., Young, R.A., and Posner, A.S. (1964). Crystal structure of hydroxyapatite. *Nature* 204: 1050–1052. https://doi.org/10.1038/2041050a0.

6 Ivanova, T.I., Frank-Kamenetskaya, O.V., Kol'tsov, A.B. et al. (2001). Crystal structure of calcium-deficient carbonated hydroxyapatite. Thermal decomposition. *Journal of Solid State Chemistry* 160: 340–349. https://doi.org/10.1006/jssc.2000.9238.

7 Matsuura, Y., Onda, A., Ogo, S. et al. (2014). Acrylic acid synthesis from lactic acid over hydroxyapatite catalysts with various cations and anions. *Catalysis Today* 226: 192–197. https://doi.org/10.1016/j.cattod.2013.10.031.

8 Lee, K.Y., Houalla, M., Hercules, D.M. et al. (1994). Catalytic oxidative decomposition of dimethyl methylphosphonate over cu-substituted hydroxyapatite. *Journal of Catalysis* 145: 223–231. https://doi.org/10.1006/jcat.1994.1026.

9 Bianco, A., Cacciotti, I., Lombardi, M. et al. (2010). F-substituted hydroxyapatite nanopowders: thermal stability, sintering behaviour and mechanical properties. *Ceramics International* 36: 313–322. https://doi.org/10.1016/j.ceramint.2009.09.007.

10 Okazaki, M., Miake, Y., Tohda, H. et al. (1999). Functionally graded fluoridated apatites. *Biomaterials* 20: 1421–1426. https://doi.org/10.1016/S0142–9612(99)00049-6.

11 Yamashita, K., Kitagaki, K., and Umegaki, T. (1995). Thermal instability and proton conductivity of ceramic hydroxyapatite at high temperatures. *Journal of the American Ceramic Society* 78: 1191–1197. https://doi.org/10.1111/j.1151–2916.1995.tb08468.x.

12 Ellis, D.E., Terra, J., Warschkow, O. et al. (2006). A theoretical and experimental study of lead substitution in calcium hydroxyapatite. *Physical Chemistry Chemical Physics* 8: 967–976. https://doi.org/10.1039/B509254J.

13 Mori, K., Yamaguchi, K., Hara, T. et al. (2002). Controlled synthesis of hydroxyapatite-supported palladium complexes as highly efficient heterogeneous catalysts. *Journal of the American Chemical Society* 124: 11572–11573. https://doi.org/10.1021/ja020444q.

14 Bett, J.A.S., Christner, L.G., and Hall, W.K. (1967). Hydrogen held by solids. XII. Hydroxyapatite catalysts. *Journal of the American Chemical Society* 89: 5535–5541. https://doi.org/10.1021/ja00998a003.

15 Liu, Y., Tsunoyama, H., Akita, T. et al. (2011). Aerobic oxidation of cyclohexane catalyzed by size-controlled au clusters on hydroxyapatite: size effect in the sub-2 nm regime. *ACS Catalysis* 1: 2–6. https://doi.org/10.1021/cs100043j.

16 Mori, K., Yamaguchi, K., Mizugaki, T. et al. (2001). Catalysis of a hydroxyapatite-bound Ru complex: efficient heterogeneous oxidation of primary amines to nitriles in the presence of molecular oxygen. *Chemical Communications* 461–462. https://doi.org/10.1039/B009944I.

17 Kibby, C.L. and Hall, W.K. (1973). Studies of acid catalyzed reactions: XII. Alcohol decomposition over hydroxyapatite catalysts. *Journal of Catalysis* 29: 144–159. https://doi.org/10.1016/0021-9517(73)90213-3.

18 Venugopal, A. and Scurrell, M.S. (2003). Hydroxyapatite as a novel support for gold and ruthenium catalysts: behaviour in the water gas shift reaction. *Applied Catalysis A: General* 245: 137–147. https://doi.org/10.1016/S0926-860X(02)00647-6.

19 Yin, A., Margolis, H.C., Grogan, J. et al. (2003). Physical parameters of hydroxyapatite adsorption and effect on candidacidal activity of histatins. *Archives of Oral Biology* 48: 361–368. https://doi.org/10.1016/S0003-9969(03)00012-8.

20 Feng, Y., Gong, J.-L., Zeng, G.-M. et al. (2010). Adsorption of cd (II) and Zn (II) from aqueous solutions using magnetic hydroxyapatite nanoparticles as adsorbents. *Chemical Engineering Journal* 162: 487–494. https://doi.org/10.1016/j.cej.2010.05.049.

21 Nakamura, S., Takeda, H., and Yamashita, K. (2001). Proton transport polarization and depolarization of hydroxyapatite ceramics. *Journal of Applied Physics* 89: 5386–5392. https://doi.org/10.1063/1.1357783.

22 Wang, D., Paradelo, M., Bradford, S.A. et al. (2011). Facilitated transport of cu with hydroxyapatite nanoparticles in saturated sand: effects of solution ionic strength and composition. *Water Research* 45: 5905–5915. https://doi.org/10.1016/j.watres.2011.08.041.

23 Wang, D., Chu, L., Paradelo, M. et al. (2011). Transport behavior of humic acid-modified nano-hydroxyapatite in saturated packed column: effects of Cu, ionic strength, and ionic composition. *Journal of Colloid and Interface Science* 360: 398–407. https://doi.org/10.1016/j.jcis.2011.04.064.

24 Ogo, S., Onda, A., and Yanagisawa, K. (2011). Selective synthesis of 1-butanol from ethanol over strontium phosphate hydroxyapatite catalysts. *Applied Catalysis A: General* 402: 188–195. https://doi.org/10.1016/j.apcata.2011.06.006.

25 Hu, A., Li, M., Chang, C. et al. (2007). Preparation and characterization of a titanium-substituted hydroxyapatite photocatalyst. *Journal of Molecular Catalysis A: Chemical* 267: 79–85. https://doi.org/10.1016/j.molcata.2006.11.038.

26 Sun, H., Su, F.-Z., Ni, J. et al. (2009). Gold supported on hydroxyapatite as a versatile multifunctional catalyst for the direct tandem synthesis of imines and Oximes. *Angewandte Chemie International Edition* 48: 4390–4393. https://doi.org/10.1002/anie.200900802.

27 Matsumura, Y., Sugiyama, S., Hayashi, H. et al. (1995). Lead-calcium hydroxyapatite: Cation effects in the oxidative coupling of methane. *Journal of Solid State Chemistry* 114: 138–145. https://doi.org/10.1006/jssc.1995.1020.

28 Lee, K.-Y., Han, Y.-C., Suh, D.J. et al. (1998). Pb-substituted hydroxyapatite catalysts prepared by coprecipitation method for oxidative coupling of methane. In: *Studies in Surface Science and Catalysis*, vol. 119 (ed. A. Parmaliana, D. Sanfilippo, F. Frusteri, et al.), 385–390. Elsevier.

29 Park, J.H., Lee, D.-W., Im, S.-W. et al. (2012). Oxidative coupling of methane using non-stoichiometric lead hydroxyapatite catalyst mixtures. *Fuel* 94: 433–439. https://doi.org/10.1016/j.fuel.2011.08.056.

30 Kwak, J.H., Lee, S.Y., Nam, S.-W. et al. (2008). Partial oxidation of n-butane over ceria-promoted nickel/calcium hydroxyapatite. *Korean Journal of Chemical Engineering* 25: 1309–1315. https://doi.org/10.1007/s11814-008-0214-z.

31 Wei, X. and Yates, M.Z. (2012). Yttrium-doped hydroxyapatite membranes with high proton conductivity. *Chemistry of Materials* 24: 1738–1743. https://doi.org/10.1021/cm203355h.

32 Lomonosov, V.I. and Sinev, M.Y. (2016). Oxidative coupling of methane: mechanism and kinetics. *Kinetics and Catalysis* 57: 647–676. https://doi.org/10.1134/S0023158416050128.

33 Hammond, C., Conrad, S., and Hermans, I. (2012). Oxidative methane upgrading. *ChemSusChem* 5: 1668–1686. https://doi.org/10.1002/cssc.201200299.

34 Lunsford, J.H. (1995). The catalytic oxidative coupling of methane. *Angewandte Chemie International Edition in English* 34: 970–980. https://doi.org/10.1002/anie.199509701.

35 Kiyoshi, O., Qin, L., Masaharu, H. et al. (1986). The catalysts active and selective in oxidative coupling of methane. Alkali-doped samarium oxides. *Chemistry Letters* 15: 467–468. https://doi.org/10.1246/cl.1986.467.

36 Maiti, G.C. and Baerns, M. (1995). Dehydration of sodium hydroxide and lithium hydroxide dispersed over calcium oxide catalysts for the oxidative coupling of methane. *Applied Catalysis A: General* 127: 219–232. https://doi.org/10.1016/0926-860X(95)00071-2.

37 Dissanayake, D., Lunsford, J.H., and Rosynek, M.P. (1993). Oxidative coupling of methane over oxide-supported barium catalysts. *Journal of Catalysis* 143: 286–298. https://doi.org/10.1006/jcat.1993.1273.

38 Dedov, A.G., Loktev, A.S., Moiseev, I.I. et al. (2003). Oxidative coupling of methane catalyzed by rare earth oxides: unexpected synergistic effect of the oxide mixtures. *Applied Catalysis A: General* 245: 209–220. https://doi.org/10.1016/S0926-860X(02)00641-5.

39 Chan, T.K. and Smith, K.J. (1990). Oxidative coupling of methane over cobalt—magnesium and manganese—magnesium mixed oxide catalysts. *Applied Catalysis* 60: 13–31. https://doi.org/10.1016/S0166-9834(00)82169-7.

40 Gao, Z. and Shi, Y. (2010). Suppressed formation of CO_2 and H_2O in the oxidative coupling of methane over La_2O_3/MgO catalyst by surface modification. *Journal of Natural Gas Chemistry* 19: 173–178. https://doi.org/10.1016/S1003-9953(09)60047-5.

41 Matsumura, Y. and Moffat, J.B. (1993). Catalytic oxidative coupling of methane over hydroxyapatite modified with lead. *Catalysis Letters* 17: 197–204. https://doi.org/10.1007/BF00766142.

42 Hadrich, A., Lautié, A., and Mhiri, T. (2001). Vibrational study and fluorescence bands in the FT–Raman spectra of $Ca_{10-x}Pb_x(PO_4)_6(OH)_2$ compounds. *Spectrochimica Acta Part A: Molecular and Biomolecular Spectroscopy* 57: 1673–1681. https://doi.org/10.1016/S1386-1425(01)00402-4.

43 Reyes, S.C., Iglesia, E., and Kelkar, C.P. (1993). Kinetic-transport models of bimodal reaction sequences—I. Homogeneous and heterogeneous pathways in oxidative coupling of methane. *Chemical Engineering Science* 48: 2643–2661. https://doi.org/10.1016/0009-2509(93)80274-T.

44 Korf, S.J., Roos, J.A., Vreeman, J.A. et al. (1990). A study of the kinetics of the oxidative coupling of methane over a Li/Sn/MgO catalyst. *Catalysis Today* 6: 417–426. https://doi.org/10.1016/0920-5861(90)85035-M.

45 Miro, E., Santamaria, J., and Wolf, E.E. (1990). Oxidative coupling of methane on alkali metal-promoted nickel titanate: I. Catalyst characterization and transient studies. *Journal of Catalysis* 124: 451–464. https://doi.org/10.1016/0021-9517(90)90192-M.

46 Amorebieta, V.T. and Colussi, A.J. (1988). Kinetics and mechanism of the catalytic oxidation of methane over lithium-promoted magnesium oxide. *The Journal of Physical Chemistry* 92: 4576–4578. https://doi.org/10.1021/j100327a003.

47 Lehmann, L. and Berns, M. (1992). Kinetic studies of the oxidative coupling of methane over a NaOH/CaO catalyst. *Journal of Catalysis* 135: 467–480. https://doi.org/10.1016/0021-9517(92)90048-M.

48 Beck, B., Fleischer, V., Arndt, S. et al. (2014). Oxidative coupling of methane—a complex surface/gas phase mechanism with strong impact on the reaction engineering. *Catalysis Today* 228: 212–218. https://doi.org/10.1016/j.cattod.2013.11.059.

49 Wolf, E.E. (2014). Methane to light hydrocarbons via oxidative methane coupling: lessons from the past to search for a selective heterogeneous catalyst. *The Journal of Physical Chemistry Letters* 5: 986–988. https://doi.org/10.1021/jz500197h.

50 Thybaut, J.W., Sun, J., Olivier, L. et al. (2011). Catalyst design based on microkinetic models: oxidative coupling of methane. *Catalysis Today* 159: 29–36. https://doi.org/10.1016/j.cattod.2010.09.002.

51 Iwamatsu, E. and Aika, K.-I. (1989). Kinetic analysis of the oxidative coupling of methane over Na+-doped MgO. *Journal of Catalysis* 117: 416–431. https://doi.org/10.1016/0021-9517(89)90352-7.

52 Taheri, Z., Seyed-Matin, N., Safekordi, A.A. et al. (2009). A comparative kinetic study on the oxidative coupling of methane over LSCF perovskite-type catalyst. *Applied Catalysis A: General* 354: 143–152. https://doi.org/10.1016/j.apcata.2008.11.017.

53 Su, Y.S., Ying, J.Y., and Green, W.H. (2003). Upper bound on the yield for oxidative coupling of methane. *Journal of Catalysis* 218: 321–333. https://doi.org/10.1016/S0021-9517(03)00043-5.

54 Takanabe, K. and Iglesia, E. (2009). Mechanistic aspects and reaction pathways for oxidative coupling of methane on $Mn/Na_2WO_4/SiO_2$ catalysts. *The Journal of Physical Chemistry C* 113: 10131–10145. https://doi.org/10.1021/jp9001302.

55 Sadjadi, S., Jašo, S., Godini, H.R. et al. (2015). Feasibility study of the Mn–Na2WO4/SiO2 catalytic system for the oxidative coupling of methane in a fluidized-bed reactor. *Catalysis Science & Technology* 5: 942–952. https://doi.org/10.1039/C4CY00822G.

56 Roos, J.A., Korf, S.J., Veehof, R.H.J. et al. (1989). Kinetic and mechanistic aspects of the oxidative coupling of methane over a Li/MgO catalyst. *Applied Catalysis* 52: 131–145. https://doi.org/10.1016/S0166-9834(00)83377-1.

57 Liu, Z.-W., Jun, K.-W., Roh, H.-S. et al. (2002). Partial oxidation of methane over nickel catalysts supported on various aluminas. *Korean Journal of Chemical Engineering* 19: 735–741. https://doi.org/10.1007/BF02706961.

58 Diskin, A.M. and Ormerod, R.M. (2000). Partial oxidation of methane over supported nickel catalysts. In: *Studies in Surface Science and Catalysis*, vol. 130 (ed. A. Corma, F.V. Melo, S. Mendioroz and J.L.G. Fierro), 3519–3524. Elsevier.

59 Dissanayake, D., Rosynek, M.P., and Lunsford, J.H. (1993). Are the equilibrium concentrations of carbon monoxide and hydrogen exceeded during the oxidation of methane over a nickel/ytterbium oxide catalyst? *The Journal of Physical Chemistry* 97: 3644–3646. https://doi.org/10.1021/j100117a002.

60 Horn, R. and Schlögl, R. (2015). Methane activation by heterogeneous catalysis. *Catalysis Letters* 145: 23–39. https://doi.org/10.1007/s10562-014-1417-z.

61 Christian Enger, B., Lødeng, R., and Holmen, A. (2008). A review of catalytic partial oxidation of methane to synthesis gas with emphasis on reaction mechanisms over transition metal catalysts. *Applied Catalysis A: General* 346: 1–27. https://doi.org/10.1016/j.apcata.2008.05.018.

62 Pantaleo, G., La Parola, V., Deganello, F. et al. (2015). Synthesis and support composition effects on CH4 partial oxidation over Ni–CeLa oxides. *Applied Catalysis B: Environmental* 164: 135–143. https://doi.org/10.1016/j.apcatb.2014.09.011.

63 Diskin, A.M., Cunningham, R.H., and Ormerod, R.M. (1998). The oxidative chemistry of methane over supported nickel catalysts. *Catalysis Today* 46: 147–154. https://doi.org/10.1016/S0920-5861(98)00336-8.

64 Chen, Y., Hu, C., Gong, M. et al. (2000). Partial oxidation and chemisorption of methane over Ni/Al2O3 catalysts. In: *Studies in Surface Science and Catalysis*, vol. 130 (ed. A. Corma, F.V. Melo, S. Mendioroz and J.L.G. Fierro), 3543–3548. Elsevier.

65 Lu, Y., Liu, Y., and Shen, S. (1998). Design of stable Ni catalysts for partial oxidation of methane to synthesis gas. *Journal of Catalysis* 177: 386–388. https://doi.org/10.1006/jcat.1998.2051.

66 Nakagawa, K., Ikenaga, N., Teng, Y. et al. (1999). Transient response of catalyst bed temperature in the pulsed reaction of partial oxidation of methane to synthesis gas over supported rhodium and iridium catalysts. *Journal of Catalysis* 186: 405–413. https://doi.org/10.1006/jcat.1999.2576.

67 Jun, J.H., Lee, T.-J., Lim, T.H. et al. (2004). Nickel–calcium phosphate/hydroxyapatite catalysts for partial oxidation of methane to syngas: characterization and activation. *Journal of Catalysis* 221: 178–190. https://doi.org/10.1016/j.jcat.2003.07.004.

68 Choudhary, V.R., Uphade, B.S., and Mamman, A.S. (1997). Oxidative conversion of methane to syngas over nickel supported on commercial low surface area porous catalyst carriers Precoated with alkaline and rare earth oxides. *Journal of Catalysis* 172: 281–293. https://doi.org/10.1006/jcat.1997.1838.

69 Boukha, Z., Gil-Calvo, M., de Rivas, B. et al. (2018). Behaviour of Rh supported on hydroxyapatite catalysts in partial oxidation and steam reforming of methane: on the role of the speciation of the Rh particles. *Applied Catalysis A: General* 556: 191–203. https://doi.org/10.1016/j.apcata.2018.03.002.

70 Jun, J.H., Lim, T.H., Nam, S.-W. et al. (2006). Mechanism of partial oxidation of methane over a nickel-calcium hydroxyapatite catalyst. *Applied Catalysis A: General* 312: 27–34. https://doi.org/10.1016/j.apcata.2006.06.020.

71 Jin, R., Chen, Y., Li, W. et al. (2000). Mechanism for catalytic partial oxidation of methane to syngas over a Ni/Al2O3 catalyst. *Applied Catalysis A: General* 201: 71–80. https://doi.org/10.1016/S0926-860X(00)00424-5.

72 Takarroumt, N., Kacimi, M., Bozon-Verduraz, F. et al. Characterization and performance of the bifunctional platinum-loaded calcium-hydroxyapatite in the one-step synthesis of methyl isobutyl ketone. *Journal of Molecular Catalysis A: Chemical* 377: 42–50. https://doi.org/10.1016/j.molcata.2013.04.017.

73 Ho, C.R., Zheng, S., Shylesh, S. et al. (2018). The mechanism and kinetics of methyl isobutyl ketone synthesis from acetone over ion-exchanged hydroxyapatite. *Journal of Catalysis* 365: 174–183. https://doi.org/10.1016/j.jcat.2018.07.005.

74 Waters, G., Richter, O., and Kraushaar-Czarnetzki, B. (2006). Gas-phase conversion of acetone to MIBK over Bifunctional metal/carbon catalysts. I. Investigation of the acid−base sites. *Industrial & Engineering Chemistry Research* 45: 5701–5707. https://doi.org/10.1021/ie060184b.

75 Shylesh, S., Hanna, D., Gomes, J. et al. (2015). The role of hydroxyl group acidity on the activity of silica-supported secondary amines for the self-condensation of n-Butanal. *ChemSusChem* 8: 466–472. https://doi.org/10.1002/cssc.201402443.

76 Talwalkar, S. and Mahajani, S. (2006). Synthesis of methyl isobutyl ketone from acetone over metal-doped ion exchange resin catalyst. *Applied Catalysis A: General* 302: 140–148. https://doi.org/10.1016/j.apcata.2006.01.004.

77 Suzuki, T., Hatsushika, T., and Hayakawa, Y. (1981). Synthetic hydroxyapatites employed as inorganic cation-exchangers. *Journal of the Chemical Society, Faraday Transactions 1: Physical Chemistry in Condensed Phases* 77: 1059–1062. https://doi.org/10.1039/F19817701059.

78 Ho, C.R., Shylesh, S., and Bell, A.T. (2016). Mechanism and kinetics of ethanol coupling to Butanol over hydroxyapatite. *ACS Catalysis* 6: 939–948. https://doi.org/10.1021/acscatal.5b02672.

79 Jin, C., Yao, M., Liu, H. et al. (2011). Progress in the production and application of *n*-butanol as a biofuel. *Renewable and Sustainable Energy Reviews* 15: 4080–4106. https://doi.org/10.1016/j.rser.2011.06.001.

80 Savage, N. (2011). Fuel options: the ideal biofuel. *Nature* 474: S9–S11. https://doi.org/10.1038/474S09a.

81 Keitaro, K., Toyomi, M.-u., Yasushi, O. et al. (2009). Guerbet reaction of ethanol to *n*-Butanol catalyzed by iridium complexes. *Chemistry Letters* 38: 838–839. https://doi.org/10.1246/cl.2009.838.

82 Weissermel, K. and Arpe, H.J. (2003). Syntheses involving carbon monoxide. *Industrial Organic Chemistry* 127–144. https://doi.org/10.1002/9783527619191.ch6.

83 Kozlowski, J.T. and Davis, R.J. (2013). Heterogeneous catalysts for the Guerbet coupling of alcohols. *ACS Catalysis* 3: 1588–1600. https://doi.org/10.1021/cs400292f.

84 Gabriëls, D., Hernández, W.Y., Sels, B. et al. (2015). Review of catalytic systems and thermodynamics for the Guerbet condensation reaction and challenges for biomass valorization. *Catalysis Science & Technology* 5: 3876–3902. https://doi.org/10.1039/C5CY00359H.

85 Ueda, W., Ohshida, T., Kuwabara, T. et al. (1992). Condensation of alcohol over solid-base catalyst to form higher alcohols. *Catalysis Letters* 12: 97–104. https://doi.org/10.1007/BF00767192.

86 Birky, T.W., Kozlowski, J.T., and Davis, R.J. (2013). Isotopic transient analysis of the ethanol coupling reaction over magnesia. *Journal of Catalysis* 298: 130–137. https://doi.org/10.1016/j.jcat.2012.11.014.

87 Carvalho, D.L., de Avillez, R.R., Rodrigues, M.T. et al. (2012). Mg and Al mixed oxides and the synthesis of *n*-butanol from ethanol. *Applied Catalysis A: General* 415–416: 96–100. https://doi.org/10.1016/j.apcata.2011.12.009.

88 Earley, J.H., Bourne, R.A., Watson, M.J. et al. (2015). Continuous catalytic upgrading of ethanol to *n*-butanol and >C4 products over cu/CeO_2 catalysts in supercritical CO_2. *Green Chemistry* 17: 3018–3025. https://doi.org/10.1039/C4GC00219A.

89 Yang, C. and Meng, Z.Y. (1993). Bimolecular condensation of ethanol to 1-Butanol catalyzed by alkali Cation zeolites. *Journal of Catalysis* 142: 37–44. https://doi.org/10.1006/jcat.1993.1187.

90 Silvester, L., Lamonier, J.-F., Faye, J. et al. (2015). Reactivity of ethanol over hydroxyapatite-based Ca-enriched catalysts with various carbonate contents. *Catalysis Science & Technology* 5: 2994–3006. https://doi.org/10.1039/C5CY00327J.

91 Tsuchida, T., Kubo, J., Yoshioka, T. et al. Reaction of ethanol over hydroxyapatite affected by Ca/P ratio of catalyst. *Journal of Catalysis* 259: 183–189. https://doi.org/10.1016/j.jcat.2008.08.005.

92 Hanspal, S., Young, Z.D., Prillaman, J.T. et al. Influence of surface acid and base sites on the Guerbet coupling of ethanol to butanol over metal phosphate catalysts. *Journal of Catalysis* 352: 182–190. https://doi.org/10.1016/j.jcat.2017.04.036.

93 Ogo, S., Onda, A., Iwasa, Y. et al. (2012). 1-Butanol synthesis from ethanol over strontium phosphate hydroxyapatite catalysts with various Sr/P ratios. *Journal of Catalysis* 296: 24–30. https://doi.org/10.1016/j.jcat.2012.08.019.

94 Silvester, L., Lamonier, J.-F., Lamonier, C. et al. (2017). Guerbet reaction over strontium-substituted hydroxyapatite catalysts prepared at various (Ca+Sr)/P ratios. *ChemCatChem* 9: 2250–2261. https://doi.org/10.1002/cctc.201601480.

95 Scalbert, J., Thibault-Starzyk, F., Jacquot, R. et al. (2014). Ethanol condensation to butanol at high temperatures over a basic heterogeneous catalyst: how relevant is acetaldehyde self-aldolization? *Journal of Catalysis* 311: 28–32. https://doi.org/10.1016/j.jcat.2013.11.004.

96 Tsuchida, T., Sakuma, S., Takeguchi, T. et al. (2006). Direct synthesis of *n*-Butanol from ethanol over nonstoichiometric hydroxyapatite. *Industrial & Engineering Chemistry Research* 45: 8634–8642. https://doi.org/10.1021/ie0606082.

97 Ndou, A.S., Plint, N., and Coville, N.J. (2003). Dimerisation of ethanol to butanol over solid-base catalysts. *Applied Catalysis A: General* 251: 337–345. https://doi.org/10.1016/S0926-860X(03)00363-6.

98 Gines, M.J.L. and Iglesia, E. (1998). Bifunctional condensation reactions of alcohols on basic oxides modified by copper and potassium. *Journal of Catalysis* 176: 155–172. https://doi.org/10.1006/jcat.1998.2009.

99 Diallo-Garcia, S., Osman, M.B., Krafft, J.-M. et al. (2014). Identification of surface basic sites and acid–base pairs of hydroxyapatite. *The Journal of Physical Chemistry C* 118: 12744–12757. https://doi.org/10.1021/jp500469x.

100 Kibby, C.L. and Hall, W.K. (1973). Dehydrogenation of alcohols and hydrogen transfer from alcohols to ketones over hydroxyapatite catalysts. *Journal of Catalysis* 31: 65–73. https://doi.org/10.1016/0021-9517(73)90271-6.

101 Lovón-Quintana, J.J., Rodriguez-Guerrero, J.K., and Valença, P.G. (2017). Carbonate hydroxyapatite as a catalyst for ethanol conversion to hydrocarbon fuels. *Applied Catalysis A: General* 542: 136–145. https://doi.org/10.1016/j.apcata.2017.05.020.

102 Tsuchida, T., Yoshioka, T., Sakuma, S. et al. (2008). Synthesis of biogasoline from ethanol over hydroxyapatite catalyst. *Industrial & Engineering Chemistry Research* 47: 1443–1452. https://doi.org/10.1021/ie0711731.

103 Eagan, N.M., Lanci, M.P., and Huber, G.W. (2020). Kinetic modeling of alcohol oligomerization over calcium hydroxyapatite. *ACS Catalysis* 10: 2978–2989. https://doi.org/10.1021/acscatal.9b04734.

104 Shen, Z., Jin, F., Zhang, Y. et al. (2009). Effect of alkaline catalysts on hydrothermal conversion of glycerin into lactic acid. *Industrial & Engineering Chemistry Research* 48: 8920–8925. https://doi.org/10.1021/ie900937d.

105 Hisanori, K., Fangming, J., Zhouyu, Z. et al. (2005). Conversion of glycerin into lactic acid by alkaline hydrothermal reaction. *Chemistry Letters* 34: 1560–1561. https://doi.org/10.1246/cl.2005.1560.

106 Xu, J., Zhang, H., Zhao, Y. et al. (2013). Selective oxidation of glycerol to lactic acid under acidic conditions using AuPd/TiO_2 catalyst. *Green Chemistry* 15: 1520–1525. https://doi.org/10.1039/C3GC40314A.

107 Purushothaman, R.K.P., van Haveren, J., van Es, D.S. et al. (2014). An efficient one pot conversion of glycerol to lactic acid using bimetallic gold-platinum catalysts on a nanocrystalline $CeO2$ support. *Applied Catalysis B: Environmental* 147: 92–100. https://doi.org/10.1016/j.apcatb.2013.07.068.

108 Roy, D., Subramaniam, B., and Chaudhari, R.V. (2011). Cu-based catalysts Show Low temperature activity for glycerol conversion to lactic acid. *ACS Catalysis* 1: 548–551. https://doi.org/10.1021/cs200080j.

109 Yang, G.-Y., Ke, Y.-H., Ren, H.-F. et al. (2016). The conversion of glycerol to lactic acid catalyzed by $ZrO2$-supported CuO catalysts. *Chemical Engineering Journal* 283: 759–767. https://doi.org/10.1016/j.cej.2015.08.027.

110 Yin, H., Zhang, C., Yin, H. et al. (2016). Hydrothermal conversion of glycerol to lactic acid catalyzed by Cu/hydroxyapatite, Cu/MgO, and cu/ZrO_2 and reaction kinetics. *Chemical Engineering Journal* 288: 332–343. https://doi.org/10.1016/j.cej.2015.12.010.

111 Shen, L., Yu, Z., Zhang, D. et al. (2019). Glycerol valorization to lactic acid catalyzed by hydroxyapatite-supported palladium particles. *Journal of Chemical Technology & Biotechnology* 94: 204–215. https://doi.org/10.1002/jctb.5765.

112 Liptáková, B., Hronec, M., and Cvengrošová, Z. (2000). Direct synthesis of phenol from benzene over hydroxyapatite catalysts. *Catalysis Today* 61: 143–148. https://doi.org/10.1016/S0920-5861(00)00359-X.

113 Liptáková, B., Báhidský, M., and Hronec, M. (2004). Preparation of phenol from benzene by one-step reaction. *Applied Catalysis A: General* 263: 33–38. https://doi.org/10.1016/j.apcata.2003.12.002.

114 Niwa, S.-i., Eswaramoorthy, M., Nair, J. et al. (2002). A one-step conversion of benzene to phenol with a palladium membrane. *Science* 295: 105–107. https://doi.org/10.1126/science.1066527.

115 Goettmann, F., Fischer, A., Antonietti, M. et al. (2006). Chemical synthesis of mesoporous carbon nitrides using hard templates and their use as a metal-free catalyst for Friedel–Crafts reaction of benzene. *Angewandte Chemie International Edition* 45: 4467–4471. https://doi.org/10.1002/anie.200600412.

116 Chen, X., Zhang, J., Fu, X. et al. (2009). Fe-g-C_3N_4-catalyzed oxidation of benzene to phenol using hydrogen peroxide and visible light. *Journal of the American Chemical Society* 131: 11658–11659. https://doi.org/10.1021/ja903923s.

117 Jazie, A.A., Pramanik, H., and Sinha, A.S.K. (2013). Transesterification of peanut and rapeseed oils using waste of animal bone as cost effective catalyst. *Materials for Renewable and Sustainable Energy* 2: 11. https://doi.org/10.1007/s40243-013-0011-4.

118 Jindapon, W., Jaiyen, S., and Ngamcharussrivichai, C. (2016). Seashell-derived mixed compounds of Ca, Zn and Al as active and stable catalysts for the transesterification of palm oil with methanol to biodiesel. *Energy Conversion and Management* 122: 535–543. https://doi.org/10.1016/j.enconman.2016.06.012.

119 Sahu, H. and Mohanty, K. (2015). Al grafted natural hydroxyapatite for neem oil transesterification: kinetic study at optimal point. *Chemical Engineering Journal* 280: 564–574. https://doi.org/10.1016/j.cej.2015.06.040.

120 Chen, G., Shan, R., Shi, J. et al. (2015). Biodiesel production from palm oil using active and stable K doped hydroxyapatite catalysts. *Energy Conversion and Management* 98: 463–469. https://doi.org/10.1016/j.enconman.2015.04.012.

121 Essamlali, Y., Amadine, O., Larzek, M. et al. (2017). Sodium modified hydroxyapatite: highly efficient and stable solid-base catalyst for biodiesel production. *Energy Conversion and Management* 149: 355–367. https://doi.org/10.1016/j.enconman.2017.07.028.

122 Gupta, J., Agarwal, M., and Dalai, A.K. (2018). Marble slurry derived hydroxyapatite as heterogeneous catalyst for biodiesel production from soybean oil. *The Canadian Journal of Chemical Engineering* 96: 1873–1880. https://doi.org/10.1002/cjce.23167.

123 Obadiah, A., Swaroopa, G.A., Kumar, S.V. et al. (2012). Biodiesel production from palm oil using calcined waste animal bone as catalyst. *Bioresource Technology* 116: 512–516. https://doi.org/10.1016/j.biortech.2012.03.112.

124 Rubio-Caballero, J.M., Santamaría-González, J., Mérida-Robles, J. et al. (2009). Calcium zincate as precursor of active catalysts for biodiesel production

under mild conditions. *Applied Catalysis B: Environmental* 91: 339–346. https://doi.org/10.1016/j.apcatb.2009.05.041.
125 Zięba, A., Pacuła, A., Serwicka, E.M., and Drelinkiewicz, A. (2010). Transesterification of triglycerides with methanol over thermally treated $Zn_5(OH)_8(NO_3)_2.2H_2O$ salt. *Fuel* 89: 1961–1972. https://doi.org/10.1016/j.fuel.2009.11.013.

6

Aerobic Selective Oxidation of Alcohols and Alkanes over Hydroxyapatite-Based Catalysts
Guylène Costentin and Franck Launay

Sorbonne Université, CNRS, Laboratoire de Réactivité de Surface (LRS), 4 place Jussieu, 75005 Paris, France

6.1 Introduction

Selective oxidation catalysis plays a crucial role in the current chemical industry for the production of many key intermediates. In particular, aldehydes and ketones can be obtained from primary or secondary alcohols. Despite oxygen activation remains challenging, selective aerobic alcohol oxidation is very attractive. In addition, in the context of decay of oil supplies, there is a renewed interest in light C1–C3 alkane valorization. The proven reserves in natural gas are abundant, which makes this energy source available and quite cheap. Alkanes are also formed in large quantities by coal-to-liquid (CTL) and gas-to-liquid (GTL) plants. Partial oxidation of methane (POM) is a promising route for syngas production ($CH_4 + 1/2\,O_2 \rightarrow CO + 2H_2$). Similarly, given the increasing demand in olefins, there is a renewed interest for oxidative dehydrogenation (ODH) of $C \geq 2$ alkanes ($C_nH_{2n+2} + 1/2\,O_2 \rightarrow C_nH_{2n} + H_2O$).

For all these classes of selective oxidation reactions operating in the presence of oxygen, the optimization of selectivity should contribute to the establishment of novel green and sustainable chemical processes. It remains however quite challenging to avoid overoxidation, and the development of selective heterogeneous catalysts remains an important issue. Their formulations are mainly based on transition metal ions (TMIs). Depending on the nature of the molecule to be selectively oxidized, different TMIs are used either in their cationic or zerovalent form. Cu, Ru, Pd, Au, Fe, V, Ir, Os, or Co are commonly used for the selective oxidation of alcohols [1]. Zerovalent Rh and Ni have been studied for the POM [2–4]. In the case of $C \geq 2$ alkane oxidative dehydrogenation reactions (ODHs), numerous TMIs were used, alone or in combination [5] to form reducible oxides [6–8], nanostructured oxides [9], or isolated species [10], and V-based catalysts are the most extensively investigated for propane activation [11]. Beside the nature of TMI, its dispersion is a key parameter to control the selectivity. One can discriminate two classes of catalysts, bulk catalysts containing the TMI in its host lattice or metal-supported catalysts. In the latter case, the nature of the carrier has been found to influence the performance of the system, not only due to its impact on the metal dispersion but also

Design and Applications of Hydroxyapatite-Based Catalysts, First Edition. Edited by Doan Pham Minh.
© 2022 WILEY-VCH GmbH. Published 2022 by WILEY-VCH GmbH.

due to the role played by the acid–base properties on the adsorption–desorption of reactants and products. It was also found that the addition of phosphorus to Cr/TiO_2 and V_2O_5/TiO_2 improves the performance of various catalysts in the oxidative dehydrogenation of ethane (ODHE) or propane (ODHP) [12–14]. Several explanations for such beneficial role of phosphorus were proposed, such as appearance on the carrier of a new phosphate with adequate structure of active sites, isolation of active site (in contrast to promotion of total oxidation over oxide-like catalysts), and/or the tuning of the surface acid–base properties [15–18]. Among the investigated phosphate compounds, $(VO)_2P_2O_7$ was identified as the active phase for the selective oxidation of n-butane to maleic anhydride [19], and many other iron [15, 20], molybdenum [21], and vanadium phosphate phases [22–24] were also investigated. In this context, the hydroxyapatite (HA) system was also studied from the 1990s, since, depending on the strategy of synthesis, it can be advantageously used as a tunable carrier to get metal supported HA or directly as metal-modified bulk catalyst, thanks to the various substitutions afforded by its flexible network. As developed in Chapter 3 of this book, the stability of HA both in solution and at high temperature makes this system ideal for use in selective oxidation of alcohols operating in liquid phase and for high temperature gas phase reactions required for alkane activation. Its acid–base properties can also be tuned both by controlling the stoichiometry and by modifying its composition, especially going from calcium HA to strontium HA to favor basicity, eventually counterbalancing the acidity induced by TMI deposition. The great modulation of composition by substitution on both its cationic and anionic sites offers endless opportunities to get metal-modified systems. Several strategies are used to incorporate TMIs, one-pot coprecipitation syntheses likely to modify the bulk (and surface) properties to favor high metal dispersion, or more classical metal surface deposition onto the HA carrier. Such two-step approach includes impregnation or cationic exchanges. Both are performed in excess of solution and mainly differ by the absence of a washing step in the former case. As a result, the metal loading and the metal dispersion may differ to a large extent.

Given their key impact for the control of reactivity in the selective oxidation reactions, these aspects will be considered in this chapter. In a first section, the performances and key parameters governing the oxidation of alcohols will be reviewed. In a second section dedicated to gas phase reactions, results and main conclusions obtained from methane partial oxidation then oxidative dehydrogenation of propane (ODHP), ethane, heptane, and n-octane will be developed.

6.2 Liquid Phase Reactions: Selective Aerobic Oxidation of Alcohols

Aldehydes and ketones are key compounds that can be obtained from primary and secondary alcohols. Oxidants, such as oxides (Cr, Mn, Ru), halides, peroxides, or ozone, have been traditionally used for these reactions, but they are often pointed out to be not environment friendly [25]. Guided by sustainable development, efforts have been made over the years to replace Cr reagents, particularly in medicinal chemistry, and the most popular oxidants right now are either stoichiometric reagents such as Dess–Martin periodinane (IBX), the Swern reagent, or combinations of a terminal oxidant with a catalyst such as N-methyl morpholine-N-oxide (NMO) and tetrapropylammonium perruthenate (TPAP) or sodium hypochlorite and TEMPO [26]. In parallel, recent researches have led to an exponential increase in the number of publications dealing with combinations of O_2 (or H_2O_2) with homogeneous, heterogeneous catalysts, or biocatalysts that afford a good atom economy and only water as a by-product, but none of these systems has replaced yet those mentioned before in synthetic organic chemistry [27–29]. Dioxygen deserves special attention due to its natural availability, but its activation is more challenging than that of H_2O_2 [30]. Catalytic systems working at low pressure as well as under mild reaction conditions and with a good selectivity and functional group tolerance are needed. Low catalyst loadings and the avoidance of costly or toxic additives are required too.

In this context, liquid phase apatite-based heterogeneous catalysts working with oxygen and involving mainly Ru or Pd have been explored. The different available systems tested in batch conditions will be compared considering benzyl alcohol (BA) as a model substrate. Their performance toward other primary and secondary alcohols, activated or not, will also be discussed. Information on the mechanism will be given using, if possible, in situ characterization of catalysts or experiments, allowing the determination of Hammett constants or of the kinetic isotope effect (KIE).

6.2.1 Apatite-Based Catalysts Efficient in the Aerobic Oxidation of Alcohols

Apatite is not an oxidation catalyst by itself. It must be combined with a transition metal. The main catalytic systems based on apatite developed for the oxidation of alcohols by oxygen reported since 2000 are presented in Table 6.1 [31–40]. At best, we have tried to highlight the quantities of metal involved in the catalytic tests in order to be able to compare the performance of the catalysts with each other. The situation varies greatly from one article to another. Some authors do not give any information on the turnover frequency (TOF) values, or, if they do, the TOFs are measured either over the whole reaction or at the lowest conversion values, which is normally more favorable. Historically, the first example dates back to the work of Kaneda and co-workers (Table 6.1, entry 1) [31]. One of the most common reasons emphasized for using apatite as a support for heterogeneous catalysts is its natural availability. However, very few authors use natural apatite (NAp), probably because the composition and physicochemical properties are not well controlled. Most of the research teams work with apatites synthesized following the protocol of Sugiyama et al. [41]. Regarding NAp, Shaabani et al. have used cow bones (Table 6.1, entries 2 and 3) [32, 33]. The latter was combined, for example, with 15 wt% of manganese (IV) oxide [32] or anionic complexes of cobalt(II) phthalocyanine (0.4 wt% Co) [33], affording benzaldehyde with yields of 97 and 95% under air in p-xylene using *8.5% mol of Mn and 0.35% mol of Co per BA*, respectively. In both cases, Shaabani et al. introduced nonnegligible amounts of base (K_2CO_3 or KOH were added with up to 50% mol vs.

the alcohol), which makes these systems not very attractive from a Green Chemistry point of view. Indeed, as in many cases in the aerobic oxidation of alcohols, bases are promoting the deprotonation of the alcohol and thus the anchoring of the derived alkoxide to the metal in order to allow its subsequent activation. In addition to the catalyst recycling proved by repeated runs, the use of supports such as apatite would avoid the deactivation of the cobalt complexes (Table 6.1, entry 3), which are prone to yield nonactive μ-oxo dimers. It has to be pointed out here that this is not specific to the apatite material. In the case of MnO_2/NAp (Table 6.1, entry 2), the idea was to increase the area-to-volume ratio of MnO_2, which is detrimental for the bulk oxide. It should be noted that in these studies, little information is given on the anchoring mode of the active phases on the apatite. Apparently, detailed studies of those catalytic materials were difficult to carry out due to the use of NAp. In the case of cobalt phthalocyanine, the authors only mention the advantage of working with a ligand with anionic functions to enhance the deposition of cobalt on the support.

The literature in the field of aerobic oxidation of alcohols includes many more papers dealing with combinations of apatite with either Ru(III) salts (Table 6.1, entries 1 and 4–7) or Pd(II) salts or complexes (Table 6.1, entries 8–9). Like in the previous examples (Table 6.1, entries 2 and 3), the resulting heterogeneous catalysts are generally efficient at temperatures between 60 and 100 °C and under atmospheric pressure of pure oxygen or air.

The use of Ru(III) or Pd(II) precursors combined with apatite allows working without further addition of a base, which is a particularly attractive advantage in the context of green chemistry. Utilization of apatite in that context was initiated by Kaneda's group [31]. Baiker's group also deeply studied catalysts based on apatite and Ru(III) (Table 6.1, entries 4–6) [34–36]. It is not surprising to find these two metals combined with apatite. Indeed, they are often associated with more classical inorganic supports such as zeolites, alumina, hydrotalcite, or carbon [37–40]. Solids like HA, montmorillonite, and hydrotalcite [42] would allow controlling the location of catalytically active metal species (either in the crystal lattice, or on the surface, or in the interlayer space, respectively), to create well-defined active metal sites on the solid surface or understand the molecular basis of heterogeneous catalysis.

In the following, we will describe in detail the studies involving Ru(III) and Pd(II), trying to highlight the impact of apatite on the catalytic activity when it is possible to make comparisons. These can only be made a priori if the methods of deposition and activation of the active phases and the catalysis conditions of the resulting catalysts are similar. Finally, it should be noted that, surprisingly, gold HA-based systems have not been explored yet for alcohol oxidation reactions, although gold has been used extensively elsewhere in combination with other supports [43].

6.2.2 Apatite/Ru(III) Catalysts

In 2000, Yamaguchi et al. reported the first application of apatite-based catalysts using an Ru(III) precursor as the active phase in the partial oxidation of alcohols in liquid phase, with molecular oxygen as an oxidant [31, 44]. According to these authors, the 17 wt% Ru-HA catalyst was prepared following an ion exchange (IE) procedure involving stoichiometric HA (HA characterized by a molar Ca/P ratio

Table 6.1 Aerobic oxidation of benzyl alcohol (BA) with apatite-based catalysts. Some data have been recalculated from those provided by the authors and are thus reported in italics.

Entry	Support	Loading	Preparation method	Conditions	Yield	References
1	Stoich. HA	17.1 wt% Ru *1.7 mmol g^{-1}*	I.E. with aq Ru^{III}/24 h drying 110 °C O.N.	Atm O$_2$/80 °C/100 mg cat mmol ROH^{-1}/toluene/ *17 mol% cat*	> 99% conv. 98–95% Ald (3 h) TOF = 2 h^{-1}	[31]
2	NHA from cow bone	15 wt% MnO$_2$ *1.7 mmol g^{-1}*	KMnO$_4$ 74 h RT + washing + drying 100 °C	Air/80 °C/50 mg cat mmol ROH^{-1}/p-xylene/8.5 mol% cat + 50 mol% K$_2$CO$_3$	97% Ald (3 h)	[32]
3		0.4 wt% Co *0.07 mmol g^{-1}*	CoTSPc 48 h RT + washing + drying 70 °C O.N.	Air flow/100 °C/50 mg cat mmol ROH^{-1}/p-xylene/ 0.35 mol% cat + 50 mol% KOH	95% Ald (4 h) TOF = 69.8 h^{-1}	[33]
4	HA calcined at 500 °C + aq. CoCl$_2$ 20 min + wash. treat.	c.a. 0.35 mmol g^{-1}	aq Ru^{III}/24 h drying 80 °C vacuum 8 h	Atm O$_2$/60 °C/60 mg cat mmol ROH^{-1}?/toluene/*2.1 mol% cat*	– (Kinetic study at low conv.)	[34]
5	Stoich. HA calcined at 500 °C + CoCl$_2$ 20 min or Pb^{2+}, + wash. treat.	0.4 mmol g^{-1}	aq Ru^{III}/10 min or 24 h drying 80 °C vacuum 8 h (also 25, 110 and 180 °C)	Atm O$_2$/90 °C/50 mg cat mmol ROH^{-1}?/toluene/ *1.7 mol% cat*	>99% Ald (0.75 h) TOF 78 h^{-1}	[35]
6	Method I – Stoich. HA modified by Hacid or Phos (drying vacuum RT, 24 h)	0.28 mmol g^{-1} (0.37 w/o modif)	aq Ru^{III}/10 min drying RT vacuum 6 h	Atm O$_2$/60 °C/60 mg cat mmol ROH^{-1}/toluene/ 1.7–2.2 mol% cat	> 99% Ald (0.7–2.6 h) TOF from 21 to 70 h^{-1} *242 h^{-1}*	[36]
	Method II – HA synthesized in the presence of prolinol, proline, or benzoic acid (drying O.N. 90 °C)	0.31–0.35 mmol g^{-1}				

(continued)

Table 6.1 (Continued)

Entry	Support	Loading	Preparation method	Conditions	Yield	References
7	HA-γ-Fe$_2$O$_3$	1 wt% Ru *0.1 mmol g^{-1}*	aq Ru^{III} RT/24 h drying vacuum O.N.	1 bar O$_2$ flow/90 °C/50 mg cat mmol ROH^{-1}/toluene/0.5 mol% cat	> 99% conv. 98% Ald (1 h) TOF = 196 h^{-1}	[37]
8	Stoich. HA	0.02 mmol g^{-1}	Impregnation with PdCl$_2$(PhCN)$_2$ in acetone	O$_2$/90 °C/100 mg cat mmol ROH^{-1}/trifluorotoluene/0.2 mol% cat	99% Ald (1 h) TOF = 495 h^{-1}	[38, 39]
9	HA	1.1 mmol g^{-1}	Impregnation with Pd(OAc)$_2$ 3 h + N$_2$H$_4$ + washing at reflux	Atm air/100 °C/100 mg cat mmol ROH^{-1}/water/0.11 mol% cat	95% Ald (7 h) TOF = 123 h^{-1}	[40]

Ald, aldehyde; Wash. Treat., washing treatment; Stoich, stoichiometric; I.E., ion exchange; O.N., Overnight.

of 1.67) and an aqueous RuCl$_3$ solution for an extended period of 24 h (Table 6.1, entry 1). The resulting material, dried overnight at 110 °C, was also characterized by a (Ru+Ca)/P of 1.67, thus validating the exchange of Ca^{2+} by Ru(III)X^{2+} species. Extended X-ray absorption fine structure (EXAFS) data confirmed the presence of oxygen atoms in the coordination sphere of Ru^{3+} as well as of one chloro ligand (X = Cl). As a result, the authors considered that the active sites of those catalysts are probably monomeric RuCl^{2+} cations surrounded by O atoms of the phosphate matrix. Aromatic (including heterocyclic ones), allylic, and aliphatic primary alcohols such as 1-octanol were tested as substrates at 80 °C in toluene under 1 atm of O$_2$ with *17 mol% of Ru per substrate*. In those conditions, very good yields of aldehydes were obtained with the activated alcohols. The oxidation of 1-octanol was the slowest one. Interestingly, neither carboxylic acids, nor their esters (through their reaction with the starting alcohol) were observed. In the case of BA, three recycling experiments were conducted successfully. From the test of different benzyl alcohols varying by the substitution of the aromatic ring, the authors could establish that the value of the Hammett slope is negative. According to the authors, its value equal to −0.43 is in fact very close to that obtained in homogeneous catalysis with RuCl$_2$(PPh$_3$)$_3$ (7), thus validating monomeric forms of Ru. Similarly, the selectivities (especially primary > secondary alcohols, e.g. 1,7-octanediol was chemoselectively oxidized to 7-hydroxyoctanal in 80% yield) reached in the presence of Ru-HA turned out to be much closer to that of RuCl$_2$(PPh$_3$)$_3$ than those of other heterogeneous catalysts such as RuO$_2$ or Ru/Al$_2$O$_3$ [45]. From this result, the authors suggested that, as for RuCl$_2$(PPh$_3$)$_3$, the alcohol oxidation would be initiated by the formation of a Ru-alcoholate species and that the carbonyl compound would result from β-hydride elimination in the rate-determining step as evidenced by the strong KIE (k_H/k_D = 7) evaluated in the competitive oxidation of BA and its deuterated form. A homolytic mechanism was excluded. Ru-HA catalyst does not involve oxo-ruthenium species as the active oxidant. In the last step of the catalytic cycle (Figure 6.1a), the Ru-hydride species produced with the aldehyde would be converted into a Ru(III)—OOH species by molecular oxygen that would release H$_2$O$_2$ or its decomposition products (H$_2$O and O$_2$). According to Kaneda and co-workers, [46] apatite would behave as an inorganic ligand that is robust and tolerant of severe reaction conditions: e.g. strongly oxidizing or reducing environments at high temperatures.

One important drawback of this system is the very low TOF value measured (2 h^{-1}) by the authors because of the high metal content and the relatively large amount of catalyst used. Opre et al. reinvestigated this type of catalyst in order to better control its structure and thus optimize its performance [34–36]. The first observation made by these authors was that it is not useful to substitute Ca located in the most internal sites by Ru, because the latter cannot be accessible to the substrates. Hence, TOF value obtained by the Yamaguchi's group in the case of BA is probably higher than 2 h^{-1}. Opre et al. [34–36] also paid close attention to the conditions of introduction of the Ru(III) species. They were particularly concerned about the mechanism of the metal introduction in Yamaguchi's original work. Was it just adsorption to the surface? Was it ion exchange (IE) alone or accompanied by

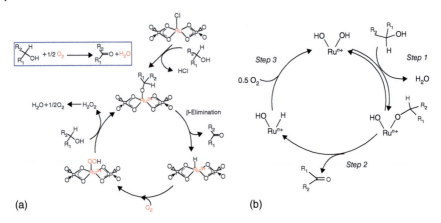

Figure 6.1 (a) Plausible mechanism of Ru-HA-catalyzed alcohol oxidation. Source: Kaneda and Mizugaki [46]. / With permission of American Chemical Society. (b) Proposed reaction mechanism for alcohol oxidation over Ru-CoHA. Source: Opre et al. [34]. / With permission of Elsevier.

apatite dissolution–precipitation mechanisms or combinations of these processes, in particular in the presence of acidic aqueous solutions such as those of Ru(III)? Is it also possible that some restructuration occurs leading, for example, to smaller HA crystallites when long contact times are used? One way to better control these points was to shorten the contact time of the aqueous solution of Ru(III) with apatite from 24 hours to 20 minutes or/and use a modified apatite. Shorter contact time would allow reducing the amount of Ru(III) introduced but also to avoid the dissolution/restructuration of apatite (lot of Ca and P leaching as well as increase of the pH of the contacted solution). Indeed, for longer contact times, Opre et al. [35] showed that the Ca+Ru/P molar ratio decreased, but this was not the case in the work reported by Yamagachi et al. [31]. Opre et al. [34–36] also worked to minimize the dehydration of the support by softening the drying conditions of the material recovered after Ru(III) incorporation. They were able to establish that the higher the drying temperature, the lower the activity. Two types of approaches based on modified apatites have been implemented. The first consisted in using apatite previously substituted with divalent cations such as Co^{2+} (Table 6.1, entries 4 and 5) and Pb^{2+} [35], the second being based on the use of apatites synthesized in the presence of polar organic molecules (Table 6.1, entry 6); one of the aims of which is to reduce the size of the crystals formed and thus increase the specific surface area.

Using Co-HA as a support [34, 35] contacted with aqueous Ru(III) solution during 10 minutes followed by drying under vacuum at 80 °C for 8 hours afforded catalysts exhibiting TOF values up to 78 h^{-1} in the conversion of BA to benzaldehyde under 1 atm of O_2 at 60 °C in toluene (Table 6.1, entry 4, *1.7 mol% of Ru per BA*). It is not clear yet if such improvement of the TOF value for Ru-CoHA is related to an optimized location of Ru on the surface (as shown by a higher Ru/Ca determined by X-ray photoelectron spectroscopy (XPS) compared to the bulk analysis) or/and to indirect electronic effects of Co^{2+} (it was shown that the incorporation of Co increased the reducibility of Ru for samples prepared at short times). According to XANES and

Figure 6.2 Proposed structure of the Ru species in Ru-CoHA. Source: Opre et al. [34]. / With permission of Elsevier.

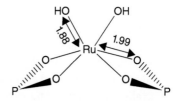

EXAFS, Ru would be highly dispersed. Simple replacement of Ca^{2+} was excluded, and similar structures were established for promoted and unpromoted materials. A point of divergence between the groups of K. Kaneda and A. Baiker also concerned the presence of a chloro ligand in the coordination sphere of ruthenium once the latter was incorporated into the apatite. Clearly, no Cl^- could be detected by XPS, and the occurrence of an $Ru(OH)_2^+$ species (Figure 6.2) with the reaction mechanism of Figure 6.1b was preferred in the work of Opre et al. [34]. The catalytic activity in alcohol oxidation was partially lost by drying the catalyst at temperatures above 80 °C, presumably, due to irreversible dehydration of the active species.

Excellent yields of aldehydes with reasonable TOF values ($>20\,h^{-1}$) were also obtained with substituted BA as well as heterocyclic alcohols. The oxidation of allylic alcohols was also complete but at a slower rate. Octanol-1 oxidation was reasonably fast without octanoic acid formation at short reaction times (less than one hour), but it can be noticed that, for longer reaction times, the formation of the carboxylic acid could not be avoided, maybe as the result of the oxidation of some Co^{2+} into Co^{3+}. Like for simple Ru-HA, competitive reactions between primary and secondary alcohols showed that the primary alcohols are more reactive. Interestingly, these authors performed a "hot filtration" test as recommended by Sheldon [47]. No further reaction occurred when the solid was removed after 40% BA conversion. In situ EXAFS study of the Ru-CoHA catalyst (Co-HA with $RuCl_3$ for 24 hours) during BA oxidation emphasized that no reduction of Ru^{3+} in Ru^0 occurs and agreed with the possible formation of ruthenium alcoholate species. Carrying out the reaction at 40 °C under argon, the conversion of BA stopped at 4.7% after 10 minutes, meaning that the reaction obeys a kind of Mars and Van Krevelen mechanism [28, 48]. In the absence of O_2, the active sites are "reduced" during the reaction and thus become inactive for further reaction. Moreover, considering the amount of catalyst (Ru) used in these conditions, the authors concluded that half of the Ru sites would be accessible.

The other approach initiated by Opre et al. consisted in using apatites with surfaces modified by the introduction of water-soluble polar organic molecules [36, 49] (alkyl phosphates, alkenoic acids, or amino acids) able to form strong hydrogen bonds with the OH and phosphate functions of HA (Table 6.1, entry 6). These molecules were used either during (case of proline, prolinol, benzoic acid (BAcid), drying at 90 °C) or after (case of hexanoic acid (HAcid) or 2-ethylhexylphopshate (Phos), drying 24 hours/RT) the synthesis of the HA support. Ru was later introduced at room temperature and for short contact time followed by a filtration step. Materials prepared by incorporation of the molecules in the synthesis gel afforded catalysts with the best TOF values. Considering the oxidation of BA at 60 °C under 1 bar O_2 in mesitylene, instead of toluene, Ru-HA-BAcid was characterized by a TOF of 242 h^{-1} at 100% conversion (with no further oxidation of benzaldehyde to benzoic acid), which is three times better than Ru-CoHA [28] and more than two order of magnitude than Ru-HA [31]. Again, Ru-HA turned out to be a good and highly selective catalyst for aromatic and allylic alcohol, but sluggish transformations were obtained with aliphatic primary and secondary alcohols. The authors think that higher intrinsic activity of Ru related to some modification of its coordination sphere is a more plausible explanation of the good catalytic performances of the resulting materials than the increase of the accessible active sites. Indeed, based on diffuse reflectance infrared Fourier transform (DRIFT), SEM, STEM-EDX, ICP-OES, and BET measurements, it could be concluded that the location and coordination of Ru is different in organically modified HA. Upon the introduction of Ru, no exchange of Ca^{2+} could be observed. In the resulting materials, the incorporation of Ru would not result from an IE process but from an adsorption process controlled by the polar organic molecules. It has to be noticed that, according to the authors, those modifiers are removed during Ru incorporation (washing and drying at room temperature for six hours). The method using prolinol, proline, or benzoic acid in the synthesis gel of HA led systematically to nonstoichiometric materials and also to a reduced crystallite size, giving rise to better specific surface areas.

Regarding kinetic studies, Opre et al. [34–36] have underlined the importance of operating conditions, where there are no mass transport limitations. In these reactions, the removal of water from the surface of the catalysts is slow and can be rate limiting. That is the reason why nonvolatile solvents are used at temperatures close to their boiling point in order to favor the elimination of water. However, if those solvents are used at temperatures too close to their boiling point, there is a risk of decreasing the oxygen partial pressure making the reoxidation step of Ru rate-limiting instead of the β-elimination. Using solvents with very high boiling point in lesser amounts or less catalyst is recommended. Those authors confirmed that under truly kinetically controlled conditions, electron-withdrawing substituents of benzyl alcohol have a decelerating effect [36], thus agreeing with a rate-limiting β-hydride elimination step (Figure 6.3). This was exactly the opposite in the case of cobalt phthalocyanine/apatite [33] where an autoxidation mechanism was proposed by Shaabani et al. In the work of Opre et al. [34], the K.I.E. value measured at 5% BA conversion in the case of Ru-CoHA was equal to 1.6 under kinetically controlled conditions and only 1.2 under transport limited conditions, which is rather small, but also agrees with β-hydride elimination as the rate-determining step. According to these authors, reoxidation of the ruthenium site should definitely not limit the overall reaction rate in the oxidation of aliphatic alcohols, which are less reactive than BA.

Later, Mori et al. used HA-γ-Fe_2O_3 materials made of magnetic γ-Fe_2O_3 nanocrystallites dispersed in an HA matrix instead of HA as a support for Ru-based oxidation catalysts (Table 6.1, entry 7) [37]. After Ru^{3+} introduction at a much lower loading than in their initial work (1 wt% instead of 17) using an impregnation at room temperature, they showed that highly dispersed monomeric Ru species could be obtained. These prepared catalysts afforded a significant improvement in activity

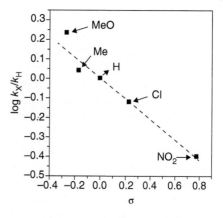

Figure 6.3 Hammett plot of p-substituted benzyl alcohol oxidation. Reaction conditions: 1 mmol p-X-benzylalcohol (X = H, Cl, NO$_2$, Me, or MeO), 10 mg RuHA-BAcid (0.0031 mmol Ru), 4 ml mesitylene, 60 °C, 1 bar O$_2$, t = 2 minutes. Source: Opre et al. [36]. / With permission of Elsevier.

in comparison with Ru-HA [31] with very interesting separation properties with an external permanent magnet, thus facilitating recycling experiments. One of the reasons was the increase of the BET surface area (115 m^2 g^{-1} instead of 49 m^2 g^{-1} for conventional Ru-HA) that would result from the suppression of the crystal growth. In this work, involving EXAFS measurements also, the authors agreed with Opre et al. and proposed the occurrence of Ru(OH)$_2$$^+$ with six coordinated oxygen atoms (two from hydroxo and four from phosphato ligands). By analogy with the use of Co-HA [34, 35], HA-γ-Fe$_2$O$_3$ would promote some electron transfer from Ru(III) to Fe(III), affording Ru(IV) oxidizing species. Catalysis tests with benzylic (or heterocyclic), allylic, and aliphatic primary alcohols as well as with primary alcohols with cyclopropyl substituents at 90 °C under O$_2$ flow in toluene with *0.5 or 1 mol% of Ru per substrate* showed that the double bonds or the cyclopropyl rings are not touched. In contrast to the systems described before, Ru/HA-γ-Fe$_2$O$_3$ led to the aerobic oxidation of aliphatic alcohols (e.g. 1-dodecanol) into their corresponding carboxylic acids. According to the authors, those catalysts were as efficient with sterically hindered substrates as their soluble analogs (homogeneous catalysis), thus emphasizing the location of most of Ru sites on the external surface of the apatite crystals. Secondary alcohols could be converted almost completely into ketones in the same conditions, especially when using 5 atm of O$_2$. The authors have claimed TOF values of 196 h^{-1} for BA instead of 2 h^{-1} in their original work (17 wt% Ru/HA [31]) and 78 h^{-1} for Ru-CoHA described by Opre et al. [34, 35]. The addition of radical traps or clocks led to the conclusion that no radical intermediates would be involved. Apparently, some oxidation tests with BA could be performed even at room temperature for 24 hours, but the amount of catalyst used in these examples was quite important. However, it can be noticed that other groups describing tests in so mild conditions in the literature proceeded with the addition of a base [50–52]. No structural changes upon the catalysis test could be detected.

6.2.3 Apatite/Pd(0) Catalysts

Palladium is also a transition metal widely used in aerobic oxidation catalysis [53, 54]. Heterogeneous catalysts based on Pd(0) and supports like activated carbon, pumice, hydrotalcite, oxides, or even organic polymers have been reported [55]. Usually, those materials suffer from low catalytic activities and a limited substrate scope. In the case of primary alcohols, the selectivity in aldehyde is rather low. Support effects on alcohol conversion and product selectivity have been mentioned many times [1]. Carbonaceous supports are notorious for their gas adsorption capacity, which may have some impact on the catalytic performance. Alumina with fairly large specific surface areas are known for their ability to disperse Pd and to stabilize single-site Pd^{2+} catalytic centers, but selectivity and conversion values are rather low. Optimum Pd(0) particle sizes have been reported for benzyl alcohol oxidation, which implies that such reaction is structure sensitive, with edge and corner Pd atoms more active than terrace sites [56].

In 2002, Kaneda et al. reported a pioneering work on the use of apatite for the development of supported Pd catalysts that turned out to be very successful, among the best, in the selective oxidation of alcohols using O_2 at atmospheric pressure (Table 6.1, entry 8) [38, 39]. Other supports such as Al_2O_3, SiO_2, and C are known to lead to less aldehyde and more acid. In this study, the authors investigated the influence of the stoichiometry of the apatite support, working either with HA-0 (stoichiometric apatite, $Ca_{10}(PO_4)_6X_2$, where X is a monovalent anion such as OH^-, F^-, or Cl^-) or HA-1 (nonstoichiometric, calcium-deficient HA, $Ca_{10-n}(HPO_4)_n(PO_4)_{6-n}(OH)_{2-n}$ ($n = Ca^{2+}$ vacancies) with Ca/P ratio <1.67). Both types of materials were impregnated by an acetone solution of $PdCl_2(CH_3CN)_2$, leading to two kinds of bound Pd complexes with similar low Pd loadings (0.02 mmol g^{-1}) despite the presence of up to 0.42 mmol g^{-1} of PO_4^{3-} on the apatite surface. Thereafter, monomeric Pd(II) species could be pointed out in both cases, but Pd to Cl was 1 : 2 for PdHA-0, and chlorine was not detected in the case of PdHA-1, leading to the proposals shown in Figure 6.4.

PdHA-0 gave rise to a very efficient catalyst in the aerobic oxidation of BA, while PdHA-1 did not. Almost complete conversion of BA into benzaldehyde was reached in one hour with *0.2% mol of Pd per BA* at 90 °C under 1 atm of O_2 in trifluorotoluene (Table 6.1, entry 8). Oxidation of a wide variety of other benzylic and allylic alcohols giving the corresponding ketones and aldehydes was reported in the same conditions. Aliphatic and heterocyclic alcohols needed more Pd catalyst (*0.6% mol Pd per substrate*) to reach completion. With BA and 1-phenylethanol, oxidation tests could be even performed in water with only *0.04% mol Pd per BA* but at 110 °C. Last but not least, a large-scale reaction was carried out with pure 1-phenyl

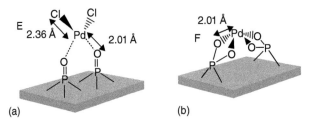

Figure 6.4 Proposed surface structures around Pd(II) center of (a) PdHA-0 and (b) PdHA-1. Source: Mori et al. [38]. / With permission of American Chemical Society.

ethanol at 160 °C during 24 hours with $4 \times 10^{-4}\%$ mol Pd per mol of substrate, giving rise to a TON of 236 000 (TOF of 9800 h^{-1}) [39]. The lack of activity of PdHA-1 might be explained by the coordination of heteroatoms to the Pd(II) centers, thus preventing their reduction to Pd(0). According to the authors, the behavior of Pd-HA0 was analogous to that of Pd nanoclusters such as Pd$_{561}$phen$_{60}$-(OAc)$_{180}$ and Pd$_{2060}$(NO$_3$)$_{360}$(OAc)$_{360}$O$_{80}$. Hence, Pd(II) would be converted in situ into Pd(0) during alcohol oxidation as demonstrated by the presence of an induction period in the time profile for PdHA-catalyzed oxidation of 1-phenylethanol at 90 °C in trifluorotoluene under O$_2$ atmosphere. This was confirmed by transmission electron microscopy (TEM) (average diameter of 3.5 nm) and Pd K-edge XAFs analyses. Using normalized TOF values, the authors suggested that the oxidation of the alcohols would occur at the low-coordination sites of the nanoparticles.

One mechanism that is generally proposed for Pd(II)X$_2$-catalyzed aerobic oxidation of alcohols is based on the formation of a Pd(II) alcoholate intermediate assisted by the deprotonation of the alcohol in the presence of an exogenous base, B, leading to the formation of BH$^+$X$^-$. Then, a Pd(II) hydride intermediate is formed upon a β-hydride elimination mechanism that is followed by a reductive elimination on the Pd(II)HX complex affording Pd(0) and HX. Pd(0) is finally re-oxidized to Pd(II) and the two X ligands reintegrated in the coordination sphere of Pd as shown in Figure 6.5a. Here, no additional base is required. The authors propose instead the occurrence of an oxidative addition step of the O—H bond of the alcohol to coordinately unsaturated Pd(0) atoms at the edge of the nanoparticles. Those Pd(0) atoms would be regenerated in two steps. First, a β-elimination would afford the carbonyl compound and a dihydride Pd(II) intermediate, then the latter would react with O$_2$, leading to H$_2$O$_2$ (or H$_2$O + ½ O$_2$) and Pd(0) (Figure 6.5b).

Aerobic oxidation of benzyl alcohols in water was also reported by Paul and co-workers [40]. Those authors proceeded by the impregnation of HA with a Pd(II)

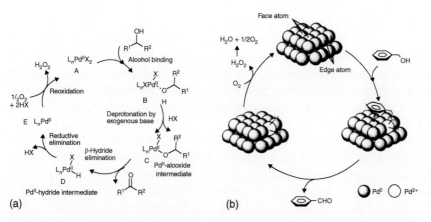

Figure 6.5 (a) Generally accepted mechanism of the aerobic Pd(II)-catalyzed oxidation of alcohols. Source: Parmeggiani and Cardona [1]. / With permission of the Royal Society of Chemistry. (b) Possible reaction mechanism for the oxidation of alcohols on the surface of Pd nanocluster. Source: Mori et al. [39]. / With permission of American Chemical Society.

complex, i.e. Pd(OAc)$_2$, but unlike Kaneda and co-workers, a reducing agent, i.e. hydrazine, was used to obtain Pd(0) nanoparticles of 20 nm average diameter (Table 6.1, entry 9). Excess of Pd(OAc)$_2$ was removed prior to the catalysis tests. It should be noted that this catalyst was much less efficient than the one reported by Kaneda and co-workers. Indeed, Paul and co-workers had to introduce *c.a.* 50 times more Pd (*0.11 vs. 0.02 mol% of Pd per substrate*) for the oxidation of BA in order to obtain a 95% yield of benzaldehyde in seven hours at 100 °C instead of one hour in the work of Kaneda et al. [38, 39]. This strong difference in reactivity may be related to the use of air instead of O$_2$. It is also likely that the size of the nanoparticles is too large. The oxidation of an aliphatic primary alcohol (1-pentanol), not described in the work of Kaneda and co-workers [38, 39], showed that the reaction was much difficult (72% yield after 48 hours), but, importantly, pentanal was shown to be selectively formed (no pentanoic acid). The absence of added base with apatite-supported Pd(0) catalysts would be responsible for such selectivity in aldehydes in water [40]. Indeed, the oxidation of aldehydes to acids is known to be favored in alkaline water because aldehydes undergo hydration to gem diols that are easily transformed into carboxylic acids.

6.3 Gas Phase Reactions

6.3.1 Partial Oxidation of Methane

Methane is an important alternative for the replacement and the diversification of the conventional sources of energy. HA-based catalysts have been investigated for its various valorization routes through oxidative coupling, steam or dry reforming, or partial oxidation of methane (POM) [2, 41, 57–64]. Only the studies related to the POM reaction (CH$_4$ + 1/2 O$_2$ → CO + 2H$_2$) will be considered in this chapter that is devoted to the selective oxidation in the presence of oxygen. Steam reforming of methane (SRM) is the dominant commercial process and is of interest for hydrogen for fuel cells applications since it provides H$_2$/CO ratio ≥3. However, compared with POM of methane, SRM exhibits some drawbacks: it is highly endothermic, it is operated under high pressure, and CO$_2$ is significantly produced by water–gas shift reaction. POM is comparatively more energy efficient since mildly exothermic and may be suitable for Fischer–Tropsch synthesis, which requires H$_2$/CO ratio close to 2 [2, 65].

Supported zerovalent Ni and Rh are accepted to be active catalysts for POM. Rhodium is known as the most active for POM, but Ni was the most extensively studied metal because it is effective and cheaper [66]. The related supported catalyst studies emphasized that strategies to avoid sintering and coke formation should to be developed [66, 67] and suggested that high dispersion of metal species and/or use of alkaline earth metals may be beneficial to improve the catalytic performance for the POM.

In this context, the Ni strontium HA system has been first investigated by Yoon et al. [65, 68, 69]. SrNiA(a) catalysts (with a = 100 × Ni/(Sr + Ni) molar percentage

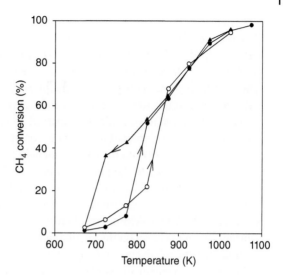

Figure 6.6 Methane conversion with sequential temperature increase and decrease over SrNiA(10), (●) first increase, (▲) decrease, (○) second increase. Source: Lee et al. [68]. / With permission of Elsevier.

varying from 1 to 20%) were prepared by a coprecipitation preparation route, but, from X-ray diffraction (XRD), mixtures of nickel substituted strontium HA and of strontium phosphate ($Sr_3(PO_4)_2$) were rather obtained. Their relative proportions were reported to depend on the Ni content, and the final global (Sr + Ni)/P molar ratios were found to vary in a large extent, from 1.5 to 3. Although metallic Ni was not detected on the calcined samples, no prereduction was required to get excellent catalytic performance. Ni addition significantly improved the results compared with Ni free HA, and at 800 °C, with $P(CH_4)/P(O_2) = 2$ and W/F = 10 mg min^{-1} ml^{-1} (W and F being the catalyst weight and the total gas flow rate, respectively). The best CO and H_2 yields were achieved for Ni/(Sr + Ni) molar ratio of 0.15, with excellent H_2 yield of 90%, which is even slightly above the equilibrium, probably due to the existence of hot spots. They observed a hysteresis upon sequential increase and decrease of the reaction temperature (Figure 6.6) and proposed that the catalyst was activated under the reducing reaction environment. This was ascribed to the extraction of Ni out of the phosphate matrix that generates highly dispersed Ni metallic active species under the reactional reductive conditions.

Nickel calcium HA system was further investigated [65]. Coprecipitation was carried out at high pH but varying the Ca/P and Ni/P precursor molar ratios in a quite large range. In these conditions, the authors obtained mixtures of calcium phosphate, calcium HA, calcium nickel phosphate, calcium nickel HA, and NiO (for high Ni loadings) (Figure 6.7a). Measuring the performance at lower contact time ($P(CH_4)/P(O_2) = 2$, W/F = 2 mg min^{-1} ml^{-1}) than in their former study dealing with the strontium HA [68], they reported excellent performance. At 800 °C, the methane conversion reached 93%, with CO and H_2 selectivities of 96 and 90%, respectively [65]. They once more concluded to the extraction of nickel out of the matrix during the POM reaction. Following the system upon sequential increase and decrease of the reaction temperature (see Figure 6.6) with related temperature programmed reduction (TPR), TEM, energy dispersive X-ray spectroscopy (EDS),

XPS characterizations, they described the evolution for the catalyst (Figure 6.7). They reported that the conversion abruptly rose from a given reaction temperature, which is explained by the reduction of NiO by the produced H_2 and CO (Figure 6.7b). In addition, in these operating conditions, the nickel present in the phosphate or apatite structures gradually got out of the phosphate matrices, formed fine particles and was reduced (Figure 6.7b). It was inferred that most of the catalytic activity should be attributed to such highly dispersed metal particles. Upon decreasing the reaction temperature, gradual reoxidation occurred, resulting in the formation of a thin layer of NiO layer onto the surface metallic nickel entities, while the extracted nickel did not go back to the phosphate or the apatite structure (Figure 6.7c). The metal state and high catalytic performance could finally be recovered upon simple increase of the reaction temperature to 450 °C (Figure 6.7b). Compared to the other Ni catalysts, Ni from calcium phosphate/HA catalysts was found to be more easily reduced, which may explain the peculiar ability of this system to be activated under reaction feed at a relatively low temperature (at or below 650 °C).

The reaction mechanism over this system was also investigated, thanks to pulse experiments [69]. It was shown that the reaction occurs primarily via the direct dissociation of methane. In addition, while metallic nickel was accepted to be an essential active component in POM, given that fully reduced catalyst exhibited lower activity, it was suggested that both metallic Ni and partially oxidized nickel might be required to get high activity and selectivity [69].

A different reaction mechanism, a two-step mechanism (complete methane oxidation followed by reforming of the residual methane with the primarily produced CO_2 and H_2O) was assumed by Boukha et al. who investigated rhodium impregnated onto non-stoichiometric HA (Ca/P = 1.5). Various Rh weight loadings ($x = 0.5$, 1, and 2 wt%) have been considered (Rh(x)/HA catalysts). From XRD, TEM, TPR, XPS, and diffuse reflectance spectroscopy (DRS) characterizations, it was found that Rh existed in three different forms in the samples as: (i) large crystallites of Rh_2O_3 deposited on the surface, (ii) RhO_x in small particles exhibiting strong interaction with the support, and (iii) Rh^{2+} species incorporated in the calcium vacancy site of the HA. Upon reduction at high temperature, the appearance of a new diffraction line ascribed to calcium triphosphate (TCP) showed that the incorporated Rh favors the thermal transformation of HA to TCP, which may be responsible for the decrease of specific surface areas observed as Rh loading increases. The POM reaction was carried out over the Rh(x)/HA catalysts after in situ reduction at 727 °C in the following conditions: $P(CH_4)/P(O_2) = 2$, $W/F = 0.3125$ mg min^{-1} ml^{-1}. Comparison of the three Rh catalysts pointed out the structure sensitivity of the Rh particles spread over the HA carrier. The best performances were obtained for Rh(1)/HA, where methane conversion, 53% at 600 °C, exceeds 46%, obtained for Rh/SiO$_2$ in similar operating conditions. In addition, the H_2/CO ratio was always higher than 2, and the value obtained at 700 °C (2.1) appears suitable for Fischer–Tropsch downstream application. The conversion and products yields were found to be very stable at least for 30 hours at 700 °C (76% conversion for Rh(1)/HA), which was explained by excellent resistance against coking. Note that, despite the limited specific surface area of its carrier (here 33 m^2 g^{-1} in the synthesis conditions applied) that was detrimental

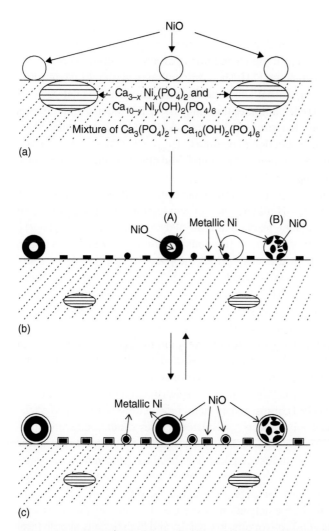

Figure 6.7 Proposed scheme of changes of Ni state in POM reaction. (a) Before the reaction. (b) During the reaction. (c) After the reaction (below 673 K) and exposure to air. Source: Jun et al. [65]. / With permission of Elsevier.

to the dispersion of rhodium, this catalyst was also found excellent for the SRM reaction. The best performance obtained for the Rh(1)/HA composition for the two reactions was explained by its higher thermal stability ascribed to its lower fraction of incorporated Rh^{2+} species that hindered its thermal transformation into TCP.

6.3.2 Alkane Oxidative Dehydrogenation Reactions

In the context of the increasing gap between the production and the demand in olefins on the world market, especially in the cases of ethylene and propylene, their production from corresponding alkanes is of great interest [4, 8, 11, 70–73]. Up to now, commercial productions of alkenes from alkanes were carried out in the absence of oxygen. However, the dehydrogenation reactions suffer from high endothermicity and thermodynamic limitations, which can be overcome by operating at high temperature. In these conditions, carbon deposition on the active sites is responsible for deactivation [74]. The introduction of oxygen in the reaction feed appears more appealing since it allows performing exothermic oxidative dehydrogenation reactions (ODH), eliminates thermodynamic limitations, and prevents the growth of carbonaceous deposits over the catalysts [11, 75]. However, such alternative ODH routes for the production of olefins ($C_nH_{2n+2} + 1/2\ O_2 \rightarrow C_nH_{2n} + H_2O$) still remain challenging. Yet no catalyst has been found efficient enough to produce competitive ethylene or propylene yields [11, 72].

The main challenge in the gas phase mild oxidation reactions is to achieve the selective activation of the C—H alkane bond at temperature low enough to limit overoxidation of both the alkane reactant and the alkene targeted product [11, 71]. Although the limiting step of the ODH reactions is commonly believed to be the abstraction of hydrogen from the adsorbed alkane [18, 74, 76, 77], there are still open questions about this activation step [71] and the factors governing the alkene selectivity [11, 78]. The proposed mechanisms for oxidative dehydrogenation of light alkanes over catalyst containing a reducible metal are essentially based on the Mars and Van Krevelen mechanism [48], which involves the cleavage of C—H bond, participation of lattice oxygen, and its in situ regeneration by gaseous oxygen present in the feed. Catalysts are, therefore, based on supported or bulk TMI, but their compositions depend on the nature of the alkane considered. It is well accepted that highly polymerized oxide-like catalytic species are detrimental to the control of the alkene selectivity, but it is still controversial whether metal units with limited nuclearity (involvement of ensemble effects) or isolated species should be favored [79–82]. In addition, the influence of acid–base properties of the support and of the nature of its interactions with the active phase have been recognized for many years [83]. Acid–base properties impact both conversion and selectivity of ODH reactions [12, 83]. On the one hand, basic properties, although less often studied than acidic ones may play a key role in the activation of the C—H bond. On the other one, according to Grzybowska and co-workers, low acidity and high basicity should favor the desorption of the olefin [77, 84].

In this context, metal-modified HA have been considered as potential candidates for the oxidative dehydrogenation of alkanes. HA should not be considered only as a simple carrier to disperse the TMI active phase, since its tunable acid–base properties make it an original cocatalyst. In addition, the numerous opportunities offered by the modulation of its composition by either cationic or anionic metal ions make it a unique catalyst system. The conditions of metal deposition (coprecipitation, IE, or impregnation as well as metal loading) are thus key parameters impacting the metal dispersion and its surface availability.

6.3.2.1 Catalytic Performance of the Metal-Modified Hydroxyapatite in the ODH Reactions

Table 6.2 gathers some results reported in literature for metal-modified HA used in the ODH of several alkanes. For the sake of comparison, the results are presented for the same reaction temperature for a given alkane, and the operating conditions have been reported using a homogeneous presentation for the reaction parameters. In line with the high demand in propylene, the ODHP was the most investigated up to now (Table 6.2, entries 1–8). Several metals, such as vanadium (Table 6.2, entries 1–5), cobalt (Table 6.2, entry 6), chromium (Table 6.2, entry 7), and iron (Table 6.2, entry 8), have been considered. Those metals have been introduced with different loadings using several preparation methods (Table 6.2). One-pot coprecipitation synthesis was achieved in the case of vanadium, the metal precursor being introduced in solution with P and Ca (or other alkaline earth metal) (Table 6.2, entries 1, 4, 5) [85, 86]. In these conditions, despite only a very limited fraction of the introduced metal is accessible on the surface, the modification of the bulk apatite could induce some benefit on the surface reactivity [87, 88]. Cation-exchange and wet impregnation were also commonly achieved for the ODHP reaction. In practice, the protocols are very similar except there is no washing step in the case of impregnation, and this later method also allows reaching higher metal loading. As shown in Table 6.2, in most cases (entries 1, 2, 5, 7–10, 13, 14), the optimal performance was obtained for low or intermediate metal loading, emphasizing that an optimal surface dispersion should be achieved. Although greatly dependent on the synthesis conditions, the highest propylene selectivity was obtained for the vanadium system. The vanadium impregnated system was also used for the oxidative dehydrogenation of n-heptane (Table 6.2, entry 12) and n-octane (Table 6.2, entries 13 and 14) that were taken as models for medium chain length linear paraffins [89–92]. Beside octenes, other mild oxidation products were formed in increasing proportions as the vanadium loading increases. The ODHE reaction was also studied on cobalt modified apatites (Table 6.2, entries 9 and 10). Despite the activation of ethane is known to be very challenging due to the very high stability of its C—H bond, promising results were reported for this system that remains stable even for high reaction temperature.

The results obtained for the various metal are detailed in the next section. Common general features observed for the investigated systems will finally allow to discuss how the HA structural properties assist the transition metal in the activation of the alkane.

6.3.2.2 Metal Ion Modifications
6.3.2.2.1 Modifications by Metal Cations
Divalent Cobalt Ions Several studies emphasized the advantageous role of cobalt in redox catalysis [100–103]. Both cobalt-modified calcium (Co-HA) (Table 6.2, entry 6) [10, 97] and strontium (Co-SrHA) (Table 6.2, entries 9 and 10) [95] HAs have been studied for ODHP and ODHE reactions, respectively. Since the influence of the metal loading was investigated for both impregnation and IE deposition methods, these Co-apatite systems provide appropriate model systems to discuss the influence of the speciation and dispersion of this redox center that can be present as isolated centers or polymeric entities.

In the case of Co-SrHA samples obtained by the IE procedure [95], the formation of supported $CoO/Sr_{10}(PO_4)_6(OH)_2$ was discarded by XANES characterization [104]. Based on the unicity of the PO_4 ^{31}P NMR signal, the authors concluded to the

Table 6.2 Explored metal-modified hydroxyapatite systems (calcium HA, except otherwise mentioned) and related properties and catalytic performance (alkane conversion and alkene selectivity) reported for the optimal composition in the mentioned conditions and at a fixed reaction temperature T_R for a given alkane.

Entry	Alkane (reaction temperature)	Metal	Metal deposition method and the M loading range explored	Composition properties	M (wt%)	SSA (m² g⁻¹)	Operating conditions	Conv. (%)	Sel. (%)	References
1	**Propane** ($T_R = 450°C$)	V	One-pot coprecipitation V/P = 0 → 0.15	V/P = 0.10 $Ca_{10}(PO_4)_{5.4}(VO_4)_{0.6}(OH)_2$	2.6[a)]	62	$P(C_3H_8)/P(O_2) = 3.5$ W/F = 16.7 mg min⁻¹ ml⁻¹	16.5	54.2	[86]
2			Wet impregnation 0–10 wt%VO_x/CaHA 0–10 wt%VO_x/SrHA		2.5 5	53 34	$P(C_3H_8)/P(O_2) = 3.5$ W/F = 16.7 mg min⁻¹ ml⁻¹	10.8 12.7	48.1 62.5	[93]
3			Wet impregnation 0–10 wt% VO_x/BaHA		5	12	$P(C_3H_8)/P(O_2) = 3.5$ W/F = 16.7 mg min⁻¹ ml⁻¹	4.4	26.5	[94]
4			One-pot coprecipitation at constant pH = 9 $Ca_{10}(PO_4)_{6-x}(VO_4)_x(OH)_2$ ($x0 → 1$)	Ca/(V + P) = 1.66 V/P = 6.69 $Ca_{10}(PO_4)_{0.78}VO_4)_{5.22}(OH)_{2-y0y}$	24*	20	$P(C_3H_8)/P(O_2) = 3.7$ W/F = 8 mg min⁻¹ ml⁻¹	5	27	[87]
5			One-pot coprecipitation with periodic base addition	$Ca_4V_4O_{14}$: V-HA = 5 : 95 wt% Epitaxial growth	18.7*	28	$P(C_3H_8)/P(O_2) = 3.7$ W/F = 8 mg min⁻¹ ml⁻¹	8.5	44	[88]
6		Co	Ion exchange from SrHA	Sr/P = 1.53 1000Co/Sr (at) = 55	2.1	56.1	$P(C_3H_8)/P(O_2) = 3.5$ W/F = 16.7 mg min⁻¹ ml⁻¹	23	48.6	[95]
7		Cr	Ion exchange 0–3.7 wt%	(Cr + Ca)/P ~ 1.5	~1.5	53–61	$P(C_3H_8)/P(O_2) = 2$ W/F = 0.66 mg min⁻¹ ml⁻¹	13	~30	[5]

#	Feed (T_R)	Metal	Preparation	Support/Notes			Conditions			Ref
8		Fe	Ion exchange 0.5–5.4 wt%	(Fe + Ca)/P = 1.63	0.5	42	$P(C_3H_8)/P(O_2) = 2$ W/F = 0.83 mg min^{-1} ml^{-1}	~4	~32–35	[96]
9	Ethane ($T_R = 550\,°C$)	Co	Ion exchange 0–1.35 wt%		1.1	34–40	$P(C_2H_6)/P(O_2) = 2$ W/F adjusted[b]	35	60	[10]
10			Wet impregnation 0–14 wt%		0.96	48–52	$P(C_2H_6)/P(O_2) = 2$ W/F adjusted[b]	16	53	[97]
11	Isobutane ($T_R = 450\,°C$)	Cr	Wet impregnation 0–7 wt%	Ca/P = 1.67	7	52.5	$P(iC_4H_{10})/P(O_2) = 1.17$ W/F = 16.67 mg min^{-1} ml^{-1}	24.4	31.1	[98]
12	n-heptane ($T_R = 500\,°C$)	V	Wet impregnation	M/P = 1.65–1.63	15	9–12	$P(n\text{-}C_7H_{16})/P(O_2) = 2$ GHSV = 4000 h^{-1}			[89]
			15 wt%/CaHA					23	42	
			15 wt%/SrHA,					25	43.5	
			15 wt%/MgHA					30	39	
			15 wt%/BaHA					26	47	
13	n-octane ($T_R = 450\,°C$)	V	Wet impregnation 2.5–15 V$_2$O$_5$ wt%	Ca/P = 1.51, SS of the support 72 m^2 g^{-1}	0.7(2.5)	51	[O$_2$] not provided GHSV = 4000 h^{-1}	29	66 (7)	[90]
				V$_2$O$_5$ ≥10 wt%, additional pyrovanadate phase	3 (10)	13		22	58 (14)	
14		V	Wet impregnation	Sr/P = 1.64 SS of the support 81.2 m^2 g^{-1}			$P(n\text{-}C_8H_{18})/P(O_2) = 1$ GHSV = 4000 h^{-1}			[99]
			5 V$_2$O$_5$ wt%/SrHA	V$_2$O$_5$	1.45	45.7		25	75 (10)	
			7.5 V$_2$O$_5$ wt%/SrHA		2.2	38		30	60 (25)	
			15 V$_2$O$_5$ wt%/SrHA	(+pyrovanadate for 15 wt%)	4.7	9.5		33	56 (29)	

In the case of n-octane, beside selectivity in aromatics and mild oxygenated compounds is additionally provided in parentheses. To facilitate the comparison, metal loading M wt% and operating conditions are reported in a homogeneous way. Some data have been recalculated from those provided by the authors and are thus reported in italics.

a) For one-pot coprecipitation syntheses, the metal loading gathers metal distributed both in the bulk and on the surface.
b) The weight has been varied to compare the system in the same conversion range.

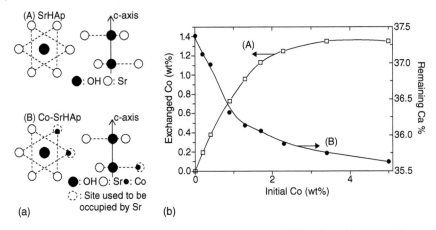

Figure 6.8 (a) Schematic illustration of local structure of SrHA (A) and Co-SrHA (B). Source: Sugiyama et al. [95]. / With permission of Elsevier. (b) Amounts of exchanged cobalt (A) and remaining calcium (B) versus initial Co content in the solution. Source: Elkabouss et al. [10] / With permission of Elsevier.

formation of $Sr_{10-x}Co_x(PO_4)_6(OH)_2$ rather than a mixture of $Sr_{10}(PO_4)_6(OH)_2$ and $Co_{10}(PO_4)_6(OH)_2$ (Figure 6.8a). They did not precise if this $Sr_{10-x}Co_x(PO_4)_6(OH)_2$ formulation refers to the formation of a bulk solid solution or if the cationic substitution (achieved by IE) is limited to the surface. This is questionable, since, given the acidic pH of the exchange solution, dissolution–precipitation may possibly occur, as clearly evidenced in the case of iron exchanged HA [96]. Although not discussed in this study, in the absence of detectable modification of the XRD patterns, the increasing broadness of the nuclear magnetic resonance (NMR) signal with increasing Co loading might be induced by long range perturbations induced by the presence of surface paramagnetic Co^{2+} ions. Using a similar preparation method starting from calcium HA, Bozon Verduraz and co-workers concluded to surface cation exchange and estimated that the Co^{2+}/Ca^{2+} level of exchange was limited to 1.35 wt% Co(Figure 6.8b) [10].

Given that no oxidation to Co^{3+} was observed even upon thermal treatment up to 550 °C in air (the maximum reaction temperature), a marked support effect was emphasized: the apatite matrix impedes the oxidation of Co^{2+} [10]. Indeed, according to magnetic studies, the cobalt remains highly dispersed as paramagnetic isolated Co^{2+} ions. Upon impregnation method, cobalt was also found to be first exchanged with Ca^{2+} as isolated Co^{2+} for cobalt loading ≤0.9 wt% [97]. The additional formation of Co_xO_y clusters, then of Co_3O_4 nanocrystals, was induced as cobalt loading increases as summarized in Figure 6.9 [97].

Sugiyama reported first that over Co-SrHA, the conversion of propane, the selectivity to propylene, and the reaction rate per unit of the corresponding surface area increased upon increasing incorporation of Co^{2+} as isolated exchanged ions (Table 6.2, entry 6) [95]. Such beneficial impact of isolated Co^{2+} species was confirmed for the ODHE reaction on Co-HA [10, 97]. In the case of exchanged samples, the ethylene yield reaches a maximum (21%) for ~1 wt% Co at 550 °C (Table 6.2,

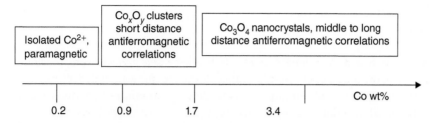

Figure 6.9 Co species appearing upon increasing the Co content deposited by impregnation method. Source: El Kabouss et al. [97]. / With permission of the Royal Society of Chemistry.

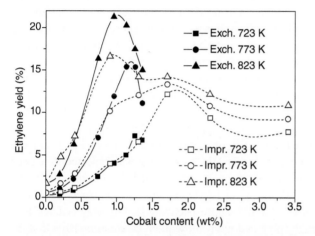

Figure 6.10 Comparison of ethylene yields measured at various reaction temperatures on cobalt-hydroxyapatite samples prepared by cation exchange and impregnation methods. Source: El Kabouss et al. [97]. / With permission of the Royal Society of Chemistry.

entry 9) [10]. Consistently, high cobalt loaded catalysts prepared by impregnation methods (involving Co_xO_y then Co_3O_4, Figure 6.9) were found much less active than those involving isolated Co^{2+} ions only prepared at lower loadings [97]. Note that the maximum yield of ethylene observed for a given reaction temperature was higher for the exchanged samples than for the impregnated ones (21.3 against 16.6% at 550 °C, Figure 6.10). This was ascribed to the lower exchange capacity when depositing surface cobalt species in impregnation conditions. The maxima observed were explained by the involvement of different Co^{2+} sites depending on Co content, the isolated cobalt cations (as Co^{2+}—O—Ca^{2+}) being found more efficient than polymeric ones.

Interestingly, the occurrence of similar maxima was observed in the butan-2-ol model reaction that almost exclusively led to butanone upon dehydrogenation route. The involvement of active basic oxygen derived from the abstraction of hydrogen from OH groups was thus proposed (see Section 3.2.3). The maximum observed was explained by a compensation of two parameters, the intrinsic dehydrogenating activity first provided by cobalt and the decrease in basicity of apatite induced by the replacement of Ca^{2+} by Co^{2+} (decrease of basicity of bridging oxygen atoms).

Trivalent Iron and Chromium Cations High alkene selectivities were previously reported in the presence of iron-containing phosphates for various dehydrogenation reactions [15, 105, 106]. From previous studies of chromium-supported catalysts for the ODHP reaction, the "Cr^{3+} clusters" were proposed to be involved as active centers for the ODHP reaction, although this point still remained controversial [107–110]. This question prompted Zyiad and co-workers to consider chromium and iron exchanged HA for the ODHP reaction (Table 6.2, entries 7, 8). The impregnation method was only tested in the case of the isobutane oxidative dehydrogenation (Table 6.2, entry 11) [98]. Although not tested in comparable conditions for the ODHP reaction, the iron or chromium exchanged systems appeared less performant than the cobalt modified one (Table 6.2, entry 6). However, these studies also clearly emphasized the beneficial role of isolated metal cations [5, 96].

In line with the acidity of the metal solution, the occurrence of a dissolution–precipitation process during the IE procedure was clearly evidenced for iron HA, which explains the significant modification of the specific surface areas of the resulting materials compared with that of the support [95]. Moreover, in the context of exchange of Ca^{2+} ions by trivalent cations, the question of the charge balance is a key issue. In fact, the extent of the Ca^{2+}/M^{3+} exchange was expected to depend on the nature and charge of the metal species predominant in solution. While formally exchanged as $[Fe(H_2O)_5(OH)]^{2+}$ or $[Cr(H_2O)_5(OH)]^{2+}$ complexes present in the precursor solutions at pH 2.3 and 4.8, respectively [5, 96], only isolated Fe^{3+} and Cr^{3+} cations were finally obtained upon subsequent calcination. Note that the oxidability of the exchanged Cr^{3+} ions was found to be limited to the lowest Cr content [5], and Cr^{6+} species, detected at very low metal loading, were finally found to be rapidly reduced by the reaction feed. For both metals, the charge balance associated with Ca^{2+}/M^{3+} exchange in the calcined samples was proposed to be adjusted by the removal of protons. As Fe loading increases, the increasing intensity of the visible absorption band at 850–910 nm shows there is an agglomeration of iron species, which is detrimental for propene yield (Figure 6.11). In addition, due to the correlation observed between the ODHP and the butanol conversion to butanone, the basicity was considered to be a key parameter to rationalize the performance. It was once again concluded that the M^{3+}—O—Ca^{2+} sites generated at low metal loading provide a reasonable performance in propane ODH, most likely because they maintain the needed basicity for the hydrogen abstraction from the propane. On the contrary, the limitation observed for higher metal content was explained by the decreased basicity induced by additional fixation of M^{3+} as M^{3+}—O—M^{3+} [5, 96].

In the case of isobutane ODH, undoped HAs were reported to be moderately active and selective. The higher performance obtained for stoichiometric HA than for under-stoichiometric ones once more supports that basic sites are essential to catalyze the ODH reaction [98]. Upon deposition of increasing loadings of chromium by impregnation, despite improvement of isobutane conversion, the selectivity to isobutene was greatly decreased, probably due the formation of

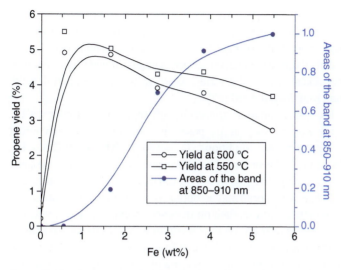

Figure 6.11 Comparison of yields obtained for Fe(x)/HA with the areas of the band at 850–910 nm versus Fe loading (blue curve). Source: Khachani et al. [96]. / With permission of Elsevier.

chromium oxide [98]. Such a tendency confirms that isolated cationic species are favorable to this class of reaction and that this configuration is better controlled upon cation exchange.

6.3.2.2.2 Modifications by Vanadate Anions

Vanadium-based systems have been extensively studied for the ODHP, and VMgO is a well-known reference catalyst for this reaction [8, 11, 72]. Among magnesium vanadates, $Mg_2V_2O_7$ is generally accepted as the most active phase. It was suggested that the V=O and/or V—O—V bonds participate to the activation of propane and that the activity may be related to the easy abstraction of lattice oxygen [11, 79].

Modifications of HAs by vanadate anions have been achieved by both one-pot coprecipitation (Table 6.2, entries 1, 4, 5) and impregnation in excess of solution (Table 6.2, entry 2, 3). Sugiyama and co-workers first investigated the coprecipitation synthesis, but in their operating conditions, they reported that the limit of vanadium incorporation by substitution was quite low (V/P $0 \rightarrow 0.15$) [86]. Promising performance, close to those reached for the $Mg_2V_2O_7$ phase under the same reaction conditions (14.0% conversion and 51% propylene selectivity), was reported for the V/P = 0.1 composition (Table 6.2, entry 1). The authors proposed that the combination of VO_4 groups and active sites of HA (see Section 3.2.3) may be involved. They also emphasized the interesting redox properties of vanadate-substituted calcium HA: the reduction of V^{5+} to V^{4+} in the used catalysts was illustrated by the drastic decrease of the ^{51}V solid-state NMR signal (that only probes V^{5+} species), and the initial signal could be easily regenerated in presence of oxygen at 450 °C [86]. Costentin and co-workers recently showed that thanks to peculiar attention devoted to the constant control of the pH during precipitation, vanadium could be successfully incorporated in the full range of the solid solution, $Ca_{10}(VO_4)_x(PO_4)_{6-x}(OH)_2$

(x $0 \rightarrow 6$) with Ca/(P+V) = 1.66 [85]. Within this series of compounds, an optimal propylene yield was observed for $x = 5.22$ (Table 6.2, entry 4). Unfortunately, despite such high bulk vanadium loading, XPS analyses and adsorption of CO probe molecule (likely to detect surface acidic vanadium species) showed that vanadate tetrahedra are deeply relaxed inside the surface layer, which may explain that the performance obtained remained limited [87].

In contrast with the coprecipitation approach, wet impregnation over CaHA, SrHA, and BaHA allowed to vary at purpose the surface vanadium content [93, 94]. Under the same reaction conditions, $T_R = 450\,°C$, $P(C_3H_8)/P(O_2) = 3.5$, W/F = 16.7 mg min^{-1} ml^{-1}, the order of both propane conversion and propene selectivity was the following VO$_x$/SrHA > VO$_x$/CaHA > VO$_x$/BaHA (Table 6.2, entries 2, 3) [93, 94]. The poor performance obtained for the barium system was ascribed too weak interactions of vanadate on the corresponding support whereas the peculiar redox properties of the vanadium strontium system would be responsible for its better properties [94]. From Table 6.2 (entries 2, 3), both the nature of the alkaline earth cation and the vanadium content greatly impact the specific area [93, 94]. It is found to decrease as V loadings increase. It is questionable if dissolution–precipitation, likely to hinder the accessibility of vanadium on the surface, might be favored as the vanadium loading increased. In addition, Friedrich and co-workers observed similar decrease of specific surface areas on the vanadium impregnated HAs (Ca, Sr, Mg, Ba), and they mentioned that corresponding pyrovanadate phase was formed in the as prepared samples for V_2O_5 weight loading higher than 10% (corresponding to 3 wt% V loading, Table 6.2, entries 12, 13) [89–91]. They evaluated the performance of these catalysts for n-heptane or n-octane ODHs (Table 6.2, entries 12–14) [89–92]. The vanadium pentoxide, which is the dominant dispersed phase in the lower loading (2.5 wt% in V_2O_5, corresponding to 0.7 wt% in vanadium), gave high selectivity toward octenes and low selectivity toward aromatics [90]. At higher loading, due to the presence of pyrovanadate phase, the octene selectivity was decreased at benefit of the formation of secondary mild oxidation products, C8 aromatics and oxygenates [90, 99]. The nature of the alkaline earth cation influences the products distributions, and highest selectivity to octane, aromatics, or oxygenates was obtained for calcium, strontium, and magnesium HAs, respectively. A possible synergistic effect due to the coexistence between the V_2O_5 and the pyrovanadate phases was proposed [99] since the highly reactive surface oxygen species of the isolated tetrahedra of the pyrovanadate phase should promote the secondary oxidation of octenes [91, 99].

Synergistic effects between vanadium-substituted HA and $Ca_4V_4O_{14}$ phases were also shown to be responsible for the significant enhancement of the propylene selectivity compared to that obtained with pure vanadium-substituted HA (Table 6.2, entry 5) [87, 88]. The $Ca_4V_4O_{14}$ phase alone led to very poor propane conversion. While continuous adjustment of the pH of the coprecipitation medium to 9 during the preparation led to the formation of the pure substituted HA, periodic base addition [85] led to the epitaxial growth of the $Ca_4V_4O_{14}$ phase onto the vanadium substituted phase [88]. Combination of HRTEM, solid-state NMR, and electron paramagnetic resonance (EPR) studies allowed to precise the structure of the interface. The best phase cooperation obtained for the phase composition VHA : $Ca_4V_4O_{14}$: = 95/5 wt% corresponds to complementary properties between the catalytic functions provided by each phase: the intrinsic ability of vanadium-substituted HA to activate propane (see Section 3.2.3) and the easier electron delocalization on the vanadium tetramers located at the phase's boundaries compared with monomeric vanadate present in V-HA that is beneficial to propene selectivity [88].

6.3.2.3 Activation Site and Mechanism

From the former section, the modification of HA by exchanged Co, Cr, or Fe cations concluded that isolated cations should be favored to achieve selective activation of alkanes. On the other hand, except for vanadium impregnated samples where oxygen of the VO_x formed at the surface is able to activate the propane, isolated VO_4 tetrahedra in $Ca_{10}(VO_4)_x(PO_4)_{6-x}(OH)_2$ solid solutions incorporated by coprecipitation were found to be too poorly accessible for efficient activation of propane. This raises the question of the nature of the active site able to activate light alkane over metal-modified HA. Given that the limiting step is H abstraction, a beneficial influence of basicity on the activation process was proposed [84, 87, 96, 111, 112]. It is well supported by the established relationships between the alkane conversion and basic reactivity probed by model reactions [10]. The nature of basic entities able to activate alkane deserves discussion.

Due to their electrophilic properties, adsorbed active O^- and O_2^- oxygen possibly present on HA surface are expected to be associated with nonselective oxidation [70, 78]. At reverse, nucleophilic lattice O^{2-} oxide ions should be preferentially involved in the selective process. In the HA structure, among the two types of lattice oxygen species present in phosphate and hydroxyl groups, the involvement of O^{2-} from PO_4^{3-} was discarded on the basis on the much lower activity of $Ca_3(PO_4)_2$ compound compared with that of HA [113]. On the contrary, based on the progressive deactivation associated with the conversion of HA into corresponding chloroapatites in the presence of tetrachloromethane in the feed stream and to the rather low activities observed on chloroapatite [113, 114], it was inferred that hydroxyl groups must somehow participate to the activation process of light alkanes [58, 89, 95, 104]. However, hydroxyl groups being weak bases [115, 116], it was also underlined that it would be surprising that OH^- groups of HA directly contribute to hydrogen abstraction of alkanes [95, 104]. However, Sugiyama et al. rather assumed that OH^- of HA may be a source of active oxygen species [104]. Described as similar to lattice oxygen in oxide catalysts, based on results obtained for SrHA with and without Co^{2+} doping, they proposed that this active oxygen species would derive from OH^- by hydrogen desorption [86, 95, 104]. They studied the gas phase H—D exchangeability of hydroxyl groups of calcined HA by D_2O at 450 °C. Solid state ^1H MAS NMR characterizations of SrHA and Co-SrHA allowed to evaluate the H—D exchangeability of OH groups following the signal at ~0 ppm [95, 104]. Apart a progressive decrease of the chemical shift of this peak, its intensity decreased with increasing Co^{2+} content. Although not considering some misleading effect due to the paramagnetism induced by Co^{2+}, the authors concluded to the enhancement of H—D exchangeability [95, 104]. Hydrogen abstraction ability could also be achieved with CH_3OD [117]. Although these H—D exchange experiments could not provide direct evidence for the stable presence of oxygen species that would be involved in the activation of propane, the authors underlined that the tendencies observed strongly support that active oxygen species should be formed during the catalytic reaction and that the modification of HA by cobalt would favor this process [95].

The generation of active O^{2-} from OH^- was later discussed in relation with intrinsic ionic conduction properties of HA [10, 87, 118, 119]. In the case of cation exchange of Ca^{2+} by trivalent cations, the OH deprotonation required to ensure the charge balance was proposed to generate $Ca^{2+}-O-M^{3+}$, which is able to activate propane [5, 96]. In the case of divalent Co^{2+} ions modification (introduced by cation exchange), it was hypothesized that the formation of active O^{2-} may be favored by the mobility of neighboring lattice oxygen or OH^- groups and may thus influence the activation process [10]. The authors proposed a mechanism described by Eqs. (6.1–6.2), where O_o is a lattice oxygen and V_o a neutral oxygen vacancy. This oxygen present in $Co^{2+}-O-Ca^{2+}$ entities would both activate ethane and act as labile oxygen. The neutral oxygen vacancy formed upon its departure would be further replenished by gaseous oxygen [10].

$$C_2H_6 + O_o \rightarrow C_2H_4 + H_2O + V_o \qquad (6.1)$$

$$\tfrac{1}{2}O_2 + V_o \rightarrow O_o \qquad (6.2)$$

Such key role played by the bulk anisotropic conduction onto the surface reactivity was later unambiguously demonstrated by Costentin and co-workers to rationalize the behavior of solid solution of vanadium-substituted apatite for ODHP, thanks to combination of advanced in situ and operando characterization techniques [87]. First, ^1H NMR evidenced that there was a lack of protons inside the OH columns, while the vanadium was increasingly incorporated in the bulk lattice by coprecipitation achieved maintaining the pH constant to 9 value. In addition, the formation of the vanadium oxy-hydroxy-apatite $Ca_{10}(PO_4)_{6-x}(VO_4)_x(OH)_{2-2y}O_y$ solid solution was favored upon thermal treatment, thanks to partial dehydration along the OH columns generating additional O^{2-} oxide ions. Their formation was also supported by in situ DRIFT with the appearance of a characteristic shoulder at ~3530 cm^{-1} that was ascribed to H-bonding interaction between the generated O^{2-} oxide ions and the remaining adjacent O—H groups inside the channels. These bulk O^{2-} species were found to be stable at high temperature. According to impedance spectroscopy, they initiate the activation of proton anisotropic ionic conductivity along the columns running along the c-axis. This proton migration process eventually resulted in the exposure of surface O^{2-} species (Figure 6.12) that could act as strong basic sites. This is supported by the ability of surface $Ca^{2+}-O^{2-}$ pairs to dissociate H_2 [87]. The key role played in the activation of propane by these strong basic O^{2-} ions was supported by the unique synergistic effect between the propane to propene catalytic reaction and the proton mobility process evidenced by operando impedance spectroscopy. It is shown that the transformation of propane to propene proceeds both upon ODH (O_2 is required for propene formation) and dehydrogenation reaction (minor route).

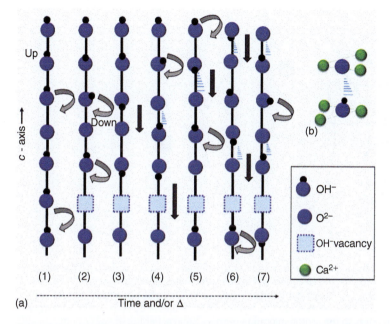

Figure 6.12 (a) Schematic representation of the thermally activated proton migration mechanism in the oxy-hydroxy-apatite system. This schematic representation shows (1) the stacking of OH^- along the c-axis including O^{2-} native defects present in vanadium-modified apatites and the proton rotation activated at low temperature that are favored by the polarization induced by the defective stacking, (2) O^{2-} species generated upon thermal dehydration together with a OH vacancy site, (3–7) and the proton transportation along the c-axis activated in the medium range temperature. The proton migration occurs from the upper layers to the lower ones with time, resulting in O^{2-} climbing up from the bulk to the surface. Slight displacement of the O^{2-} ion from the initial position of the OH^- group allows the formation of O^{2-}...HO^- H-bonding interaction. (b) Ca^{2+} triangles coordination sphere of the OH^- and O^{2-} anions present in the channels. Source: Petit et al. [87]. / With permission of John Wiley & Sons.

Finally, based on all the collected data, a catalytic cycle for the ODHP reaction was proposed (Figure 6.13). It requires two types of surface oxygen ions: a basic O^{2-} species emerging from the OH channels of the apatite framework and a labile oxygen from the VO_4 tetrahedra. The former species is formed and regenerated upon the assistance of the peculiar proton conduction ability of the apatite material. The latter species can be protonated by proton diffusion and can contribute to the formation/desorption of a water molecule. This oxygen species is eventually regenerated by oxidation with gaseous oxygen.

6.4 Conclusions and Perspectives

Metal-modified HAs have been studied since the 1990s as heterogeneous catalysts for selective oxidation reactions. Very interesting results have already been reported for the oxidation of primary alcohols in the liquid phase under aerobic conditions

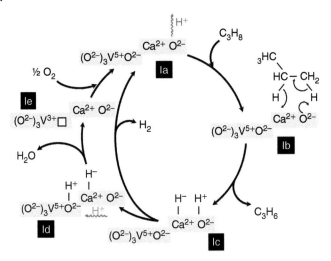

Figure 6.13 Schematized representation of the catalytic transformation of propane to propene involving dehydrogenation and oxidative dehydrogenation routes on V-HA samples exhibiting $Ca^{2+}-O^{2-}$ surface acid–base pairs generated, thanks to proton migration process that are close to VO_4 tetrahedra. The square box represents an oxygen vacancy and refers to the proton migration process. Source: Petit et al. [87] / With permission of John Wiley & Sons.

and the partial methane oxidation operating in gas phase at very high temperature (up to 800 °C). This illustrates the large application scope of the HA system due to its stability in a very large range of conditions.

From the analysis of the literature dealing with the selective oxidation of primary alcohols in the liquid phase under aerobic conditions over metal-modified HA, Pd-based systems are quite interesting. Despite the higher price of Pd compared with Ru, the latter has to be used at a higher loading. Significant improvements have however been achieved in reducing the amount of Ru while maintaining good catalytic performance but complicating the catalyst synthesis procedure (addition of another metal or of organic additives). The reference substrate for aerobic oxidations of primary alcohol is clearly benzyl alcohol. It should be noted that its conversion into benzaldehyde by aerobic and heterogeneous catalysis in liquid phase is of interest because of the use of this compound as a flavoring agent (7000 tons produced per year in 1999) [118]. Other aldehydes of industrial importance are cinnamaldehyde and vanillin, but the one that is the most industrially produced is formaldehyde because of its involvement in the synthesis of polymeric materials. Gas phase aerobic oxidation of methanol was not examined in this chapter, but it can be noted that there are some reports dealing with the use of HA-supported molybdenum oxide catalysts for this reaction [119].

Rhodium-supported HA shows the best performance for the POM. Although also very promising, the nickel catalyst obtained from coprecipitation synthesis is a complex multiphasic sample. It is thus unclear if the highly dispersed active Ni species that were extracted from the bulk to the surface during the POM reaction comes from the Ni-substituted HA and/or from the other Ni phosphate phase. Optimization of the precipitation conditions to get single phase Ni-HA would be useful to investigate this point. In addition, comparisons with the highly dispersed nickel or even isolated single Ni atoms obtained upon IE and reported to be very efficient for the dry reforming of methane [64] would also be of interest.

The oxidative dehydrogenation of light alkanes into alkenes remains more challenging since the lower stability of alkenes compared with alkanes and CO_2 makes the control of alkene selectivity difficult in the presence of oxygen. Nevertheless, quite promising results have been reported with the vanadium system for propane conversion to propene, with performances very comparable to those of a well-known reference catalyst. Cobalt exchanged HA also shows good selectivity for the oxidative dehydrogenation of ethane.

For all the investigated systems, further optimization requires better control of the synthesis and advanced characterizations, allowing to establish fine structure–reactivity relationships. In particular, better description of the role played by the HA itself in the selective oxidation reactions is of peculiar interest.

In most studies, HA was mainly envisaged as a support for the dispersion of the active phase. The formation of highly dispersed metal sites is favored due to the peculiar ability of the HA structure for substitution, and it can be achieved either by one-step coprecipitation procedure or by post-synthesis cation exchange in the excess of solution. Wet impregnation approach may also lead to highly dispersed species, but nonselective oxide like active phase could be preferentially obtained for high metal loadings. This may explain that an optimum metal loading was often identified, namely, for Co impregnated system for the ODHE reaction. Better attention should also be devoted to the control of dissolution–precipitation of the support possibly occurring in acid conditions imposed by the metal salt containing solution since such process may be detrimental to the specific surface area and to the metal accessibility onto the surface. Although reported efficient for vanadium-substituted HA, coprecipitation is a priori less favorable in terms of economy of metal atoms. Moreover, as mentioned previously, depending on the metal, the formation of single phased metal-HA is not easy to monitor. Note, however, that, as sometimes described for other systems also governed by the Mars and Van Krevelen mechanism, beneficial synergistic effects resulting from the complementary functions provided by each phase were finally observed for a vanadate tetrameric phase grown onto a vanadium-substituted HA.

Beside the control of the metal dispersion, HA is an original support with modulating acid–base properties that can be monitored by varying its stoichiometry and/or by substituting calcium by other alkaline earth cations. As far as aerobic oxidation of alcohol reaction is concerned, although the authors preferentially used stoichiometric HA, known to exhibit basic properties, there is no mention of the possible involvement of the basic sites of apatite itself. Note, however, that the authors point out that the Ru or Pd/apatite catalytic systems avoid the use of any additional base. It could be interesting to investigate if the basic properties of the HA itself take part in the liquid phase oxidation reactions. The direct involvement of the HA as part of the active phase was more directly discussed in the case of alkane ODHs. Similar tendencies were reported between the basic reactivity evaluated by a model reaction and the ODH activity. Rather than OH groups that are too weak bases, their derived O^{2-} form, which are in situ generated via the thermally activated proton ionic conduction process, has been shown to be much stronger basic sites and to be involved in the activation of the C—H bond of propane. Such role played by basic sites of HA in the ODH mechanism illustrates that metal-modified HA system is original and can be very efficient bifunctional redox and acid–base catalysts.

References

1 Parmeggiani, C. and Cardona, F. (2012). Transition metal based catalysts in the aerobic oxidation of alcohols. *Green Chem.* 14: 547–564.
2 Boukha, Z., Gil-Calvo, M., de Rivas, B. et al. (2018). Behaviour of Rh supported on hydroxyapatite catalysts in partial oxidation and steam reforming of methane: on the role of the speciation of the Rh particles. *Appl Catal A.* 556: 191–203.
3 Jiménez-González, C., Boukha, Z., de Rivas, B. et al. (2013). Catalytic performance of Ni/CeO$_2$/X-ZrO$_2$ (X = Ca, Y) catalysts in the aqueous-phase reforming of methanol. *Appl Catal A.* 466: 9–20.
4 Dong, S., Altvater, N.R., Mark, L.O., and Hermans, I. (2021). Assessment and comparison of ordered & non-ordered supported metal oxide catalysts for upgrading propane to propylene. *Appl Catal A.* 617: 118121.
5 Boucette, C., Kacimi, M., Ensuqye, A. et al. (2009). Oxidative dehydrogenation of propane over chromium-loaded calcium-hydroxyapatite. *Appl Catal A.* 256: 201–210.
6 Cavani, F. and Trifiro, F. (1999). Selective oxidation of light alkanes: interaction between the catalyst and the gas phase on different classes of catalytic materials. *Catal Today.* 51: 561–580.
7 Centi, G., Cavani, F., and Trifiro, F. (2001). *Selective Oxidation by Heterogeneous Catalysis*. New York: Publishers KAP.
8 Cavani, F. (2010). Catalytic selective oxidation: the forefront in the challenge for a more sustainable chemical industry. *Catal Today.* 157 (1): 8–15.
9 Corberan, V.C. (2009). Nanostructured oxide catalysts for oxidative activation of alkanes. *Top Catal.* 52: 962–969.
10 Elkabouss, K., Kacimi, M., Ziyad, M. et al. (2004). Cobalt-exchanged hydroxyapatite catalysts: magnetic studied, spectroscopic investigations, performance in 2-butanol and ethane oxidative dehydrogenations. *J Catal.* 226: 16–24.
11 Carrero, C.A., Schloegl, R., Wachs, I.E., and Schomaecker, R. (2014). Critical literature review of the kinetics for the oxidative dehydrogenation of propane over well-defined supported vanadium oxide catalysts. *ACS Catal.* 4: 3357–3380.
12 Concepción, P., Blasco, T., López Nieto, J.M. et al. (2004). Preparation, characterization and reactivity of V- and/or Co-containing AlPO-18 materials (VCoAPO-18) in the oxidative dehydrogenation of ethane. *Micropor Mesopor Mater.* 67 (2): 215–227.

13 Savary, L., Saussey, J., Costentin, G. et al. (1996). Propane oxydehydrogenation reaction on a VPO/TiO$_2$ catalyst. Role of the nature of acid sites determined by dynamic in situ IR studies. *Catal Today.* 32: 57–61.

14 Savary, L., Saussey, J., Costentin, G. et al. (1996). Role of the nature of the acid sites in the oxydehydrogenation of propane on a VPO/TiO$_2$ catalyst. An in situ FT-IR spectroscopy investigation. *Catal Lett.* 38: 197–201.

15 Millet, J.M.M. and Védrine, J.C. (2001). Importance of site isolation in the oxidation of isobutyric acid to methacrylic acid on iron phosphate catalysts. *Top Catal.* 15 (2–4): 139–144.

16 Grasselli, R.K. (2014). Site isolation and phase cooperation: two important concepts in selective oxidation catalysis: retrospective. *Catal Today.* 238: 10–27.

17 Callahan, J.L. and Grasselli, R.K. (1963). A selectivity factor in vapor-phase hydrocarbon oxidation catalysis. *AIChE J.* 9 (6): 755–760.

18 Singh, R.P., Bañares, M.A., and Deo, G. (2005). Effect of phosphorous modifier on V$_2$O$_5$/TiO$_2$ catalyst: ODH of propane. *J Catal.* 233 (2): 388–398.

19 Centi, G., Trifirò, F., Busca, G. et al. (1989). Nature of active species of (VO)$_2$P$_2$O$_7$ for selective oxidation of *n*-butane to maleic anhydride. *Faraday Disc Chem Soc.* 87 (0): 215–225.

20 Wang, Y. and Otsuka, K. (1997). Partial oxidation of ethane by reductively activated oxygen over iron phosphate catalyst. *J Catal.* 171: 106–114.

21 Savary, L., Costentin, G., Maugé, F. et al. (1997). Characterization of AgMo$_3$P$_2$O$_{14}$ catalyst active in propane mild oxidation. *J Catal.* 169 (1): 287–300.

22 Savary, L., Costentin, G., Bettahar, M.M. et al. (1996). Structure-sensitivity study of partial propene oxidation over AV$_2$P$_2$O$_{10}$ vanadium phosphate compounds. *J Chem Soc Faraday Trans.* 92 (8): 1423–1428.

23 Costentin, G., Savary, L., Lavalley, J.C. et al. (1998). Structural effects on propane mild oxidation from comparative performances of molybdenum and vanadium phosphate model catalysts. *Chem Mater.* 10 (1): 59–64.

24 Savary, L., Costentin, G., Bettahar, M.M. et al. (1996). Effects of the structural and cationic properties of AV$_2$P$_2$O$_{10}$ solids on propane selective oxidation. *Catal Today.* 32: 305–309.

25 Sheldon, R.A. (2017). The E factor 25 years on: the rise of green chemistry and sustainability. *Green Chem.* 19: 18–43.

26 Alfonsi, K., Colberg, J., Dunn, P.J. et al. (2008). Green chemistry tools to influence a medicinal chemistry and research chemistry based organization. *Green Chem.* 10: 31–36.

27 Gunasekaran, N. (2015). Aerobic oxidation catalysis with air or molecular oxygen and ionic liquids. *Adv Synth Catal.* 357: 1990–2010.

28 Cornell, C.N. and Sigman, M.S. (2007). Recent progress in Wacker oxidations: moving toward molecular oxygen as the sole oxidant. *Inorg Chem.* 46: 1903–1909.

29 Centi, G., Cavani, F., and Trifirò, F. (2012). *Selective Oxidation by Heterogeneous Catalysis*. New York: Media SSB.

30 Woo Lee, H., Nam, H., Han, G.H. et al. (2021). Solid-solution alloying of immiscible Pt and Au boosts catalytic performance for H_2O_2 direct synthesis. *Acta Mater.* 205: 116563.

31 Yamaguchi, K., Mori, K., Mizugaki, T. et al. (2000). Creation of a monomeric Ru species on the surface of hydroxyapatite as an efficient heterogeneous catalyst for aerobic alcohol oxidation. *J Am Chem Soc.* 122: 7144–7145.

32 Shaabani, A., Afaridoun, H., and Shaabani, S. (2016). Natural hydroxyapatite-supported MnO_2: a green heterogeneous catalyst for selective aerobic oxidation of alkylarenes and alcohols. *Appl Organomet Chem.* 30: 772–776.

33 Shaabani, A., Shaabani, S., and Afaridoun, H. (2016). Highly selective aerobic oxidation of alkyl arenes and alcohols: cobalt supported on natural hydroxyapatite nanocrystals. *RSC Adv.* 6 (54): 48396–48404.

34 Opre, Z., Grunwaldt, J.D., Mallat, T., and Baiker, A. (2005). Selective oxidation of alcohols with oxygen on Ru-Co-hydroxyapatite: a mechanistic study. *J Mol Catal.* 242 (1–2): 224–232.

35 Opre, Z., Grunwaldt, J.D., Maciejewski, M. et al. (2005). Promoted Ru–hydroxyapatite: designed structure for the fast and highly selective oxidation of alcohols with oxygen. *J Catal.* 230: 406–419.

36 Opre, Z., Ferri, D., Krumeich, F. et al. (2006). Aerobic oxidation of alcohols by organically modified ruthenium hydroxyapatite. *J Catal.* 241 (2): 287–295.

37 Mori, K., Kanai, S., Hara, T. et al. (2007). Development of ruthenium-hydroxyapatite-encapsulated superparamagnetic $\gamma\text{-}Fe_2O_3$ nanocrystallites as an efficient oxidation catalyst by molecular oxygen. *Chem Mater.* 19: 1249–1256.

38 Mori, K., Yamaguchi, K., Hara, T. et al. (2002). Controlled synthesis of hydroxyapatite-supported palladium complexes as highly efficient heterogeneous catalysts. *J Am Chem Soc.* 124 (39): 11572–11573.

39 Mori, K., Hara, T., Mizugaki, T. et al. (2004). Hydroxyapatite-supported palladium nanoclusters: a highly active heterogeneous catalyst for selective oxidation of alcohols by use of molecular oxygen. *J Am Chem Soc.* 126 (34): 10657–10666.

40 Jamwal, N., Gupta, M., and Paul, S. (2008). Hydroxyapatite-supported palladium (0) as a highly efficient catalyst for the Suzuki coupling and aerobic oxidation of benzyl alcohols in water. *Green Chem.* 10 (9): 999–1003.

41 Sugiyama, S., Minami, T., Hayashi, H. et al. (1996). Enhancement of the selectivity to carbon monoxide with feedstream doping by tetrachloromethane in the oxidation of methane on stoichiometric calcium hydroxyapatite. *J Chem Soc Faraday Trans.* 92 (2): 293–299.

42 Kaneda, K. and Mizugaki, T. (2009). Development of concerto metal catalysts using apatite compounds for green organic syntheses. *Energy Environ Sci.* 2: 655–673.

43 Abad, A., Almela, C., Corma, A., and García, H. (2006). Efficient chemoselective alcohol oxidation using oxygen as oxidant. Superior performance of gold over palladium catalysts. *Tetrahedron Lett.* 62: 6666–6672.

44 Kaneda, K., Mori, K., Hara, T. et al. (2004). Design of hydroxyapatite-bound transition metal catalysts for environmentally-benign organic syntheses. *Catal Surv Asia.* 8: 231–239.

45 Matsumoto, M. and Watanabe, N. (1984). Oxidation of allylic alcohols to unsaturated carbonyl compounds by ruthenium dioxide and dioxygen/ruthenium dioxide. *J Org Chem.* 49: 3435–3436.

46 Kaneda, K. and Mizugaki, T. (2017). Design of high-performance heterogeneous catalysts using apatite compounds for liquid-phase organic syntheses. *ACS Catal.* 7: 920–935.

47 Lempers, H.E.B. and Sheldon, R.A. (1998). The stability of chromium in CrAPO-5, CrAPO-11, and CrS-1 during liquid phase oxidations. *J Catal.* 175: 62–69.

48 Mars, P. and van Krevelen, D.W. (1954). Oxidations carried out by means of vanadium oxide catalysts. *Chem Eng Sci.* 3: 41.

49 El Shafei, G.M.S. and Moussa, N.A. (2021). Adsorption of some essential amino acids on hydroxyapatite. *J Colloid Interface Sci.* 238: 160–166.

50 Moody, C.J. and Palmer, F.N. (2002). Dirhodium(II) carboxylate-catalysed oxidation of allylic and benzylic alcohols. *Tetrahedron Lett.* 43: 139–141.

51 Muldoon, J. and Brown, S.N. (2002). Practical Os/Cu-cocatalyzed air oxidation of allyl and benzyl alcohols at room temperature and atmospheric pressure. *Org Lett.* 4: 1043–1045.

52 Martín, S.E. and Suárez, D.F. (2002). Catalytic aerobic oxidation of alcohols by $Fe(NO_3)_3$—$FeBr_3$. *Tetrahedron Lett.* 43: 4475–4479.

53 Vinod, C.P., Wilson, K., and Lee, A.F. (2011). Recent advances in the heterogeneously catalysed aerobic selective oxidation of alcohols. *J Chem Technol Biotechnol.* 86: 161–171.

54 Kaneda, K., Ebitani, K., Mizugaki, T., and Mori, K. (2006). Design of high-performance heterogeneous metal catalysts for green and sustainable chemistry. *Bull Chem Soc Jpn.* 79: 981–1016.

55 Davis, S.E., Ide, M.S., and Davis, R.J. (2013). Selective oxidation of alcohols and aldehydes over supported metal nanoparticles. *Green Chem.* 15: 17–45.

56 Chen, J., Zhang, Q., Wang, Y., and Wan, H. (2008). Size-dependent catalytic activity of supported palladium nanoparticles for aerobic oxidation of alcohols. *Adv Synth Catal.* 350: 453–464.

57 Matsumura, Y., Sugiyama, S., Hayashi, H., and Moffat, J.B. (1995). An apparent ensemble effect in the oxidative coupling of methane on hydroxyapatites with incorporated lead. *Catal Lett.* 30 (1–4): 235–240.

58 Sugiyama, S., Iguchi, Y., Minami, T. et al. (1997). Partial oxidation of methane to methyl chloride with tetrachloromethane on strontium hydroxyapatites ion-exchanged with lead. *Catal Lett.* 46 (3–4): 279–285.

59 Sugiyama, S., Minami, T., Moriga, T. et al. (1996). Surface and bulk properties, catalytic activities and selectivities in methane oxidation on near-stoichiometric calcium hydroxyapatites. *J Mater Chem.* 6 (3): 459–464.

60 Oh, S.C., Xu, J., Tran, D.T. et al. (2018). Effects of controlled crystalline surface of hydroxyapatite on methane oxidation reactions. *ACS Catal.* 8 (5): 4493–4507.

61 Boukha, Z., Kacimi, M., Ziyad, M. et al. (2007). Comparative study of catalytic activity of Pd loaded hydroxyapatite and fluoroapatite in butan-2-ol conversion and methane oxidation. *J Mol Catal A.* 270 (1): 205–213.

62 Zhang, Y., Zhu, J., Li, S. et al. (2020). Rational design of highly H_2O- and CO_2-tolerant hydroxyapatite-supported Pd catalyst for low-temperature methane combustion. *ChemEng J.* 396: 125225.

63 Boukha, Z., Kacimi, M., Pereira, M.F.R. et al. (2007). Methane dry reforming on Ni loaded hydroxyapatite and fluoroapatite. *Appl Catal A.* 317 (2): 299–309.

64 Akri, M., Shu Zhao, S., Li, X. et al. (2019). Atomically dispersed nickel as coke-resistant active sites for methane dry reforming. *Nat Commun.* 10 (1): 1–10.

65 Jun, J.H., Lee, T.-J., Lim, T.H. et al. (2004). Nickel–calcium phosphate/hydroxyapatite catalysts for partial oxidation of methane to syngas: characterization and activation. *J Catal.* 221 (1): 178–190.

66 Choudhary, V.R., Uphade, B.S., and Mamman, A.S. (1997). Oxidative conversion of methane to syngas over nickel supported on commercial low surface area porous catalyst carriers precoated with alkaline and rare earth oxides. *J Catal.* 172: 281–293.

67 Liu, Z.W., Jun, K.W., Roh, H.S. et al. (2002). Partial oxidation of methane over nickel catalysts supported on various aluminas. *Korean J Chem Eng.* 19 (5): 735–741.

68 Lee, S.J., Jun, J.H., Lee, S.-H. et al. (2002). Partial oxidation of methane over nickel-added strontium phosphate. *Appl Catal A.* 230 (1): 61–71.

69 Jun, J.H., Lim, T.H., Nam, S.W. et al. (2006). Mechanism of partial oxidation of methane over a nickel–calcium hydroxyapatite catalyst. *Appl Catal A.* 312: 27–34.

70 Blasco, T. and Lopez Nieto, J.M. (1997). Oxidative dehydrogenation of short chain alkanes on supported vanadium oxide catalysts. *Appl Catal A.* 157: 117–142.

71 Grabowski, R. (2006). Kinetics of oxidative dehydrogenation of C2–C3 alkanes on oxide catalysts. *Catal Rev.* 48 (2): 199–268.

72 Cavani, F., Ballarini, N., and Cericola, A. (2007). Oxidative dehydrogenation of ethane and propane: how far from commercial implementation? *Catal Today.* 127: 113–131.

73 Kube, P., Frank, B., Wrabetz, S. et al. (2017). Functional analysis of catalysts for lower alkane oxidation. *ChemCatChem.* 9: 573–585.

74 Airaksinen, S.M.K., Bañares, M.A., and Krause, A.O.I. (2005). In situ characterisation of carbon-containing species formed on chromia/alumina during propane dehydrogenation. *J Catal.* 230 (2): 507–513.

75 Fuchs, S., Leveles, L., Seshan, K. et al. (2001). Oxidative dehydrogenation and cracking of ethane and propane over LiDyMg mixed oxides. *Top Catal.* 15 (2–4): 169–174.

76 Kung, H.H. (1994). Oxidative dehydrogenation of light (C2 to C4) alkanes. In: *Advances in Catalysis* (ed. D.D. Eley, H. Pines and W.O. Haag), 1–38. New York: Academic Press.

77 Klisinska, A., Samson, K., Gressel, I., and Grzybowska, G.B. (2006). Effect of additives on properties of V_2O_5/SiO_2 and V_2O_5/MgO catalysts I. Oxidative dehydrogenation of propane and ethane. *Appl Catal A.* 309: 10–16.

78 Banares, M.A. (1999). Supported metal oxide and other catalysts for ethane conversion: a review. *Catal Today.* 51: 319–348.

79 Barman, S., Maity, N., Bhatte, K. et al. (2016). Single-site VO_x moieties generated on silica by surface organometallic chemistry: a way to enhance the catalytic activity in the oxidative dehydrogenation of propane. *ACS Catal.* 6: 5908–5921.

80 Chalupka, K., Thomas, C., Millot, Y. et al. (2013). Mononuclear pseudo-tetrahedral V species of VSiBEA zeolite as the active sites of the selective oxidative dehydrogenation of propane. *J Catal.* 305: 46–55.

81 Cheng, M.J., Chenoweth, K., Oxgaard, J. et al. (2007). Single-site vanadyl activation, functionalization, and reoxidation reaction mechanism for propane oxidative dehydrogenation on the cubic V_4O_{10} cluster. *J Phys Chem C.* 111: 5115–5127.

82 Corma, A., Lopez-Nieto, J.M., and Paredes, N. (1993). Influence of the preparation methods of V–Mg–O catalysts on their catalytic properties for the oxidative dehydrogenation of propane. *J Catal.* 144: 425–433.

83 Védrine, J.C. (2002). The role of redox, acid–base and collective properties and of crystalline state of heterogeneous catalysts in the selective oxidation of hydrocarbons. *Top Catal.* 21 (1–3): 97–106.

84 Klisinska, A., Loridant, S., Grzybowska, B. et al. (2006). Effect of additives on properties of V_2O_5/SiO_2 and V_2O_5/MgO catalysts II. Structure and physicochemical properties of the catalysts and their correlations with oxidative dehydrogenation of propane and ethane. *Appl Catal A.* 309: 17–26.

85 Petit, S., Gode, T., Thomas, C. et al. (2017). Incorporation of vanadium in the framework of hydroxyapatites: importance of the vanadium content and pH conditions during the precipitation step. *Phys Chem Chem Phys.* 19: 9630–9640.

86 Sugiyama, S., Osaka, T., Hirata, Y., and Sotowa, K.I. (2006). Enhancement of the activity for oxidative dehydrogenation of propane on calcium hydroxyapatite substituted with vanadate. *Appl Catal A.* 312: 52–58.

87 Petit, S., Thomas, C., Millot, Y. et al. (2020). Activation of C—H bond of propane by strong basic sites generated by bulk proton conduction on V-modified hydroxyapatites for the formation of propene. *ChemCatChem.* 12: 2506–2521.

88 Petit S, Thomas C, Millot Y, Averseng F, Brouri D, Krafft JM, et al. (2021). Synergistic effect between $Ca_4V_4O_{14}$ and vanadium-substituted hydroxyapatite in the oxidative dehydrogenation of propane. *ChemCatChem.* 13(18): 3995–4009. http://dx.doi.org/https://doi.org/10.1002/cctc.202100807.

89 Dasireddy, V.D.B.C., Singh, S., and Friedrich, H.B. (2015). Effect of the support on the oxidation of heptane using vanadium supported on alkaline earth metal hydroxyapatites. *Catal Lett.* 145: 668–678.

90 Dasireddy, V.D.B.C., Singh, S., and Friedrich, H.B. (2012). Oxidative dehydrogenation of *n*-octane using vanadium pentoxide-supported hydroxyapatite catalysts. *Appl Catal A.* 421–422: 58–69.

91 Dasireddy, V.D.B.C., Friedrich, H.B., and Singh, S. (2013). Studies towards a mechanistic insight into the activation of *n*-octane using vanadium supported on alkaline earth metal hydroxyapatites. *Appl Catal A.* 467: 142–153.

92 Dasireddy, V.D.B.C., Singh, S., and Friedrich, H.B. (2013). Activation of *n*-octane using vanadium oxide supported on alkaline earth hydroxyapatites. *Appl Catal A.* 456: 105–117.

93 Sugiyama, S., Osaka, T., Ueno, Y., and Sotowa, K.I. (2008). Oxidative dehydrogenation of propane over vanadate catalysts supported on calcium and strontium hydroxyapatites. *J Jpn Petrol Inst.* 51 (1): 50–57.

94 Sugiyama, S., Sugimoto, N., Hirata, Y. et al. (2008). Oxidative dehydrogenation of propane on vanadate catalysts supported on various metal hydroxyapatites. *Phosphorus Res Bull.* 22: 13–16.

95 Sugiyama, S., Shono, T., Makino, D. et al. (2003). Enhancement of the catalytic activities in propane oxidation and H—D exchangeability of hydroxyl groups by the incorporation with cobalt into strontium hydroxyapatite. *J Catal.* 214: 8–14.

96 Khachani, M., Kacimi, M., Ensuque, A. et al. (2010). Iron–calcium–hydroxyapatite catalysts: Iron speciation and comparative performances in butan-2-ol conversion and propane oxidative dehydrogenation. *Appl Catal A.* 388: 113–123.

97 El Kabouss, K., Kacimi, M., Ziyad, M. et al. (2006). Cobalt speciation in cobalt oxide-apatite materials: structure–properties relationship in catalytic oxidative dehydrogenation of ethane and butan-2-ol conversion. *J Mater Chem.* 16: 2453–2463.

98 Ehiro, T., Misu, H., Nitta, S. et al. (2017). Effects of acidic–basic properties on catalytic activity for the oxidative dehydrogenation of isobutane on calcium phosphates, doped and undoped with chromium. *J Chem Eng Jpn.* 50 (2): 122–131.

99 Dasireddy, V.D.B.C., Singh, S., and Friedrich, H.B. (2014). Vanadium oxide supported on non-stoichiometric strontium hydroxyapatite catalysts for the oxidative dehydrogenation of *n*-octane. *J Mol Catal A.* 395: 398–408.

100 Cortés Corberán, V., Jia, M.J., El-Haskouri, J. et al. (2004). Oxidative dehydrogenation of isobutane over Co-MCM-41 catalysts. *Catal Today.* 91–92: 127–130.

101 Kim, H., Park, J., Park, I. et al. (2015). Coordination tuning of cobalt phosphates towards efficient water oxidation catalyst. *Nat Commun.* 6 (1): 8253.

102 Yoon, Y.S., Fujikawa, N., Ueda, W. et al. (1995). Propane oxidation over various metal molybdate catalysts. *Catal Today.* 24: 327–333.

103 Lee, S., Halder, A., Ferguson, G.A. et al. (2019). Subnanometer cobalt oxide clusters as selective low temperature oxidative dehydrogenation catalysts. *Nat Commun.* 10 (1): 954.

104 Sugiyama, S. and Hayashi, H. (2003). Role of hydroxide groups in hydroxyapatite catalysis for the oxidative dehydrogenation of alkanes. *Int J Mod Phys B.* 17: 1476–1481.

105 Wei, W., Moulijn, J.A., and Mul, G. (2009). FAPO and Fe-TUD-1: promising catalysts for N_2O mediated selective oxidation of propane? *J Catal.* 262 (1): 1–8.
106 Ai, M. (2003). Oxidation activity of iron phosphate and its characters. *Catal Today.* 85 (2): 193–198.
107 Jiménez-López, A., Rodriguez-Castellón, E., Maireles-Torres, P. et al. (2001). Chromium oxide supported on zirconium- and lanthanum-doped mesoporous silica for oxidative dehydrogenation of propane. *Appl Catal A.* 218: 295–330.
108 Liu, Y.M., Feng, W.L., Wang, L.C. et al. (2006). Chromium supported on mesocellular silica foam (MCF) for oxidative dehydrogenation of propane. *Catal Lett.* 106 (3–4): 145–152.
109 Cherian, M., Rao, M.S., Hirt, A.M. et al. (2002). Oxidative dehydrogenation of propane over supported chromia catalysts: influence of oxide supports and chromia loading. *J Catal.* 211 (2): 482–495.
110 Jibril, B.Y. (2004). Propane oxidative dehydrogenation over chromium oxide-based catalysts. *Appl Catal A.* 264 (2): 193–202.
111 Mamedov, E.A., Vislovskii, V.P., Talyshinskii, R.M., and Rizayev, R.G. (1992). Role of oxide catalysts basicity in selective oxidation. *Stud Surf Sci.* 72: 379–386.
112 Klisinska, A., Haras, A., Samson, K. et al. (2004). Effect of additives on properties of vanadia-based catalysts for oxidative dehydrogenation of propane. Experimental and quantum chemical studies. *J Mol Catal A.* 210: 87–92.
113 Sugiyama, S., Nitta, E., Hayashi, H., and Moffat, J.B. (1999). Alkene selectivity enhancement in the oxidation of propane on calcium-based catalysts. *Catal Lett.* 59: 67–72.
114 Sugiyama, S., Nitta, E., Hayashi, H., and Moffat, J.B. (2000). The oxidation of propane on nonstoichiometric calcium hydroxyapatites in the presence and absence of tetrachloromethane. *Appl Catal A.* 198: 171–178.
115 Chizallet, C., Bailly, M.L., Costentin, G. et al. (2006). Thermodynamic bronsted basicity of clean MgO surfaces determined by their deprotonation ability: Role of Mg^{2+}-O^{2-} pairs. *Catal Today.* 116 (2): 196–205.
116 Petitjean, H., Krafft, J.M., Che, M. et al. (2010). Basic reactivity of CaO: investigating active surface sites under realistic conditions. *Phys Chem Chem Phys.* 12: 14740–14748.
117 Sugiyama, S. and Moffat, J.B. (2002). Cation effects in the conversion of methanol on calcium, strontium, barium and lead hydroxyapatites. *Catal Lett.* 81 (1–2): 77–81.
118 Passos, M.L. and Ribeiro, C.P. (2010). *Innovation in Food Engineering: New Techniques and Products*, 87. Boca Raton, FL: CRC Press.
119 Thrane, J., Mentzel, U.V., Thorhauge, M. et al. (2021). Hydroxyapatite supported molybdenum oxide catalyst for selective oxidation of methanol to formaldehyde: studies of industrial sized catalyst pellets. *Catal Sci Technol.* 11: 970–983.

7

Selective Hydrogenation and Dehydrogenation Using Hydroxyapatite-Based Catalysts

Vijay K. Velisoju[1], Hari Padmasri Aytam[2], and Venugopal Akula[1]

[1] CSIR-Indian Institute of Chemical Technology, Department of Catalysis and Fine Chemicals, Hyderabad 500 007, Telangana State, India
[2] Osmania University, University College of Science, Department of Chemistry, Hyderabad 500 007, Telangana State, India

7.1 Introduction

Hydroxyapatite (HA; $Ca_{10}(PO_4)_6(OH)_2$), the main component of bones and teeth, is of considerable interest in many areas because of its typical physicochemical properties [1–3]. Owing to its thermal stability, it has been examined as a suitable catalyst and as a support for many catalytic reactions, such as hydrogenation, dehydrogenation, reforming, and organic transformation in the liquid as well as in the gas phase [4]. Due to its amphoteric nature, HA was considered to serve as a suitable support for various metals (Ni, Ru, Au, etc.) in both hydrogenation–dehydrogenation reactions in the gas phase as well as the liquid phase applications. As mentioned in previous chapters of this book, the following features make HA a suitable material in many catalytic reactions [3, 5–8]:

1) Amphoteric nature of the material (simultaneous presence of both acid and base nature).
2) High stability under oxidizing and reducing conditions.
3) High surface area suitable for metal deposition.
4) High ion exchange capacity with various metals.
5) Control over the synthesis methods to tune the surface properties by changing Ca/P ratio and synthesis parameters.
6) Good structural flexibility and high adsorption capacity.

A variety of methods have been reported to synthesize HA, with the majority employing either wet chemical, precipitation, sol–gel, hydrolysis, microwave irradiation, or hydrothermal methods, as discussed in Chapter 2 of this book. A detailed overview of the synthesis methods for HA was also reviewed by Fihri et al. [2]. Despite its importance as suitable support for different metals in various reactions, this chapter would mainly focus on the hydrogenation and dehydrogenation reactions. As depicted in Figure 7.1, both stoichiometric and

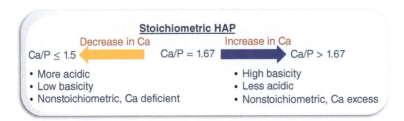

Figure 7.1 Differences in acid–base properties of HA with variation in Ca/P ratio. Source: Vijay K Velisoju.

nonstoichiometric (Ca excess) HAs are suitable for dehydrogenation processes. In contrast, Ca-deficient HA is better suited for dehydration reactions due to changes in acid–base properties of HA upon varying Ca/P molar ratio. However, HA possesses both acid and base sites when its Ca/P ratio is found between 1.30 and 1.67 [8, 9].

7.2 HA as Catalyst Support in Hydrogenation Reactions

Catalytic hydrogenation is of great importance to the petrochemical and fine chemical industries, which allows converting unsaturated hydrocarbons, i.e. alkenes and aromatics, into saturated alkanes and cycloalkanes. During the last decades, this process is intensively investigated to convert carbon dioxide to valuable products such as methanol, ethanol, methane, and higher hydrocarbons via CO_2 Fischer–Tropsch (FT) synthesis [10]. Various processes that investigate catalytic applications of HA are thoroughly reviewed in the following sections.

7.2.1 Hydrogenation of Biomass-Derived Compounds to Fuels and Fine Chemicals

Several research groups have used HA as a support to design heterogeneous catalysts for hydrogenation reactions. Sudhakar et al. investigated the catalytic activity of HA-supported Ru, Pt, Pd, and Ni catalysts in the hydrogenation of levulinic acid (LA) to γ-valerolactone (GVL) in both liquid and gas phases using water as solvent (Table 7.1) [7, 11]. Among the evaluated catalysts, the Ru/HA exhibited good catalytic activity and selectivity to GVL at 70 °C under 0.5 MPa H_2 pressure in the liquid phase. Recycling studies have shown that the Ru/HA catalyst can be readily recovered and reused up to four times without significant loss in activity. In the gas phase, authors have identified that HA-supported Ru catalyst also exhibited better performance than the other supported metal catalysts (Table 7.1). They concluded that the presence of weak Brønsted and moderate Lewis acid sites is a possible reason for the high selectivity toward the desired product GVL as indicated by the pyridine adsorbed FTIR spectra of Ru/HA catalyst. Mizugaki et al. also developed an HA-supported Pt—Mo catalyst for LA hydrogenation to produce 1,4-pentanediol in aqueous phase at 130 °C under 30 bar pressure after 24 hours of reaction time.

Table 7.1 Aqueous phase and gas phase levulinic acid to γ-valerolactone hydrogenation over HA-supported metal (M = Pt, Pd, Ru, Cu, and Ni) catalysts.

Levulinic acid →(Catalyst, Δ; +H_2, –H_2O)→ γ-Valerolactone

Metal (wt%) (HA)	LA conversion (mol%) (aqueous phase)	GVL selectivity (mol%) (aqueous phase)	LA conversion (mol%) (vapor phase)	GVL selectivity (mol%) (vapor phase)
Pd	25.3	66.5	26	90
Pt	88.2	78.6	42	88
Ru	92.2	99.8	99	99
Ni	21.0	65.0	18	65
Cu	32.2	74.8	–	–

Source: Sudhakar et al. [7]/with permission of Elsevier.

Authors also evaluated and compared various conventional support materials such as SiO_2, TiO_2, MgO, Al_2O_3, and CeO_2 and found better performance over the HA-supported Pt catalyst for this reaction. Pt–Mo/HA catalyst was recoverable and reusable while maintaining its high activity and selectivity as compared with other catalysts investigated in the same study [12]. The dramatically improved selectivity to 1,4-pentanediol (~93% yield) was explained due to a strong interaction between Pt nanoparticles and MoO_x species supported on HA.

In many cases, the catalytic activity of HA-supported metal (e.g. Cu and Ru) catalysts showed better performances as compared with the metal-based catalysts supported on conventional supports like SiO_2 and TiO_2 [13, 14]. For instance, Xu et al. studied the hydrogenation of stearic acid and compared the catalytic activity of Ru/HA catalyst with those of Ru catalysts supported on activated carbon, TiO_2, SiO_2, and ZrO_2 [14]. Ru/HA catalyst was not only more active but also showed better recyclability with consistent activity as compared with the Ru/TiO_2 catalyst (Figure 7.2), indicating its robust nature under the reaction conditions adopted.

Jia et al. explored HA as support for cobalt catalysts in the direct selective hydrogenation of fatty acids (stearic acid) and jatropha oil to fatty alcohols [15]. Simple wet impregnation method was used for the deposition of cobalt nanoparticles on the surface of HA and other conventional supports like CeO_2, ZrO_2, Al_2O_3, SiO_2, HZSM-5, and TiO_2. In the selective hydrogenation of stearic acid, HA-supported cobalt catalyst was found to be more active and selective than the counterparts prepared from other catalyst supports, with >97% yield toward 1-octadecanol at 190 °C reaction temperature under 40 bar pressure in water. In addition, it was also found that the Co/HA was efficient in the direct hydrogenation of the natural oil, jatropha oil, to fatty alcohols without any further preprocessing. Catalyst characterizations revealed

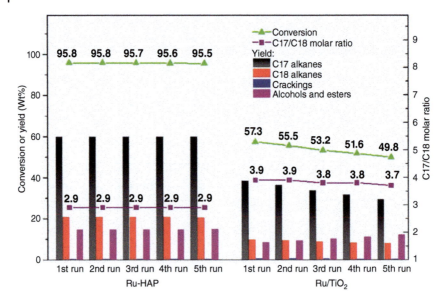

Figure 7.2 Recycling experiments on the hydrogenation of stearic acid over Ru/HA and Ru/TiO$_2$ catalysts at 180 °C under 2 MPa H$_2$ pressure. Source: Xu et al. [14]/Reproduction with permission of American Chemical Society.

that the controlled acidity and high cobalt dispersion on HA support along with presence of strong metal–support interaction were the key reasons for high catalytic performance of the Co/HA catalyst. Additionally, it was also claimed that the high absorption of fatty acids on the surface of the Co/HA catalyst was found to promote hydrogenation of fatty acids over the Co species as revealed by FTIR results, and finally a plausible reaction mechanism was also proposed over the Co/HA catalyst based on the results obtained in the study.

Sun et al. investigated the HA-supported Ru catalysts for quinoline hydrogenation to decahydroquinoline (DHQ) in cyclohexane and found that the catalyst was active at 150 °C and under hydrogen pressure of 40 bar [16]. The influence of the Ca/P molar ratio in the range of 1.30–1.50 was also investigated in this reaction. The optimum Ca/P molar ratio was found at 1.38, and the corresponding catalyst showed the best performance in quinoline hydrogenation with 100% yield at 150 °C under 40 bar pressure in 3 hours reaction time. The authors believed that the high performance of this catalyst is attributed to the presence of high crystallinity of the material and cooperation between the active metal species and hydroxyl groups on the surface of HA, as highlighted by the characterization of the catalysts before and after quinoline adsorption using X-ray diffraction (XRD) and Fourier transform infrared (FTIR) techniques. Other research group also demonstrated the catalytic application of HA in the quinoline hydrogenation process using different active metals and concluded that the HA is a suitable support for this reaction, owing to its desired and suitable physicochemical properties required for this reaction [17, 18].

Recently, Zhang's group developed HA-supported γ-Fe$_2$O$_3$@HA catalyst for the selective hydrogenation of furfural into furfuryl alcohol with high yield of furfuryl alcohol (91.7%) [19]. After the reaction, the γ-Fe_2O_3@HA catalyst was able to be easily collected with the use of a magnet and was reused for six runs with an insignificant loss in activity at 180 °C under 10 bar pressure in isopropanol as solvent. In addition, kinetic investigations also revealed that the value of activation energy for the transfer hydrogenation of furfural was high (47.69 kJ mol^{-1}), indicating that reaction was sensitive to the temperature of the reaction.

Furthermore, Li et al. demonstrated Pd/HA catalyst for the selective hydrogenation of furfural with 100% yield of furfuryl alcohol under relatively mild reaction conditions (at 40 °C under 10 bar pressure in 2-propanol as solvent) [20]. In both cases, the authors found that the HA-supported metal catalysts were recyclable, indicating the robustness of HA support under hydrogenation conditions. Authors also claimed that the activity of other Pd-based catalysts (Pd–MgO, Pd–CeO_2, Pd–ZrO_2, Pd–TiO_2, Pd–Al_2O_3, and Pd–SiO_2) was lower compared with that of HA-supported Pd catalyst, due to quasicoordination effect between HA support and Pd species that allowed a high dispersion and a good stabilization of Pd nanoparticles that are key in activating hydrogen. For the same reaction, recently, Putrakumar et al. also explored HA as support for Cu and found that a 5 wt% Cu supported on HA catalyst was active, selective, and stable with 96% furfural conversion and 100% selectivity toward the desired product at 140 °C in 4 hours of reaction time [21]. The characterization data emphasized that the HA provided highly dispersed Cu species on the surface of the catalyst with required basicity at 5 wt% Cu loading.

In addition, HA was also widely explored as catalytic support material for different metals in the hydrogenation of aromatics, an important transformation in the petroleum and chemical industries [22, 23]. Zahmakıran et al. reported the synthesis of HA-supported Ru and Rh catalysts and investigated them in the hydrogenation of aromatics to their corresponding cyclohexane derivatives under mild reaction conditions. The authors achieved high turnover frequencies over both the catalysts at room temperature and 3 bar pressure. They claimed that the high catalytic activity, easy preparation, isolability, and reusability of HA-supported Ru and Rh catalysts make them very attractive for the hydrogenation of aromatics in industrial applications and in small-scale organic synthesis procedures.

7.2.2 Hydrogenation of Olefins and Nitro Compounds

Different research groups also investigated and developed various HA-supported metal-based catalysts for the hydrogenation of nitroarenes and olefins in different solvents under mild reaction conditions [24–27]. Huang and coworkers developed a highly active and selective Rh/HA catalyst for the hydrogenation of olefins and nitroarenes under mild reaction conditions [27]. Various nitroarenes were selectively converted to corresponding anilines with hydrazine as reducing agents without modification of reducible groups (e.g. alkene and cyano). The authors also found that the same Rh/HA catalyst was effective in hydrogenating olefins in presence of external hydrogen gas. This Rh/HA catalyst was not only active and selective but also stable and recyclable up to five cycles of the reaction without a significant loss in activity, indicating the robust nature of the HA support. Furthermore, Sreedhar and coworkers reported a facile and green route to synthesize Gum Acacia assisted hydroxyapatite (GA-HA) nanostructures and explored its potential application in the catalytic hydrogenation of nitrobenzene after in situ deposition of gold nanoparticles on this support [26]. The catalytic activity of the GA-HA toward nitrobenzene was compared with two different GA-HA-supported Au (1 and 5 wt%) catalysts. Higher catalytic activity was observed for 1 wt% Au/GA-HA catalyst, and at higher loadings of Au (5 wt%), significant agglomeration of Au particles was observed that resulted in decreased BET surface area of the catalyst at 230 °C (WHSV = 36.12 hours^{-1}; H$_2$/nitrobenzene = 4; residence time: 0.0276 hours). Recently, Pashaei et al. developed an ionic-tagged magnetic mesoporous HA-supported silver catalyst for the hydrogenation of nitroarenes. The catalyst was active and relatively stable (98% yield in the first cycle and 90% yield after five cycles under reflux conditions in H$_2$O and NaBH$_4$ as reducing agent) and easily recoverable after the reaction due to its supermagnetic nature via application of magnetic field [24]. Finally, Shokouhimehr et al. [25] found that the deposition of Pd nanoparticles on HA serves as an efficient catalyst in the hydrogenation of nitroarenes and in green oxidation of alcohols with excellent activity (98% yield) under mild conditions (at room temperature and water as solvent). Additionally, the catalyst was also found to be active and selective even after six recycles of the reaction. From the characterization results, it was concluded that the HA support material allowed the fine dispersion of metal nanoparticles, which is a key parameter desirable for the hydrogenation of nitroarenes to their corresponding derivatives.

In another study, Jiang and Zhang developed HA-based catalyst (γ-Fe$_2$O$_3$@HA-Pd) in the transfer hydrogenation of nitro compounds to produce aniline with HCOONH$_4$ as hydrogen donors [28]. Maximum yield (>99%) of aniline was obtained in water at room temperature with six equivalents of HCOONH$_4$. Authors have successfully demonstrated the activity of catalyst for a wide range of nitro compounds and achieved excellent yields of respective hydrogenating products. A synergistic effect between HA support and Pd nanoparticles was found to be key for the hydrogenation process with good stability. Owing to the varied acid–base properties of the HA support material, the reaction also did not require the use of any additives and ligands in the air as opposed to the reported studies confirming the excellent catalytic properties of HA as support material of Pd for this process [28].

In addition to the previously discussed catalytic applications, HA and modified HA materials are found to be suitable for various other industrially important processes. Depending on the application, different metals were used to modify HA, and they were found to have excellent catalytic activity under mild reaction conditions in liquid phase processes such as benzene and phenol hydrogenation [14, 29] and in gas-phase applications such as NH$_3$-SCR [30, 31] which will be discussed in the following sections. The bare HA also found to exhibit good catalytic activity with optimized ratio of Ca/P in Guerbet coupling reaction of ethanol to butanol through transfer hydrogenation [32, 33].

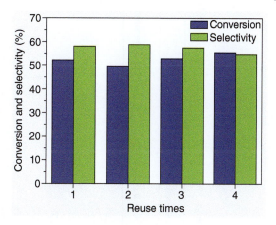

Figure 7.3 The performance of the catalyst in the recycling experiment. Reaction conditions: benzene, 0.5 ml; reaction temperature, 150 °C; H_2 pressure, 50 bar; catalyst Ru–Zn/HA-1, 20 mg; NaOH concentration, 0.5 M; water, 1.5 ml. Source: Zhang et al. [34]. Reproduction with permission of Royal Society of Chemistry.

7.2.3 Hydrogenation of Benzene, Phenol, and Diols

Zhang et al. explored HA as support for Ru–Zn in the partial hydrogenation of benzene to cyclohexene and investigated the influences of reaction temperature, Ru/Zn molar ratio, pressure, and amount of NaOH modifier on hydrogenation activity [34]. Metallic nanoparticles smaller than 2 nm and dispersed uniformly on the surface of the HA were observed, which exhibited high activity and selectivity toward the desired product. Compared with the monometallic Ru/HA catalyst, Zn addition to Ru on the HA surface was found to increase the selectivity toward cyclohexene. In addition, it was also found that the added NaOH in the solution could improve the yield of cyclohexene by retarding its hydrogenation. Cyclohexene yield of 33% was achieved over Zn-modified Ru/HA catalyst (having 1 : 1 M ratio of Ru : Zn) at 150 °C reaction temperature under 50 bar pressure of H_2 (Figure 7.3). Authors successfully demonstrated the same catalyst for four consecutive cycles of the reaction without significant loss of activity or selectivity toward cyclohexene.

In another study, HA was used as support for Pd in the selective hydrogenation of phenol to cyclohexanone in aqueous media under mild reaction conditions [29]. The prepared Pd/HA catalyst exhibited excellent catalytic activity (100% yield) toward cyclohexanone from phenol at room temperature under ambient H_2 pressure in water. The characterization data proved a high and uniform dispersion of Pd nanoclusters on the surface of the HA support material. Moreover, basic sites were mainly identified as active sites on catalyst's surface. Additionally, the catalyst was also found to be stable that could be recycled without deactivation or morphology change. In further investigation, Perez et al. studied the selective hydrogenation of phenol into cyclohexanone over palladium supported on different supports such as HA, carbon (C), and different types of alumina (Al_2O_3) [14]. Table 7.2 shows the main results obtained. Pd/HA exhibited excellent catalytic performance with 100% conversion after 50 minutes of reaction time and 97% of selectivity into cyclohexanone. This performance is equivalent to Pd catalysts supported on synthesized alumina supports and is much higher than that of commercial Pd/Al_2O_3 and carbon-supported Pd catalyst.

Table 7.2 Comparative catalytic performance of different palladium catalyst in selective hydrogenation of phenol into cyclohexanone.

Catalyst	Pd[b] (wt%)	$m_{catalyst}$ (g)	Conversion[a] (%)					Selectivity (%)	
			10 min	30 min	50 min	90 min	120 min	C=O cyclohexanone	C–OH cyclohexanol
Pd/Al$_2$O$_3$-CWE	13	0.188	70	96	100	100	100	98	3
Pd/γ-Al$_2$O$_3$	0.2	1.41	26	30	37	40	58	100	–
Pd/γ-Al$_2$O$_3$	1.5	0.188	35	89	100	100	100	97	3
Pd/Al$_2$O$_3$-com	10	0.028	8	31	60	100	100	78	2
Pd/HA	1.5	0.188	21	87	100	100	100	97	2
Pd/C	1	0.282	17	46	73	100	100	98	2
Pd/C-com	5	0.056	5	13	28	80	100	74	26
Pd/C-com	10	0.028	9	21	35	61	100	86	14

Reaction conditions: phenol (100 mg and 1.06 mmol) dissolved in water (5 ml) in the presence the solid catalyst, stirred in a glass reactor (8.0 ml), pressurized with H$_2$ (5 bars), and heated to 100 °C.
a) The conversion was determined by GC analysis and the products identified by GC-MS with the external standard method.
b) wt% of palladium measured by inductively coupled plasma (ICP) chemical analysis.
Source: Pérez et al. [14]. Reproduced with permission of Elsevier.

In another study, Gama-Lara et al. studied HA-supported Pt catalyst in the hydrogenation of 2-butyne-1,4-diol to 2-butene-1,4-diol [35]. Authors carried out biosynthesis of Pt-nanoparticles (Pt NPs) supported on bovine bone powder without any calcination procedure and applied in the hydrogenation process. The synthesized nanocomposite material exhibited >96% yield toward the desired product with 100% conversion and 96% selectivity. The characterization of the investigated HA-based Pt catalyst revealed uniformly dispersed Pt nanoparticles (average particle size of 7.1 nm from TEM), and the presence of both Pt° (48%) and PtO (51%) on the surface (from XPS) was the key reason for the high activity of the catalyst.

7.2.4 Selective Catalytic Reduction of Nitric Oxide

Tounsi and coworkers developed HA-supported Cu catalyst for the selective catalytic reduction of nitric oxide with ammonia prepared by ion exchange method [31]. The catalysts were characterized using Brunauer, Emmett and Teller (BET), XRD, H_2-Temperature programmed reduction (H_2-TPR), and Ultraviolet–visible (UV–vis) spectroscopy to correlate the structure of the catalyst with their performance in the selective catalytic reduction of NO by NH_3 under oxidizing atmosphere. The authors found that the highly dispersed CuO clusters, without libethenite phase ($Cu_2(OH)(PO_4)$) were active with 85% conversion for the desired product formation in the low temperature range (328–350 °C) for this reaction. The presence of H_2 (2.5%) in the stream significantly reduced NO conversion (47% at 375 °C), and this negative impact was more pronounced at high temperatures (400 and 425 °C).

Further, Campisi et al. studied Fe-functionalized HA catalysts for the selective catalytic reduction of NO_x to N_2 by the use of ammonia (NH_3-SCR) [30]. The catalysts were prepared by an ion exchange procedure (with Fe content of 2–7 wt%) and investigated in the NO_x reduction by ammonia. A 6 wt%Cu/HA was also prepared by ion exchange method, which served as a reference catalyst. The activity measurements on Fe/HA catalysts revealed that these catalysts required high temperatures (around 450 °C) to convert NO_x to N_2. Also Fe/HA catalysts were less active than the reference Cu/HA catalyst. The latter reached its highest activity around 250 °C. The authors claimed that both the catalysts were active, and differences in the activity in the reaction were further corroborated by the different reduction properties of Cu^{2+} (about 250 °C) and Fe^{3+} (about 450 °C) for Cu/HA and Fe/HA catalysts, respectively. Molecular nitrogen was found as a major product of the reaction over these catalysts.

7.2.5 Higher Alcohol Synthesis by Simultaneous Dehydrogenation and Hydrogenation Reactions

Higher or heavier alcohols (e.g. *n*-butanol) synthesis from ethanol (Guerbet reaction) is also one of the most studied processes over the modified HA-based catalysts [9, 36–39]. In a typical reaction mechanism of Guerbet alcohol synthesis, ethanol first undergoes dehydrogenation followed by dehydration/coupling and hydrogenation. This is detailed in the Chapter 5 of this book. Considering many studies explored using HA-based catalysts for this reaction and the hydrogenation of enal, which is the rate-limiting step that could strongly influence the product

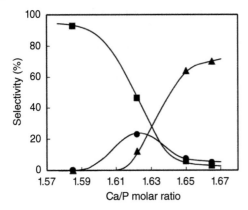

Figure 7.4 Selectivity of main products at 50% conversion of ethanol: (▲) total Guerbet alcohols, (●) 1,3-butadiene, (■) ethylene. Reaction conditions: catalyst: 4 ml; GHSV: 2000 hours^{-1}. Source: Tsuchida et al. [9] Reproduced with permission of Elsevier.

distribution, some of the relevant examples are briefly reviewed and discussed in this section. Further discussions on the Guerbet reaction over HA-based catalysts are also reported in the Chapter 10 of this book. The produced alcohols, for example, 1-butanol is an important precursor for the production of polymer raw materials (butyl acrylate and butyl methacrylate) and also used as solvent in organic transformations [36]. For this reaction, Ueda and coworkers developed different HA and modified HA catalysts to produce higher alcohols or biogasoline directly from ethanol or bioethanol [9, 36, 37]. Nonstoichiometric HA with different Ca/P molar ratios were investigated in the production of n-butanol from ethanol at different temperatures (300–450 °C) [36]. They found that the HA with Ca/P ratio of 1.64 achieved the highest selectivity (76.3%) toward n-butanol at a contact time of 1.78 seconds at 300 °C. Authors concluded and explained the performance of this catalyst by its large number of both acid sites and base sites that plays a key role in this process. In another study, the ethanol conversion to higher or total Guerbet alcohols over these HA catalysts was also investigated, and the catalytic performance of HA catalysts was compared with those of CaO and MgO catalysts [9]. As shown in Figure 7.4, ethylene was the main product at low ratio of Ca/P, while the selectivity of Guerbet alcohols increased with the increase in Ca/P ratio at a constant conversion (50%) of ethanol. Authors claimed that yields of Guerbet alcohols over the HA catalysts can be expressed quantitatively as a function of the probability of activation of the raw material that is function of basic site density of the material.

The same group also explored these HA catalysts with different ratios of Ca/P in the synthesis of biogasoline from plant-derived ethanol (bioethanol) in one-step catalytic conversion [37]. The ratio of oxygenates in the biogasoline could be adjusted by changing the reaction conditions (mainly reaction temperature). However, no sufficient efforts were made to characterize the catalysts to correlate the structure of catalyst with its performance toward any desired product. The authors proposed a mechanism for the synthesis of biogasoline from ethanol over HA, where the reactions to propagate carbon number occurred by the Guerbet reaction between mainly normal alcohols, forming normal and branched alcohols, whereas branched aldehydes and olefins were formed by dehydrogenation and dehydration.

Recently, Silvester et al. also investigated the HA catalysts with different Ca/P ratios ranged between 1.62 and 2.39 using various carbonate contents in ethanol conversion [38]. Acid–base properties were correlated with the reactivity of the solids; particularly ethanol conversion rate was higher as acidity/basicity ratio is >5, and reverse trend was observed, while the ratio was <5. The best performance was accordingly obtained over the HA-CO$_3$ catalyst, which gave a yield of 30% of heavier alcohols at 40% ethanol conversion. In contrast, the catalysts with high ratios (>5) of acidity/basicity exhibited higher ethanol conversion but were less selective to total alcohols due to an increase in ethylene selectivity, thereby decreasing total alcohol selectivity. It was confirmed that fine-tuning of acid–base sites (both their nature and amount) in HAs is necessary to have higher ethanol conversion without losing Guerbet alcohol selectivity.

Recently, Ogo and coworkers synthesized strontium hydroxyapatite (Sr-HA, where Sr is used instead of Ca) having different molar ratio of Sr/P (1.58–1.70) by hydrothermal method and investigated them as catalytic materials in the synthesis of 1-butanol from ethanol [39]. XRD analysis revealed that these catalysts had the apatitic structure. NH$_3$-TPD analysis shows that the strength of acid sites is the same because desorption temperature is similar, but the density of acid sites followed a volcano shape with the highest value at the Sr/P molar ratio of 1.67 (Figure 7.5a). On the other hand, both the strength and the density of the basic sites of these materials depend on the Sr/P ratio: the higher the Sr/P ratio is, the higher the strength and density of the basic sites are (Figure 7.5b). In the conversion of ethanol in the gas phase, the Sr/P ratio strongly influenced the ethanol conversion and the selectivity into 1-butanol (Table 7.3). The Sr-HA catalyst with high Sr/P molar ratio exhibited high activity toward 1-butanol at 300 °C. The presence of high proportion of strong basic and acidic sites (with relatively high density of basic sites in comparison to acidic sites) was key for the observed activity over the investigated catalysts. Moreover, the selectivity into by-product not only depended on the Sr/P ratio but also on the ethanol conversion, with 2-buten-1-ol and aldehydes

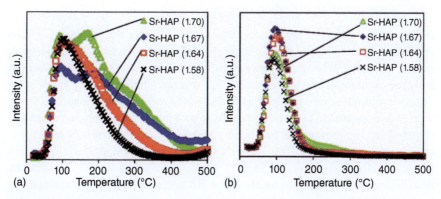

Figure 7.5 TPD profiles of CO$_2$ (a) and NH$_3$ (b) adsorbed onto Sr-HA catalysts with various Sr/P molar ratios and (c) changes in 1-butanol selectivity with increasing ethanol conversion over Sr-HA catalysts with various Sr/P molar ratios. Source: Ogo et al. [39]. Reproduction with permission of Elsevier.

Table 7.3 Catalytic conversion of ethanol over Sr-HAP catalysts with various Sr/P molar ratios.[a]

Catalyst	Conv. (C-%)	Selectivity (C-%)				Specific reaction rate ($\mu mol\ h^{-1}\ m^{-2}$)
		Acetaldehyde	1-Butanol	2-Buten-1-ol	C_6, C_8–OH	
No catalyst	Trace	n.d.[b]	n.d.[b]	n.d.[b]	n.d.[b]	–
Sr-HAP (1.58)	1.1	1.5	69.0	22.3	2.6	2.7
Sr-HAP (1.64)	5.9	0.7	78.1	6.5	10.7	15.5
Sr-HAP (1.67)	7.9	0.5	81.7	3.0	12.0	20.6
Sr-HAP (1.70)	11.3	0.7	86.4	3.0	7.8	23.3
Sr-HAP-NH$_3$ (1.71)[c]	4.4	0.4	79.2	5.8	8.8	8.4

a) Catalyst 2.0 g, temperature 300 °C, W/F$_{ethanol}$ 130 h g mol^{-1}, ethanol 16.1 mol% (Ar balanced), time on stream 3 hours.
b) Not detected.
c) This catalyst was prepared by almost the same method as the Sr-HAP (1.70) except for using NH$_3$ instead of NaOH as the alkaline source. The Sr/P molar ratio and specific surface area were 1.71 and 40.0 m^2 g^{-1}.
Source: Ogo et al. [39]. Reproduced with permission of Elsevier.

as the main by-products below 7% ethanol conversion and C$_6$ and C$_8$ alcohol as main by-products above 9% ethanol conversion. The rate-determining step was considered to be dimerization process in ethanol conversion over Sr-HA catalysts as aldol condensation is mainly accelerated by base catalysis that would explain why the Sr-HA catalysts with higher basic site density showed higher catalytic activity and 1-butanol selectivity.

7.2.6 Hydrogenation of Carbon Dioxide

Directly utilizing the greenhouse gas CO_2 as a feedstock for the production of fuels and chemicals can open up a major strategy, contributing to a closed anthropogenic carbon cycle [40]. Many researchers put up a lot of efforts in synthesizing various fuel-type compounds and fine chemicals over various supported metal catalysts. In this reference, HA-based catalysts were also widely explored and found to be efficient in hydrogenating CO_2 with green H_2 to produce various fuels and chemicals (Figure 7.6). The presence of basic Ca^{2+} cations on the HA surface is reported to chemisorb molecular CO_2 onto the surface even at room temperature; hence, HA is a potentially suitable catalyst support worth investigating for CO_2 conversions. In addition, the weakly basic nature of the HA surface and high thermal stability has been utilized and reported for various high temperature catalytic reactions [41]. Among different routes, FT (for liquid fuels), methanation (for CH_4), and water–gas shift reaction (WGS) (for syngas) are the major routes studied in the open domain and are mainly discussed in the following sections.

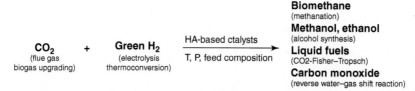

Figure 7.6 General illustration of selective catalytic hydrogenation of CO_2 into valuable energy carriers.

7.2.6.1 CO_2 Methanation

CO_2 methanation has attracted greater attention of researchers in both academia and industry during the last decade in order to mitigate the global warming resulted from CO_2 emissions and to cut the dependency on the use of fossil fuels. CH_4 is an energy-rich carrier, which can be stored in liquified form under cryogenic condition with current infrastructure and facilities. Sustainable methane production on a commercial scale while reducing CO_2 emissions may help to ensure energy security for countries, which imports huge amount of natural gas for its domestic consumption [41, 42]. Thus, despite many catalytic systems already being explored for this process, still researchers use various strategies to find out more efficient catalysts in terms of activity, selectivity, stability, and cost. In this reference, HA-based catalysts have been recently reported to exhibit excellent catalytic activity, owing to its desired catalytic properties needed for this conversion [41].

In fact, Wai and coworkers [41] used HA-supported Ni catalyst to convert CO_2 into CH_4 and also studied the influence of formate species on methane selectivity. Ni/HA catalysts were prepared by ion exchange method, with and without the presence of oleic acid (OA). Thus, three catalysts were prepared: NiHAP (0), NiHAP (0.5), and NiHAP (2.0); the first one did not use OA; the second and the last ones used OA with the molar ratio of OA to Ni equal to 0.5 and 2.0, respectively. The use of OA during ion exchange procedure is to modify chemical state of surface nickel species. The catalytic performance and stability of the prepared catalysts were illustrated in Figure 7.7a–c. The three catalysts demonstrate increasing CO_2 conversion under increasing temperature from 200 to 350 °C. Beyond 350 °C, the CO_2 conversion decreases due to the emergence of reverse water–gas shift (RWGS) reaction, which results in more CO production as compared with CH_4. The temperatures corresponding to 50% conversion for NiHAP (0.5), NiHAP (2.0), and NiHAP (0) are around 291, 328, and 363 °C, respectively, and the increasing order of catalytic activity goes in this order: NiHAP (0.5) > NiHAP (2.0) > NiHAP (0). Hence, NiHAP (0.5) catalyst was found to be catalytically more active and selective for CH_4 production than the other two catalysts. The authors attributed this difference to the size distribution of nickel nanoparticles: small Ni nanoparticles favor CO production, while large Ni nanoparticles favor CH_4 production.

7.2.6.2 CO_2 Fisher–Tropsch (FT) Synthesis

Mounting interest to valorize biomass and wastes for sustainable environment has augmented the production of syngas, which can be further processed by FT

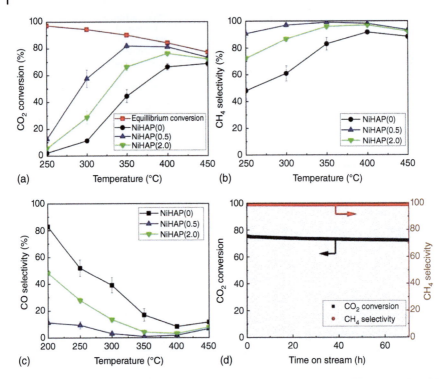

Figure 7.7 Effect of OA addition on Ni supported on HAP catalysts. (a) CO_2 conversion, (b) CH_4 selectivity, and (c) CO selectivity: 4 ml min^{-1} of CO_2, 16 ml min^{-1} of H_2, 30 ml min^{-1} of He with 100 mg catalyst. (d) Stability test of NiHAP (0.5) at 350 °C under same reaction condition. Source: Wai et al. [41]. Reproduction with permission of John Wiley and Sons.

synthesis to produce value added biofuels [43]. The direct hydrogenation of CO_2 into liquid fuels (called: modified or CO_2 FT synthesis) has not been done yet with HA-based catalysts, but has been done with other families of catalysts. However, HA was explored as a suitable support for cobalt in FT synthesis [43], and this could be a starting point to further investigate HA in CO_2-FT synthesis. In this regard, Munirathnam et al., for the first time, reported Co/HA catalyst for FT process (CO + H_2 → hydrocarbons) showing a very good activity in CO conversion in comparison with Co/alumina catalyst [43]. The catalytic performance at 20 bar, 220 °C, and H_2/CO = 2.1 showed that the CO conversion was higher 10 wt%Co/HA than 15 wt%Co/Al_2O_3 (Figure 7.8). Co/HA catalyst also exhibited very good C_{5+} selectivity (~83%) with low methane selectivity (10%). In contrast, Co/Al_2O_3 catalyst showed lower C_{5+} selectivity (73%) and higher methane selectivity (14%) even at lower CO conversion. HA support was found to exhibit considerably less acid site density, consequently, reducing detrimental interactions of the support with Co precursors, leading to hardly reducible Co species that are generally observed with its alumina counterpart. In addition, Co/HA catalyst consisted relatively larger Co particles (~9 vs. ~6 nm) and better reducibility of ionic cobalt species when compared with the reduction of Co species on alumina support.

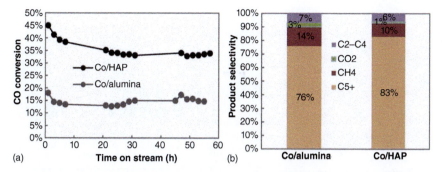

Figure 7.8 (a) CO conversions and (b) product selectivity in FT synthesis using Co/HAP and Co/Alumina catalysts reduced at 400 °C under reactions conditions of 20 bar pressure with a gas hourly space velocity (GHSV) of 770 ml g^{-1} h^{-1}) and 220 °C temperature. Source: Munirathinam et al. [43]. Reproduction with permission of Elsevier.

7.2.6.3 Alcohol Synthesis

Alcohols (C_1 or C_2) produced industrially from a mixture of $CO/CO_2/H_2$ is in high demand over the world as a promising alternative fuel and industrially useful bulk chemical [44]. There are numerous reports on heterogeneous catalysts (typically Cu-based catalysts) used for the thermocatalytic hydrogenation of CO_2 into alcohols [44, 45]. In this reference, Sans et al. explored the permanently polarized HA (p-HA) obtained from a thermal stimulated polarization process as a highly selective catalyst for the production of ethanol from CO_2 and CH_4. Authors demonstrated the selective catalytic activity of p-HA to convert gaseous CO_2 into ethanol. The yield of ethanol achieved in this work was ~90% and is among the highest reported in the literature. On the other hand, the bioceramic catalyst showed lower CO_2 conversion (~0.08%) than those reported for metallic-based catalysts, even though these usually worked at higher pressures. For instance, Cu^I@Zr_{12}-bpdc exhibited a conversion of 0.5% at 20 bar, while Zr_{12}-bpdc-CuCs showed a conversion of 52% at 40 bar. Both theoretical calculations and experiments under different reaction conditions provide important insights about the mechanism pathway.

7.2.6.4 Water–Gas Shift and Reverse Water–Gas Shift

In this sense, to avoid the poisoning effect of the residual carbon monoxide on the efficiency of the fuel cell electrodes and to produce syngas and further processing of syngas to higher hydrocarbons, WGS/ RWGS and CO preferential oxidation catalytic processes are traditionally applied [6, 46, 47]. Though many materials have been studied for use in the RWGS reaction, some offer particular advantages over others. Mainly, Cu- and Fe-based catalysts were explored for these important processes. However, there is still room for the improvements in this area given the faster deactivation and selectivity issues with the already existing catalytic systems. In this reference, HA-based materials were also explored in WGS, preferential oxidation reactions, and not many reported studies exist in the open domain that use

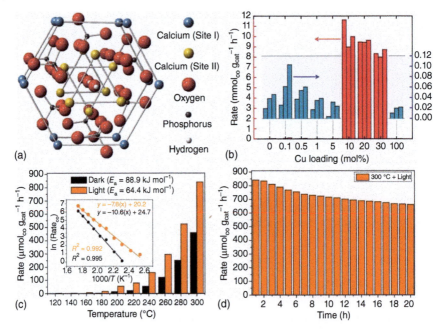

Figure 7.9 (a) Crystal structure of the HAP lattice, (b) photocatalytic RWGS activity of calcined Cu-HAP samples as a function of Cu content. Experiments were performed using three different samples obtained from the same synthesis, thereby confirming the reproducibility of this catalytic performance. Reaction conditions: H_2/CO_2 ratio = 1:1, light intensity = 40 sun, no external heating and measurement time = 1 h. (c) Capillary flow reactor results for the calcined 10 mol% Cu-HAP catalyst. (d) Catalytic stability results for the calcined 10 mol% Cu-HAP sample over 20 sequential hour-long runs. Reaction conditions for flow measurements (c, d): atmospheric pressure, light intensity = ~2.0 W cm^{-2}, H_2/CO_2 ratio = 1:1, flow rate = 2 sccm. Source: Guo et al. [47]. Reproduction with permission of American Chemical Society.

HA as support material for RWGS process. About WGS reaction over HA-based catalysts, Chapter 9 of this book will discuss in more details this reaction. Recently, Guo et al. extensively studied the Cu-HA catalyst in photothermal CO_2 reduction to CO [47]. Authors found Cu^{2+} ions first substituted preferentially into the HAP lattice at lower copper loadings (Figure 7.9). As the Cu content was increased above 5 mol% Cu, Cu^{2+} began to form surface Cu^{2+} hydrates outside the HAP lattice, enabling the formation of CuO nanosheets during subsequent calcination. Catalytic activity testing for the RWGS reaction revealed that only these surface Cu^{2+} species could be reduced to form metallic Cu nanoparticles, thereby greatly enhancing CO_2 reduction to CO via a strong photothermal effect. Consequently, light-enhanced catalytic RWGS rates of up to 12 mmol$_{CO}$ g$_{cat}^{-1}$ h^{-1} and selectivities exceeding 99 mol% CO were attained by samples containing between 10 and 30 mol% Cu. Remarkably, evidence of metal hydride stretching modes suggested that H_2 could be heterolytically activated even at room temperature. Finally, the low cost, low toxicity, stability, and scalability of Cu-HAP make it a singularly promising candidate for large-scale CO_2 remediation.

7.2.7 Partial Conclusions

In summary, HA as catalyst support prepared under controlled synthesis conditions to obtain desired Ca/P molar ratio, which in turn allows tuning the acid–base properties of the support, in combination with an appropriate active phase, is proved to be efficient in various hydrogenation processes with high activity, stability, recyclability, and reusability under mild conditions. In general, the hydrogenation processes were occurred on the metal species, while the hydrogen-transfer processes were accelerated in the presence of acid sites on HA support surface.

7.3 HA as Support in Dehydrogenation Reactions

HA was also widely explored in various dehydrogenation processes due to its possible surface basic characters that are key in dehydrogenation reactions (Figure 7.1). Among various routes to produce olefins in the industry, catalytic dehydrogenation processes are attractive methods with low greenhouse gas emissions, reduced cost, and various other advantages [48, 49]. Extensive and systematic investigations were carried out on this industrially important transformation over various HA-based catalysts [50–56]. Sugiyama and coworkers developed various HA-based materials in the oxidative dehydrogenation of alkanes with different metals (e.g. strontium, barium, and vanadium) and studied the effects of the density of hydroxyl groups, gas-phase additive (tetrachloromethane) and solid-phase additives (copper and lead) in the reaction [50, 55–57]. Authors investigated the Sr-HA (which was synthesized only from Sr and phosphate salts, without Ca presence) and Co-exchanged HA (Co-HA) catalysts in the oxidative dehydrogenation of propane and compared their H–D exchangeability of OH groups in those HAs with D_2O. It was claimed that the activity and H–D exchangeability of OH groups in those HAs with D_2O were enhanced by the increased of Co^{2+} content. Therefore, it was further suggested that the hydrogen abstraction was directly contributed from oxygen species derived from the abstraction of hydrogen of OH groups. In a specific example, authors investigated Sr-HA and Ba-HA catalysts in the oxidative dehydrogenation of propane with and without support treatment with Pb^{2+} and Cu^{2+} [56]. Sr-HA and Ba-HA were synthesized from Sr, Ba, and phosphate salts, and did not contain calcium. These materials were doped with Pb^{2+} and Cu^{2+} by ion exchange method. In the dehydrogenation of propane at 450 °C, Sr-HA catalyst was much more active than Ba-HA. However, Ba-HA showed higher propene and ethylene selectivity than Sr-HA. The incorporation of Pb^{2+} at the place of Sr^{2+} on the surface of Sr-HA was not favorable for the propane conversion in comparison with the undoped material. Sr-HA doped with Cu^{2+} was also less active than undoped Sr-HA in the absence of tetrachloromethane but more active in the presence of tetrachloromethane (Figure 7.10). However, the understanding of the origin of this improvement remains an open question as tetrachloromethane could also reduce Cu^{2+} to metallic Cu.

Boucetta et al. prepared Cr-loaded HA catalysts ($Cr(x)/Ca-HA$ ($0.1 \leq x \leq 3.7$ wt% Cr)) by ion exchange method and examined these catalysts in the dehydrogenation

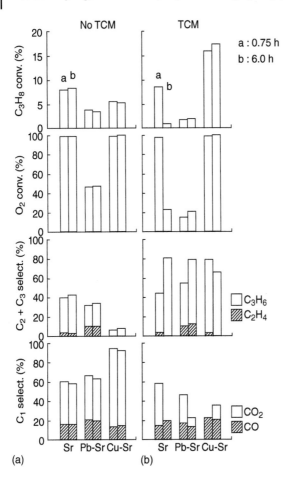

Figure 7.10 Influence of tetrachloromethane (TCM) on the dehydrogenation activity of Sr-HA, Cu treated Sr-HA and Pb treated Sr-HA catalysts measured at 450 °C with 0.5 g catalyst. Source: Sugiyama et al. [56]. Reproduced with permission of Elsevier.

of propane at 300–550 °C [54]. By using various characterization techniques such as Raman spectroscopy, FTIR, UV–vis–NIR, EPR, XPS, and TPR, different states of ionic Cr species (Cr^{2+}, Cr^{3+}, Cr^{5+}, and Cr^{6+}) at different Cr loadings could be determined. It was concluded that the Cr^{6+} species present on the catalyst surface would initiate the (initially present) reaction of propane cracking, owing to their relatively high acidic nature that also results in the significant increase in the conversion of propane. In contrast, because the reduction took place during the reaction, simultaneously the Cr^{2+} ions are formed along with a drop in conversion at 550 °C. In addition, the isolated Cr^{3+} species were found to be responsible for the propylene formation that was ascribed to the decrease of surface basicity (induced by the fixation of Cr^{3+}). Furthermore, Dasireddy and coworkers explored HA substituted with V_2O_5 with different V_2O_5 loadings from 2.5 to 15 wt% in dehydrogenation of n-octane in a continuous flow fixed bed reactor at 350–550 °C [52]. Undoped HA showed much lower n-octane conversion in comparison with modified HA within the temperature range studied. The selectivity into octenes was impacted by both the V_2O_5 loading and the reaction temperature. In all cases, increase of the reaction temperature led to a decrease of the selectivity into octenes and to an increase of the selectivity into

7.3 HA as Support in Dehydrogenation Reactions | 259

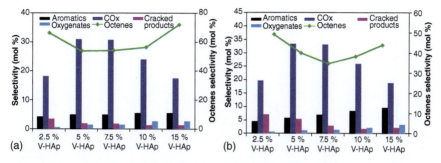

Figure 7.11 Selectivity of the product stream at iso-conversion (24%), temperature = 450 °C (a) and iso-conversion (36%), temperature = 550 °C (b). Source: Dasireddy et al. [52]. Reproduced with permission of Elsevier.

cracked products and aromatics. The increase of V_2O_5 loading led to a decrease of the selectivity into octenes and to an increase of the selectivity into aromatics. Authors also compared the product selectivities at two different iso-conversion levels by conducting the experiments at 450 and 550 °C reaction temperatures (Figure 7.11). As shown in Figure 7.11, in both cases, a gradual decrease in octene selectivity was observed until 7.5% loading of vanadium, and further vanadium loading increase led to significant increase in octenes selectivity. In contrast, an opposite trend was observed in case of aromatics, suggesting that octenes are precursors for the formation of aromatic compounds by cyclization. From CO_2-TPD results, the authors found that the basicity (key for dehydrogenation) of the catalysts increased with the increase in V_2O_5 content. The characterization data of the used catalysts showed minimal changes in the catalyst structure and morphology, explaining the robust nature of these HA-based catalysts under the reaction conditions used.

Studies from other research groups reported the efficient use of HA-based catalysts, which were modified with iron and cobalt in the dehydrogenation of 2-butanol and oxidative dehydrogenation of ethane and propane [14, 51, 52]. In case of Co exchange, the high activity observed was explained due to the partial compensation of the intrinsic dehydrogenating activity of Co by the decrease in basicity of HA, which is induced by the replacement of Ca^{2+} species by Co and the involvement of two types of active sites in the reaction [51]. After introducing Co to Ca-HA, by 1.35 wt% Co, even after calcination in air at 550 °C, the presence of cobalt in its divalent oxidation state confirmed the effect of the support material. With increasing Co loading, higher yield of ethylene was observed at 500 and 550 °C (22%).

The modification of HA by Fe^{3+} via ion exchange led to the formation of $Fe^{3+}-O-Ca^{2+}$ sites at low Fe content (0.52 wt%) and $Fe^{3+}-O-Fe^{3+}$ sites at high Fe loadings (> 0.52 wt%) [53]. The loaded Fe^{3+} could only be reduced into Fe^{2+} under H_2 atmosphere. The Fe-loaded HA catalysts were studied in the dehydrogenation (without oxygen) and oxydehydrogenation (with the presence of oxygen) of butan-2-ol. Under dehydrogenation conditions at 170–230 °C, the conversion of butan-2-ol and the selectivity in butenes increased with the reaction temperature and with iron loading. Under the oxydehydrogenation conditions, the effect of reaction temperature was similar, but the effect of iron loading on the conversion of

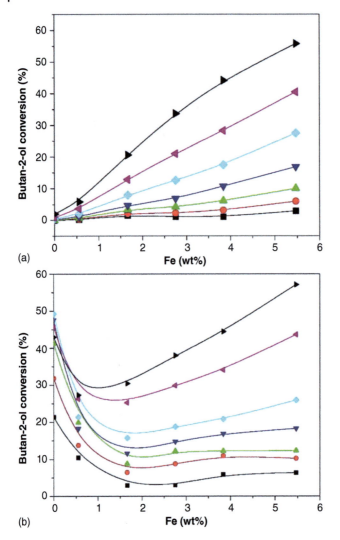

Figure 7.12 Butan-2-ol conversion versus Fe content at various temperatures: 170 °C (♦); 180 °C (●); 190 °C (▲); 200 °C (▼); 210 °C (◆); 220 °C (◄); and 230 °C (►): (a) in the absence of dioxygen and (b) in the presence of dioxygen. Source: Khachani et al. [53]. Reproduced with permission of Elsevier.

butan-2-ol showed a minimum. This minimum shifted from c. 2 to 1 wt% Fe when the reaction temperature increased from 170 to 230 °C (Figure 7.12). This catalytic behavior under oxydehydrogenation conditions was explained by the fact that the conversion of butan-2-ol is due to both the dehydrogenation and oxydehydrogenation reactions. These two reactions evolve in the opposite direction when iron content increases. These catalysts were also investigated in the oxydehydrogenation of propane (theoretical reaction: $2C_3H_8 + O_2 \rightarrow 2C_3H_6 + 2H_2O$) at 350–550 °C. Pure HA showed very low propane conversion at 550 °C. Among the doped catalysts, the one containing the lowest Fe content (0.52 wt%) showed the highest propane conversion and propene selectivity. The results were ascribed to the compensation of two parameters: the dehydrogenating ability of iron and the impact of iron on the basicity of the HA support. In fact, the addition of iron is favorable for dehydrogenation reaction. But in parallel, the addition of iron decreases the basicity of the HA support, which is not favorable for the dehydrogenation. This work showed the important contribution of the HA support in the oxydehydrogenation of propane.

In addition to oxidative dehydrogenation of alkanes, various HA-based catalysts were also found active and efficient in dehydrogenation of alcohols and other compounds (e.g. glycerol, indoles, and ammonia borane) [58–63]. Various modifications were proposed to tune the surface acid–base properties of HA (in both stoichiometric and nonstoichiometric forms) in order to achieve high activity and selectivity toward the desired products. For instance, Lu et al. investigated Cu/HA catalyst and compared its performance with Cu/SiO$_2$ (low basic support; SiO$_2$) and Cu/MgO (highly basic support; MgO) catalysts in the dehydrogenation of methanol. It was concluded that the surface basicity of the final catalysts significantly affected the catalytic activity in methanol dehydrogenation. The SiO$_2$-supported catalyst found to favor the methanol dehydrogenation to produce methyl formate due to low basicity of the catalyst. In contrast, HA- and MgO-supported catalysts produced CO and H$_2$ due to the rapid degradation of methyl formate over the strong and more basic sites present on the surface of these catalysts [49]. In other studies, different surface modifications were done to HA with various metals such as Au, Pd, and Ru. The modified catalysts were found to be active in the dehydrogenation of glycerol to lactic acid (Au/HA), indolines to indoles (Pd/HA), and ammonia borane dehydrogenation (Ru/HA) [32, 51–55]. Notable activity and stability were achieved over these catalysts, which was mainly attributed to high dispersion of metal nanoparticles on HA support. In a specific case, as shown in the Figure 7.13, highly dispersed Ru metallic nanoparticles with particle size in the range of 3.0–5.5 nm (mean diameter: 4.7 ± 0.7 nm) were observed on the surface of HA by TEM analysis. In addition, ion exchange of ionic Ru species followed by reduction to metallic Ru species caused no change in the framework lattice of HA. The HA-supported Ru nanoparticles show

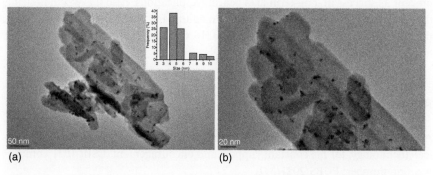

Figure 7.13 TEM images of Ru/HA catalyst at different magnifications (a) 50 nm and (b) 20 nm. Source: Akbayrak et al. [63], figure 05 (p. 05)/with permission of Elsevier.

high catalytic activity in H_2 generation from the hydrolysis of ammonia borane, providing a release of 3.0 equivalent of H_2 gas per mole of ammonia borane with a turnover frequency of up to 137 minutes^{-1} at 25 °C. Authors found that the Ru/HA catalyst showed good stability and reusable providing 87 000 turnovers for hydrogen generation from the hydrolysis of ammonia borane and maintaining 92% of their initial catalytic activity even after the fifth run.

In summary, due to tunable basic properties of HA support by varying Ca/P molar ratio, various metals in combination with HA were proved to be active and stable in various gas- and liquid-phase catalytic applications. Fine dispersion of the metals, desired ratio of Ca/P, and appropriate method for catalyst preparation were key factors in obtaining high activity and stability of HA-based catalysts in dehydrogenation reactions.

7.4 Summary and Conclusions

As demonstrated in this chapter, tremendous progress has been made during the last few decades pertaining to the synthesis and application of HA in hydrogenation and dehydrogenation reactions. This solid material offers numerous advantages such as high thermal stability and weak-moderate acid–base characters (which are undesirable for certain reactions) that may prevent curtail/minimize the side reactions in order to obtain desired compounds at high rates. The use of HA opens up possibility of producing modified chemical compositions and transitioning its structure from stoichiometric to the nonstoichiometric (Ca excess and/or deficient) composition with wide range of Ca/P ratios. Consequently, one could create desired range of acid–base properties required for a given chemical transformation. In addition, replacement of Ca^{2+} sites in Ca-HA by other cations leads to stable material formation with enhancement in catalytic performance. However, a number of challenges remain untouched dealing with structural evaluation of modified HA and thus opens up a wide scope on its exploration. For instance, the textural properties and the mesoporosity of such material are still difficult to control. As stated in the Chapter 1 of this book, a drawback of HA is its poor mechanical properties that are not adequate to prevent its success in long-term load bearing applications. Also it is worth noticing that HA may decompose in a strongly acid medium. For all aforementioned reasons, the interest on studying the earth-abundant HA will continue to be a fast-progressing topic in the coming years.

Acknowledgments

All the authors thank Dr. S Chandrasekhar, Director, CSIR-Indian Institute of Chemical Technology, for his constant support and encouragement. CSIR-IICT communication number: PUBS/2021/291.

References

1 Mori, K., Yamaguchi, K., Hara, T. et al. (2002). Controlled synthesis of hydroxyapatite-supported palladium complexes as highly efficient heterogeneous catalysts. *Journal of the American Chemical Society* 124 (39): 11572–11573. https://doi.org/10.1021/JA020444Q.

2 Fihri, A., Len, C., Varma, R.S. et al. (2017). Hydroxyapatite: a review of syntheses, structure and applications in heterogeneous catalysis. *Coordination Chemistry Reviews* 347: 48–76. https://doi.org/10.1016/J.CCR.2017.06.009.

3 Ibrahim, M., Labaki, M., Giraudon, J.M. et al. (2020). Hydroxyapatite, a multifunctional material for air, water and soil pollution control: a review. *Journal of Hazardous Materials* 383: 121139. https://doi.org/10.1016/J.JHAZMAT.2019.121139.

4 Venugopal, A. and Scurrell, M.S. (2003). Hydroxyapatite as a novel support for gold and ruthenium catalysts: behaviour in the water gas shift reaction. *Applied Catalysis A: General* 245 (1): 137–147. https://doi.org/10.1016/S0926-860X(02)00647-6.

5 Kamieniak, J., Bernalte, E., Foster, C.W. et al. (2016). High yield synthesis of hydroxyapatite (HAP) and palladium doped HAP via a wet chemical synthetic route. *Catalysts* 6 (8): 119. https://doi.org/10.3390/CATAL6080119.

6 Iriarte-Velasco, U., Ayastuy, J.L., Boukha, Z. et al. (2018). Transition metals supported on bone-derived hydroxyapatite as potential catalysts for the water-gas shift reaction. *Renewable Energy* 115: 641–648. https://doi.org/10.1016/J.RENENE.2017.08.086.

7 Sudhakar, M., Kumar, V.V., Naresh, G. et al. (2016). Vapor phase hydrogenation of aqueous Levulinic acid over hydroxyapatite supported metal (M=Pd, Pt, Ru, Cu, Ni) catalysts. *Applied Catalysis B: Environmental* 180: 113–120. https://doi.org/10.1016/j.apcatb.2015.05.050.

8 Silvester, L., Lamonier, J.-F., Vannier, R.-N. et al. (2014). Structural, textural and acid–base properties of carbonate-containing hydroxyapatites. *Journal of Materials Chemistry A* 2 (29): 11073–11090. https://doi.org/10.1039/C4TA01628A.

9 Tsuchida, T., Kubo, J., Yoshioka, T. et al. (2008). Reaction of ethanol over hydroxyapatite affected by Ca/P ratio of catalyst. *Journal of Catalysis* 259 (2): 183–189. https://doi.org/10.1016/J.JCAT.2008.08.005.

10 Zell, T. and Langer, R. (2018). From ruthenium to iron and manganese—a mechanistic view on challenges and design principles of base-metal hydrogenation catalysts. *ChemCatChem* 10 (9): 1930–1940. https://doi.org/10.1002/CCTC.201701722.

11 Sudhakar, M., Lakshmi Kantam, M., Swarna Jaya, V. et al. (2014). Hydroxyapatite as a novel support for Ru in the hydrogenation of levulinic acid to γ-valerolactone. *Catalysis Communications* 50: 101–104. https://doi.org/10.1016/J.CATCOM.2014.03.005.

12 Mizugaki, T., Nagatsu, Y., Togo, K. et al. (2015). Selective hydrogenation of levulinic acid to 1,4-pentanediol in water using a hydroxyapatite-supported Pt–Mo bimetallic catalyst. *Green Chemistry* 17 (12): 5136–5139. https://doi.org/10.1039/C5GC01878A.

13 Wen, C., Cui, Y., Chen, X. et al. (2015). Reaction temperature controlled selective hydrogenation of dimethyl oxalate to methyl glycolate and ethylene glycol over copper-hydroxyapatite catalysts. *Applied Catalysis B: Environmental* 162: 483–493. https://doi.org/10.1016/J.APCATB.2014.07.023.

14 Xu, G., Zhang, Y., Fu, Y. et al. (2017). Efficient hydrogenation of various renewable oils over Ru-HAPcatalyst in water. *ACS Catalysis* 7 (2): 1158–1169. https://doi.org/10.1021/acscatal.6b03186.

15 Jia, W., Xu, G., Liu, X. et al. (2018). Direct selective hydrogenation of fatty acids and jatropha oil to fatty alcohols over cobalt-based catalysts in water. *Energy & Fuels* 32 (8): 8438–8446. https://doi.org/10.1021/ACS.ENERGYFUELS.8B01114.

16 Sun, Y.P., Fu, H.Y., Zhang, D.L. et al. (2010). Complete hydrogenation of quinoline over hydroxyapatite supported ruthenium catalyst. *Catalysis Communications* 12 (3): 188–192. https://doi.org/10.1016/J.CATCOM.2010.09.005.

17 Elazarifi, N., Chaoui, M.A., El Ouassouli, A. et al. (2004). Hydroprocessing of dibenzothiophene, 1-methylnaphthalene and quinoline over sulfided NiMo-hydroxyapatite-supported catalysts. *Catalysis Today* 98 (1–2): 161–170. https://doi.org/10.1016/J.CATTOD.2004.07.030.

18 Norifumi, H., Yusuke, T., Takayoshi, H. et al. (2010). Fine tuning of Pd0 nanoparticle formation on hydroxyapatite and its application for regioselective quinoline hydrogenation. *Chemistry Letters* 39 (8): 832–834. https://doi.org/10.1246/CL.2010.832.

19 Wang, F. and Zhang, Z. (2016). Catalytic transfer hydrogenation of furfural into furfuryl alcohol over magnetic γ-Fe2O3@HAP catalyst. *ACS Sustainable Chemistry and Engineering* 5 (1): 942–947. https://doi.org/10.1021/ACSSUSCHEMENG.6B02272.

20 Li, C., Xu, G., Liu, X. et al. (2017). Hydrogenation of biomass-derived furfural to tetrahydrofurfuryl alcohol over hydroxyapatite-supported pd catalyst under mild conditions. *Industrial and Engineering Chemistry Research* 56 (31): 8843–8849. https://doi.org/10.1021/ACS.IECR.7B02046.

21 Putrakumar, B., Seelam, P.K., Srinivasarao, G. et al. (2020). High performance and sustainable copper-modified hydroxyapatite catalysts for catalytic transfer hydrogenation of furfural. *Catalysts* 10 (9): 1045. https://doi.org/10.3390/CATAL10091045.

22 Zahmakıran, M., Tonbul, Y., and Özkar, S. (2010). Ruthenium(0) nanoclusters supported on hydroxyapatite : highly active, reusable and green catalyst in the hydrogenation of aromatics under mild conditions with an unprecedented catalytic lifetime. *Chemical Communications* 46 (26): 4788–4790. https://doi.org/10.1039/C0CC00494D.

23 Zahmakıran, M., Román-Leshkov, Y., and Zhang, Y. (2011). Rhodium(0) nanoparticles supported on nanocrystalline hydroxyapatite: highly effective

catalytic system for the solvent-free hydrogenation of aromatics at room temperature. *Langmuir* 28 (1): 60–64. https://doi.org/10.1021/LA2044174.

24 Pashaei, M. and Mehdipour, E. (2018). Silver nanoparticles supported on ionic-tagged magnetic hydroxyapatite as a highly efficient and reusable nanocatalyst for hydrogenation of nitroarenes in water. *Applied Organometallic Chemistry* 32 (4): e4226. https://doi.org/10.1002/AOC.4226.

25 Shokouhimehr, M., Yek, S.M.-G., Nasrollahzadeh, M. et al. (2019). Palladium nanocatalysts on hydroxyapatite: green oxidation of alcohols and reduction of nitroarenes in water. *Applied Sciences* 9 (19): 4183. https://doi.org/10.3390/APP9194183.

26 Sreedhar, B., Devi, D.K., Neetha, A.S. et al. (2015). Green synthesis of gum-acacia assisted gold-hydroxyapatite nanostructures – characterization and catalytic activity. *Materials Chemistry and Physics* 153: 23–31. https://doi.org/10.1016/J.MATCHEMPHYS.2014.12.031.

27 Huang, L., Luo, P., Pei, W. et al. (2012). Selective hydrogenation of nitroarenes and olefins over rhodium nanoparticles on hydroxyapatite. *Advanced Synthesis & Catalysis* 354 (14–15): 2689–2694. https://doi.org/10.1002/ADSC.201200330.

28 Jiang, L. and Zhang, Z. (2016). Efficient transfer hydrogenation of nitro compounds over a magnetic palladium catalyst. *International Journal of Hydrogen Energy* 41 (48): 22983–22990. https://doi.org/10.1016/J.IJHYDENE.2016.09.182.

29 Xu, G., Guo, J., Zhang, Y. et al. (2015). Selective hydrogenation of phenol to cyclohexanone over pd–HAP catalyst in aqueous media. *ChemCatChem* 7 (16): 2485–2492. https://doi.org/10.1002/CCTC.201500442.

30 Campisi, S., Galloni, M.G., Bossola, F. et al. (2019). Comparative performance of copper and iron functionalized hydroxyapatite catalysts in NH3-SCR. *Catalysis Communications* 123: 79–85. https://doi.org/10.1016/J.CATCOM.2019.02.008.

31 Tounsi, H., Djemal, S., Petitto, C. et al. (2011). Copper loaded hydroxyapatite catalyst for selective catalytic reduction of nitric oxide with ammonia. *Applied Catalysis B: Environmental* 107 (1–2): 158–163. https://doi.org/10.1016/J.APCATB.2011.07.009.

32 Young, Z.D. and Davis, R.J. (2018). Hydrogen transfer reactions relevant to Guerbet coupling of alcohols over hydroxyapatite and magnesium oxide catalysts. *Catalysis Science & Technology* 8 (6): 1722–1729. https://doi.org/10.1039/C7CY01393K.

33 Hanspal, S., Young, Z.D., Prillaman, J.T. et al. (2017). Influence of surface acid and base sites on the Guerbet coupling of ethanol to butanol over metal phosphate catalysts. *Journal of Catalysis* 352: 182–190. https://doi.org/10.1016/J.JCAT.2017.04.036.

34 Zhang, P., Wu, T., Jiang, T. et al. (2012). Ru–Zn supported on hydroxyapatite as an effective catalyst for partial hydrogenation of benzene. *Green Chemistry* 15 (1): 152–159. https://doi.org/10.1039/C2GC36596K.

35 Gama-Lara, S.A., Natividad, R., Vilchis-Nestor, A.R. et al. (2019). Ultra-small platinum nanoparticles with high catalytic selectivity synthesized by an eco-friendly method supported on natural hydroxyapatite. *Catalysis Letters* 149 (12): 3447–3453. https://doi.org/10.1007/S10562-019-02919-Z.

36 Tsuchida, T., Sakuma, S., Takeguchi, T. et al. (2006). Direct synthesis of N-butanol from ethanol over nonstoichiometric hydroxyapatite. *Industrial and Engineering Chemistry Research* 45 (25): 8634–8642. https://doi.org/10.1021/IE0606082.

37 Tsuchida, T., Yoshioka, T., Sakuma, S. et al. (2008). Synthesis of biogasoline from ethanol over hydroxyapatite catalyst. *Industrial and Engineering Chemistry Research* 47 (5): 1443–1452. https://doi.org/10.1021/IE0711731.

38 Silvester, L., Lamonier, J.-F., Faye, J. et al. (2015). Reactivity of ethanol over hydroxyapatite-based Ca-enriched catalysts with various carbonate contents. *Catalysis Science & Technology* 5 (5): 2994–3006. https://doi.org/10.1039/C5CY00327J.

39 Ogo, S., Onda, A., Iwasa, Y. et al. (2012). 1-Butanol synthesis from ethanol over strontium phosphate hydroxyapatite catalysts with various Sr/P ratios. *Journal of Catalysis* 296: 24–30. https://doi.org/10.1016/J.JCAT.2012.08.019.

40 Peters, M., Köhler, B., Kuckshinrichs, W. et al. (2011). Chemical technologies for exploiting and recycling carbon dioxide into the value chain. *ChemSusChem* 4 (9): 1216–1240. https://doi.org/10.1002/CSSC.201000447.

41 Wai, M.H., Ashok, J., Dewangan, N. et al. (2020). Influence of surface Formate species on methane selectivity for carbon dioxide methanation over nickel hydroxyapatite catalyst. *ChemCatChem* 12 (24): 6410–6419. https://doi.org/10.1002/CCTC.202001300.

42 Wai, M.H. (2018). Development of nickel hydroxyapatite catalyst for CO_2 hydrogenation to methane. Master thesis, National University of Singapore. https://scholarbank.nus.edu.sg/handle/10635/146863.

43 Munirathinam, R., Pham Minh, D., and Nzihou, A. (2020). Hydroxyapatite as a new support material for cobalt-based catalysts in Fischer-Tropsch synthesis. *International Journal of Hydrogen Energy* 45 (36): 18440–18451. https://doi.org/10.1016/J.IJHYDENE.2019.09.043.

44 Kanega, R., Onishi, N., Tanaka, S. et al. (2021). Catalytic hydrogenation of CO_2 to methanol using multinuclear iridium complexes in a gas–solid phase reaction. *Journal of the American Chemical Society* 143 (3): 1570–1576. https://doi.org/10.1021/JACS.0C11927.

45 Sans, J., Revilla-López, G., Sanz, V. et al. (2021). Permanently polarized hydroxyapatite for selective electrothermal catalytic conversion of carbon dioxide into ethanol. *Chemical Communications* 57 (42): 5163–5166. https://doi.org/10.1039/D0CC07989H.

46 Boukha, Z., González-Velasco, J.R., and Gutiérrez-Ortiz, M.A. (2020). Platinum supported on Lanthana-modified hydroxyapatite samples for realistic WGS conditions: on the nature of the active species, kinetic aspects and the resistance to shut-down/start-up cycles. *Applied Catalysis B: Environmental* 270: 118851. https://doi.org/10.1016/J.APCATB.2020.118851.

47 Guo, J., Duchesne, P.N., Wang, L. et al. (2020). High-performance, scalable, and low-cost copper hydroxyapatite for photothermal CO_2 reduction. *ACS Catalysis* 10 (22): 13668–13681. https://doi.org/10.1021/ACSCATAL.0C03806.

48 Airaksinen, S.M.K., Bañares, M.A., and Krause, A.O.I. (2005). In situ characterisation of carbon-containing species formed on chromia/alumina during propane dehydrogenation. *Journal of Catalysis* 230 (2): 507–513. https://doi.org/10.1016/J.JCAT.2005.01.005.

49 Airaksinen, S.M.K., Kanervo, J.M., and Krause, A.O.I. (2001). Deactivation of CrOx/Al2O3 catalysts in the dehydrogenation of i-butane. *Studies in Surface Science and Catalysis* 136: 153–158. https://doi.org/10.1016/S0167-2991(01)80296-2.

50 Sugiyama, S., Osaka, T., Hirata, Y. et al. (2006). Enhancement of the activity for oxidative dehydrogenation of propane on calcium hydroxyapatite substituted with vanadate. *Applied Catalysis A: General* 312 (1–2): 52–58. https://doi.org/10.1016/J.APCATA.2006.06.018.

51 Elkabouss, K., Kacimi, M., Ziyad, M. et al. (2004). Cobalt-exchanged hydroxyapatite catalysts: magnetic studies, spectroscopic investigations, performance in 2-butanol and ethane oxidative dehydrogenations. *Journal of Catalysis* 226 (1): 16–24. https://doi.org/10.1016/J.JCAT.2004.05.007.

52 Dasireddy, V.D.B.C., Singh, S., and Friedrich, H.B. (2012). Oxidative dehydrogenation of N-octane using vanadium pentoxide-supported hydroxyapatite catalysts. *Applied Catalysis A: General* 421–422: 58–69. https://doi.org/10.1016/J.APCATA.2012.01.034.

53 Khachani, M., Kacimi, M., Ensuque, A. et al. (2010). Iron–calcium–hydroxyapatite catalysts: iron speciation and comparative performances in Butan-2-Ol conversion and propane oxidative dehydrogenation. *Applied Catalysis A: General* 388 (1–2): 113–123. https://doi.org/10.1016/J.APCATA.2010.08.043.

54 Boucetta, C., Kacimi, M., Ensuque, A. et al. (2009). Oxidative dehydrogenation of propane over chromium-loaded calcium-hydroxyapatite. *Applied Catalysis A: General* 356 (2): 201–210. https://doi.org/10.1016/J.APCATA.2009.01.005.

55 Sugiyama, S., Osaka, T., Hashimoto, T. et al. (2005). Oxidative dehydrogenation of propane on calcium hydroxyapatites partially substituted with vanadate. *Catalysis Letters* 103 (1): 121–123. https://doi.org/10.1007/S10562-005-6513-7.

56 Sugiyama, S., Shono, T., Nitta, E. et al. (2001). Effects of gas- and solid-phase additives on oxidative dehydrogenation of propane on strontium and barium hydroxyapatites. *Applied Catalysis A: General* 211 (1): 123–130. https://doi.org/10.1016/S0926-860X(00)00864-4.

57 Sugiyama, S. and Hayashi, H. (2012). Role of hydroxide groups in hydroxyapatite catalysts for the oxidative dehydrogenation of alkanes. *International Journal of Modern Physics B* https://doi.org/10.1142/S0217979203019186 17 (8-9 I): 1476–1481. https://doi.org/10.1142/S0217979203019186.

58 Matsumura, Y. and Moffat, J.B. (1996). Methanol adsorption and dehydrogenation over stoichiometric and non-stoichiometric hydroxyapatite catalysts. *Journal of the Chemical Society, Faraday Transactions* 92 (11): 1981–1984. https://doi.org/10.1039/FT9969201981.

59 Kibby, C.L. and Hall, W.K. (1973). Dehydrogenation of alcohols and hydrogen transfer from alcohols to ketones over hydroxyapatite catalysts. *Journal of Catalysis* 31 (1): 65–73. https://doi.org/10.1016/0021-9517(73)90271-6.

60 Lu, Z., Gao, D., Yin, H. et al. (2015). Methanol dehydrogenation to methyl Formate catalyzed by SiO2-, hydroxyapatite-, and MgO-supported copper catalysts and reaction kinetics. *Journal of Industrial and Engineering Chemistry* 31: 301–308. https://doi.org/10.1016/J.JIEC.2015.07.002.

61 Bharath, G., Rambabu, K., Hai, A. et al. (2020). Development of au and 1D hydroxyapatite nanohybrids supported on 2D boron nitride sheets as highly efficient catalysts for dehydrogenating glycerol to lactic acid. *ACS Sustainable Chemistry & Engineering* 8 (19): 7278–7289. https://doi.org/10.1021/ACSSUSCHEMENG.9B06997.

62 Hara, T., Mori, K., Mizugaki, T. et al. (2003). Highly efficient dehydrogenation of indolines to indoles using hydroxyapatite-bound pd catalyst. *Tetrahedron Letters* 44 (33): 6207–6210. https://doi.org/10.1016/S0040-4039(03)01550-8.

63 Akbayrak, S., Erdek, P., and Özkar, S. (2013). Hydroxyapatite supported ruthenium(0) nanoparticles catalyst in hydrolytic dehydrogenation of ammonia borane: insight to the nanoparticles formation and hydrogen evolution kinetics. *Applied Catalysis B: Environmental* 142–143: 187–195. https://doi.org/10.1016/J.APCATB.2013.05.015.

8

Reforming Processes Using Hydroxyapatite-Based Catalysts

Zouhair Boukha, Rubén López-Fonseca, and Juan R. González-Velasco

Chemical Technologies for Environmental Sustainability Group, Department of Chemical Engineering, Faculty of Science and Technology, University of the Basque Country UPV/EHU, P.O. Box 644, E-48080 Bilbao, Spain

8.1 Introduction

The development of clean and efficient solutions for fuels and chemicals production from fossil resources is considered one of the most important challenges facing scientific and technological research since the turn of the century. In this sense, the proven large reserves of natural gas (NG) and the development of biogas production technologies allow implementing relatively long-term strategies for their efficient utilization. For instance, the transformation of the available natural resources into products of high added-value, such as ammonia, alcohols, olefins, and higher hydrocarbons, among others, via synthesis gas (H_2 + CO), is a major challenge of the chemical industry. Moreover, owing to its clean combustion, emitting only water, and its high calorific value (120.7 kJ g^{-1}), hydrogen is considered as the most promising fuel to be integrated in the future energy strategies [1]. Currently, over 50 million tons of hydrogen are produced by global chemical industry for many applications [2].

The synthesis gas is mainly produced from fossil fuels by different reforming processes [1–8]. Among them, the endothermic steam reforming of methane (SRM) (8.1) is the most used one [8]. This technology has largely been applied for the production of large amounts of H_2, providing a high H_2/CO ratio (≥3). However, in many cases, the exothermic methane partial oxidation (POM) (8.2) is considered a good alternative, since it offers some interesting advantages, especially when the posterior use requires an H_2/CO ratio close to 2, as in the case of Fischer–Tropsch synthesis [8]. In the last decades, an increasing interest has also been devoted to the DRM (8.3), which appears to be a suitable strategy for long chain hydrocarbon production, which requires an H_2/CO ratio close to unity [9]. For environmental concerns, this is an attractive strategy fulfilling one of the constant challenges, because it allows the use of greenhouse gases (CH_4 and CO_2) as reactants [4–9].

$$\text{SRM: } CH_4 + H_2O \rightarrow CO + 3H_2 \quad \Delta H° = +206 \text{ kJ mol}^{-1} \quad (8.1)$$

$$\text{POM: } CH_4 + \tfrac{1}{2}O_2 \rightarrow CO + 2H_2 \quad \Delta H° = -36 \text{ kJ mol}^{-1} \quad (8.2)$$

Design and Applications of Hydroxyapatite-Based Catalysts, First Edition. Edited by Doan Pham Minh.
© 2022 WILEY-VCH GmbH. Published 2022 by WILEY-VCH GmbH.

$$\text{DRM: } CH_4 + CO_2 \rightarrow 2CO + 2H_2 \quad \Delta H° = +247 \text{ kJ mol}^{-1} \quad (8.3)$$

Due to its high symmetry, the activation of the tetravalent methane molecule through the dissociation of CH_3—H bond requires a relatively high energy (439.3 kJ mol^{-1}) [3]. Therefore, the use of active catalysts is essential for all strategies aiming its efficient conversion by reforming processes. Noble metal catalysts, such as Rh, Pt, Pd, and Ru, exhibit a high reforming activity and good resistance to coke formation. Regarding economic considerations, current research is focused on other promising alternatives based on cheaper transition metals. Among them, judging from their high activity, Co- and Ni-based catalysts are the most investigated ones. However, the major drawback that limits their use is their sensitivity to the structural changes under the severe conditions of the reductive pretreatment and/or during the reforming process that can in turn affect the efficiency of their active species. According to previous reports, these changes result in sintering and coking, leading to a severe drop in the activity and the resistance during a long-term operation [4, 7, 8]. To overcome these issues considerable efforts have been devoted to develop robust catalysts by optimizing the preparation route and finding suitable supports to obtain highly dispersed and sintering-resistant metallic particles.

Though high surface area alumina-supported active metals are by far the most investigated catalysts [10–13], a number of studies dealing with the use of calcium hydroxyapatite (HA, $Ca_{10}(PO_4)_6(OH)_2$) as a support for the methane reforming reactions are already available [4–8, 14–19]. Actually, the reactivity of the HA surface in the methane oxidation and oxidative coupling of methane reactions was first demonstrated by Moffat's group [20–23]. These studies pointed out that the advantage of the use of HA as a modulable catalyst support for the selectivity optimization lies in its capacity to retain its structure when it undergoes important changes in composition. Since then, the application of HA has fascinated more researchers due to its potential to achieve more challenging goals, such as the stabilization of the metallic active phase under severe reaction conditions. The good performance of HA catalyst support is also found to be connected to its high thermal stability, textural, and acid–base properties and the fact that it could provide beneficial synergistic effects with the active phases. Moreover, the surface properties of HA can be tuned by varying the Ca/P atomic ratio and/or the ion exchange with a wide number of cations and anions [4, 5, 8, 15–23]. When optimized, this set of properties does help to achieve a high efficiency in the methane reforming processes. For instance, according to the literature data, HA has been successfully used as a suitable support for Ni catalysts in DRM, a process in which it has been mostly investigated. The suitability of HA support for this reforming process was first reported by Boukha et al. [5] in their study on the activity of Ni/HA catalysts. This system resulted highly active showing 100% methane conversion at 700 °C.

The standard HA crystal is a hexagonal prism belonging to $P6_3/m$ space group. Its lattice comprises an assemblage of Ca^{2+} and tetrahedral PO_4^{3-} groups (Figure 8.1). The Ca^{2+} ions are hosted in two different sites, denoted columnar Ca (I) and screw axis Ca (II), respectively. The Ca (I) sites are surrounded by nine oxygen atoms belonging to six PO_4 tetrahedrons. The Ca (II) sites form a channel configuration

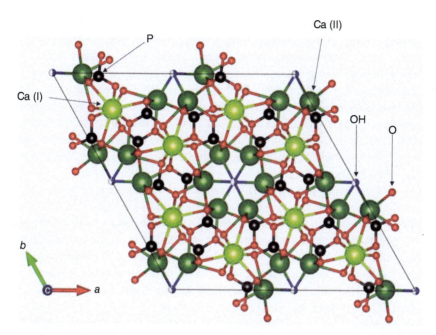

Figure 8.1 Crystallographic structure of HA (JCPDS: 01-082-2956): Projection on the plane (001). Source: Zouhair Boukha.

with a larger diameter oriented along the c-axis, which hosts monovalent anions (OH^-) for the charge compensation. The Ca (II) sites play an important role in tuning the chemical properties of HA, since they are more prone to be exchanged with cationic species, including the lanthanides and transition metals. The hydroxyl groups hosted in the HA channels can also be exchanged by different monovalent anions, such as F^-, I^-, Cl^-, and Br^-. Details on the HA structure has been presented in Chapter 3 of this book.

Commercial and/or synthesized materials with a composition close to that of a stoichiometric HA (Ca/P = 1.67) have been by far the most investigated as catalyst supports in the reforming reactions. A number of the available studies addressed the effect of the preparation method, the nature of the active phase, metal loading, promoter effect, and the HA composition, among others [4–8, 18–25]. However, there is no detailed review on the performance of the HA-based catalysts in the methane reforming reactions. In this sense, this work summarizes the recent results published up to now and could help researchers to design and improve even more the current achievements.

8.2 Overview on the Nature of the Interactions of HA with Transition Metal Catalysts

The type of metal–support interactions is of great importance for transition metal-supported catalysts. It is widely accepted that the nature of the support influences the strength and the distribution of the supported active phases. In turn, this would mainly affect (i) the size of the metallic particles and the (ii) structural, (iii) chemical, and (iv) textural properties, proved to be key parameters in controlling the efficiency, selectivity, and stability of the reforming catalysts [9]. Though the active phases for the reforming reactions are generally reduced transition metal species, the preparation of the catalyst formulation usually involves a previous oxidative treatment. This induces the deposition of oxidized forms with a distribution that depends on the nature of the used support. However, after reduction, these precursor phases should be capable to generate metallic species presenting suitable interactions with the support.

In view of its distinct properties, HA materials provide different types of interactions with the supported metallic phase. Irrespective of the preparation method used to deposit the corresponding metal oxide, often cationic species readily integrate the HA framework by exchanging Ca^{2+} sites [4, 5, 8, 15]. It is noteworthy that the degree of this type of substitution is controlled by the nature of the transition metal. Hence, it was found that on a stoichiometric HA the total amount of exchanged Co^{2+} did not exceed 1.35 wt% [15], while up to 3.2 wt% of Cu^{2+} could exchange the Ca^{2+} sites [26]. On the other hand, the lattice parameters of HA can be remarkably modified depending on the ionic radius of the exchanged metal and its electronegativity with respect to those of Ca^{2+}. For instance, after calcination of a Pt/HA catalyst at 500 °C, a shrinkage of the HA lattice was noticed, owing to the incorporation of Pt^{2+} ions [27]. This was explained by the smaller ionic radius and the high electronegativity of Pt^{2+} (94 pm and 2.28, respectively) compared with Ca^{2+} (114 pm and 1, respectively). Likewise, the surface vacant sites occurring on a Ca-deficient HA support favor the incorporation of transition metal cations. This range of ion exchange possibilities caused by the flexible structure of HA allows tuning the distribution of the supported metallic species and the acid–base properties of the catalyst.

Boukha and coworkers [15] identified three types of cobalt species after a simple impregnation of Co over HA followed by calcination at 500 °C: (i) Co^{2+} ions integrating the HA framework, (ii) Co_xO_y clusters exhibiting a relatively small size (<10 nm), and (iii) a well crystallized Co_3O_4 phase. This distribution significantly affected the reducibility of the catalyst. Thus, Co^{2+} ion exchanged species reduced at relatively higher temperatures (660 °C) with respect to those spread on the catalyst surface (<500 °C). They also observed that the addition of large amounts of Co ($\geq$8.5%) decreased the number of the surface acid sites. This was explained by the achievement of a critical Co surface density, which covered a large number of the surface acid sites. However, irrespective of the Co loading, the density of the basic sites increased by more than twice. A similar metallic distribution had been observed on Ni/HA catalysts prepared by impregnation method and calcined at 550 °C [5]. The characterization of the samples by H_2-TPR, XPS, and UV–Vis–NIR techniques revealed that Ni occupied three distinct reducible sites: (i) exchanged Ni^{2+}, (ii) small NiO particles interacting with the support, and (iii) large NiO particles, which were only formed at high loadings.

On the other hand, the calcination temperature of the HA support seems to influence the distribution of the supported Ni phases. This can be due to structural evolution affecting chiefly its lattice parameters and degree of crystallinity. For instance, when a stoichiometric HA support was calcined at a higher temperature (750 °C) before the impregnation of Ni, the TPR profile of the resulting Ni/HA catalyst did not exhibit the feature corresponding to exchanged Ni^{2+} species [4]. This sample showed only two reduction peaks in the range of 350–550 °C, associated with NiO species presenting different interactions with the support. According to a study by Rêgo de Vasconcelos and coworkers [28], an increase in the HA calcination temperature, up to 1000 °C, weakened drastically the interaction of Ni with HA. In this case, only one reduction peak at 400 °C, attributed to a surface NiO phase, was observed. TEM images, moreover, revealed that the oxide particles were sized between 100 and 200 nm.

For so long, the concept of strong metal–support interactions (SMSI) has been referred to oxide supports only. To our knowledge, the occurrence of SMSI between noble metals and non-oxide supports was first demonstrated by Tang et al. [29] in their study on HA-supported Au nanoparticles. Unlike the oxide supports, they confirmed that SMSI between Au and HA took place through an oxidative treatment at 800 °C. The resulting encapsulated Au nanoparticles proved to be highly sintering resistant and exhibited high stability in liquid-phase catalysis. The same research group [30] extended their investigation to examine the strength of the interaction of HA with Pd and Pt, respectively. They found that the two systems (Pd/HA and Pt/HA) also generated metallic species encapsulated by a thin support layer, under oxidative conditions. They attributed this reversible effect to a loss in the surface hydroxyl groups. Moreover, the resulting active species did not undergo aggregation and/or leaching, and, thus, they proved to be highly resistant under various cycles of the Suzuki cross-coupling reaction. We observed a similar behavior in our previous work on the influence of the calcination temperature (500–800 °C) on the properties and the activity of HA-supported Pd catalysts in the methane oxidation reaction [17]. In addition, we paid a special attention to analyzing the evolution of the structure of Pd^{2+} species by using UV–Vis–NIR techniques. We reported, besides the Pd encapsulation derived from the dehydroxylation of the support occurring at $T \geq 700\,°C$, a change in the coordination of the Pd^{2+} species with the calcination temperature from tetrahedral (Td) to square planar (D_{4h}) geometry. H_2-TPR results revealed that this distribution affected the reducibility of PdO, where the encapsulated particles required higher temperatures for reduction (Figure 8.2). The occurrence of these embedded species was also evidenced by FTIR experiments, which suggested a suppression of CO adsorption at room temperature. These properties seemed to be suitable to provide a relatively high specific activity in the methane oxidation reaction. Moreover, in contrast to the catalysts calcined at lower temperatures ($T \leq 700\,°C$), the stability tests revealed a progressive activation of the catalyst calcined at 800 °C with time on stream (TOS), resulting from a de-encapsulation of the PdO phase under the reducing atmosphere of the reaction mixture.

Unlike the Au, Pd, and Pt supported on HA catalysts, up to now, no evidence was reported on the occurrence of SMSI in the case of Rh/HA system. The distribution of the deposited Rh phases seems to follow a similar model as for Ni and Co catalysts. In this sense, Boukha et al. [8] identified three different forms of Rh species after

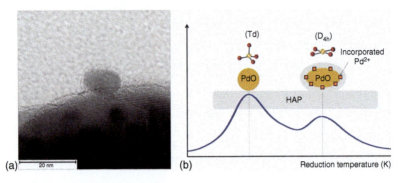

Figure 8.2 (a) HRTEM image of the Pd/HA-1073 sample, reduced at 473 K for 2 hours and (b) proposed scheme summarizing the distribution of the Pd species. Source: Boukha et al. [17], pg. 277. / With permission of Elsevier.

calcination of a series of Rh/HA samples at 727 °C: (i) large crystallites of Rh_2O_3, (ii) small RhO_x particles, and (iii) Rh^{2+} substituting Ca^{2+} sites of HA. After reduction at 727 °C, the structural and textural properties were found to be dependent on the distribution of Rh among the different sites. Thus, the reduction of the incorporated Rh species provoked a phase transformation from HA to tricalcium phosphate (TCP), which significantly lowered the specific surface area. Nevertheless, the analysis of the reduced samples by TEM microscopy did not evidence the presence of embedded Rh particles. We provide here another proof that the Rh/HA samples reduced at 727 °C did not manifest SMSI. As can be deduced from their analysis, the FTIR CO adsorption spectra recorded at 35 °C over a reduced Rh(1)/HA sample (Figure 8.3) revealed the presence of a doublet peaked at 2020 and 2095 cm^{-1}, respectively, associated with the gem-dicarbonyl species and a single feature at 2020 cm^{-1} assigned to linearly bonded CO [31]. These features, adsorbed on metallic Rh species, seem to resist evacuation under N_2 in the temperature range of 35–100 °C.

8.3 HA-Supported Non-noble Metal Catalysts for Methane Reforming Reactions

8.3.1 Suitability of the HA-Based Catalysts

As aforementioned, the current issues that limit the application of non-noble metal catalysts such as Ni and Co in the reforming processes are mainly related to their low resistance to deactivation during a long-term operation [4–9]. Many reports attributed this undesirable behavior to the carbon deposition, which blocks and/or encapsulates the active metallic phases. This was essentially caused by the occurrence of methane cracking (8.4) and Boudouard (8.5) side reactions as follows:

$$CH_4 \leftrightarrow C + 2H_2 \quad \Delta H° = +74.8 \text{ kJ mol}^{-1} \qquad (8.4)$$

$$2CO \leftrightarrow C + CO_2 \quad \Delta H° = -173 \text{ kJ mol}^{-1} \qquad (8.5)$$

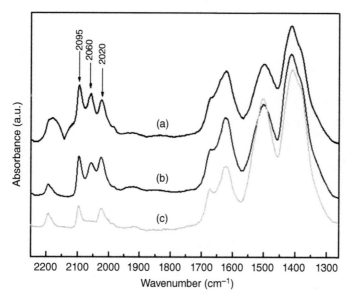

Figure 8.3 FTIR spectra of the CO adsorption over a Rh(1)/HA catalyst, reduced and evacuated at 727 °C, recorded after (a) adsorption of 750 ppm of CO at 35 °C and after subsequent evacuations under N_2 for 30 minutes at (b) 35 °C and (c) 100 °C, respectively. All spectra were recorded at 35 °C. Source: Zouhair Boukha.

To address this shortcoming, various supports have been investigated in order to optimize suitable formulations, exhibiting not only a high efficiency but also a reasonable stability. For instance, it is known that increasing the strength of the metal–support interactions improves markedly the dispersion of the metallic active phase and their resistance to sintering and carbon deposition under the severe conditions of the reforming reactions [5, 8, 9, 32, 33]. However, previous studies pointed out that undesirable metal–support interactions can lead to the formation of inactive metallic phases. For example, it is widely accepted that at high temperatures the Ni/Al_2O_3 and Ni/SiO_2 systems are susceptible to the formation of a large fraction of the inactive and hardly reducible nickel aluminate and nickel silicate spinel phases, respectively. Interestingly, the demonstrated suitability of HA support for Ni reforming catalysts can be adjusted by controlling the metal–support interactions. Accordingly, previous studies showed that a fraction lower than 1 wt% Ni could incorporate the HA structure [4, 5].

Promising results were reported when HA was used as a support for Co and Ni catalysts for DRM [4, 5, 34]. The presence of phosphate groups in the standard composition of HA results in advantageous acid properties, leading to the stabilization of the active phases in the reforming reactions [35–39]. Park et al. [35] studied the effect of the phosphate groups on the stabilization of the Co active phases in DRM. They found that the addition of P to Al_2O_3 markedly enhanced the stability of the Co catalyst (Figure 8.4). The superiority of the P-modified catalyst was assigned to the formation of $AlPO_4$ at the expense of the inactive $CoAl_2O_4$ phases. Ibrahim et al.

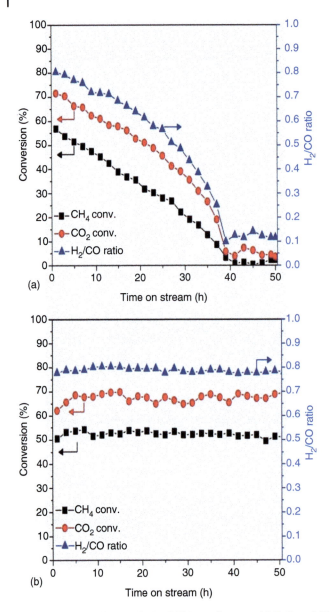

Figure 8.4 Stability test in the DRM reaction over (a) CoAl and (b) CoP(2)Al catalysts. Reaction conditions: 0.05 g catalyst, $T = 750\,°C$, SV = 60 000 cm^3 g^{-1} h^{-1}, and CH$_4$/CO$_2$/N$_2$ (vol%) = 40/40/20. Source: Park et al. [35]. Reproduced with permission of Elsevier.

[36] reported that phosphate modification enhanced the Ni—Zr interaction in their Ni/ZrO$_2$ DRM catalyst, which limited the formation of coke.

The literature has claimed a positive effect of the basicity on the performance of the catalysts in methane reforming [40, 41]. This was mainly associated with the role of the surface basic sites in the rapid activation of CO$_2$, which help to suppress the coke formation from CH_4 decomposition. Note that there is a wide consensus on the bifunctionality of the DRM mechanism, where the activation of CH_4 occurs on the metal and that of CO_2 on the chemically active support surface [9]. In this mechanism, the surface hydroxyl groups play a crucial role over both acidic and basic supports, when they react with CO_2 to form formates and oxy-carbonates intermediates, respectively. By contrast, the surface of inert supports such as SiO_2 does not provide these features for CO_2 activation. This makes them less active and less resistant to deactivation compared with the acidic or basic supports [9].

In order to provide insight into their suitability in DRM, Table 8.1 compares the performance, in terms of activity and durability, of HA-based catalysts with those obtained with other catalyst supports reported in the literature. It should be remarked that it was difficult to make a precise comparison due to the differences in the experimental conditions applied over each formulation. Nevertheless, the results obtained on the HA-based catalysts globally evidence their competitiveness with respect to those of conventional catalyst supports.

Jun et al. [14] studied the catalytic behavior of HA-supported Ni samples in the POM reaction. They prepared Ni–Ca–HA catalysts by coprecipitation of Ni, Ca, and P salts in a basic medium. After calcination at 800 °C, in addition to the exposed NiO particles exhibiting sizes in the range of 5–20 nm, two solid solutions were also formed: $Ca_{3-x}Ni_x(PO_4)_2$ and $Ca_{10-y}Ni_y(OH)_2(PO_4)_6$. The H_2-TPR data indicated that NiO phase was the most easily reducible species followed by the Ni incorporated into the apatite. These catalysts exhibited an excellent performance under a stoichiometric POM mixture, yielding values close to the thermodynamic equilibrium (Figure 8.2). Furthermore, the catalyst could be easily activated under the reaction mixture during the first light-off run; thereafter, no hysteresis of catalyst activity was observed. According to the authors, this behavior was in contrast with that shown by traditional oxide supports. They suggested that the Ni species present in their catalysts were more easily reducible under the reaction mixture.

In our previous study, we studied the behavior of HA-supported Ni catalysts in DRM. Under a diluted DRM feed mixture ($CH_4/CO_2/He$ = 2/2/60), the Ni/HA catalysts (1 < Ni wt% < 10) proved to be highly active in the temperature range of 600–800 °C, where the CH_4 and CO_2 conversion values were found to be close to the thermodynamic equilibrium. The CO_2 conversion (X_{CO2}) values were in all cases greater than those of CH_4 conversion (X_{CH4}), thereby suggesting the contribution of the reverse water–gas shift reaction (8.6). Carbon deposition on the catalysts seemed to increase with nickel loading but without provoking any significant decay of the activity. These promising achieved results were attributed to the synergy among the basic features of HA, its ability to chemisorb CO_2, and the catalytic properties of the supported Ni. Interestingly, a total fluoridation of HA did not influence significantly the performances of the investigated Ni catalysts, which questions the intrinsic role of the crystallographic hydroxyl groups.

$$CO_2 + H_2 \leftrightarrow CO + H_2O \quad \Delta H° = 41 \text{ kJ mol}^{-1} \tag{8.6}$$

Rêgo de Vasconcelos et al. [7] compared the activity of HA- and 30MgAl-supported Ni catalysts in DRM. Despite the observed similarities in their Ni particle sizes,

Table 8.1 Comparison of the performance of Ni/HA and Co/HA catalysts with Ni and Co catalysts reported in the literature.

Catalyst	Ni (wt%)	Co (wt%)	S_{BET} (m² g⁻¹)	T (°C)	Feed composition $CH_4/CO_2/N_2$	WHSV (ml h⁻¹ g⁻¹)	TOS (h)	X_{CH4}	X_{CO2}	H_2/CO	Coke (%)	References
Ni/HA-(1.57)	3.4	0	24	750	50/50/0	60 000	24	75	83	0.78	0.6	[4]
Ni/HA-(1.62)	3.4	0	26	750	50/50/0	60 000	24	78	86	0.79	0.8	
Ni/HA-(1.67)	3.4	0	17	750	50/50/0	60 000	24	70	80	0.75	0.7	
Ni/HA-(1.73)	3.4	0	9	750	50/50/0	60 000	24	67	78	0.72	0.9	
Co/HA-(1.40)	0	8.2	7	700	20/20/60	31 764	50	52	63	0.71	n.d.	[6]
Ni/HA-(1.60)	8.6	0	5	700	20/20/60	31 764	50	37	45	0.57	n.d.	
Ni—Co/HA-(1.43)	4.4	4.7	8	700	20/20/60	31 764	50	45	53	0.63	n.d.	
Ni/HA—Ce (*)	2.6	0	124	750	20/20/60	60 000	100	80	95	0.80	n.d.	[18]
Ni—Co/HA	2.8	2.8	46	750	20/20/60	15 882	24	73	79	0.83	n.d.	[34]
Ni/CYSZ	5.1	0	23	750	50/50/0	60 000	70	60	70	0.75	11	[42]
Ni/Al₂O₃	10	0	108	750	40/40/20	60 000	20	55	58	0.83	25.4	[43]
Ni/P—Al₂O₃	10	0	133	750	40/40/20	60 000	20	63	61	0.86	8	
Ni/H-ZSM5-700	5	0	327	700	50/50/0	3600	9	76	79	0.70	7	[44]
Ni/CeO₂/Al₂O₃	5	0	146	700	50/50/0	18 000	24	80	80	n.d.	4.6	[45]
Ni—Mg—Al—8La	20	0	202	750	50/50/0	240 000	10	67	72	n.d.	n.d.	[46]
Ni—Co—CeZrO₂/S	1.2	2.8	17	750	50/50/0	60 000	550	68	73	0.8	n.d.	[47]
Ni/Al₂O₃	4.3	0	10	750	50/50/0	36 000	6	66	73	0.65	9	[48]
Co/Al₂O₃	0	4.4	13	750	50/50/0	36 000	6	75	82	0.70	4	
Ni/CeO₂-700	8	0	24	700	20/20/60	30 000	6	62	65	0.80	20	[49]
Co/CeO₂-700	0	8	19	700	20/20/60	30 000	6	<10	<10	—	n.d.	

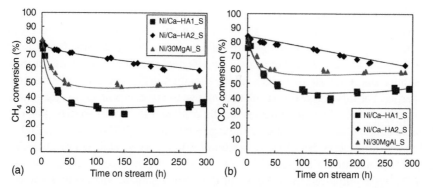

Figure 8.5 CH_4 and CO_2 conversions at high contact time of the catalysts prepared with calcined supports. Reaction conditions: 700 °C, 160 000 Pa, m_{cat} = 340 mg, WHSV = 20.3 h^{-1}. Source: Rego de Vasconcelos et al. [7]. Reproduced with permission of John Wiley and Sons.

owing to its large specific surface area (94 m² g^{-1}) and improved surface basicity, the Ni/30MgAl catalyst was the most active sample, showing X_{CH4} and X_{CO2} values close to 85% during 50 hours TOS. Over the Ni/HA (57 m² g^{-1}), however, conversion of both reactants did not exceed 79%. An annealing treatment at 1200 °C under air provoked a significant loss in the specific surface area (<3 m² g^{-1}) and the number of the basic sites. Though all catalysts exhibited a comparable initial activity, their stability tests at 700 °C for 300 hours TOS suggested that their deactivation proceeded via distinct pathways. As can be deduced from Figure 8.5, over the Ni/HA catalyst the deactivation process was relatively slower than that of Ni/30MgAl. The observed deactivation was assigned to the carbon deposition in core–shell form, which covered the active phase. Nevertheless, they pointed out that despite its lower activity compared to a high surface area sample, an increase in the contact time could open a new perspective for the use of the optimized Ni/HA catalyst on a large scale.

The size of supported metal particles is also a critical parameter, which affects deeply the performance of the catalysts. Many reports correlated the efficiency and the stability of the supported catalysts with the particle sizes spread on the catalyst surface. Hence, the deposition of small metal particles offers a large specific metallic surface area and prevents the carbon formation. Nevertheless, at high temperature and under a long-term operation, they may sinter, thereby provoking a dramatic decay in their activity. Recently, Akri et al. [19] deposited Ni over HA support using a strong electrostatic adsorption method. The reduction of the resulting Ni/HA sample presenting 0.5 wt% Ni induced the formation of individual and uniformly isolated Ni single atoms. At higher loadings (5–10 wt%) nanoparticles of Ni of 2.1–2.6 nm were observed. Their DRM kinetic data, estimated at 750 °C, revealed that the Ni single atom sample exhibited a specific activity (816.5 mol g_{Ni}^{-1} h^{-1}) four times higher than that of the Ni nanoparticle catalyst (185.2 mol g_{Ni}^{-1} h^{-1}). Moreover, due to its excellent resistance to carbon deposition, the former exhibited a good stability during 16 hours TOS.

On the other hand, the regeneration of the reforming catalysts has been extensively investigated as well. From an economic point of view, a successful regeneration process can allow the reutilization of the catalyst keeping its efficiency intact. Essentially the goal is to achieve an effective cleaning of the active surface by suppressing the coke deposits without alteration of the textural and structural properties. For this purpose, controlled oxidative treatments with O_2 or steam have been proposed as efficient strategies for the oxidation of carbon species. Rêgo de Vasconcelos and coworkers studied the regeneration of a Ni/HA catalyst at 700 °C submitted to three cycles of DRM reaction (30 hours TOS) and two regeneration steps for 90 minutes in between [28]. Two different regeneration atmospheres (air or CO_2) were employed. Irrespective of the gasifying agent, the catalyst showed a significant decrease in its initial DRM activity, probably due to the oxidation of Ni species. However, after 30 hours TOS the regenerated catalyst presented a good reproducibility of the results shown by its freshly reduced counterpart. A very small loss in the reactants' conversion was observed, but the selectivity toward synthesis gas products was not affected by the regeneration step. They attributed the absence of a marked irreversible deactivation between cycles to the formation of core–shell carbon and/or densification of Ni particles under the reaction conditions.

It is known that the textural properties of the support play a crucial role in the nucleation pathway and the final structure of the deposited metallic phase [50–54]. This, in turn, impacts the performance of the heterogeneous catalyst. Accordingly, the presence of macropores facilitates the mass transfer of reactants and products, while the mesopores are helpful in increasing the dispersion of the metallic phases. The influence of the distribution of the pore sizes on the activity and stability of Ni/HA catalysts in DRM was studied by Li and coworkers [50]. By adjusting the coprecipitation temperature of their HA preparation (20–120 °C), they obtained various support samples with different pore distributions. Specifically, the relative contribution of macropores decreased following the order: HA-20 > HA-50 > HA-80 > HA-120, while the abundance of the mesopores increased from 33 to 49%. Moreover, when the Ni particles were preferentially dispersed on the mesopore channels, they exhibited a stronger interaction with the support and relatively small sizes. However, small-size mesopores (2–5 nm) were completely blocked since the fine active nanoparticles were not accessible to the gas phase. They observed that the abundance of large-size mesopores (5–50 nm) provided highly dispersed and sintering-resistant active Ni particles in DRM reaction.

Though Ni catalysts have been the most investigated in DRM, their specific activity compared with Co catalysts has not been yet clarified. It seems that the efficiency/performance of both of them depends on the nature of the support. For instance, Li et al. [48] found that Co/Al_2O_3 was more active than a Ni/Al_2O_3 catalyst (Table 8.1). However, the addition of Y_2O_3 as a promoter improved significantly the performance of the nickel catalyst, which became more active than the promoted Co/Al_2O_3 catalyst. The promoting effect was attributed to the deposition of smaller Ni particles and an increase in the density of surface basic sites compared with the unpromoted catalyst. Ay and coworkers reported that Ni/CeO_2 catalyst had a higher performance compared with a Co/CeO$_2$ catalyst [49]. From these results, it is reasonable to conclude that over acid supports, such as alumina, Co is more active than Ni. By contrast, over a support with a basic character, such as ceria, Ni catalysts exhibit a superior activity. The difference could be associated with the type of interactions generated between the acid or basic support and the active phase (Co or Ni). Then, tuning the acid–base properties of the used support can improve the performance of the active phase. In line with the role of alumina as an acid support, the use of an HA support with improved acid properties was found to be more suitable for Co than Ni catalysts. Thus, Tran and coworkers compared the activity of Co and Ni catalysts supported on a Ca-deficient HA in DRM [6]. The results obtained at 700 °C evidenced that the former clearly outperforms the latter in terms of activity and stability.

8.3.2 Effect of the Composition on the Performance of HA in the Reforming Reactions

As described in earlier works, the composition of HA is an important property that influences the distribution and the density of the basic and acid sites and its reactivity in the catalytic processes [4, 6, 23, 25, 55–58]. The modification of these properties has been principally correlated with the Ca/P atomic ratio value. Hence, the effect of composition on the activity of HA bare support in methane oxidation reaction at 700 °C was studied by Sugiyama and coworkers [23]. They reported that the composition change did not induce significant effect on the methane conversion and selectivity to CO$_2$ production. By contrast, over the near stoichiometric sample (Ca/P = 1.68), the selectivity to C2 compounds reached a maximum, while it gave the lowest value of selectivity to carbon monoxide (S_{CO}). However, the highest S_{CO} values were obtained over a substoichiometric HA sample (1.53).

Jun et al. [25] varied the Ca/P and Ni/P molar ratios in their preparation of a series of Ni—Ca—PO$_4$/HA catalysts for the POM reaction. These catalysts were prepared by coprecipitation of calcium nitrate, nickel nitrate, and dibasic ammonium phosphate at pH of 10–11. The optimum catalysts presented a Ca/P ratio close to 10/6 and Ni/P ratios ranging between 1/6 and 3/6. Over the latter, methane conversions and CO and H$_2$ yield values were close to those of the thermodynamic equilibrium. At lower Ca/P ratios, however, a significant loss in activity was noticed, owing to the dominant formation of a hardly reducible nickel phosphate phase. When the Ca/P ratio was too high, excess calcium totally suppressed the reduction of Ni. They concluded that the HA phase is catalytically more beneficial to the activity than the phosphate phase. A long-term test performed at 750 °C for 85 hours revealed that both X_{CH4} and S_{CO} slowly decreased with TOS (from 90 to 80% and from 94 to 88%, respectively), but the regeneration of the catalyst was possible after O$_2$ treatment for one hour.

By contrast, it seems that the HA composition effect on the DRM differs from that observed in the POM reaction. Recently, the influence of Ca/P ratio on the catalytic performance of Ni/ HA catalysts in DRM has been studied using severe experimental conditions (Table 8.1) [4]. The characterization of the catalysts revealed the presence

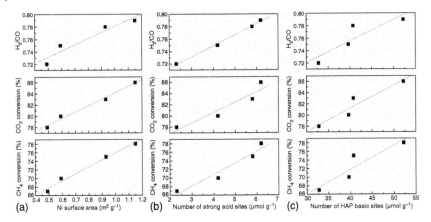

Figure 8.6 Dependence of DRM activity over Ni/HA catalysts on (a) the metallic surface area, (b) the number of strong acid sites, and (c) the number of basic sites. Source: Boukha et al. [4]. Reproduced with permission of Elsevier.

of NiO, Ni^{2+} ion exchanged species, and HA phases over the Ca-deficient samples calcined at 750 °C, while their reduction induced the formation of Ni, TCP, and HA phases. However, in the stoichiometric and Ca-enriched samples, in addition to Ni and HA, a considerable contribution of the CaO phase was detected. Ca-deficient HA catalysts (Ca/P = 1.62) provided suitable properties, which gave the highest performance in DRM compared with the stoichiometric (Ca/P = 1.67) and Ca-surplus (Ca/P = 1.73) HA samples. Particularly, by increasing the Ca/P ratio the catalysts suffered from a decrease in the surface area (from 26 to 9 $m^2 g^{-1}$), and a loss in the number of both strong acid sites and basic sites was evident. These changes resulted in an increase in the Ni particle sizes from 18.3 nm, for the Ca-deficient sample, to 42.8 nm, for the Ca-surplus sample (Ca/P = 1.73). Interestingly, it has found a strong relationship between the abundance of strong acid centers, which might act as anchoring sites for stabilizing Ni particles, and the improved dispersion of Ni particles (Figure 8.6). The TPO results for the spent catalysts evidenced a good resistance against coke deposition, with coke contents lower than 1 wt% (Table 8.1). HAADF-STEM analyses of the used samples did not show significant changes in the Ni particle size. Moreover, the carbon deposits were mainly filament shaped, which explained the high resistance of the catalysts to deactivation.

Nevertheless, as observed in Figure 8.7, which includes images corresponding to a Ca-surplus spent sample, some analyzed zones evidenced the presence of other forms of carbonaceous species, which encapsulated Ni particles. In accordance with this observation, XRD analysis of this spent sample evidenced the formation of a graphitic carbon phase.

Another study by Tran and coworkers also pointed out that the Ca/P atomic ratio is a parameter to take into consideration in order to design efficient HA-supported catalysts for DRM [6]. The comparison of the activity of cobalt supported on two HA samples presenting Ca/P molar ratios of 1.43 and 1.6, respectively, pointed out that the former exhibited higher CH_4 and CO_2 conversion and higher H_2 and CO

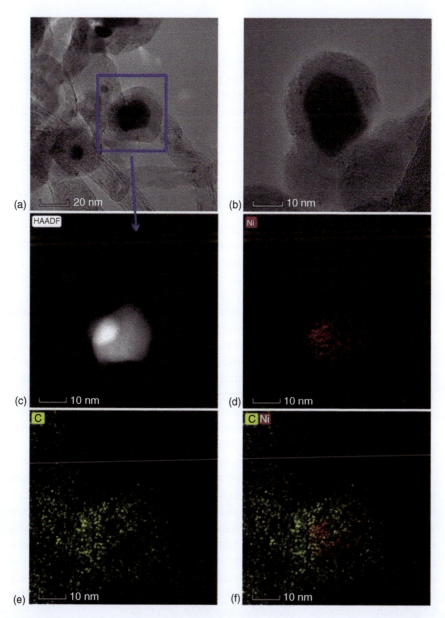

Figure 8.7 (a), (b), and (c) HAADF-STEM images of the Ni/HA-E spent catalyst (DRM at 750 °C) evidencing the encapsulation of Ni particles. Images (d), (e), and (f) display the color maps for Ni (red) and C (green). Source: Boukha et al. [4], pg. 580: 34–45. / With permission of Elsevier.

selectivity values during 50 hours TOS. However, the catalyst with a Ca/P atomic ratio close to 1.6 showed a higher selectivity toward C and H_2O production.

8.3.3 Bimetallic Catalysts

The inclusion of metal oxides into the reforming catalyst formulations has proved to be an effective strategy for the activity enhancement as well as for reducing carbon formation and improving their stability [59–65]. Among them, CeO_2 and its derived mixed oxides have been extensively investigated, owing to their reversible redox properties [18, 59–63, 66, 67]. Their reduction generally leaves oxygen vacancies, which increase the density of surface defects. When located at the vicinity of the metal–support interface, these defects can generate very reactive and dynamic sites.

Kim et al. [66] studied the influence of ceria addition on the performance of Ni/HA catalysts in DRM. They found that this promoter was efficient for the removal of formed carbon. They attributed this behavior to the oxidative property of lattice oxygen in ceria. Furthermore, ceria addition significantly enhanced the stability of the catalyst. The optimum Ce/Ni atomic ratio was found to be equal to 0.3/2.5. At higher CeO_2 contents, the activity decreased due to the blocking of the metallic active surface.

The positive effect of ceria promoter on the activity of a Ni/HA catalyst was also reported in the POM reaction by Ki and coworkers [67]. At 600 °C, the modified catalysts showed significantly higher conversion, CO selectivity, and H_2 yield, and a good stability during 15 hours TOS, compared with the unpromoted counterpart. The optimum Ce/Ni molar ratio was found to be ranging between 0.1/2.5 and 0.2/2.5. The examination of the spent catalysts by SEM revealed the presence of carbon nanotubes with large diameters (60–130 nm) over the unpromoted catalyst, while these were hardly observed on the ceria-promoted catalysts. It was considered that the positive effect of ceria originated from its high oxygen storage capacity.

Akri et al. [18] deposited Ni single atoms on HA and Ce-substituted HA support, respectively, using a strong electrostatic adsorption method. These catalysts were tested in DRM at 750 °C using a mixture composed of 20 vol% CH_4 and 20 vol% CO_2 with a WHSV of 60 000 ml g^{-1} h^{-1}. As can be seen in Figure 8.8, the authors observed a significant improvement of the Ni catalysts performance in terms of activity and stability when doping with 5% of ceria. The optimum catalysts corresponded to those comprising Ni single atoms with Ni loadings ranging between 1 and 2%. Moreover, their experimental and computational studies evidenced that isolated Ni atoms were intrinsically coke resistant due to their ability to only activate the first C—H bond in CH_4, which prevents the methane decomposition into carbon.

Lee et al. [68] examined the activity of Ni-added strontium phosphate catalysts in the POM reaction. For the preparation of these catalysts, they adopted an unusual method based on the complete inclusion of Sr^{2+} and Ni^{2+} into the cationic sites of HA, by coprecipitation of Ni, Sr, and P precursors and a subsequent calcination at 800 °C. The (Sr + Ni)/P atomic ratio was fixed at 10/6 whereas the Ni/(Sr + Ni) ratio was varied from 0.01 to 0.2. After an oxidative treatment the optimum

8.3 HA-Supported Non-noble Metal Catalysts for Methane Reforming Reactions | 285

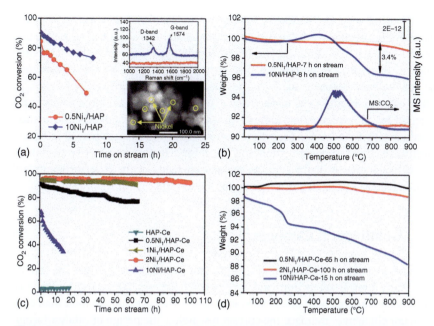

Figure 8.8 DRM performance and carbon deposition analysis over Ni/HA and Ni/HA−Ce samples. (a) CO_2 conversion during DRM over HA and c HA−Ce supported different Ni catalysts. Conditions: $T = 750\,°C$, $CH_4/CO_2/He = 10/10/30$, total flow = 50 ml min^{-1} (GHSV = 60 000 ml h^{-1} g$_{cat}^{-1}$), inset is Raman spectra of spent 0.5Ni$_1$/HA and 10Ni/HA after reaction at 750 °C and STEM image of 0.5Ni$_1$/HA after 7 h reaction; (b–d) TGA-MS and TGA profiles of various catalysts after different reaction times onstream at 750 °C. Source: Akri et al. [18], 10(1), 1–10 / Springer nature / CC BY 4.0.

sample, presenting a Ni/(Sr + Ni) molar ratio of 0.058, contained three different crystalline phases: $Sr_{10}(PO_4)_6(OH)_2$, $Sr_3(PO_4)_2$, and $Sr_{3-x}Ni_x(PO_4)_2$. However, when activated under hydrogen at 750 °C, only the $Sr_3(PO_4)_2$ species were still present as a crystalline phase. The catalytic tests in POM were carried out with sequential increase and decrease of the reaction temperature, without pretreatment, using partial pressures of 0.16 and 0.08 atm for CH_4 and O_2, respectively, and a WHSV of 6000 ml h^{-1} g^{-1}. The optimum catalyst exhibited an excellent performance at temperatures higher than 600 °C, reaching X_{CH4} and X_{O2} values close to those of the thermodynamic equilibrium and the H_2/CO molar ratio close to 2. Furthermore, the catalyst was activated under the reaction mixture during the first light-off run, and its performance was enhanced during a subsequent decrease of the reaction temperature.

HA-supported bimetallic Ni—Co catalysts were first evaluated in DRM by Phan et al. [34]. Regardless of the preparation method (successive impregnation or co-impregnation), Ni—Co bimetallic nanoparticles with size of dozens nm were formed. Moreover, all catalysts contained two fractions of Co^{2+} and Ni^{2+} species exchanged HA, forming $Ca_{10-x}Co_x(PO_4)_6(OH)_2$ and $Ca_{10-y}Ni_y(PO_4)_6(OH)_2$ phases, respectively. The in situ reduced catalysts (700 °C) were tested in DRM under a WHSV of 15 882 ml h^{-1} g^{-1} and a mixture composed of 20% CH_4 and 20%

CO_2 balanced with He. A long-term test at 700 °C showed that the order of the successive impregnation of the two active phases did not affect the efficiency and the resistance of the catalysts. The initial activity was around 73–78%, and then a deactivation process within the first 10 hours TOS was observed. Thereafter, the activity was relatively stable (up to 50 hours TOS) achieving X_{CH4} and X_{CO2} values close to 60 and 68%, respectively. However, the Ni—Co bimetallic sample prepared by co-impregnation method showed the poorer catalytic performance. Its initial activity did not exceed 60%. Nevertheless, this catalyst proved to be less selective toward carbon formation and the most selective toward CO formation. This behavior was explained by the main occurrence of the reverse Boudouard reaction, where carbon deposits react with CO_2 to form CO.

8.4 Noble Metal Catalysts

Noble metal catalysts such as Rh, Pt, Pd, Ir, and Ru exhibit a high specific activity in methane reforming reactions and are claimed to be much more resistant to coke deposition compared with other transition metals. However, due to their high cost, the major challenges that face the current research is the design of catalyst formulations presenting a high dispersion of the active phase. Furthermore, the catalysts should resist the severe conditions of the reforming reactions as well as the reductive treatments at high temperatures. For these purposes, precious metals have been supported on a variety of chemically active or inert oxides presenting a high surface area, namely, Al_2O_3, TiO_2, CeO_2, MgO, La_2O_3, SiO_2, ZrO_2, and ZSM-5, among others, with the objective of finding suitable metal–support interactions, leading to a high dispersion of the active phase [69–74]. Besides the metal dispersion, a wide range of works pointed out that the nature of the support also affected its redox properties and its sintering resistance. Wang and coworkers established that the specific activity of a Rh/MgO-Al_2O_3 catalyst in the SRM reaction decreased by increasing the Rh particle size [70]. Using ZrO_2, CeO_2, $CeZrO_2$, and SiO_2 for the dispersion of the active phase, Ligthart and coworkers studied the effect of Rh particle size on the catalytic performance in SRM [71]. They concluded that the nature of the support influenced the dispersion and the reduction degree of Rh. The reduction degree of the CeO_2-supported catalyst was found to be lower than that supported on $CeZrO_2$, ZrO_2, and SiO_2. This difference was explained by the occurrence of strong metal-ceria interactions, which hindered the reduction of the metal oxide. It was also observed that catalysts with Rh particles smaller than 2.5 nm underwent a rapid deactivation compared with those of larger sizes. These results were attributed to the oxidation of these fine metallic particles under the SRM reaction mixture.

Unlike Ni and Co catalysts, very few studies have been carried out on noble metal catalysts supported on HA materials. Some results have indicated that the growth of Pd, Pt and Ru, respectively, could be a serious drawback negatively affecting the activity and the durability of the HA-based catalysts. Rêgo De Vasconcelos et al. [75] investigated the activity of Ru/HA and Pt/HA catalysts in DRM. Prior to the catalytic tests, the samples were heated in inert atmosphere at 400 °C. The light-off runs revealed that Ru catalyst was more active in the temperature range of 400–600 °C, whereas at a higher temperature (700 °C) the Pt catalyst was more active and more selective toward CO and H_2 production. The stability test at 700 °C performed over the latter pointed out that the CO_2 conversion values decreased from 67 to 40% within 65 hours TOS, which entailed an activity loss of about 40%. Actually, data from the available literature evidence the superiority of a number of noble metal catalysts (such as Pt/Al_2O_3, Pt/La_2O_3, and Pt/ZrO_2 [76, 77]) with respect to the Pt/HA system investigated by Rêgo De Vasconcelos and coworkers [75]. The low activity of the latter could be associated with the structures of the Pt particles deposited on the HA surface in view of the wide size distribution from some nm to hundreds of nm.

Kamieniak and coworkers [78] compared the activity of Ni/HA and Pd/HA catalysts in DRM. The catalysts were reduced under pure hydrogen at 300 °C. Interestingly, the two formulations displayed a notable and comparable activity reaching X_{CH4} values close to 87% at 650 °C. Nevertheless, the Ni catalyst was still more efficient since it contained a lower metallic loading (6 wt%) compared with the Pd catalyst (13.6 wt%). It was also noticed that the most active Pd catalyst exhibited the smallest Pd particle size (16.1 nm) and the largest surface area (46 $m^2\,g^{-1}$). The same research group reported that the catalytic properties of Pd/HA catalysts could be improved by the optimization of the Pd distribution [79]. They observed that an ultrasound-assisted incipient wetness impregnation preparation carried out in alkaline media led to the deposition of highly dispersed and active Pd particles in DRM. Hence, at 650 °C the methane conversion increased from 57% for the sample prepared at pH 2 to 97% for the sample prepared at pH 10.

Boukha et al. [8] examined a series of Rh/HA catalysts in POM and SRM reactions. The catalysts exhibited good activity and excellent stability at 700 °C during 30 hours TOS. This behavior was explained by the high dispersion of Rh particles and their high coke resistance. In the SRM reaction, for instance, the activity of the catalyst with the optimum Rh loading (1 wt%) and the average Rh particle size of 11.6 nm was very close to that achieved over a commercial Rh/Al_2O_3 catalyst (Figure 8.9). However, the Rh/HA catalysts presenting a large contribution of the TCP phase was significantly less active. It was concluded that the interaction of Rh with TCP had a negative impact on the activity.

8.5 Reforming of Other Hydrocarbons

Hydrogen production by alcohols reforming processes has attracted great interest in the last decades. This strategy allows the use of easy-to-store renewable resources with high hydrogen contents and large power density [80–83]. Owing to these interesting advantages, steam reforming of renewable sources such as methanol (8.7), ethanol (8.8), and glycerol (8.9) presents promising alternative solutions for fuel cell technologies to those based on NG.

$$CH_3OH + H_2O \rightarrow CO_2 + 3H_2 \quad \Delta H° = +49.5 \text{ kJ mol}^{-1} \qquad (8.7)$$

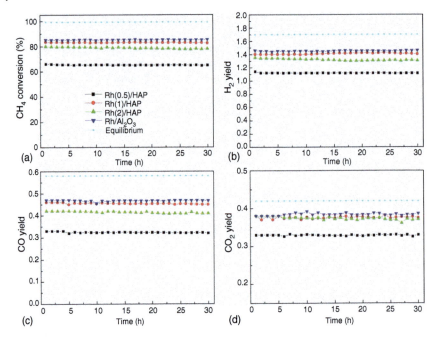

Figure 8.9 Methane conversion (a) and H_2 (b), CO (c), and CO_2 (d) yields over reduced Rh(x)/HA catalysts versus time on stream in the SRM reaction. Reaction conditions: 19200 cm³ CH4 g⁻¹ h⁻¹; W = 0.250 g; T = 973 K. Gas mixture: 10%CH_4/30%H_2O/N_2. Data corresponding to the commercial Rh/Al_2O_3 catalyst and thermodynamic equilibrium were also included. Source: Boukha et al. [8]. Reproduced with permission of Elsevier.

$$C_2H_5OH + 3H_2O \rightarrow 2CO_2 + 6H_2 \quad \Delta H° = +174 \text{ kJ mol}^{-1} \quad (8.8)$$

$$C_3H_5(OH)_3 + 3H_2O \rightarrow 3CO_2 + 7H_2 \quad \Delta H° = +128 \text{ kJ mol}^{-1} \quad (8.9)$$

In recent years, some studies have been devoted to investigate the applicability of HA-based catalysts for the hydrogen generation by alcohol reforming processes [81–87]. Cichy et al. [81] examined the activity of calcium deficient HA (Ca/P = 1.57) supported Ni catalysts (2.5 ≤ Ni wt% ≤ 10) in the glycerol steam reforming reaction. The experiments were carried out under a mixture with steam to carbon ratio of 1 : 3 and a WHSV of 1000 ml g⁻¹ min⁻¹. Prior to the catalytic tests, the samples were pre-reduced in H_2 stream at 800 °C. The most promising results were obtained over the catalyst with 7.5 wt% Ni, which provided glycerol conversions higher than 94% at temperatures above 750 °C. However, the catalyst showed a low selectivity toward hydrogen production, which did not exceed 40% at 800 °C (vs. 70% for the selectivity into CO, S_{CO}). Nevertheless, the assayed system was considered efficient for the preparation of synthesis gases with low H_2/CO ratio, suitable for the synthesis of liquid fuels. It should be highlighted that the catalyst was significantly less selective toward the formation of CO_2 (<11%), CH_4 (12–20%), and C_2H_4 (<6%) carbonaceous compounds. The same research group studied the influence of ceria addition on the activity of Co/HA catalysts (2.5 ≤ Co wt% ≤ 10) in the steam reforming

Figure 8.10 Evaluation of (a) glycerol conversion and (b) H_2 selectivity at 800 °C. Source: Dobosz et al. [82]. Reproduced with permission of Elsevier.

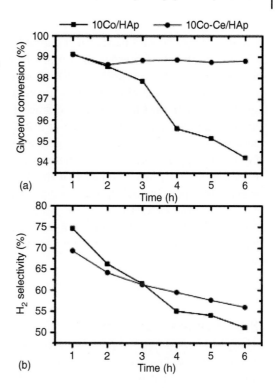

of glycerol [82]. In this study the catalytic tests were performed under a lower WHSV (100 ml g^{-1} min^{-1}). The ceria-modified samples resulted more active and selective and exhibited a higher stability compared with the free promoter catalysts (Figure 8.10). At 650 °C, the optimum catalyst (7.5%Co—Ce/HA) showed a glycerol conversion close to 96% and H_2 selectivity around 81%. The high performance of the Co—Ce/HA catalysts was attributed to the strong Co—Ce interactions, which seemed to prevent the growth of Co particles. It was also observed that the selectivity toward H_2 production increased with increasing the pre-reduction temperature of the catalysts. The optimum performance was reached when the catalysts were reduced at 800 °C.

The catalytic properties of Co/HA system were also studied in the ethanol steam reforming reaction [87]. The activity of the catalysts seemed to decrease with increasing the Co loading, from 2.5 to 7.5 wt%. By contrast, a significant increase in the H_2 selectivity was observed. Two possible reactions occurred on the assayed catalysts, namely, dehydrogenation of ethanol and steam reforming. Moreover, it was found that the former was promoted by Co^{2+} species whereas the latter was promoted by the metallic Co species. Ogo et al. [83] prepared four catalyst supports, using a hydrothermal method, presenting the following chemical formula: $Ca_{10}(PO_4)_6(OH)_2$, $Ca_{10}(VO_4)_6(OH)_2$, $Sr_{10}(PO_4)_6(OH)_2$, and $Sr_{10}(VO_4)_6(OH)_2$ (namely, Ca—P, Ca—V, Sr—P, and Sr—V, respectively). The resulting supports were impregnated with 5 wt% Co and tested in the ethanol steam reforming reaction at 550 °C. Among all tested catalysts, the Co/Sr—P was the most active (ethanol conversion: X_{EtOH} = 37%) and selective toward H_2 production (hydrogen selectivity: S_{H2} = 43%), which was linked to its improved surface basicity compared with the rest of the catalysts. Moreover, the Co/Sr—P catalyst proved to be more effective for the coke formation suppression when compared with a Co/α-Al_2O_3 reference catalyst. The observed difference was associated with the low reducibility of Co species spread on the Co/Sr—P surface, since it contained relatively large amounts of Co^{2+} species, even after reduction at 550 °C. Dobosz et al. [84] compared the activity of Co—Ce/HA catalysts in ethanol steam reforming reaction starting from HA supports prepared by microwave-assisted hydrothermal and sol–gel routes, respectively. It was found that the hydrothermally prepared HA gave the best catalytic performance in terms of activity and selectivity. Over the optimized catalyst (5%Co/10%Ce/HA) hydrogen yield achieved 3.5 mol H_2 (mol $C_2H_5OH)^{-1}$. The superiority of the latter was attributed to its improved textural properties, its higher reducibility and oxygen mobility, and the presence of larger amounts of basic sites.

8.6 Summary and Remarks

This chapter summarizes the relevant results from previous studies on the application of HA-based catalysts in the methane reforming reactions. The main conclusions are listed as follows:

- The structural properties of the HA allow the control of the metal–support interactions. Despite its high capacity to ion exchange a wide number of transition metals, the extent of these substitutions is not too large to provoke a significant loss in the active metallic phase.
- The presence of phosphate groups confers to the HA moderate acid properties essential for the stability of the metallic active phases.
- The surface basic properties provided by calcium and hydroxyl groups seem to play a key role in the reaction mechanism and help to suppress the carbon formation during the reforming reactions.
- The results obtained over Ni/HA catalysts in DRM evidence their competitiveness with respect to those based on conventional supports.
- Recent studies pointed out that Ca-deficient HA-based catalysts are more active and more resistant in DRM than those based on stoichiometric and Ca-surplus HA supports.
- The use of promoters, such as ceria, and the development of Ni—Co bimetallic catalysts has proved to be good strategies to noticeably improve the performance of Ni/HA systems.
- Though less detailed and numerous, studies on the activity of HA-supported noble metal catalysts (Rh and Pd) also reported good results in DRM, POM, and SRM reactions. However, because they manifest SMSI the activity of the HA-supported Pt catalysts was lower than that of Pt catalysts supported on conventional supports.

- Future efforts should be encouraged with the objective of revealing the kinetic aspects of HA-based catalysts in reforming reactions and the intrinsic role of the support.
- Most of the investigated HA-based catalysts have been focused on the DRM reaction. We encourage their application in alternative/other methane reforming reactions such as SRM, POM, autothermal reforming of methane and, in tri-reforming of methane (TRM).
- In view of the decrease in the surface area of the HA-based catalysts, when operated at high temperatures, the use of textural promoters, such as lanthana, is highly recommended.

Acknowledgments

The financial support for this work provided by Ministerio de Economía y Competitividad (CTQ2015-73219-JIN [AEI/FEDER/UE]), Spanish ministry of Science and Innovation (PID2019-107105RB-I00 AEI/FEDER, UE), and Basque Government (GIC IT-1297-19) is gratefully acknowledged. Likewise, the technical support provided by SGIker (UPV/EHU) is gratefully acknowledged.

References

1 Haryanto, A., Fernando, S., Murali, N., and Adhikari, S. (2005). Current status of hydrogen production techniques by steam reforming of ethanol: a review. *Energy & Fuels* 19: 2098–2106.
2 Saavedra, J., Whittaker, T., Chen, Z. et al. (2016). Controlling activity and selectivity using water in the Au-catalysed preferential oxidation of CO in H_2. *Nature Chemistry* 8: 584–589.
3 Christian Enger, B., Lødeng, R., and Holmen, A. (2008). A review of catalytic partial oxidation of methane to synthesis gas with emphasis on reaction mechanisms over transition metal catalysts. *Applied Catalysis A: General* 346: 1–27.
4 Boukha, Z., Yeste, M.P., Cauqui, M.Á., and González-Velasco, J.R. (2019). Influence of Ca/P ratio on the catalytic performance of Ni/hydroxyapatite samples in dry reforming of methane. *Applied Catalysis A: General* 580: 34–45.
5 Boukha, Z., Kacimi, M., Pereira, M.F.R. et al. (2007). Methane dry reforming on Ni loaded hydroxyapatite and fluoroapatite. *Applied Catalysis A: General* 317: 299–309.
6 Tran, T.Q., Pham Minh, D., Phan, T.S. et al. (2020). Dry reforming of methane over calcium-deficient hydroxyapatite supported cobalt and nickel catalysts. *Chemical Engineering Science.* 228: 115975.
7 Rego de Vasconcelos, B., Pham Minh, D., Martins, E. et al. (2020). A comparative study of hydroxyapatite- and alumina-based catalysts in dry reforming of methane. *Chemical Engineering & Technology* 43: 698–704.

8 Boukha, Z., Gil-Calvo, M., de Rivas, B. et al. (2018). Behaviour of Rh supported on hydroxyapatite catalysts in partial oxidation and steam reforming of methane: on the role of the speciation of the Rh particles. *Applied Catalysis A: General* 556: 191–203.

9 Pakhare, D. and Spivey, J. (2014). A review of dry (CO_2) reforming of methane over noble metal catalysts. *Chemical Society Reviews* 43: 7813–7837.

10 Boukha, Z., Jiménez-González, C., Gil-Calvo, M. et al. (2016). MgO/NiAl2O4 as a new formulation of reforming catalysts: Tuning the surface properties for the enhanced partial oxidation of methane. *Applied Catalysis B: Environmental* 199: 372–383.

11 Jiménez-González, C., Boukha, Z., De Rivas, B. et al. (2013). Structural characterisation of Ni/alumina reforming catalysts activated at high temperatures. *Applied Catalysis A: General* 466: 9–20.

12 Jiménez-González, C., Boukha, Z., de Rivas, B. et al. (2015). Behaviour of nickel–alumina spinel ($NiAl_2O_4$) catalysts for isooctane steam reforming. *International Journal of Hydrogen Energy* 40: 5281–5288.

13 Jiménez-González, C., Boukha, Z., De Rivas, B. et al. (2014). Behavior of coprecipitated $NiAl_2O_4/Al_2O_3$ catalysts for low-temperature methane steam reforming. *Energy Fuels* 28: 7109–7121.

14 Jun, J.H., Lee, T.-J., Lim, T.H. et al. (2004). Nickel-calcium phosphate/hydroxyapatite catalysts for partial oxidation of methane to syngas: characterization and activation. *Journal of Catalysis* 221: 178–190.

15 Boukha, Z., González-Prior, J., Rivas, B. et al. (2016). Synthesis, characterisation and behaviour of Co/hydroxyapatite catalysts in the oxidation of 1,2-dichloroethane. *Applied Catalysis B: Environmental* 190: 125–136.

16 Gruselle, M. (2015). Apatites: a new family of catalysts in organic synthesis. *Journal of Organometallic Chemistry* 793: 93–101.

17 Boukha, Z., Choya, A., Cortés-Reyes, M. et al. (2020). Influence of the calcination temperature on the activity of hydroxyapatite-supported palladium catalyst in the methane oxidation reaction. *Applied Catalysis B: Environmental* 277: 119280.

18 Akri, M., Zhao, S., Li, X. et al. (2019). Atomically dispersed nickel as coke-resistant active sites for methane dry reforming. *Nature Communications* 10: 1–10.

19 Akri, M., El Kasmi, A., Batiot-Dupeyrat, C., and Qiao, B. (2020). Highly active and carbon-resistant nickel single-atom catalysts for methane dry reforming. *Catalysts* 10: 1–20.

20 Matsumura, Y. and Moffat, J.B. (1994). Partial oxidation of methane to carbon-monoxide and hydrogen with molecular-oxygen and nitrous-oxide over hydroxyapatite catalysts. *Journal of Catalysis* 148: 323–333.

21 Matsumura, Y., Sugiyama, S., Hayashi, H., and Moffat, J.B. (1994). An apparent ensemble effect in the oxidative coupling of methane on hydroxyapatites with incorporated lead. *Catalysis Letters* 30: 235–240.

22 Matsumura, Y., Sugiyama, S., Hayashi, H. et al. (1994). Strontium hydroxyapatites: catalytic properties in the oxidative dehydrogenation of methane to carbon oxides and hydrogen. *Journal of Molecular Catalysis* 92: 81–94.

23 Sugiyama, S., Minami, T., Moriga, T. et al. (1996). Surface and bulk properties, catalytic activities and selectivities in methane oxidation on near-stoichiometric calcium hydroxyapatites. *Journal of Material Chemistry* 6: 459–464.

24 Brunton, P.A., Davies, R.P.W., Burke, J.L. et al. (2013). Treatment of early caries lesions using biomimetic self-assembling peptides – a clinical safety trial. *British Dental Journal* 215.

25 Jun, J.H., Jeong, K.S., Lee, T.-J. et al. (2004). Nickel–calcium phosphate/hydroxyapatite catalysts for partial oxidation of methane to syngas: effect of composition. *Korean Journal of Chemical Engineering* 21: 140–146.

26 Tounsi, H., Djemal, S., Petitto, C., and Delahay, G. (2011). Copper loaded hydroxyapatite catalyst for selective catalytic reduction of nitric oxide with ammonia. *Applied Catalysis B: Environmental* 107: 158–163.

27 Boukha, Z., González-Velasco, J.R., and Gutiérrez-Ortiz, M.A. (2020). Platinum supported on lanthana-modified hydroxyapatite samples for realistic WGS conditions: on the nature of the active species, kinetic aspects and the resistance to shut-down/start-up cycles. *Applied Catalysis B: Environmental* 270: 118851.

28 Rego de Vasconcelos, B., Pham Minh, D., Sharrock, P., and Nzihou, A. (2018). Regeneration study of Ni/hydroxyapatite spent catalyst from dry reforming. *Catalysis Today* 310: 107–115.

29 Tang, H., Wei, J., Liu, F. et al. (2016). Strong metal-support interactions between gold nanoparticles and nonoxides. *Journal of the American Chemical Society* 138: 56–59.

30 Tang, H., Su, Y., Guo, Y. et al. (2018). Oxidative strong metal-support interactions (OMSI) of supported platinum-group metal catalysts. *Chemical Science* 9: 6679–6684.

31 Bernal, S., Blanco, G., Calvino, J.J. et al. (1997). Influence of the preparation procedure on the chemical and microstructural properties of lanthana promoted Rh/SiO$_2$ catalysts: a FTIR spectroscopic study of chemisorbed CO. *Journal of Alloys and Compounds* 250: 461–466.

32 Boukha, Z., Jiménez-González, C., de Rivas, B. et al. (2014). Synthesis, characterisation and performance evaluation of spinel-derived Ni/Al$_2$O$_3$ catalysts for various methane reforming reactions. *Applied Catalysis B: Environmental* 158–159: 190–201.

33 Bian, Z., Wang, Z., Jiang, B. et al. (2020). A review on perovskite catalysts for reforming of methane to hydrogen production. *Renewable & Sustainable Energy Reviews* 134: 110291.

34 Phan, T.S., Sane, A.R., Rêgo de Vasconcelos, B. et al. (2018). Hydroxyapatite supported bimetallic cobalt and nickel catalysts for syngas production from dry reforming of methane. *Applied Catalysis B: Environmental* 224: 310–321.

35 Park, J., Yeo, S., Kang, T. et al. (2018). Enhanced stability of Co catalysts supported on phosphorus-modified Al$_2$O$_3$ for dry reforming of CH$_4$. *Fuel* 212: 77–87.

36 Ibrahim, A.A., Al-Fatesh, A.S., Khan, W.U. et al. (2019). Enhanced coke suppression by using phosphate-zirconia supported nickel catalysts under dry methane reforming conditions. *International Journal of Hydrogen Energy* 44: 27784–27794.

37 Cimino, S., Lisi, L., and Mancino, G. (2017). Effect of phosphorous addition to Rh-supported catalysts for the dry reforming of methane. *International Journal of Hydrogen Energy* 42: 23587–23598.

38 Pelletier, L. and Liu, D.D.S. (2007). Stable nickel catalysts with alumina-aluminum phosphate supports for partial oxidation and carbon dioxide reforming of methane. *Applied Catalysis A: General* 317: 293–298.

39 Diallo-Garcia, S., Osman, M.B., Krafft, J.-M. et al. (2014). Identification of surface basic sites and acid-base pairs of hydroxyapatite. *Journal of Physical Chemistry C* 118: 12744–12757.

40 Abdulrasheed, A., Jalil, A.A., Gambo, Y. et al. (2019). A review on catalyst development for dry reforming of methane to syngas: recent advances. *Renewable & Sustainable Energy Reviews* 108: 175–193.

41 Aziz, M.A.A., Jalil, A.A., Wongsakulphasatch, S., and Vo, D.-.N. (2020). Understanding the role of surface basic sites of catalysts in CO_2 activation in dry reforming of methane: a short review. *Catalysis Science and Technology* 10: 35–45.

42 Muñoz, M.A., Calvino, J.J., Rodríguez-Izquierdo, J.M. et al. (2017). Highly stable ceria-zirconia-yttria supported Ni catalysts for syngas production by CO_2 reforming of methane. *Applied Surface Science* 426: 864–873.

43 Bang, S., Hong, E., Baek, S.W., and Shin, C.-H. (2018). Effect of acidity on Ni catalysts supported on P-modified Al_2O_3 for dry reforming of methane. *Catalysis Today* 303: 100–105.

44 Fakeeha, A.H., Khan, W.U., Al-Fatesh, A.S., and Abasaeed, A.E. (2013). Stabilities of zeolite-supported Ni catalysts for dry reforming of methane. *Chinese Journal of Catalysis* 34: 764–768.

45 Wang, S. and Lu, G.Q. (2000). Effects of promoters on catalytic activity and carbon deposition of Ni/γ-Al_2O_3 catalysts in CO_2 reforming of CH_4. *Journal of Chemical Technology & Biotechnology* 75: 589–595.

46 Kalai, D.Y., Stangeland, K., Li, H., and Yu, Z. (2017). The effect of La on the hydrotalcite derived Ni catalysts for dry reforming of methane. *Energy Procedia* 142: 3721–3726.

47 Djinović, P. and Pintar, A. (2017). Stable and selective syngas production from dry CH_4—CO_2 streams over supported bimetallic transition metal catalysts. *Applied Catalysis B: Environmental* 206: 675–682.

48 Li, B., Su, W., Wang, X., and Wang, X. (2016). Alumina supported Ni and Co catalysts modified by Y_2O_3 via different impregnation strategies: Comparative analysis on structural properties and catalytic performance in methane reforming with CO_2. *International Journal of Hydrogen Energy* 41: 14732–14746.

49 Ay, H. and Üner, D. (2015). Dry reforming of methane over CeO_2 supported Ni, Co and Ni—Co catalysts. *Applied Catalysis B: Environmental* 179: 128–138.

50 Li, B., Yuan, X., Li, B., and Wang, X. (2020). Impact of pore structure on hydroxyapatite supported nickel catalysts (Ni/HAP) for dry reforming of methane. *Fuel Processing Technology* 202: 106359.

51 Li, J., Wu, H., and Ma, B. (2017). Preparation of bimodal mesoporous silica containing cerious salt and its application as catalyst for the synthesis of biodiesel by esterification. *Journal of Porous Materials* 24: 1279–1288.

52 Meinusch, R., Ellinghaus, R., Hormann, K. et al. (2017). On the underestimated impact of the gelation temperature on macro- and mesoporosity in monolithic silica. *Physical Chemistry Chemical Physics* 19: 14821–14834.

53 Zhang, Y., Liu, Y., Yang, G. et al. (2007). Effects of impregnation solvent on Co/SiO_2 catalyst for Fischer-Tropsch synthesis: a highly active and stable catalyst with bimodal sized cobalt particles. *Applied Catalysis A: General* 321: 79–85.

54 Zhang, Y., Nagamori, S., Yamaguchi, M. et al. (2008). Hetero-atom combination bimodal pore structure supports prepared by inside-pore organized oxide precursors from advanced sol-gel solution. *Catalysis Communications* 9: 902–906.

55 Boukha, Z., Kacimi, M., Ziyad, M. et al. (2007). Comparative study of catalytic activity of Pd loaded hydroxyapatite and fluoroapatite in butan-2-ol conversion and methane oxidation. *Journal of Molecular Catalysis A: Chemical* 270: 205–213.

56 Resende, N.S., Nele, M., and Salim, V.M.M. (2006). Effects of anion substitution on the acid properties of hydroxyapatite. *Thermochimica Acta* 451: 16–21.

57 Mori, K., Yamaguchi, K., Hara, T. et al. (2002). Controlled synthesis of hydroxyapatite-supported palladium complexes as highly efficient heterogeneous catalysts. *Journal of the American Chemical Society* 124: 11572–11573.

58 Tsuchida, T., Kubo, J., Yoshioka, T. et al. (2008). Reaction of ethanol over hydroxyapatite affected by Ca/P ratio of catalyst. *Journal of Catalysis* 259: 183–189.

59 Boaro, M., Colussi, S., and Trovarelli, A. (2019). Ceria-based materials in hydrogenation and reforming reactions for CO_2 valorization. *Frontiers in Chemistry* 7: 28.

60 Devaiah, D., Reddy, L.H., and Park, S. -., Reddy, B. M. (2018). Ceria–zirconia mixed oxides: synthetic methods and applications. *Catalysis Reviews – Science and Engineering* 60: 177–277.

61 Montini, T., Melchionna, M., Monai, M., and Fornasiero, P. (2016). Fundamentals and catalytic applications of CeO_2-based materials. *Chemical Reviews* 116: 5987–6041.

62 Trovarelli, A. (1996). Catalytic properties of ceria and CeO_2-containing materials. *Catalysis Reviews – Science and Engineering* 38: 439–520.

63 Xie, S., Wang, Z., Cheng, F. et al. (2017). Ceria and ceria-based nanostructured materials for photoenergy applications. *Nano Energy* 34: 313–337.

64 Rodriguez-Gomez, A. and Caballero, A. (2018). Bimetallic Ni—Co/SBA-15 catalysts for reforming of ethanol: How cobalt modifies the nickel metal phase and product distribution. *Molecular Catalysis* 449: 122–130.

65 San-José-Alonso, D., Juan-Juan, J., Illán-Gómez, M.J., and Román-Martínez, M.C. (2009). Ni, Co and bimetallic Ni—Co catalysts for the dry reforming of methane. *Applied Catalysis A: General* 371: 54–59.

66 Kim, K.H., Lee, S.Y., and Yoon, K.J. (2006). Effects of ceria in CO_2 reforming of methane over Ni/calcium hydroxyapatite. *Korean Journal of Chemical Engineering* 23: 356–361.

67 Ki, K.H., Sang, Y.L., Nam, S.-W. et al. (2006). Promotion effects of ceria in partial oxidation of methane over Ni-calcium hydroxyapatite. *Korean Journal of Chemical Engineering* 23: 17–20.

68 Lee, S.J., Jun, J.H., Lee, S. et al. (2002). Partial oxidation of methane over nickel-added strontium phosphate. *Applied Catalysis A: General* 230: 61–71.

69 Buyevskaya, O.V., Wolf, D., and Baerns, M. (1994). Rhodium-catalyzed partial oxidation of methane to CO and H_2. Transient studies on its mechanism. *Catalysis Letters* 29: 249–260.

70 Wang, Y., Chin, Y.H., Rozmiarek, R.T. et al. (2004). Highly active and stable Rh/MgO—Al_2O_3 catalysts for methane steam reforming. *Catalysis Today* 98: 575–581.

71 Ligthart, D.A.J.M., Van Santen, R.A., and Hensen, E.J.M. (2011). Influence of particle size on the activity and stability in steam methane reforming of supported Rh nanoparticles. *Journal of Catalysis* 280: 206–220.

72 Tian, Z., Dewaele, O., and Marin, G.B. (1999). The state of Rh during the partial oxidation of methane into synthesis gas. *Catalysis Letters* 57: 9–17.

73 Varga, E., Baán, K., Samu, G.F. et al. (2016). The effect of Rh on the interaction of Co with Al_2O_3 and CeO_2 supports. *Catalysis Letters* 146: 1800–1807.

74 Varga, E., Pusztai, P., Oszkó, A. et al. (2016). Stability and temperature-induced agglomeration of Rh nanoparticles supported by CeO_2. *Langmuir* 32: 2761–2770.

75 Rêgo De Vasconcelos, B., Zhao, L., Sharrock, P. et al. (2016). Catalytic transformation of carbon dioxide and methane into syngas over ruthenium and platinum supported hydroxyapatites. *Applied Surface Science* 390: 141–156.

76 Ghelamallah, M. and Granger, P. (2014). Supported-induced effect on the catalytic properties of Rh and Pt—Rh particles deposited on La_2O_3 and mixed α-Al_2O_3—La_2O_3 in the dry reforming of methane. *Applied Catalysis A: General* 485: 172–180.

77 Özkara-Aydınoğlu, S., Özensoy, E., and Aksoylu, A.E. (2009). The effect of impregnation strategy on methane dry reforming activity of Ce promoted Pt/ZrO_2. *International Journal of Hydrogen Energy* 34: 9711–9722.

78 Kamieniak, J., Kelly, P.J., Banks, C.E., and Doyle, A.M. (2017). Methane emission management in a dual-fuel engine exhaust using Pd and Ni hydroxyapatite catalysts. *Fuel* 208: 314–320.

79 Kamieniak, J., Bernalte, E., Doyle, A.M. et al. (2017). Can ultrasound or pH influence Pd distribution on the surface of HAP to improve its catalytic properties in the dry reforming of methane? *Catalysis Letters* 147: 2200–2208.

80 Du, C., Mo, J., and Li, H. (2015). Renewable hydrogen production by alcohols reforming using plasma and plasma-catalytic technologies: challenges and opportunities. *Chemical Reviews* 115: 1503–1542.

81 Cichy, M., Dobosz, J., Borowiecki, T., and Zawadzki, M. (2017). Glycerol steam reforming over calcium deficient hydroxyapatite supported nickel catalysts. *Reaction Kinetics, Mechanisms and Catalysis* 122: 69–83.

82 Dobosz, J., Cichy, M., Zawadzki, M., and Borowiecki, T. (2018). Glycerol steam reforming over calcium hydroxyapatite supported cobalt and cobalt-cerium catalysts. *Journal of Energy Chemistry* 27: 404–412.

83 Ogo, S., Maeda, S., and Sekine, Y. (2017). Coke resistance of Sr-hydroxyapatite supported Co catalyst for ethanol steam reforming. *Chemistry Letters* 46: 729–732.

84 Dobosz, J., Hull, S., and Zawadzki, M. (2016). Catalytic activity of cobalt and cerium catalysts supported on calcium hydroxyapatite in ethanol steam reforming. *Polish Journal of Chemical Technology* 18: 59–67.

85 Goula, M.A., Charisiou, N.D., Pandis, P.K., and Stathopoulos, V.N. (2016). A Ni/apatite-type lanthanum silicate supported catalyst in glycerol steam reforming reaction. *RSC Advances* 6: 78954–78958.

86 Hakim, L., Yaakob, Z., Ismail, M. et al. (2013). Hydrogen production by steam reforming of glycerol over Ni/Ce/Cu hydroxyapatite-supported catalysts. *Chemical Papers* 67: 703–712.

87 Dobosz, J., Małecka, M., and Zawadzki, M. (2018). Hydrogen generation via ethanol steam reforming over Co/HAp catalysts. *Journal of the Energy Institute* 91: 411–423.

9

Hydroxyapatite-Based Catalysts for the Production of Energetic Carriers

Othmane Amadine[1], Karim Dânoun[1], Younes Essamlali[1], Said Sair[1], and Mohamed Zahouily[1,2]

[1] MASCIR Foundation, Rabat Design, Rue Mohamed El Jazouli, Madinat El Irfane 10100 Rabat, Morocco
[2] Laboratory of Materials, Catalysis and Valorization of Natural Resources, FSTM, University Hassan II, B.P. 146, 20650 Casablanca, Morocco

9.1 Introduction

Calcium hydroxyapatite (HA) has been widely studied for use in a variety of applications [1, 2], because of its high chemical and thermal stability [3], ion exchangeability [4], acid–base proprieties, and affinity for organic compounds [5]. This is mainly due to its unique structure and intrinsic properties including its high flexibility, ability to accept various substitutions and ion exchanges in its framework, and easy surface modification [6, 7]. The HA shows a high exchangeability of Ca^{2+}, OH^-, and PO_4^{3-} ions by other cations and anions [8, 9]. More details on the properties of HA and HA-based materials have been presented in Chapters 2 and 3 of this book. The partially substituted HAs by many ions were investigated as catalysts for the dehydration and dehydrogenation of alcohols [10, 11], the direct synthesis of *n*-butanol from ethanol [12, 13], the oxidative dehydrogenation of propane [14], the methane dry reforming to produce synthesis gas [15], the decomposition of volatile organic compounds (VOC) [16], etc. Furthermore, the HA has both acidic and basic sites in the crystal lattice, and so it is expected to perform as bifunctional catalyst [17].

Generally, calcium HA is produced using different methods such as coprecipitation [18], sol–gel [19], hydrothermal [20], and microwave assisted [21], as previously detailed in Chapter 2 of this book. By varying the synthesis method and by controlling the associated parameters, one can tune the structure, morphology, and texture of HA (Chapter 2). Therefore, these remarkable features along with the acid–basic properties and the ability to modify the behavior of the underlying HA by the use of ions, conferred to the material several applications in the domain of energy [22, 23], environmental protection [24, 25], biotechnology [26], optics [27], biological or physical sensors [28], high temperature fuel cells [29], and catalysis [30, 31].

In this context, the purpose of this chapter is to review different applications of HA in the energy area via catalytic applications and to offer an in-depth discussion of the advantages and drawbacks of its use while striving to present and highlight evidence

Design and Applications of Hydroxyapatite-Based Catalysts, First Edition. Edited by Doan Pham Minh.
© 2022 WILEY-VCH GmbH. Published 2022 by WILEY-VCH GmbH.

of the contribution of its substitution and/or modification. Thus, this chapter aims to provide insights into the applications of HA for the production of different energetic carriers such as biodiesel, hydrogen, and energy additives. This chapter is specifically focused on synthesized HA. Works related to HA derived from animal bones will be treated in Chapter 13.

9.2 Biodiesel Production

Biodiesel or, more accurately, fatty acid methyl esters (FAME) has attracted considerable interest all over the world as a renewable, ecologically, and green fuel alternative to fossil-based diesel [32]. It shares similar physicochemical properties as petroleum-based diesel, reflected in terms of kinematic viscosity, specific gravity, calorific value, and cetane number [33–35], and even better in terms of flash point (>130 °C) and of course greenhouse gas emissions. Others include increased lubricity and biodegradability as well as ease of handling, transport, and storage. Biodiesel is conventionally produced by transesterification or alcoholysis, in which triglycerides moieties react with alcohols (typically methanol), in the presence of suitable catalyst, either homogeneous or heterogeneous, to produce FAME [36]. Figure 9.1 demonstrates the overall transesterification reaction for biodiesel production. The development of green and sustainable chemical processes for biodiesel production is one of the most challenges that both academics and industrials are facing today. In this way, the design of new catalysts for heterogeneous processes is of prime importance and plays a key role to reduce the environmental impact of biodiesel production. In this context, this section of the chapter reviews various HA-based catalysts used for biodiesel production to date according to the state of the art and compares their suitability in the transesterification and/or esterification processes.

In the literature, few studies have been devoted to the use of HA-based catalysts for biodiesel production by transesterification and esterification reactions. A literature survey [37] revealed that the most widely used HA-based catalysts are derived from natural HA, and only few studies are reported on the use of synthetic HA as catalyst or catalyst support for biodiesel production. From a catalytic point of view, HA-based catalysts used for biodiesel production can be divided in four major classes, mainly natural HA derived from waste bone materials, alkaline earth and alkali metal-doped HA catalysts, mixed metal-based HA catalysts, and supported HA catalysts (Figure 9.2). The natural HA is a class of material in which the HA is the main component along with other secondary phases such as CaO, $CaCO_3$, $Ca(OH)_2$, and $Ca_3(PO_4)_2$. The alkaline earth and alkali metal-doped HA along with mixed metal-based HA catalysts includes modified apatite in which the calcium atoms are replaced by monovalent alkaline earth and alkali metal, resulting in the generation of new catalytic phases generally assigned to calcium deficient HA and metal phosphate phases. In the last class of material, the HA serves as a support on which actives species were dispersed on its surface.

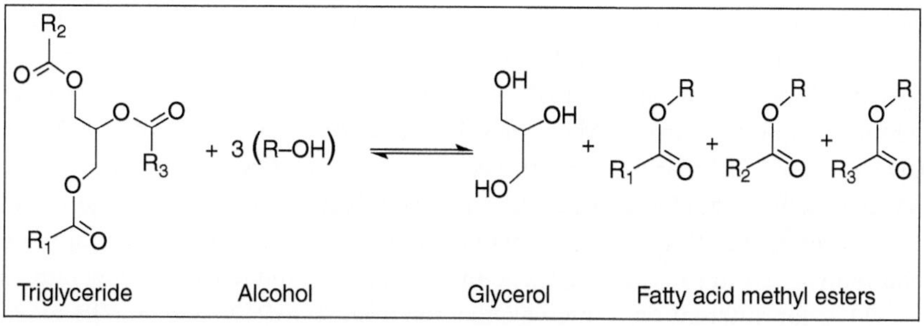

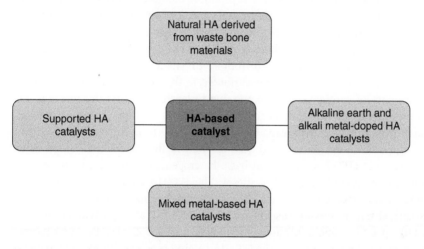

Figure 9.1 Transesterification reactions of triglycerides with an alcohol. Source: Othmane AMADINE and Mohamed ZAHOUILY.

Figure 9.2 Typology of HA-based catalysts for biodiesel production. Source: Othmane AMADINE and Mohamed ZAHOUILY.

9.2.1 Transesterification Reactions

The unique physicochemical properties of the material such as structural stability, ion substitution ability, acid–base properties, and adsorption capacity support the recent exploitation of HA as a catalyst or catalyst support. The heterogeneous catalyst derived from HA material has demonstrated potential in catalyzing transesterification reaction for the production of biodiesel. Some investigations on the use of HA-based catalysts to produce biodiesel from different bio-feedstocks will be described hereafter.

Abukhadra et al. [38] synthesized spongy carbonate fluorapatite loaded with nickel and iron, with its catalytic activity tested for the conversion of waste cooking oil (WCO) into biodiesel. The metal loaded spongy carbonate fluorapatite was prepared by dissolution of natural phosphorite in nitric acid followed by the addition of nickel chloride in the filtrate containing Ca^{2+} and H_3PO_4. Then the resulted solution was precipitated using liquid ammonia solution (20%) and hydrothermally treated in a Teflon-lined stainless steel autoclave for 12 hours at 150 °C. The modified apatite was mainly consisting of a unique highly crystalline carbonate-fluorapatite phase having an average crystallite size of 8.27 nm. Elemental analysis by energy dispersive X-ray spectroscopy (EDX) analysis showed the presence of oxygen, calcium, phosphorous, iron, carbon, nickel, and fluorine with mass% of 36.8, 30.03, 10.69, 8.54, 7.36, 3.34, and 3.26%, respectively. The catalytic performance of the produced catalyst was investigated with transesterification reaction of fifty grams of WCO in a three-neck flask (150 ml) under the following conditions: methanol-to-oil molar ratio of 2 : 1 to 15 : 1, catalyst amount of 4–12 wt%, and reaction temperatures of 50–90 °C for different time intervals ranging from 30 to 240 minutes. The maximum biodiesel yield of 98% was obtained under the following optimal conditions: reaction time of 120 minutes, temperature of 70 °C, catalyst amount of 10 wt%, and a methanol-to-oil molar ratio of 8 : 1. The produced biodiesel exhibited characteristics that match the requirement of American Society for Testing and Materials (ASTM) D-6571 and most of EN 14214 biodiesel standards. The reusability of the modified apatite was also studied through three runs, and the results demonstrated that the efficiency of the catalyst decreased from 97 to 54% after three consecutive runs. The drop in biodiesel yield can be attributed to the decrease in the Ca^{2+} ions in the structure of carbonate fluorapatite due to leaching during the reaction. The concentration of calcium ions leached in the reaction medium gradually increased from 25 to 83 ppm with increasing the number of runs from run 1 to run 3.

On the other hand, Brasil et al. [39] showed in their work that hydrotalcite–hydroxyapatite (HT–HA) is a good basic heterogeneous catalyst for the transesterification of soybean oil into biodiesel. The HT–HA catalyst was prepared by coprecipitation method, in which HA was the predominant phase. The synthesis was carried out by mixing suspension of HA and HT at room temperature and pH ($10 \leq pH \leq 11$) under ultrasonic homogenization, followed by hydrothermal treatment for 24 hours at 80 °C. The HT–HA catalyst produced 80.4% yield of FAME through the transesterification of soybean oil at 240 °C, 4 hours reaction, and 12 : 1 molar ratio of methanol/soybean oil. However, the proposed transesterification process suffers from one limitation related to the high reaction temperature required to achieve a quantitative yield, which limits large-scale application.

Similarly, Vilas-Bôas et al. [40] prepared HT–HA using the coprecipitation method at constant pH of 10. The efficacy of the individual hydrotalcite (HT) and HA catalysts and the HT–HA hybrid catalyst was determined through transesterification reactions using different lipid raw materials, namely, commercial soybean oil and vegetable oil deodorization distillate (VODD). In this study, commercial soybean oil was used for comparative purposes as a model of feedstock and the VODD as a by-product of the vegetable oil processing industry. VODD was first pretreated through simple filtration and then dehydrated with the addition of sodium sulfate (25% w/w) to reduce the water content to 1%. The transesterification experiments were carried out in a 125 ml round-bottom flask coupled to a reflux condenser and under fixed mechanical agitation (400 rpm). The reaction was performed under fixed conditions of temperature of 70 °C, mass ratio of 40 g of the studied oil to 150 g of ethanol, catalyst concentration of 5 wt%, and a maximum duration of six hours. It was discovered that the individual HT and HA catalysts exhibited a low activity toward the transesterification of commercial soybean oil, leading to low conversions in ethyl esters (about 39–65% after six hours of reaction). When the HT–HA hybrid catalyst, calcined at 700 °C, was used, high soybean oil conversion of about 98% was obtained, under conditions of 45/1 ethanol/oil molar ratio, 70 °C, catalyst concentration of 5 wt%, and six hours of reaction. It was reported that under the same conditions, the catalyst was also able to transesterify the VODD with 98% conversion. This confirmed that the hybrid catalyst developed from HT and HA was efficient for the transesterification reaction.

Essamlali et al. [31] used sodium-modified hydroxyapatite (NaHA) as a highly efficient solid base catalyst for biodiesel production from rapeseed oil. The virgin HA was prepared via the pseudo sol–gel method using cetyl trimethylammonium bromide (CTAB) as cationic surfactant. The NaHA catalyst was prepared by the incipient wetness impregnation method using sodium nitrate ($NaNO_3$) as sodium precursor, followed by drying at 100 °C overnight and calcination at 300–800 °C for 2 hours in air. A series of catalysts were prepared by varying the loading level of the $NaNO_3$ precursor from 10 to 50 wt%. The catalysts were designated as X-NaHA, where X refers to the $NaNO_3$ loading (wt%). These catalysts were then used to produce biodiesel from soybean oil with the following reaction conditions: 100 °C, 8 hours reaction time, 6/1 methanol/oil molar ratio, and 4 wt% catalyst loading. It was found that biodiesel conversion was highly dependent on the sodium content and on the catalyst calcination temperature (300–800 °C), which controls the catalyst basicity. The 50-NaHA catalyst calcined at 800 °C showed the highest basicity (121 µmol g^{-1}), leading to the highest conversion of 99%. The increase in the basicity of the catalyst was assigned to calcium-deficient HA and $NaCaPO_4$ formed after impregnation and calcination steps. However, a small catalytic deactivation occurred due to the Na^+, Ca^{2+}, and PO_4^{3-} ions leaching into the biodiesel product. However, the catalyst exhibited excellent reusability with over 90% yield even after five recycling runs. The authors also showed that synthetic HA without any modification is not able to catalyze the transesterification or the esterification reactions under the previously mentioned conditions [31]. Similar results were also reported by Ngamcharussrivichai et al. [41] where HA of analytical reagent (AR grade) was used as catalyst for the transesterification of palm kernel oil (free fatty acids [FFAs] of 0.14 mg KOH g^{-1}) at 60 °C, 30/1 methanol/oil molar ratio, and 6 wt% catalyst loading for 3 hours. In the study, it was observed that HA alone was not active enough to promote the transesterification reaction because the FAME yield was within 2.6%.

Xie et al. [42] studied the catalytic activity of magnetic separable solid catalyst based on biguanide-functionalized HA-encapsulated-γ-Fe_2O_3 nanoparticles (HA-γ-Fe_2O_3) as heterogeneous catalyst for the synthesis of soybean oil biodiesel via methanolysis. The HA-γ-Fe_2O_3 was prepared by using the coprecipitation method followed by calcination at 200 °C for 3 hours. Then the biguanide was bonded to the HA-γ-Fe_2O_3 surface via covalent attachments. The basic strength of the biguanide-functionalized HA-γ-Fe_2O_3 catalyst determined by the Hammett indicator method was 9.8–15.0. The loading of biguanide on the magnetic support, estimated from nitrogen content in the catalysts obtained by elemental analysis, was about 0.03 mmol g^{-1}. By using the methanol/oil molar ratio of 25/1 and catalyst loading of 3 wt%, the maximum oil conversion of 99.6% was achieved over the solid base catalyst under reflux at 65 °C for 3 hours. Interestingly, the solid catalyst could be easily recovered by simple magnetic separation. The used catalyst could maintain its catalytic activity after five reuses.

Gupta et al. [43] studied the catalytic activity of marble slurry-derived HA as heterogeneous catalyst prepared by the coprecipitation method in the synthesis of soybean oil biodiesel via methanolysis. For HA synthesis, the marble slurry ($CaCO_3$) reacted with concentrated nitric acid under continuous stirring to produce calcium nitrate. This latter was then mixed with phosphate source precursors (H_3PO_4, Na_2HPO_4, or $(NH_4)_3PO_4$), and the mixture was precipitated using ammonium hydroxide until pH = 10.5, to produce HA, which was finally sintered at 700 and 950 °C in an oven for 2 hours. The reaction was carried out at reaction time (1–6 hours), methanol/oil molar ratio (3/1 to 15/1), and catalyst loading (2–10 wt%), while the reaction temperature and the stirring speed were maintained constant at 65 °C and 600 rpm, respectively. Under optimized reaction conditions (reaction temperature: 65 °C; reaction time: 3 hours; methanol/oil molar ratio: 9/1; and catalyst concentration: 6 wt%), the HA calcined at 800 °C for 3 hours had the highest basicity (13.30 mmol g^{-1}) and achieved 94 ± 1% biodiesel yield, which is better than that obtained with marble slurry calcined at temperature lower than 800 °C. Recycling study demonstrated that the produced HA was almost stable and exhibited slight fluctuations in its catalytic activity during five reuses.

Biodiesel production via transesterification reaction has also been intensively studied using HA-based catalysts derived from animal bones. These catalysts will be presented in more details in Chapter 13. In this view, Widayat et al. [22] explored the suitability of a KI/KIO_3-loaded HA catalyst derived from natural phosphate rocks for biodiesel production from WCO. The catalyst was prepared in two steps: first, phosphate rocks was dissolved in nitric acid solution (HNO_3 1 M, pH 2) to produce the Ca^{2+} and PO_4^{3-} ions. The solution was then precipitated using NH_4OH until the pH reached 10 and left for 24 hours until a white precipitate was formed. Calcination was performed at 1050 °C for 4 hours to produce the HA catalyst. Second, the calcined HA was impregnated using a solution containing KI and KIO_3 in the range of 1–6 wt%. Biodiesel was produced from WCO under simultaneous esterification–transesterification reactions at 60 °C for 6 hours. The authors showed that the biodiesel yield increased by increasing impregnation ratio and the maximum yield (92%) was achieved at an impregnation ratio of 5 wt%. They also showed that the KI/HA catalyst exhibited better catalytic performance (92% biodiesel yield) as compared with the KIO_3/HA catalyst (90% biodiesel yield). The developed catalysts were qualified as bifunctional catalysts since they possess Lewis's acid as well as conjugated basic sites, with the basic sites allowing them to promote both esterification of FFAs and transesterification of triglyceride simultaneously. The reusability of the catalyst was also investigated, and the results indicated that after three times of use, the catalyst was still active with about 72% biodiesel yield.

Nisar et al. [44] used potassium hydroxide catalyst supported on HA derived from animal bones as an active catalyst for the transesterification reaction of *Jatropha curcas* oil to produce pure biodiesel of ASTM specifications. The catalyst was prepared by soaking calcined animal bone in KOH solution of different concentration followed by the drying and calcinating at temperature ranging from 300 to 900 °C. The transesterification reaction was conducted at 70 ± 3°C and methanol-to-oil molar ratio of 9/1 using 6 wt% (based on oil weight) of the catalyst loaded with 10 wt% of potassium hydroxide and calcined at 900 °C. Under these optimum conditions, 96% production yield was achieved within three hours. The catalytic activity of the KOH/HA catalyst was attributed to the generation of new catalytic phase mainly $KCaPO_4$ and CaO along with HA as a predominant phase.

Sahu and Mohanty [45] studied the transesterification of neem (*Azadirachta indica*) oil to biodiesel using Al-grafted HA derived from cooked waste fish bone (CWFB). The waste fish bones were washed with tap water, air dried for two days, washed with acetone for 24 hours, dried again at 60 °C, and finally the CWFB was crushed to yield fine powder biological hydroxyapatite (CWFB-HA). To improve the catalytic activity of the CWFB-HA, aluminum was added by wet chemical precipitation technique using $Al(NO_3)_3 \cdot 9H_2O$ as precursor, followed by calcination at 650 °C. Three catalysts with weight proportions of 0.5/1, 1/1, and 1/0.5 of CWFB-HA to aluminum nitrate were prepared and designated as 0.5HA, 1HA, and 2HA, respectively. The X-ray powder diffraction (XRD) results revealed the presence of well resolved characteristic peak for HA along with crystalline Al_2O_3. The authors proposed a mechanism for the reaction of biodiesel production, which involves both Al and PO_4 species as acidic and basic sites, thereby making it a bifunctional catalyst. The results showed that biodiesel yield was around 95% below 70 °C, methanol-to-oil molar ratio of 30/1, reaction time of 120 minutes, and catalyst amount of 3 wt% based on oil weight. A maximum acid conversion of 97% was obtained at 90 °C, catalyst dosage of 1 wt%, methanol-to-oil molar ratio of 30/1, reaction time of four hours, and stirring speed of 400 rpm. The catalyst performance could be maintained even after seven cycles without significant yield deterioration.

Yan et al. [46] studied a series of $CaO-CeO_2/HA$ catalysts produced by impregnating calcined waste bone-derived HA by an aqueous solution containing both $Ca(NO_3)_2$ and $Ce(NO_3)_3$ at Ca/Ce molar ratio of 1/1 followed by calcination at 500, 650, and 800 °C. The produced catalysts possessed the following loadings of $CaO-CeO_2$ as 10, 20, 30, and 40 wt%. The prepared catalysts were used to catalyze the methanolysis of palm oil. It was found that the $CaO-CeO_2$ loading and the calcination temperature significantly affected biodiesel yield. Thus, 30% $CaO-CeO_2/HA$ calcined at 650 °C catalyst exhibited the best catalytic activity toward the transesterification of palm oil with 92% FAME yield at the following reaction parameters: catalyst dosage of 11 wt%, methanol-to-oil molar ratio of 9/1, reaction temperature of 65 °C, and reaction time of three hours. Moreover, the catalyst exhibited good performance after eight cycles as the yield of FAME remained almost high, around 84%.

In 2019, Kowthaman and Varadappan [47] studied the catalytic activity of new and highly efficient catalyst based on sodium-doped nanohydroxyapatite (nHA) for the production of FAME from Schizochytrium algae oil via methanolysis. The nHA was prepared from cow bones by successive steps of washing, drying, grounding, and calcination. The bones were hydraulically crushed and ground into fine powder using planetary ball mill at 1500 rpm for 3 hours. The obtained fine powder was then calcinated at 900 °C for 5 hours. Furthermore, the fine nHA obtained was then impregnated with sodium nitrate ($NaNO_3$) by incipient wetness impregnation method and calcinated at 500–900 °C for 3 hours under static air. By varying the sodium nitrate loading from 10 to 50 wt%, different catalysts, denoted as X/Na-nHA, were synthesized. XRD characterization showed the formation of orthorhombic $NaCaPO_4$ as new phase. However, typical diffraction peaks of other compounds such as CaO and Na_2O were not observed. It was discovered that the 50/Na-nHA catalyst calcined at 900 °C had the highest catalytic activity with 96% yield at an optimized reaction condition of 12 : 1 methanol-to-oil molar ratio, catalyst loading of 9.5 wt%, reaction temperature of 65 °C, and reaction time of 121 minutes. The basicity density of the fresh nHA support was about 60 $\mu mol\,g^{-1}$, which increased to 119 $\mu mol\,g^{-1}$ upon impregnation in 50/Na-nHA catalyst calcined at 900 °C. This increase in basicity was mainly ascribed to the formation of new phosphate compounds generated by the solid-state chemical reaction between $NaNO_3$ and nHA, forming Lewis's base sites (—O—) on the outer surface of the catalyst. This synthesized phase was responsible for catalyzing the reaction.

Chen et al. [48] developed a solid base catalyst from pig bones and K_2CO_3. The HA-based catalyst support was obtained by calcining pig bones at 600 °C for 4 hours. The K-doped counterpart was prepared by wet impregnation method using a solution of K_2CO_3 with different K_2CO_3 loadings from 10 to 40 wt%. The synthesized catalysts were calcined in the temperature range of 500–750 °C for 5 hours. The results revealed that catalyst impregnated with 30 wt% K_2CO_3 and calcinated at 600 °C (30K/HA-600) demonstrated the highest activity in transesterification of refined palm oil into FAME. At the optimized reaction conditions (9 : 1 ratio of methanol to oil, 8 wt% catalyst loading, and reaction temperature of 65 °C for 90 minutes), the FAME yield reached 96.4%. The XRD results of this catalyst revealed that the HA structure was modified with the concomitant formation of new crystalline phases assigned to $KCaPO_4$ and CaO generated by solid-state reaction between HA and K_2CO_3 during calcination process. Moreover, potassium addition increased both basicity strength and density. In fact, the respective base strength (H) and basic site density were in the range of $9.8 < H < 15.0$ and 2.2 $mmol\,g^{-1}$ for the initially calcined support and $15.0 < H < 18.4$ and 13.2 $mmol\,g^{-1}$ for the synthesized K/HA catalyst. This explains the high catalytic activity of K/HA catalyst in the transesterification reaction. The 30K/HA-600 catalyst maintained high activity with an ester yield above 90% after eight reuse cycles. A small catalyst deactivation occurred due to the K^+ ions leaching into the product during reaction. Consequently, with the same support, Sr/HA and Li/HA catalysts were prepared under similar conditions, which demonstrated comparable catalytic activity with that of K/HA catalyst. This proves the high potential of pig bones-derived HA as supports for the preparation of high performing catalysts for transesterification reactions.

9.2.2 Esterification Reaction for Biodiesel Production

In addition to the results presented previously, selected HA-based catalysts derived from animal waste and bones will be presented and discussed. Muria et al. [49] reported various metal-supported nano-HA synthesized hydrothermally from eggshells (ES) or shells as catalysts for the esterification of palm fatty acid distillate (PFAD) into biodiesel (Figure 9.3). The metal-HA catalysts were prepared by impregnating 4.85 g of HA with 0.15 g of metal precursors such as $Ni(NO_3)_2{\cdot}6H_2O$, $CuSO_4{\cdot}5H_2O$, and $Co(NO_3)_2{\cdot}6HO$, followed by the calcination of the impregnated dried material at 500 °C for 3 hours. The produced catalysts were denoted as Cu-HA, Co-HA, and Ni-HA, respectively. The XRD analysis showed no significant differences between HA and its supported catalysts Cu-HA, Co-HA, and Ni-HA for various concentrations of the metal loadings (3, 6, and 12%) synthesized. The esterification reaction was performed under the following conditions: 25 g PFAD, 10 : 1 methanol-to-PFAD ratio, 0.25 g catalyst, reaction temperature of 60 °C, and reaction time ranging from 2 to 4 hours. After four hours of reaction time, Cu-HA catalyst showed the highest biodiesel yield (94.4%), followed by Ni-HA (34.33%) and Co-HA (26%), while the initial HA support without metal only produced 10.67% biodiesel yield.

HA derived from waste animal bone was used as catalyst by Corro et al. [50], in a two-step catalytic process for the production of biodiesel from waste frying oil (WFO). In the preparation of the catalyst, the animal bones were first washed in a heating cooker to remove attached tissues and fats, mechanically crushed, and sieved (2–5 mm) and then calcined at 400–800 °C for 8 hours. The authors found that HA derived from waste animal bone can effectively promote the esterification reaction of FFAs into their corresponding FAMEs. The catalytic activity was retained after 10 successive reuses. This is because in spite of the low specific surface, the calcined animal bone displayed a high basic site density. The basicity of the catalyst originated from the oxygen atoms surrounding the phosphorous atom in the PO_4^{3-} moiety in $Ca_5(PO_4)_3OH$. These PO_4^{3-} groups could probably act as Lewis basic sites (electron donor species) during the FFA esterification reaction.

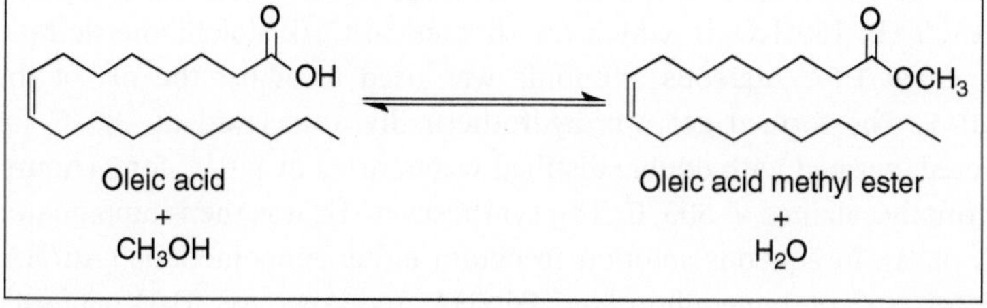

Figure 9.3 Esterification reaction of oleic acid by methanol as a reaction model. Source: Othmane AMADINE and Mohamed ZAHOUILY.

Conversely, Chakraborty and RoyChowdhury [51] reported a novel fish bone-derived natural hydroxyapatite (NHA)-supported copper solid acid catalyst in the esterification of oleic acid with ethanol. The Cu–NHA catalyst was prepared through wet impregnation method involving tungsten–halogen-irradiation-assisted freeze drying at different temperatures (70, 80, and 90 °C) followed by calcination in air at 500 °C for 2 hours. Three weight ratios of copper nitrate ($Cu(NO_3)_2 \cdot 5H_2O$) to NHA were used (0.5 : 1, 0.75 : 1, and 1 : 1), resulting in three catalysts denoted as C0.5, C0.75, and C1.0. The acidity of the Cu–NHA catalysts increased by increasing the copper nitrate loading as well as the freeze-drying temperature. The highest acidity (11.22 mmol KOH (g catalyst)$^{-1}$) was obtained for C1.0 catalyst prepared at 90 °C freeze-drying temperature. The esterification reaction was conducted in a laboratory semibatch reactor, and reaction parameters governing oleic acid esterification were optimized by the Taguchi design method (L9 orthogonal array). The maximum oleic acid conversion reached 92% over C1.0 catalyst. The catalytic activity was attributed to the presence of Cu^{2+} species, originated from CuO and $Cu_2(OH)(PO_4)$ phases, which are present on catalyst surface and act as Lewis acid in the esterification reaction studied.

9.2.3 Other Esterification Reactions

The conversion of lignocellulose biomass into fuels and value-added fine chemicals is considered as one of the most explored subjects in the prospective vision of biorefinery. In this context, a lot of interest has been driven toward the conversion of furfural, a lignocellulose biomass, into fuels and value-added fine chemicals. The development of highly efficient heterogeneous catalysts for the synthesis of esters through environmentally benign and cost-effective procedures is attracting much interest [52].

In this light, Radhakrishna et al. [53] have prepared a series of monometallic supported porous HA nanorods (Au/HA, Ag/HA, and Pd/HA) and bimetallic supported HA (Au_1Pd_{1-x}/HA and $Au_{1-x}Ag_x$/HA, $x = 0.2, 0.4, 0.6, 0.8$) and explored their catalytic activities toward the oxidative esterification of furfural. The HA support was prepared by hydrothermal method using cetyl trimethyl ammonium bromide (CTAB) as a template and calcium nitrate ($Ca(NO_3)_2 \cdot 4H_2O$, 99.9%) and di-ammonium hydrogen orthophosphate (($NH_4)_2 \cdot HPO_4$, 99%) as a source of calcium and phosphorus. The calcium source, phosphorus source, and CTAB were mixed to prepare a gel having the following molar composition: $1Ca^{2+}$: $0.66PO_4^{3-}$: $0.3CTAB$: $120H_2O$, in which the theoretical Ca/P stoichiometric ratio was maintained at 1.67. Aqueous ammonia was used to adjust the pH of the solution to 10.5. The formed gel was hydrothermally autoclaved at 180 °C for 24 hours, filtered, washed with double distilled water, dried at 110 °C for 24 hours, ground, and finally calcined at 500 °C. The synthesized HA was then impregnated with Au, Pd, or Ag in aqueous solution to obtain either monometallic (Au/HA, Pd/HA, and Ag/HA) or bimetallic ($Au_{1-x}Pd_x$/HA and $Au_{1-x}Ag_x$/HA) catalysts. The amount of the metal or mixed metals was defined so as to support 5 wt% of the monometallic or bimetallic system on HA. The impregnated catalyst was then reduced using sodium borohydride (NaBH$_4$), washed with deionized water and with ethanol twice, dried at 70 °C for 6 hours, calcined at 350 °C under the air atmosphere for 2 hours, and completely reduced under the flow of hydrogen (0.2LMP) at 350 °C for 5 hours. By varying Au and Ag or Au and Pd contents, various bimetallic catalysts such as Au$_{0.8}$Ag$_{0.2}$/HA, Au$_{0.6}$Ag$_{0.4}$/HA, Au$_{0.4}$Ag$_{0.6}$/HA, Au$_{0.2}$Ag$_{0.8}$/HA, Au$_{0.8}$Pd$_{0.2}$/HA, Au$_{0.6}$Pd$_{0.4}$/HA, Au$_{0.4}$Pd$_{0.6}$/HA, and Au$_{0.2}$Pd$_{0.8}$/HA were prepared. Structural analysis by means of Fourier Transform Infra-Red (FTIR), XRD, and X-Ray photoelectron spectrometry (XPS) analyses confirmed that both the crystallinity and stability are retained even after the impregnation of metals and reduction under H$_2$ atmosphere. The XRD patterns of Au/HA, Pd/HA, Ag/HA, Au$_{1-x}$Pd$_x$/HA catalysts ($x = 0.2$–0.8), and Au$_{1-x}$Ag$_x$/HA showed the presence of diffraction peaks of HA phase together with well separable peak for the metal species. Moreover, the specific surface area and the pore volume of the HA support calculated from the nitrogen adsorption–desorption isotherm were found to be 44 m^2 g^{-1} and 0.2485 cm^3 g^{-1}. The pore size distribution of the synthesized HA was centered at 2.4, 3.1, and 5.4 nm. The HR-TEM (Transmission Electron Microscopy) micrograph revealed that the Au—Pd bimetallic system adopted core–shell morphology, while the Au—Ag system showed alloy morphology over HA nanorods. The catalytic activities of the monometallic (Au/HA, Pd/HA, and Ag/HA) and bimetallic (Au$_{1-x}$Pd$_x$/HA and Au$_{1-x}$Ag$_x$/HA) catalysts were evaluated at atmospheric pressure toward the oxidative esterification of furfural to methyl 2-furoate using methanol as solvent and *tert*-butyl hydroperoxide (TBHP) as an oxidant. Catalytic experiments were carried in liquid phase using 300 μl of furfural, 3 eq. of TBHP (70%) and 50 mg of catalyst (Au-20 mol%) and 20 ml of methanol. The optimized reaction temperature for the oxidative esterification of furfural was found to be 120 °C. Among the monometallic system, the Pd–HA catalyst shows higher conversion of furfural (86.6%) than Au/HA (76%). When considering selectivity toward methyl 2-furoate, the Au/HA showed higher selectivity of (94.2%) than Pd/HA (2.3%) and Ag/HA (25%). Among the bimetallic HA system, the Au$_{0.8}$Pd$_{0.2}$/HA showed the highest conversion of 94.2% of furfural and the highest selectivity (99%) toward the desired product of methyl 2-furoate. According to this study, it was found that in the Pd-rich bimetallic system, a core–shell structure was formed, which reduce the exposition of the Au species to reactant molecules and hence results in poor conversion and selectivity. The stability and recyclability of the Au$_{0.8}$Pd$_{0.2}$/HA catalyst was investigated, and the results obtained revealed that the drop in the conversion at the end of the fifth cycle did not exceed 4%, while the selectivity toward methyl 2-furoate remained around 99%. The Au catalysts were stable and quite easily recovered and represent a feasible and promising route to efficiently convert furfural to methyl-2-furoate to be scaled up at industrial level.

In another study, the oxidative esterification of acetol with methanol to methyl pyruvate was reported by Wan et al. [54] using Au/HA with appropriate strength and balanced ratio of acid and base sites. In this study, various Au/HA catalysts were prepared by varying the Ca/P molar ratios of the HA support (1.50, 1.67, 1.80). The supported gold catalysts were prepared by using the deposition–precipitation (DP) method. The powder XRD patterns of HAP and Au/HAP confirm that the

Au ($2\theta = 38.2°$ (111)) was successfully deposited on the HA surface and that the crystal structure of HA was well maintained after the loading of Au nanoparticles. The acid/base properties of synthesized Au/HAP catalysts as determined by ammonia temperature-programmed desorption (NH_3-TPD) and carbon dioxide temperature-programmed desorption (CO_2-TPD) showed that the density of acid sites decreased (from 3.34 to 2.21 µmol g^{-1}) with the increase in the Ca/P ratio from 1.59 to 1.65, while a reverse trend was found in the density of base sites, which was found to increase from 2.87 to 4.00 µmol g^{-1}. Moreover, elemental analysis showed that the measured Au content was approximately 0.98 wt%, close to the nominal loading for all catalysts. TEM micrograph revealed that the Au particles were homogeneously distributed with uniform particle sizes. The mean particle sizes of Au were in the range 3.4–3.6 nm for all analyzed catalysts. The catalytic activity of the Au/HA-1.59, Au/HA-1.62, and Au/HA-1.65 was studied in the oxidative esterification of acetol in a batch-type Teflon-lined stainless steel autoclave. The results obtained showed that the Au/HA catalyst exhibited superior catalytic activity and high methyl pyruvate (MPA) selectivity (S = 77%) than Au/HT (S = 8%), Au/γ-Al_2O_3 (48%), Au/ZnO (S = 53%), Au/SiO_2 (S = 0%), and Au/TiO_2 (S = 8%). Among the tested catalysts, the Au/HA-1.65, which contains the strongest acid but weakest base sites, exhibited the best performance among the catalysts studied, achieving 87% selectivity into MPA at 62% conversion of acetol. The selectivities toward side products such as methyl lactate (MLA) and methyl acetate (MAA) were all kept below 5%, suggesting that the aldol condensation or Cannizzaro side reactions do not favorably occur over Au/HA-1.65 catalyst. The stability and recyclability of the optimized Au/HA-1.65 catalyst were investigated over five cycles. The recovered Au/HA-1.65 catalyst exhibited similar catalytic activity after each run, demonstrating excellent catalyst stability and durability.

Novel Au/HA-based catalysts have been employed for the direct oxidative esterification of methacrolein to the corresponding methyl methacrylate in methanol in the presence of molecular oxygen as a benign oxidant [55]. The Au/HA catalysts were prepared by DP method. The authors explored the catalytic activity of several Au/HA-based catalysts wherein the HA support adopts different morphology (needlelike or N-HA, lamella-like or L-HA, and rodlike HA or R-HA) but with similar gold particle size in the direct oxidative esterification of methacrolein to the methyl methacrylate. They found that the morphology of the HA support plays an important role and affects the catalytic activity of the Au/HA catalyst. Remarkably, the Au/N-HA having a needlelike morphology allows obtaining a high MMA selectivity of 98% at 68% conversion under mild conditions (i.e. ambient pressure, low reaction temperature of 70 °C, and low methanol/aldehyde ratio of 8 : 1). However, for Au/L-HA and Au/R-HA, the conversions only reach 48 and 36%, respectively, together with low MMA selectivity (76% with Au/L-HA and 43% with Au/R-HA). The N-HA structure could facilitate the higher dispersion of Au species, owing to its high specific surface area, and the stronger interaction between Au NPs and the support resulted in the formation of more surface defects, due to the existence of partially encapsulated Au particles by the needlelike HA structure. Furthermore, the superior catalytic performance of the Au/N-HA catalyst was also attributed to a cooperative effect between abundant acid–base sites of N-HA for the preferential chemisorption of methacrolein and highly dispersed active Au species for the favorable formation of β-hydride and oxygen activation.

Recently, in a study on the esterification of acetic acid with 2-ethyl-1-hexanol to produce 2-ethylhexyl acetate, Mukhopadhyay et al. [56] reported on the exploration of the catalytic performance of nano-tin oxide grafted natural HA (SnO_2/HA) as a highly efficient nano-photocatalyst. The nano-SnO_2 grafted natural HA photocatalyst was prepared by hydrothermal impregnation method using different kinds of energy, i.e. conventional energy, solar-type energy, and sequential application of solar-type and ultrasound wave energy. The authors found that the catalyst prepared under application of sequential solar-type and ultrasound wave energy allowed to obtain a catalyst with higher surface area (174.57 $m^2\,g^{-1}$) and uniform dispersion of nano-SnO_2 particles on the natural HA support surface within only 1 h. The XRD results showed prominent HA and cassiterite (SnO_2) phases. Additionally, scanning electron microscopy (SEM) and TEM micrographs showed that the catalyst prepared by conventional method has greater grain size along with less homogeneity in nano SnO_2 distribution compared with the catalyst prepared under ultrasound irradiation, suggesting that the sequential application of solar-type and ultrasound wave energy inhibited the agglomeration of SnO_2 nanoparticles. The XPS analysis showed that no signal of Sn(0) was observed, validating XRD analyses and indicating that Sn existed only in its oxidized form with a valence state of +2/+4 (Sn(II) and Sn(IV)). The photoactivity of the SnO_2/HA catalysts was evaluated for the synthesis of 2-ethylhexyl acetate by esterifying acetic acid with ethyl-1-hexanol in a solar-type/ultrasound batch reactor. Among the tested catalysts, the SnO_2/HA catalyst prepared by sequential application of solar-type and ultrasound wave energy exhibited the highest photocatalytic activity by achieving 98 ± 1% acetic acid conversion in only two hours under mild process condition, i.e. 60 °C and atmospheric pressure. The superiority of this catalyst was ascribed to its higher surface area, which allow better contact between the catalyst actives sites and the reactant molecules. More interestingly, this photocatalyst presented the slowest deterioration in its performance even after the eighth cycle when compared with the conventionally prepared photocatalyst.

Ranu and coworkers [57] described the application of HA-supported Pd(II) catalyst (Pd/HA) for the synthesis of (E)-2-alkene-4-ynecarboxylic esters by the coupling vicinal-diiodoalkenes with conjugated carboxylic esters. The use of Pd/HA catalyst in presence of K_2CO_3 was found to give the desired product in 75% yield. The procedure was applied to several substituted vicinal diiodoalkenes and acrylic esters under the catalysis of Pd/HA, and the obtained results demonstrated that this procedure produce the corresponding enynecarboxylic esters in high yields (up to 90%) depending on the nature of the substituent with a turnover number (TON) based on Pd up to 16 000. Pd/HA also showed a good recyclability without significant loss of efficiency after three recycling tests.

The use of nanocrystalline HA as a recoverable heterogeneous catalyst for acid-catalyzed esterification of methacrylic acid (MA) with ethylene glycol (EG) was reported by Sandhyarani and coworkers [58]. The authors have prepared various HA samples at different Ca/P ratios (1.67, 1.62, 1.57, and 1.5) and evaluated their catalytic activity toward this esterification. The characterization by XRD, FT-IR, and TEM confirmed the formation of nano-sized HA particles with desired Ca/P ratio. The catalytic activity of the catalysts was highly affected by the Ca/P ratio and reached the highest value with the HA sample having the Ca/P molar ratio of 1.5. The kinetic study allowed determining the first reaction order with respect to acid. Interestingly, it was demonstrated that the esterification reaction does not require the addition of hydroquinone to prevent polymerization in this kind of reaction, suggesting that the catalyst exhibits mild acidic sites.

9.3 Hydrogen Production

Hydrogen, a very light gas (density = 0.09 g cm^{-3} at 0 °C and 1 atm), is the simplest of all the gaseous elements and represents the third most abundant element on our planet. It has a higher energy density (higher calorific value – HHV: 120 MJ kg^{-1}) than that of natural gas (HHV: 50 MJ kg^{-1}). It is more flammable than natural gas or conventional hydrocarbons, and its flammability limit is 4% by volume in air, while it is 5.3% for natural gas [59]. The annual global production of hydrogen is more than 30 Mt and places this gas among the first major industrial gases. Hydrogen is mainly used in refineries (hydrocracking, hydrodesulfurization, etc.) and in the synthesis of several fine chemicals and fuels such as methanol, ammonia, and in the Fischer–Tropsch process [60, 61]. From an environmental point of view, it serves as a desulfurization agent for hydrocarbons in order to reduce sulfur content and thus the emission of sulfur oxides during hydrocarbon combustion [62]. In transport sector, hydrogen is considered as a clean fuel since water is the only by-product of hydrogen combustion. Over the past decades, research has focused on the use of hydrogen as an energy source for prototype hydrogen vehicles and stationary power generators. Hydrogen can also be used to power a fuel cell. The latter is an electrochemical system that provides electrical energy and heat from a redox reaction.

9.3.1 Water–Gas Shift Reactions

Industrially, hydrogen is mostly produced from coal, natural gas, and hydrocarbons, which account for more than 95%, the rest being produced by water electrolysis (Figure 9.4) [63]. Coal is gasified with steam, while natural gas and hydrocarbons are either reformed with steam or partially oxidized to produce the synthesis gas (syngas), which is a mixture rich in hydrogen and carbon monoxide.

This carbon monoxide can be converted with water vapor into hydrogen and carbon dioxide according to the water–gas shift (WGS) reaction:

$$CO + H_2O \leftrightarrow CO_2 + H_2 \quad \Delta H°_{298} = -41.09 \text{ kJ mol}^{-1}.$$

After the WGS step, pure hydrogen is separated from the gas mixture by using a separation process such as pressure swing adsorption (PSA), allowing the production of H_2 of very high purity (up to 99.99%). This hydrogen production route implies the utilization of solid catalysts for both reforming and WGS steps, which is identified

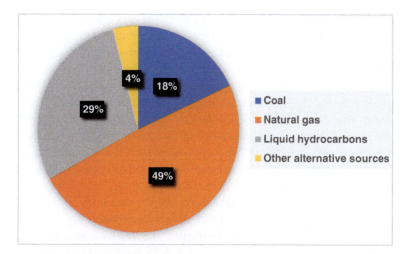

Figure 9.4 Hydrogen production from various source. Source: [63]. Reproduced with permission of Elsevier.

as key parameters. To this end, the choice of the catalyst support is a major factor for a successful chemical transformation. Among the various supports used for the production of hydrogen, particular interest was given to the HA as an excellent alternative to conventional supports, thanks to its (i) structural stability, (ii) acid–base properties, and (iii) ion exchange ability and substitution property.

Details of reforming processes can be found in Chapter 8 of this book. Thus, this section is only focused on the WGS process. This WGS reaction was discovered more than two centuries ago, by the Italian physicist Felice Fontana in 1780. However, its exploitation was realized much later. The WGS reaction is moderately exothermic, and its equilibrium constant decreases with increasing temperature. The reaction is favored thermodynamically at low temperatures and kinetically at elevated temperatures but is unaffected by changes in pressure [64]. For this reason and at the commercial level, the WGS reaction is achieved in two consecutive steps consisting of a high temperature WGS (HT-WGS) operated in the range of 350–500 °C and then a low temperature WGS (LT-WGS, 150–250 °C) to achieve high conversions of CO (generally less than 0.3 vol% of CO remains in the gas mixture from the LT-WGS outlet).

In view of the significance of WGS reaction in several industrial areas, it is of great interest to develop highly efficient catalytic systems. Generally, the availability of oxygen vacancies and activity for water dissociation are the main characteristics of WGS catalyst systems [65]. In the literature, a large number of noble metal-based catalysts, such as palladium (Pd), ruthenium (Ru), platinum (Pt), rhodium (Rh), and gold (Au), have been extensively explored in WGS reaction in the last few years [66–75]. Likewise, the transition metals such as Ni, Co, Cu, Fe, etc. have also been investigated as active phase in WGS reaction.

In this light, Dengyun et al. [76] explored the catalytic activity of platinum loaded onto synthetic hydroxyapatite (Pt/HA) for WGS reaction. It was demonstrated that the efficiency of this catalytic system is strongly affected by the Ca/P until a value of 1.75. The dependence of the catalytic on the Ca/P ratio suggests that the stoichiometry of the HA plays a crucial role in defining the characteristic of the surface of the catalyst support and particularly the abundance of the Lewis acidic (Ca^{2+} species) and base (PO_4^{3-}) sites on its surface. In fact, the authors used quantum chemical calculations to show that H_2O molecule is strongly activated on apatite surfaces via simultaneous coordination to Lewis acidic sites (Ca^{2+} species) and H-bonding to the oxygen atoms of PO_4^{3-} groups. In another study, the same authors [77] showed that the activity of Pt/HA was significantly improved when Ca^{2+} was exchanged with Sr^{2+}, and that the produced Pt/SrHA did not exhibit any methanation activity up to 450 °C. They also showed that the catalytic performance of the Pt/SrHA catalyst had a volcano behavior, which increased with the increase of Pt content up to 1 wt% and then decreased above that content due to apatite surface area limitations.

Correspondingly, Boukha et al. [78] investigated the catalytic activity of Pt supported on lanthanum-modified HA samples for WGS reaction. In the catalyst, lanthanum was incorporated into HA structure as promoter due to its capacity to achieve good ionic conductivity. Catalysts having different La loading levels were prepared and designated as Pt/La(x)/HA where x stands for La loading ($x = 1, 4$, and 8 wt%). WGS reaction experiments were conducted in a tubular flow reactor operating at atmospheric pressure. Prior to the catalytic test, the catalysts were reduced under a flow of 20% H_2/He at 450 °C for one hour and cooled to 200 °C, under a flow of He. Experimental results revealed that the addition of La to the Pt/HA catalyst enhanced its oxygen storage capacity, blocked the incorporation of Pt species into the HA framework, promoted its reducibility, decreased the Pt particle size, and markedly improved activity. It is noteworthy to mention that the catalytic activity of the Pt/La(4)/HA is quite similar to that of Pt/Al_2O_3 and Pt/CeO_2.

Palladium (Pd = 0.5–2 wt%) supported on the calcium-deficient HA (Ca/P = 1.50) was investigated by Boukha and coworker as a heterogeneous catalyst for WGS reaction to produce hydrogen [79]. Experimental results showed that to reach the same CO conversion of 10%, the 0.5 wt% Pd/HA catalyst required lower temperature (300 °C) than that of 1 wt% Pd/HA (345 °C) and 2 wt% Pd/HA (385 °C) catalysts. Increasing Pd loading up to 2 wt% led to a drop in the catalytic activity because of the agglomeration tendency of Pd particles, which lowers the density of active sites. Likewise, the authors showed that the 0.5 wt%Pd/HA catalyst exhibited the best catalytic performance and achieved about 26% CO conversion at 400 °C, which was much higher than that obtained by Lupescu et al. using Pd/Al_2O_3 catalyst [80] (below 11% at 400 °C). The poor activity of the Pd/Al_2O_3 catalyst was due to the lack of oxygen mobility in the support material.

In another study, the influence of cobalt (Co) addition on the catalytic activity of the Pd/HA toward the WGS reaction was investigated [81]. The added cobalt acts as a promoter because of its highly anisotropic redox properties of the oxide. Considering the relatively low mobility of oxygen on the Pd/HA catalyst, cobalt addition probably increases the surface oxygen mobility and induces an improvement in the catalytic performance. The HA-supported bimetallic Pd-Co catalyst was prepared by two consecutive impregnation steps. First, Co(x)/HA (with $x = 1, 4,$ and 8 wt%) was prepared by incipient wetness impregnation using different amounts of Co $(NO_3)_2 \cdot 6H_2O$ and then calcined at 500 °C for 4 hours. Afterward, the Co(x)/HA was impregnated with an aqueous solution of $Pd(NH_3)_4Cl_2 \cdot H_2O$, dried at 120 °C, and calcined at 500 °C for 4 hours to obtain Pd/Co(x)/HA catalysts having 0.5 wt% Pd loading. From a catalytic point of view, it was found that the catalytic activity was strongly dependent on the Co loading and the Pd/Co(8)/HA catalyst exhibited the highest activity with CO conversion about 11% at 250 °C, 64% at 300 °C, 84.5% at 350 °C, and 82% at 400 °C. The increased Co loading improved the intimate contact between Pd and Co species, which is responsible of the improvement of the CO conversion. Consequently, high Co loading also favored the methanation activity. Furthermore, it was observed that Co-rich catalysts (Pd/Co (4)/HA and Pd/Co(8)/HA) were insensitive to the presence of oxygen in the feeding mixture. On the other hand, the addition of O_2 to the feed seemed to enhance the activity of the Pd/Co(1)/HA catalyst. To underline the role of the catalyst support in promoting the WGS reaction, additional experiments were conducted with Pd-Co(1)/Al_2O_3 containing the same Pd and Co loading in comparison to Pd—Co(1)/HA. Surprisingly, Pd/Co(1)/HA catalyst was four times more efficient than Pd—Co(1)/Al_2O_3 catalyst, demonstrating the potential of HA support in WGS reaction.

In the study of Venugopal et al. [82], they compared the catalytic performance of gold-supported HA (Au/HA) and ruthenium-supported HA (Ru/HA) toward the WGS reaction to produce hydrogen. The HA support was prepared by chemical precipitation, and the active species were deposited on the support surface by wet impregnation using a dilute aqueous solution of hydrogen tetra-chloroaurate or ruthenium trichloride. They showed that the Au/HA catalyst displayed superior catalytic activity compared with Ru/HA, particularly at low reaction temperature ranging from 110 to 120 °C. They attributed the superiority of the Au/HA catalyst to the role of acid–base surface properties. It should be noted that the catalytic activity of the Au/HA and Ru/HA catalysts become quite similar (conversion about 90%) at higher reaction temperature of about 320 °C. However, the catalytic system involving the Ru/HA as catalyst suffers from the methanation as a side reaction.

Nevertheless, the high cost and low availability of noble metals limit their large-scale application. As an alternative, nickel (Ni), iron (Fe), cobalt (Co), and copper (Cu) metals supported on ZrO_2, Al_2O_3, SiO_2, and CeO_2 have also been employed as catalysts for the WGS reaction [83–86]. Yet, up to date, there is relatively little work reported on the use of HA as support for non-noble metals catalysts for the WGS reaction. In 2018, Boukha et al. [87] developed a novel catalytic system based on the CuO—ZnO/HA (designated as Cu—Zn(x)/HA, with x = 1, 3, and 5, which refers to the Cu/Zn molar ratio), which were prepared by co-impregnation method and evaluated its catalytic performance for WGS reaction. The reaction was performed in a tubular flow reactor with an internal diameter of 9 mm operating at atmospheric pressure. Among the tested catalysts, Cu—Zn(3)/HA exhibited the best catalytic performance and allowed obtaining a CO conversion of 26% at 250 °C, 29% at 350 °C, and 35% at 400 °C. The catalytic activity follows the order Cu—Zn(1)/HA < Cu—Zn(5)/HA < Cu—Zn(3)/HA. The superiority of the CuZn(3)/HA can be assigned to the suitable surface chemistry developed at the catalyst surface, which plays an important role in stabilizing the active species. Furthermore, the CO-TPD, H_2-TPD, and CO_2-TPD analyses revealed that the CuZn (3)/HA catalyst is the most active and displayed the highest density of basic sites (1.7 µmol CO_2 m^{-2}). In this light, several studies reclaimed that the presence of weak–medium strength basic sites could be favorable for the catalytic activity of the Cu-based catalyst in WGS reaction [88].

Iriarte-Velasco et al. [89] investigated the catalytic activity of NHA derived from pork bone as catalytic support for various transition metals in WGS reaction using a fixed-bed stainless steel reactor. The M/HA (with M: Ni, Co, Cu, and Fe) catalysts were prepared by impregnation method and calcined at 450 °C for 2 hours before being used. The results obtained showed that the catalytic activity of the prepared catalyst followed the order Ni/HA > Co/HA ≫ Cu/HA ≫ Fe/HA. At 400 °C, Ni/nHA achieved the highest H_2 yield of about 85% and a small CH_4 yield of 0.3%. Under the similar condition, Ni-loaded synthetic HA (sHA) led only to 60% H_2 yield of 60% but 9% CH_4 yield. This difference was explained by the fact that Ni/sHA contains abundant medium strength basic sites, which favor CH_4 production, while Ni/nHA displayed a deficit in basic sites, which could inhibit the methanation. Moreover, on the basis of the chemical elemental analysis (EDX and inductively coupled plasma [ICP]), the authors suggested that the presence of ubiquitous species such as Mg, Na, and K in the nHA could have inhibited the methanation side reaction, showing the advantage of nHA versus sHA in this reaction. As comparative study regarding the effectiveness of HA versus the conventional supports used in WGS reaction by using Ni as transition metal, A. Haryanto et al. [90] used Ni/CeO$_2$—Al$_2$O$_3$ at temperature of 450 °C with a CO/steam molar ratio of 1 : 3. The result of this study showed that the Ni/nHA catalyst displayed a better catalytic activity (85% H_2 yield) than Ni/CeO$_2$—Al$_2$O$_3$ (52% H_2 yield). Additionally, the authors concluded that the alumina-supported catalysts (with or without ceria promotion) produced the highest CH_4 yield, around 5 vol% on average. Therefore, the remarkable WGS catalytic performance of Ni/HA catalysts appears to be strongly depended on the structural properties of HA supports.

9.3.2 Borohydride Hydrolysis Reaction

It is known that the hydrogen can be stored as solid form through physisorption in different carbon materials, or molecular hydrogen in pressurized vessels, and cryogenic liquid [91–100]. It can also be stored in the form of atomic hydrogen in different metal hydrides (MH$_x$), or in the form of hydride ion in protide compounds (MAlH$_4$), and under borohydrides form (MBH$_4$); M being a metal. The generation of hydrogen using metal hydrides is carried out by thermal decomposition, and they are considered reproducible in reversible ways. Nevertheless, their hydrogen amounts are lower than protide compounds as shown in Table 9.1.

Among these different compounds, NaBH$_4$ is considered as one of the most potential materials for hydrogen storage. Hydrogen generation rate from NaBH$_4$ hydrolysis could be controlled by using heterogeneous catalysts. Hence, during the past 10 years, several heterogeneous catalytic systems have been reported for

Table 9.1 Hydrogen capacity in metal hydride and protide compound.

	Typical metal hydride	Example of metal hydride	Hydrogen storage capacity (wt%)
Metal hydride	AB_5H_6	$LaNi_5H_6$	1.4
	AB_2H_4	$ZrMn_2H_{3.46}$	1.7
	ABH_2	$TiFeH_{1.9}$	1.8
	A_2BH_4	Mg_2NiH_4	3.6
	AH_X	MgH_2	7.6
Protide compound	$M^I AlH_4$	$NaAlH_4$	7.4
	$M^I BH_4$	$LiBH_4$	18.4
		$NaBH_4$	10.6
	$M^{III}(BH_4)_3$	$Al(BH_4)_3$	16.8

Source: [101]. Reproduced with permission of Elsevier.

sodium borohydride hydrolysis; many of them based on cobalt and ruthenium as active phases.

Jaworski et al. [102] investigated the $NaBH_4$ hydrolysis by using cobalt immobilized onto HA support (CoHA). Different HA-based supports were synthesized and investigated, which were Ca-deficient HA (cd_CoHA), amorphous HA (a_CoHA), and single crystal HA (s_CoHA). For each HA support, cobalt deposition was carried out by ion exchange method using aqueous solutions of cobalt nitrate of different cobalt concentration in order to obtain different cobalt loadings. For each hydrolysis experiment, 0.2 g of $NaBH_4$ dissolved in 20 ml 1% NaOH (0.25 M) was added to a Teflon-coated flask at c. 22 °C and 1 bar under stirring at 1000 rpm. First, blank test without catalyst showed that no hydrolysis took place. Second, the addition of HA-based catalyst to the reaction accelerated the hydrolysis reaction, and by increasing Co loading, the amount of hydrogen generated from the hydrolysis increased. Third, cd_CoHA and a_CoHA had the similar crystalline structure, but cd_CoHA facilitated the incorporation of Co to its surface in comparison with a_CoHA. Thus, cd_CoHA showed higher catalytic activity than a_CoHA in the hydrolysis reaction. Finally, even with the lowest specific surface area, s_CoHA showed the highest catalytic activity. The authors evoked that the predominant presence of [98] surface of s_CoHA could have been the reason for this high catalytic performance, but this needs to be further studied and confirmed. About the catalytic stability, all three catalysts showed a similar catalytic behavior during 20-day test, with a consecutive loss of activity, which reached nearly 75% during this period. Various characterizations were done for the used catalysts, but the reason of the deactivation could not be revealed.

Similarly, another research group [103] demonstrated that the Co(0)HA nanoclusters obtained by reduction of cobalt (II) was very effective to produce hydrogen toward $NaBH_4$ hydrolysis. Co(0)HA catalyst was prepared by ion exchange method using a commercial HA support and cobalt nitrate aqueous solution, followed by the reduction of Co(II) fixed on HA surface in $NaBH_4$ solution. The hydrolysis of

NaBH$_4$ was also carried out in a sealed flask coated with Teflon, which was filled with 50 ml NaBH$_4$ aqueous solution (149.4 mM) and 818.5 mg of catalyst (0.72 wt% Co) at 25 °C, 1 bar, and 900 rpm. The studied catalyst was considered as long-lived catalyst providing in the hydrolysis of basic NaBH$_4$ about 256 000 turnovers, before deactivation. Further recycling tests also confirmed the finding of Jaworski et al. [102] about the catalyst deactivation. In fact, during the five successive recycling tests, Co(0)HA could only conserve c. 81% of its activity in comparison with the previous test. Recently, Nakazato et al. [104] converted ES and scallop shells (SS) as calcium sources with phosphoric acid into stoichiometric HA under different conditions of stirring (120 and 300 rpm) and ageing (variation of temperature: 25 and 100 °C and duration: 1 and 72 hours). Pure commercial CaO was also used as reference for HA synthesis. HA powders were then used as catalyst supports to prepare Ru/HA catalysts by ion exchange method. In the hydrolysis of NaBH$_4$, the HA supports synthesized with 300 rpm, 100 °C ageing temperature, and one hour ageing duration led to the most active Ru/HA catalysts. The analysis of physicochemical properties/catalytic performance revealed that Ru/HA catalysts having an average pore size below 20 nm and a molar ratio Ca/P of approximately 1.60 are the most favorable factors in this reaction.

9.3.3 Ammonia Borane Hydrolysis Reaction

The ammonia borane (H$_3$NBH$_3$) appears to be a suitable hydrogen storage material, particularly for portable applications, since it displayed numerous advantages such as high hydrogen content (19.6 H$_2$ wt%), high water solubility (33.6 g H$_3$NBH$_3$/100 g H$_2$O), high water stability at room temperature, nontoxicity, and more importantly, it is able to release hydrogen by means of hydrolysis reaction at room temperature in the presence of a suitable catalyst [100, 105]. Several catalysts have demonstrated effectiveness toward the ammonia borane hydrolysis reaction, including various transition metal salts, noble metal nanoclusters, and non-noble metals supported on different catalyst supports such as carbon, SiO$_2$, Al$_2$O$_3$, Fe(0) nanoparticles, intra-zeolite cobalt(0) nanoclusters, and HA.

In 2011, Rakap et al. [106] investigated HA-supported palladium (0) nanoclusters (Pd(0)/HA) as catalyst in the hydrolysis of ammonia borane. This catalyst was prepared in situ from the reduction of palladium (II) ion exchanged HA (Pd^{2+}-HAP) by ammonia borane during the hydrolysis reaction. The ICP-OES analysis showed that the palladium content of the sample was around 1.04 wt%. After evaluating the catalytic activity of HA-supported palladium (0) nanoclusters in hydrogen generation from the hydrolysis of ammonia borane, the authors found that the Pd(0)/HA nanoclusters were extremely active and exhibited a long lifetime span by providing a TON of 12 300 with a maximum hydrogen generation rate about 1425 ml H$_2$ min^{-1} (g Pd)$^{-1}$ at 25 °C. More importantly, this catalyst showed great potential to be isolated and reused several times, with only a slight decrease in its catalytic activity about 88% of its initial activity after five cycles. The slight decrease in catalytic activity can be attributed to the passivation of nanoclusters' surface by increasing the concentration of boron products, e.g. metaborate, which decreases the accessibility of active sites. To highlight the role of HA as support in the hydrolysis of H_3NBH_3, several studies consistently showed that HA played a significant role in this reaction in comparison with Pd/conventional support system. For instance, M. Chandra et al. [107] used Pd-γ-Al_2O_3 (2.0 wt% Pd) with the average particle size of 3.6 nm to release H_2 at the molar ratio of $H_2/H_3NBH_3 = 2.9$ in 120 minutes, while Pd(0)/HA nanoclusters (1.04 wt%) with average particle size of 3.3 nm allowed releasing H_2 at the molar ratio $H_2/H_3NBH_3 = 3.0$ only in 30 minutes. Durk et al. [108] reported the synthesis of ruthenium(0) nanoparticles supported on the surface of nano-HA (RuNPs@nano-HA) by ion exchange of Ru^{3+} ions with Ca^{2+} ions of the HA matrix followed by the ammonia borane reduction of the resulting Ru^{3+}@nano-HA precatalyst during the hydrolytic dehydrogenation of ammonia borane. After a short induction time, the RuNPs@nano-HA (0.51 wt% Ru loading) exhibited remarkable catalytic activity with TOF = 205 min^{-1} in hydrogen generation from the hydrolysis of ammonia borane at 25 °C, while in the same reaction, ruthenium (0) nanoparticles supported on micro-HA showed a TOF = 137 min^{-1}. The authors attribute this difference to the large surface area of nano-HA spheres than that of the micro-HA particles and to the smaller size of ruthenium (0) nanoparticles supported on nano-HA (2.56 nm) than that of the nanoparticles supported on the micro-HA particles (4.70 nm). It was important to note that the ruthenium (0) nanoparticles show high stability against the sintering and leaching, which makes them importantly reusable catalyst in the hydrolytic dehydrogenation of ammonia borane. In comparative study between HA and Al_2O_3 as catalyst supports for ruthenium dispersion in hydrogen generation from the hydrolysis of ammonium borane, the RuNPs@nano-HA catalyst showed higher catalytic activity than Ru@Al_2O_3, with TOF about 205 and 39.6 min^{-1}, respectively. The catalytic performance of RuNPs@nano-HA can be attributed to the well dispersion of nanoparticles with a high accessible active site on the nanoparticles surface [109, 110]. These results clearly confirmed the advantage of using HA as catalyst support in hydrogen generation from the hydrolysis of ammonia borane.

9.4 Catalytic Production of High Value-Added Energy Additives

The diesel market contributes to the prosperity of the worldwide economy; it is essential for transport sector, power generation, and heavy-duty engines working [111]. However, the conventional diesel fuel suffers from several issues related to its incompatibility with some engines [112], incomplete combustion [113], and contribution to greenhouse gas emissions affecting the environment and human health [114, 115]. Hence, the development of new additives allowing the clean combustion of diesel has become an urgent necessity [116–118]. These new alternative chemicals are actually used to assess the quality of diesel in order to meet up the different specified standards, enhance the engine performance, and reduce the harmful emissions (Figure 9.5) [111, 119].

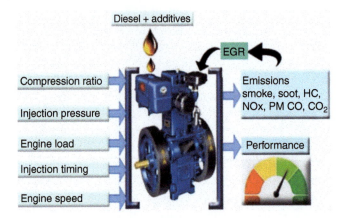

Figure 9.5 Affecting parameters for engine performance and emission. Source: [120]. Kumar et al. (2018). pg 25: 22475–22498/with permission of Springer Nature.

The oxygenated additives are the most widely used additives. This family includes alcohols (ethanol, methanol, propanol, butanol, etc.), esters (acetoacetic esters, dicarboxylic acid esters, 2-hydroxy-ethyl esters, EG, etc.), ethers (ethyl tertiary butyl ether, methyl *tert*-butyl ether, dimethyl ether, etc.), organic carbonate (diethyl carbonate (DEC), dimethyl carbonate (DMC), propylene carbonate (PC), glycerol carbonate, etc.), and others (Table 9.2).

Several research attempts have been devoted to the development of new heterogeneous catalyst for the synthesis of high added-value energy additives that are able to assess the diesel quality [142–145]. Different materials, in particular metal oxides [146], zeolites [147], and HA [148] have been reported in literature to be effective catalytic materials for the synthesis of these additives. This section reports the latest advances in energy additives production using catalysts derived from HA. More precisely, this section will focus on the *n*-butanol synthesis, furfural synthesis, and Guerbet reaction.

9.4.1 *n*-Butanol and Its Derivative Chemicals

n-Butanol is recognized as the most promising chemical platform for the synthesis of energy additives. It is often produced by the Guerbet reaction in which ethanol was first converted into ethanal, which is transformed via self-condensation reaction into crotonaldehyde and underwent hydrogenation reaction yielding *n*-butanol product (Figure 9.6) [149]. Besides, the ethanol could be converted to some *n*-butanol derivatives that are also used as energy additives such as butadiene [150] and isobutene [151]. The chemical transformation process of ethanol into high added-value energy additives such as *n*-butanol requires an efficient heterogeneous [13]. Various heterogeneous catalysts have been previously reported in the literature like basic zeolites, HTs, mixed oxides, alkaline earth, zinc oxide, and HA [152–156]. However, the majority of these catalysts led to a nonselective reaction and deactivate during the

Table 9.2 Energy additives and their specific characters in energy sector.

Additives classes	Family	Additive name	Specific characteristic in energy sector	References
Oxygenated additives	**Alcohols**	Ethanol	Gasoline alternative and additive	[121]
		Methanol	Octane booster	[122]
		Propanol	Improvement of spark ignition engine	[123]
		n-Butanol	Gasoline alternative and additive	[124]
		Iso-butanol	Gasoline additive	[125]
		Tert-butanol	Gasoline additive and octane booster	[126]
		3 1-Octylamino-3-octyloxy-2-propanol	Improve the viscosity index	[117]
	Esters	Acetoacetic esters	Improve ignition and flash point	[127]
		Dicarboxylic acid esters	Reduce the wear	[128]
		2-Hydroxy-ethyl esters	Improve the lubricity	[128]
	Ether	Diethylene glycol	Improve the combustion efficiency increase the cetane number	[129]
		Dimethoxy methane	Improve miscibility and reduce knock	[130]
		Dimethoxy ethane	Flame additives	[131]
		Tripropylene glycol	Improve the viscosity index	[132]
		Monomethyl ether	Reduce the viscosity deviation caused by temperature	[133]

(continued)

Table 9.2 (Continued)

Additives classes	Family	Additive name	Specific characteristic in energy sector	References
	Organic carbonate	Diethyl carbonate	Improve ignition and flash point	[134]
		Dimethyl carbonate	Improve the viscosity index	[135]
		Propylene carbonate	Enhance ignition of delay time	[136]
		Glycerol carbonate	Enhance ignition of delay time and increase the cetane number	[136]
	Other products	Sorbitan monooleate	Improve ignition	[137]
		Polyoxyethylene sorbitan monooleate	Reduce the wear and enhance the stability of emulsified blend	[138]
		Dibutyl maleate	Improve the lubricity	[139]
		N-octyl nitramine	Enhance the stability and combustion efficiency by increasing the cetane number	[119]
		Di-*tert* butyl glycerol (DTBG)	Dispersants and emulsifiers agents	[140]
		Tri-*tert* butyl glycerol (TTBG)	Flame additives	
		Diacetin and triacetin	Gasoline alternative and additive	[141]

Source: Othmane AMADINE and Mohamed ZAHOUILY.

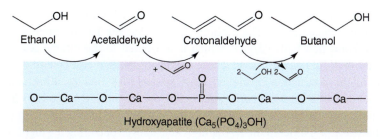

Figure 9.6 Mechanism of the formation of butanol over the HA catalyst. Source: [158]. Reproduced with permission of American Chemical Society.

reaction except the use of HA, which demonstrates the highest selectivity toward *n*-butanol synthesis [157].

Tsuchida et al. [13] reported that unmodified HA is a good catalyst for the highly selective synthesis of *n*-butanol and its derivatives such as 1.3-butadiène by Lebedev's reaction. They demonstrated that the Ca/P ratio is the controlling factor of the acidic and basic character of the HA catalyst and significantly affects the catalytic activity. The maximum selectivity for 1,3-butadiene was found over HA with Ca/P = 1.62. Moreover, Scalbert et al. [12] investigated the condensation of ethanol to butanol over HA catalysts in a temperature ranging from 350 to 410 °C. The authors demonstrated that the major fraction of butanol was formed by the direct condensation reaction between two ethanol molecules whereas the indirect pathway remains less selective and led to the formation of acetaldehyde and H_2 as by-products. A detailed catalytic mechanism for the production of butanol over the HA catalyst has been described by Christopher et al. [158] using spectroscopy analysis and in situ titration method (Figure 9.6). The authors identified two types of active sites, the first one with a basic character ≪Ca—O≫ responsible for the dehydrogenation step, and the second one, CaO/PO_4^{3-}, allows the aldol condensation of acetaldehyde. The reaction mechanism occurs in three main phases in which (i) the acetaldehyde was first formed over the Ca—O basic active sites, (ii) then the acetaldehyde was converted into crotonaldehyde over the CaO/PO_4^{3-} groups, and finally (iii) the formed crotonaldehyde was subsequently hydrogenated via hydrogen transfer reaction from an ethanol molecule over the Ca—O basic sites. All these intermediates have been observed when the reaction was conducted at 330 °C. The HA has a high capacity for substitution of Ca^{2+} and PO_4^{3-} ions in the framework by other cations and anions. This has a strong impact on its catalytic selectivity and allowed the conversion of methanol [159] and aldol condensation [160]. The effects of acid–base properties of palladium-modified HA on butan-2-ol conversion have been reported by Boukha et al. [159]. The Pd(z)/HA (with z = 1%) has been prepared by impregnation method, then calcined at 550 °C for 12 hours. The obtained catalyst has been used for the catalytic conversion of butan-2-ol into methyl ethyl ketone under a continuous airflow reactor and temperature range of 120–260 °C. A total conversion up to 90% has been achieved using Pd/HA at 180 °C. A comparative study between the catalytic performances of Pd/HA and another catalyst based on Pd supported fluorapatite has been conducted [159]. The authors showed that the Pd/HA catalyst is more efficient compared with Pd/FAP, which is attributed to the differences in surface acidity and PdO particle size. These findings highlight the potential of HA support in the production of methyl ethyl ketone[161]. Besides, Khachani et al. [162] studied the catalytic performance of iron modified HA (Fe-HA) in the butan-2-ol conversion and showed that the content of incorporated iron enhanced the acidic properties of the HA catalyst and allowed obtaining the highest butan-2-ol conversion (butene yield of 55%) under oxygenated conditions. The authors highlighted that a good propene selectivity (35–90%) was obtained using Fe-HA catalyst. However, at low Fe content (0.52 wt%), propene yield did not exceed 6.2%, which could be related to the presence of Fe^{3+}—O—Ca^{2+} species (single iron species deposited on the HA surface) that maintain the basicity needed for the hydrogen abstraction from propane compared with the acidic sites of Fe^{3+}—O—Fe^{3+} (iron oxides) formed at higher iron loadings. These results highlight the critical role of the acid–base properties of the catalyst in the dehydrogenation process.

The synthesis of n-butanol from ethanol over strontium phosphate hydroxyapatite (Sr-HA) has been reported by Ogo et al. [163]. According to the authors [163], Sr-HA catalysts with various Sr/P molar ratios (analytical value varied from 1.58 to 1.70) were synthesized using hydrothermal methods at 110 °C for 14 hours in an autoclave. Ethanol conversion into n-butanol was performed in a fixed-bed reactor at 300 °C. Blank test without catalyst showed nearly zero ethanol conversion at this temperature. On the other hand, the use of Sr-HA catalyst allowed obtaining 1.1–11.3% ethanol conversion when the Sr/P molar ratio raises from 1.58 to 1.70. In parallel, the selectivity into n-butanol as the main product was also favored by increasing Sr/P molar ratio (69% at Sr/P = 1.58 and 86.4% at Sr/P = 1.70). Other products of the reaction were 2-buten-1-ol (3–22.3%), C_6—OH and C_8—OH (2.6–12.0%), and some traces of acetaldehyde. A mechanism study was also conducted to determine the rate-determining step (RDS) of the reaction. This RDS was attributed to the aldol condensation, including both the condensation process itself and the formation of two aldehyde adsorbates on neighboring basic sites just before condensation. High Sr/P molar ratio allowed increasing the density of strong basic sites, which favored both ethanol conversion and n-butanol selectivity [163].

9.4.2 Fuel Additives from Furfural

Furfural is one of the furan derivatives with the molecular formula of $C_5H_4O_2$. There are many synonyms for furfural such as 2-furancarboxaldehyde, 2-furanaldehyde, furaldehyde, 2-furfuraldehyde, furfurol, and fural. Generally, this chemical compound can be easily produced from cheap straw and wastes containing a furan ring and aldehyde group [164]. The furfural and its derivatives have been widely applied in various chemical fields [165–167]. It is not only employed as feedstock for producing high-grade gasoline [168], diesel [169], or jet fuel [170] but can also be efficiently converted into energy additive by oxidation [171], hydrogenation [172], acetalization [173], decarboxylation [174], condensation [175], etc. Particularly, efficient aldol condensation of furfural with acetone has been subjected as an intermediate step into second-generation biofuels synthesis from lignocellulosic biomass [176]. More generally, the conversion of furfural into fuels and high added-value chemicals has gained much attention during the last decades, and many researchers investigated the use of oxidative esterification catalytic reactions using HA as a catalyst or a catalyst support.

Mishra et al. [177] used HA-supported Au nanocatalyst (1 wt% Au) for oxidative esterification reaction of 5-hydroxymethyl-2-furfural (HMF) to 2,5-furan dimethyl carboxylate (FDMC) in oxygenated conditions. An excellent yield of FDMC up to 89.3% was recorded at 130°C, under an air pressure of 2.4 MPa during 6 hours of reaction. The used catalyst keeps its performance without any drop in the catalytic activity after five consecutive cycles, thus highlighting the good reusability and the high stability of the considered catalyst. Moreover, Tong et al. [178] reported the oxidative condensation of furfural-n-propanol system catalyzed by Au—Fe_2O_3—HA + K_2CO_3 at 140°C for 4 hours under an oxygen pressure of 0.3 MPa. They reported a successful synthesis of 3-(furan-2-yl)-2-methylacryaldehyde with 91.4% conversion and 97.2% selectivity. Under similar conditions, oxidative esterification reaction of the furfural-methanol system in the presence of O_2 at 140°C, 0.3 MPa, and 5 wt% of Au—Fe_xO_y—HA + K_2CO_3 was reported. The obtained results confirmed a good yield of 91.8% and high selectivity of 98.7% into methyl 2-furoate. Hence, the produced compound (methyl 2-furoate) is identified as one of the most promising renewable chemicals for the sustainable production of commodity chemicals and fuels. The synthesis of the same product was reported by Radhakrishnan et al. [53] using HA nanorods impregnated (HA-T) with various monometallic (Au, Pb, Ag) and bimetallic ($Au_{1-x}Pd_x$, $Au_{1-x}Ag_x$) species. The catalytic activities of the studied catalysts were evaluated under atmospheric pressure using tert-butyl hydroperoxide (TBHP) as an oxidant and methanol as solvent. The obtained results proved that $Au_{0.8}Pd_{0.2}$/HA-T with core–shell morphology achieved the highest furfural conversion (94.2%) along with very high selectivity into methyl-2-furoate (99%) and good reusability. This finding highlighted the fact that the addition of second metal into the HA structure enhances the catalytic activity and stability due to the change in their electronic and geometric properties. Other investigations have reported the use of magnetic catalysts based on HA supported γ-Fe_2O_3 to produce furfural [179]. Zhang et al. [180] have reported γ-Fe_2O_3 dispersed in HA as a new catalyst for the oxidation of HMF to 2,5-diformylfuran (DFF). In the study, a modified HA has been applied by immobilizing monomeric Ru^{4+} species on the outer surface of the HA. The RuHA-γFe_2O_3 catalyst achieved a high DFF yield of 89.1% and an HMF conversion of 100% under the optimal reaction conditions (90°C, 4 hours). On the other hand, catalytic transfer hydrogenation of C=O is one of the most important reactions for fuels and fine chemicals synthesis. For this purpose, Wang et al. tested the catalytic performance of HA-γFe_2O_3 for metal-free transfer hydrogenation of furfural into furfuryl and other valuable derivatives such as 5-hydroxymethyl-2-furfural (HMF).

The use of HA-γFe$_2$O$_3$ produced excellent furfural conversion of 96.2% along with a high yield up to 91.7% after 10 hours of reaction at 180 °C using iso-propanol as a hydrogen donor. In another study, HA-bounded palladium (Pd-HA) catalyst was used in furfural hydrogenation reaction. This catalyst showed high catalytic performance compared to others, reaching approximately 100% selectivity into tetrahydrofurfuryl alcohol (THFAL) using 2-propanol solvent at 40 °C for 3 hours under a hydrogen pressure of 1 MPa. For the hydrogenation reaction of furfural to THFAL, the authors proposed two mechanisms: in the first one, furfural was partially converted to furfuryl alcohol and followed by hydrogenation to form THFAL, and in the second one, 2-(isopropoxymethyl) furan (2-IPMF) was first formed by reaction between the furfural and the 2-propanol via esterification reaction, then the intermediate compound was converted into THFAL. As a conclusion, 2,5-furan dimethyl carboxylate, 3-(furan-2-yl)-2-methylacryaldehyde, methyl-2-furoate, and the 5-hydroxymethyl-2-furfural are some of the high added value energy additives that have been obtained by chemical catalytic processes involving the use of HA-based catalysts.

9.4.3 Organic Carbonates Agents

Organic carbonates as green chemicals are used in several industrial sectors. In energy sector, they are widely used as energy additives to improve the antiknock characteristics of gasoline, to reduce CO and NO$_x$ emissions, to reduce Reid vapor pressure (RVP), to improve the octane number, and to replace some products already used for this purpose such as ethyl tertiary butyl ether (ETBE), methyl-*tert*-butyl ether (MTBE), and other oxygenated additives [134, 181–183]. These chemicals are characterized by a high oxygen content [182], the presence of a single bond between carbon and oxygen atoms [184], ecological character [185], and immiscibility with water, which make organic carbonates an important class of energy additives [181, 182]. For example, the addition of 5, 10, and 15 wt% DMC to gasoline strongly improves the thermal efficiency of the engines in the range of, respectively, 15, 32, and 50% [186]. The family of organic carbonates is large and includes a variety of compounds and high potency molecules such as glycerol carbonate [136], which recently have been evaluated as a diesel promoter additive. This compound can decompose exclusively during the ignition of diesel to produce CO$_2$ and 3-hydroxypropanal, which is known for its beneficial impact on soot reduction. Organic carbonate compounds are generally obtained by the reaction between one or more alcohol function(s) and a carbonate or carbonyl groups. The phosgenation remained the most used synthetic homogeneous route to produce organic carbonate at the industrial scale [187]. However, the toxicity of phosgene and the inherent problems of storage and transport have prompted researchers to develop other alternative routes. Green synthesis of DEC has been reported by Shukla et al. [188] using a non-phosgene process based on a transesterification reaction between PC and ethanol in the presence of modified HA, Mg/HA, and Zn/HA. The reaction was conducted at 160 °C for 5 hours. Mg/HA was classified as the best catalyst among the studied catalysts, owing to its high basicity. A maximum

DEC yield of 52.1% was obtained at 160 °C after 5 hours of reaction using Mg/HA catalyst.

The synthesis of glycerol carbonate was also investigated via the transesterification reaction between DMC and glycerol over modified HA at 78 °C, 1 bar, under 2000 rpm in a batch reactor [189]. HA support was synthesized by the conventional precipitation method using $(NH_4)_2 \cdot HPO_4$ and $Ca(NO_3)_2 \cdot 4H_2O$ as P and Ca precursors at 70 °C, pH 10.5 for 2 hours. Then M/HA catalysts (M = K, La, Zr, Li, Ce, KF), containing 15 wt% M, were prepared by wet impregnation. Among the different M/HA catalysts, KF/HA was found to be the best one, which allowed reaching more than 99% glycerol conversion and more than 99% glycerol carbonate selectivity under the investigated conditions (78 °C; 2/1 DMC/glycerol molar ratio; 3 wt% catalyst amounts with respect to the glycerol amount; 50–120 minutes). KF/HA catalyst also showed an excellent reusability with only slight activity loss (2% after four runs) and unchanged selectivity [189].

Catalysts based on metals impregnated HA were prepared, and their catalytic activities were evaluated in the reaction of the alcoholysis of urea and 1,2-propanediol to produce PC[190]. HA support was first prepared by a co-precipitation method. Then it was subsequently impregnated with different metal precursors to obtain M/HA-supported catalyst (M = LiOH, K_2O, MgO, SrO, BaO, La_2O_3, and CeO). The prepared catalysts were used for urea alcoholysis reaction under nitrogen gas at 170 °C for 2 hours. The catalytic activity of M/HA catalysts was much higher compared with that of the undoped HA support. La/HA showed the highest activity, and PC yield up to 91.5% was achieved. The high activity of La/HA catalysts was attributed to the change in the basic properties of the catalyst surface. Besides, M/HA (M = Mg, Sr, and Ba) provides lower catalytic activity compared with La/HA, Li/HA, and K/HA, but an acceptable yield of PC up to 70% was also recorded using these catalysts.

9.4.4 Energy Additives from Alcohols via Guerbet Reaction

Alcohol coupling reaction, also known as Guerbet reaction with reference to Marcel Guerbet, is an important class of reactions for the synthesis of many important branched saturated alcohols used as potential oxygenated additives in the energy sector. The Guerbet reaction process consists of a successive series of reactions of dehydrogenation, aldolization, crotonization, and hydrogenation. The transformation of ethanol to heavier products via Guerbet reaction has been widely reported by using HA and modified HA to produce oxygenated additives such as acetaldehyde, ethylene, diethyl ether (DEE), hexanol, butadiene, 2-ethyl-1-butanol, 3-butene-2-ol, 3-butene-1-ol, and 2-methyl-2-propenol. Takashi et al. [152] studied the utilization of Sr-substituted HA having different (Ca + Sr)/P molar ratio, and they found that the change in the stoichiometric ratio affected the structural and the morphological character as well as the acidic and basic properties of the resulting catalyst. They concluded that the physical and chemical properties including the specific surface area, the stoichiometry, and the acid–base properties strongly affected the ethanol conversion into heavier alcohols [152]. The ethanol conversion at 450 °C on a nonstoichiometric HA (Ca/P = 1.64) has been reported by the same research team [191]. The authors demonstrated that the obtained liquid was a mixture of several compounds containing olefins, dienes, alcohols, aldehydes, and aromatics of 12 carbon atoms. The obtained chromatographic spectrum was similar to that of commercial gasoline. Lovón-Quintana et al. [192] reported the use of carbonated HA obtained by chemical precipitation for the synthesis of a hydrocarbon fuel via the conversion of ethanol by the Guerbet reaction. Structural analysis of carbonated HA showed that carbonate ions (CO_3^{2-}) substituted PO_4^{3-} ions in the apatite structure, forming a B-type HA [$Ca_{10-x/2}(PO_4)_{6-x}(CO_3)_x(OH)_2$]. The morphological analysis revealed the existence of textural mesopores, while the surface chemistry analysis confirmed the bifunctional character of the material: acid function promoting ethanol dehydration reactions and basic function promoting ethanol dehydrogenation reactions, which simultaneously provoked the formation or cleavage of C—C, C—H, and C—O bonds. Catalytic experiments have proven the ability of the catalyst to convert ethanol to hydrocarbon fuels with a conversion rate of 97% at 500 °C. Chemical analysis of the produced fuel revealed that it is mainly composed of oxygenated and non-oxygenated hydrocarbons including aromatics, dienes, olefins, paraffins, alcohols, aldehydes, esters, ethers, and ketones with various molecular carbon chains from 1 to more than 18 carbon atoms. Similarly, Sylvester et al. have also studied the Guerbet reaction of ethanol to produce heavier products using carbonated HA catalyst [10]. A yield of 30% in heavier alcohols at a 40% ethanol conversion was obtained at 400 °C at 5000 ml h^{-1} g^{-1} gas hourly space velocity (GHSV) [10]. The combined catalytic activity of HA and Cu—ZnO has also been investigated to convert ethanol to methyl benzaldehydes by the Guerbet reaction followed by cyclization. The selectivity of the reaction into the desired product was about 30% with a slow rate. Besides, using a tailored bifunctional cobalt–hydroxyapatite (Co/HA) catalyst, aromatic alcohols can be synthesized from ethanol according to the Guerbet reaction followed by cyclization [193]. HA support was synthesized by the conventional preparation method using $Ca(NO_3)_2$ and $(NH_4)_2HPO_4$ as precursors and NH_4OH as precipitant. Co deposition was performed by incipient wetness impregnation followed by drying at 50 °C, calcination at 350 °C, and reduction under H_2/N_2 at 400 °C. Catalytic test was carried out in a fixed-bed reactor. The obtained results showed that an ethanol conversion of 35% and a selectivity of 54% into methylbenzyl alcohols (MB-OH) can be achieved over a single bed of Co–HA catalyst at 325 °C. A significant enhancement in selectivity achieving 71% at a temperature of 225 °C was recorded after the combination of Cu/C with Co-HA in a dual-bed catalyst system. Effort was devoted to the determination of the reaction pathway for the formation of MB–OH and aliphatic alcohols from ethanol, as represented in Figure 9.7 [193]. Based on the kinetic reaction results and spectroscopic analysis, Wang et al. [193] have proposed a plausible pathway for 2-MB–OH formation from ethanol over Co–HAP. The proposed pathway consists of successive reactions of ethanol dehydrogenation to acetaldehyde that occurs on the Co^{2+} sites followed by a fast condensation conversion of acetaldehyde to 2-butenal, which then transformed to aromatic alcohols by coupling, dehydrocyclization over basic sites such as those on HA surface, and hydrogenation process.

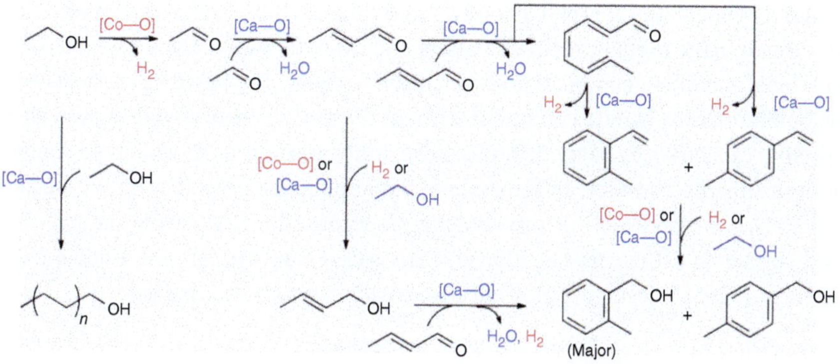

Figure 9.7 Reaction pathway of ethanol condensation into aliphatic and aromatic alcohols over Co/HA catalysts. Source: [193]. Reproduced with permission of American Chemical Society.

9.4.5 Other Value-Added Chemicals

The DEE can be considered as a renewable oxygenated fuel when it is synthesized from the continuous dehydration of bioethanol over a nonstoichiometric aluminum phosphate–HA [194]. HA and $AlPO_4$/HA having the (Ca + Al)/P molar ratio of 1.62 were prepared by precipitation method using $Ca(NO_3)_2 \cdot 4H_2O$, $(NH_4)_2HPO_4$, and $Al(NO_3)_3 \cdot 9H_2O$ as precursors. Ethanol conversion was performed at 300–400 °C and 1–200 bars. The results showed that 78% ethanol conversion and high selectivity to DEE up to 96% were obtained under the optimum conditions (340 °C, 200 bars). The developed catalyst has proved an excellent stability without any loss in activity and selectivity for 41 hours [194]. Methanol conversion into methyl formate (MF), which is an effective gasoline additive and an important intermediate in the synthesis of other energy additives, was investigated using Cu-based catalysts [195]. Under the investigated conditions (200–280 °C), Cu/SiO_2 was found as effective catalyst for the formation of MF due to its low basicity and weak strength basic sites, while Cu/HA and Cu/MgO degraded the resultant MF into CO and H_2 because of their strong strength basic sites [195]. The results would be improved by preparing Cu supported catalyst with a calcium-deficient HA support. Methyl glycolate (MG) and EG could be obtained by the selective hydrogenation of dimethyl oxalate (DMO) over copper-supported HA catalyst (Cu/HA) [196]. Cu/HA catalyst containing 20 wt% Cu was the most active, achieving a high MG selectivity and 90% DMO conversion at 310 °C. Furthermore, this catalyst showed a high stability without any catalytic activity loss during 120 hours of time-on-stream [196].

9.5 Conclusion

As we described in this chapter, tremendous progress has been made in recent years in the development and application of HA in the energy field. HA and HA-based catalysts present different advantages in comparison with other solid materials, such as high chemical and thermal stability, ion exchangeability, acid–base proprieties, etc., which make them attractive catalysts and catalytic supports onto which other functional materials can be deposited. In many cases, HA-based catalysts can be used under relatively mild conditions without the use of cosolvent and show a good recycling capacity. The catalytic properties mainly arise from the acid–base character of the apatite's surface. In some cases, the impregnated moieties on the surface of the HA are responsible of the catalytic properties, the HA playing the role of a solid support. Further investigations on HA-based catalysts for renewable energy are encouraged to develop highly effective catalytic systems in this field.

References

1 Hendi, A.A. (2017). Hydroxyapatite based nanocomposite ceramics. *Journal of Alloys and Compounds* 712: 147–151.
2 Sadat-Shojai, M., Khorasani, M.-T., Dinpanah-Khoshdargi, E. et al. (2013). Synthesis methods for nanosized hydroxyapatite with diverse structures. *Acta Biomaterialia* 9: 7591–7621.
3 Javadinejad, H.R., Saboktakin Rizi, M., Aghababaei Mobarakeh, E. et al. (2017). Thermal stability of nano-hydroxyapatite synthesized via mechanochemical treatment. *Arabian Journal for Science and Engineering* 42: 4401–4408.
4 Cazalbou, S., Eichert, D., Ranz, X. et al. (2005). Ion exchanges in apatites for biomedical application. *Journal of Materials Science: Materials in Medicine* 16: 405–409.
5 Daniels, Y. and Alexandratos, S.D. (2010). Design and synthesis of hydroxyapatite with organic modifiers for application to environmental remediation. *Waste and Biomass Valorization* 1: 157–162.
6 Borum, L. and Wilson, O.C. (2003). Surface modification of hydroxyapatite. Part II. Silica. *Biomaterials* 24: 3681–3688.
7 Li, Y. and Weng, W. (2008). Surface modification of hydroxyapatite by stearic acid: characterization and in vitro behaviors. *Journal of Materials Science: Materials in Medicine* 19: 19–25.
8 Amedlous, A., Amadine, O., Essamlali, Y. et al. (2019). Aqueous-phase catalytic hydroxylation of phenol with H_2O_2 by using a copper incorporated apatite nanocatalyst. *RSC Advances* 9: 14132–14142.
9 Yu, W., Sun, T.-W., Ding, Z. et al. (2017). Copper-doped mesoporous hydroxyapatite microspheres synthesized by a microwave-hydrothermal method using creatine phosphate as an organic phosphorus source: application in drug delivery and enhanced bone regeneration. *Journal of Materials Chemistry B* 5: 1039–1052.
10 Silvester, L., Lamonier, J.-F., Faye, J. et al. (2015). Reactivity of ethanol over hydroxyapatite-based Ca-enriched catalysts with various carbonate contents. *Catalysis Science and Technology* 5: 2994–3006.

11 Young, Z.D. and Davis, R.J. (2018). Hydrogen transfer reactions relevant to Guerbet coupling of alcohols over hydroxyapatite and magnesium oxide catalysts. *Catalysis Science and Technology* 8: 1722–1729.

12 Scalbert, J., Thibault-Starzyk, F., Jacquot, R. et al. (2014). Ethanol condensation to butanol at high temperatures over a basic heterogeneous catalyst: how relevant is acetaldehyde self-aldolization? *Journal of Catalysis* 311: 28–32.

13 Tsuchida, T., Kubo, J., Yoshioka, T. et al. (2008). Reaction of ethanol over hydroxyapatite affected by Ca/P ratio of catalyst. *Journal of Catalysis* 259: 183–189.

14 Boucetta, C., Kacimi, M., Ensuque, A. et al. (2009). Oxidative dehydrogenation of propane over chromium-loaded calcium-hydroxyapatite. *Applied Catalysis A: General* 356: 201–210.

15 Rego de Vasconcelos, B., Pham Minh, D., Martins, E. et al. (2020). Highly-efficient hydroxyapatite-supported nickel catalysts for dry reforming of methane. *International Journal of Hydrogen Energy* 45: 18502–18518.

16 Xin, Y., Ando, Y., Nakagawa, S. et al. (2020). New possibility of hydroxyapatites as noble-metal-free catalysts towards complete decomposition of volatile organic compounds. *Catalysis Science and Technology* 10: 5453–5459.

17 Diallo-Garcia, S., Osman, M.B., Krafft, J.-M. et al. (2014). Identification of surface basic sites and acid–base pairs of hydroxyapatite. *The Journal of Physical Chemistry C* 118: 12744–12757.

18 Luo, J., Chen, J., Li, W. et al. (2015). Temperature effect on hydroxyapatite preparation by co-precipitation method under carbamide influence. *MATEC Web of Conferences* 26: 01007.

19 Nazeer, M.A., Yilgor, E., Yagci, M.B. et al. (2017). Effect of reaction solvent on hydroxyapatite synthesis in sol–gel process. *Royal Society Open Science* 4: 171098.

20 Yang, Y., Wu, Q., Wang, M. et al. (2014). Hydrothermal synthesis of hydroxyapatite with different morphologies: influence of supersaturation of the reaction system. *Crystal Growth and Design* 14: 4864–4871.

21 Hassan, M.N., Mahmoud, M.M., El-Fattah, A.A. et al. (2016). Microwave-assisted preparation of nano-hydroxyapatite for bone substitutes. *Ceramics International* 42: 3725–3744.

22 Widayat, W., Hadiyanto, H., Wardani, P.W.A. et al. (2020). Preparation of KI/hydroxyapatite catalyst from phosphate rocks and its application for improvement of biodiesel production. *Molecules* 25: 2565.

23 Yaemsunthorn, K. and Randorn, C. (2017). Hydrogen production using economical and environmental friendly nanoparticulate hydroxyapatite and its ion doping. *International Journal of Hydrogen Energy* 42: 5056–5062.

24 Piccirillo, C. and Castro, P.M.L. (2017). Calcium hydroxyapatite-based photocatalysts for environment remediation: characteristics, performances and future perspectives. *Journal of Environmental Management* 193: 79–91.

25 Wang, M., Zhang, K., Wu, M. et al. (2019). Unexpectedly high adsorption capacity of esterified hydroxyapatite for heavy metal removal. *Langmuir* 35: 16111–16119.

26 Jyotsna and Vijayakumar, P. (2020). Synthesis and characterization of hydroxyapatite nanoparticles and their cytotoxic effect on a fish vertebra derived cell line. *Biocatalysis and Agricultural Biotechnology* 25: 101612.

27 Szmacinski, H., Hegde, K., Zeng, H.-H. et al. (2020). Imaging hydroxyapatite in sub-retinal pigment epithelial deposits by fluorescence lifetime imaging microscopy with tetracycline staining. *Journal of Biomedical Optics* 25: 047001.

28 Zhang, Q., Liu, Y., Zhang, Y. et al. (2015). Facile and controllable synthesis of hydroxyapatite/graphene hybrid materials with enhanced sensing performance towards ammonia. *Analyst* 140: 5235–5242.

29 Yang, C.-C., Li, Y.J., Chiu, S.-J. et al. (2008). A direct borohydride fuel cell based on poly(vinyl alcohol)/hydroxyapatite composite polymer electrolyte membrane. *Journal of Power Sources* 184: 95–98.

30 Amadine, O., Essamlali, Y., Amedlous, A. et al. (2019). Iron oxide encapsulated by copper-apatite: an efficient magnetic nanocatalyst for *N*-arylation of imidazole with boronic acid. *RSC Advances* 9: 36471–36478.

31 Essamlali, Y., Amadine, O., Larzek, M. et al. (2017). Sodium modified hydroxyapatite: highly efficient and stable solid-base catalyst for biodiesel production. *Energy Conversion and Management* 149: 355–367.

32 Pinzi, S., Garcia, I.L., Lopez-Gimenez, F.J. et al. (2009). The ideal vegetable oil-based biodiesel composition: a review of social, economical and technical implications. *Energy and Fuels* 23: 2325–2341.

33 Ramadhas, A.S., Jayaraj, S., and Muraleedharan, C. (2005). Biodiesel production from high FFA rubber seed oil. *Fuel* 84: 335–340.

34 Karmee, S.K. and Chadha, A. (2005). Preparation of biodiesel from crude oil of Pongamia pinnata. *Bioresource Technology* 96: 1425–1429.

35 Ma, F. and Hanna, M.A. (1999). Biodiesel production: a review Journal Series 12109, Agricultural Research Division, Institute of Agriculture and Natural Resources, University of Nebraska–Lincoln.1. *Bioresource Technology* 70: 1–15.

36 Mahlia, T.M.I., Syazmi, Z.A.H.S., Mofijur, M. et al. (2020). Patent landscape review on biodiesel production: technology updates. *Renewable and Sustainable Energy Reviews* 118: 109526.

37 Rizwanul Fattah, I.M., Ong, H.C., Mahlia, T.M.I. et al. (2020). State of the art of catalysts for biodiesel production. *Frontiers in Energy Research* 8: 101.

38 Abukhadra, M.R., Dardir, F.M., Shaban, M. et al. (2018). Spongy Ni/Fe carbonate-fluorapatite catalyst for efficient conversion of cooking oil waste into biodiesel. *Environmental Chemistry Letters* 16: 665–670.

39 Brasil, H., Pereira, P., Corrêa, J. et al. (2017). Preparation of hydrotalcite–hydroxyapatite material and its catalytic activity for transesterification of soybean oil. *Catalysis Letters* 147: 391–399.

40 Vilas-Bôas, R.N., da Silva, L.L.C., Fernandes, L.D. et al. (2020). Study of the use of hydrotalcite–hydroxyapatite as heterogeneous catalysts for application in biodiesel using by-product as raw material. *Catalysis Letters* 150: 3642–3652.

41 Ngamcharussrivichai, C., Nunthasanti, P., Tanachai, S. et al. (2010). Biodiesel production through transesterification over natural calciums. *Fuel Processing Technology* 91: 1409–1415.

42 Xie, W., Han, Y., and Tai, S. (2017). Biodiesel production using biguanide-functionalized hydroxyapatite-encapsulated-γ-Fe$_2$O$_3$ nanoparticles. *Fuel* 210: 83–90.

43 Gupta, J., Agarwal, M., and Dalai, A.K. (2018). Marble slurry derived hydroxyapatite as heterogeneous catalyst for biodiesel production from soybean oil. *Canadian Journal of Chemical Engineering* 96: 1873–1880.

44 Nisar, J., Razaq, R., Farooq, M. et al. (2017). Enhanced biodiesel production from Jatropha oil using calcined waste animal bones as catalyst. *Renewable Energy* 101: 111–119.

45 Sahu, H. and Mohanty, K. (2015). Al grafted natural hydroxyapatite for neem oil transesterification: Kinetic study at optimal point. *Chemical Engineering Journal* 280: 564–574.

46 Yan, B., Zhang, Y., Chen, G. et al. (2016). The utilization of hydroxyapatite-supported CaO—CeO$_2$ catalyst for biodiesel production. *Energy Conversion and Management* 130: 156–164.

47 Kowthaman, C.N. and Varadappan, A.M.S. (2019). Synthesis, characterization, and optimization of Schizochytrium biodiesel production using Na$^+$-doped nanohydroxyapatite. *International Journal of Energy Research* 43: 3182–3200.

48 Chen, G., Shan, R., Shi, J. et al. (2015). Biodiesel production from palm oil using active and stable K doped hydroxyapatite catalysts. *Energy Conversion and Management* 98: 463–469.

49 Muria, S.R., Azis, Y., Khairat et al. (2020). Biodiesel synthesis from palm fatty acid distillate (PFAD) by palm oil industry product using metal-hydroxyapatite catalyst. *Journal of Physics: Conference Series* 1655: 012030.

50 Corro, G., Sánchez, N., Pal, U. et al. (2016). Biodiesel production from waste frying oil using waste animal bone and solar heat. *Waste Management* 47: 105–113.

51 Chakraborty, R. and RoyChowdhury, D. (2013). Fish bone derived natural hydroxyapatite-supported copper acid catalyst: Taguchi optimization of semi-batch oleic acid esterification. *Chemical Engineering Journal* 215–216: 491–499.

52 Manzoli, M., Menegazzo, F., Signoretto, M. et al. (2016). Biomass derived chemicals: furfural oxidative esterification to methyl-2-furoate over gold catalysts. *Catalysts* 6: 107.

53 Radhakrishnan, R., Kannan, K., Kumaravel, S. et al. (2016). Oxidative esterification of furfural over Au—Pd/HAP-T and Au—Ag/HAP-T bimetallic catalysts supported on mesoporous hydroxyapatite nanorods. *RSC Advances* 6: 45907–45922.

54 Wan, Y., Zheng, C., Lei, X. et al. (2019). Oxidative esterification of acetol with methanol to methyl pyruvate over hydroxyapatite supported gold catalyst: Essential roles of acid-base properties. *Chinese Journal of Catalysis* 40: 1810–1819.

55 Gao, J., Fan, G., Yang, L. et al. (2017). Oxidative esterification of methacrolein to methyl methacrylate over gold nanoparticles on hydroxyapatite. *ChemCatChem* 9: 1230–1241.

56 Mukhopadhyay, P. and Chakraborty, R. (2020). Energy-efficient 2-ethylhexyl acetate synthesis with a nano-Sn-hydroxyapatite photocatalyst. *Chemical Engineering & Technology* 43: 531–539.

57 Ranu, B.C., Adak, L., and Chattopadhyay, K. (2008). Hydroxyapatite-supported palladium-catalyzed efficient synthesis of (*E*)-2-alkene-4-ynecarboxylic esters. Intense fluorescene emission of selected compounds. *Journal of Organic Chemistry* 73: 5609–5612.

58 Sandhyarani, M., Prabhakarn, A., Rameshbabu, N., and Subrahmanyam, C. Nanosized non-stoichiometric hydroxyapatite: synthesis, characterization and evaluation as a catalyst for esterification reaction. *Current Topics in Catalysis* 8: 81–90.

59 Jonchère, J. P. (2013). Mémento de l'Hydrogène, AFHYPAC, Fiche 3-2-1 mai 2013.

60 Lu, H., Tong, J., Cong, Y., and Yang, W. (2005). Partial oxidation of methane in $Ba_{0.5}Sr_{0.5}Co_{0.8}Fe_{0.2}O_3-\delta$ membrane reactor at high pressures. *Catalysis Today* 104: 154–159.

61 Dong, H., Shao, Z., Xiong, G. et al. (2001). Investigation on POM reaction in a new perovskite membrane reactor. *Catalysis Today* 67: 3–13.

62 Granovskii, M., Dincer, I., and Rosen, M.A. (2006). Environmental and economic aspects of hydrogen production and utilization in fuel cell vehicles. *Journal of Power Sources* 157: 411–421.

63 Prakash, P. and Sheeba, N. (2014). Hydrogen production from steam gasification of biomass: influence of process parameters on hydrogen yield – a review. *Renewable Energy* 66: 570–579.

64 Wei-Hsin, C., Mu-Rong, L., Jau-Jang, L. et al. (2010). Thermodynamic analysis of hydrogen production from methane via autothermal reforming and partial oxidation followed by water gas shift reaction. *International Journal of Hydrogen Energy* 35: 11787–11797.

65 Salai, C.A. and Andreas, H. (2013). Origin of the unique activity of Pt/TiO_2 catalysts for the water–gas shift reaction. *Journal of Catalysis* 306: 78–90.

66 Chandra, R. and Jon, P.W. (2009). Water gas shift catalysis. *Catalysis Reviews: Science and Engineering* 51: 325–440.

67 Idakiev, V., Tabakova, T., Naydenov, A. et al. (2006). Gold catalysts supported on mesoporous zirconia for low-temperature water–gas shift reaction. *Applied Catalysis B: Environmental* 63: 178–186.

68 Idakiev, V., Tabakova, T., Yuan, Z.-Y. et al. (2004). Gold catalysts supported on mesoporous titania for low-temperature water–gas shift reaction. *Applied Catalysis A: General* 270: 135–141.

69 González-Castano, M., Reina, T.R., Ivanova, S. et al. (2016). O_2-assisted water gas shift reaction over structured Au and Pt catalysts. *Applied Catalysis B: Environmental* 185: 337–343.

70 Pedro, A., Mario, M., and Eduardo, E.M. (2005). Monolithic reactors for environmental applications: a review on preparation technologies. *Chemical Engineering Journal* 109: 11–36.

71 Wheeler, C., Jhalani, A., Klein, E.J. et al. (2004). The water–gas-shift reaction at short contact times. *Journal of Catalysis* 223: 191–199.

72 Palma, V. and Pisano, D. (2018). Structured noble metal-based catalysts for the WGS process intensification. *International Journal of Hydrogen Energy* 43: 11745–11754.

73 Jaclyn Seok, K.T., Siew, P.T., William, P.A. et al. (2011). Monolithic gold catalysts: Preparation and their catalytic performances in water gas shift and CO oxidation reactions. *International Journal of Hydrogen Energy* 36: 5763–5774.

74 Quiney, A.S., Germani, G., and Schuurman, Y. (2006). Optimization of a water–gas shift reactor over a Pt/ceria/alumina monolith. *Journal of Power Sources* 160: 1163–1169.

75 Wolfgang, R., Oleg, I., and Robert, J.F. (2003). A new generation of water gas shift catalysts for fuel cell applications. *Journal of Power Sources* 118: 61–65.

76 Dengyun, M., Andreas, G., and Hengyong, X. (2016). Platinum/apatite water–gas shift catalysts. *ACS Catalysis* 6: 775–783.

77 Dengyun, M., Gulperi, C., Henning, L. et al. (2017). Water–gas shift reaction over platinum/strontium apatite catalysts. *Applied Catalysis B: Environmental* 202: 587–596.

78 Zouhair, B., González-Velasco, J.R., and Miguel, A.G. (2020). Platinum supported on lanthana-modified hydroxyapatite samples for realistic WGS conditions: on the nature of the active species, kinetic aspects and the resistance to shut-down/start-up cycles. *Applied Catalysis B: Environmental* 270: 118851.

79 Zouhair, B., Ayastuy, J.L., González-Velasco, J.R., and Gutiérrez-Ortiz, M.A. (2017). CO elimination processes over promoter-free hydroxyapatite supported palladium catalysts. *Applied Catalysis B: Environmental* 201: 189–201.

80 Lupescu, J.A., Schwank, J.W., Dahlberg, K.A. et al. (2016). Pd model catalysts: effect of aging environment and lean redispersion. *Applied Catalysis B: Environmental* 183: 343–360.

81 Zouhair, B., Ayastuy, J.L., Gutierrez-Ortiz, M.A. et al. (2018). Catalytic properties of cobalt-promoted Pd/HAP catalyst for CO-cleanup of H_2-rich stream. *International Journal of Hydrogen Energy* 43: 16949–16958.

82 Akula, V. and Mike, S.S. (2003). Hydroxyapatite as a novel support for gold and ruthenium catalysts behaviour in the water gas shift reaction. *Applied Catalysis A: General* 245: 137–147.

83 Ali, R.S.R., Maryam, B.K., Soheil, R. et al. (2012). Study of Cu—Ni/SiO_2 catalyst prepared from a novel precursor, $[Cu(H_2O)_6][Ni(dipic)_2]\cdot 2H_2O/SiO_2$, for water gas shift reaction. *Fuel Processing Technology* 96: 9–15.

84 Jiann-Horng, L. and Vadim, V.G. (2012). Alumina-supported Cu@Ni and Ni@Cu core–shell nanoparticles: synthesis, characterization, and catalytic activity in water–gas-shift reaction. *Applied Catalysis A: General* 445: 187–194.

85 Dae-Woon, J., Won-Jun, J., Jae-Oh, S. et al. (2014). Low-temperature water-gas shift reaction over supported Cu catalysts. *Renewable Energy* 65: 102–107.

86 Chongqi, C., Chunxiao, R., Yingying, Z. et al. (2014). The significant role of oxygen vacancy in Cu/ZrO$_2$ catalyst for enhancing water gas-shift performance. *International Journal of Hydrogen Energy* 39: 317–324.

87 Zouhair, B., José, L., Ayastuy, J.R. et al. (2018). Water–gas shift reaction over a novel Cu-ZnO/HAP formulation: enhanced catalytic performance in mobile fuel cell applications. *Applied Catalysis A, General* 566: 1–14.

88 Kunimasa, S., Imazu, N., and Hidenori, Y. (2013). Study on factors controlling catalytic activity for low-temperature water–gas-shift reaction on Cu-based catalysts. *Catalysis Today* 201: 145–150.

89 Unai, I.-V., Jose, L.A., Zouhair, B. et al. (2018). Transition metals supported on bone-derived hydroxyapatite as potential catalysts for the water–gas shift reaction. *Renewable Energy* 115: 641–648.

90 Agus, H., Sandun, D., Fernando, S.D. et al. (2009). Hydrogen production through the water–gas shift reaction: thermodynamic equilibrium versus experimental results over supported Ni catalysts. *Energy and Fuels* 23: 3097–3102.

91 Joan, M., Ogden, T.G.K., and Margaret, M.S. (2000). Fuels for fuel cell vehicles. *Fuel Cells Bulletin* 3: 5–13.

92 Aceves, S.M., Berry, G.D., and Rambach, G.D. (1998). Insulated pressure vessels for hydrogen storage on vehicles. *International Journal of Hydrogen Energy* 23: 583–591.

93 Aceves, S.M., Martinez-Frias, J., and Garcia-Villazana, O. (2000). Analytical and experimental evaluation of insulated pressure vessels for cryogenic hydrogen storage. *International Journal of Hydrogen Energy* 25: 1075–1085.

94 Dillon, A.C., Jones, K.M., Bekkedahl, T.A. et al. (1997). Storage of hydrogen in single-walled carbon nanotubes. *Nature* 386: 377–379.

95 Pehr, K. (1996). Aspects of safety and acceptance of LH2 tank systems in passenger cars. *International Journal of Hydrogen Energy* 21: 387–395.

96 Chahine, R. and Bose, T.K. (1994). Low-pressure adsorption storage of hydrogen. *International Journal of Hydrogen Energy* 19: 161–164.

97 Scott, H., Ware, F., and Jeffrey, B. (1997). Hydrogen storage by carbon sorption. *International Journal of Hydrogen Energy* 22: 601–610.

98 Liu, C., Fan, Y.Y., Liu, M. et al. (1999). Hydrogen storage in single-walled carbon nanotubes at room temperature. *Science* 286: 1127–1129.

99 Ye, Y., Ahn, C.C., Witham, C. et al. (1999). Hydrogen adsorption and cohesive energy of single-walled carbon nanotubes. *Applied Physics Letters* 74: 2307–2309.

100 Yusuke, Y., Kentaro, Y., Qiang, X. et al. (2010). Cu/Co$_3$O$_4$ Nanoparticles as catalysts for hydrogen evolution from ammonia borane by hydrolysis. *Journal of Physical Chemistry C* 39: 16456–16462.

101 Liu, B.H. and Li, Z.P. (2009). A review: hydrogen generation from borohydride hydrolysis reaction. *Journal of Power Sources* 187: 527–534.

102 Justyn, W.J., Sunghwa, C., Yeoung, K. et al. (2013). Hydroxyapatite supported cobalt catalysts for hydrogen generation. *Journal of Colloid and Interface Science* 394: 401–408.

103 Murat, R. and Özkar, S. (2012). Hydroxyapatite-supported cobalt(0) nanoclusters as efficient and cost-effective catalyst for hydrogen generation from the hydrolysis of both sodium borohydride and ammonia–borane. *Catalysis Today* 183: 17–25.

104 Nakazato, T., Murata, Y., and Kai, T. (2016). Hydroxyapatite prepared from biomineral calcium carbonate resources: a Ru-catalyst support for hydrogen generation. *Journal of Ecotechnology Research* 18: 11–16.

105 Bo, P. and Jun, C. (2008). Ammonia borane as an efficient and lightweight hydrogen storage medium. *Energy & Environmental Science* 1: 479–483.

106 Murat, R. and Saim, O. (2011). Hydroxyapatite-supported palladium(0) nanoclusters as effective and reusable catalyst for hydrogen generation from the hydrolysis of ammonia-borane. *International Journal of Hydrogen Energy* 36: 7019–7027.

107 Manish, C. and Qiang, X. (2007). Room temperature hydrogen generation from aqueous ammonia-borane using noble metal nano-clusters as highly active catalysts. *Journal of Power Sources* 168: 135–142.

108 Halil, D., Mehmet, G., Mehmet, Z. et al. (2014). Hydroxyapatite-nanosphere supported ruthenium(0) nanoparticle catalyst for hydrogen generation from ammonia-borane solution: kinetic studies for nanoparticle formation and hydrogen evolution. *RSC Advances* 4: 28947–28955.

109 Hasan, C. and Önder, M. (2012). A facile synthesis of nearly monodisperse ruthenium nanoparticles and their catalysis in the hydrolytic dehydrogenation of ammonia borane for chemical hydrogen storage. *Applied Catalysis B: Environmental* 125: 304–310.

110 Serdar, A., Pelin, E., and Saim, Ö. (2013). Hydroxyapatite supported ruthenium(0) nanoparticles catalyst in hydrolytic dehydrogenation of ammonia borane: Insight to the nanoparticles formation and hydrogen evolution kinetics. *Applied Catalysis B: Environmental* 142: 187–195.

111 Ribeiro, N.M., Pinto, A.C., Quintella, C.M. et al. (2007). The role of additives for diesel and diesel blended (ethanol or biodiesel) fuels: a review. *Energy & Fuels* 21: 2433–2445.

112 Chandran, D. (2020). Compatibility of diesel engine materials with biodiesel fuel. *Renewable Energy* 147: 89–99.

113 Scheepers, P.T.J. and Bos, R.P. (1992). Combustion of diesel fuel from a toxicological perspective. *International Archives of Occupational and Environmental Health* 64: 149–161.

114 Reşitoğlu, İ.A., Altinişik, K., and Keskin, A. (2015). The pollutant emissions from diesel-engine vehicles and exhaust after treatment systems. *Clean Technologies and Environmental Policy* 17: 15–27.

115 İlkiliç, C. and Aydin, H. (2012). The harmful effects of diesel engine exhaust emissions. *Energy Sources, Part A: Recovery, Utilization, and Environmental Effects* 34: 899–905.

116 Bidita, B.S., Suraya, A.R., Shazed, M.A. et al. (2014). Influence of fuel additive in the formulation and combustion characteristics of water-in-diesel nanoemulsion fuel. *Energy & Fuels* 28: 4149–4161.

117 De Caro, P.S., Mouloungui, Z., Vaitilingom, G. et al. (2001). Interest of combining an additive with diesel–ethanol blends for use in diesel engines. *Fuel* 80: 565–574.

118 Ndiaye, M., Arhaliass, A., Legrand, J. et al. (2020). Reuse of waste animal fat in biodiesel: biorefining heavily-degraded contaminant-rich waste animal fat and formulation as diesel fuel additive. *Renewable Energy* 145: 1073–1079.

119 Imdadul, H.K., Masjuki, H.H., Kalam, M.A. et al. (2015). A comprehensive review on the assessment of fuel additive effects on combustion behavior in CI engine fuelled with diesel biodiesel blends. *RSC Advances* 5: 67541–67567.

120 Kumar, C., Rana, K.B., Tripathi, B. et al. (2018). Properties and effects of organic additives on performance and emission characteristics of diesel engine: a comprehensive review. *Environmental Science and Pollution Research* 25: 22475–22498.

121 Jafari, H., Idris, M.H., Ourdjini, A. et al. (2010). Effect of ethanol as gasoline additive on vehicle fuel delivery system corrosion. *Materials and Corrosion* 61: 432–440.

122 Wang, C., Li, Y., Xu, C. et al. (2019). Methanol as an octane booster for gasoline fuels. *Fuel* 248: 76–84.

123 Li, B., Sun, Q., Li, A. et al. (2020). Effects of propanol isomers enrichment on in-cylinder thermochemical fuel reforming (TFR) in spark ignition natural gas engine. *International Journal of Hydrogen Energy* 45: 10932–10950.

124 Atmanlı, A., Ileri, E., and Yüksel, B. (2015). Effects of higher ratios of n-butanol addition to diesel–vegetable oil blends on performance and exhaust emissions of a diesel engine. *Journal of the Energy Institute* 88: 209–220.

125 Zaharin, M.S.M., Abdullah, N.R., Masjuki, H.H. et al. (2018). Evaluation on physicochemical properties of iso-butanol additives in ethanol-gasoline blend on performance and emission characteristics of a spark-ignition engine. *Applied Thermal Engineering* 144: 960–971.

126 da Silva Trindade, W.R. and dos Santos, R.G. (2017). Review on the characteristics of butanol, its production and use as fuel in internal combustion engines. *Renewable and Sustainable Energy Reviews* 69: 642–651.

127 Anastopoulos, G., Lois, E., Zannikos, F. et al. (2001). Influence of aceto acetic esters and di-carboxylic acid esters on diesel fuel lubricity. *Tribology International* 34: 749–755.

128 Rezende, M.J.C., Perruso, C.R., de Azevedo, D.A. et al. (2005). Characterization of lubricity improver additive in diesel by gas chromatography–mass spectrometry. *Journal of Chromatography A* 1063: 211–215.

129 Bertoli, C., Del Giacomo, N., and Beatrice, C. (1997). Diesel combustion improvements by the use of oxygenated synthetic fuels. *SAE Transactions* 106: 1557–1567.

130 Kocis, D., Song, K., Lee, H. et al. (2000). Effects of dimethoxymethane and dimethylcarbonate on soot production in an optically-accessible DI diesel engine. *SAE Transactions* 2299–2308.

131 Jiang, T., Liu, C.-J., Rao, M.-F. et al. (2001). A novel synthesis of diesel fuel additives from dimethyl ether using dielectric barrier discharges. *Fuel Processing Technology* 73: 143–152.

132 Smith, B.L., Ott, L.S., and Bruno, T.J. (2008). Composition-explicit distillation curves of diesel fuel with glycol ether and glycol ester oxygenates: fuel analysis metrology to enable decreased particulate emissions. *Environmental Science and Technology* 42: 7682–7689.

133 McCormick, R.L., Ross, J.D., and Graboski, M.S. (1997). Effect of several oxygenates on regulated emissions from heavy-duty diesel engines. *Environmental Science and Technology* 31: 1144–1150.

134 Arteconi, A., Mazzarini, A., and Di Nicola, G. (2011). Emissions from ethers and organic carbonate fuel additives: a review. *Water, Air, and Soil Pollution* 221: 405.

135 González, D.M.A., Piel, W., Asmus, T. et al. (2001). Oxygenates screening for advancedpetroleum-based diesel fuels: part 2. The effect of oxygenate blending compounds on exhaust emissions. *SAE Transactions* 2246–2255.

136 Szőri, M., Giri, B.R., Wang, Z. et al. (2018). Glycerol carbonate as a fuel additive for a sustainable future. *Sustainable Energy and Fuels* 2: 2171–2178.

137 Vellaiyan, S. and Amirthagadeswaran, K.S. (2016). The role of water-in-diesel emulsion and its additives on diesel engine performance and emission levels: a retrospective review. *Alexandria Engineering Journal* 55: 2463–2472.

138 Liu, T. and Feng, L. (2012). Effect of additives on the stability of emulsified diesel collector. *Research and Exploration in Laboratory* 12: 19–23.

139 Layton, D.W., and Marchetti, A.A., (2002). Comparative environmental performance of two diesel-fuel oxygenates: dibutyl maleate (DBM) and tripropylene glycol monomethyl ether (TGME). United States: N. p., 2001. Web. doi:https://doi.org/10.4271/2002-01-1943.

140 Izquierdo, J.F., Montiel, M., Palés, I. et al. (2012). Fuel additives from glycerol etherification with light olefins: state of the art. *Renewable and Sustainable Energy Reviews* 16: 6717–6724.

141 Veluturla, S., Narula, A., and Shetty, S.P. (2017). Kinetic study of synthesis of bio-fuel additives from glycerol using a hetropolyacid. *Resource-Efficient Technologies* 3: 337–341.

142 Trifoi, A.R., Agachi, P.Ş., and Pap, T. (2016). Glycerol acetals and ketals as possible diesel additives. A review of their synthesis protocols. *Renewable and Sustainable Energy Reviews* 62: 804–814.

143 Cao, Q., Liang, W., Guan, J. et al. (2014). Catalytic synthesis of 2, 5-bis-methoxymethylfuran: A promising cetane number improver for diesel. *Applied Catalysis A: General* 481: 49–53.

144 Baranowski, C.J., Bahmanpour, A.M., and Kröcher, O. (2017). Catalytic synthesis of polyoxymethylene dimethyl ethers (OME): a review. *Applied Catalysis B: Environmental* 217: 407–420.

145 Mallesham, B., Sudarsanam, P., and Reddy, B.M. (2014). Eco-friendly synthesis of bio-additive fuels from renewable glycerol using nanocrystalline SnO_2-based solid acids. *Catalysis Science and Technology* 4: 803–813.

146 Ndou, A.S., Plint, N., and Coville, N.J. (2003). Dimerisation of ethanol to butanol over solid-base catalysts. *Applied Catalysis A: General* 251: 337–345.

147 Nandiwale, K.Y., Patil, S.E., and Bokade, V.V. (2014). Glycerol etherification using *n*-butanol to produce oxygenated additives for biodiesel fuel over H-beta zeolite catalysts. *Energy Technology* 2: 446–452.

148 Tsuchida, T., Sakuma, S., Takeguchi, T. et al. (2006). Direct synthesis of *n*-butanol from ethanol over nonstoichiometric hydroxyapatite. *Industrial and Engineering Chemistry Research* 45: 8634–8642.

149 Larina, O.V., Valihura, K.V., Kyriienko, P.I. et al. (2019). Successive vapour phase Guerbet condensation of ethanol and 1-butanol over Mg-Al oxide catalysts in a flow reactor. *Applied Catalysis A: General* 588: 117265.

150 Angelici, C., Weckhuysen, B.M., and Bruijnincx, P.C.A. (2013). Chemocatalytic conversion of ethanol into butadiene and other bulk chemicals. *ChemSusChem* 6: 1595–1614.

151 Sun, J., Zhu, K., Gao, F. et al. (2011). Direct conversion of bio-ethanol to isobutene on nanosized $Zn_xZr_yO_z$ mixed oxides with balanced acid–base sites. *Journal of the American Chemical Society* 133: 11096–11099.

152 Silvester, L., Lamonier, J., Lamonier, C. et al. (2017). Guerbet reaction over strontium-substituted hydroxyapatite catalysts prepared at various (Ca + Sr)/P ratios. *ChemCatChem* 9: 2250–2261.

153 Kulkarni, N.V., Brennessel, W.W., and Jones, W.D. (2018). Catalytic upgrading of ethanol to *n*-butanol via manganese-mediated Guerbet reaction. *ACS Catalysis* 8: 997–1002.

154 Sun, Z., Couto Vasconcelos, A., Bottari, G. et al. (2017). Efficient catalytic conversion of ethanol to 1-butanol via the Guerbet reaction over copper-and nickel-doped porous. *ACS Sustainable Chemistry & Engineering* 5: 1738–1746.

155 Hernández, W.Y., De Vlieger, K., Van Der Voort, P. et al. (2016). Ni—Cu hydrotalcite-derived mixed oxides as highly selective and stable catalysts for the synthesis of β-branched bioalcohols by the guerbet reaction. *ChemSusChem* 9: 3196–3205.

156 Hanspal, S., Young, Z.D., Prillaman, J.T. et al. (2017). Influence of surface acid and base sites on the Guerbet coupling of ethanol to butanol over metal phosphate catalysts. *Journal of Catalysis* 352: 182–190.

157 Han, X., An, H., Zhao, X. et al. (2020). Influence of acid-base properties on the catalytic performance of Ni/hydroxyapatite in *n*-butanol Guerbet condensation. *Catalysis Communications* 146: 106130.

158 Ho, C.R., Shylesh, S., and Bell, A.T. (2016). Mechanism and kinetics of ethanol coupling to butanol over hydroxyapatite. *ACS Catalysis* 6: 939–948.

159 Boukha, Z., Choya, A., Cortés-Reyes, M. et al. (2020). Influence of the calcination temperature on the activity of hydroxyapatite-supported palladium catalyst in the methane oxidation reaction. *Applied Catalysis B: Environmental* 277: 119280.

160 Young, Z.D., Hanspal, S., and Davis, R.J. (2016). Aldol condensation of acetaldehyde over titania, hydroxyapatite, and magnesia. *ACS Catalysis* 6: 3193–3202.

161 Boukha, Z., Kacimi, M., Ziyad, M. et al. (2007). Comparative study of catalytic activity of Pd loaded hydroxyapatite and fluorapatite in butan-2-ol conversion and methane oxidation. *Journal of Molecular Catalysis A: Chemical* 270: 205–213.

162 Khachani, M., Kacimi, M., Ensuque, A. et al. (2010). Iron–calcium–hydroxyapatite catalysts: Iron speciation and comparative performances in butan-2-ol conversion and propane oxidative dehydrogenation. *Applied Catalysis A: General* 388: 113–123.

163 Ogo, S., Onda, A., Iwasa, Y. et al. (2012). 1-Butanol synthesis from ethanol over strontium phosphate hydroxyapatite catalysts with various Sr/P ratios. *Journal of Catalysis* 296: 24–30.

164 Amiri, H., Karimi, K., and Roodpeyma, S. (2010). Production of furans from rice straw by single-phase and biphasic systems. *Carbohydrate Research* 345: 2133–2138.

165 Anthonia, E.E. and Philip, H.S. (2015). An overview of the applications of furfural and its derivatives. *International Journal of Advanced Chemistry* 3: 42–47.

166 Lewkowski, J. (2001). Synthesis, chemistry and applications of 5-hydroxymethyl-furfural and its derivatives. *Archive for Organic Chemistry* 2001: 17–54.

167 Lange, J., Van Der Heide, E., van Buijtenen, J. et al. (2012). Furfural – a promising platform for lignocellulosic biofuels. *ChemSusChem* 5: 150–166.

168 Zhang, L., He, Y., Zhu, Y. et al. (2018). Camellia oleifera shell as an alternative feedstock for furfural production using a high surface acidity solid acid catalyst. *Bioresource Technology* 249: 536–541.

169 Li, G., Li, N., Wang, Z. et al. (2012). Synthesis of high-quality diesel with furfural and 2-methylfuran from hemicellulose. *ChemSusChem* 5: 1958–1966.

170 Xu, J., Li, N., Yang, X. et al. (2017). Synthesis of diesel and jet fuel range alkanes with furfural and angelica lactone. *ACS Catalysis* 7: 5880–5886.

171 Taarning, E., Nielsen, I.S., Egeblad, K. et al. (2008). Chemicals from renewables: aerobic oxidation of furfural and hydroxymethylfurfural over gold catalysts. *ChemSusChem* 1: 75–78.

172 Yan, K., Wu, G., Lafleur, T. et al. (2014). Production, properties and catalytic hydrogenation of furfural to fuel additives and value-added chemicals. *Renewable and Sustainable Energy Reviews* 38: 663–676.

173 Elias, E., Costa, R., Marques, F. et al. (2015). Oil-spill cleanup: the influence of acetylated curaua fibers on the oil-removal capability of magnetic composites. *Journal of Applied Polymer Science* 132: 41732.

174 Meszaros, L. (1960). Conversion of furfural into furan by vapor phase oxidative decarboxylation on metal oxide catalysts. *Acta Universitatis Szegediensis Acta Physico-Chimica* 6: 97–98.

175 Kikhtyanin, O., Ganjkhanlou, Y., Kubička, D. et al. (2018). Characterization of potassium-modified FAU zeolites and their performance in aldol condensation of furfural and acetone. *Applied Catalysis A: General* 549: 8–18.

176 Bendeddouche, W., Bedrane, S., Zitouni, A. et al. (2021). Highly efficient catalytic oNE-POT biofuel production from lignocellulosic biomass derivatives. *International Journal of Energy Research* 45: 2148–2159.

177 Mishra, D.K., Cho, J.K., Yi, Y. et al. (2019). Hydroxyapatite supported gold nanocatalyst for base-free oxidative esterification of 5-hydroxymethyl-2-furfural to 2, 5-furan dimethylcarboxylate with air as oxidant. *Journal of Industrial and Engineering Chemistry* 70: 338–345.

178 Tong, X., Liu, Z., Yu, L. et al. (2015). A tunable process: catalytic transformation of renewable furfural with aliphatic alcohols in the presence of molecular oxygen. *Chemical Communications* 51: 3674–3677.

179 Wang, F. and Zhang, Z. (2017). Catalytic transfer hydrogenation of furfural into furfuryl alcohol over magnetic γ-Fe_2O_3@ HAP catalyst. *ACS Sustainable Chemistry and Engineering* 5: 942–947.

180 Zhang, Z., Yuan, Z., Tang, D. et al. (2014). Iron oxide encapsulated by ruthenium hydroxyapatite as heterogeneous catalyst for the synthesis of 2,5-diformylfuran. *ChemSusChem* 7: 3496–3504.

181 Pacheco, M.A. and Marshall, C.L. (1997). Review of dimethyl carbonate (DMC) manufacture and its characteristics as a fuel additive. *Energy & Fuels* 11: 2–29.

182 Rounce, P., Tsolakis, A., Leung, P. et al. (2010). A comparison of diesel and biodiesel emissions using dimethyl carbonate as an oxygenated additive. *Energy & Fuels* 24: 4812–4819.

183 Schifter, I., González, U., Díaz, L. et al. (2017). Comparison of performance and emissions for gasoline-oxygenated blends up to 20 percent oxygen and implications for combustion on a spark-ignited engine. *Fuel* 208: 673–681.

184 Huang, S., Yan, B., Wang, S. et al. (2015). Recent advances in dialkyl carbonates synthesis and applications. *Chemical Society Reviews* 44: 3079–3116.

185 Mehta, B.H., Mandalia, H.V., and Mistry, A.B. (2011). A review on effect of oxygenated fuel additive on the performance and emission characteristics of diesel engine. *National Conference on Recent Trends in Engineering & Technology*, Gujarat, India (13–14 May 2011).

186 Gopinath, D. and Sundaram, E.G. (2012). Experimental investigation on the effect of adding di methyl carbonate to gasoline in a SI engine performance. *International Journal of Scientific & Engineering Research* 3: 1–5.

187 Shukla, K. and Srivastava, V.C. (2016). Diethyl carbonate: critical review of synthesis routes, catalysts used and engineering aspects. *RSC Advances* 6: 32624–32645.

188 Shukla, K. and Srivastava, V.C. (2020). Efficient synthesis of diethyl carbonate by Mg, Zn promoted hydroxyapatite via transesterification reaction. *International Journal of Chemical Reactor Engineering* 18: 2019–0205.

189 Bai, R., Wang, S., Mei, F. et al. (2011). Synthesis of glycerol carbonate from glycerol and dimethyl carbonate catalyzed by KF modified hydroxyapatite. *Journal of Industrial and Engineering Chemistry* 17: 777–781.

190 Du, Z., Liu, L., Yuan, H. et al. (2010). Synthesis of propylene carbonate from alcoholysis of urea catalyzed by modified hydroxyapatites. *Chinese Journal of Catalysis* 31: 371–373.

191 Tsuchida, T., Yoshioka, T., Sakuma, S. et al. (2008). Synthesis of biogasoline from ethanol over hydroxyapatite catalyst. *Industrial & Engineering Chemistry Research* 47: 1443–1452.

192 Lovón-Quintana, J.J., Rodriguez-Guerrero, J.K., and Valença, P.G. (2017). Carbonate hydroxyapatite as a catalyst for ethanol conversion to hydrocarbon fuels. *Applied Catalysis A: General* 542: 136–145.

193 Wang, Q.-N., Weng, X.-F., Zhou, B.-C. et al. (2019). Direct, selective production of aromatic alcohols from ethanol using a tailored bifunctional cobalt–hydroxyapatite catalyst. *ACS Catalysis* 9: 7204–7216.

194 Rahmanian, A. and Ghaziaskar, H.S. (2013). Continuous dehydration of ethanol to diethyl ether over aluminum phosphate–hydroxyapatite catalyst under sub and supercritical condition. *The Journal of Supercritical Fluids* 78: 34–41.

195 Lu, Z., Gao, D., Yin, H. et al. (2015). Methanol dehydrogenation to methyl formate catalyzed by SiO_2^-, hydroxyapatite-, and MgO-supported copper catalysts and reaction kinetics. *Journal of Industrial and Engineering Chemistry* 31: 301–308.

196 Wen, C., Cui, Y., Chen, X. et al. (2015). Reaction temperature controlled selective hydrogenation of dimethyl oxalate to methyl glycolate and ethylene glycol over copper-hydroxyapatite catalysts. *Applied Catalysis B: Environmental* 162: 483–493.

10

Hydroxyapatite-Based Catalysts in Organic Synthesis

Michel Gruselle[1], Kaia Tõnsuaadu[2], Patrick Gredin[3,4], and Christophe Len[5,6]

[1]*CNRS, UMR 8232, Parisian Institute for Molecular Chemistry, Sorbonne University, 4 place Jussieu, F75005, Paris, France*
[2]*Laboratory of Inorganic materials, Institute of Materials and Environmental Technology, Tallinn University of Technology, Ehitajate tee 5, 19086 Tallinn, Estonia*
[3]*Institut de Recherche de Chimie Paris, ChimieParisTech, PSL, Research University, CNRS, 11 rue Pierre et Marie Curie, F-75005 Paris, France*
[4]*Faculté des Sciences et Ingénierie, Sorbonne Université, 4 place Jussieu, F-75005 Paris, France*
[5]*Centre de recherche Royallieu, Sorbonne université, Université de Technologie de Compiègne, CS 60 319, F-60203 Compiègne, Cedex, France*
[6]*Institute of Chemistry for Life and Health Sciences, ChimieParisTech, PSL Research University, CNRS, 11 rue Pierre et Marie Curie, F-75005 Paris, France*

10.1 Introduction

From the laboratory level to the industrial, many processes in organic synthesis involve heterogeneous catalysis, for both fine chemicals and drug molecules [1–3]. The major interest of catalysis is to make organic reactions easier and more selective. In addition, environmental requirements lead to the search for energy-saving and minimum waste-generating processes [4, 5].

Catalysts must have at least four qualities: high activity, selectivity, stability, and reusability and be as inexpensive as possible. The control of the surface composition, structure and properties, the specific surface area, and porosity as well as the thermal and mechanical stability is of highest importance in the design of heterogeneous catalysts [6].

Apatites, natural or synthetic, inherently have such properties and can exhibit a large spectrum of composition, due to their ability to form solid solutions, which can accept a large number of cationic and anionic substitutions in their structures [7–9]. In addition, because of their surface properties, they can play the role of macro-ligand or solid support for various catalytic moieties anchored to the surface. The use of apatites as catalysts in organic synthesis is widely described in the literature and is the subject of several reviews [10–15].

The main feature of apatitic surfaces is related to their acid–base properties, which are essential in organic synthesis catalysis. Modulation of the synthesis parameters of apatites, as well as their ability toward substitution, allows their versatility in terms of composition, stoichiometry, and morphology to be used to highlight the

Design and Applications of Hydroxyapatite-Based Catalysts, First Edition. Edited by Doan Pham Minh.
© 2022 WILEY-VCH GmbH. Published 2022 by WILEY-VCH GmbH.

relationships that exist between structure and acid–base properties. Indeed, at a macroscopic level, stoichiometry, represented by the Ca/P ratio, and morphology are the two key parameters controlling the base properties of those materials. For example, lowering the Ca/P ratio leads to a decrease in the basicity and increases the acidic properties of the surface. With a low nonstoichiometric Ca/P ratio of 1.50, hydroxyapatite (HA) acts as an acid catalyst with the existence of basic sites. In contrast, with a stoichiometric ratio of 1.67, HA acts as a basic catalyst, while acid sites are still present [16]. Therefore, when dealing with the catalytic properties of a material, it is necessary to investigate the surface composition, taking into consideration the fact that the composition of the bulk and the surface are not the same. With regard to apatites, this problem has been reported by several authors [17–21].

The surface hydrated layer is also a key factor influencing the interaction with molecules present in the fluids. In fact, since the reaction takes place at the surface, the nature of the interaction between the catalyst and the reactants is of considerable importance for the course of the reaction [22]. This has been detailed in Chapter 3 of this book.

10.2 Synthesis and Characterization of HA and HA-Based Catalysts

10.2.1 Synthesis

10.2.1.1 Stoichiometric and Nonstoichiometric Apatites

The synthesis, physical properties, and structure of apatites, mainly calcium HA, are extensively described in the literature [23–25] and in Chapter 2 of this book. There is a large variety of synthetic methods depending on the purpose of HA use, which can be classified as wet precipitation and solid-state or thermal methods. The apatites obtained in solid-state reactions (sol–gel synthesis, mechanical activation) are stable at higher temperatures [26]. Most of the apatites used as catalysts are prepared by wet methods, which allow the modification of the Ca/P ratio of HA and, therefore, the determination of the surface acid/base properties [16], which have significant importance in catalytic processes.

10.2.1.2 Apatites as Catalyst Supports

To introduce other cations, having catalytic properties, into the HA structure, different methods can be used. The introduction of divalent cations such as Cu^{2+}, Zn^{2+}, Co^{2+}, Ni^{2+}, Mg^{2+}, Pb^{2+}, and Pd^{2+} into the CaHA structure is achieved by coprecipitation [27, 28] or by ion exchange in a corresponding solution [29, 30]. In the case of coprecipitation, the content of the substitute depends on the ability of the cation to take place into the HA structure; for that, the ionic radius of the incoming cation, compared with the Ca^{2+} radius, is of prime importance [31]. In ion exchange processes, the apatite surface properties and the sorption conditions are also important [8, 32, 33]. The impregnation of metal salts or complexes on HA surfaces is widely used to obtain catalytic materials in which HA serves as much as a support as an auxiliary in the reaction process. In general, the technique used consists of a reaction where the apatite is suspended in water or an organic solvent, while the part to be impregnated is in solution. Such a procedure is, for example, widely used for palladium starting from $PdCl_2(PhCN)_2$ in acetone [34, 35] or $Pd(Oac)_2$ in ethanol [36]. The concentration of the metal ions in such modified HA varies greatly (e.g. Pd-0.002–0.007 mol% [35, 37], Cu 0.1 mol% [38]). Unfortunately, often the content of metal ions introduced into the apatite structure or bound at the surface of HA is not given or detected, which makes *comparison/characterization* of the catalysts complicated. In addition, one must be very attentive to catalyst modifications during the reaction, insofar as it can release the supported entity. In certain cases, it is possible to transform the catalytic moiety supported on the apatite surface, for example, to obtain nanoparticles of Pd_0 [37]. Impregnation of HA in aqueous solutions of rare earth metals (La, Ru, Sc, Y) often leads to its partial dissolution due to the low pH of the solution and the formation of a new phase [39, 40].

A higher content of the active metals can be achieved through the formation of other phases (hydroxide or phosphate) during the loading process, and polyphase catalysts are obtained [39–41].

10.2.1.3 HA as Macro-ligands for Catalytic Moieties

The concept of HA as "macro-ligands for organometallic moieties" has been developed by several authors [42, 43]. This concept is based on the ability of either phosphate groups located on the surface to bind a transition metal ion or for surface calcium ions to bind the anionic part of an organic ligand complexing a transition metal ion.

10.3 Apatites as Catalysts in C–C Bond Formation

Carbon–carbon bond formation is one of the most important reaction in organic synthesis, allowing complex molecules to be built [44–48]. It covers a broad spectrum of reaction types: cross-coupling reactions, nucleophilic carbon–carbon bond forming reactions, and multicomponent reactions (MCRs). For the sake of clarity, this chapter focuses on the main carbon–carbon bond formation reactions mainly reported from 2015 to 2020.

10.3.1 Cross-coupling Reactions

A carbon–carbon cross-coupling reaction is achieved when two identical (homo-coupling) or nonidentical (hetero-coupling) starting materials endowed with an activating group are connected together, using a transition metal as catalyst. Among them, palladium as noble metal is the most widely used atom, and it catalyzes cross-coupling reactions such as Suzuki–Miyaura, Mizoroki–Heck, Sonogashira, Stille, Hiyama, Negishi, Kumada, Murahashi, and Buchwald–Hartwig using either homogeneous or heterogeneous catalysts.

The Suzuki–Miyaura cross-coupling reaction is one of the most versatile and frequently utilized methods for C—C bond formation and involves coupling an organoboron compound with aryl, alkenyl, or alkynyl halides. Various works have reported Suzuki–Miyaura cross-coupling reaction using palladium-modified HA [34, 35, 49–51]. In 2018, an efficient Suzuki–Miyaura cross-coupling reaction was reported [37] using a new nanocatalyst Pd^{II}NPs@nano-HA containing palladium nanoparticles (0.75 wt% Pd) supported on HA nanospheres (50 nm). Obviously, the precatalyst Pd^{II}NPs@nano-HA was transformed to the catalyst Pd^{0}NPs@nano-HA in situ in the presence of the boronic acid. Starting from p-bromoacetophenone (1.0 mmol), phenylboronic acid (1.5 mmol), and K_2CO_3 (2.0 mmol) in the presence of the supported catalyst (Pd 0.1%) in N,N-dimethylformamide (DMF)-water (1 : 4, v/v) mixture at 100 °C under air atmosphere for 15 seconds, the target biphenyl derivatives were obtained with an excellent turnover frequency (TOF: 3×10^5 h^{-1}) (Scheme 10.1). Application of this methodology was realized with a range of various aryl bromides bearing either electron-donating or electron-withdrawing substituents (Figure 10.1). It is noteworthy that the use of water as sole solvent gave lower TOF and longer reaction times, and the activation of the aryl chloride was more difficult than usual. The host HA framework did not change during the reaction, indicating that the palladium nanoparticles are located on the apatitic surface and the catalyst can be reused at least ten times without significant loss in activity.

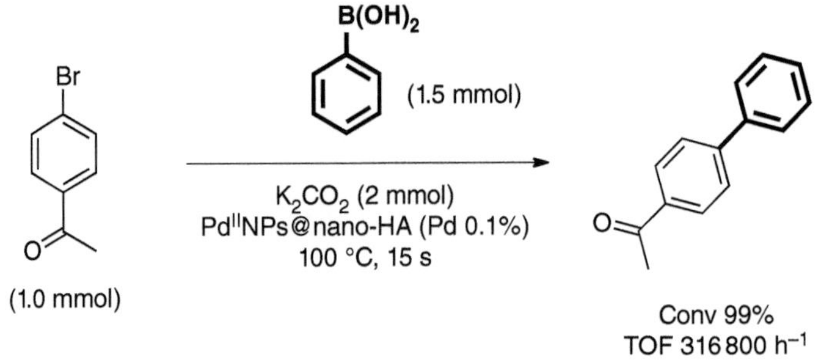

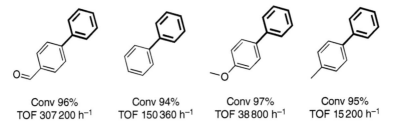

Scheme 10.1 Suzuki–Miyaura cross-coupling sequence using Pd^{II}NPs@nano-HA as supported precatalyst [37]. Source: Michel Gruselle.

The Mizoroki–Heck cross-coupling reaction is one of the most efficient methods for the vinylation of aryl/vinyl halides or triflates to form a C—C bond. Only a

Figure 10.1 Substrate scope of Suzuki–Miyaura cross-coupling sequence using Pd^{II}NPs@nano-HA as supported precatalyst [37]. Source: Michel Gruselle.

few works have been reported using palladium-modified HA [52, 53]. In 2019, a mesoporous ionic-modified γ-Fe_2O_3@HA-DABCO-Pd containing palladium nanoparticles were prepared and used as a sustainable nanocatalyst for the Mizoroki–Heck cross-coupling reaction in aqueous media for the production of intermediates of pharmaceuticals and fine chemicals [36]. According to the authors, the appropriate surface area and pore size of the starting mesoporous nanocomposite γ-Fe_2O_3@HA-DABCO provide a good template for immobilization of Pd nanoparticles (DABCO means 1,4-diazabicyclo[2.2.2]octane). Starting from iodobenzene (1 mmol), styrene (1 mmol) and K_2CO_3 (2.5 mmol) in the presence of γ-Fe_2O_3@HA-DABCO-Pd (0.05 g) in refluxing water for 45 minutes furnished the stilbene in 98% yield (Scheme 10.2). The reaction of different aryl iodides and bromides with different vinyl derivatives gave excellent yields even if the aryl bromides are less reactive (Figure 10.2). Using the concept of magnetic nanocatalyst, γ-Fe_2O_3@HA-DABCO can easily be recovered from the reaction mixture through magnetic trapping and reused with no appreciable loss of activity after six runs.

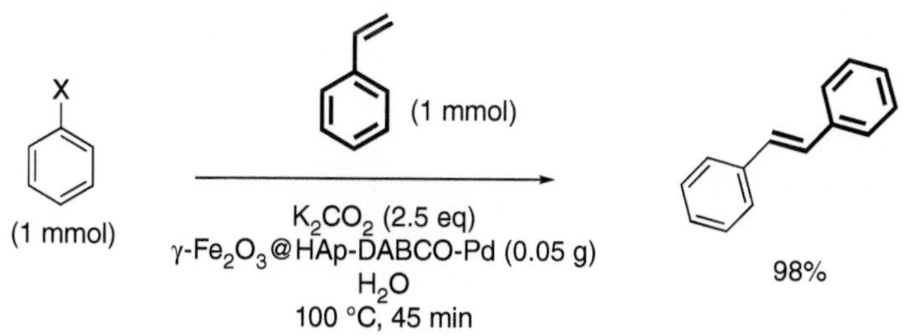

Scheme 10.2 Mizoroki–Heck cross-coupling sequence using γ-Fe_2O_3@HA-DABCO-Pd as supported catalyst [36]. Source: Michel Gruselle.

The Sonogashira cross-coupling reaction permits the coupling of a terminal alkyne with aryl or vinyl halides in the presence of a palladium catalyst, a copper

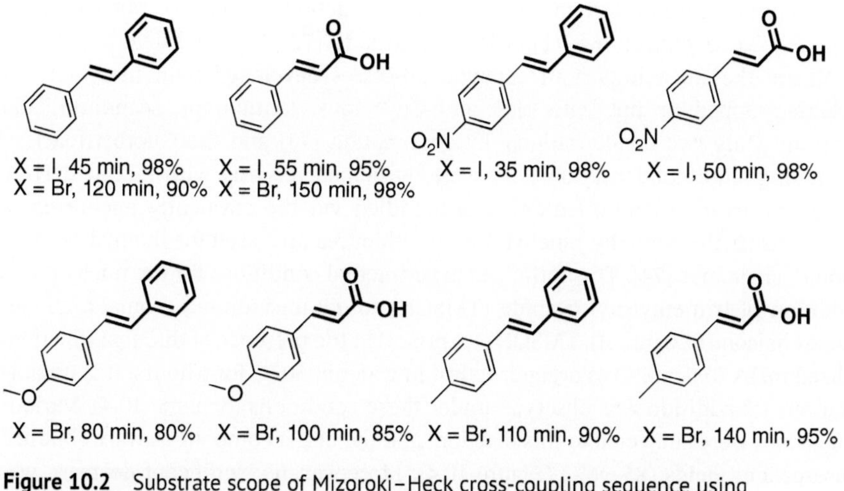

Figure 10.2 Substrate scope of Mizoroki–Heck cross-coupling sequence using γ-Fe_2O_3@HA-DABCO-Pd as supported catalyst [36]. Source: Michel Gruselle.

cocatalyst, and an amine base. Only two reports described this C—C coupling using HA [54, 55]. HA-supported palladium with 2 wt% of metal showed an excellent activity for the Sonogashira cross-coupling reaction under copper and phosphine conditions for the synthesis of pharmaceutical ingredients [54]. Starting from iodobenzene (0.5 mmol), phenylacetylene (0.5 mmol), and K_2CO_3 (0.5 mmol) in the presence of 2 wt% Pd@HA-12 catalyst (20 mg) in a mixture of CH_3OH-H_2O (1 : 1, v/v) at 95 °C for 150 minutes, the corresponding alkyne was obtained in 96% yield (Scheme 10.3). Recycling of the catalyst was carried out for four recycles with consistent activity and selectivity. The basic sites of HA-12 support were the main parameter in the high catalytic activity. The scope of the reaction showed that aryl iodides with electron-withdrawing groups gave a higher yield than those with electron-donating groups, and the aryl bromide was less active (Figure 10.3). It is notable that the optimized conditions produced the corresponding target in moderate yields starting from heterocyclic alkynes.

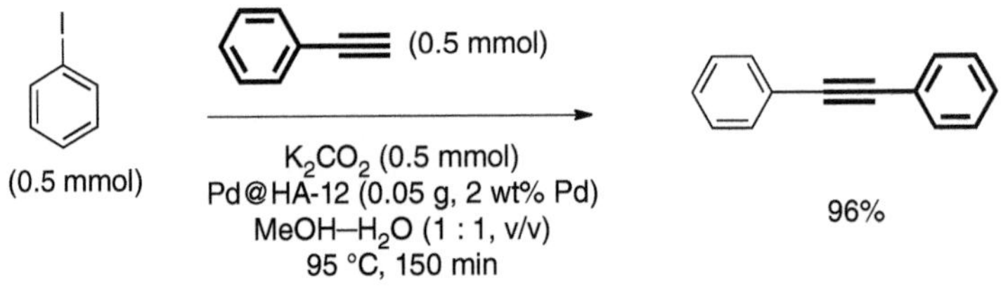

Scheme 10.3 Sonogashira cross-coupling sequence using 2 wt% Pd@HA-12 as catalyst [54]. Source: Michel Gruselle.

It is notable that the Glaser–Hay reaction, which allows the production of 1,3-diyne derivatives, was studied using HA catalyst [38].

10.3.2 Nucleophilic Carbon–Carbon Bond Forming Reactions

Nucleophilic carbon–carbon bond forming reactions, such as Michael [39], [56–63], Knoevenagel [64–71], Claisen–Schmidt [72, 73], and hydrocyanation [74], are the most important catalytic processes developed with modified HA whether doped or not with Na^+ and Zn^{2+} ions, lanthanum, scandium, and yttrium. Only two condensations: hydrocyanation [74] and the Guerbet reaction including aldol condensation [75], were developed recently with modified HAs. Magnetic hydroxyapatite (mHA) was modified via the covalently anchoring of 1-(3,5-bis-(trifluoromethyl)phenyl-3-propyl)thiourea and used for the hydrocyanation of chalcones [74]. The optimized experimental conditions for the nucleophilic addition of trimethylsilyl cyanide (TMSCN) to α,β-unsaturated aromatic enones were chalcone (1.0 mmol), TMSCN (1.1 mmol) in the presence of thiourea functionalized mHA (0.5 mol%) as organocatalyst in water at 50 °C for 6 hours. It is notable that no 1,2-addition was observed under these conditions (Scheme 10.4). Various chalcones were subjected to this reaction, and the target compounds were obtained in excellent yields (85–96%) (Figure 10.4). Moreover, no significant decrease was observed in the magnetic saturation and in the activity after ten cycles.

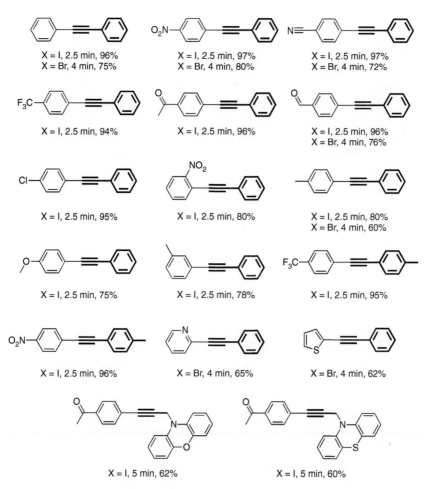

Figure 10.3 Substrate scope of Sonogashira cross-coupling sequence using 2 wt% Pd@HA-12 as catalyst [54]. Source: Michel Gruselle.

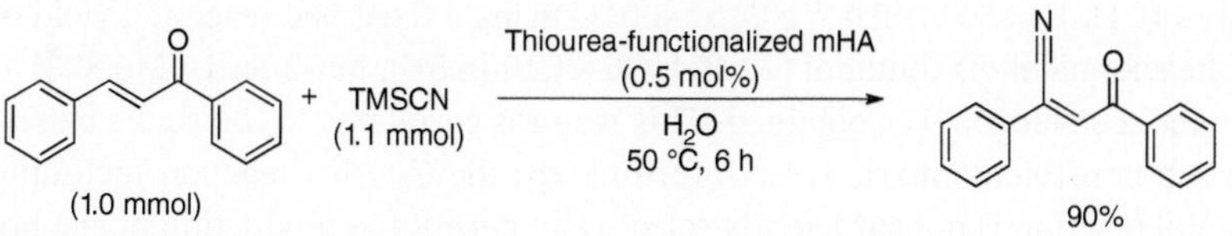

Scheme 10.4 Synthesis of β-ketonitriles catalyzed by thiourea functionalized mHA (0.5 mol%) as organocatalyst [74]. Source: Michel Gruselle.

Aldol condensation as a major method of C—C bond formation has been developed over modified HA as an intermediate reaction of the Guerbet coupling. The Guerbet reaction is a well-known industrial C—C coupling for the production of alcohol from two short-chain alcohols without gain or loss of carbon atoms. Conventionally, ethanol is converted to butanol as a higher energy density fuel additive,

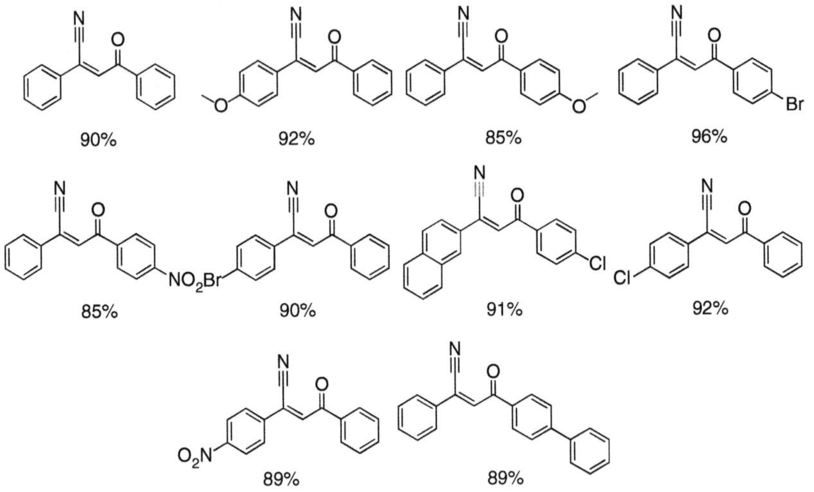

Figure 10.4 Substrate scope of the hydrocyanation of chalcones catalyzed by thiourea functionalized mHA (0.5 mol%) as organocatalyst [74]. Source: Michel Gruselle.

an intermediate in perfumes and also a solvent for paints and dyes. The sequence involves successive reaction: dehydrogenation (step 1), aldol condensation reaction (step 2), and hydrogenation (step 3), furnishing the corresponding higher alcohol. In this regard, different works were reported for the transformation of ethanol to n-butanol as a biofuel additive [76–80]. Different applications of this process have been studied [81]; one of which is the production of methylisobutylketone starting from acetone [82]. Concerning the production of n-butanol from bio-based ethanol, various HA having different Ca/P ratios showed the importance of the nature of the active acid/base pairs: among Ca^{2+}—OH^- and POH—OH^- acid–base pairs, the former is involved in the dehydrogenation of ethanol to acetaldehyde and the latter in the aldol condensation [77, 83]. Using carbonated apatites with a Ca/P ratio ranging from 1.69 to 2.39 with or without the presence of Na^+ ions in the structure [75] under the optimized conditions of 20 vol% ethanol in helium gas, 0.60 g of catalyst (GHSV = 5000 ml h^{-1} g^{-1}) at 400 °C using a fixed-bed reactor, a yield of 30% of heavier alcohols (butanol [22.4%], C6—C10 [6.8%], but-2-en-1-ol [0.8%]) at 40% ethanol conversion was obtained. This result is correlated to the acidic behavior of such nonstoichiometric HA. Unfortunately, the Guerbet reaction including C—C aldol reaction is not sufficiently selective to permit the production of alcohol for specific applications.

10.3.3 Multicomponent Reaction

MCRs are one-pot convergent syntheses involving a cascade of elementary chemical reactions starting from three or more materials. Among the main MCRs, Biginelli, Passerini, Ugi, Mannich, Bucherer–Bergs, Gewald, Hantzsch, Kabachnik–Fields, and Strecker reactions allow the formation aliphatic and aromatic heterocycles.

Recently, different modified HAs were used for the development of MCRs, producing new heterocycles containing nitrogen and oxygen atoms.

Pyrano[2,3-d]pyrimidine derivatives have a wide range of pharmaceutical activities and are produced via MCRs including Knoevenagel reaction as the first component condensation reaction. Ni^{2+} that supported on HA–core–shell–γ-Fe_2O_3 magnetic nanoparticles was synthesized and characterized [84]. A one-pot three-component synthesis using benzaldehyde (15 mmol), malononitrile (15 mmol), barbituric acid (15 mmol), and γ-Fe_2O_3@HA-Ni^{2+} (10 mg) in ethanol at room temperature for 25 minutes afforded the corresponding pyrano[2,3-d] pyrimidinone derivative in 95% yield (Scheme 10.5). Obviously, the catalyst activated the Knoevenagel reaction in the first step and then the rearrangement in the third step. Starting from aromatic aldehydes having electron-withdrawing and electron-donating groups, application of the optimized conditions produced various pyrano[2,3-d]pyrimidinone derivatives in good to excellent yields (85–95%) (Figure 10.5). Following the conventional procedure, the magnetic catalyst γ-Fe_2O_3@HA-Ni^{2+} could be used for six consecutive runs. The recyclability of the heterogeneous catalyst and the synthesis at room temperature in ethanol as bio-based solvent give important advantages in terms of green chemistry and sustainable development.

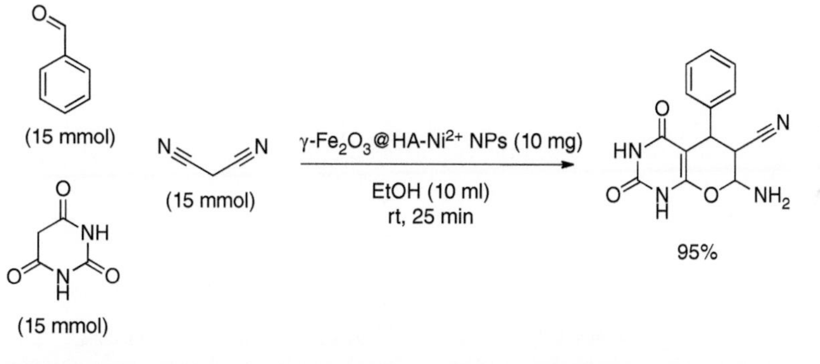

Scheme 10.5 One-pot three-component synthesis of pyrano[2,3-d]pyrimidinones catalyzed by γ-Fe_2O_3@HA-Ni^{2+} [84]. Source: Michel Gruselle.

Another team reported similar results for the production of novel pyrido[2,3-d] pyrimidine derivatives as new pyrans having pharmacological and biological activities [85, 86]. A four-component reaction between malononitrile (1.1 mmol), 4-methoxybenzaldehyde (1 mmol), dimethylbarbituric acid (1 mmol), and ammonium acetate (1.5 mmol) catalyzed by CeO_2-doped HA (50 mg) in ethanol at room temperature for 45 minutes furnished the target compound in 96% yield (Scheme 10.6) [85]. In contrast with the previous work [84], it was noticeable that the barbituric acid was permethylated. In this sequence, the reusability and stability of the heterogeneous catalyst was good after six catalytic runs. The scope of the reactions was investigated by varying the nature of the substituents of the aldehyde (Figure 10.6).

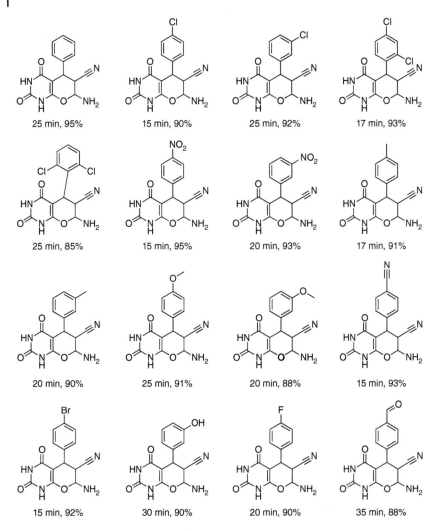

Figure 10.5 Substrate scope of MCRs with successive Knoevenagel and Michael condensations using γ-Fe_2O_3@HA-Ni^{2+} [84]. Source: Michel Gruselle.

The same group published the synthesis of novel 2,6-diamino-pyran-3,5-dicarbonitriles with a symmetric structure via a one-pot three-component synthetic protocol. The optimized conditions with malononitrile (1.1 mmol), 2-methoxyaldehyde (1.0 mmol), and cyanoacetamide (1 mmol) in the presence of chitosan-doped HA (CS/HA, 30 mg) in ethanol (10 ml) at room temperature for 30 minutes gave the corresponding target compound in 96% yield (Scheme 10.7) [86]. Application of this sequence furnished different analogs with good yields (Figure 10.7).

The Biginelli condensation reaction was studied using modified HA for the production of 3,4-dihydropyrimidin-2(1H)-one [87, 88]. Starting from benzaldehyde (10 mmol), ethylacetoacetate (10 mmol), and urea (10 mmol) in the presence of Ca_7Mg_3HA (2 mol%) obtained by the double decomposition method in either

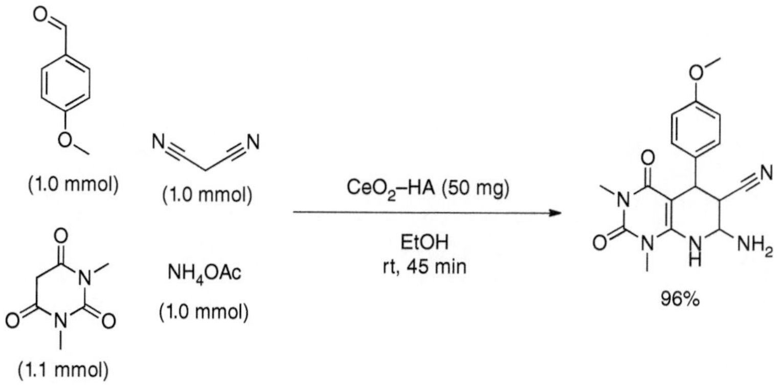

Scheme 10.6 One-pot four-component synthesis of pyrido[2,3-*d*]pyrimidines catalyzed by CeO$_2$-doped HA [85]. Source: Michel Gruselle.

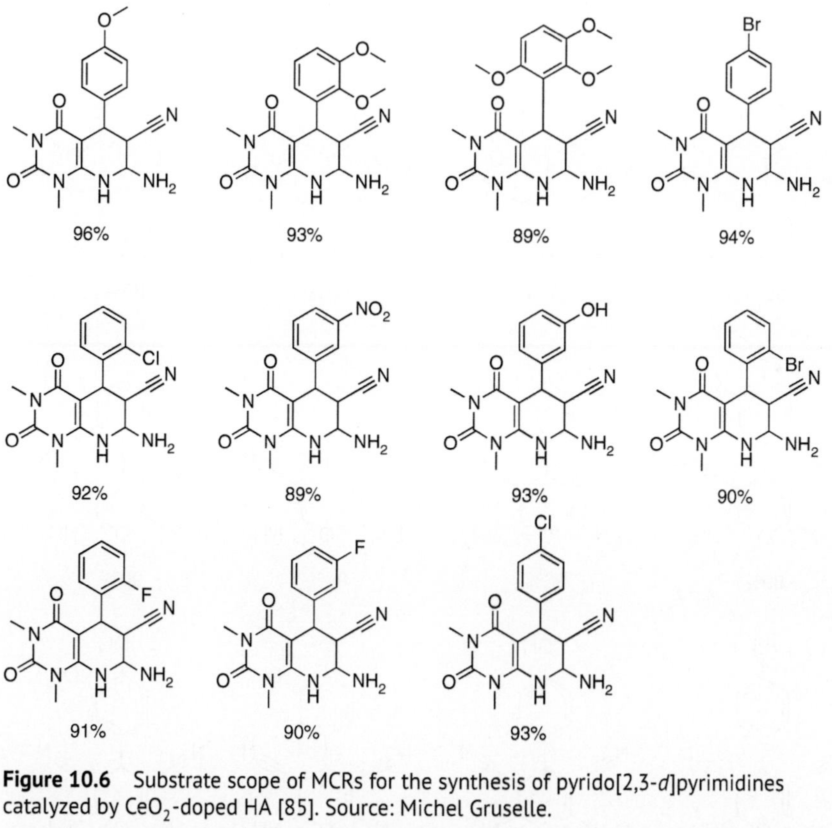

Figure 10.6 Substrate scope of MCRs for the synthesis of pyrido[2,3-*d*]pyrimidines catalyzed by CeO$_2$-doped HA [85]. Source: Michel Gruselle.

refluxing ethanol or refluxing toluene produced, the target heterocycle in 85% yield after 24 hours (Scheme 10.8) [88]. Variation of the nature of the substituent in the para-position of the benzaldehyde produced the desired compounds in similar yields with electron-donating groups (OCH$_3$, N(CH$_3$)$_2$) and in lower yields with electron-withdrawing groups (NO$_2$, F). Substitution of calcium ions with

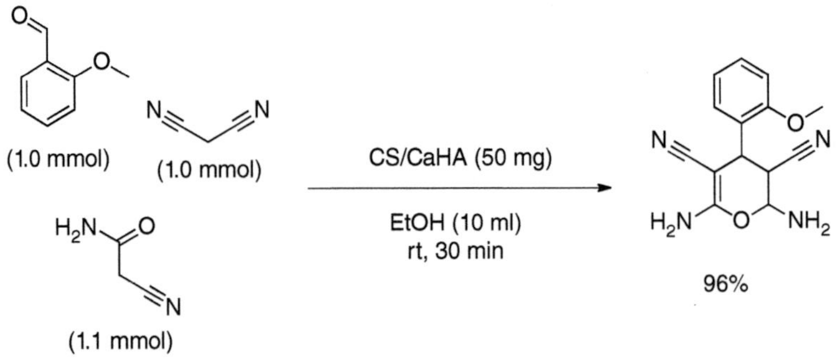

Scheme 10.7 One-pot three-component synthesis of 2,6-diamino-pyran-3,5-dicarbonitriles catalyzed by CS/CaHA [86]. Source: Michel Gruselle.

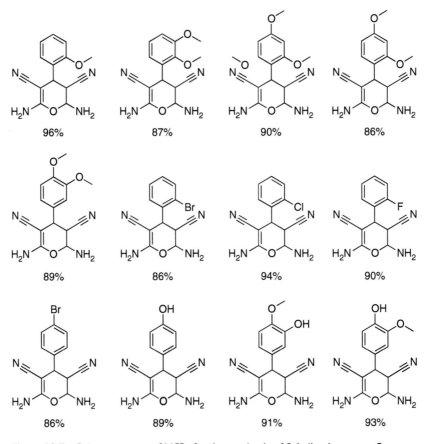

Figure 10.7 Substrate scope of MCRs for the synthesis of 2,6-diamino-pyran-3,5-dicarbonitriles catalyzed by CS/CaHA [86]. Source: Michel Gruselle.

magnesium cations increased the yield due to the enhancement of the OH⁻ activity at the surface of the catalyst.

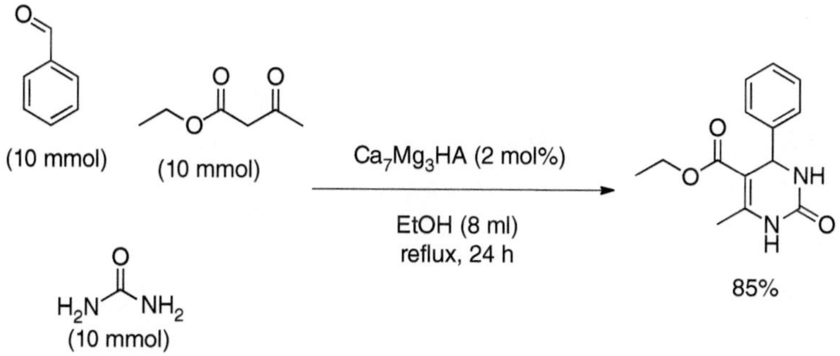

Scheme 10.8 One-pot three-component synthesis of ethyl 6-methyl-2-oxo-4-phenyl-1,2,3,4-tetrahydropyrimidine-5-carboxylate catalyzed by Ca_7Mg_3HA [88]. Source: Michel Gruselle.

The Hantzsch condensation reaction using γ-Fe_2O_3@HA@melamine as magnetic nanocatalyst was studied for the production of 1,4-dihydropyridine and polyhydroquinoline with biological applications [89]. For the 1,4-dihydropyridine, a mixture of benzaldehyde (1 mmol), ethylacetoacetate (2 mmol), and ammonium acetate (1.5 mmol) in the presence of γ-Fe_2O_3@HA@melamine (0.15 g) under solvent-free conditions at 80 °C for 15 minutes afforded the corresponding aromatic ring with a symmetrical structure in 94% yield (Scheme 10.9). The conversion was higher than 83%, with no detectable loss after recycling five times using the conventional procedure. The scope of the Hantzsch condensation allowed a series of 1,4-dihydropyridines with different steric and electronic properties to be synthesized with similar yields (Figure 10.8).

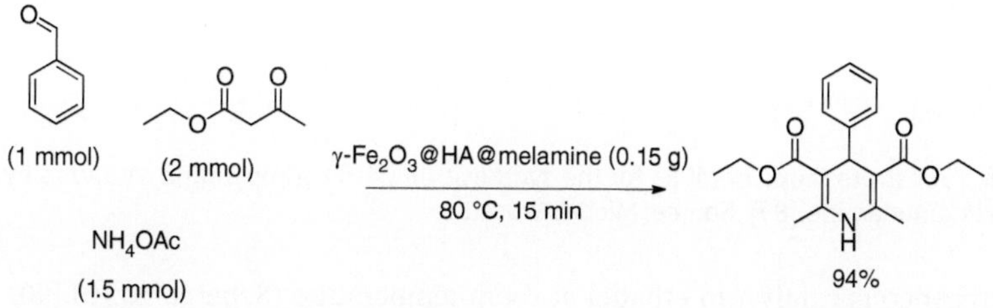

Scheme 10.9 One-pot three-component synthesis of diethyl 2,6-dimethyl-4-phenyl-1,4-dihydropyridine-3,5-dicarboxylate catalyzed by γ-Fe_2O_3@HA@melamine [89]. Source: Michel Gruselle.

Application of this process for the production of polyhydroquinoline was reported starting from 5,5-dimethylcyclohexane-1,3-dione (Figure 10.9) [89]. It was noticeable that the polyhydroquinoline derivatives have no symmetry in their structure.

The formation of furo[3,4-b]chromenes was reported starting from 4-ethylbenzaldehyde (1 mmol), 4-hydroxyfuran-2(5H)-one (1 mmol), and 5,5-dimethylcyclohexane-1,3-dione (1 mmol) in the presence of yttria (2.5 wt%) doped HA (50 mg) as

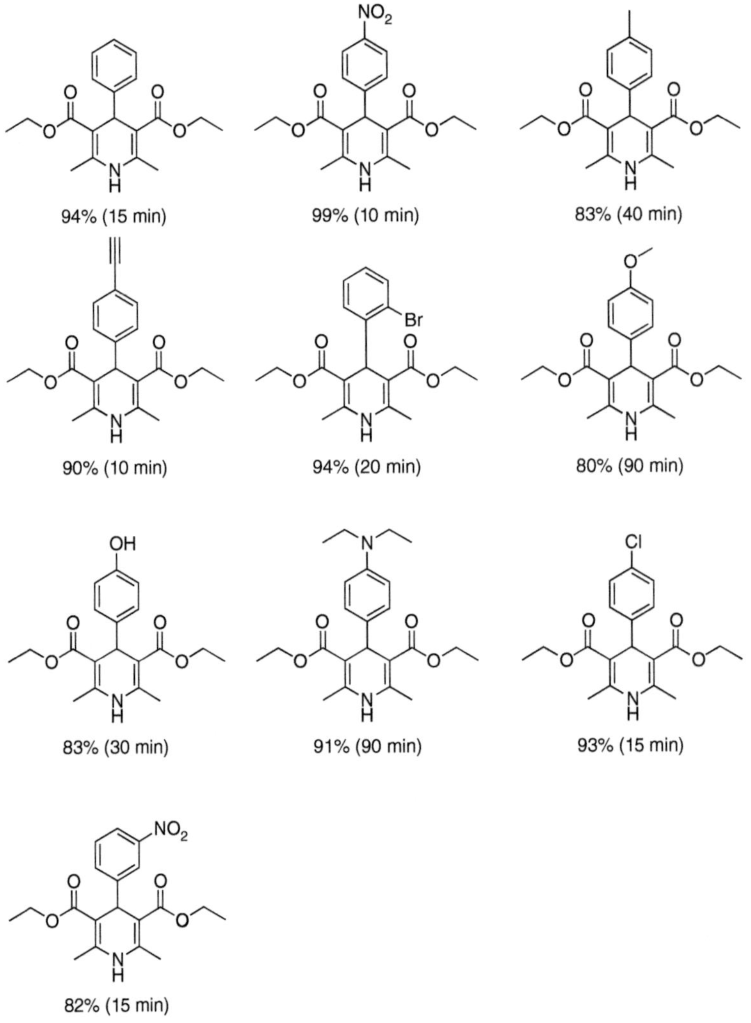

Figure 10.8 Substrate scope of MCRs for the synthesis of 1,4-dihydropyridine catalyzed by γ-Fe$_2$O$_3$@HA@melamine [89]. Source: Michel Gruselle.

solid heterogeneous catalyst in ethanol at room temperature (Scheme 10.10) [90]. The application of this approach with different aldehydes having electron-donating and electron-withdrawing substituents led to the production of various furo[3,4-b] chromenes with more than 92% yield (Figure 10.10). The synthesis of the chromene derivatives at room temperature in the presence of an efficient heterogeneous catalyst in a bio-based solvent suggests various benefits in terms of green chemistry, sustainable development, and catalysis.

A one-pot three-component A^3-coupling reaction for the synthesis of racemic acyclic propargylamine was reported using Zn(II) anchored onto magnetic natural HA. Starting from benzaldehyde (1 mmol), morpholine (1 mmol), phenylacetylene (1 mmol), and γ-Fe$_3$O$_4$@HA@Zn(II) (8 mol%) at 110 °C under solvent-free

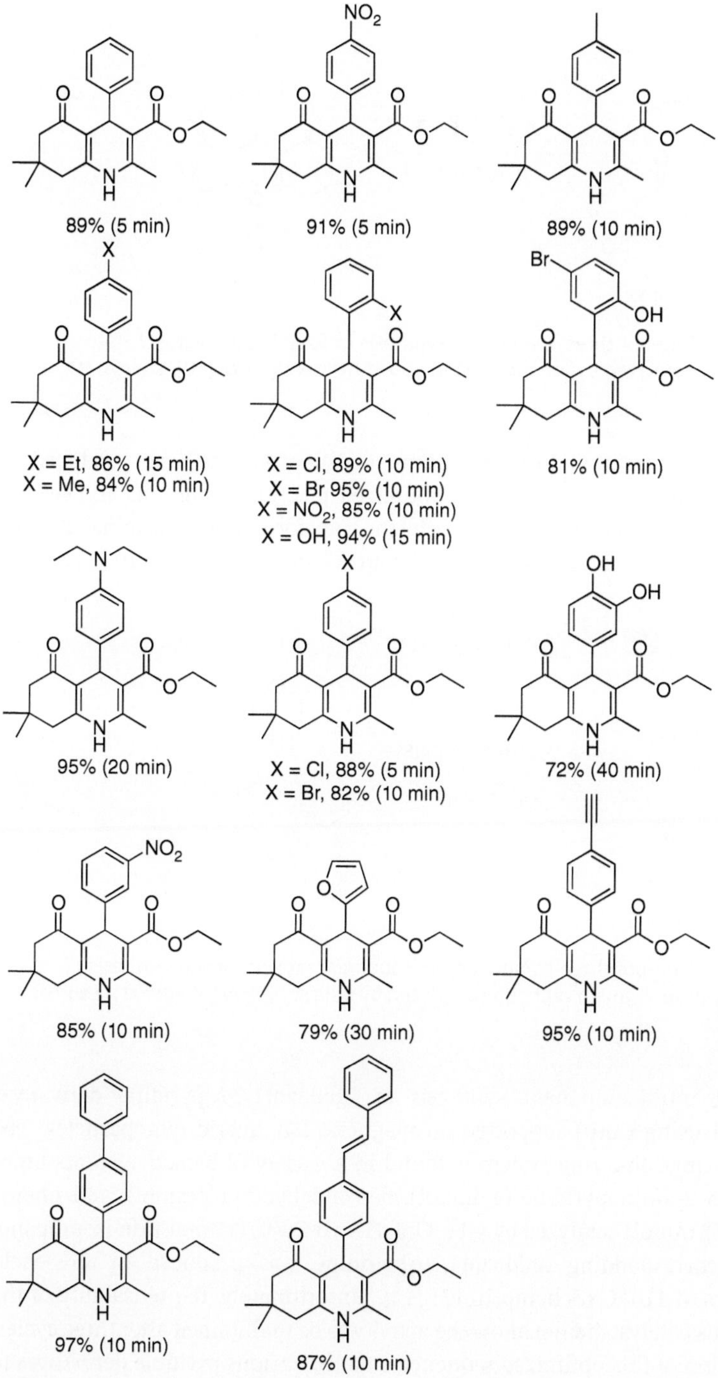

Figure 10.9 Substrate scope of MCRs for the synthesis of polyhydroquinoline catalyzed by γ-Fe$_2$O$_3$@HA@melamine [89]. Source: Michel Gruselle.

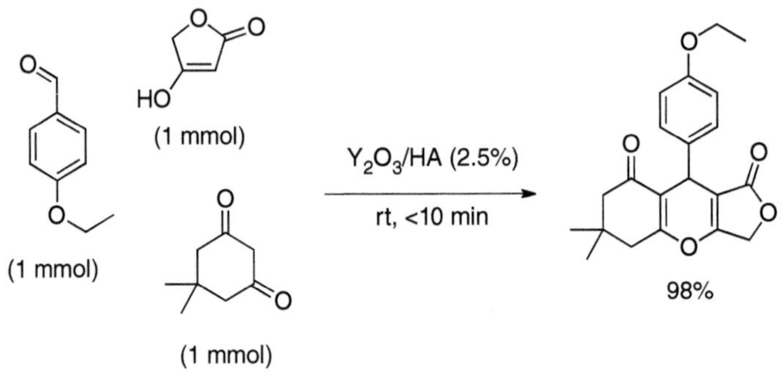

Scheme 10.10 One-pot three-component synthesis of 5,6,7,9-tetrahydro-1*H*-furo[3,4-*b*]chromene catalyzed by yttria (2.5 wt%) doped hydroxyapatite [90]. Source: Michel Gruselle.

conditions, the pure 4-(1,3-diphenylprop-2-ynyl)morpholine was produced in 95% yield (Scheme 10.11) [91]. These optimized reaction conditions allowed the synthesis of a variety of propargyl amines by the transformation of terminal alkynes, secondary amines, and aldehydes with different substituents (Figure 10.11).

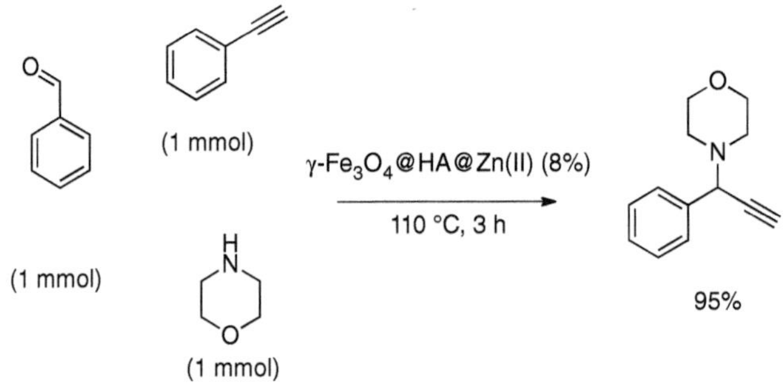

Scheme 10.11 One-pot three-component A^3-coupling reaction for the synthesis of 4-(1,3-diphenylprop-2-ynyl)morpholine catalyzed by γ-Fe$_3$O$_4$@HA@Zn(II) [91]. Source: Michel Gruselle.

A one-pot multicomponent synthesis of imidazo[1,2-*a*]pyridine derivatives was reported using Cu(0) supported on magnetic HA hybrid nanoparticles. The imidazo[1,2-*a*]pyridine ring system is found in a variety of bioactive compounds. Starting from 2-aminopyridine (1 mmol), benzaldehyde (1.2 mmol), and phenylacetylene (1.5 mmol) catalyzed by γ-Fe$_2$O$_3$@HAp@Cu(0) (10 mol%) in isopropanol (1 ml), the corresponding imidazo[1,2-*a*]pyridine was produced in 85% yield after 12 hours at 110 °C (Scheme 10.12) [92]. Unfortunately, the reusability of the heterogeneous catalyst did not allow the activity to be maintained after three cycles. The application of this optimized sequence enabled various pyridine derivatives to be synthesized by modulation of the substituents of the 2-aminopyridine and the nature of the aldehyde and the alkyne (Figure 10.12). It is noteworthy that substitutions of phenylacetylene and benzaldehyde derivatives by the corresponding

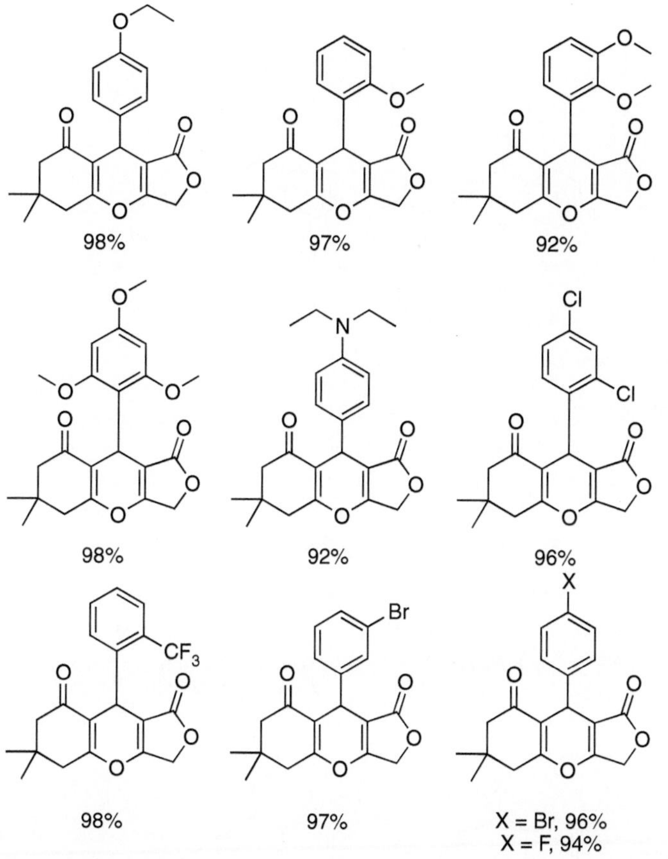

Figure 10.10 Substrate scope of multicomponent reactions for the synthesis of 5,6,7,9-tetrahydro-1H-furo[3,4-b]chromene catalyzed by yttria 2.5% doped hydroxyapatite [90]. Source: Michel Gruselle.

phenylpropiolic acid and phenylglyoxylic acid, respectively, allowed the identical chemical to be obtained in a similar yield via decarboxylative coupling.

10.4 Conclusions

Different modified HAs were used for the catalytic C—C bond formation. Depending on the type of reactions: cross-coupling, nucleophilic C—C bond forming, and MCRs, HA was modified with different noble or non-noble metal salts and/or functional groups. In some cases, $\gamma\text{-Fe}_2\text{O}_3$ forms the core of the modified HA in order to obtain synthesized magnetic inorganic–organic hybrid nanocomposite generating good recycling and catalytic properties. Recent cross-coupling reactions were studied successfully using palladium nanoparticles supported by either HA nanosphere (Suzuki–Miyaura), HA-12 (Sonogashira),

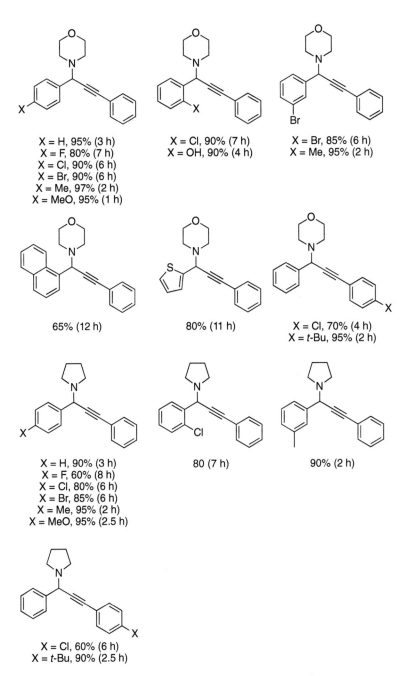

Figure 10.11 Substrate scope of one-pot three-component A³-coupling reaction for the synthesis of propargylamine catalyzed by γ-Fe₃O₄@HA@Zn(II) [91]. Source: Michel Gruselle.

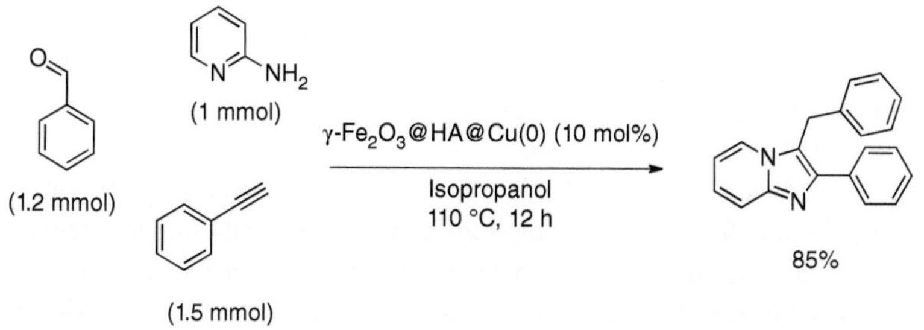

Scheme 10.12 One-pot three-component reaction for the synthesis of 3-benzyl-2-phenylimidazo[1,2-a]pyridine catalyzed by γ-Fe$_2$O$_3$@HA@Cu(0) [92]. Source: Michel Gruselle.

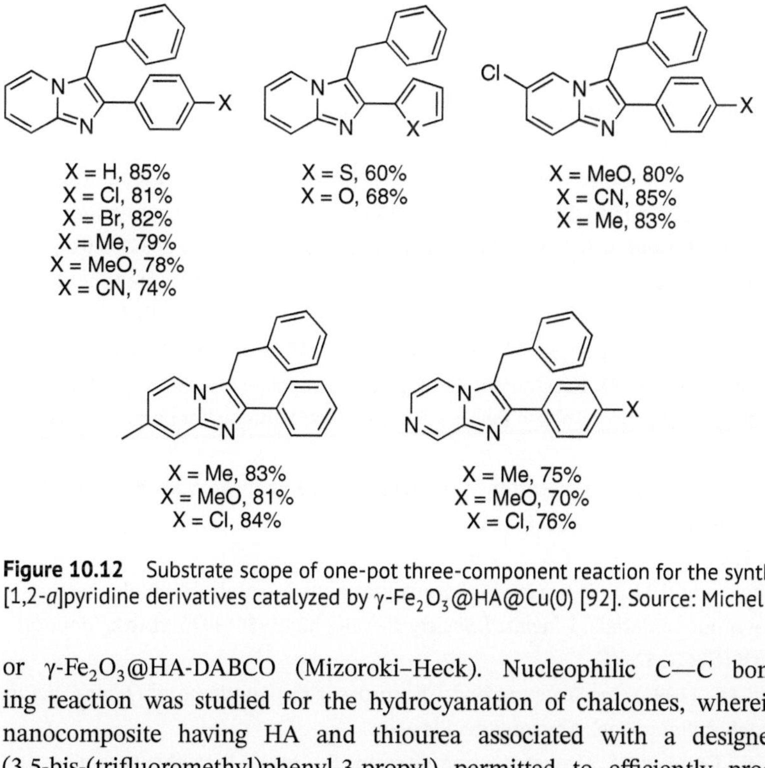

Figure 10.12 Substrate scope of one-pot three-component reaction for the synthesis of [1,2-a]pyridine derivatives catalyzed by γ-Fe$_2$O$_3$@HA@Cu(0) [92]. Source: Michel Gruselle.

or γ-Fe$_2$O$_3$@HA-DABCO (Mizoroki–Heck). Nucleophilic C—C bond forming reaction was studied for the hydrocyanation of chalcones, wherein hybrid nanocomposite having HA and thiourea associated with a designed linker (3,5-bis-(trifluoromethyl)phenyl-3-propyl) permitted to efficiently produce the target chemicals in water. Recent MCRs such as one-pot three-component and one-pot four-component reactions were reported for the synthesis of heteroaromatic derivatives. Similar strategy was developed based on HA associated with either CeO$_2$ nanoparticles, magnesium, or chitosan for the synthesis of pyrido[2,3-d]pyrimidine, 3,4-dihydropyrimidin-2(1H)-one, and 2,6-diamino-pyran-3,5-dicarbonitriles, respectively. Other modified HAs having metal salts were successfully used for the development of MCRs. In that way, yttria-doped HA was reported for the efficient synthesis of furo[3,4-b]chromenes, while non-noble metal-doped magnetic hybrid materials such as γ-Fe$_2$O$_3$@HA-Ni^{2+}, γ-Fe$_3$O$_4$@HA@Zn(II), and

γ-Fe$_2$O$_3$@HA@Cu(0) were used for the synthesis of pyrano[2,3-*d*]pyrimidine, 4-(1,3-diphenylprop-2-ynyl)morpholine, and imidazo[1,2-*a*]pyridine, respectively. When melanin was supported on γ-Fe$_2$O$_3$@HA, this basic nanocatalyst afforded 1,4-dihydropyridine/polyhydroquinoline.

In general, modified HAs have got a lot of advantages in heterogeneous catalysis; they are easily modifiable and have high thermal stability and tunable acid–base characters. Nevertheless, modified HAs have drawbacks; they have textural properties and mesoporosity, which are difficult to control, and they are highly sensible in strong acidic media. For all aforementioned reasons, the development of new modified HA will pave the way for green chemistry and sustainable system in the near future.

References

1. Satterfield, C.N. (2000). *Heterogeneous Catalysis in Industrial Practice*, 2ee. New York: McGraw Hill.
2. Martinez, C. and Corma, A. (2011). Inorganic molecular sieves: preparation, modification and industrial application in catalytic processes. *Coordination Chemistry Reviews* 225: 1558–1580. https://doi.org/10.1016/j.ccr.2011.03.014.
3. Plantard, G., Janin, T., and Goetz, V. (2012). Solar photocatalysis treatment of phytosanitary refuses: efficiency of industrial photocatalysts. *Applied Catalysis B: Environmental* 115–116: 38–44. https://doi.org/10.1016/j.apcatb.2011.11.034.
4. Sheldon, R.A. (2012). Fundamentals of green chemistry: efficiency in reaction design. *Chemical Society Reviews* 41: 1437–1451. https://doi.org/10.1039/C1CS15219J.
5. Dunn, P.J. (2012). The importance of green chemistry in process research and development. *Chemical Society Reviews* 41: 1452–1461. https://doi.org/10.1039/C1CS15041C.
6. Perego, C. and Millini, R. (2013). Porous materials in catalysis: challenges for mesoporous materials. *Chemical Society Reviews* 42: 3956–3976. https://doi.org/10.1039/C2CS35244C.
7. Elliot, J.C. (1994). In: *Structure and Chemistry of the Apatites and Other Calcium Orthophosphates*, 1ee. Amsterdam: Elsevier.
8. Rey, C., Combes, C., Drouet, C., and Grossin, D. (2011). Bioactive ceramics: physical chemistry. In: *Comprehensive Biomaterials* (ed. P. Ducheyne, K. Healy, D. Hutmacher, et al.), 187–221. Amsterdam: Elsevier ISBN978-0-08-055302-3.
9. Gomez-Morales, J., Iafisco, M., Delgado-Lopez, J.M. et al. (2013). Progress on the preparation of nanocrystalline apatites and surface characterization: overview of fundamental and applied aspects. *Progress in Crystal Growth and Characterization* 59: 1–46. https://doi.org/10.1016/j.pcrysgrow.2012.11.001.
10. Gruselle, M. and Tõnsuaadu, K. (2017). Tunable calcium-apatites as solid catalysts for classical organic reactions. *Current Organic Chemistry* 21: 688–697. https://doi.org/10.2174/1385272821666161219155302.

11 Kaneda, K. and Mizukagi, T. (2017). Design of high-performance heterogeneous catalysts using apatite compounds for liquid-phase organic syntheses. *ACS Catalysis* 7: 920–935. https://doi.org/10.1021/acscatal.6b02585.

12 Fihri, A., Len, C., Varma, R.S., and Solhy, A. (2017). Hydroxyapatite: a review of syntheses, structures and applications in heterogeneous catalysis. *Coordination Chemistry Reviews* 347: 48–76. https://doi.org/10.1016/j.ccr.2017.06.009.

13 Gruselle, M. (2015). Apatites: a new family of catalysts in organic synthesis. *Journal of Organometallic Chemistry* 793: 93–101. https://doi.org/10.1016/j.jorganchem.2015.01.018.

14 Kaneda, K. and Mizukagi, T. (2009). Development of *concerto* metal catalysts using apatite compounds for green organic syntheses. *Energy and Environmental Science* 2: 655–673. https://doi.org/10.1039/B901997A.

15 Sugyiama, S. (2007). Approach using apatite to studies on energy and environment. *Phosphorus Research Bulletin* 21: 1–8. https://doi.org/10.3363/prb.21.1.

16 Diallo-Garcia, S., Ben Osman, M., Krafft, J.M. et al. (2014). Identification of surface basic sites and acid–base pairs of hydroxyapatite. *The Journal of Physical Chemistry C* 118: 12744–12757. https://doi.org/10.1021/jp500469x.

17 Bolis, V., Busco, C., Martra, G. et al. (2012). Coordination chemistry of Ca sites at the surface of nanosized hydroxyapatite: interaction with H_2O and CO. *Philosophical Transactions of the Royal Society A* 370: 1313–1336. https://doi.org/10.1098/rsta.2011.0273.

18 Errassifi, F., Menbaoui, A., Autefage, H. et al. (2010). Adsorption on apatitic calcium phosphates: applications to drug delivery. *Ceramic Transactions Series* 218: 159–174. https://doi.org/10.1002/9780470909898.ch16.

19 Jarlbring, M., Sandströen, D.E., Antzutkin, O.N., and Forsling, W. (2006). Characterization of active phosphorus surface sites at synthetic carbonate-free fluorapatite using single-pulse ^1H, ^{31}P, and ^{31}P CP MAS NMR. *Langmuir* 22: 4787–4792. https://doi.org/10.1021/la052837j.

20 Bengtsson, A. and Sjöberg, S. (2009). Surface complexation and proton-promoted dissolution in aqueous apatite systems. *Pure and Applied Chemistry* 81: 1569–1584. https://doi.org/10.1351/PAC-CON-08-10-02.

21 Ospina, C.A., Terra, J., Ramirez, A.J. et al. (2012). Experimental evidence and structural modeling of nonstoichiometric (0 1 0) surfaces coexisting in hydroxyapatite nano-crystals. *Colloids and Surfaces B: Biointerfaces* 89: 15–22. https://doi.org/10.1016/j.colsurfb.2011.08.016.

22 Sebti, S., Zahouily, M., and Lazrek, H.B. (2006). Les Phosphates: Nouvelle famille de catalyseurs très performant. *The Second International Conference on the Valorization of Phosphates and Phosphorus Compounds (COVAPHOS II)*, Marrakech, Morocco, (Novembre 2006), 4: 19–40.

23 Koutsopoulos, S. (2002). Synthesis and characterization of hydroxyapatite crystals: a review study on the analytical methods. *Journal of Biomedical Materials Research* 4: 600–612. https://doi.org/10.1002/jbm.10280.

24 Nayak, A.K. (2010). Hydroxyapatite synthesis methodologies: an overview. *International Journal of ChemTech Research* 2: 903–907.

25 Lin, K., Wu, C., and Chang, J. (2014). Advances in synthesis of calcium phosphate crystals with controlled size and shape. *Acta Biomaterialia* 10: 4071–4102. https://doi.org/10.1016/j.actbio.2014.06.017.

26 Tõnsuaadu, K., Gross, K.A., Pluduma, L., and Veiderma, M.A. (2012). A review on the thermal stability of calcium apatites. *Journal of Thermal Analysis and Calorimetry* 110: 647–659. https://doi.org/10.1007/s10973-011-1877-y.

27 Imrie, F.E., Aina, V., Lusvardi, G. et al. (2013). Synthesis and characterisation of strontium and magnesium co-substituted biphasic calcium phosphates. *Key Engineering Materials* 529–530: 88–93. https://doi.org/10.4028/www.scientific.net/KEM.529-530.88.

28 Cox, S.C., Jamshidi, P., Grover, L.M., and Mallick, K.K. (2014). Preparation and characterisation of nanophase Sr, Mg, and Zn substituted hydroxyapatite by aqueous precipitation. *Materials Science and Engineering: C* 35: 106–114. https://doi.org/10.1016/j.msec.2013.10.015.

29 Smiciklas, I., Onjia, A., Raicevic, S. et al. (2008). Factors influencing the removal of divalent cations by hydroxyapatite. *Journal of Hazardous Materials* 152 (2): 876–884. https://doi.org/10.1016/j.jhazmat.2007.07.056.

30 Cazalbou, S., Eichert, D., Ranz, X. et al. (2005). Ion exchanges in apatites for biomedical application. *Journal of Materials Science: Materials in Medicine* 16: 405–409. https://doi.org/10.1007/s10856-005-6979-2.

31 Matsunaga, K., Inamori, H., and Murata, H. (2008). Theoretical trend of ion exchange ability with divalent cations in hydroxyapatite. *Physical Review B* 78: 094101. https://doi.org/10.1103/PhysRevB.78.094101.

32 Stötzel, C., Müller, F.A., Reinert, F. et al. (2009). Ion adsorption behaviour of hydroxyapatite with different crystallinities. *Colloids and Surfaces B: Biointerfaces* 74: 91–95. https://doi.org/10.1016/j.colsurfb.2009.06.031.

33 Viipsi, K., Sjöberg, S., Tõnsuaadu, K., and Shchukarev, A. (2013). Hydroxy- and fluorapatite as sorbents in Cd(II)–Zn(II) multi-component solutions in the absence/presence of EDTA. *Journal of Hazardous Materials* 252–253: 91–98. https://doi.org/10.1016/j.jhazmat.2013.02.034.

34 Mori, K., Yamaguchi, K., Hara, T. et al. (2002). Controlled synthesis of hydroxyapatite-supported palladium complexes as highly efficient heterogeneous catalysts. *Journal of the American Chemical Society* 124: 11572–11573. https://doi.org/10.1021/ja020444q.

35 Mori, K., Hara, T., Mizugaki, T. et al. (2005). Catalytic investigations of carbon–carbon bond-forming reactions by a hydroxyapatite-bound palladium complex. *New Journal of Chemistry* 29: 1174–1181. https://doi.org/10.1039/B506129F.

36 Pashaei, M., Mehdipour, E., and Azaroon, M. (2019). Engineered mesoporous ionic-modified γ-Fe_2O_3@hydroxyapatite decorated with palladium nanoparticles and its catalytic properties in water. *Applied Organometallic Chemistry* 33: e4622. https://doi.org/10.1002/aoc.4622.

37 Bulut, A., Aydemir, M., Durap, F. et al. (2018). Palladium nanoparticles supported on hydroxyapatite nanospheres: highly active, reusable and green catalyst for Suzuki – Miyaura cross coupling reactions under aerobic conditions. *ChemistrySelect.* 3: 1569–1576. https://doi.org/10.1002/slct.201702537.

38 Maaten, B., Moussa, J., Desmarets, C. et al. (2014). Cu-modified hydroxy-apatite as catalyst for Glaser–Hay C—C homo- coupling reaction of terminal alkynes. *Journal of Molecular Catalysis A: Chemical* 393: 112–116. https://doi.org/10.1016/j.molcata.2014.06.011.

39 Aissa, A., Debbabi, M., Gruselle, M., Thouvenot, R., Flambard, A., Gredin, P., Beaunier, P., Tõnsuaadu, K. (2009). Sorption of tartrate ions to lanthanum (III)-modified calcium fluor- and hydroxyapatite. *Journal of Colloid and Interface Science* 330: 20–28. https://doi.org/10.1016/j.jcis.2008.10.043.

40 Tõnsuaadu, K., Gruselle, M.; Villain, F.; Thouvenot, R.; Peld, M.; Mikli, V., Traksmaa, R., Gredin, P., Carrier, X. Salles, L. (2006). A new glance at ruthenium sorption mechanism on hydroxy, carbonate, and fluor apatites: analytical and structural studies. *Journal of Colloid and Interface Science* 304: 283–291. https://doi.org/10.1016/j.jcis.2006.07.079.

41 Elazarifi, N., Ezzamarty, A., Leglise, J., de Ménorval, L.-C., Moreau, C. (2004). Kinetic study of the condensation of benzaldehyde with ethylcyanoacetate in the presence of Al-enriched fluorapatites and hydroxyapatites as catalysts. *Applied Catalysis A: General A* 267: 235–240. https://doi.org/10.1016/j.apcata.2004.03.012.

42 Wang K., Kennedy, J.G., Cook, R.A. (2009). Hydroxyapatite-supported $Rh(CO)_2(acac)$ (acac = acetylacetonate): structure characterization and catalysis for 1-hexene hydroformylation. *Journal of Molecular Catalysis A: Chemical* 298: 88–93. https://doi.org/10.1016/j.molcata.2008.10.012.

43 Dessoudeix, M., Jauregui-Haza, U.J., Heughebaert, M. et al. (2002). Apatitic tricalcium phosphate as novel smart solids for supported aqueous phase catalysis (SAPC). *Advanced Synthesis & Catalysis* 344: 406–412. https://doi.org/10.1002/1615-4169(200206)344:3/4<406::AID-ADSC406>3.0.CO;2-3.

44 Devendar, P., Qu, R.Y., Kang, W.M. et al. (2018). Palladium-catalyzed cross-coupling reactions: a powerful tool for the synthesis of agrochemicals. *Journal of Agricultural and Food Chemistry* 66: 8914–8934. https://doi.org/10.1021/acs.jafc.8b03792.

45 Hong, K., Sajjadi, M., Suh, J.M. et al. (2020). Palladium nanoparticles on assorted nanostructured supports: Applications for Suzuki, Heck, and Sonogashira cross-coupling reactions. *ACS Applied Nano Materials* 3: 2070–2103. https://doi.org/10.1021/acsanm.9b02017.

46 Jana, R., Pathak, T.P., and Sigman, M.S. (2011). Advances in transition metal (Pd, Ni, Fe)-catalyzed cross-coupling reactions using alkyl-organometallics as reaction partners. *Chemical Reviews* 111: 1417–1492. https://doi.org/10.1021/cr100327p.

47 Len, C., Bruniaux, S., Delbecq, F., and Parmar, V.S. (2017). Palladium-catalyzed Suzuki–Miyaura cross-coupling in continuous flow. *Catalysts* 7: 146/1–146/23. https://doi.org/10.3390/catal7050146.

48 Polshettiwar, V., Len, C., and Fihri, A. (2009). Silica-supported palladium: sustainable catalysts for cross-coupling reactions. *Coordination Chemistry Reviews* 253: 2599–2626. https://doi.org/10.1016/j.ccr.2009.06.001.

49 Jamwal, N., Gupta, M., and Paul, S. (2008). Hydroxyapatite-supported palladium (0) as a highly efficient catalyst for the Suzuki coupling and aerobic oxidation of benzyl alcohols in water. *Green Chemistry* 10: 999–1003. https://doi.org/10.1039/B802135J.

50 Masuyama, Y., Sugioka, Y., Chonan, S. et al. (2012). Palladium(II)-exchanged hydroxyapatite-catalyzed Suzuki–Miyaura-type cross-coupling reactions with potassium aryltrifluoroborates. *Journal of Molecular Catalysis A: Chemical* 352: 81–85. https://doi.org/10.1016/j.molcata.2011.10.017.

51 Indra, A., Gopinath, C.S., Bhaduri, S., and Kumar Lahiri, G. (2013). Hydroxyapatite supported palladium catalysts for Suzuki–Miyaura cross-coupling reaction in aqueous medium. *Catalysis Science & Technology* 3: 1625–1633. https://doi.org/10.1039/C3CY00160A.

52 Climent, M.J., Corma, A., Iborra, S., and Mifsud, M. (2007). Heterogeneous palladium catalysts for a new one-pot chemical route in the synthesis of fragrances based on the Heck reaction. *Advanced Synthesis & Catalysis* 349: 1949–1954. https://doi.org/10.1002/adsc.200700026.

53 Zhang, D., Zhao, H., Zhao, X. et al. (2011). Application of hydroxyapatite as catalyst and catalyst carrier. *Progress in Chemistry* 23: 687–694.

54 Saha, D., Chatterjee, T., Mukherjee, M., and Ranu, B.C. (2012). Copper(I) hydroxyapatite catalyzed Sonogashira reaction of alkynes with styrenyl bromides. Reaction of cis-styrenyl bromides forming unsymmetric diynes. *The Journal of Organic Chemistry* 77: 9379–9383. https://doi.org/10.1021/jo3015819.

55 Bilakanti, V., Boosa, V., Velisoju, V.K. et al. (2017). Role of surface basic sites in Sonogashira coupling reaction over $Ca_5(PO_4)_3OH$ supported Pd catalyst: investigation by diffuse reflectance Infrared Fourier Transform Spectroscopy. *The Journal of Physico Chemistry C* 121: 22191–22198. https://doi.org/10.1021/acs.jpcc.7b07620.

56 Gruselle, M., Kanger, T., Thouvenot, R. et al. (2011). Calcium hydroxyapatites as efficient catalysts for the Michael C—C bond formation. *ACS Catalysis* 1: 1729–1733. https://doi.org/10.1021/cs200460k.

57 Zahouily, M., Abrouki, Y., Bahlaouan, B. et al. (2003). Hydroxyapatite: new efficient catalyst for the Michael addition. *Catalysis Communications* 4: 521–524. https://doi.org/10.1016/j.catcom.2003.08.001.

58 Zahouily, M., Bahlaouan, W., Bahlaouan, B. et al. (2005). Catalysis by hydroxyapatite alone and modified by sodium nitrate: a simple and efficient procedure for the construction of carbon-nitrogen bonds in heterogeneous catalysis. *ARKIVOC* 2005 (xiii): 150–161.

59 Tahir, R., Banert, K., Solhy, A., and Sebti, S. (2006). Zinc bromide supported on hydroxyapatite as a new and efficient solid catalyst for Michael addition of indoles to electron-deficient olefins. *Journal of Molecular Catalysis A: Chemical* 246: 39–42. https://doi.org/10.1016/j.molcata.2005.10.012.

60 Mori, K., Oshiba, M., Hara, T. et al. (2005). Michael reaction of 1,3-dicarbonyls with enones catalyzed by a hydroxyapatite-bound La complex. *Tetrahedron Letters* 46: 4283–4286. https://doi.org/10.1016/j.tetlet.2005.04.099.

61 Mori, K., Oshiba, M., Hara, T. et al. (2006). Creation of monomeric La complexes on apatite surfaces and their application as heterogeneous catalysts for Michael reactions. *New Journal of Chemistry* 30: 44–52. https://doi.org/10.1039/B512030F.

62 Hara, T., Kanai, S., Mori, K. et al. (2006). Highly efficient C—C bond-forming reactions in aqueous media catalyzed by monomeric vanadate species in an apatite framework. *The Journal of Organic Chemistry* 71: 7455–7462. https://doi.org/10.1021/jo0614745.

63 Kaneda, K., Hara, T., Hashimoto, N., Mitsudome, T., Mizukagi, T., Itsukawa, K. (2010). Creation of a monomeric vanadate species in an apatite framework as an active heterogeneous base catalyst for Michael reactions in water. *Catalysis Today* 152: 93–98. https://doi.org/10.1016/j.cattod.2009.08.018.

64 Sebti, S., Tahir, R., Nazih, R. et al. (2002). Hydroxyapatite as a new solid support for the Knoevenagel reaction in heterogeneous media without solvent. *Applied Catalysis A: General* 228: 155–159. https://doi.org/10.1016/S0926-860X(01)00961-9.

65 Smahi, A., Solhy, A., Al Badaoui, H. et al. (2003). Potassium fluoride doped fluorapatite and hydroxyapatite as new catalysts in organic synthesis. *Applied Catalysis A: General* 250: 151–159. https://doi.org/10.1016/S0926-860X(03)00254-0.

66 Alazarifi, N., Ezzamarty, A., Leglise, J. et al. (2004). Kinetic study of the condensation of benzaldehyde with ethylcyanoacetate in the presence of Al-enriched fluoroapatites and hydroxyapatites as catalysts. *Applied Catalysis A: General* 267: 235–240. https://doi.org/10.1016/j.apcata.2004.03.012.

67 Zhang, Y. and Xia, C. (2009). Magnetic hydroxyapatite-encapsulated γ-Fe_2O_3 nanoparticles functionalized with basic ionic liquids for aqueous Knoevenagel condensation. *Applied Catalysis A: General* 366: 141–147. https://doi.org/10.1016/j.apcata.2009.06.041.

68 Zhang, Y., Zhao, Y., and Xia, C. (2009). Basic ionic liquids supported on hydroxyapatite-encapsulated γ-Fe_2O_3 nanocrystallites: an efficient magnetic and recyclable heterogeneous catalyst for aqueous Knoevenagel condensation. *Journal of Molecular Catalysis A: Chemical* 306: 107–112. https://doi.org/10.1016/j.molcata.2009.02.032.

69 Mallouk, S., Bougrin, K., Laghzizil, A., and Benhida, R. (2010). Microwave-assisted and efficient solvent-free Knoevenagel condensation. A sustainable protocol using porous calcium hydroxyapatite as catalyst. *Molecules* 15: 813–823. https://doi.org/10.3390/molecules15020813.

70 Priya, K. and Buvaneswari, G. (2009). Apatite phosphates containing heterovalent cations and their application in Knoevenagel condensation. *Materials Research Bulletin* 44: 1209–1213. https://doi.org/10.1016/j.materresbull.2009.01.014.

71 Pillai, M.K., Singh, S., and Jonnalagadda, S.B. (2010). Solvent-free Knoevenagel condensation over cobalt hydroxyapatite. *Synthetic Communications* 40: 3710–3715. https://doi.org/10.1080/00397910903531714.

72 Sebti, S., Solhy, A., Tahir, R., and Smahi, A. (2002). Modified hydroxyapatite with sodium nitrate: an efficient new solid catalyst for the Claisen–Schmidt condensation. *Applied Catalysis A: General* 235: 273–281. https://doi.org/10.1016/S0926-860X(02)00273-9.

73 Solhy, A., Tahir, R., Sebti, S. et al. (2010). Efficient synthesis of chalcone derivatives catalyzed by re-usable hydroxyapatite. *Applied Catalysis A: General* 374: 189–193. https://doi.org/10.1016/j.apcata.2009.12.008.

74 Oskouie, A.A., Taheri, S., Mamani, L., and Heydari, A. (2015). Thiourea-functionalized magnetic hydroxyapatite as a recyclable inorganic–organic hybrid nanocatalyst for conjugate hydrocyanation of chalcones with TMSCN. *Catalysis Communications* 72: 6–10. https://doi.org/10.1016/j.catcom.2015.08.016.

75 Silvester, L., Lamonier, J.F., Faye, J. et al. (2015). Reactivity of ethanol over hydroxyapatite-based Ca-enriched catalysts with various carbonate contents. *Catalysis Science & Technology* 5: 2994–3006. https://doi.org/10.1039/C5CY00327J.

76 Wang, S.C., Cendejas, M.C., Hermans, I. (2020). Insights into ethanol coupling over hydroxyapatite using modulation excitation *operando* infrared spectroscopy. *ChemCatChem* 12: 4167–4175. https://doi.org/10.1002/cctc.202000331.

77 Ben Osman, M., Kraft, J.M., Thomas, C. et al. (2019). Importance of the nature of the active acid/base pairs of hydroxyapatite involved in the catalytic transformation of ethanol to *n*-butanol revealed by operando DRIFTS. *ChemCatChem* 11: 1765–1778. https://doi.org/10.1002/cctc.201801880.

78 Young, Z.D., Hanspal, S., and Davis, R.J. (2016). Aldol condensation of acetaldehyde over titania, hydroxyapatite, and magnesia. *ACS Catalysis* 6: 3193–3202. https://doi.org/10.1021/acscatal.6b00264.

79 Moteki, T., Rowley, A.T., and Flaherty, D.W. (2016). Self-terminated cascade reactions that produce methylbenzaldehydes from ethanol. *ACS Catalysis* 6: 7278–7282. https://doi.org/10.1021/acscatal.6b02475.

80 Hanspal, S., Young, Z.D., Shou, H., and Davis, R.J. (2015). Multiproduct steady-state isotopic transient kinetic analysis of the ethanol coupling reaction over hydroxyapatite and magnesia. *ACS Catalysis* 5: 1737–1746. https://doi.org/10.1021/cs502023g.

81 Hernandez-Gimenez, A.M., Ruiz-Martinez, J., Puertolas, B. et al. (2017). Operando spectroscopy of the gas-phase aldol condensation of propanal over solid base catalysts. *Topics in Catalysis* 60: 1522–1536. https://doi.org/10.1007/s11244-017-0836-7.

82 Ho, C.R., Zheng, S., Shylesh, S., and Bell, A.T. (2018). The mechanism and kinetics of methyl isobutyl ketone synthesis from acetone over ion-exchanged hydroxyapatite. *Journal of Catalysis* 365: 174–183. https://doi.org/10.1016/j.jcat.2018.07.005.

83 Dai, J. and Zhang, H. (2019). Recent advances in selective C—C bond coupling for ethanol upgrading over balanced Lewis acid-base catalysts. *Science China Materials* 62: 1642–1654. https://doi.org/10.1007/s40843-019-9454-x.

84 Rezayati, S., Abbasi, Z., Nezhad, E.R. et al. (2016). Three-component synthesis of pyrano[2,3-*d*]pyrimidinone derivatives catalyzed by Ni^{2+} supported on hydroxyapatite-core–shell-γ-Fe_2O_3 nanoparticles in aqueous medium. *Research on Chemical Intermediates* 42: 7597–7609. https://doi.org/10.1007/s11164-016-2555-2.

85 Maddila, S., Gangu, K.K., Maddila, S.N., and Jonnalagadda, S.B. (2018). A viable and efficacious catalyst, CeO_2/HAp, for green synthesis of novel pyrido[2,3-*d*]pyrimidine derivatives. *Research on Chemical Intermediates* 44: 1397–1409. https://doi.org/10.1007/s11164-017-3174-2.

86 Maddila, S., Gangu, K.K., Maddila, S.N., and Jonnalagadda, S.B. (2017). A facile, efficient, and sustainable chitosan/CaHAp catalyst and one-pot synthesis of novel 2,6-diamino-pyran-3,5-dicarbonitriles. *Molecular Diversity.* 21: 247–255. https://doi.org/10.1007/s11030-016-9708-5.

87 Ben Moussa, S., Lachheb, J., Gruselle, M. et al. (2017). Calcium, Barium and Strontium apatites: A new generation of catalysts in the Biginelli reaction. *Tetrahedron* 73: 6542–6548. https://doi.org/10.1016/j.tet.2017.09.051.

88 Ben Moussa, S., Mehri, A., and Badraoui, B. (2020). Magnesium modified calcium hydroxyapatite: An efficient and recyclable catalyst for the one-pot Biginelli condensation. *Journal of Molecular Structure* 1200: 127111. https://doi.org/10.1016/j.molstruc.2019.127111.

89 Igder, S., Kiasat, A., and Shushizadeh, M.R. (2015). Melamine supported on hydroxyapatite-encapsulated-γ-Fe_2O_3: a novel superparamagnetic recyclable basic nanocatalyst for the synthesis of 1,4-dihydropyridines and polyhydroquinolines. *Research on Chemical Intermediates* 41: 7227–7244. https://doi.org/10.1007/s11164-014-1808-1.

90 Ganja, H., Robert, A.R., Lavanya, P. et al. (2020). Y_2O_3/HAp, a sustainable catalyst for novel synthesis of furo[3,4-*b*]chromenes derivatives via green strategy. *Inorganic Chemistry Communications* 114: 107807. https://doi.org/10.1016/j.inoche.2020.107807.

91 Zarei, Z. and Akhlaghinia, B. (2016). Zn(II) anchored onto the magnetic natural hydroxyapatite (Zn^{II}/HAP/Fe_3O_4): as a novel, green and recyclable catalyst for A^3-coupling reaction towards propargylamine synthesis under solvent-free conditions. *RSC Advances* 6: 106473–106484. https://doi.org/10.1039/C6RA20501A.

92 Sun, W., Jiang, W., Zhu, G., and Li, Y. (2018). Magnetic Cu^0@HAP@γ-Fe_2O_3 nanoparticles: An efficient catalyst for one-pot three-component reaction for the synthesis of imidazo[1,2-*a*]pyridines. *Journal of Organometallic Chemistry* 873: 91–100. https://doi.org/10.1016/j.jorganchem.2018.07.039.

11

Electrocatalysis and Photocatalysis Using Hydroxyapatite-Based Materials

Eric Puzenat and Mathieu Prévot

Université de Lyon, Université Claude Bernard-Lyon 1, CNRS, IRCELYON-UMR5256, 2 av. A. Einstein, F-69626, Villeurbanne Cedex, France

11.1 Photocatalysis with Hydroxyapatite-Based Materials

In this section the basic principles of photocatalysis are first recalled. Then, the intrinsic properties and structures of hydroxyapatites (HA) are examined with regard to the principles of photocatalysis. It is clear that pure phases of HA cannot be photocatalytic. Therefore, different modification strategies are employed such as doping or preparing HA composites with TiO_2 or other photocatalytic materials. The use of such composites for the treatment of water and air shows an interest in coupling the properties of adsorption of HA to those of photocatalytic TiO_2. The scientific perspectives of the use of HA in photocatalysis are finally exposed.

11.1.1 Basic Photocatalysis Principles

As defined by IUPAC, photocatalysis deals with the change in the rate of a chemical reaction or its initiation under the action of radiation in the range from ultraviolet (UV) to visible light in the presence of a substance, called the photocatalyst that absorbs light. As any catalyst, the photocatalyst is not changed after each reaction cycle [1]. Photocatalysis lays on the semiconducting properties of the photocatalyst. Under right irradiation, meaning with photons having energy greater than the bandgap of the photocatalyst, electrons and holes are photogenerated in the photocatalyst in the conduction band and in the valence band, respectively. Holes are able to oxidize molecules having an electrochemical potential lower than the valence band and electrons to reduce molecules with potential higher than the conduction band as illustrated in Figure 11.1.

The process can be involved either in homogeneous or heterogeneous medium. In homogeneous photocatalysis, both the reactants and the photocatalyst are in the liquid phase. But it is commonly assumed that the general term "photocatalysis" deals with heterogeneous photocatalysis. In this case, the photocatalyst is a solid,

Design and Applications of Hydroxyapatite-Based Catalysts, First Edition. Edited by Doan Pham Minh.
© 2022 WILEY-VCH GmbH. Published 2022 by WILEY-VCH GmbH.

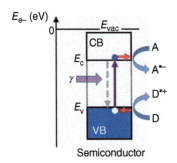

Figure 11.1 Energetic scheme of a photocatalytic reaction representing the different electron transfers; γ represents photons with energy higher than bandgap semiconductor E_G ($E_G = E_C - E_V$); E_C is the minimum energy level of the conduction band (CB); E_V is the maximum energy level of the valence band (VB); A is electron acceptor; D is electron donor (hole acceptor). Source: Eric Puzenat.

while the reactants are in gas or liquid phase. Heterogeneous photocatalysis processes have found applications for 30 years in fields ranging from water and air treatment, self-cleaning materials, and for 10 years in energy through the production of solar fuels [2–4]. Photocatalysts are inorganic semiconductors (SCs) as oxides, sulfides, or nitrides of transition metals. But the mostly known is titanium dioxide TiO_2 being the most efficient one. As explain before, under irradiation with light having energy equal or higher to that of the bandgap, an electron is promoted from the valence band to the conduction band; this will generate a hole–electron pair, according to the following reaction:

$$SC + h\nu \rightarrow e^- + h^+ \tag{11.1}$$

Electrons and holes are mobile in the all volume of the solid. First, charged species can recombine if the dielectric properties of the photocatalyst are not enough. On the contrary, electrons and holes reaching the surface of the photocatalyst can react with adsorbed chemical compounds. Electrons will reduce acceptor species, while holes will oxidize donor species. It is important to keep in mind that all electron transfers occur on the surface of the photocatalyst as well as all the reactions [5]. Surface electronic transfers with acceptors and electron donors lead to the formation of radicals and/or highly oxidizing ionic species.

A reaction with water or molecular oxygen, for instance, are reported as follows [5]:

$$H_2O + h^+ \rightarrow HO^{\cdot} + H^+ \tag{11.2}$$

$$O_2 + e^- \rightarrow O_2^{\cdot -} \tag{11.3}$$

For applications in environmental decontamination, the photogenerated radicals can then partially or totally oxidize the contaminating or toxic organic pollutants adsorbed on the surface of the solid. The efficiency of the photocatalyst is based on several characteristics; the light absorption capacity is a very important parameter for photogeneration of charges (e^-–h^+). But once generated, their recombination, which is a very rapid and very limiting phenomenon, must be minimized [6]. The different timescales of each elementary process involved for the photocatalytic water splitting, which can be generalized to all photocatalytic reactions, are illustrated in Figure 11.2.

11.1 Photocatalysis with Hydroxyapatite-Based Materials

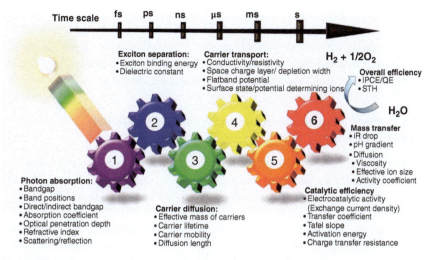

Figure 11.2 Parameters associated with photocatalysis. Overall water splitting is only successful for high efficiencies of all six gears depicted in the scheme. The different timescales of the reactions are also displayed. Source: [6]. Reproduced with permission of American Chemical Society.

Several parameters, besides the nature of the photocatalyst, can affect these two properties such as the morphology of the material, the size of the crystallites, the presence of structural defects, and the dielectric constant of the SC, which are crucial parameters. For the case of photocatalytic films, the epitaxial growth of the crystals is important [7–9]. The stability of photocatalysts in the environments in which they will be used is an important parameter to take into account for applications, so, for example, ZnO, which is an effective photocatalyst in water pollution control, cannot be applied because it photocorrodes in the water; finally, it is necessary to take into account the nontoxicity of the materials for their use in environmental decontamination. TiO_2 is the most widely used photocatalyst to date because it has all the properties previously mentioned. TiO_2 can exist in three common structures, which are anatase, rutile, and brookite. Other structures exist; TiO_2-B, ramsdellite-type TiO_2, hollandite-type TiO_2, and α-PbO_2-type TiO_2; the latter has been little studied. Although anatase is the photocatalyst that is often presented as the most effective form, all are active as well as their mixture [10, 11]. However, TiO_2 has a bandgap energy in the UV region (about 3.2 and 3.0 eV for anatase and rutile, respectively), which limits its use for heliophotocatalysis, that is, photocatalysis using direct sunlight, yet only about 4% of the light received at the Earth's surface is in the UV range [5, 12]. Also active research is being carried out to develop new photocatalytic materials active in the overall solar spectrum. Indeed, several visible light photocatalysts were developed, some examples being zinc and tungsten oxides or silver phosphate [13]. At the same time, however, composite photocatalysts were also considered; several combinations using different materials were studied; many of which were still based on TiO_2. The presence of more than one compound can increase the efficiency in the charges' generation and reduce their recombination, hence leading to materials with higher photocatalytic activity; moreover, photocatalysts with different bandgap values can be obtained [14]. HA was one of the materials studied while searching for alternative photocatalysts. The following sections will describe the structure and the most important properties of this compound; subsequently a review of the HA application as photocatalyst will be presented.

11.1.2 Hydroxyapatite Structure and Properties Implication in Photocatalysis

Hydroxyapatite (HA) is a phosphate with the formula $Ca_{10}(PO_4)_6(OH)_2$. It crystallizes in two main forms, either monoclinic or hexagonal (space groups P21/b and P63/m, respectively); the most common shape is the hexagonal shape. Calcium can be located in two cationic positions denoted Ca1 and Ca2 [15]. Hexagonal HA is biocompatible [16] and is, therefore, used to make bone implants. HA has very low solubility at neutral pH. However, very different acidity constants have been reported in literature. According to Bertazzo [17], pK_a values vary from 8.4 to 12.0. These differences may be due to the variation in the stoichiometry of the compound as well as the presence of carbonates in the HA network [18]. These adjustable properties of the surface acidity of HA will be a parameter, which must be taken into account in terms of the adsorption capacity of HA as a function of the charge of the molecules to be treated. In particular the rate of the photocatalytic reactions of discoloration of solutions of anionic and cationic dyes is dependent on the surface charge of the photocatalyst, and it is this adsorption step, which can be limiting [5, 19]. The acid–base properties of HA depend on their size and shape [20]. More precisely, HA is stable in basic solution but can be slowly dissolved in an acidic environment. HA dissolution kinetics do not seem to depend on its stoichiometry, but rather on the solubility of calcium or phosphorus in the medium [21, 22]. Due to its crystalline structure and its high specific surface area, HA can efficiently absorb metal cations, in particular divalent metals. Two mechanisms are possible, either by ion exchange with calcium, namely, the exchange of ions between calcium and divalent ions, or by a dissolution–precipitation mechanism [23]. Both mechanisms are based on the flexibility of the HA structure but above all on the surface concentration of calcium. This property of HA is used in the depollution of heavy metal water in wastewater and/or contaminated soils [24]. On the other hand, in photocatalysis, it can be either useful for facilitating the adsorption of molecules or ions such as NH_4^+ to be degraded in water treatment applications or disadvantageous inducing an adsorption competition between the molecules to be degraded and poisoning inorganic species for the photocatalytic reaction.

HA without stoichiometric defect is classified as insulators, that is to say, in photocatalysis fields, with a high bandgap (greater than 4 eV) [25]. Thus, the pure phase of HA has a bandgap value greater than 6 eV [26]. Therefore, the unmodified HA will only have photocatalytic activity in the UV–vis range, which is not useful for photocatalytic applications.

Therefore, to make HA photocatalytic, it is necessary either to introduce structural defects or to dope it in order to modify its electronic structure by reducing the value of the bandgap energy or by introducing acceptor (p doping) or donor (n doping) levels in the forbidden band inducing new electronic transitions, allowing the absorption of photons of energy in the 2–3.5 eV range and allowing the initiation of photocatalytic reactions. The other strategy is the coupling of HA with a photocatalyst to obtain a multifunctional catalyst. In the following sections, the different forms of photocatalytic HA from undoped single-phase materials to doped HA compounds and, finally, multiphase HA-based photocatalysts will be considered.

11.1.3 Single-Phase HA for Photocatalysis

Since HA without a stoichiometric defect is insulating, it has no photocatalytic activity using common UV–vis light source. However, with some forms of HA, photocatalytic properties have been observed. The most probable hypothesis put forward to explain these observations is the presence of defects in the crystal structure of HA or the modification of the electronic surface structure. The first study on photocatalytic monophasic HA was published almost 20 years ago and described the UV oxidation of methylmercaptan [27]. In this study, a first HA photocatalyst is prepared by precipitation followed by calcination at 200 °C. On this material, a 96% conversion of methylmercaptane is measured under UV light, while for HA prepared with a heat treatment at 1150 °C, no activity is detected. Electron paramagnetic resonance (EPR) measurements showed the presence of hydroxyl radicals HO· and superoxide radicals $O_2\cdot^-$ for HA treated at 200 °C, but not for that treated at 1150 °C [28]. It has been shown that heat treatment at 200 °C led to the formation of anionic vacancies in the HA network [29], modifying the electronic structure of HA and making possible the photogeneration of oxidizing radicals responsible for the degradation of methylmercaptan. Tanaka et al. [30] studied the role of phosphate groups on the photocatalytic properties of HA. They demonstrated that the P—O—H groups were the sites of adsorption of dimethyl sulfide. This was confirmed by Fourier transform infrared (FTIR) spectroscopy measurements showing a decrease in the intensity of the PO_4 peak of HA after photocatalytic experiments [31, 32]. The hypothesis put forward to explain this observation is the dissolution of phosphate groups under illumination and the formation of electronic surface defects. The photocatalytic activity for the monophasic HA is still not fully understood. Shariffudine et al. [32] observed that the HA obtained from spent mussel shells exhibited higher photocatalytic activity than a commercial material. They put forward the hypothesis according to which the excess carbonates of HA from mussel shells were at the origin of the photocatalytic activity without precisely explaining the mechanisms involved. As previously indicated, the pH value of the reaction medium plays a crucial role. Reddy et al. [33] have shown that the rate of photocatalytic degradation of an azo dye in acidic medium with monophasic HA is comparable to that of commercial TiO_2 P25 Evonik-Degussa but is slower in basic medium. The hypothesis put forward is the dependence of the surface charge (PZC) with the pH. The surface modification may also have an effect on the photocatalytic behavior. Tripathi et al. [34] prepared HA nanofibers by precipitation in the presence of yeast as a structuring agent. The HA obtained showed photocatalytic

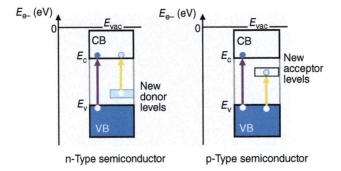

Figure 11.3 Electronic structure of an n-type and a p-type semiconductors. Source: Eric Puzenat.

activity for the degradation of methylene blue. However, it is important to keep in mind that photocatalytic tests carried out by discoloration of dyes can be distorted by parallel photochemical reactions [5]. Control of the light source is essential. Some forms of HA derived from fishbones are also believed to have photocatalytic activity. Piccirillo et al. [35] prepared HA from cod bones treated and calcined at 1000 °C showing photocatalytic properties. The hypothesis put forward to explain the photocatalytic activity is the formation of anionic oxygen vacancies in the HA network. Modeling by density functional theory (DFT) [36] would indicate that an oxygen deficiency in the phosphate groups would lead to the introduction of an electronic transition of 3.45 eV in agreement with experimental measurements. In addition, the DFT calculations also showed that an OH gap at the surface would lead to a surface electronic transition between 1.6 and 2.4 eV potentially active in visible light. These results remain theoretical, and such materials have not been able to be synthesized to date.

11.1.4 Doped Photocatalytic HA

One way to modify the properties of a photocatalyst is to dope it in the sense of SCs. The concept is to introduce defects in the electronic structure in particular in the bandgap. The dopants may be either an electron donor (n-type doping) or an electron acceptor (p-type doping). The interest of doping is to introduce new electronic transitions, allowing the absorption of photons of energy lower than that of the bandgap. It is illustrated on Figure 11.3.

Numerous studies have been carried out to spike HA with titanium. Wakamura et al. [37] were the first to publish on the doping of HA by Ti. The preparation method is the coprecipitation of calcium salts and titanium oxysulfate $TiOSO_4$ in the presence of phosphate. The material obtained showed a bandgap of 3.2 eV similar to that of TiO_2. The materials obtained showed a photocatalytic activity for the degradation of albumin in liquid phase and acetaldehyde in gas phase under UV irradiation. The Ti-doped HA materials exhibit SC properties similar to that of TiO_2 by electronic transition between 2p orbitals of the oxygen atom and the 3d orbital of titanium [26]. Different bandgap energies were measured for Ti-doped HAs, such as 3.75, 3.65, and 3.13 eV [26, 38–40]. All of these values are consistent with electronic transition energies observed with TiO_2. Structural modeling shows that Ti^{4+} would replace calcium in the Ca1 position [26]. This substitution would cause a narrowing of the volume of the HA network, due to the smaller ionic radius of titanium compared to calcium 0.68 vs. 0.99 Å, respectively [41, 42]. On the other hand, the incorporation of titanium into HA reduces its crystallinity, which is disadvantageous to the formation of SC properties [42, 43]. Under specific preparation conditions, Ti—OH groups can form on the surface of Ti-doped HA samples, promoting photocatalytic properties. The formation of such photocatalytic sites has been demonstrated by an IR band at 3737 cm^{-1} [39]. The materials obtained were tested on the photoinduced degradation of three different high molecular weight proteins, bovine serum albumin (BSA), neutral myoglobin (MGB), and basic lysozyme (LSZ). The results showed some activity with LSZ, but not with BSA or MGB. The authors conclude by proposing to use HA doped with Ti for blood purification phototherapies.

For the Ti-doped HAs, an increase in the photocatalytic activity was obtained by depositing copper on the surface of the material, the copper playing the role of cocatalyst. The EPR analysis made it possible to show a transfer of charges between the HA valence band doped toward the Cu^{2+} ions. The proposed explanation is a slowing down of recombination by electron trapping [44]. On the other hand, the incorporation of copper does not induce activity in the visible light illumination.

Wakamura et al. [45] treated Ti-doped HA in a solution containing Cr^{3+} ions. Different concentrations of chromium were considered (between 10^{-4} and 10^{-2} mol l^{-1}). An activity under irradiation in visible light was observed, but it was correlated with the formation of DMSO-d6 optical transitions of Cr^{3+} ions. The formation of a layer of Cr—OH groups on the surface of materials was demonstrated, which played a role in the activity with visible light [46].

For HA doped with vanadium, Nishikawa et al. [47] prepared materials by hydrothermal method, with VCl_3 as the source of vanadium. In the doped material, vanadium was at degree +5 as a substitute for calcium. The absorption of HA doped with vanadium goes up to 530 nm corresponding to a transition involving V^{5+}. It has been shown that the photocatalytic activity of these materials is very dependent on the calcination conditions and on the vanadium content.

11.1.5 Multiphasic HA-Containing Photocatalyst

Forming a junction between two SCs is a strategy that can improve charge separation. In the configuration illustrated in Figure 11.4, a photon is absorbed by each SC of different bandgap. The energy shift of the valence and conduction bands of the SCs leads to the transfer of the photogenerated electrons in the SC 2 into the conduction band of the SC 1 and to the transfer of the photogenerated holes in the SC 1 into the valence band of the SC 2.

As HA and TiO_2 have different bandgaps, the concept can potentially be applied to HA—TiO_2 biphasic systems.

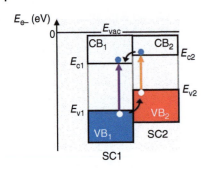

Figure 11.4 Electronic transfers at the junction between two semiconductors under illumination. Source: Eric Puzenat.

11.1.5.1 HA–TiO$_2$ Biphasic Composites

Combining HA with photocatalytic materials seems to appear to be the best strategy. The combination of TiO$_2$ with HA is based on the concept of increasing photocatalytic activity by enhancing the adsorption of molecules by the high absorbent properties of HA. Indeed, like any heterogeneous catalysis process, after the transfer of molecules from the fluid phase to the surface, the second step is the absorption of molecules at the surface of the photocatalyst [5]. Komazaki et al. [48] were the first to prepare an HA/TiO$_2$ composite. The materials show photocatalytic activity for the treatment of NOx under UV illumination. HA is combined with TiO$_2$ to allow the adsorption of pollutants on HA and increase the interaction with TiO$_2$ in a strategy of catalytic synergy by nanometric intimacy [49].

Anmin et al. prepared HA–TiO$_2$ by the hydrothermal method [50] and was tested by following the degradation of methylene blue. All three composites showed higher activity than either HA monophase or TiO$_2$ alone, which would indicate a synergistic effect.

Mitsionis et al. [51] prepared HA–TiO$_2$ by precipitation of HA in the presence of TiO$_2$ P25 commercial Evonik-Degussa. These composites have been tested for the treatment of NOx. The best composite is the one with a ratio of 50 wt%. Conversely, two-phase materials were synthesized using commercial HA and TiO$_2$ prepared in sol–gel. The best results were obtained for composites 50 and 67 wt% [52].

Yang and Zhang [53] described the performance of TiO$_2$ nanoparticles with a small amount of HA on their surface (maximum 1 wt% by weight); absorption of protein (bilirubin) is higher at the surface of composite materials and improved UV photodegradation of the protein. The material with the best performance is the one with a content of 0.75 wt%, which was not the one with the highest surface absorption. This shows that the surface diffusion of molecules and active species is the limiting step.

HA–TiO$_2$ composite films were prepared by magneton sputtering [35, 54, 55] to deposit multilayer TiO$_2$–HA coatings on a glass substrate followed by annealing at 500 °C to increase the crystallinity of the deposits. The film obtained has a formaldehyde degradation activity greater than the TiO$_2$ film. TiO$_2$ coatings with HA nanoparticles embedded in their structure have also been prepared using aerosol-assisted chemical vapor deposition (AACVD) [56]. In this work, despite the very low concentration of HA nanoparticles, their inclusion in the coatings led to increased photodegradation of methylene blue explained by an increase in adsorption.

11.1.5.2 HA–TiO$_2$ Multiphasic Composites

Other composites based on (HA—TiO$_2$), in which additional phases have been added to the HA—TiO$_2$ system, to shift the bandgap in the visible and/or to give additional properties to the material have been proposed.

HA has been associated with TiO$_2$-modified Ag nanoparticles [57]. This was done to make the material antibacterial, due to the known antimicrobial properties of Ag—TiO$_2$ systems [58]. Multiphasic Ag—TiO$_2$/HA was prepared on Al$_2$O$_3$ membrane. Membranes prepared in this way showed higher inactivation rates of the Gram-negative strain *Escherichia coli* under UV irradiation, compared with membranes with only HA or Ag—TiO$_2$. The increased antibacterial effect was mainly due to the antimicrobial action of silver, but silver nanoparticles could also reduce charge recombination, leading to higher photocatalytic performance.

Palladium (Pd) has also been added to the HA—TiO$_2$ systems; this noble metal is considered as a cocatalyst. A Pd—TiO$_2$—HA composite (Pd: 3 wt%, TiO$_2$: 25 wt%) was prepared by ultrasound-assisted precipitation [59]. The material was effective in the photodegradation induced by visible light of cyanides, showing an efficiency superior to that of the Pd—TiO$_2$ system. This better performance would be due to the increase in the adsorption by HA and the slowing down of the recombination by the presence of palladium.

Composites in HA, TiO$_2$, and β-Ca$_3$(PO$_4$)$_2$ have also been prepared [35]. The material obtained consisted of only about 2wt% TiO$_2$, the remainder being a mixture of HA and β-Ca$_3$(PO$_4$)$_2$. Methylene blue photodegradation is very rapid, despite the very low TiO$_2$ content. Again, the methylene blue test may be questionable. The authors suggest that TiO$_2$ was placed on the surface of a very porous HA-based material favoring the access of molecules to photocatalytic sites. The presence of tricalcium phosphate can improve the efficiency as observed by Ryu et al. [60] and Marquez-Brazon et al. [61].

11.1.5.3 Other Photocatalytic HA-Containing Composites

The HA was also combined with other compounds, which included either metals or other photocatalysts. A biphasic composite of HA—Pt was prepared by Vukomanovic et al. [62] by sonochemical precipitation, using either H$_2$PtCl$_6$·H$_2$O or C$_{10}$H$_{14}$O$_4$Pt as the source of platinum. The material appears to show photocatalytic activity with both UV and visible light in the degradation of methylene blue. The authors hypothesize a photocatalytic behavior with a debatable ligand-to-metal charge transfer (LMCT) mechanism. Several studies have been carried out on multiphase materials comprising iron oxides, potentially active in the visible [63]. α-Fe$_2$O$_3$—HA composites were prepared by forming HA nanoparticles on an α-Fe$_2$O$_3$ dendritic crystal, using a hydrothermal process [64]. The photocatalytic activity of the composite, tested on the degradation of methyl violet, was higher than the one for the single-phase α-Fe$_2$O$_3$ crystals. In addition, the material exhibits properties for the removal of Pb^{2+} in an aqueous medium. The interest of this type of multifunctional materials is that they can both degrade organic compounds and decontaminate water from heavy metals. The use of Fe_3O_4 in water treatment is based on its magnetic properties, facilitating its separation from the reaction medium. Yang et al. prepared Fe_3O_4/HA nanoparticles by both magnetic and photocatalytic precipitation [65].

Composites of HA with silver phosphate Ag_3PO_4 leads to a visible light photocatalyst [13]. Although composites comprising HA and Ag_3PO_4 were already known, their use was limited to biomedicine, due to the antibacterial nature of Ag_3PO_4 [66, 67]. HA nanoparticles loaded with Ag_3PO_4 were prepared by Hong et al. [68], showing good photocatalytic activity both in UV light and in visible light (methylene blue degradation test), superior to that of unmodified Ag_3PO_4. Li et al. [69] treated nanowires of HA in an $AgNO_3$ solution to obtain nanowires loaded with Ag_3PO_4. The materials, in addition to showing photocatalytic behavior in visible light, also had the ability to adsorb Pb (II) as well as to inactivate *E. coli*. An HA/Ag_3PO_4 magnetic photocatalyst was prepared by combining these two compounds with Fe_3O_4 [70]. The protocol followed to prepare the material was similar to that described previously [69], thus resulting in an HA—Ag_3PO_4—Fe_3O_4 composite with visible light catalytic activity and magnetic properties.

11.1.6 Summary and Outlook

HA-based photocatalysts are gaining interest in water and air treatment applications. The HA mainly brings its capacity of absorption of organic and inorganic species interesting in heterogeneous catalysis. By combining HA and TiO_2, it is easy to obtain a multifunctional material ideal for water treatment. The non-exhaustive work reported in the previous section indeed shows the potential but, at the same time, the critical obstacles to be removed in order to use HA as a photocatalyst. Various HA-based materials can be prepared using numerous techniques inducing varying crystal surface, shape, and size characteristics. The photocatalytic tests used to evaluate their activity are very diverse, which makes it difficult to compare performance. The most frequently used test is the discoloration of an aqueous solution containing a dye. Now in the field of photocatalysis, this type of test is controversial because the discoloration can be induced by phenomena other than a photocatalytic reaction. In many cases, this calls into question the observed photocatalytic activities. Despite this, however, some general considerations can still be made regarding HA-based photocatalysts. For single-phase HAs, for example, high pollutant degradation rates have always been achieved with a high catalyst load and/or with slow, much lower kinetics than with conventional photocatalysts like TiO_2. For the gas phase experiments, much better performance was observed. A reservation must be advanced because very rarely the dark adsorption steps are presented. An important limitation of the use of HA for the treatment of water is its possible dissolution according to the pH.

In the gas phase, the stability of the photocatalysts is hardly ever presented; as a reminder, a catalyst is always characterized by its activity, its selectivity, and its stability. Likewise, no characterization of the materials after their use is presented although this could explain the low activities observed. Overall, the experimental methodology used should be carried out on monophasic photocatalytic HAs, in a more rigorous way to better understand the phenomena involved by drawing inspiration from that carried out on TiO_2. The photoactivity of the Ti-doped HA appears to be the most effective; HA doped with Ti can in certain cases present an efficiency comparable to that of commercial with always the reserve of the heterogeneity of the experimental devices, making it difficult to compare the results. To improve the efficiency of photocatalytic reactions, multiphase compounds also appear to be of interest. Combining HA with other photocatalytic compounds in bi- or multiphase materials has been the approach that has led to highly efficient photocatalysts. The key criterion for these multicomposites is the HA/TiO_2 ratio. The combination of HA with other photocatalysts (i.e. Fe_2O_3, Ag_3PO_4), could lead to visible light photocatalysts or multifunctional materials, i.e. magnetic photocatalysts, although the results published would deserve to be reproduced. Regarding the applications of HA-based materials for environmental remediation, a special feature is surely their dual function photocatalyst/heavy metal absorber.

Nevertheless, HA is a material with an interesting potential for applications in depollution photocatalysis. In the same way that TiO_2 has been associated with activated carbon for water treatment, TiO_2 coupled with HA can find applications in water decontamination. The studies developed so far have shown that the best performing materials are obtained when the HA is combined with other photocatalysts. One of the reasons for the improvement of the photocatalytic activity is the high adsorption capacity of HA, which promotes and facilitates the degradation of pollutants. Overall, HA-based materials could be used for photocatalytic environmental remediation, due to their biocompatibility and low toxicity.

11.2 Electrocatalysis with Hydroxyapatite-Based Materials

In this section, the capability of HA ceramics to transport charges and the corresponding charge transport mechanisms are first examined. The resulting conductivity, combined with the rich surface chemistry and high surface area of HA nanocrystals, makes them an attractive material for electrocatalytic applications. First, the biocompatibility of HA has attracted considerable interest for its use as component in biosensors. Thus, a series of composite electrocatalytic HA sensors for the detection a wide range of hazardous analytes is presented. Then recently developed efficient HA-supported palladium electrocatalytic anodes for the oxidation of alcohols are detailed, as well as their application in direct alcohol fuel cell (DAFC) devices. A subsequent section will present how the ability of easily exchanging Ca^{2+} ions from the HA surface with transition metal cations has been leveraged to develop efficient HA-based oxygen evolution electrocatalysts for energy storage applications through water splitting. A scientific outlook and future research efforts are finally proposed.

11.2.1 Charge Transport Mechanism in Hydroxyapatites

The charge transport mechanism in HA ceramics has been investigated over the past 40 years but remains somewhat unclear. Early measurements evidenced polarization in HAs above 250 °C, associated with an activation energy of 30 kcal mol^{-1} and a resistivity of 4×10^9 Ω cm^{-1} at 400 °C, already suggesting ionic conductivity [71]. Following studies quickly dismissed Ca^{2+} [72] and PO$_4^{3-}$ [73] as candidate charge carriers for electrical conduction in HA. For instance, in 1986, Yamashita et al. [72] calculated that the diffusion rate of Ca^{2+} along the c-axis ($D = 1 \times 10^{-13}$ cm^2 s^{-1} at 700 °C) would yield a conductivity of 9×10^{-19} S cm^{-1}, a value far inferior to their measured HA conductivity of $\sigma = 1 \times 10^{-7}$ S cm^{-1} at 700 °C. Instead, researchers proposed protons (H$^+$), oxide ions (O^{2-}), or hydroxyl ions (OH$^-$) as majority charge carriers in HAs. Measurements from a 1978 study from Takahashi et al. [73] suggested that hydroxyl transport would be easier than proton transport because of weak hydrogen bonding in HAs. Another work from 1987 by Nagai and Nishino [74] suggested that room temperature conduction was driven by migration of protons in adsorbed water molecules bound to the polarizing phosphorus atoms of the ceramic, while high temperature conduction was still attributed to hydroxyl transport. However, this interpretation was later questioned because high temperature measurements were performed on nondense samples. In contrast, a 1995 study of Yamashita et al. [75] presented high-temperature conductivity measurements on 99% dense sintered HA samples. In this work, HA specimens were sintered at 1250 °C in dry air or in presence of water vapor, and the authors conducted impedance measurements on them to evaluate their conductivity at 700 and 800 °C. They observed that an optimal value of ambient vapor partial pressure was necessary to enable high conductivities (as high as 10^{-3} S cm^{-1} at 800 °C for ceramics sintered in the presence of vapor). The authors suggested that OH$^-$ vacancies in the HA structure were in equilibrium with the ambient vapor according to the following equation:

$$OH^- + OH^- \rightarrow O^{2-} + V_{OH^-} + H_2O_{(g)}$$

where V_{OH^-} indicates a vacancy of lattice OH$^-$. This equilibrium also indicates that conduction would be induced by proton transport in the HA structure. While the authors agreed with the previous assertion that lattice OH$^-$ was too far apart (0.344 nm) for efficient hydrogen bonding and thus efficient proton transport, they also pointed out that conductivity dependency on vapor pressure was not compatible with ionic conduction driven by the transport of lattice OH$^-$ charge carriers. Rather, the authors suggested that the distance between adjacent OH$^-$ and PO$_4^{3-}$ sites was short enough (0.307 nm) to enable the efficient formation of hydrogen bonds and that proton transport could explain the observed conductivity according to the following mechanism:

$$OH^- + OPO_4^{3-} + OH^- \rightarrow O^{2-} + H \cdots OPO_4^{2-} + OH^-$$
$$\rightarrow O^{2-} + OPO_4^{3-} + H \cdots OH$$
$$\rightarrow O^{2-} + OPO_4^{3-} + V_{OH^-} + H_2O_{(g)}$$

Furthermore, this model supported an observed aging of the ionic conductivity, whereby the conductivity initially increased as more OH⁻ vacancies (and, therefore, more free protons) were created through dehydration at high temperature, followed by a decrease in conductivity triggered by the disruption of conduction paths through the increase of vacancy density. Finally, the conductivity reached a plateau corresponding to the equilibrium density of vacancies (typically around 10^{-5} S cm^{-1}). Activation energies associated with ionic conduction were found to be in the 0.7–1.3 eV range, values consistent with proton condition. Interestingly, Laghzizil et al. [76] later reported similar activation energy and conduction mechanism.

In 2001, Nakamura et al. [77] alternatively used thermally stimulated depolarization current (TSDC) measurements – measuring the polarization current density as a function of applied temperature – to investigate the polarization and depolarization of HA ceramics sintered at 1250 °C for 2 hours in saturated water vapor. The authors studied the depolarization of samples polarized at different temperatures and under different electric fields. The peak temperature (corresponding to the highest current density) in the depolarization TSDC spectra was found to vary from 300 to 450 °C and to be highly dependent on the polarization temperature and electric field, suggesting that the polarization was due to ionic transport in the lattice rather than to a ferroelectric–paraelectric phase transition. HA crystals present a lattice of hydroxyl anions (OH⁻) situated at the center of calcium ion (Ca^{2+}) triangles and assembled in columns along the c-axis of the unit cell. Some Ca^{2+} ions thus define a screw axis around the hydroxyl columns, while additional Ca^{2+} is organized in columns parallel to the columnar OH⁻. Possible ionic charge carrier in this structure can be either protons (H$^+$), oxide ions (O^{2-}), or lattice hydroxyl anions (OH⁻). Based on the activation energies (0.84–0.89 eV) and relaxation times (0.077–0.43 ms) extracted from the TSDC analysis, the authors concluded that HA ceramics were more likely polarized by proton transport inside the OH⁻ columnar channels of the structure, rather than by the migration of OH⁻. More precisely, protons were suggested to migrate through vacancies (O^{2-}) in the columnar hydroxyl sites.

In 2007 and 2008, Gittings et al. [78, 79] studied the polarization of HA ceramics and set out to compare conductivity measurements between porous and dense samples. At low temperature (below 200 °C), porous samples displayed higher conductivities than dense samples. This was attributed to the higher surface area of porous samples, leading to more efficient surface conduction. On the contrary, at high temperature (700–1000 °C), dense samples displayed higher conductivities, which was associated to bulk ionic conduction being dominant. At intermediate temperature (200–700 °C), the conductivity of HA was dependent on its thermal history and degree of hydration and in particular the amount of surface-bound water. Furthermore, the relatively high measured activation energy of ~2 eV for conduction at high temperature was ascribed to ionic conduction by hydroxyl ions, contradicting the previously discussed reports by Yamashita et al. and Nakamura et al. that reported lower activation energies associated with proton conduction.

11.2.2 Electrocatalytic Sensors

Despite their relatively poor conduction properties at room temperature, HA ceramics have been researched as components in electrocatalytic biosensors, usually in combination with a more conductive material (e.g. metal nanoparticles or reduced graphene oxide). Indeed, their excellent adsorption properties and high surface area make them prime candidates for the anchoring of electrocatalytic species capable of a fast electrochemical response to concentration changes of a designated substrate. Furthermore, the biocompatibility of HAs makes them attractive for biomedical devices, particularly in combination with enzymes that can act as electrocatalytic biological molecules. In this context, the adsorption mechanism of electrocatalytic enzymes on HA is commonly believed to involve electrostatic interaction between the negatively charged carboxylate moieties of the protein and the Ca^{2+} sites of the ceramic and/or the positively charged NH_3^+ moieties and the phosphate groups of the ceramic surface [80]. A wide range of HA-based electrocatalytic sensors has been developed for a variety of analytes over the last decade, as illustrated in the following paragraphs, and for each sensor, a few important parameters are studied to benchmark its performance as follows:

1) *Linear range*: This corresponds to the range of analyte concentrations over which the electrocatalytic current produced by the sensor is directly proportional to the analyte concentration.
2) *Sensitivity*: This corresponds to the slope of the current-concentration line over this linear range. It provides a standardized way of expressing how much output (current density) change is produced by a change in input (analyte concentration).
3) *Detection limit or limit of detection (LOD)*: This provides an estimation of the lowest detectable amount of analyte. It is calculated from the standard deviation of the current-concentration line (σ) and the sensitivity (S) according to the following formula:

$$\text{LOD} = 3.3 \times \frac{\sigma}{S}$$

In 2009, You et al. studied the electrocatalytic detection of hydrogen peroxide [81] on a sensor composed of redox-active hemoglobin (Hb) immobilized on an Au—HA nanocomposite. Gold nanoparticles are traditionally employed in biosensors because of their ability to efficiently stabilize a wide range of biomolecules, stemming from a high surface energy, and their high surface-to-volume ratio, allowing for the binding of a large number of molecules. In this particular study, the authors synthesized HA nanotubes from the dropwise addition of a H_3PO_4 solution to a $Ca(OH)_2$ solution at pH 7.0. The gold nanoparticles were prepared by a citrate colloidal route and subsequently attached to the HA nanotubes by simply mixing the two materials in solution and under sonication, resulting in an Au/HA nanocomposite suspension at 0.2 mg ml^{-1}. Then 5 μl of this nanocomposite suspension was cast on a glassy carbon electrode (GCE), along with 10 μl of a 5 mg ml^{-1} Hb solution to afford, after drying, the desired electrode. Scanning electron microscopy

(SEM) revealed intertwined HA nanotubes forming a porous structure onto which the gold nanoparticles were well dispersed. A cyclic voltammogram of this electrode revealed well-defined peaks associated to a reversible electrochemical process at a formal potential of -340 ± 2 mV vs. the saturated calomel electrode (SCE) at pH 7.0, consistent with the signal of the electroactive Fe^{III}/Fe^{II} center of the hemoglobin molecule, and thus confirming the efficient adsorption of the biomolecule onto the nanocomposite. The composite electrode was then tested for the reduction of H_2O_2, a process for which Hb is well known to be electrocatalytic. As expected, after addition of the H_2O_2, the voltammogram displayed a drastic enhancement of the reduction current, due to the electrocatalytic behavior of the adsorbed hemoglobin. Furthermore, the sensor exhibited a relatively fast response to concentration changes in H_2O_2. Indeed, upon successive additions of H_2O_2, a stable current value could always be reached within 15 seconds, and a linear response was observed in the 0.5–25 µM H_2O_2 range with a detection limit of 0.2 µM, the lowest reported value for H_2O_2 biosensor at the time of the study. Finally, the proposed sensor retained 82.9% of its initial response to 20 µM H_2O_2 after being kept at 4 °C for 30 days and lost 6.6% of its initial response to 0.1 mM H_2O_2 after 100 consecutive measurements, demonstrating adequate stability. Interestingly, the same year, a different group proposed a different electrocatalytic H_2O_2 sensor based on the immobilization of horseradish peroxidase (HRP) on SiO_2—HA thin films. In this work by Wang et al. [82], a sol of HA nanoparticles was prepared by slowly dripping a 0.01 M solution of phosphoric acid into a saturated calcium hydroxide solution, and a desired amount of HRP was added to this sol to form an HRP–HA solution at 2.0 and 10 mg mL^{-1} in HRP and HA, respectively. Finally, this solution was drop-casted onto a GCE, followed by the drop-casting of a silica sol to afford a SiO_2/HRP–HA/GCE electrode. Likewise, to what was previously observed for hemoglobin, a cyclic voltammogram of this electrode revealed a reversible redox signal at -0.35 V vs. SCE, consistent with the Fe^{III}/Fe^{II} couple of the heme in HRP, confirming the electrochemical accessibility of the electrocatalytic center. Based on the surface area of the corresponding redox peaks, the authors calculated a high surface coverage of 2.13×10^{-11} mol cm^{-2} of HRP on the HA. When exposed to H_2O_2, the sensor displayed the expected electrocatalytic activity, characterized by a drastic increase in reduction current and the disappearance of the Fe(III)/Fe(II) oxidation wave. The authors observed a significantly higher electrocatalytic reduction current for the SiO_2/HRP–HA/GCE than for a SiO_2/HRP/GCE, a high number of adsorbed HRP enzymes due to the high density of adsorption sites on the surface of HA. The necessity of the SiO_2 layer was justified as a protection against HA leaching under operating conditions. The response of the sensor was tested at pH 7.0 and -0.4 V vs. SCE and was found to be fast (five seconds to reach 95% of the steady-state current), and the catalytic current was found to be proportional to the H_2O_2 concentration in the 1.0–100 µM range, with a slope equation of $I(\mu A) = 0.2770 + 0.0506C$ (µM), resulting in a detection limit of 0.35 µM and a sensitivity of 50.56 mA M^{-1}. The sensor showed good stability in 0.1 M phosphate buffer (pH 7.0) under continuous cyclic voltammetry between -0.7 and 0 V vs. SCE for 2 hours. After this stability measurement, the peak current decreased by 4.1% only. Additionally, after storage in 0.1 M phosphate buffer for two weeks at 4 °C, the sensor retained 95% of its initial performance. A third and final HA-based H_2O_2 biosensor decorated with hemoglobin was proposed by Zhao et al. [83]. In this work, multiwall carbon nanotubes (MWCNTs) were employed to generate a HA/MWCNTs composite. To synthesize this material, 10–30 nm MWCNTs were oxidized by nitric acid to generate carboxylate groups on their surface. They were then reacted with calcium chloride and Na_2HPO_4. The resulting suspension of HA/MWCNTs composite was drop-casted on a GCE, and then a phosphate buffer containing hemoglobin was casted on top of the composite to form the resulting Hb/HA/MWCNTs/GCE electrode. Once again, the reversible redox signal of the hemoglobin Fe^{III}/Fe^{II} site was observed at −0.335 V vs. SCE, indicating the electrochemical accessibility of the electrocatalytic site. By plotting the peak-to-peak separation of the voltammogram vs. the scan rate, the authors calculated an apparent electron transfer rate constant of 5.05 ± 0.41 s^{-1}, which was found to be higher than values calculated for Hb immobilized on mesoporous TiO_2, MWCNTs, or nanocrystalline SnO_2. Therefore, HA was found to perform as a better substrate for this electrocatalytic application. The surface coverage of Hb was calculated to be $(9.25 \pm 0.69) \times 10^{-10}$ mol cm^{-2}, and the sensor was found to reduce H_2O_2 electrocatalytically, although with a lower performance and with linearity over a smaller concentration range than the previously mentioned studies (0.5–2 µM) and with a detection limit of 0.09 µM. This sensor was also found to retain 90% of its initial activity after storage in a phosphate buffer for three weeks at 4 °C. Overall, these seminal works on H_2O_2 sensing using HA-based electrodes demonstrated the viability of HA as a substrate for the creation of biocompatible high surface area, fast-responding electrocatalytic sensors for the detection of H_2O_2. They demonstrated that the high density of adsorption sites of the HA surface allowed for the immobilization of a high number of electrocatalytic species and that the conductivity of the material was sufficient enough to generate a measurable response to small change in substrate concentration.

HA nanoparticles were subsequently applied to a variety of electrocatalytic sensors toward a wide range of analytes and particularly for the detection of harmful pollutants or medically relevant biomolecules. Thus, in 2010, Yin et al. designed an electrocatalytic sensor for the detection of 4-nitrophenol (4-NP), a hazardous waste with high environmental impact due to its long persistence [84]. Interestingly, the sensor was only composed of HA nanoparticles dispersed on a GCE. While no oxidation signal was observed in the absence of 4-NP for GCE and HA/GCE electrodes, a clear oxidation wave appeared when the electrodes were exposed to 4-NP. It appeared that the presence of HA slightly increased the oxidation current density of 4-NP. Once again, pH 7.0 was found to be optimal in terms of oxidation current density. The catalytic rate constant of HA for the oxidation of 4-NP was determined using chronoamperometry measurements and according to the equation:

$$\frac{I_{cat}}{I_L} = \pi^{1/2}(k_{cat}c_0 t)^{1/2}$$

where I_{cat} was the measured catalytic current, I_L was the measured current in the absence of 4-NP, k_{cat} (M^{-1} s^{-1}), c_0(M) was the concentration of 4-NP, and t (s) was the elapsed time. Based on this equation, k_{cat} was estimated to be 8.78×10^3 M^{-1} s^{-1}, indicating a significant electrocatalytic activity of HA toward 4-NP oxidation. Furthermore, the authors calculated an adsorption capacity of 7.12×10^{-10} mol cm^2 of the HA electrode toward 4-NP, stemming from the excellent adsorption properties of the material. A linear response was observed for 4-NP in the 1–300 µM range, with a detection limit of 0.6 µM, which was found to be average when compared to existing electrocatalytic sensors. Moreover, the HA sensor retained 82% of its initial performance after 30 days of storing, and it was found that a wide number of other substrates did not interfere with the detection of 4-NP. Indeed, 500 µM of Na$^+$, Ca^{2+}, Mg^{2+}, Fe^{3+}, Ba^{2+}, Al^{3+}, Zn^{2+}, Ni^{2+}, Cu^{2+}, F$^-$, Cl$^-$, SO$_4^{2-}$, CO$_3^{2-}$, PO$_4^{3-}$, NO$_3^-$, or 200 µM of 2,4-dinitrophenol, pyrocatechol, hydroquinone, o-nitrobenzoic acid, m-nitrobenzoic acid, and p-nitrobenzoic acid did not interfere with the oxidation signal of 10 µM of 4-NP. However, it was found that phenol, 2-nitrophenol, and hydroxyphenol interfered with the detection of 4-NP. Finally, although no 4-NP could be detected in water samples collected in different environment, the authors found that they could reach 95.86–104.38% recovery rates for 4-NP added by the standard addition method in water samples from river, lake, waste, and tap. This study suggests that HA, due to its numerous adsorption sites, can be used to produce a measurable electrochemical response toward a pollutant such as 4-NP, which could be leveraged for the analysis of water samples.

Wang et al. proposed around the same time to utilize the inhibition of HRP by cyanide ions to create an HA-based composite cyanide biosensor involving chitosan and the HRP as sensing element [85]. In this work, HA was electrodeposited onto a GCE covered by a track-etched porous polycarbonate template from a solution containing 42 mM of Ca(NO$_3$)$_2$ and 25 mM of NH$_4$H$_2$PO$_4$. The polycarbonate template was subsequently dissolved in chloroform, affording a nanowire array of HA on the electrode surface. The nanowires were regular and ordered, with an average diameter of 200 nm and an average length of 1 µm, with a density of 5×10^8 cm^{-2}. Then, a chitosan (CHIT, a polysaccharide biopolymer)/HRP thin film was cast on top of this HA array to afford the final CHIT-HRP/HA/GCE sensor. When submitted to cyclic voltammetry in phosphate buffer (pH 7.4), the sensor did not display the reversible redox waves expected from the HRP Fe^{III}/Fe^{II} center, suggesting a limitation in charge transfer from the HRP to the GCE, although this was not commented on by the authors. Regardless, an increase in reductive current density was expectedly observed in the presence of H$_2$O$_2$. On the contrary, when cyanide is added, the H$_2$O$_2$ reduction current is inhibited as CN$^-$ ions bind to the HRP active sites, causing their deactivation. This decrease in reduction current density is close to proportional to the amount of CN$^-$ in a 2–80 ng ml^{-1} range, with a detection limit of 0.6 ng ml^{-1}, with no interference from other common ions (Cl$^-$, CO$_3^{2-}$, PO$_4^{3-}$, SO$_4^{2-}$, S^{2-}, Na$^+$, K$^+$, Hg^{2+}, Pb^{2+}, Cd^{2+}). The sensor was regenerated by a simple 60-second rinse by tap water and performed well on the detection of cyanide in river water, distilled wine, and cassava starch. Furthermore, the biosensor was stable for 20 days, being tested twice a day, and then started to show a decline in performance and stability. The same group proposed another HA-chitosan biosensor for the detection of phenolic compounds [86]. In this case, they functionalized the HA with tyrosinase, an enzyme that generates quinones from phenolic compounds in the presence of O_2. More precisely, HA nanoparticles were prepared by a hydrothermal method and mixed with a chitosan solution, to which a tyrosinase solution was added. This solution was subsequently drop-casted onto a gold electrode to form the sensor. Cyclic voltammograms were acquired on HA/chitosan/tyrosinase/Au, chitosan/tyrosinase/Au, and HA/chitosan/Au electrodes, and it appeared that while all three electrodes exhibited a similar voltammogram in the absence of catechol, only HA/chitosan/tyrosinase/Au displayed an enhancement of reduction current upon the addition of catechol. This enhancement was attributed to the following process:

$$\text{cathechol} + \text{tyrosinase}(O_2) \rightarrow o\text{-quinone} + H_2O$$

$$o\text{-quinone} + 2H^+ + 2e^- \rightarrow \text{cathechol}$$

In this situation, the reduction of the quinone is the process monitored at the electrode. Interestingly this suggested that not only the tyrosinase was necessary to trigger the electrocatalysis, but the presence of HA was also crucial to generate a measurable current upon the addition of 0.1 mM catechol. The sensor responded quickly to the addition of catechol (reaching 95% of the steady-state current within 8 seconds) and further responded to the presence of different phenolic compounds with an increasing sensitivity for the molecules of m-cresol, phenol, and catechol. The linear detection range was found to be 10 nM–5 µM for m-cresol, 70 nM–5 µM for phenol, and 10 nM–7 µM for catechol with a detection limit of 5 nM. The sensor was stable for 18 days and could provide repeatable readings with standard deviation of 2.5% over 10 successive determinations, and did not suffer from the interference of ascorbic acid, uric acid, caffeine, H_2O_2, and glucose. Finally, it was found that the use of HA allowed reaching a much higher sensitivity, a wider linear range, a lower detection limit, and a longer stability than other previously reported substrates (poly-3,4-ethylenedioxythiophene, colloidal gold on graphite, mercaptopropionic acid-modified gold, and Nafion-coated ZnO).

Another example of HA-supported tyrosinase was proposed by Kanchana et al. for the sensing of L-tyrosine [87]. The authors doped HA nanoparticles with iron to improve electron transfer through the material. These nanoparticles were produced by microwave heating (600 W for 35 minutes) of calcium hydroxide and hydrogen orthophosphate following the reaction:

$$10Ca(OH)_2 + 6(NH_4)_2HPO_4 \rightarrow Ca_{10}(PO_4)_6(OH)_2 + 6H_2O + 12NH_4OH$$

To this solution, a 1 M% concentration of iron chloride was added to achieve the doping of HA nanoparticles whereby Fe^{3+} replaces Ca^{2+} in the structure, resulting in a light orange powder. Subsequently, the nanoparticle suspension was mixed with a solution of tyrosinase and then cast onto a GCE. Electrochemical impedance spectroscopy was used to measure the charge transfer resistance of each electrode: $128\,\Omega\,cm^{-2}$ for GCE, $4319\,\Omega\,cm^{-2}$ for tyrosinase/GCE, $2700\,\Omega\,cm^{-2}$ for tyrosinase/HA/GCE, and $1973\,\Omega\,cm^{-2}$ for tyrosinase/Fe-HA/GCE. This demonstrated that doping the HA substrate resulted in an improved charge transfer to the electrolyte compared with the undoped substrate. When L-tyrosine was put in the presence of the tyrosinase/Fe-HA/GCE sensor in phosphate buffer (pH 7.0), an electrocatalytic oxidation wave was observed in the voltammogram, corresponding to the oxidation of L-tyrosine into L-dopa by the enzyme. The electrocatalytic response was linear in the 0.1–11 µM range, with a detection limit of 0.24 µM. Moreover, urea, sucrose, galactose, boric acid, and oxalic acid were found not to interfere with the detection of L-tyrosine, and the sensor retained 80% of its initial performance after two weeks of storage in a pH 7 buffer solution at 4°C. The same research group further modified their microwave synthesis and added ethylenediaminetetraacetic acid (EDTA) as a means to control the size of the resulting HA nanoparticles [18]. Impedance spectroscopy revealed that using EDTA resulted in a decrease in the charge transfer resistance from 2889 to 970 $\Omega\,cm^{-2}$, and cyclic voltammograms revealed an increase in electrochemically active surface area (ECSA) from 4.641×10^9 to 5.772×10^9 mol cm^{-2}, indicating a beneficial effect of EDTA-assisted HA (E-HA) synthesis for electrochemical applications. The E-HA-modified GCE electrode was then tested for uric acid detection. The electrode modified with E-HA showed slightly improved current response compared with bare GCE (27.1 vs. 25 µA), probably due to increased surface area. The response was found to be linear in a 0.1–30 µM concentration range, with a detection limit of 142 nM, and to be very fast (typically within five seconds). This HA-based sensor was then used to detect uric acid in biological samples (human urine and human serum) with a 98.5–102.1% recovery rate.

In 2015, Bharath et al. developed a reduced grapheme oxide (rGO) and HA-based composite prepared by hydrothermal growth of 1D HA nanorods directly on graphene oxide [88]. To do so, $CaCl_2\cdot2H_2O$ and $(NH_4)_2(HPO_4)$ were added to a solution of graphene oxide in deionized water and heated to 180°C for 12 hours. The resulting composite consisted in 60–85 nm long HA nanorods with a diameter of 32 nm, dispersed on sheets of rGO, resulting in a Brunauer-Emmet-Teller (BET) surface area of 120 $m^2\,g^{-1}$. The suggested growth mechanism was the following: Ca^{2+} ions were electrostatically adsorbed to epoxy, hydroxyl, and carboxyl groups of the GO and subsequently reacted with phosphate ions to trigger the growth of HA nanocrystals, while GO is simultaneously reduced to rGO. The resulting composite was then drop-cast onto a GCE and functionalized with a glucose oxidase (GOx) solution to generate a GOx/HA/rGO/GCE sensor. When subjected to cyclic voltammetry in phosphate buffer (pH 7.0), the electrode displayed a reversible redox signal at −0.3 V vs. Ag/AgCl. This redox process was associated with the reversible conversion of the flavin adenine dinucleotide (FAD) cofactor of the GOx:

$$GOx - FAD + 2e^- + 2H^+ \leftrightarrow GOx - FADH_2$$

This confirmed the electrochemical accessibility of the enzymatic sites of the sensor. Moreover, the authors confirmed that in the presence of dissolved oxygen, an increase in reduction current was observed, due to the following process:

$$GOx - FADH_2 + O_2 \rightarrow GOx - FAD + H_2O_2$$

When glucose is added to the electrolyte, the concentration of oxygen decreases according to the following reaction:

$$GOx - FAD + glucose \rightarrow GOx - FADH_2 + gluconolactone$$

$$GOx - FADH_2 + O_2 \rightarrow GOx - FAD + H_2O_2$$

Therefore, with increasing glucose concentration, a lower catalytic reduction current for O_2 is observed, allowing for the indirect electrocatalytic detection of glucose by the sensor. The amperometric determination of glucose was performed with a rotating disk electrode at 1500 rpm to reduce mass transport limitations. The oxygen content expectedly decreased with each addition of glucose, and it was found that the GOx/HA/rGO/GCE electrode presented a linear steady-state response in the 0.1–19.2 mM glucose concentration range, with a detection limit of 0.03 mM and a sensitivity of 16.9 μA mM^{-1} cm^{-2}. Moreover, the sensor did not suffer from the interference of dopamine, uric acid, ascorbic acid, and L-cysteine, and retained 89.4% of its initial performance after storage for 25 days at 4 °C in phosphate buffer. Finally, the sensor could be employed to measure glucose levels in urine samples of both healthy and diabetic patients, with a recovery rate of 96.3–103.6%, and overall was found to perform better than previously reported biosensors. More recently, the same group developed an HA-based sensor for the detection of hydrazine, a hazardous industrial pollutant [89]. In this study, the authors synthesized 1D HA nanorods using adenosine 5′-triphosphate (ATP) as an organic phosphate source and $CaCl_2$ as a calcium source under hydrothermal conditions (180 °C for 12 hours). Colloidal gold nanoparticles were then prepared by the citrate route and were adsorbed on the HA nanorods, through the interaction of the negatively charged citrate ligands with the Ca^{2+} sites of the HA. The resulting HA/Au nanocomposites consisted of 10 × 65 nm HA nanorods decorated with Au nanoparticles exhibiting a diameter of 8 nm. A suspension of the Au/HA composite was drop-cast onto a GCE to generate an Au/HA/GCE sensor. This sensor was tested, with a rotating disk electrode setup, for its electrocatalytic properties toward hydrazine oxidation:

$$N_2H_4 \rightarrow N_2 + 4H^+ + 4e^-$$

Cyclic voltammetry revealed that the HA-supported sensor generated a more intense current at a lower potential (0.47 V vs. Ag/AgCl) than a bare Au/GCE reference. The amperometric oxidation of hydrazine revealed that the sensor showed a linear response to hydrazine concentration in the 0.5 μM–1.43 mM concentration range, in phosphate buffer (pH 7.0), with a detection limit of 0.017 μM and a sensitivity of 0.05 μA μM^{-1} cm^{-2}. The sensor also displayed a fast response to changes in hydrazine concentration (within 10 seconds). These performances were found to be better than previously reported hydrazine sensors based on gold nanoparticles, ZnO nanorods, or Pt–TiO_2. Moreover, the sensor did not suffer any interference from ascorbic acid, dopamine, uric acid, $ZnCl_2$, $CoNO_3$, $NiCl_2$, $CaCl_2$, Na_2SO_3, and KCl, and was found to retain 91.2% of its initial performance after 10 days of storage in air. Finally, the Au/HA/GCE sensor was able to detect 50 μM of added hydrazine in various water samples from lake, river, and tap, with a ~100% recovery rate.

An alternative HA-based hydrazine sensor was proposed in 2017 by Gao et al. [90], who developed an rGO/HA composite in an attempt to overcome the poor conductivity of HA by mixing it with a conductive material. Similarly to what was observed by Bharath et al. [88], the coprecipitation of HA from $Ca(OH)_2$ and H_3PO_4

on GO resulted on the in situ reduction of this substrate to form rGO. The authors further mixed their HA/rGO composite with chitosan and subsequently drop-cast this solution onto GCE to form an HA/rGO/chitosan/GCE sensor. This sensor showed a linear response in the 2.5 µM–0.26 mM range and a different one in the 0.26–1.16 mM range with associated sensitivities of 4.21 and 1.61 µA µM^{-1} cm^{-2}, respectively. Interference studies showed that hydrazine could be well detected in the presence of H_2O_2, glucose, glutathione, urea, ethanol, methanol, ascorbic acid, acetic acid, formic acid, citric acid, and uric acid. Finally, the sensor was stable, with an attenuation in current of only 8% after 30 days of storage in phosphate buffer.

Recently, Iyyappan et al. set out to conduct a more in-depth analysis of the influence of the morphology and surface properties of HA in its performance as component of electrocatalytic sensors [91]. They prepared HA powder through a standard precipitation of calcium nitrate and diammonium hydrogen phosphate at pH 10. After centrifugation, washing with distilled water, and redispersion of the colloid, the powder was dried and calcined at 600 °C for 2 hours in air, resulting in sample HA-0T. Furthermore, the authors varied the synthesis with several treatments before drying and calcinating the colloid to afford a series of different HA materials: refluxing for one hour (sample HA-0T$_{R1}$), using a hydrothermal treatment at 100 °C for 1 hour (sample HA-0T-373$_{H1}$), or subjecting the powder to ultrasonication for one hour (sample HA-0T$_{U1}$). Furthermore, other samples were prepared by adding 4 wt% of Triton X-100, a widely used surfactant, to the calcium nitrate solution before precipitating the HA powder. The same series of treatments were applied to the resulting colloidal suspension, resulting in samples HA-4T, HA-4T$_{R1}$, HA-4T-373$_{H1}$, and HA-4T$_{U1}$. Here, the use of Triton X-100 was expected to increase the porosity of the final material, according to previous works [92, 93]. X-ray diffraction (XRD) patterns of all the materials confirmed that they were all composed of crystalline HA. Furthermore, the density of acidic sites on each material was explored through NH_3 temperature-programmed desorption (TPD) experiments (see Table 11.1). In general, the use of Triton X-100 did not have a clear effect on the acidity of HA, whereas HA treated by reflux showed significantly higher surface acidity than the other samples. Interestingly, this higher acidity was well correlated with a higher carbonate/phosphate ratio on the HA surface, as confirmed by FTIR and XPS analysis. To quantify the amount of each species on the material surface, the authors used XPS peaks with binding energies of 288.6 eV (subpeak of the carbon 1s signal) for the carbonate and peaks of binding energies 134.1 and 133.1 eV, corresponding to P=O and P—O bonds, respectively, for the phosphate. The observed surface carbonate was assumed to be formed from atmospheric CO_2 during calcination, by filling of OH$^-$ vacancies in the HA crystal structure, suggesting that refluxing the colloid caused an increase in OH$^-$ vacancies. The morphology of each sample was further observed by SEM, revealing that all samples were composed of HA nanoparticles and that refluxed HA presented a higher estimated percentage of rodlike particles (60–80%) compared with other samples (30–50%) as presented in Table 11.1. Transition electronic microscopy (TEM) further revealed that HA formed in the presence of Triton X-100 resulted in nanopowders, exhibiting significantly less aggregates than their

Table 11.1 Correlation between uric acid sensing capability and chemical, morphological, and structural properties of HA.

Electrode	I_p (µA)	Acidity (µmol g^{-1})[a]	Fraction of rodlike particles (%)[b]	Mesopore volume (cm^3 g^{-1})[c]
CPE	43.3	n/a	n/a	n/a
HA-0T-CPE	56.7	151	30.5	0.291
HA-4T-CPE	68.9	125	42.5	0.352
HA-0T-373$_{H1}$-CPE	64.6	125	43.0	0.383
HA-4T-373$_{H1}$-CPE	65.4	158	49.0	0.437
HA-0T$_{R1}$-CPE	51.0	176	62.8	0.201
HA-4T$_{R1}$-CPE	96.0	207	82.1	0.235
HA-0T$_{U1}$-CPE	55.8	102	43.0	0.328
HA-4T$_{U1}$-CPE	65.7	129	28.2	0.375

CPE: carbon paste electrode; I_p: peak uric acid oxidation current.
a) Calculated from NH$_3$ temperature-programmed desorption.
b) Estimated from scanning electron microscopy images.
c) Calculated from N$_2$ adsorption isotherms.
Source: [91]. Reproduced with the permission of American Chemical Society.

surfactant-free counterparts, and this observation was consistent regardless of the other synthesis parameters. Additionally, N$_2$ sorption measurements revealed that the use of Triton X-100 resulted in an increase in pore volume of 20% for HA-4T vs. HA-0T and of ca. 15% for the other samples (see Table 11.1). Each HA powder was then grinded with graphite powder and silicon oil to form a modified carbon paste electrode (CPE). Each electrode was subsequently contacted with a copper wire and polished before electrochemical testing. Electrochemical impedance spectroscopy was employed to measure the charge transfer resistance of each electrode, and it appeared that HA samples prepared using Triton X-100 possessed a lower resistance, which was linked to the previously mentioned lower aggregation and consistent with previous measurements by Kanchana et al. on HA particles synthesized in the presence of EDTA [94]. Overall, it appeared that HA-4T$_{R1}$ presented the best properties with a charge transfer resistance of 6.2 Ω only.

The electrocatalytic properties of each sample were then investigated for the sensing of uric acid, and the corresponding peak oxidation currents (I_p) are reported in Table 11.1. First, all samples containing HA showed better performance than a bare CPE, confirming the beneficial effect of HA on the electrochemical sensing of uric acid previously reported by Kanchana et al. [94]. Then it appeared that using Triton X-100 in the electrode preparation always resulted in an increase of peak current density for uric acid sensing (as monitored by cyclic voltammetry), which was again attributed to better particle distribution and lower aggregation resulting in higher accessible surface area for the analyte. Finally, HA-4T$_{R1}$, the material synthesized using Triton X-100 and posttreated with reflux, showed the highest current density for uric acid oxidation of all the electrocatalysts surveyed in the study. However, among the materials synthesized in the absence of Triton X-100, HA-0T-373$_{H1}$, which underwent a hydrothermal posttreatment, produced the highest current density, while the refluxed powder HA-0T$_{R1}$ produced the lowest. This indicated that the parameters determining uric acid sensing performance were drastically different depending whether Triton X-100 was employed during the synthesis, i.e. depending whether the resulting HA material was agglomerated or well dispersed. The authors correlated the current associated with uric acid detection of dispersed HA powders (HA-4T, HA-4T$_{R1}$, HA-4T$_{U1}$, HA-4T-373$_{H1}$) with the percentage of rodlike nanoparticles they exhibited and with surface acidity. They inferred that surface acidity was higher in rodlike structures grown along the c-axis of the HA crystal structure because they exhibited a high density of exposed Ca^{2+} Lewis acid sites on the faces of the crystals. On the contrary, the oxidation currents generated by materials synthesized in the absence of Triton X-100 and, therefore, displaying a higher degree of agglomeration, was better correlated with the mesopore volume measured by N_2 adsorption. The authors rationalized that for aggregated materials, while the diffusion of small NH_3 molecules (used to measure the density of accessible acidic sites) was not controlled by the mesopore size, it was not the case for the much bulkier uric acid (or more accurately urate ion in alkaline conditions), which diffusion in the material was complicated by the high density of aggregates. On the contrary, in well-dispersed materials prepared in the presence of surfactant, a direct correlation between density of acidic sites and electrocatalytic current density, which was consistent with the notion that these acidic sites, active for urate ion adsorption and oxidation, was well exposed and easily accessible regardless of the mesopore volume. Then the best performing electrode, HA-4T$_{R1}$-CPE, was subjected to different scan rates, and a linear relationship between the logarithm of the peak current for uric acid oxidation (I_p) and the logarithm of the scan rate (v) was observed: $\log(I_p) = 0.6793 \log(v) + 1.1137$ with a correlation coefficient of 0.994. A slope of 0.6793 suggests an electrocatalytic process controlled by both diffusion and adsorption, confirming the previously inferred importance of adsorption site accessibility, although diffusion seems to still play a significant role. This electrode showed a linear response in a uric acid concentration range of 0.068–50 µM, with a detection limit of 0.05 µM, and was found to be stable after storage in phosphate buffer (pH 7.0) for 15 days. Moreover, the electrode did not suffer any detection interference from glucose, urea, ammonium chloride, and sodium chloride. Finally, the electrode was found to be efficient for the sensing of uric acid in diluted human urine. Overall, this study proposed an original approach to correlate HA electrocatalytic performance with structural and chemical properties produced by different synthetic treatments, namely, the increased amount of exposed acidic Ca^{2+} active sites in rodlike HA nanostructure and the importance of limiting aggregation to remove anolyte diffusion limitation.

The different studies reported in this section clearly showed the versatility of HA as component of composite electrocatalytic biosensors for a wide range of analytes. The relatively low conductivity of HA could be overcome by mixing it with conductive materials, doping with metallic ions, or nanostructuring approaches.

Table 11.2 Summary table of the HA-based electrocatalytic sensors reported in this section, presenting quantitative estimations of sensing properties when available.

Sensor	Analyte	Linear range	LOD	Sensitivity	References
HA/Au/Hb	H_2O_2	0.5–25 µM	0.2 µM	n/a	[81]
SiO_2/HA/HRP	H_2O_2	1.0–100 µM	0.35 µM	50.56 mA M^{-1}	[82]
MWCNTs/HA/Hb	H_2O_2	0.5–2 µM	0.09 µM	n/a	[83]
HA	4-NP	1–300 µM	0.6 µM	n/a	[84]
Chit./HA/HRP	CN-	76 nM–3.1 µM	23 nM	n/a	[85]
Chit./HA/Tyr./Au	Cresol	10 nM–5 µM	5 nM	n/a	[86]
	Phenol	70 nM–5 µM	5 nM	n/a	
	Catechol	10 nM–7 µM	5 nM	n/a	
HA/Tyr.	L-tyrosine	0.1–11 µM	0.24 µM	n/a	[87]
HA	Uric acid	0.1–30 µM	0.14 µM	n/a	[94]
rGO/HA/GOx	Glucose	0.1–19.2 mM	0.03 mM	16.9 µA mM^{-1} cm^{-2}	[88]
HA/Au	Hydrazine	0.5 µM–1.4 mM	17 nM	0.05 µA µM^{-1} cm^{-2}	[89]
rGO/HA	Hydrazine	2.5 µM–0.26 mM	0.43 µM	4.21 µA µM^{-1} cm^{-2}	[90]
		0.26–1.16 mM		1.61 µA µM^{-1} cm^{-2}	
MWCNTs/HA	Uric acid	68 nM–50 µM	0.05 µM	n/a	[91]

LOD: limit of detection; Hb: hemoglobin; HRP: horseradish peroxidase; CNT: carbon nanotubes; 4-NP: 4-nitrophenol; Chit.: chitosan; Tyr.: tyrosinase; rGO: reduced graphene oxide; GOx: glucose oxidase.

All sensors yielded performance on par or better than state-of-the-art reports at the time of their respective reports and displayed satisfying stability. Their relevant sensing properties are quantified and organized in Table 11.2.

11.2.3 Fuel Cell Application

DAFCs represent an emerging technology aimed at generating electricity from the reduction of O_2 into H_2O combined with the direct oxidation of alcohol into CO_2, skipping the prior reforming of alcohol into hydrogen necessary in conventional fuel cells:

Anode: $C_nH_{2n+1}OH + (2n-1)H_2O \rightarrow nCO_2 + 6nH^+ + 6ne^-$

Cathode: $O_2 + 4H^+ + 4e^- \rightarrow 2H_2O$

Overall reaction: $2C_nH_{2n+1}OH + 3nO_2 \rightarrow 2nCO_2 + (2n+2)H_2O$

This technology is attractive due to the high energy density of alcohols and the possibility of generating them from preexisting biomass exploitation infrastructures. One of the major challenges this technology is facing is the development of an efficient anode capable of efficiently breaking C—C bonds at low overpotential. Over the last decade, palladium and its alloys have been shown to be very active for the direct oxidation of ethanol into acetate, with higher current density at lower overpotential, and with reduced poisoning compared with platinum [95]. Although Pd-based catalysts do not allow to break the C—C bond and fully oxidize ethanol into CO_2, it represents a significant first step toward the development of potential multifunctional catalysts able to do so. In alkaline media, this process follows the following mechanism:

$$Pd + OH^- \rightarrow Pd-OH_{ads} + e^- \tag{11.4}$$

$$Pd + CH_3CH_2OH \rightarrow Pd-(CH_3CH_2OH)_{ads} \tag{11.5}$$

$$Pd-(CH_3CH_2OH)_{ads} + 3OH^- \rightarrow Pd-(CH_3CO)_{ads} + 3H_2O + 3e^- \tag{11.6}$$

$$Pd-(CH_3CO)_{ads} + Pd-OH_{ads} \rightarrow Pd-(CH_3COOH)_{ads} + Pd \tag{11.7}$$

$$Pd-(CH_3COOH)_{ads} + OH^- \rightarrow Pd + CH_3COO^- + H_2O \tag{11.8}$$

The rate-limiting step of this mechanism is step (11.7), i.e. the reaction of an adsorbed ethoxy intermediate with an adsorbed hydroxyl. Therefore, the kinetics of this reaction are governed by the surface coverage of each species, and supporting Pd on hydroxyl-rich substrates such as layered double hydroxides has been shown to favor the formation of $Pd-OH_{ads}$ species and, therefore, improve the performance of the electrocatalyst toward ethanol oxidation [96]. As a result, HA ceramics have recently been proposed as alternative biocompatible substrates for palladium-based anode for DAFCs due to their great adsorption properties and hydroxyl-rich surface. However, because of its low conductivity, pristine HA cannot be directly employed as electrode substrate in DAFCs. Instead, it is necessary to mix it with a conductive material to create synergetic composites exhibiting high hydroxyl surface coverage, as well as good electrical conductivity.

For instance, in 2014, Cui et al. proposed a mixture of high surface area carbon black (Vulcan XC-72) and HA-supported Pd nanoparticles [97]. HA was prepared by a typical coprecipitation of calcium nitrate and hydrogen phosphate, while Pd nanoparticles were generated from the reduction of palladium chloride by ethylene glycol in 1 M KOH. Electron microscopy revealed that Pd nanoparticles were evenly dispersed on the HA with an average diameter of 4 nm, while the same nanoparticles directly dispersed on carbon black showed poorer size distribution with an average diameter of 5 nm, suggesting a beneficial stabilizing ability of HA as a substrate for Pd nanoparticles. Electrochemical tests were then performed on an electrode prepared by the drop-casting of a 15 µl of a solution composed of 3 mg of the Pd/HA/C catalyst and 30 µl of 5 wt% perfluorosulfonic acid onto a GCE. The metal loading of this electrode was measured to be 386 µg cm^{-2}. The ECSA of the catalyst was further determined by the oxidative stripping of adsorbed CO, following the formula:

$$\text{ECSA (m}^2\text{ g}^{-1}) = \frac{Q\,(\mu C)}{420\,(\mu C\,cm^{-2}) \times m(\mu g)} \times 100$$

where Q is the measured charge passed for the oxidation of adsorbed CO, m represents the total amount of Pd on the electrode, and 420 µC cm^{-2} is the density of charge required to oxidize one monolayer of adsorbed CO on Pd. Using this method, an ECSA of 33.08 m² g⁻¹ was determined for the Pd/HA/C catalyst, while a reference Pd/C with the same metal loading only produced an ECSA of 15.48 m² g⁻¹. To determine the electrocatalytic abilities of the proposed material, it was subjected to cyclic voltammetry in 1 M KOH + 1 M ethanol. A peak current density of 246 mA cm⁻² (i.e. 637 mA mg$_{Pd}$⁻¹) was observed for Pd/HA/C compared to 109 mA cm⁻² (i.e. 298 mA mg$_{Pd}$⁻¹) for Pd/C, an improvement consistent with the previously observed increase in ECSA. Moreover, an onset potential of −0.65 V vs. SCE was observed on Pd/HA/C, c. 50 mV more negative than the one observed on Pd/C, further suggesting improved electrocatalytic properties for Pd deposited on HA compared to bare carbon black. Another important parameter to consider is the tolerance of the electrocatalyst toward the deposition of carbonaceous species. A way to quantify this parameter is to measure the ratio of the forward peak anodic current (I_f) to the backward anodic peak current (I_b) in the cyclic voltammogram. A better tolerance to carbonaceous species leads to a higher current density for ethanol oxidation during the forward scan and lower residual oxidation (i.e. lower current density) during the backward scan and, therefore, to a higher I_f/I_b ratio. Pd/HA/C exhibited a ratio of 1.22, while the Pd/C reference only exhibited a ratio of 0.84, further demonstrating the beneficial effect of HA on the electrocatalytic properties of Pd. Stability was tested by chronoamperometry at −0.2 V vs. SCE, and the HA-supported catalyst showed better resilience toward the degradation of current density over time, although it still clearly suffered from poisoning by carbonaceous residues. The catalysts were finally tested in a DEFC configuration, with a membrane electrode assembly with an active surface area of 4.0 cm². The DEFC based on Pd/HA/C generated an open-circuit voltage of 0.73 V and a peak power density of 50 mW cm⁻², compared with 0.66 V and 36 mW cm⁻², respectively, for Pd/C. Thus, this study demonstrated the clearly beneficial effect of supporting Pd on HA for DEFC applications, in terms of increased output power and resilience toward surface poisoning.

Following this first report, Safavi et al. reported a Pd-loaded HA/MWCNTs composite for the electrocatalytic oxidation of alcohols in 2015 [98]. In this work, MWCNTs were flash oxidized in air at 700 °C to functionalize them with carboxylic groups, and then HA was deposited on their surface by a solventless solid-state metathesis developed in another work [99]: in essence, powders of $CaCl_2$ and $Na_3PO_4 \cdot 12H_2O$ were finely mixed with a Ca/P stoichiometric ratio of 1.66, as well as with a desired amount of MWCNTs, and then subjected to a microwave treatment at 900 W for 3 minutes. The deposition of Pd nanoparticles was then achieved by the sequential adsorption of Pd^{2+} ions from a $PdCl_2$ aqueous solution and their reduction by treatment with $NaBH_4$. This resulted in MWCNTs wrapped by a uniform shell of HA with a thickness of about 30 nm, with well-dispersed 3 nm Pd nanoparticles on its surface. This composite was loaded on a GCE by drop-casting, with the subsequent deposition of a Nafion membrane for mechanical stability. The resulting Pd/HA/MWCNT/GCE was found to be loaded with 7.9 μg cm⁻², while a reference Pd/MWCNTs/GCE electrode was loaded with 5.5 μg cm⁻². The ECSA of each electrode was calculated from the reduction of the PdO monolayer present at the surface of the catalyst, according to the formula:

$$\text{ECSA (m}^2\text{ g}^{-1}) = \frac{Q\,(\mu C)}{405\,(\mu C\,cm^{-2}) \times m\,(\mu g)} \times 100$$

where Q is the measured charge passed for the reduction of PdO, m represents the total amount of Pd on the electrode, and 405 µC cm^{-2} is the density of charge required to reduce one monolayer of adsorbed CO on Pd. These calculations resulted in an ECSA of 43.56 m^2 g^{-1} for Pd/HA/MWCNTs and 3.8 m^2 g^{-1} for Pd/MWCNTs, highlighting the excellence of HA as a high-surface area substrate for the adsorption of Pd. Using cyclic voltammetry, a peak current density of 1810.89 mA mg$_{Pd}^{-1}$ was measured for the electrocatalytic oxidation of ethylene glycol on Pd/HA/MWCNTs, with an onset potential of −0.38 V vs. Ag/AgCl in 1 M KOH + 1 M ethylene glycol. Furthermore, the Pd/HA/MWCNTs was tested for the electrocatalytic oxidation of various alcohols, resulting in high current densities and I_f/I_b ratios, confirming the good electrocatalytic properties of the composite and its tolerance toward poisoning by carbonaceous species. Finally, the electrocatalyst was subjected to a 20-hour stability testing by chronoamperometry at −0.25 V vs. Ag/AgCl. A sharp initial decrease in current was ascribed to the formation of a diffusion layer, while a later degradation was attributed to changes in electrolyte composition and notably its pH. Importantly, no significant current was observed over the Pd/MWCNTs reference, showing again the good performance of HA as a promoting substrate for the electrocatalytic oxidation of alcohols.

In a follow-up study from 2016, Safavi et al. studied an optimized version of their Pd/HA/MWCNTs composite specifically for ethanol oxidation [100]. This study set out to compare the composite obtained by the previously described microwave-assisted solventless metathesis approach, labeled HA/MWCNT(MW), and a composite obtained from the standard coprecipitation of CaCl$_2$ and Na$_3$PO$_4$ in alkaline medium on the MWCNTs, labeled HA/MWCNT(SP). Pd was loaded on each composite by impregnation of PdCl$_2$, followed by a reductive NaBH$_4$ treatment. A Pd/MWCNTs reference was also prepared. The authors reported that the standard precipitation route resulted in a nonuniform coverage of the MWCNTs by the HA, whereas the microwave route produced a uniform 30-nm-thick shell of HA around the MWCNTs. After impregnation, a uniform distribution of Pd nanoparticles of 3 nm was observed on the HA shell (see Figure 11.5a and b). The composites were then loaded onto a GCE, with a loading of 15.79 and 15.22 µg cm^{-2} in Pd for Pd/HA/MWCNTs(MW) and Pd/HA/MWCNTs(SP), respectively. Finally, a reference electrode was produced by directly loading Pd on MWCNTs with a loading of 11.05 µg cm^{-2}. ECSA were calculated by the reduction of a PdO monolayer, according to the previously mentioned formula. This calculation resulted in ECSA values of 4.42, 37.85, and 49.74 m^2 g$_{Pd}^{-1}$ for Pd/MWCNTs, Pd/HA/MWCNTs(SP), and Pd/HA/MWCNTs(MW), respectively, highlighting the excellence of HA as a high-surface area substrate for the adsorption of Pd. The higher ECSA obtained by the microwave route could result from the improved morphology previously discussed. Cyclic voltammetry experiments in 0.8 M NaOH + 0.2 NaClO$_4$ + 1 M ethanol displayed in Figure 11.5d, resulted in a peak

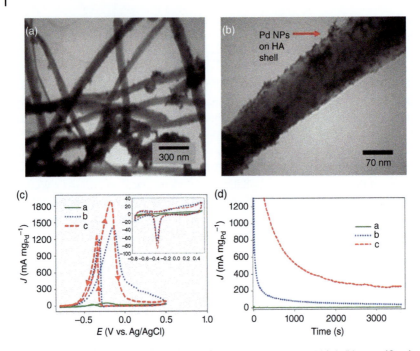

Figure 11.5 TEM pictures of Pd/HA/MWCNT(MW) at low (a) and high (b) magnifications. (c) Cyclic voltammograms recorded at 50 mV s^{-1} in 1 M EtOH + 0.8 M NaOH + 0.2 M NaClO$_4$ at GCEs coated with Pd/MWCNTs (a), Pd/HA/MWCNTs(SP) (b), and Pd/HA/MWCNTs(MW) (c). Inset: electrochemical behavior of the same electrodes in the absence of EtOH. (d) Chronoamperograms of the same electrodes at −0.25 V vs. Ag/AgCl in 1 M EtOH + 0.8 M NaOH + 0.2 M NaClO$_4$. Source: [100] Safavi, A. et al. (2016), Figure 02 (p. 02) / with permission of John Wiley and Sons.

current density of 1412.3 mA mg$_{Pd}^{-1}$ and an onset potential of −0.6 V vs. Ag/AgCl for Pd/HA/MWCNTs(SP) and a current density of 1874.6 mA mg$_{Pd}^{-1}$ and an onset potential of −0.65 V vs. Ag/AgCl for Pd/HA/MWCNTs(MW), values 36 times greater and 0.12 V lower than their respective counterparts for Pd/MWCNTs. Therefore, Pd/HA/MWCNTs(MW) displayed significantly better performance than the other electrocatalysts, in good agreement with the optimized morphology observed by electron microscopy. Chronoamperometry measurements further revealed that Pd/HA/MWCNTs(MW) possessed a much better resilience to current density degradation over time, although a significant decrease in performance was still observed after one hour (see Figure 11.5d). Overall, these two studies highlight the crucial effect of synthesis conditions on the morphology and, therefore, the performance of HA-supported electrocatalysts by presenting an original microwave-assisted preparation method, leading to excellent electrocatalytic performances for ethanol oxidation.

The studies reported in this section demonstrate the excellent properties of HA as a substrate for supported electrocatalytic Pd nanoparticles for the oxidation of alcohols. The high density of adsorption sites on the surface of HA favors the generation of a uniform distribution of Pd nanoparticles, while its hydroxyl-rich surface improves the adsorption rate of ethoxy intermediates, which is the rate-limiting process in alcohol oxidation on Pd.

11.2.4 Electrocatalytic Water Oxidation

Water oxidation is an important electrochemical process in the context of the storage of energy through water splitting. Indeed, this process is employed to generate hydrogen as an energy vector following the following electrochemical reaction:

$$\text{Anode:} \quad 2H_2O \rightarrow O_2 + 4H^+ + 4e^-$$
$$\text{Cathode:} \quad 2H^+ + 2e^- \rightarrow H_2$$
$$\text{Overall reaction:} \quad 2H_2O \rightarrow O_2 + 2H_2$$

For this reaction to proceed as efficiently as possible, it is necessary to develop efficient electrocatalysts to lower the necessary overpotential, especially for the oxygen evolution reaction (OER) that requires the difficult breaking of four covalent bonds and creation of a double bond. One well-known and extensively studied efficient oxygen evolution electrocatalyst is a (photo)electrodeposited amorphous cobalt phosphate, usually labeled Co—Pi [101]. Due to the easy surface functionalization of HA with metal ions through the replacement of Ca^{2+} ions, researchers have, therefore, generated cobalt-decorated HA electrocatalyst to emulate the Co—Pi electrocatalyst. Indeed, Pyo et al. published a study in 2019 in which they report a cobalt-incorporated HA examined for electrocatalytic oxygen evolution [102]. HA powder was prepared by a hydrothermal method and subsequently immersed in a 0.1 M Co^{2+} aqueous solution. After this treatment, the color of the HA powder changed from white to light blue. Inductively coupled plasma (ICP) and energy-dispersive X-ray (EDX) analyses were in agreement and provided an estimated 1 wt% Co content for the modified HA powder. XRD indicated no change in crystal structure upon cobalt incorporation, but also no diffraction pattern corresponding to a crystalline cobalt species was observed. Furthermore, no cobalt nanoparticles were observed by high-resolution TEM, where only the HA lattice diffraction was observed. The Co—HA was then drop-casted with Nafion onto a GCE and subjected to cyclic voltammetry in 0.1 M phosphate buffer (pH 7.0). The voltammogram exhibited an oxidation wave at 1.10 V vs. the normal hydrogen electrode (NHE) corresponding to the precatalytic oxidation of Co^{2+}, followed by the catalytic water oxidation wave at 1.25 V vs. NHE (corresponding to an overpotential η of ~0.43 V). A comparative voltammogram was acquired for a Co—Pi film electrodeposited from 0.5 mM Co^{2+} in 0.1 M phosphate buffer. Despite a more well-defined Co^{2+} oxidation peak, the electrocatalytic behavior of Co—Pi was found to be akin to the one observed for Co—HA. Stability testing over time at 1.3 vs. NHE (i.e. $\eta = 0.48$ V) resulted in a sharp initial decrease in current over the first hour, ascribed to leaching of cobalt ions from the HA structure due to the absence of material sintering, resulting in a relatively stable steady-state current after 100 minutes. A turnover frequency (TOF) of 6.81×10^{-3} s^{-1} was measured at $\eta = 0.48$ V, a value comparable with the one measured for Co—Pi (2×10^{-3} s^{-1}

at $\eta = 0.41$ V). Furthermore, a Tafel plot (overpotential vs. logarithm of current density) exhibited a slope of ~80 mV dec^{-1}, comparable with other oxygen evolution catalysts (OECs) typically ranging from 60 to 100 mV dec^{-1}. Finally, the Faradic efficiency of the Co—HA electrocatalyst was estimated using gas chromatography and resulted in a value of 94%. Overall, this study proposed a novel approach to generate cobalt-phosphate electrocatalysts through the incorporation of cobalt ions on the surface of phosphate-containing HA.

A 2020 study from Safavi et al., who previously developed HA/MWCNTs for the electrocatalytic oxidation of alcohols, described the application of the same solventless HA/MWCNTS synthesis technique to the production of oxygen evolution electrocatalysts [103]. To this end, the HA/MWCNT composite was soaked in 50 mM of Co(NO$_3$)$_2$ and Ni(NO$_3$)$_2$ for 24 hours to incorporate cobalt and nickel on the surface of the material and generate Co—HA, Ni—HA, and Co—Ni—HA. Transition electron microscopy revealed that the resulting material retained the previously reported morphology of MWCNTs coated with a thin layer of HA (see Figure 11.6a). Moreover, it appeared that nickel and cobalt impregnation resulted in the formation of nanoparticles with 13–23 nm diameter on the composite surface (see Figure 11.6b). XRD further revealed the formation of Ni$_3$(PO$_4$)$_2$ and Co$_3$(PO$_4$)$_2$ after impregnation, and EDX showed that Co was preferentially adsorbed over Ni, with a Co/Ni ratio of 3 in the Co—Ni—HA material. The materials were then deposited on a GCE with Nafion to evaluate their electrocatalytic properties for water oxidation. Cyclic voltammograms performed in 1.0 M KOH revealed the expected Co^{2+} redox signature (at 1.15 V vs. RHE) and Ni^{2+} redox signature (at 1.38 V vs. RHE) in Co—HA and Ni—HA, respectively, while Co—Ni—HA exhibited a combination of both. Linear sweep voltammograms recorded in 0.1 M KOH are displayed in Figure 11.6c and were used to monitor the electrocatalytic properties of different materials for the OER. As a result, bimetallic Co—Ni—HA was found to perform significantly better than the other catalysts, with the production of a lower onset potential and higher current densities. This result was partially attributed to an increase in surface area by 20% compared to the monometallic compounds, as evaluated by capacitance measurements; however, this increase was not sufficient to explain the observed improved electrocatalytic performance. The authors further measured that the Co—Ni—HA/MWCNTs possessed a lower charge transfer resistance, as measured by electrochemical impedance spectroscopy. Unfortunately, the authors did not comment on the fundamental underlying reasons for this difference in performance. Finally, the Co—Ni—HA electrocatalyst was found to be stable for the electrocatalytic oxygen evolution at 10 mA cm^{-2} for almost 3 hours, with no significant increase in the potential required to generate this current density (see Figure 11.6d). Furthermore, compared with Co—Ni particles directly deposited on MWCNTs, the Co—Ni—HA//MWCNTs electrocatalyst needed a 60 mV lower potential to generate 10 mA cm^{-2} and showed much better stability.

These studies demonstrated the possibility of taking advantage of the easy functionalization of HA surfaces with transition metal ions to generate efficient electrocatalysts for water oxidation. While the resulting HA-based catalysts did not show any major improvement compared to state-of-the-art OEC, they produced

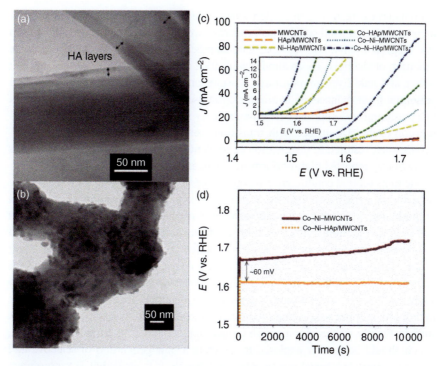

Figure 11.6 (a) TEM picture of HA/MWCNTs composite. (b) TEM picture of the Co−Ni−HA/MWCNTs electrocatalyst. (c) Linear sweep voltammetry measurements of MWCNTs, HA/MWCNTs, Ni−HA/MWCNTS, Co−HA/MWCNTS, Co−Ni−MWCNTs, and Co−Ni−HA/MWCNTs electrodes, acquired at 50 mV s^{-1} in 1.0 M KOH. Inset: voltammograms zoomed-in around the onset of electrocatalytic oxygen evolution current. (d) Stability testing of Co−Ni−MWCNTs and Co−Ni−HA/MWCNTs electrodes by chronopotentiometry at 10 mA cm^{-2} over 10 000 seconds. Source: [103] Safavi, A. et al. (2020), Figure 01 (p. 03) / with permission from Springer.

performances comparable to efficient metal phosphate-based electrodes. This paves the way for further modifications of HA surfaces with metal ions to generate novel electrocatalysts.

11.2.5 Summary and Outlook

In this chapter, the conductivity, charge transport mechanism, and electrocatalytic applications of HA ceramics were examined. Despite a relatively low conductivity at room temperature, most likely originating from surface transport, HA has been successfully integrated into electrocatalytic devices. First, as component of various biosensors, HA was combined with better conducting materials, such as MWCNTs or rGO, and coated with electrocatalytic species, such as enzymes or metal nanoparticles. The resulting HA composite biosensors usually showed better properties than the reference unsupported electrocatalyst in terms of linear range, LOD, sensitivity, reproducibility, and stability, when reported. This was generally attributed to the excellent adsorption properties and high surface area of HA nanomaterials, resulting in a high surface density of electrocatalytic sites. Furthermore, the same advantageous properties were leveraged to successfully design HA-supported palladium nanoparticles as efficient anodes for the electrocatalytic oxidation of alcohols. These anodes were then employed in DAFCs, which output power, and stability were monitored. Despite long-term instability due to palladium poisoning by carbonaceous byproducts, the use of HA was demonstrated to limit this poisoning and improve performances over short period of times. Finally, modification of the HA surface with transition metal cations, by replacement of Ca^{2+} surface sites, resulted in the creation of efficient OECs, matching the performances of previously reported high-performing cobalt and nickel-based OECs.

Due to its limited conductivity, it is probable that in order to obtain high electrocatalytic performances with HA, further controlling the deposition of thin layers of the material on conductive substrates will remain a successful approach. This will produce potentially biocompatible composite electrocatalysts combining good conductivity and a high density of electrocatalytic sites. Furthermore, a few recent studies demonstrate the importance of morphology control of HA nanocrystals for electrochemical applications, and additional work would be welcomed to clearly study the properties of different HA crystal facets in different aqueous environments in order to optimize adsorption phenomena and electrocatalysis kinetics at the HA surface. Finally, surface doping with transition metal ion is an intriguing approach, which should be further explored to generate a new generation of selective and efficient robust HA-based electrocatalysts. Overall, it is clear that HA presents a number of desirable features for electrocatalytic applications, although its conductivity remains a limiting factor that needs to be efficiently overcome to truly exploit the full potential of this material in electrocatalytic devices.

References

1 Braslavsky, S.E., Braun, A.M., Cassano, A.E. et al. (2011). Glossary of terms used in photocatalysis and radiation catalysis. *Pure and Applied Chemistry* 83 (4): 931–1014.
2 Herrmann, J.M. (1999). Heterogeneous photocatalysis: fundamentals and applications to removal of various types of aqueous pollutants. *Catalysis Today* 53 (1): 115–129.
3 Matsuoka, M., Kitano, M., Takeuchi, M. et al. (2007). Photocatalysis for new energy production: recent advances in photocatalytic water splitting reactions for hydrogen production. *Catalysis Today* 122 (1–2): 51–61.
4 Peruchon, L., Puzenat, E., Girard-Ergot, A. et al. (2008). Characterization of self-cleaning glasses using Langmuir-blodgett technique to control thickness of stearic acid multilayers – importance of spectral emission to define standard test. *Journal of Photochemistry and Photobiology A: Chemistry* 197 (2–3): 170–176.
5 Herrmann, J.M. (2010). Fundamentals and misconceptions in photocatalysis. *Journal of Photochemistry and Photobiology A: Chemistry* 216 (2–3): 85–93.

6 Takanabe, K. (2017). Photocatalytic water splitting: quantitative approaches toward photocatalyst by design. *ACS Catalysis* 7: 8006–8022.

7 Hu, D., Xie, Y., Liu, L. et al. (2016). Constructing TiO_2 nanoparticles patched nanorods heterostructure for efficient photodegradation of multiple organics and H_2 production. *Applied Catalysis B* 188: 207–216.

8 Lutrell, T., Halpegamage, S., Sutter, E. et al. (2014). Photocatalytic activity of anatase and rutile TiO_2 epitaxial thin film grown by pulsed laser deposition. *Thin Solid Films* 564: 146–155.

9 Lyandres, O., Finkelstein-Shapiro, D., Chakthranont, P. et al. (2012). Preferred orientation in sputtered TiO_2 thin films and its effect on the photooxidation of acetaldehyde. *Chemistry of Materials* 24 (17): 3355–3362.

10 Kandiel, T.A., Robben, L., Alkaim, A. et al. (2013). Brookite versus anatase TiO_2 photocatalysts: phase transformations and photocatalytic activities. *Photochemical and Photobiological Sciences* 12 (4): 602–609.

11 Ding, Z., Lu, G.Q., Greenfield, P.F. et al. (2000). Role of the crystallite phase in heterogeneous photocatalysis for phenol oxidation in water. *Journal of Physical Chemistry B* 104 (19): 4815–4820.

12 Bui, T.H., Karkmaz, M., Puzenat, E. et al. (2007). Solar purification and potabilization of water containing dyes. *Research on Chemical Intermediates* 33 (3–5): 421–431.

13 Martin, D.J., Liu, G., Moniz, S.J.A. et al. (2015). Efficient visible driven photocatalyst, silver phosphate: performance, understanding and perspective. *Chemical Society Reviews* 44: 7808–7828.

14 Marschall, R. (2014). Semiconductor composites: strategies for enhancing charge carrier separation to improve photocatalytic activity. *Advanced Functional Materials* 24: 2421–2440.

15 Ivanova, T.I., Frank-Kamenetskaya, O.V., Kol'tsov, A.B. et al. (2001). Crystal structure of calcium-deficient carbonated hydroxyapatite. Thermal decomposition. *Journal of Solid State Chemistry* 160: 340–349.

16 Figueiredo, M., Fernando, A., Martins, G. et al. (2010). Effect of the calcination temperature on the composition and microstructure of hydroxyapatite derived from human and animal bone. *Ceramics International* 36: 2383–2393.

17 Bertazzo, S., Zambuzzi, W.F., Campos, D.D.P. et al. (2010). Hydroxyapatite surface solubility and effect on cell adhesion. *Colloids and Surfaces B* 78: 177–184.

18 Bengtsson, A., Shchukarev, A., Persson, P. et al. (2009). A solubility and surface complexation study of a non-stoichiometric hydroxyapatite. *Geochimica and Cosmochimica Acta* 73: 257–267.

19 Lachheb, H., Puzenat, E., Houas, A. et al. (2002). Photocatalytic degradation of various dyes (Alizarine S, Crocein Orange G, Methyl Red, Congo Red, Methylene Blue) in water by UV-irradiated titania. *Applied Catalysis B: Environmental* 39 (1): 75–90.

20 Bose, S. and Tarafder, S. (2012). Calcium phosphate ceramic systems in growth factor and drug delivery for bone tissue engineering: a review. *Acta Biomaterialia*. 8: 1401–1421.

21 Hankermeyer, C.R., Ohashi, K.L., Delaney, D.C. et al. (2002). Dissolution rates of carbonated hydroxyapatite in hydrochloric acid. *Biomaterials* 23: 743–750.

22 Piccirillo, C., Pullar, R.C., Tobaldi, D.M. et al. (2014). Hydroxyapatite and chloroapatite derived from sardine by-products. *Ceramics International* 40: 13231–13240.

23 Nzihou, A. and Sharrock, P. (2010). Role of phosphate in the remediation and reuse of heavy metal polluted wastes and sites. *Waste and Biomass Valorization* 1: 163–174.

24 Viipsi, K., Sjoberg, S., Tonsuaadu, K. et al. (2013). Hydroxy- and fluorapatite as sorbents in Cd(II)—Zn(II) multicomponent solutions in the absence/presence of EDTA. *Journal of Hazardous Materials* 252: 91–98.

25 Dubey, A.K., Mallik, P.K., Kundu, S. et al. (2013). Dielectric and electrical conductivity properties of multi-stage spark plasma sintered HA-$CaTiO_3$ composites and comparison with conventionally sintered materials. *Journal of European Ceramic Society* 33: 3445–3453.

26 Tsukada, M., Wakamura, M., Yoshida, N. et al. (2011). Band gap and photocatalytic properties of Ti-substituted hydroxyapatite: comparison with anatase-TiO_2. *Journal of Molecular Catalysis A* 338: 18–23.

27 Nishikawa, H. and Omamiuda, K. (2002). Photocatalytic activity of hydroxyapatite for methyl mercaptane. *Journal of Molecular Catalysis A* 179 (1–2): 193–200.

28 Nishikawa, H. (2003). Radical generation on hydroxyapatite by UV irradiation. *Materials Letters* 58: 14–16.

29 Nishikawa, H. (2007). Photo-induced catalytic activity of hydroxyapatite based on photo-excitation. *Phosphorous Research Bulletin* 21: 97–102.

30 Tanaka, H., Tsuda, E., Nishikawa, H. et al. (2012). FTIR studies of adsorption and photocatalytic decomposition under UV irradiation of dimethyl sulphide on calcium hydroxyapatite. *Advanced Powder Technology* 23: 115–119.

31 Nishikawa, H. (2003). Surface changes and radical formation on hydroxyapatite by UV irradiation for inducing photocatalytic activation. *Journal of Molecular Catalysis A* 206: 231–238.

32 Shariffudin, J.H., Jones, M.I., and Patterson, D.A. (2013). Greener photocatalysts: hydroxyapatite derived from waste mussel shells for the photocatalytic degradation of a model azo dye wastewater. *Chemical Engineering Research and Design* 91: 1693–1704.

33 Reddy, M.P., Venugopal, A., and Subrahmanyam, M. (2007). Hydroxyapatite photocatalytic degradation of calmagite (an azo dye). *Applied Catalysis B* 69: 164–170.

34 Tripathi, R.M., Kumar, N., Bhadwal, A.S. et al. (2015). Facile and rapid biomimetic approach for synthesis of HA nanofibers and evaluation of their photocatalytic activity. *Materials Letters* 140: 64–67.

35 Piccirillo, C., Dunnill, C.W., Pullar, R.C. et al. (2013). Calcium phosphate-based materials of natural origin showing photocatalytic activity. *Journal of Molecular Catalysis A* 1: 6452–6461.

36 Bystrov, V.S., Piccirillo, C., Tobaldi, D.M. et al. (2016). Oxygen vacancies, optical band gap (Eg) and photocatalysis of hydroxyapatite: comparing modelling with measured data. *Applied Catalysis B* 196: 100–107.

37 Wakamura, M., Hashimoto, K., and Watanabe, T. (2003). Photocatalysis by calcium hydroxyapatite modified with Ti(IV): albumin decomposition and bactericidal effect. *Langmuir* 19: 3428–3431.

38 Tsuruoka, A., Isobe, T., Matsushita, S. et al. (2015). Comparison of photocatalytic activity and surface friction force variation on Ti-doped hydroxyapatite and anatase under UV illumination. *Journal of Photochemistry and Photobiology A* 311: 160–165.

39 Kandori, K., Oketani, M., Sakita, Y. et al. (2012). FTIR studies on photocatalytic activity of Ti(IV)-doped calcium hydroxyapatite particles. *Journal of Molecular Catalysis A* 360: 54–60.

40 Lin, H., Huang, C.P., Li, W. et al. (2006). Size dependency of nanocrystalline TiO_2 on its optical property and photocatalytic reactivity exemplified by 2-chlorophenol. *Applied Catalysis B* 68: 1–11.

41 Ergun, C. (2008). Effect of Ti ion substitution on the structure of hydroxylapatite. *Journal of European Ceramic Society* 28: 2137–2149.

42 Ribeiro, C.C., Gibson, I., and Barbosa, M.A. (2006). The uptake of titanium ions by hydroxyapatite particles structural changes and possible mechanism. *Biomaterials* 27: 1749–1761.

43 Anmin, H., Ming, L., Chengkang, C. et al. (2007). Preparation and characterization of a titanium-substituted hydroxyapatite photocatalyst. *Journal of Molecular Catalysis A* 267: 79–85.

44 Nishikawa, M., Yang, W., and Nosaka, Y. (2013). Grafting effects of Cu^{2+} on the photocatalytic activity of titanium-substituted hydroxyapatite. *Journal of Molecular Catalysis A* 378: 314–318.

45 Wakamura, M., Tanaka, H., Naganuma, Y. et al. (2011). Surface structure and visible light photocatalytic activity of titanium-calcium hydroxyapatite modified with Cr (III). *Advanced Powder Technology* 22: 498–503.

46 Ould-Chikh, S., Proux, O., Afanasiev, P. et al. (2014). Photocatalysis with chromium-doped TiO_2: bulk and surface doping. *Chemsuschem* 7 (5): 1361–1371.

47 Nishikawa, M., Tan, L.H., Nakabayashi, Y. et al. (2015). Visible light responsive vanadiumsubstituted hydroxyapatite photocatalysts. *Journal of Photochemistry and Photobiology A* 311: 30–34.

48 Komazaki, Y., Shimizu, H., and Tanaka, S. (1999). A new measurement for nitrogen oxides in the air using an annular diffusion scrubber coated with titanium dioxide. *Atmospheric Environment* 33: 4363–4371.

49 Niimi, M., Masuda, T., Kaihatsu, K. et al. (2014). Virus purification and enrichment by hydroxyapatite chromatography on a chip. *Sensors Actuators B* 201: 185–190.

50 Anmin, H., Tong, L., Ming, L. et al. (2006). Preparation of nanocrystals hydroxyapatite/TiO_2 compound by hydrothermal treatment. *Applied Catalysis B* 63: 41–44.

51 Mitsionis, A., Vaimakis, T., Trapalis, C. et al. (2011). Hydroxyapatite/titanium dioxide nanocomposites for controlled photocatalytic NO oxidation. *Applied Catalysis B* 106: 398–404.

52 Giannakopoulou, T., Todorova, N., Romanos, G. et al. (2012). Composite hydroxyapatite/TiO_2 materials for photocatalytic oxidation of NOx. *Materials Science and Engineering B* 117: 1046–1052.

53 Yang, Z. and Zhang, C. (2008). Photocatalytic degradation of bilirubin on hydroxyapatitemodified nanocrystalline titania coatings. *Catalysis Communication* 10: 351–354.

54 Ozeki, K., Janurudin, J.M., Aoki, H. et al. (2007). Photocatalytic hydroxyapatite/titanium dioxide multilayer thin film deposited onto glass using an rf magnetron sputtering technique. *Applied Surface Science* 253: 3397–3401.

55 Ye, F.X., Ohmori, A., Tsumura, T. et al. (2007). Microstructural analysis and photocatalytic activity of plasma-sprayed titania-hydroxyapatite coatings. *Journal of Thermal Spray Technology* 16: 776–782.

56 Piccirillo, C., Denis, C., Pullar, R.C. et al. (2017). Aerosol assisted chemical vapour deposition of hydroxyapatite-embedded titanium dioxide composite thin films. *Journal of Photochemistry and Photobiology A* 332: 45–53.

57 Ma, N., Fan, X., Quan, X. et al. (2009). Ag-TiO_2/HA/Al_2O_3 bioceramic composite membrane: fabrication, characterization and bactericidal activity. *Journal of Membrane Science* 336: 109–117.

58 Tobaldi, D.M., Piccirillo, C., Seabra, M.P., and et al. (2014). A green synthesis route for Ag-modified nano-titania as an antibacterial agent and photocatalyst. *Journal of Physical Chemistry C* 118: 4751–4766.

59 Mohamed, R.M. and Baeissaa, E.S. (2013). Preparation and characterisation of Pd-TiO_2-hydroxyapatite nanoparticles for the photocatalytic degradation of cyanide under visible light. *Applied Catalysis A* 464: 218–224.

60 Ryu, J., Kim, K.Y., Hahn, B.D. et al. (2009). Photocatalytic nanocomposite thin films of TiO_2-*b*-calcium phosphate by aerosol deposition. *Catalysis Communication* 10: 596–599.

61 Marquez-Brazon, E., Piccirillo, C., Moreira, I.S. et al. (2016). Photodegradation of pharmaceutical persistent pollutants using hydroxyapatitebased materials. *Journal of Environmental Management* 182: 485–495.

62 Vukomanovic, M., Zunic, V., Otonicar, M. et al. (2012). Hydroxyapatite/platinum bio-photocatalyst: a biomaterial approach to self cleaning. *Journal of Materials Chemistry* 22: 10571–10580.

63 Zeng, J., Li, J., Zhong, J. et al. (2015). Improved sun light photocatalytic activity of α-Fe_2O_3 prepared with the assistance of CTAB. *Materials Letters* 160: 526–528.

64 Bharath, G. and Ponpandian, N. (2015). Hydroxyapatite nanoparticles on dendritic α-Fe_2O_3 hierarchical architectures for a heterogeneous photocatalyst and ab-sorption of Pb(II) ions from industrial wastewater. *Royal Society Chemistry Advances* 5: 84685–84693.

65 Yang, Z.P., Gong, X.Y., and Zhang, C.J. (2010). Recyclable Fe_3O_4/hydroxyapatite composite nanoparticles for photocatalaytic applications. *Chemical Engineering Journal* 165: 117–121.

66 Buckley, J.J., Lee, A.F., Olivi, L. et al. (2010). Hydroxyapatite supported anti-bacterial Ag_3PO_4 nanoparticles. *Journal of Materials Chemistry* 20: 8056–8063.

67 Suwanprateeb, J., Thammarakcharoen, F., Wasoontararat, K. et al. (2012). Preparation and characterization of nanosized silver phosphate loaded hydroxyapatite by single step co-conversion process. *Materials Science and Engineering C* 32: 2122–2128.

68 Hong, X., Wu, X., Zhang, O. et al. (2012). Hydroxyapatite supported Ag_3PO_4 nanoparticles with higher visible light photocatalytic activity. *Applied Surface Science* 258 (10): 4801–4805.

69 Li, Y., Zhou, H., Zhu, G. et al. (2015). High efficient multifunctional Ag_3PO_4 loaded hydroxyapatite nanowires for water treatment. *Journal of Hazardous Materials* 299: 379–387.

70 Valizadeh, S., Rasoulifard, M.H., and Seyed Dorraji, M.S. (2014). Modified Fe_3O_4-hydroxyapatite nanocomposites as heterogeneous catalysts in three UV, Vis and Fenton like degradation systems. *Applied Surface Science* 319: 358–366.

71 Jarcho, M., Bolen, C.H., Thomas, M.B. et al. (1976). Hydroxylapatite synthesis and characterization in dense polycrystalline form. *Journal of Materials Science* 11 (11): 2027–2035.

72 Yamashita, K., Owada, H., Nakagawa, H. et al. (1986). Trivalent-cation-substituted calcium oxyhydroxyapatite. *Journal of the American Ceramic Society* 69 (8): 590–594.

73 Takahashi, T., Tanase, S., and Yamamoto, O. (1978). Electrical conductivity of some hydroxyapatites. *Electrochimica Acta* 23 (4): 369–373.

74 Nagai, M. and Nishino, T. (1988). Surface conduction of porous hydroxyapatite ceramics at elevated temperatures. *Solid State Ionics* 28–30: 1456–1461.

75 Yamashita, K., Kitagaki, K., and Umegaki, T. (1995). Thermal instability and proton conductivity of ceramic hydroxyapatite at high temperatures. *Journal of the American Ceramic Society* 78 (5): 1191–1197.

76 Laghzizil, A., El Herch, N., Bouhaouss, A. et al. (2001). Comparison of electrical properties between fluorapatite and hydroxyapatite materials. *Journal of Solid State Chemistry* 156 (1): 57–60.

77 Nakamura, S., Takeda, H., and Yamashita, K. (2001). Proton transport polarization and depolarization of hydroxyapatite ceramics. *Journal of Applied Physics* 89 (10): 5386–5392.

78 Gittings, J.P., Bowen, C.R., Turner, I.G. et al. (2007). Characterisation of ferroelectric-calcium phosphate composites and ceramics. *Journal of the European Ceramic Society* 27 (13): 4187–4190.

79 Gittings, J.P., Bowen, C.R., Dent, A.C.E. et al. (2009). Electrical characterization of hydroxyapatite-based bioceramics. *Acta Biomaterialia* 5 (2): 743–754.

80 Zhou, H., Wu, T., Dong, X. et al. (2007). Adsorption mechanism of BMP-7 on hydroxyapatite (001) surfaces. *Biochemical and Biophysical Research Communications* 361 (1): 91–96.

81 You, J., Ding, W., Ding, S. et al. (2009). Direct electrochemistry of hemoglobin immobilized on colloidal gold-hydroxyapatite nanocomposite for electrocatalytic detection of hydrogen peroxide. *Electroanalysis* 21 (2): 190–195.

82 Wang, B., Zhang, J.-J., Pan, Z.-Y. et al. (2009). A novel hydrogen peroxide sensor based on the direct electron transfer of horseradish peroxidase immobilized on silica–hydroxyapatite hybrid film. *Biosensors and Bioelectronics* 24 (5): 1141–1145.

83 Zhao, H.Y., Xu, X.X., Zhang, J.X., and at al. (2010). Carbon nanotube–hydroxyapatite–hemoglobin nanocomposites with high bioelectrocatalytic activity. *Bioelectrochemistry* 78 (2): 124–129.

84 Yin, H., Zhou, Y., Ai, S. et al. (2010). Electrochemical oxidative determination of 4-nitrophenol based on a glassy carbon electrode modified with a hydroxyapatite nanopowder. *Microchimica Acta* 169 (1): 87–92.

85 Wang, S., Lei, Y., and Zhang, at al. (2010). Hydroxyapatite nanoarray-based cyanide biosensor. *Analytical Biochemistry* 398 (2): 191–197.

86 Lu, L., Zhang, L., Zhang, X. et al. (2010). A novel tyrosinase biosensor based on hydroxyapatite–chitosan nanocomposite for the detection of phenolic compounds. *Analytica Chimica Acta* 665 (2): 146–151.

87 Kanchana, P., Lavanya, N., and Sekar, C. (2014). Development of amperometric L-tyrosine sensor based on Fe-doped hydroxyapatite nanoparticles. *Materials Science and Engineering C* 35: 85–91.

88 Bharath, G., Madhu, R., Chen, S.M. et al. (2015). Enzymatic electrochemical glucose biosensors by mesoporous 1D hydroxyapatite-on-2D reduced graphene oxide. *Journal of Materials Chemistry B* 3 (7): 1360–1370.

89 Bharath, G., Naldoni, A., Ramsait, K.H. et al. (2016). Enhanced electrocatalytic activity of gold nanoparticles on hydroxyapatite nanorods for sensitive hydrazine sensors. *Journal of Materials Chemistry A* 4 (17): 6385–6394.

90 Gao, F., Wang, Q., Gao, N. et al. (2017). Hydroxyapatite/chemically reduced graphene oxide composite: environment-friendly synthesis and high-performance electrochemical sensing for hydrazine. *Biosensors and Bioelectronics* 97: 238–245.

91 Iyyappan, E., Samuel Justin, S.J., Wilson, P. et al. (2020). Nanoscale hydroxyapatite for electrochemical sensing of uric acid: roles of mesopore volume and surface acidity. *American Chemical Society Applied Nano Materials* 3 (8): 7761–7773.

92 Iyyappan, E. and Wilson, P. (2013). Synthesis of nanoscale hydroxyapatite particles using triton X-100 as an organic modifier. *Ceramics International* 39 (1): 771–777.

93 Iyyappan, E., Wilson, P., Sheela, K. et al. (2016). Role of triton X-100 and hydrothermal treatment on the morphological features of nanoporous hydroxyapatite nanorods. *Materials Science and Engineering C* 63: 554–562.

94 Kanchana, P. and Sekar, C. (2014). EDTA assisted synthesis of hydroxyapatite nanoparticles for electrochemical sensing of uric acid. *Materials Science and Engineering C* 42: 601–607.

95 Wang, F., Lu, Z., Yang, L. et al. (2013). Palladium nanoparticles with high energy facets as a key factor in dissociating O_2 in the solvent-free selective oxidation of alcohols. *Chemical Communication* 49 (59): 6626–6628.

96 Zhao, J., Shao, M., Yan, D. et al. (2013). A hierarchical heterostructure based on Pd nanoparticles/layered double hydroxide nanowalls for enhanced ethanol electrooxidation. *Journal of Materials Chemistry A* 1 (19): 5840–5846.

97 Cui, Q., Chao, S., Bai, Z. et al. (2014). Based on a new support for synthesis of highly efficient palladium/hydroxyapatite catalyst for ethanol electrooxidation. *Electrochimica Acta* 132: 31–36.

98 Safavi, A., Abbaspour, A., and Sorouri, M. (2015). Hydroxyapatite wrapped multiwalled carbon nanotubes composite, a highly efficient template for palladium loading for electrooxidation of alcohols. *Journal of Power Sources* 287: 458–464.

99 Safavi, A. and Sorouri, M. (2013). Multiwalled carbon nanotube wrapped hydroxyapatite, convenient synthesis via microwave assisted solid state metathesis. *Materials Letters* 91: 287–290.

100 Safavi, A., Abbaspour, A., Sorouri, M. et al. (2016). Highly efficient ethanol electrooxidation on a synergistically active catalyst based on a Pd-loaded composite of hydroxyapatite. *ChemElectroChem* 3 (4): 558–564.

101 Kanan, M.W. and Nocera, D.G. (2008). In situ formation of on oxygen evolving catalyst in neutral water containing phosphate and Co^{2+}. *Science* 321: 1072–1075.

102 Pyo, E., Lee, K., Jang, M.J. et al. (2019). Cobalt incorporated hydroxyapatite catalyst for water oxidation. *ChemCatChem* 11 (22): 5425–5429.

103 Safavi, A., Mohammadi, A., and Sorouri, M. (2020). Cobalt–nickel wrapped hydroxyapatite carbon nanotubes as a new catalyst in oxygen evolution reaction in alkaline media. *Electrocatalysis* 11 (2): 226–233.

12

Magnetic Structured Hydroxyapatites and Their Catalytic Applications

Tasnim Munshi, Smriti Rawat, Ian J. Scowen, and Sarwat Iqbal

University of Lincoln, School of Chemistry, Joseph Banks Laboratories, Lincoln, LN6 7DL, UK

12.1 Introduction

Catalysis is an important functional tool of chemistry because it enables researchers to develop a number of synthetic pathways to access desired products using sustainable resources and to develop less polluting chemical processes. One of the most important and required conditions of catalysts is their reusability for industrial applications. While homogeneous catalysts have proved to be superior in activity and selectivity, still the major industrially preferred and applicable catalysts are heterogeneous catalysts, because of their simplicity and reusability [1]. Recently, magnetic nanoparticles have become popular catalysts because of their high activity and selectivity. The properties that make these magnetic nanoparticles popular are their size, crystallinity, monodispersity, large surface area, simple recovery processes, and composition. The focus of this chapter is about the synthesis of magnetic hydroxyapatite (mHA) and their applicability in catalysis. In this particular class of materials, the catalytic activity depends on the surface composition of mHA in a heterogeneous manner. The active species can either be the magnetic material on its own or another metal that is embedded in or attached onto the apatite surface.

Hydroxyapatite, $Ca_{10}(PO_4)_6(OH)_2$ (HA), a member of calcium apatite family, is a well-studied material with various applications, for example, as a drug and protein delivery agent, antimicrobial agent, and catalyst [2]. Chemically, HA is a naturally occurring mineral form of calcium apatite with stoichiometric ratio of $Ca/P = 1.67$. Thermodynamically, it is the most stable derivative of calcium phosphate at room temperature [3]. In the nineteenth century, attempts were made toward investigating the chemical structure and compositions of HA-type materials, with not much success. However, in the twentieth century Hausen et al. [4] proposed the existence of various phases of calcium phosphate. HA has recently become a popular choice out of all the other calcium phosphate derivatives, for its practical application in a wide range of disciplines [5–8]. One of the main reasons that HA is so widely used is its structural property related with the presence of two different crystallographic positions of Ca, which can help in tuning its material properties by specific site modification for a wider range of application – catalysis being the major one. In the HA

Design and Applications of Hydroxyapatite-Based Catalysts, First Edition. Edited by Doan Pham Minh.
© 2022 WILEY-VCH GmbH. Published 2022 by WILEY-VCH GmbH.

unit cell, Ca and phosphates are arranged in such a manner that ten Ca atoms possess two different crystallographic positions (i.e. M_1 and M_2). Four Ca atoms occupy M_1 position along threefold screw axis where they are surrounded by nine O atoms of phosphate moieties. M_2 is occupied by the other six Ca atoms along sixfold screw axis, which are surrounded by six O atoms of phosphate moieties and one of the two OH^- groups. These OH^- groups are hosted along the c-axis and balances positive charge in the matrix [9].

Two main phases of HA have been reported in literature [10, 11]. The most frequently encountered HA structure has a hexagonal crystal system and $P6_3/m$ space group, generally associated with nonstoichiometric HA containing impurities such as F^- or Cl^-. The second phase is a monoclinic crystal system with a $P2_1/b$ space group, a low temperature stable phase, generally associated with synthetic HA or stoichiometric HA [11]. The first crystal structure of HA was reported by Kay et al. [10]; they reported that ion substitution can affect the lattice parameters and as a result alters the inherent properties of the matrix. These substitutions induce alterations in biological and physicochemical properties of material. The possibility of substitution in nonstoichiometric HA has proved to be really advantageous because it can provide vacancy defects in the OH^- and cationic sites, the effect on the state of crystallinity, and an enhancement in solubility of HA.

The ability to control and manipulate the material stoichiometry with defined ion substitution has made HA very interesting for structure–property investigations, particularly in catalyst development with desired properties, while the chemistry and origin of HA gives it interesting properties and, therefore, makes it popular in a range of applications from multiple disciplines; to name a few are ion exchangeability, biocompatibility, nontoxicity, a high adsorption capacity, and availability [12].

The application of HA alone, i.e. not in conjunction with any other compound, is mostly not preferred. This is because of its brittleness, low tensile strength, and fracture toughness. If used as a coating on different types of materials, its advantages can be exploited and can help overcome its limitation of mechanical strength. To address this issue, nanophase HA coatings, especially nanocomposite HA, are used as a coating material. Coatings of HA proved to be helpful because of its crystallographic structure resembles to that of natural bone, which eliminated the biocompatibility issues [13]. These coatings possess a greater number of atoms at the surface, higher surface area, increased electron delocalization, and greater number of defects at the surface. It provides advantage over the conventional bigger size biological material in terms of higher surface reactivity. Literature has widely acknowledged the use of HA nanocomposite coatings over a variety of materials, but a recent developing interest is combining concept of magnetization with this bioactive coating [14]. Coating HA over the magnetic particles is the recent emerging idea, which has paved a path for a wide range of applications in a variety of disciplines.

12.2 Magnetic HA

Magnetic nanoparticles fascinate researchers from a wide range of disciplines, involving biomedical/biotechnology magnetic fluids [15] (Figure 12.1), catalysis

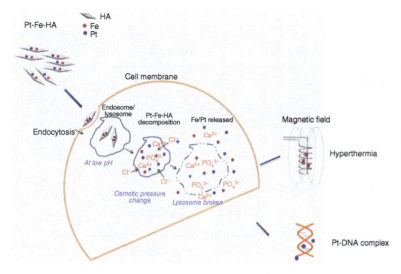

Figure 12.1 Incorporation of HA with Fe^{2+} and Pt^{2+} (Pt–Fe–HA) for lung cancer treatment. Source: Trandafir et al. [22]. Reproduced with permission of Elsevier.

[16], environmental remediation [17], data storage [18], and magnetic resonance imaging [19].

Magnetic HA (mHA) offers the use of the materials as catalysts to be used as safe, stable, and recyclable materials and offers an alternative to move away from toxic options. mHA has been extracted from cow bones, and then magnetic nanoparticles are incorporated into the scaffold. The studies show that successful application of these magnetic nanoparticles depends on their size as each nanoparticle behaves as a single magnetic domain and exhibits superparamagnetic behavior. However, they have large constant magnetic moments and show instant magnetization in the applied magnetic field with negligible resonance on removing the external magnetic force. Nanoparticles are not stable over longer period of time; due to their high surface-to-volume ratio, they tend to agglomerate to decrease their energy and become stable. Furthermore, simple metallic nanoparticles generally loose magnetism and the ability to disperse after getting oxidized in air due to their high chemical activity. Sometimes, a protective layer or coat can provide chemical stability and functionalization, suited to their specific application. Literature reports widely state that magnetic coupling in nano-HA provides external control of particle action and delivery using a magnetic field [21]. Previously HA has been coupled with various magnetic particles and used for a variety of applications. Some of these applications are listed in Table 12.1.

12.2.1 Synthesis Route of Magnetic HA

Synthesis of mHAs offers an excellent alternative to more traditional catalysts, leading to a great advantage with ease of separation of the catalyst from the products of the reaction. Production of mHA nanomaterials has become popular

Table 12.1 Applications of various magnetic HA materials.

HA	Applications	Drawbacks	References
Ag nanoparticles decorated	Catalysis	Poor reusability	[2]
Magnetite	Tissue engineering, biomedicine, parabens removal	Multistep synthesis of core–shell structure	[22–24]
Fe doped	Drug delivery	Low crystallinity and low thermal stability	[25]
Pb, Cu-doped biochar nanocomposite	Heavy metal removal	pH dependent – multistep synthesis	[26]
Fe and Pt doped	Catalysis, high stability, and sensor applications	Toxic properties limit their applications in biology	[27]
Fe_3O_4 doped	Removal of U(VI) in wastewater	Effects in aqueous medium are not known	[28]
Mn and Fe doped	MRI contrast agents, hyperthermia applications	Composite materials do not have influence of Mn substitution in terms of magnetic properties	[29]
$CoFe_2O_4$ doped	MRI contrast agent, bioimaging property	Metal toxicity in biometabolism	[30]
Gd-Nd doped	Luminescent materials for bioengineering, enhanced electrical conductivity	Requirement of large external fields and controlled environment	[31]
Fe and Cu doped	Drug delivery	Variable solubility in aqueous media	[32]
Magnetic zeolite-HA adsorbent	Protein adsorption	Possible metal leeching at high temperature	[33]

Source: Tasnim Munshi.

for adsorption, catalysis, and various other applications. Numerous techniques have been reported for their synthesis of which include simulated biofluid method [23], ultrasonic irradiation [2], precipitation combined with mechanical grinding [28], chemical doping with nanoparticles [34], nanoparticle infiltration [35], micro arc oxidation combined with hydrothermal and electrochemical treatments [36], biometric method [37], coprecipitation [38], microwave-assisted coprecipitation [20], stepwise synthesis combined with mechanical milling and subsequent heat treatments and reactions in solid state [39], hydrothermal processing [20], electrospinning, microwave irradiation-assisted precipitation [40], and emulsion and microemulsion [41]. The most common criteria for a highly efficient and best performing catalyst are their narrow size distribution and high dispersion on oxide supported (in metal-supported catalysts). This criterion is kept in mind

Table 12.2 Preparation techniques used for the synthesis of magnetic HA and their properties.

Preparation method	Size (µm)	Crystallinity	Size distribution	Phase purity	References
Chemical doping	>0.1	Low	Variable	Variable	[34]
Hydrothermal processing	>0.05	High	Wide	High	[20, 43]
Electrospinning	10×10^{-30}	Low	Variable	Variable	[44]
Microwave irradiation	100×25 nm	High	Narrow	High	[40]
Sol–gel	>0.001	Variable	Variable	Variable	[45]
Electrospraying	75×45 nm	Low	Variable	Variable	[43]
Solid state	>2.0	High	Wide	Low	[39]
Self-propagating combustion	>0.45	Variable	Wide	High	[46]
Biofluid method	>5 nm	Variable	Narrow	Low	[23]

Source: Tasnim Munshi.

when designing the active catalysts, and various innovative preparation approaches are adapted accordingly. Among the most common methods, impregnation and precipitation methods are the simplest and most wide spread. Another most widely used metric of the efficiency of catalyst preparation is dispersion of metals, which is defined as "ratio of exposed metal surface to the total number of metal atoms present in the catalyst." Dispersion has a reciprocal relation with size of metal particles, and it is of interest to compare the particle size and their distribution for materials that are obtained from various preparation methods [42]. A summary of various preparation techniques and their effect on morphology of HA is presented in Table 12.2.

There are some dry methods employed for the preparation of mHA materials. Dry methods are based on heat treatment of finely ground and mixed metal precursors. In order to have a robust and efficiently applicable HA material, they must have a perfectly homogenous mixture. Other important parameter that can influence the dry route synthesis is the weighing procedures used during synthesis. An identification of various optimized synthesis parameters has been done by Tromel et al. [47]. They have indicated temperature as the most important parameter to produce nanostructured HA material. They have shown a successful synthesis route of nano-sized HA catalytic material by preparing a homogeneous mixture of calcium and phosphorus precursors, followed by hydrothermal reaction at varied hydrothermal temperature and different hydrothermal reaction time. Temperature is the key to form nanostructure HA, e.g. the 180 °C-hydrothermal-treated HA produced needle-like shape with a diameter of 10–20 nm [47].

From the reported studies on HA synthesis, it can be concluded that particle size and morphology play an important role. The most common method adapted for their synthesis is chemical precipitation [48], microwave irradiation [40], and hydrothermal processing [49]. It is reported by Haider et al. [12] that maintaining the right stoichiometry, high crystallinity, and high aspect ratio is challenging. They have reported addition of various additives such as allylamine hydrochloride and poly(acrylic acid) during the synthesis process to produce tailored particle sizes. Similarly, the wet mechanochemical methods allow control over the stoichiometry of the final product, but if the metal ratio is not adjusted to the desired requirement at the precursor stage, then it would show the presence of various phases. Moreover, it becomes very hard to control nucleation, crystal growth, and agglomeration occurring simultaneously during the precipitation step [39].

During chemical synthesis, special care is made in tuning the surface properties of HA materials. Various aspects of interfacial behavior are related with their electrochemical properties. These electrochemical properties are determined by pH of the precipitating solutions and concentration of potential determining ions, for example, calcium, phosphate, and hydroxide ions. The surface charge of HA is an important property as it influences the adhesion process.

The point of zero charge (PZC) is the pH at which the net charge at the surface is zero. As HA has three lattice ions, its PZC can be obtained by changing the concentration of two of the ions independently. These experimental manipulations lead to a set of points with zero surface charge and thus give rise to a line of zero charge on the surface. This line of zero helps in the characterization of electrical double layer properties of HA in solution conditions. Bell et al. [50] calculated the PZC of HA in different electrolytes and reported a range of PZC from pH 4.35 to 7. This variation in PZC reflects the effects of choice of electrolyte, impurities, Ca/P ratio, and crystal lattice defects on the surface properties of HA. All these studies along with HA's ion substitution properties have indicated toward the more positive (less negative) nature of HA surface at lower pH where the potential determining ions are H^+ and Ca^+ along with HPO_4^-. On contrary the HA surface shows more negative (less positive) nature at higher pH where the potential determining ions are OH^-, HPO_4^-, and PO_4^{3-} in addition to Ca^{2+}. This has paved a path for surface modified HAs. Functionalization of HA surface via adsorption of compounds like synthetic macromolecules, surfactants, etc. depends on the surface properties and charge of HA [51].

Researchers have shown an increased interest in following a much greener approach for the synthesis of desired materials. Bahadorikhalili et al. [2] successfully used facile ultrasonic method for synthesis of AgNPs imprinted on thiourea-functionalized mHA (i.e. Ag@mHA-Si-(S)) nanocatalyst. Thiourea-functionalized mHA was proposed as a stabilizer and support for embedding AgNPs because of silver's strong affinity toward sulfur. The critical part of the synthesis was the adsorption of AgNP on thiourea-functionalized mHA. This was followed by reduction under ultrasonic agitation. $NaBH_4$ and $AgNO_3$ are used as reducing agent and Ag ion source, respectively. ICP-AES analysis showed loading of about 0.425 mM AgNPs per gram of functionalized mHA. This ultrasonic approach provides better catalytic ability to the heterogeneous catalyst in terms of providing more active sites, leading to higher yield in shorter reaction time.

This prepared nanostructure proved to be an efficient heterogeneous catalyst in partial oxidation of primary amines to related N-monoalkylated hydroxylamines as well as the selective reduction of nitro compounds [2]. Moreover, this green catalyst immobilized on biocompatible functionalized mHA could be recovered easily from the reaction mixture in the presence of an external magnetic field and can be reused at least 10 times without showing any significant decrease in catalytic activity.

A similar approach was used by Pashaei and Mehdipour in their study [52]. They also used an ultrasonic synthesis pathway to develop a novel AgNP loaded 1,4-diazabicyclo[2.2.2] octane (DABCO) grafted mHA (i.e. mHA DABCO-Ag). DABCO grafted on mHA creates ionic sites as hydrophilic domain and unhindered amines form an organophilic environment. The zeta potential values recorded for ionic-tagged mHA (i.e. IT-mHA-Ag) and mHA-Ag are −11.76 and −8.69 mV, respectively. As high zeta potential indicates greater stability in a colloidal system, the recorded values show that DABCO provides stability to this nanocomposite in aqueous medium. AgNPs loaded on IT-mHA are stabilized by amines against agglomeration. This enhances its catalytic property. A series of experiments were carried out to check the catalytic activity by using these synthesized IT-mHA-Ag nanostructures as heterogeneous catalyst in hydrogenation reaction of nitroarenes. It was found that reaction gave almost no yield when mHA and IT-MHA were individually used as catalysts whereas it showed 86% yield in the presence of mHA-Ag. This indicates that DABCO is an auxiliary catalyst and AgNPs in IT-mHA-AG are the main site of reaction; mHA only provides its mesoporous surface increasing possibilities of functionalization. Along with that it has been used because of its nontoxic and biocompatible nature. Furthermore, IT-mHA-Ag catalyst showed maximum product yield and better efficiency when compared with other catalysts reported in the literature [2, 53]. This experiment proposed a facile inexpensive green synthesis method of producing a reusable, highly efficient magnetic heterogeneous catalyst.

Wu et al. [54] adapted two-step wet chemical process for the synthesis of HA, where they reacted calcium hydroxide with orthophosphoric acid under argon atmosphere for 20 hours. In the next step, $FeCl_2 \cdot 4H_2O$ was added and stirred at 80 °C and pH 8.0–8.5 for 10 hours, and at the end they obtained molar ratio of 0.75 for Ca/Fe. The obtained material was washed several times in order to achieve neutral pH before applying it for catalytic applications. Similar approach was adapted by Zuo et al. [55] where solution intercalation method was applied. In this study, 1.0 g sodium dodecyl sulfonate was dissolved in 15 ml deionized (DI) water with 30 ml ethanol, and the mixture was heated at 60 °C. In next step, 15 ml of 3.3 M calcium nitrate and 30 ml of 1 M diammonium hydrogen phosphate were added to 30 ml of ethanol. Then 20 ml of 2.5 M NaOH and 20 ml of ethanol were added to the mixture and stirred for 30 minutes. To synthesize magnetic nanoparticles, 7.5 ml of 2.4 M iron(III) chloride, 7.5 ml of 1.2 M iron(II) chloride, and 30 ml of 3 M NaOH were mixed, after which 45 ml of ethanol was added. The mixture was refluxed for 14 hours at 85 °C and left at room temperature for 21 days to develop laminated HA.

The scope of various methodologies adapted for surface modifications and their consequent geometric orientation on the nanoparticles significantly affect the sizes and shapes of HA nanoparticles. So far, most studies have focused on improving the biocompatibility of materials, but only a few scientific studies have discussed the refining of magnetic particles after formation of a composite structure. Different facile synthetic routes and their mode of action, as well as close integration with surrounding molecules, need to be evaluated in a more comprehensive study. More critical studies are warranted to formulate more cost-effective but good-quality HA nanoparticles without any protracted purification stages.

12.3 Catalysis

As stated earlier, HA materials have an interesting surface chemistry and ion exchange ability. They owe their catalytic behavior to their distinctive size and incredible physical and chemical properties. These properties provide various advantages to magnetic nanoparticles over others [56], for example:

1. Small size (nanometer scale) provides high surface area-to-volume ratio and thus offers more surface for the reaction to take place.
2. Easy recovery and recycling with the help of external magnetic field.
3. Good biocompatibility makes it a better choice as a catalyst as it offers reduced toxicity and less harmful waste products.

12.3.1 Magnetic HA Nanoparticles as Active Catalysts for Organic Reactions

HA have proved to be active heterogeneous catalysts for a number or organic reactions. For example, Aziz et al. [9] have compiled a range of reactions where modified HA has been used as a heterogeneous catalyst. Similar to this, Das et al. [57] reported the synthesis of a nickel/HA/cobalt ferrite (Ni/HA/CoFe$_2$O$_4$) novel nanocomposite catalyst, which showed enhanced degradation of about 90% methyl orange and 99.1% methylene blue within 90 minutes in the presence of H$_2$O$_2$. A recyclable, in situ prepared, catalytic, natural HA-supported MnO$_2$ was investigated for aerobic oxidation of alkylarenes and alcohols. It showed highly efficient catalytic conversion of alkylarenes and alcohols to their corresponding carbonyl compounds without the presence of any other oxidizing agent [58]. All these studies show the significance of HA as a catalyst, owing to its strong adsorption ability, surface acidity or basicity, and ion exchange ability.

Iron oxides, i.e. magnetite (Fe$_3$O$_4$) and maghemite (γ-Fe$_2$O$_3$) are the most popular magnetic nanoparticles because of their relatively low toxicity, high magnetic saturation, easy fabrication, and superparamagnetic behavior [27]. Surface modification of their hydrophilicity/hydrophobicity is usually performed to adjust their dispersion stability and prevent agglomeration in organic or aqueous media. As mentioned earlier, HA serves as an effective solution for this; it enhances their chemical stability, prevents them from agglomerating, and provides an extra surface for further functionalization, thus making magnetic nanoparticles even better catalysts. Numerous studies show the use of HA encapsulated magnetic nanoparticles as a catalyst in various types of reactions and synthesis processes.

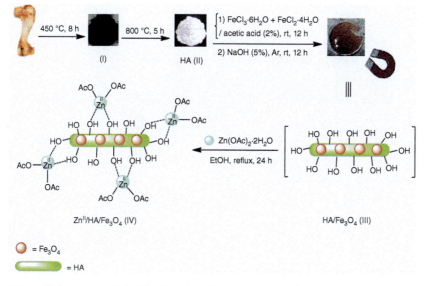

Figure 12.2 Preparation of Zn(II) anchored onto the magnetic natural hydroxyapatite. Source: Zarei et al. [61], pg. 6 (108): 106473–106484. / With permission of Royal Society of Chemistry.

Degradation of harmful industrial waste is a recent emerging and attractive problem for many researchers, and many of them have proposed mHA as a highly efficient and recyclable catalyst. Studies have also witnessed a larger scale synthesis and modification of these mHAs for their catalytic applications. Recently, Das et al. [57] suggested a three-step synthesis of magnetic Zn/HA/MgFe$_2$O$_4$ along with its characterization (Figure 12.2). The researchers were able to synthesize an efficient, recyclable, and easy to recover HA-coated magnetic nanocomposite catalyst for degradation of contaminating dyes from polluted water. They evaluated its catalytic performance in degradation of malachite green (a carcinogenic dye present in industrial wastewater) in aqueous medium with the presence of H$_2$O$_2$. Their reported data showed an exponential increase in the degradation rate, and further increase in catalytic activity was observed in the presence of H$_2$O$_2$. The proposed explanation for this was the Fenton and Fenton-like mechanism [57]. H$_2$O$_2$ acted as an oxidant providing HO$^\cdot$ and HOO$^\cdot$ radicals, which reacts with malachite green, leading to its degradation (Figure 12.3). They also studied the effects of pH and concentration of catalyst, which revealed 100% catalytic performance in the 3–7 pH range at 10 mol% H$_2$O$_2$ oxidant. Moreover, the superparamagnetic behavior of the catalyst helped in its easy separation from the reaction mixture and allowed reusability. The catalyst was used for seven consecutive cycles, and the catalyst showed maintenance of its efficiency even after seventh cycle. The powder XRD pattern of the catalyst before and after its applications shows no change in its structure and supports the previous observation.

Various other studies similar to the ones earlier have previously been reported, supporting enhanced degradation by application of magnetic HA catalysts.

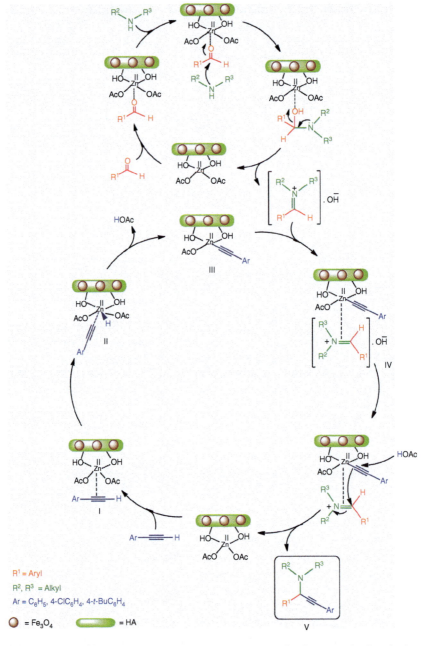

Figure 12.3 Proposed reaction mechanism for the synthesis of propargylamines in the presence of Zn(II)/HA/Fe$_3$O$_4$. Source: Zarei and Akhlaghinia [61]. Reproduced with permission of Royal Society of Chemistry.

Recyclable, nonfunctionalized Fe_3O_4-HA magnetic nanorods were synthesized by Yang et al. [59] using a homogeneous precipitation method and used them as a heterogeneous photocatalyst for effective degradation of diazinon (insecticide) under UV radiation. The structure of this study was similar to previously mentioned one, but unlike it, this study gave it more insights on the effects of pH and catalyst concentration on activity of catalyst. The acid–base equilibria controlling the surface chemistry of the catalyst was suggested as a possible explanation for this. The PZC of HA is reported to be ~6.8 indicating positively charged behavior below pH 6.8. pK_a for diazinon is 2.6, which indicates its negatively charged behavior above pH 2.6. The electrostatic interactions between the catalyst surface and diazinon are largely affected by pH change, and keeping in mind the PZC of HA and pK_a of diazinon, 5.5 pH seems to be an environment for optimum interactions. In this study, Yang et al. observed photodegradation rate of diazinon (at pH 5.5) within $2.0–6.0 g l^{-1}$ concentration range of Fe_3O_4-HA catalyst. As it is a photocatalytic reaction, presence of light is essential for the process. Relating to this fact, they found decreased catalytic activity even after an increase in concentration with the increased number of adsorbed diazinon on Fe_3O_4-HA surface. This leads to light scattering and screening effect, which hinders the penetration of light, further resulting in reduced specific activity of the catalyst. This reason is very specific to the process and especially to photocatalytic processes. Other reason for such behavior may be associated with the difficulty in maintaining homogeneous suspension at high catalyst concentrations. It may show particle agglomeration, leading to reduced catalytic activity. This is a more general explanation, which is applicable for all the types of reactions and gives a broader understanding about catalyst activity of magnetic HA materials.

The ease of preparation, robust structure, reusable nature, and easy recoverability in the presence of external magnetic field (owing it to superparamagnetic property of γ-Fe_2O_3) make it an efficient heterogeneous magnetic nanocatalyst well suited for industrial applications. Mizugaki et al. [60] successfully synthesized magnetic γ-Fe_2O_3 dispersed in HA matrix as a new catalyst support. The ion exchangeable property of HA provides it the ability of equimolar substitution of Ru and Ca, leading to the formation of active catalytic site (RuHA-γ-Fe_2O_3) (Figure 12.4). This catalyst has been employed to enhance the oxidation of various alcohols to their corresponding carbonyl compounds in the presence of oxygen as primary oxidant.

In the interest of Green Chemistry, Zarei et al. [61] successfully synthesized a novel Zn(II)-anchored magnetic HA nanocatalyst. The HA used in this catalyst is a natural form extracted from cow's bone. This nanocatalyst has been tested for the synthesis of structurally different propargylamines (4-(1,3-diphenylprop-2-ynyl)) via one-pot A^3-coupling (three-component coupling of aldehydes, secondary amine, and alkynes). A mixture of benzaldehyde, morpholine, and phenylacetylene along with Zn^{II}/HA/Fe_3O_4 as a nanocatalyst was used for its synthesis. After a series of experiments with varying reactant ratios, different catalyst concentration, and changing solvents, a reaction scheme was proposed.

The catalytic activity of Zn^{II}/HA/Fe_3O_4 in this propargylamine synthesis process was established when low yield of product was obtained in the absence of

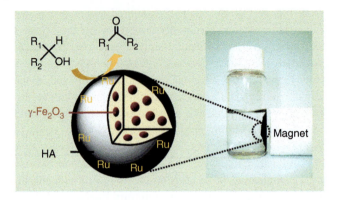

Figure 12.4 Development of ruthenium–hydroxyapatite-encapsulated superparamagnetic γ-Fe$_2$O$_3$ nanocrystallites as an efficient oxidation catalyst by molecular oxygen. Source: Mori et al. [60], pg. 19 (6): 1249–1256. / With permission of American Chemical Society.

catalyst and in the presence of Fe$_3$O$_4$ or HA or HA/Fe$_3$O$_4$, whereas presence of Zn^{II}/HA/Fe$_3$O$_4$ nanoparticle gave an excellent yield (95%). Optimization study showed that 8 mol% of nanocatalyst (Zn^{II}/HA/Fe$_3$O$_4$) in solvent-free condition at 110 °C with 1 : 1 : 1 molar ratio of benzaldehyde/phenylacetylene/morpholine results into desired A^3-coupling product with 95% isolated yield. This synthesis did not show any leaching of Zn(II) throughout the process, which proves the heterogeneous nature of Zn^{II}/HA/Fe$_3$O$_4$ nanocatalyst.

As reusability of the catalyst is one of the major needs for commercial applications, a reusability test was conducted by using the same catalyst for seven uninterrupted runs. Elemental analysis by ICP-OES showed that freshly prepared catalyst contained 0.066 g Zn(II) (g of catalyst)$^{-1}$ and the seventh reused catalyst contained 0.061 g Zn(II) (g of catalyst)$^{-1}$. According to these results, 93% of Zn(II) could still be found after seven successive runs. This confirmed that even after seven consecutive runs, the catalytic property was retained without any significant loss. Furthermore, the magnetic core of the catalyst allows easy separation after every cycle in the presence of external magnetic field. This easy to synthesize, reusable magnetic nanocatalyst (Zn^{II}/HA/Fe$_3$O$_4$) provided an inexpensive, facile green process of synthesis with excellent yield without requiring any additives/cocatalysts and eliminating toxic solvents and hazardous materials.

There are various similar studies, and all of them show agreement with the increased degradation rate on application of magnetic HA as a catalyst material. Functionalized magnetic HAs have been reportedly used as a phase transfer catalyst. Khosravinia et al. [62] successfully synthesized β-cyclodextrin conjugated γ-Fe$_2$O$_3$/HA solid–liquid phase transfer catalyst for the nucleophilic ring opening of the epoxide in water for the preparation of β-azido alcohol, β-nitro-alcohol, and β-cyanohydrin. They simply observed the increased rate of reaction with 89% yield in the presence of catalyst, the catalytic property was associated with the complex formation between epoxide oxygen and β-cyclodextrin's outer OH$^-$ group via hydrogen bonds. It provided high yield, short reaction time, and recyclable catalyst with easy separation from reaction mixture with the help of external magnetic field.

Porous HA has been reported as reusable heterogeneous catalyst, prepared by microwave irradiation method and calcined at varied temperatures. The catalyst calcined at 300 °C showed reasonably good yield in Knoevenagel condensation for a range of aldehyde substrates that incorporated electron-donating and electron-withdrawing substituents, with malonitrile and ethyl cyanoacetate. The catalyst was reported to have exhibited far better catalytic activity and selectivity compared with the catalysts that were calcined at higher temperatures [63].

The introduction of a second metal into lattice structure expands the catalytic applications of magnetic HA materials [64]. An alternative method uses the concept of metal deposition on the surfaces of HA. This strategy is more successfully reported by use of a ligand that attached itself with an incoming metal on surface of magnetic nanoparticles. Metal oxides such as Si and Ru islands can be decorated on nanoparticles of Fe_2O_3 or Al_2O_3 and have been reported to be successful catalysts for hydrogenation reactions [65].

12.3.2 HA Analogs and Their Catalytic Applications

Analogs to HA, for example, are $NaLaCa_3(PO_4)_3OH$ and $NaLaSr_3(PO_4)$; they have been reported as possible catalysts in the Knoevenagel condensation [66]. They are reported to be active, but not stable. The addition of water leads to an enhanced activity of these catalysts; an increase in basic sites on the surface may have been the cause of enhanced catalytic activity. Unfortunately, the authors have not reported recycling data, and also a more detailed investigation in terms of substrates and reaction conditions will be required to determine further scope of these catalysts. Another detailed investigation report has been published for the same reaction using HA at room temperature in the absence of any solvents. The catalyst was able to promote the reaction, although reaction times are reported to be longer compared with homogeneous catalysts. Significantly improved catalytic activity was observed by adding water and benzyl-triethylammonium chloride separately, and high yields of products were reported. It is concluded that the addition of water dissolved the ammonium salt and facilitated its interaction with HA, which, therefore, allowed the activation of catalyst. The catalyst was recoverable and subsequently calcined at 700 °C before being reused nine times without any loss in catalytic activity [67].

Further to this, another study reported Knoevenagel condensation using HA prepared by the coprecipitation method [68]. They found methanol to be the best solvent among various other organic solvents. The catalyst showed superior activity, and doping of HA with potassium fluoride further improved the activity rate and product yield. The catalyst was reported to be recyclable for up to three cycles. In another study Al-enriched HA catalyst was used for Knoevenagel condensation of benzene with ethyl cyanoacetate at ambient conditions. The catalyst showed an enhanced activity that was related with a high surface area of magnetic HA and also the presence of HPO_4^{2-} species that leads to the formation of acid–base pair sites by an interaction between Al of $AlPO_4$ and OH^- of HPO_4^{2-} [69]. Another interesting report has come from Liu et al. [65] who described a series of imidazole-based ionic liquid with different alkyl chain lengths deposited on γ-Fe_2O_3 encapsulated in HA. These basic magnetic nanoparticles were used as recyclable and active catalysts for Knoevenagel condensation reaction between different types of aldehydes and mononitrile under mild reaction conditions. The characterization data reported in this study confirms 1–3 nm γ-Fe_2O_3 particles in HA material. They also linked an enhanced catalytic activity with a synergistic cooperation between basic sites of HA framework and supported basic ionic liquids. The catalyst was recovered by magnetic decantation using a permanent magnet, and the catalyst was reported to have consistent activity up to four cycles.

Aklaghima et al. [70] have used magnetic natural HA (HA/Fe_3O_4 NPs) for the synthesis of bis-coumarin derivatives under mild green and solvent-free conditions. The bis-coumarin derivatives were synthesized in excellent yields, and importantly the magnetic HA does not need labor intensive chromatography to separate the products. The bis-coumarins can be easily filtered of after separating the HA/Fe_3O_4 NP by means of an external magnetic field. The mechanism to synthesize the bis-coumarin involves a Knoevenagel condensation followed by a Michael addition in the presence of HA/HA/Fe_3O_4 NPs as a catalyst. The use of a catalyst is the reusability, in this case after completion of the reaction, the magnetic catalyst can be separated via an external magnetic field. No significant changes were observed in the recyclable process, and the catalyst was stable up to six cycle runs. The comparison of this catalysis with other available catalysis for the synthesis of bis-coumarin derivatives shows that the HA/Fe_3O_4 NP acts better especially from the ease of separation using an external magnetic field.

In a similar study by Abbasi et al. [71], Ag-loaded mHA (i.e. γ-Fe_2O_3@HA-Ag NPs) was synthesized and used as an efficient catalyst for a simple procedure of Pechmann reaction under solvent and halogen free conditions at 80 °C. Pechmann reaction is a one-pot condensation reaction of various phenols with ethyl acetoacetate, which is considered the best procedure for preparing coumarin (2H-chromene-2-ones) derivatives (a significant compound for pharmaceutical and biological applications). In past, Pechmann condensation has been performed in the presence of variety of acids, but they suffered from at least one of the following disadvantages: low, unsatisfactory yield; tedious procedure; extreme reaction conditions; lengthy process; use of toxic catalysts; and generation of hazardous waste. Use of the proposed γ-Fe_2O_3@HA-Ag nanocatalyst served all these disadvantages and provided an additional advantage with reusability of catalyst and ease of recovery using an external magnet.

12.3.3 HA Catalysts and Green Chemistry

With increasing environmental concerns, researchers are also focusing on the development of products and procedures, which reduce or eliminate the use and generation of harmful substances. Because of its biocompatible and nontoxic nature, literature has reported the use of mHA as a catalyst in many of these green synthesis approaches. Naming a few, Mahire et al. [21] proposed a green approach for water-mediated synthesis of pyrimidinobenzimidazoles (bioactive compound) using sulfonated chitosan-encapsulated HA-Fe_3O_4. In another study Shaabani et al. [72] carried out the same synthesis process in water without the use of catalyst. A comparative analysis of both showed that the presence of catalyst provided 88% yield within 15 minutes, whereas in its absence it only provided 70% yield in 7–12 hours. Here, Fe_3O_4 provided catalytic property along with increased surface area; HA and chitosan provided adsorption surface and served as a bioactive coating to prevent Fe_3O_4 from getting oxidized in atmosphere, and finally sulfonated functional layer provided additional active sites for the reaction to take place. The catalyst synthesized was reusable and easily recoverable in the presence of magnetic field. This study articulates the significance of catalyst in promoting the rapid synthesis with maximum yield of product.

The main barriers for the implementation of enzyme synthesis is the difficulty of recovery from the reaction medium after use; however, magnetic HA provides an easy means of separation. Saire-Saire et al. [73] have recently synthesized $CoFe_2O_4$ magnetic oxide nanoparticles that have been incorporated into HA and used for the synthesis of lipase. A magnetic core of $CoFe_2O_4$ has been synthesized and coated with HA as a biocompatible interface for enzymatic functionalities. Lipase was attached either by adsorption or covalent bonding mediated by 3-aminopropyl)triethoxysilane (APTES) and glutaraldehyde (GLU). The influence of $CoFe_2O_4$/HA and lipase interface on the catalytic response was evaluated. Adsorption on the surface is through interaction between the functional groups of the $CoFe_2O_4$ nanocomposite surface and the enzyme. The HA has three different interacting groups: (i) chelating centers of Ca^{2+}, (ii) electrostatic interaction with phosphate sites, and (iii) hydroxyl groups [74]. Meanwhile, lipase enzyme has amino ($-NH_2$) and carboxylate groups ($-COO^-$). The surface adsorption occurs through chelation of carboxylate groups with Ca^{2+} metallic centers, hydrogen bonds with surface hydroxyl groups, and electrostatic interaction between enzyme positively charged amino groups and support negatively charged phosphate groups [75].

Lipase catalysts offer a major advantage such as mild conditions and environmentally friendly processes; however, the major drawbacks are the stability and recyclability [76]. The innovation in both the fabrications of the new immobilized lipase with both magnetic behaviors, core–shell structure, and its recyclability in its use as a solid-state catalyst in the interesterification of vegetable oils offers advantages over the traditional methods. The greatest advantage lies in the magnetic separability of the catalyst to the conventional immobilized lipase, and, therefore, it can easily be separated.

The mHA has also been seen in studies related to food technology [77]. The vegetable oil we use is chemically modified to produce structured lipids with desirable properties. As partial hydrogenation is one of the popular methods to do so, unfortunately the process leads to the formation of trans-fatty acid as a final product. These trans-fatty acids are known to have an adverse effect on human health. In the need of an approach to reduce the trans isomer content in modified fat, interesterification has attracted much attention. Enzymatic interesterification is a great method for production of trans-free modified fats with desired properties and nutritional values.

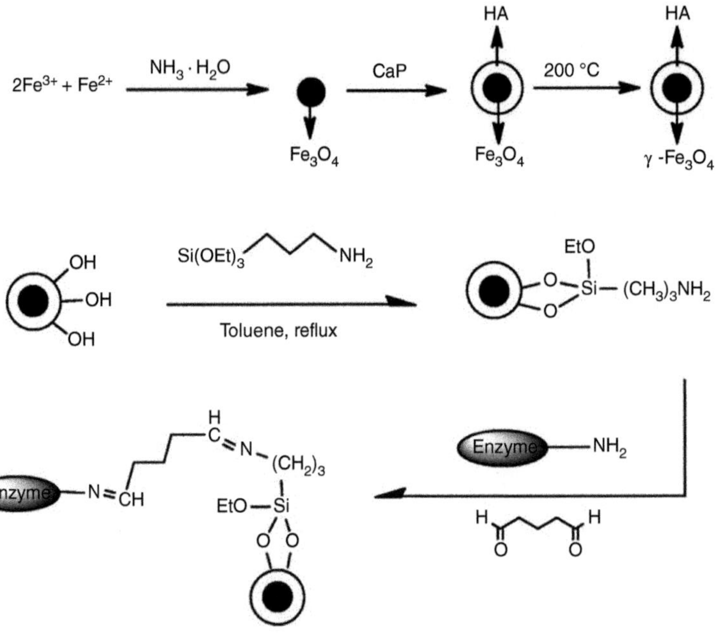

Figure 12.5 Synthesis of the magnetic immobilized lipase. Source: Xie and Zang [76]. Reproduced with permission of Elsevier.

Xie and Zang [76] successfully prepared HA/γ-Fe$_2$O$_3$ nanoparticles and employed them as magnetic carriers for lipase immobilization with covalent bonds (Figure 12.5). The lipase immobilized HA/γ-Fe$_2$O$_3$ biocatalyst showed excellent catalytic performance. The immobilized lipase allowed interesterification to take place. The synthesized magnetic biocatalyst could be easily separated from the reaction mixture with simple magnetic decantation. This can then be reused without showing any significant degradation in biocatalytic activity.

Immobilization by covalent bonding is attained by surface modification in two steps. First, the nanocomposite CoFe$_2$O$_4$/HA is silanized with APTES, an organofunctional alkoxysilane molecule with an aminofunctional group. Afterward, GLU is used to covalently bond amino-terminated APTES with amino functional groups of the enzyme through a Schiff base cross-linking reaction that forms imine bonds (Figure 12.6). The catalytic performance and selectivity of the immobilized enzymes (magnetic bio-nanocomposites) toward the transesterification reaction of (R,S)-1-phenylethanol was investigated. Evaluation of the catalytic activity showed high selectivity on the transesterification for racemic mixtures of (R,S)-1-phenylethanol with complete conversion of (R)-1-phenylethanol enantiomer. This enhances the enantiomeric excess of (S)-1-phenylethanol that can be separated from the media to obtain a 100% pure enantiomer of high added value in synthesis. Finally, magnetic recovery and reuse was evidenced by high stability of magnetic bio-nanocomposite catalysts (CoFe$_2$O$_4$/HA-Lipase and CoFe$_2$O$_4$/HA-APTES-Lipase) that maintain their catalytic activity over several cycles (Figure 12.5).

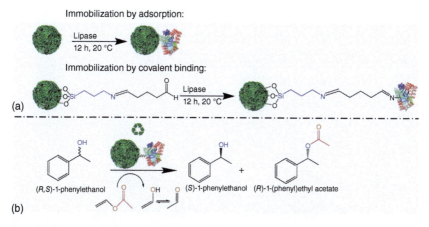

Figure 12.6 (a) Enzyme immobilization indicating temperature and time of reaction. (b) Enantioselective transesterification reaction of (R,S)-1-phenylethanol catalyzed by the bio-nanocomposite $CoFe_2O_4$/HA/lipase and $CoFe_2O_4$/HA-APTES-lipase. The green spikes the needle-shaped hydroxyapatite coating. Source: Saire-Saire et al. [73]. Reproduced with permission of Elsevier.

In the interest of oxidation of 5-hydroxymethylfurfural (HMF) into 2,5-diformylfuran (DFF), Wang et al. [78] synthesized magnetic Mo-HA@γ-Fe_2O_3 nanocatalyst. DDF is an important oxidation product of HMF as it can be used as a versatile precursor in preparation of furan polymers, porous organic frameworks, antifungal agents, and pharmaceuticals. A noble metal-based heterogeneous catalyst provides high catalytic activity, but it is too expensive. This non-noble metal-based heterogeneous catalyst (Mo-HA@γ-Fe_2O_3) provided remarkable efficiency, showing high HMF conversion of 95.8% and DFF yield of 67.5% in 4 chlorotoluene at optimum reaction temperature of 120 °C for 12 hours. The catalyst had added advantage of reusability and easy recovery from the reaction mixture in the presence of external magnetic field.

The idea of grafting desirable compound on the surface of HA has been around for a long time now. Oskouie et al. [79] in their study were able to synthesize thiourea-functionalized mHA as a recyclable inorganic–organic hybrid nanocatalyst for selective and highly efficient 1,4-addition reaction of chalcones and related enones with TMSCN (trimethylsilyl cyanide – a cyanide anion source) (Figure 12.7). First of all, organosilanes were prepared and then grafted on mHA to obtain the desired material. The basic structure of mHA-based nanocatalyst usually constitutes a superparamagnetic core coated with a layer of HA whose surface is further functionalized with desired motifs. The coating along with the functionalization affects the magnetic saturation of the whole material. The thiourea-functionalized mHA also shows the characteristic superparamagnet hysteresis curve, but it shows decreased magnetic saturation. Important thing to note is that it did not matter if the magnetic saturation was decreased; if the possessed magnetization is sufficient enough for magnetic separation using conventional magnet, then the catalyst is efficient enough.

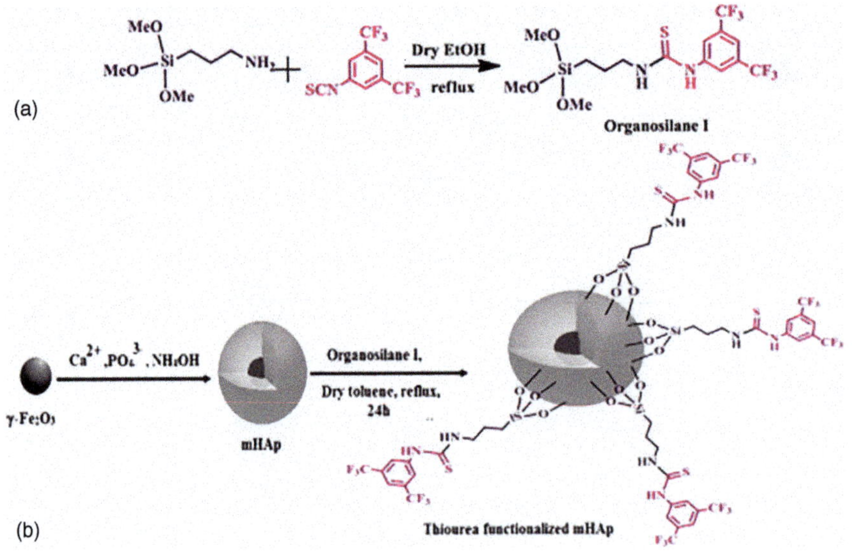

Figure 12.7 (a) Synthesis of organosilane I and (b) thiourea-functionalized mHAp. Source: Oskouie et al. [79]. Reproduced with permission of Elsevier.

Versatile mHAs have also been reported in electrochemical sensor studies. In a study by Mirzajani et al. [80], copper(II) nanomagnetic HA composite was synthesized and introduced as an electrode modifier. A highly sensitive, selective, and robust electrochemical sensor was constructed with $\gamma\text{-Fe}_2\text{O}_3/\text{HA}/\text{Cu(II)}$-modified carbon paste electrode for detection of metformin (medication for treatment of type II diabetes) in urine and pharmaceutical samples. This study highlights the major electrocatalytic characteristics of mHA. The nanoporosity and weak acid base surface nature are the reasons for the catalytic character of HA. HA along with magnetic Fe_2O_3 nanoparticles as a composite provides large specific surface area for successful adsorption of the target or any functionalization that agrees with the observations made in previous section. Here mHA surface has been functionalized by Cu, which shows an increase in the catalytic performance of the composite. With functionalization, Cu rapidly forms complex on the P—OH sites, which is followed by partial dissolution of calcium, leading to precipitation of Cu-substituted HA. Overall, the synergistic effect of Cu(II) ions and Fe_2O_3-HA composite leads to good catalytic activity, indicating the significance of modifier in electron transfer kinetics on the electrode. The application of electrocatalytic properties of mHAs holds huge scope in discipline of sensor development.

12.4 Summary and Conclusions

Tremendous progress has been made in last decade or so, pertaining to the preparation and catalytic applications of magnetic HA. This solid material has seen a number of wide range applications due to their high thermal stability and weak acid–base character that may have helped in hindering the routes to side reactions and as a result demonstrated a high yield/selectivity to desired products. The superparamagnetic behavior contributes to its significant use in the wide range of applications in catalysis and the development of environmentally friendly and sustainable synthesis routes.

References

1. Thomas, J.M. and Harris, K.D.M. (2016). Some of tomorrow's catalysts for processing renewable and non-renewable feedstocks, diminishing anthropogenic carbon dioxide and increasing the production of energy. *Energy and Environmental Sciences* 9 (3): 687–708.
2. Bahadorikhalili, S., Arshadi, H., Afrouzandeh, Z. et al. (2020). Ultrasonic promoted synthesis of Ag nanoparticle decorated thiourea-functionalized magnetic hydroxyapatite: a robust inorganic-organic hybrid nanocatalyst for oxidation and reduction reactions. *New Journal of Chemistry* 44 (21): 8840–8848.
3. Cao, H., Zhang, L., Zheng, H. et al. (2010). Hydroxyapatite nanocrystals for biomedical applications. *Journal of Physical Chemistry C* 114 (43): 18352–18357.
4. McConnel, D. (1973). *Apatite Its Crystal Chemistry, Mineralogy, Utilization, and Geologic and Biologic Occurrences*. New York: Springer.
5. Cichy, B., Kużdżał, E., and Krztoń, H. (2019). Phosphorus recovery from acidic wastewater by hydroxyapatite precipitation. *Journal of Environmental Management* 232: 421–427.
6. Cojocaru, F.D., Balan, V., Popa, M.I. et al. (2019). Biopolymers – calcium phosphates composites with inclusions of magnetic nanoparticles for bone tissue engineering. *International Journal of Biological Macromolecules* 125: 612–620.
7. Anwar, A., Rehman, I.U., and Darr, J.A. (2016). Low-temperature synthesis and surface modification of high surface area calcium hydroxyapatite nanorods incorporating organofunctionalized surfaces. *Journal of Physical Chemistry C* 120 (51): 29069–29076.
8. Song, Z., Liu, Y., Shi, J. et al. (2018). Hydroxyapatite/mesoporous silica coated gold nanorods with improved degradability as a multi-responsive drug delivery platform. *Materials Science and Engineering C* 83: 90–98.
9. Fihri, A., Len, C., Varma, R.S. et al. (2017). Hydroxyapatite: a review of syntheses, structure and applications in heterogeneous catalysis. *Coordination Chemistry Reviews* 347: 48–76.
10. Kay, M.I., Young, R.A., and Posner, A.S. (1964). Crystal structure of hydroxyapatite. *Nature* 204 (4963): 1050–1052.
11. Boda, S.K., Anupama, A.V., Basu, B. et al. (2015). Structural and magnetic phase transformations of hydroxyapatite-magnetite composites under inert and ambient sintering atmospheres. *Journal of Physical Chemistry C* 119 (12): 6539–6555.
12. Haider, A., Haider, S., Han, S.S. et al. (2017). Recent advances in the synthesis, functionalization and biomedical applications of hydroxyapatite: a review. *RSC Advances* 7 (13): 7442–7458.

13 Family, R., Solati-Hashjin, M., Nik, S.N. et al. (2012). Surface modification for titanium implants by hydroxyapatite nanocomposite. *Caspican Journal of Internal Medicine* 3 (3): 460–465.

14 Turon, P., del Valle, L.J. et al. (2017). Biodegradable and biocompatible systems based on hydroxyapatite nanoparticles. *Applied Sciences* 7 (1): 60–87.

15 Yerin, C.V. (2017). Particles size distribution in diluted magnetic fluids. *Journal of Magnetism and Magnetic Materials* 431: 27–29.

16 Abu-Dief, A.M. and Abdel-Fatah, S.M. (2018). Development and functionalization of magnetic nanoparticles as powerful and green catalysts for organic synthesis. *Beni-Seuf University Journal of Basic and Applied Sciences* 7 (1): 55–67.

17 Sanchez, L.M., Actis, D.G., Gonzalez, J.S. et al. (2019). Effect of PAA-coated magnetic nanoparticles on the performance of PVA-based hydrogels developed to be used as environmental remediation devices. *Journal of Nanoparticle Research* 21 (3): 64–80.

18 Frey, N.A. and S.S. (2011). *Magnetic Nanoparticle for Information Storage Applications*. Boca Raton, FL: Taylor & Francis.

19 Xu, Y.J., Dong, L., Lu, Y. et al. (2016). Magnetic hydroxyapatite nanoworms for magnetic resonance diagnosis of acute hepatic injury. *Nanoscale* 8 (3): 1684–1690.

20 Sangeetha, K., Vidhya, G., Girija, E.K. et al. (2019). Fabrications of magnetic responsive hydroxyapatite platform: in vitro release of chemo drug for cancer therapy. *Materials Today Proceedings* 26 (3): 3579–3582.

21 Mahire, V.N., Patil, G.P., Deore, A.B. et al. (2018). Sulfonated chitosan-encapsulated HAp@Fe_3O_4: an efficient and recyclable magnetic nanocatalyst for rapid eco-friendly synthesis of 2-amino-4-substituted-1,4-dihydrobenzo[4,5] imidazo[1,2-*a*]pyrimidine-3-carbonitriles. *Research on Chemical Intermediates* 44 (10): 5801–5815.

22 Tseng, C.L., Kuo-Chi Chang, K., Yeh, M. et al. (2014). Development of a dual-functional Pt–Fe-HAP magnetic nanoparticles application for chemo-hyperthermia treatment of cancer. *Ceramics International* 40 (4): 5117–5127.

23 Bhatt, A., Sakai, K., Madhyastha, R. et al. (2020). Biosynthesis and characterization of nano magnetic hydroxyapatite (nMHAp): an accelerated approach using simulated body fluid for biomedical applications. *Ceramics International* 46 (17): 27866–27876.

24 Ghasemi, E. and Sillanpää, M. (2019). Ultrasound-assisted solid-phase extraction of parabens from environmental and biological samples using magnetic hydroxyapatite nanoparticles as an efficient and regenerable nanosorbent. *Microchimica Acta* 186 (9): 622–629.

25 Tampieri, A., D'Alessandro, T., Sandri, M. et al. (2012). Intrinsic magnetism and hyperthermia in bioactive Fe-doped hydroxyapatite. *Acta Biomaterialia* 8 (2): 843–851.

26 Wang, Y.Y., Liu, Y.X., Lu, H.H. et al. (2018). Competitive adsorption of Pb(II), Cu(II), and Zn(II) ions onto hydroxyapatite-biochar nanocomposite in aqueous solutions. *Journal of Solid State Chemistry* 261: 53–61.

27 Tseng, C.L., Chang, K.C., Yeh, M.C. et al. (2014). Development of a dual-functional Pt–Fe–HAP magnetic nanoparticles application for chemo-hyperthermia treatment of cancer. *Ceramics International* 40 (4): 5117–5127.

28 Zeng, D., Dai, Y., Zhang, Z. et al. (2020). Magnetic solid-phase extraction of U(VI) in aqueous solution by Fe_3O_4@hydroxyapatite. *Journal of Radioanalytical Nuclear Chemistry* 324 (3): 1329–1337.

29 Pon-On, W., Meejoo, S., and Tang, I.M. (2008). Substitution of manganese and iron into hydroxyapatite: Core/shell nanoparticles. *Materials Research Bulletin* 43 (8–9): 2137–2144.

30 Petchsang, N., Pon-On, W., Hodak, J.H. et al. (2009). Magnetic properties of Co-ferrite-doped hydroxyapatite nanoparticles having a core/shell structure. *Journal of Magnetism and Magnetic Materials* 321 (13): 1990–1995.

31 Syamchand, S.S. and Sony, G. (2015). Multifunctional hydroxyapatite nanoparticles for drug delivery and multimodal molecular imaging. *Microchimica Acta* 182 (9–10): 1567–1589.

32 Liu, Y., Sun, Y., Cao, C. et al. (2014). Long-term biodistribution in vivo and toxicity of radioactive/magnetic hydroxyapatite nanorods. *Biomaterials* 35 (10): 3348–3355.

33 Piri, F., Mollahosseini, A., Khadir, A. et al. (2020). Synthesis of a novel magnetic zeolite–hydroxyapatite adsorbent via microwave-assisted method for protein adsorption via magnetic solid-phase extraction. *Journal of Iranian Chemical Society* 17 (7): 1635–1648.

34 Ribeiro, T.P., Monteiro, F.J., and Laranjeira, M.S. (2020). Duality of iron (III) doped nano hydroxyapatite in triple negative breast cancer monitoring and as a drug-free therapeutic agent. *Ceramics International* 46 (10): 16590–16597.

35 Zhu, Y., Li, Z., Zhang, Y. et al. (2020). The essential role of osteoclast-derived exosomes in magnetic nanoparticle-infiltrated hydroxyapatite scaffold modulated osteoblast proliferation in an osteoporosis model. *Nanoscale* 12 (16): 8720–8726.

36 Lan, Z., Ting, Y., Yang, X. et al. (2019). Magnetic hydroxyapatite nanotubes on micro-arc oxidized titanium for drug loading. *Materials Research Express* 6 (9): 95091.

37 Yamamoto, M., Yabutsuka, T., Takai, S. et al. (2018). Biomimetic method for production of magnetic hydroxyapatite microcapsules for enzyme immobilization. *Transactions of the Materials Research Society of Japan* 43 (3): 153–156.

38 Yusoff, A.H.M., Salimi, M.N., and Gopinath, S.C.B. (2020). Catechin adsorption on magnetic hydroxyapatite nanoparticles: a synergistic interaction with calcium ions. *Materials Chemistry and Physics* 241: 122337.

39 Ignjatovic, N., Suljovrujic, E., and Budinski-Simendic, J. (2004). Evaluation of hot-pressed hydroxyapatite/poly-L-lactide composite biomaterial characteristics. *Journal of Biomedical Materials Research – Part B Applied Biomaterials* 71 (2): 284–294.

40 Lak, A., Mazloumi, M., Mohajerani, M.S. et al. (2008). Rapid formation of mono-dispersed hydroxyapatite nanorods with narrow-size distribution via microwave irradiation. *Journal of American Ceramics Society* 91 (11): 3580–3584.

41 Sugiyama, T., Akiyama, S., and Ikoma, T. (2016). Calcium phosphate with high specific surface area synthesized by a reverse micro-emulsion method. *MRS Advances* 1 (11): 723–728.

42 Jensen, H., Pedersen, J.H., Jorgensen, J.E. et al. (2006). Determination of size distribution in nanosized powders by TEM, XD, and SAXS. *Journal of Experimental Nanoscience* 1 (3): 355–373.

43 Wu, Y., Hench, L.L., Du, J. et al. (2004). Nucleation and growth of hydroxyapatite nanocrystals by hydrothermal method. *Journal of American Chemical Society* 87: 1988–1991.

44 San Thian, E., Ahmad, Z., Huang, J. et al. (2008). The role of electrosprayed apatite nanocrystals in guiding osteoblast behaviour. *Biomaterials* 29 (12): 1833–1843.

45 Agrawal, K., Singh, G., Puri, D., and Prakash, S. (2011). Synthesis and characterization of hydroxyapatite powder by sol-gel method for biomedical application. *Journals of Minerals and Materials Characterization and Engineering* 10 (8): 727–734.

46 Vijayaraghavan, R. and Sasikumar, S. (2008). Solution combustion synthesis of bioceramic calcium phosphates by single and mixed fuels – a comparative study. *Ceramics International* 34: 1373–1379.

47 Korber, F. and Tromel, G.Z. (1932). The formation of HAP through a solid state reaction between tri and tetra - calcium phosphate. *Journal of the Electrochemical Society* 38: 578–580.

48 Joshi, P., Patel, C., and Vyas, M. (2018). Synthesis and characterization of hydroxyapatite nanoparticles by chemical precipitation method for potential application in water treatment. *AIP Conference Proceedings* 1961 (1): 030037.

49 Ashok, M., Kalkura, S.N., Sundaram, N.M. et al. (2007). Growth and characterization of hydroxyapatite crystals by hydrothermal method. *Journal of Materials Science: Materials in Medicine* 18 (5): 895–898.

50 Bell, L.C., Posner, A.M., and Quirk, J.P. (1973). The point of zero charge of hydroxyapatite and fluorapatite in aqueous solutions. *Journal of Colloid and Interface Science* 42 (2): 250–261.

51 Yosry, A.A. and Douglas, W.F. (1988). The equilibrium composition of hydroxyapatite and fluorapatite-water interfaces. *Colloids and Surfaces* 34 (3): 271–285.

52 Pashaei, M. and Mehdipour, E. (2018). Silver nanoparticles supported on ionic-tagged magnetic hydroxyapatite as a highly efficient and reusable nanocatalyst for hydrogenation of nitroarenes in water. *Applied Organometallic Chemistry* 32 (4): 1–11.

53 Rautray, T. and Kim, K. (2011). Synthesis of silver incorporated hydroxyapatite under magnetic field. *Key Engineering Materials* 493–494: 181–185.

54 Wu, H.C., Wang, T.W., Sun, J.S. et al. (2007). A novel biomagnetic nanoparticle based on hydroxyapatite. *Nanotechnology* 18 (16): 165601.

55 Zuo, G.F., Wan, Y.Z., Hou, L.Y. et al. (2015). Intercalative nanohybrid of DNA in laminated magnetic hydroxyapatite. *Materials Technology* 30 (2): 86–89.

56 Hudson, R., Feng, Y., Varma, R.S. et al. (2014). Bare magnetic nanoparticles: sustainable synthesis and applications in catalytic organic transformations. *Green Chemistry* 16 (10): 4493–4505.

57 Das, K.C., Das, B., and Dhar, S.S. (2020). Effective catalytic degradation of organic dyes by nickel supported on hydroxyapatite-encapsulated cobalt ferrite (Ni/HAP/CoFe$_2$O$_4$) magnetic novel nanocomposite. *Water, Air, and Soil Pollution* 231 (2): 1–13.

58 Shaabani, A., Afaridoun, H., and Shaabani, S. (2016). Natural hydroxyapatite-supported MnO$_2$: a green heterogeneous catalyst for selective aerobic oxidation of alkylarenes and alcohols. *Applied Organometallic Chemistry* 30 (9): 772–776.

59 Yang, Z.P., Gong, X.Y., and Zhang, C.J. (2010). Recyclable Fe$_3$O$_4$/hydroxyapatite composite nanoparticles for photocatalytic applications. *Chemical Engineering Journal* 165 (1): 117–121.

60 Mori, K., Kanai, S., Hara, T. et al. (2007). Development of ruthenium-hydroxyapatite-encapsulated superparamagnetic γ-Fe$_2$O$_3$ nanocrystallites as an efficient oxidation catalyst by molecular oxygen. *Chemical Materials* 19 (6): 1249–1256.

61 Zarei, Z. and Akhlaghinia, B. (2016). Zn(II) anchored onto the magnetic natural hydroxyapatite (ZnII/HAP/Fe$_3$O$_4$): as a novel, green and recyclable catalyst for A3-coupling reaction towards propargylamine synthesis under solvent-free conditions. *RSC Advances* 6 (108): 106473–106484.

62 Khosravinia, S., Kiasat, A.R., and Saghanezhad, S.J. (2019). Synthesis and characterization of γ-Fe$_2$O$_3$@HAP@β-CD core-shell nanoparticles as a novel magnetic nanoreactor and its application in the one-pot preparation of β-azido alcohols, β-nitro alcohols, and β-cyanohydrins. *Iranian Journal of Chemical Engineering* 38 (3): 61–68.

63 Mallouk, S., Bougrin, K., Laghzizil, A. et al. (2010). Microwave-assisted and efficient solvent-free Knoevenagel condensation. A sustainable protocol using porous calcium hydroxyapatite as catalyst. *Molecules* 15 (2): 813–823.

64 Polshettiwar, V., Baruwati, B., and Varma, R.S. (2009). Nanoparticle-supported and magnetically recoverable nickel catalyst: a robust and economic hydrogenation and transfer hydrogenation protocol. *Green Chemistry* 11 (1): 127–113.

65 Liu, J., Peng, X., Sun, W. et al. (2008). Magnetically separable Pd catalyst for carbonylative sonogashira coupling reactions for the synthesis of α,β-alkynyl ketones. *Organic Letters* 10 (18): 3933–3936.

66 Keerthi Priya, G.B. (2009). Apatite phosphates containing heterovalent cations and their application in Knoevenagel condensation. *Materials Research Bulletin* 44 (6): 1209–1213.

67 Sebti, S., Tahir, R., Nazih, R. et al. (2002). Hydroxyapatite as a new solid support for the Knoevenagel reaction in heterogeneous media without solvent. *Applied Catalysis A* 228: 155–159.

68 Smahi, A., Solhy, A., El Badaoui, H. et al. (2003). Potassium fluoride doped fluorapatite and hydroxyapatite as new catalysts in organic synthesis. *Applied Catalysis A: General* 250 (1): 151–159.

69 Elazarifi, N., Ezzamarty, A., Leglise, J. et al. (2004). Kinetic study of the condensation of benzaldehyde with ethylcyanoacetate in the presence of Al-enriched fluoroapatites and hydroxyapatites as catalysts. *Applied Catalysis A: General* 267 (1–2): 235–240.

70 Akhlaghinia, B., Sanati, P., Mohammadinezhad, A. et al. (2019). The magnetic nanostructured natural hydroxyapatite (HAP/Fe$_3$O$_4$ NPs): an efficient, green and recyclable nanocatalyst for the synthesis of biscoumarin derivatives under solvent-free conditions. *Research on Chemical Intermediates* 45 (5): 3215–3235.

71 Abbasi, Z., Rezayati, S., Bagheri, M. et al. (2017). Preparation of a novel, efficient, and recyclable magnetic catalyst, γ-Fe$_2$O$_3$@HAp-Ag nanoparticles, and a solvent- and halogen-free protocol for the synthesis of coumarin derivatives. *Chinese Chemical Letters* 28 (1): 75–82.

72 Shaabani, A., Rahmati, A., and Rad, J.M. (2008). Ionic liquid promoted synthesis of 3-(2′-benzothiazolo)-2,3-dihydroquinazolin-4(1H)-ones. *Comptes Rendus Chimie* 11 (6–7): 759–764.

73 Saire-Saire, S., Garcia-Segura, S., Luyo, C. et al. (2020). Magnetic bio-nanocomposite catalysts of CoFe$_2$O$_4$/hydroxyapatite-lipase for enantioselective synthesis provide a framework for enzyme recovery and reuse. *International Journal of Biological Macromolecules* 148: 284–291.

74 Lin, Z., Hu, R., Zhou, J. et al. (2017). A further insight into the adsorption mechanism of protein on hydroxyapatite by FTIR-ATR spectrometry. *Spectrochimica Acta – Part A Molecular Biomolecular Spectroscopy* 173: 527–531.

75 Ivić, J.T., Dimitrijević, A., Milosavić, N. et al. (2016). Assessment of the interacting mechanism between *Candida rugosa* lipases and hydroxyapatite and identification of the hydroxyapatite-binding sequence through proteomics and molecular modelling. *RSC Advances* 6 (41): 34818–34824.

76 Xie, W. and Zang, X. (2017). Covalent immobilization of lipase onto aminopropyl-functionalized hydroxyapatite-encapsulated-γ-Fe$_2$O$_3$ nanoparticles: a magnetic biocatalyst for interesterification of soybean oil. *Food Chemistry* 227: 397–403.

77 Mondal, S., Manivasagan, P., Bharathiraja, S. et al. (2017). Magnetic hydroxyapatite: a promising multifunctional platform for nanomedicine application. *International Journal of Nanomedicine* 12: 8389–8410.

78 Wang, S., Liu, B., Yuan, Z. et al. (2016). Aerobic oxidation of 5-hydroxymethylfurfural into furan compounds over Mo-hydroxyapatite-encapsulated magnetic γ-Fe$_2$O$_3$. *Journal of Taiwan Institute of Chemical Engineers* 58: 92–96.

79 Oskouie, A.A., Taheri, S., Mamani, L. et al. (2015). Thiourea-functionalized magnetic hydroxyapatite as a recyclable inorganic-organic hybrid nanocatalyst for conjugate hydrocyanation of chalcones with TMSCN. *Catalysis Communications* 72: 6–10.

80 Mirzajani, R. and Karimi, S. (2018). Preparation of γ-Fe$_2$O$_3$/hydroxyapatite/Cu(II) magnetic nanocomposite and its application for electrochemical detection of metformin in urine and pharmaceutical samples. *Sensors and Actuators B Chemical-Journal* 270: 405–416.

13

Materials from Eggshells and Animal Bones and Their Catalytic Applications

Abarasi Hart and Elias Aliu

The University of Sheffield, Department of Chemical and Biological Engineering, Mappin Street, Sheffield S1 3JD, United Kingdom

13.1 Introduction

Globally, egg and meat are commonly consumed as food because they offer high nutritional benefits. After consumption, the eggshells and animal bones are discarded as waste in landfill to undergo biodegradation. The eggshell is about 11 wt% of the total egg weight [1]. An eggshell is grainy in texture and since it is a semipermeable membrane, it contains about 17 000 pores. The eggshell shields the egg contents from mechanical and microbial attacks. In addition, it controls the exchange of gasses through the porous structure. The thickness of a typical chicken eggshell is about 0.6 mm. Eggshell comes in different sizes ranging from 30 mm (quail) to 150 mm (ostrich); however, their structural components are similar. In general, eggshell is composed of water (2%) and dry matter (98%). The composition of the dry matter is 93% ash and 5% crude protein [2]. In addition, the shell is a multilayered biomaterial composite, with calcium carbonate ($CaCO_3$) as the major mineral constituent. The $CaCO_3$ component of the eggshell structure is similar to natural limestone used in cement production.

In contrast to eggshell, animal bones are heterogeneous composite biomaterial, which comprises of the mineral hydroxyapatite (HA) ($Ca_{10}(PO_4)_6(OH)_2$) and organics such as collagen (~90%), protein (~5%), lipids (~2%), and water. The inorganic components of the bones are largely calcium and phosphorus, while the minor components includes C, Na, and Mg [3]. In some ways, animal bones contain crystalline calcium phosphate with characteristics similar to geological apatite. HA is the primary component of bones, just as calcium carbonate is to the eggshell. These biomaterials have been used in the agricultural sector to correct the pH of acidic soils (pH < 4.5) and supply essential nutrients for plant growth. The calcium carbonate mineralogical component of these biomaterials enables them to adjust the soil pH.

Figure 13.1 shows samples of animal bones, fish bone, and hen eggshell. These biomaterials are normally considered as waste and useless. Consequently, the bones and shells are disposed following consumption of the egg and meat for nutrient in

Design and Applications of Hydroxyapatite-Based Catalysts, First Edition. Edited by Doan Pham Minh.
© 2022 WILEY-VCH GmbH. Published 2022 by WILEY-VCH GmbH.

Figure 13.1 Bones and eggshells: (a) chicken and animal bones, (b) fish bone, and (c) hen eggshell. Source: Abarasi Hart.

households and butcher shops. The disposed eggshell and animal bones contribute to land and air pollution. As a result of microbial activities on the discarded biomaterials, intensive odor is emitted to the surrounding. Annually, up to 150 000 tons of eggshell is generated as hazardous waste in Europe and disposed in landfill sites at a high cost [4].

The decline in the amount of natural quarry calcium carbonate mined to feed animals and provide dietary supplements and as plastics additive can be mitigated through the conversion of abundant waste eggshells and animal bones rich in calcium carbonate and calcium phosphate. As a result of this alternative path, environmental damages and pollution caused by mining natural limestone can be mitigated while reducing the environmental burden of waste eggshells. Eggshells, fish, turkey, and chicken bones are commonly considered to be domestic waste and in some cases agricultural waste product. These biomaterials are renewable, sustainable, and cheap alternatives to limestone as heterogeneous catalyst and HA material. Hence, utilizing these biomaterials will simultaneously decrease production cost as well as environmental footprint.

Figure 13.2 displays the world egg production from 2000 to 2018, while Table 13.1 shows poultry meat and chicken meat production between 2000 and 2012. It is clear from Figure 13.2 that the egg production sector is expanding rapidly. In 2010, the world egg production rate reached 6.95×10^7 ton year^{-1}; this represents 186.1% increase from the egg production rate 20 years ago [5]. The database of Food and Agricultural Organization (FAO) of United Nations shows that world egg production has increased from 61.7 million tons in 2008 to 76.7 million in 2018, which is about 24% increase in 10 years. Geographically, China is the biggest egg producer followed by the European Union, the United States, and India.

In 2018, global egg production surpassed 76.7 million metric tons. Production will continue to increase to meet nutritional demands of growing global population. In 2014, it was estimated that about 179 eggs are available per person for consumption annually. From 2010 to 2025, the world population is projected to increase from 6.9 billion to 8.0 billion, representing 16.1% increase. As a result of this growth, egg consumption will rise as well, between 2015 and 2035. According to the forecast by Rabobank [5], egg consumption will increase by 50% over this period. Interestingly, the majority of the rise will come from developing countries. Likewise, Table 13.1 shows that the production of poultry meat continues to increase, with chicken being

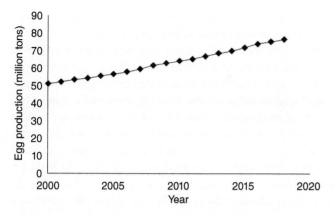

Figure 13.2 World egg production from 2000 to 2018. Source: Based on FAO, 2019.

Table 13.1 World poultry meat and chicken meat production by region between 2000 and 2012 (million tons).

Region	2000	2005	2006	2007	2008	2009	2010	2011	2012
Africa	3.0	3.6	3.6	4.0	4.2	4.4	4.6	4.8	4.9
America	30.1	35.9	37.0	38.9	41.1	40.1	41.8	42.8	43.0
Asia	22.9	27.3	28.3	30.1	31.8	32.9	34.5	36.1	37.4
Europe	11.9	13.2	13.1	14.0	14.4	15.7	16.1	16.6	16.9
Oceania	0.8	1.0	1.0	1.0	1.0	1.0	1.1	1.3	1.3
World	68.7	81.0	83.0	88.0	92.5	94.1	98.1	101.6	103.5
World chicken meat production (million tons)									
Region	2000	2005	2006	2007	2008	2009	2010	2011	2012
Africa	2.8	3.4	3.4	3.7	4.0	4.2	4.4	4.6	4.7
America	27.2	32.7	33.7	35.3	37.4	36.7	38.4	39.2	39.4
Asia	18.7	22.5	23.5	24.9	26.4	27.2	28.6	29.9	31.0
Europe	9.4	10.7	10.8	11.7	12.1	13.4	13.8	14.2	14.5
Oceania	0.7	0.9	1.0	1.0	1.0	1.0	1.1	1.3	1.3
World	58.8	70.2	72.4	76.6	80.9	82.5	86.3	89.2	90.9

Source: Based on FAO, 2019.

the most produced and consumed meat. In light of this, butchers as well as meat and fish processing industries will discard large number of bones from beef, pork, poultry, and fishes. In 2012, the FAO reported a poultry meat production of 103.5 million tons, with America, Asia, and Europe leading production (see Table 13.1). Waste eggshells and animal bones from meat production plant disposed into the environment are expected to increase annually. This will constitute environmental nuisance especially during their microbial decomposition phase. The microbial decomposition occurs because the disposed waste eggshells and animal bones are potential habitat for microorganisms thus initiate biodegradation.

The microstructures of animal bones and eggshell are nature engineered porous structures that constantly perform various functions and have many attributes superior to artificial porous structures, which are manufactured composite and complex structures [6]. The nature engineered pores like with eggshells and animal bones exchanges materials with the environment during metabolic process. Hence, they have been found suitable in HA, adsorbent, and heterogeneous catalysts production. In recent times, there has been increased awareness on recovering waste eggshells and animal bones for use in high value-added products, particularly in heterogeneous catalyst, adsorbents, fillers, and cement production. Figure 13.3 shows the typical materials that can be produced from eggshells and animal bones and their applications.

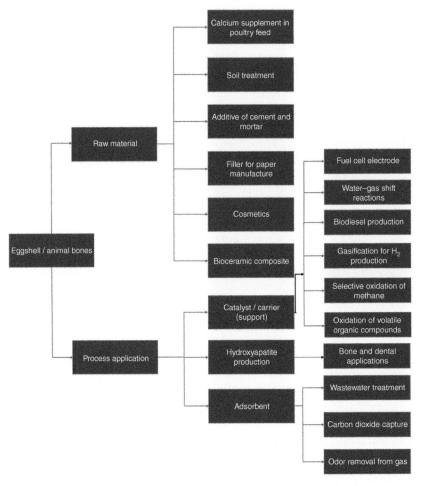

Figure 13.3 Potential applications of eggshells and animal bones as raw material and in processes. Source: Abarasi Hart.

It is well documented that eggshells and animal bones waste contain lots of valuable organic and inorganic components, which can be transformed into viable materials and create new values. For instance, the inherent pore network structure of the calcified eggshell and animal bones (i.e. bone char) makes them an attractive substitute to expensive activated carbon as a low-cost absorbent and biofilter. The difficulty in regenerating activated carbon also makes it less appealing compared to these biomaterials. In the literature, application of eggshell calcium carbonate as coating pigments for inkjet printing papers as opposed to silicon oxides (i.e. silica), which involves high production costs and demands high binder, has been reported [7]. Unlike silica, the synthesized calcium carbonate demonstrates improved rheology and lower binder demand and cost [7].

Eggshells are one of the sources of calcium carbonate ($CaCO_3$), which can partly replace limestone in cement production. The addition of eggshell powder in cement paste provided nucleation sites in the hydration and accelerated cement hydration [8]. The utilization of these solid waste powders in cement also helps to reduce the amount of carbon dioxide emitted from cement industries by reducing clinker production [9]. Consequently, ceramic tiles are mainly composed of quartz, clays, and different carbonates. The potential of eggshells and animal bones materials as source of calcium carbonate for tiles (e.g. bioceramic composite; see Figure 13.3) production has also been reported in the literature [10]. Furthermore, the powders of these solid wastes can be used as mineral fillers in thermoplastic materials such as high-density polyethylene (HDPE), to enhance their rigidity, thermal stability, and density [10]. The application of solid wastes such as eggshells and animal bones, if scaled-up, will help create a conducive and sustainable environment by minimizing impact on public health, in particular, during the decomposition stage, reducing solid wastes that are disposed into the environment, and improving solid waste management. This is possible because these solid wastes contain active components that make them highly valuable in catalytic applications, HA biomaterial for medical applications, biofilter or adsorbent material, calcium carbonate, and calcium phosphate sources. On the other hand, the use of finite natural resources to create products with a short service life that finally end up as waste is an economic sabotage in the current linear economic model. As the global population grows, this linear economic model is becoming increasingly unsustainable coupled with associated large environmental pollution. Hence, a circular economic model, which promotes waste reuse as raw material for the production of new products and energy transformation or recycled into valuable and useful material, offers remarkable business opportunities as well as sustainability. Therefore, utilization of eggshells and animal bones turn these wastes into resources, thereby preserving natural resources, the environment, while promoting waste management.

13.2 Chemical Composition and Properties of Eggshell and Animal Bones

Eggshells ($CaCO_3$) and animal bones ($Ca_{10}(PO_4)_6(OH)_2$) are mainly composed of a network of protein fibers, connected with crystals of calcium carbonate ($CaCO_3$),

Table 13.2 Elemental composition of typical eggshell and animal bone and calcined at 900 °C.

Parameter	Ca	P	S	K	Na	Mg	O	C	Al	Cl	Si	Sr	References
Raw eggshell (quail)	96.4	1.3	0.31	0.19	0.08	0.05	—	—	0.16	0.15	0.13	0.06	[11]
Calcined eggshell	98.2	0.54	—	0.11	0.05	0.04	—	—	0.10	—	—	0.05	
Raw bone	20.4	11.2	—	—	0.8	0.3	39.4	19.5	—	—	—	—	[3]
Calcined bone	29.5	14.7	—	—	0.55	0.56	35.4	5.5	—	—	—	—	

Table 13.3 Specific surface area, pore volume, and average pore size of crushed chicken eggshell, ostrich eggshell, chicken bones, and goat bones calcined at 900 °C.

Parameter	Chicken eggshell [12]	Ostrich eggshell [12]	Chicken bones [13]	Goat bones [14]
Specific surface area (m^2 g^{-1})	55	71	29	91
Pore volume (cm^3 g^{-1})	0.015	0.022	0.047	0.051
Average pore size (nm)	0.54	0.61	3.4	—

apatite carbonates, calcium phosphate $Ca_3(PO_4)_2$, magnesium carbonate ($MgCO_3$), organic materials, and water. The crystalline phases that include aragonite and calcite ($CaCO_3$) and tricalcium phosphate $Ca_3(PO_4)_2$ are the primary mineralogical component of eggshells and animal bones. The elemental composition of typical eggshell and animal bone is shown in Table 13.2.

The specific surface area, pore volume, and average pore diameter of eggshells from chicken and ostrich and animal bones from chicken and goat crushed and calcined at 900 °C are shown in Table 13.3. The specific surface area was estimated using the Brunauer–Emmett–Teller (BET) model according to ASTM C1274, while the pore volume and average pore size were determined by the t-plot model. The liberation of carbon dioxide during calcination of the crushed eggshells possibly improved the pore structure and makes it suitable as a catalyst or support material. Crushed chicken bone showed smaller specific surface area than its eggshell counterpart, while its pore volume and average pore size were higher.

The specific surface area and pore volume of waste eggshell biomaterials increases as the calcination temperature increases because the high temperature stimulates liberation of carbon dioxide. As a result of the carbon dioxide liberation, the porosity within the structure of these materials increases [15]. However, above 900 °C calcination temperature, sintering that often occurs at high temperature leads to reduction of the pore volume and specific surface area and eventually destruction of the pore structure [15]. The development of a heterogeneous catalyst from waste eggshells and animal bones would concurrently result in value addition as well as environmental pollution reduction. After calcination of animal bones, the apatite mineral $(Ca_{10}(PO_4)_6(OH)_2)$ is converted to $Ca_3(PO_4)_2$. Therefore, weight loss during calcination of animal bones between 200 and 300 °C is due to the removal of water and organic matter; whereas, 300 and 500 °C is because of the decomposition of organic materials such as collagen and apatite minerals.

Typical X-ray diffraction (XRD) pattern of powdered quail and hen eggshells, goat bone, and their calcined (heat treated) counterparts is shown in Figure 13.4. The high intensity and sharpness of both peaks of the natural and calcined waste eggshells indicates they are highly crystalline biomaterials. In Figure 13.4, the XRD pattern of the powdered natural quail eggshell showed high intensity peaks located at 2θ = 23.3°, 29.6°, 31.6°, 36.1°, 39.6°, 43.4°, 47.3°, 47.7°, 48.7°, 56.8°, 57.6°, 60.9°, 61.6°, 63.2°, 64.8°, 65.8°, and 66.0°, which have been credited to the

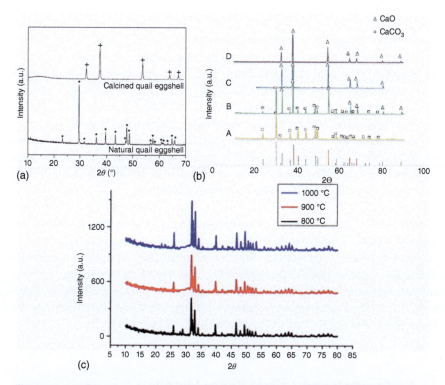

Figure 13.4 XRD pattern of crushed natural and calcined eggshells (a) quail eggshell (*$CaCO_3$, natural and + CaO, calcined at 900 °C). Source: Correia et al. [11]. https://www.hindawi.com/journals/jchem/2017/5679512/ (last accessed 23 February 2021). CC BY 4.0. (b) Chicken eggshell ((A) natural, (B) calcined at 800 °C in a closed system, (C) calcined at 1000 °C with nitrogen flow, and (D) calcined at 1000 °C with air flow). Source: Wong and Ang [16]. / Walter de Gruyter GmbH. (c) Crushed bones of goat. Source: Jazie et al. [14]. https://link.springer.com/article/10.1007/s40243-013-0011-4 (last accessed 23 February 2021). CC BY 4.0.

presence of rhombohedra calcite ($CaCO_3$). After calcination at 900 °C for 3 hours, the diffraction peaks ascribed to $CaCO_3$ disappear, with new diffraction peaks appearing at 2θ = 32.3°, 37.4°, 53.7°, 63.9°, and 67.3°. Notably, these new peaks have been reported in the XRD diffraction pattern for lime, CaO [11]. Likewise, the typical XRD pattern for chicken eggshell before and after calcination can be seen in Figure 13.4b. The discrepancies in the XRD pattern after calcination are due to the decomposition of $CaCO_3$ to CaO, releasing carbon dioxide in the process. This conversion also resulted into smaller particle size [11]. The natural eggshell shows two crystalline phases: aragonite and calcite between 400 and 700 °C calcination temperatures when the $CaCO_3$ is yet to completely change to CaO [15].

Figure 13.4b shows that the XRD patterns of chicken eggshell calcined under air and nitrogen are identical. However, for crushed bones of goat, it can be seen in Figure 13.4c that the XRD pattern of calcined bone at 900 °C demonstrates sharper peaks, signifying better crystallinity compared with those obtained at 800 and 1000 °C [14]. Therefore, beyond 900 °C calcination temperature the crystallinity decreased [17]. The common phases identified in animal bones before calcination are apatite carbonate type A ($Ca_{10}(PO_4)_6(CO_3)_2$), apatite carbonate type B ($Ca_{10}(PO_4)_3(CO_3)_3(OH)_2$), tricalcium phosphate ($Ca_3(PO_4)_2$), and octacalcium phosphate ($Ca_8H_2(PO_4)_6 \cdot 5H_2O$), while after calcination the crystal $Ca_3(PO_4)_2$ appeared [14]. The field of catalysis has shifted attention toward the exploitation and development of heterogeneous catalysts from large volume of renewable resources to boost the global sustainability of catalytic processes, owing to the high production cost and metal losses in traditional catalysts. Additionally, using renewable materials such as waste eggshells, seashells, and waste animal bones in solid catalyst production can also comparatively minimize the environmental problems and reduce the cost connected with their disposal through landfill. This will create values for these renewable biomaterials [18].

Thermogravimetric analyser (TGA) can be used to gain insight into the thermal stability and weight loss during calcination of the eggshell/animal bone-derived biomaterial. Figure 13.5 shows typical thermal analysis of a crushed raw eggshell. The TG curve represents the weight loss with temperature. The first weight loss between 50 and 530 °C corresponds to the removal of physisorbed water within the structure and decomposition of organic matter of the eggshell membrane (ESM) such as protein [11, 19]. The large weight loss between 550 and 875 °C can be attributed to the decomposition of $CaCO_3$ into CaO, resulting in CO_2 liberation [11].

13.3 Eggshell and Animal Bones Materials

The primary materials derived from eggshell and animal bones via heat treatment processes and adsorbent materials are calcium carbonate, calcium oxide, and calcium phosphate. Furthermore, they can be used to produce secondary materials such as CaO and HA material for medical applications and as a heterogeneous catalyst or its support.

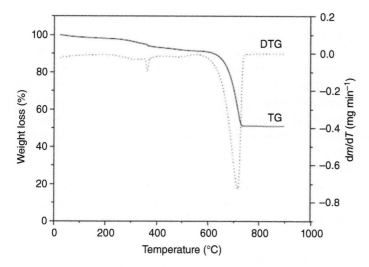

Figure 13.5 Typical TG/DTG (differential thermogravimetric) of a crushed raw quail eggshell. Source [11]. https://www.hindawi.com/journals/jchem/2017/5679512/ (last accessed 23 February 2021) *CC BY 4.0.*

13.3.1 Calcium Carbonate/Oxide/Phosphate

Calcium carbonate has many industrial applications. For instance, it is one of the major raw materials used for cement production as well as mortar in the construction sector. In the paper industry, it can be used as a filler additive to enhance brightness and smoothness of the paper [5]. Other plausible area of applications includes glass and paint production as additive [20]. Most of the calcium carbonate used in the industries is obtained from limestone. However, the process of mining limestone is energy intensive, environmentally damaging, and also exhaustible since limestones are natural resources. Animal bones and eggshells rich in calcium carbonate, apatite carbonate, and tricalcium phosphate are potentially cheap, renewable, and sustainable substitute to limestone and apatite materials. Additionally, animal bones are rich in calcium phosphate, which is one of the raw materials used in the manufacture of HA powder essential in bone regeneration, dental, and drug delivery [21]. Likewise, the calcium oxide derived from heat treatment of eggshells is a natural biomaterial or bioceramic used in HA production for biomedical applications, which is biocompatible with the human body tissues. Hence, the side effects are minimal.

13.3.2 Calcium Supplement

The Association of American Feed Control Officials has approved the use of milled or powdered eggshell biomaterial as calcium supplement for livestock animals and poultry feed [5]. Before now, calcium carbonate used in livestock is sourced generally from mined limestone to improve the strength of eggshell in poultry and promote healthy bone in livestock [22]. Eggshell, which is 94% rich in calcium carbonate, is regarded as a worthless and waste material [17]. However, compared to mined limestone, their biological and natural origin makes them eco-friendlier and more sustainable. The market value of powdered calcium carbonate commonly used in construction, pigments, fillers, poultry, and soil treatments ranges from \$60 to \$66 ton^{-1} if the calcium carbonate particles are coarse [23]. Since eggshells can be recycled as biomaterials, their use as $CaCO_3$ source is more economical when compared with the cost and intense energy associated with mining and upgrading of limestone. Additionally, the mining and upgrading operation of limestone could cause ecological damages. Presently, limestone processing accounts for about 8% carbon dioxide emissions globally [24]. Notably, milled eggshells when heated at temperatures above 650 °C also produce carbon dioxide following conversion of $CaCO_3$ to CaO but in small amount compared to limestone. Hence, it can be concluded that significant differences can be recorded between the use of eggshell and limestone as a livestock feed supplement [22].

13.3.3 Biofilter (Adsorbent) Biomaterial

Water is one of the most abundant substances and essential molecule on Earth. However, human and industrial activities continue to pollute water bodies. As a result of the pollution, domestic and industrial wastewater must be treated to remove contaminants and to meet effluent quality standards before discharge. Adsorption is one of the major means used to remove most of the pollutants in gas or water. Animal bone char is granule biomaterial produced by charring waste bones of animals in the temperature range of 400–600 °C in oxygen limited environment, controlling the quality and adsorption capacity. The char consists of small amount of carbon and mostly calcium phosphate. It can function as biofilter medium material in sugar refining industry for decolorization process to remove fluoride from water and to filter wastewater in a treatment plant [17].

In spite of the expected increase in biowaste from animal bones, it is important to recognize and understand their value in the treatment of wastewater. The removal of fluoride and arsenic can be hindered by customs and religious beliefs; for example, char originated from cow bones is not acceptable by Hindus, while those of pigs bone char are not acceptable by Muslims [25]. Hence, educating communities can be regarded as a prerequisite to bone char application in wastewater treatment. Compared with the use of zeolite or titania, the use of bone char is more environmentally friendly and cost effective. Also, bone char is considered to be an adsorbent with low negative impact on the environment and human health than alumina, clays, activated carbon, and titanium oxide [26]. Pyrolyzing conditions are critical factors that influence the produced bone char textural characteristic such as pore size distribution, surface area, and pore volume as well as its chemical composition, while the residence time and heating rate are process factors that control the quality of the bone char produced at a given temperature [25]. Depending on the bone source, pyrolysis temperature for bone char used in water treatment ranges from 500 to 700 °C. It is worth mentioning that operating the pyrolysis process at temperature below 500 °C could lead to incomplete combustion of the organic matter present in the bone structure. As a result, there is a greater possibility of leaching into water.

Table 13.4 Selected applications of animal bone char and eggshell adsorbents.

Animal bone char/eggshell	Production condition	Application	Reference
Sheep (*Ovis aries*) bone waste	500–900 °C, 1–2 h, and N_2 purge gas	Arsenic and fluoride removal	[27]
Bovine bone	500 °C at 10 °C min^{-1}, 1 h and N_2 purge gas	Copper removal	[28]
Cattle bones	350–700 °C, 2 h	Fluoride removal	[29]
Cow bone	800 °C (commercial)	Decrease mobility of hexazinone, metribuzin, and quinclorac herbicides in soil	[30]
Bovine bone	350–850 °C at 20 °C min^{-1}, 109 min, CO_2 at 130 cm^3 min^{-1}	Removal of remazol brilliant blue R (dye)	[31]
Bone char	Obtained commercially	Adsorption of some heavy metal ions such as copper (Cu^{2+}), cadmium (Cd^{2+}), and zinc (Zn^{2+})	[32]
Acetic acid-modified cattle and sheep bone char	450 °C for 4.5 h, sieved with 20–40 mesh	Removal of formaldehyde from air	[33]
Meat and bone meal char	480 °C	Sulfur and hydrogen-sulfide removal	[34]
Waste duck eggshell	Coarse particles, calcined at 500 °C and 700 °C for 2 and 4 h	Phosphorous removal from wastewater	[35]
Waste hen eggshell	Crushed eggshell, calcined at 600 °C for 3 h	Nickel (Ni) removal from wastewater	[36]
Chicken eggshell	Crushed, sieved ≤400 μm; 500–710 μm and 800–1000 μm, and dried at 80 °C	Heavy metal removal from membrane biological reactor	[37]

Consequently, purging gas such as nitrogen and carbon dioxide could be used during animal bone pyrolysis to help remove organic matter and to control bone char quality for diverse environmental applications. In general, the properties of a bone char depend on the source and the pyrolysis conditions such as pyrolysis temperature, residence time, heating rate, and purging gas. Table 13.4 shows some of animal bone char and eggshell adsorbents applications and production conditions.

Waste eggshells such as duck, chicken, and others signify a large volume of untapped calcium carbonate-rich biomaterial, which when valorized, can be utilized as an adsorbent for wastewater treatment through a heat treatment process, which converts the calcium carbonate into calcium oxide (CaO). They can be used in powder form or coarse particles as a cheap and renewable alternative to activated

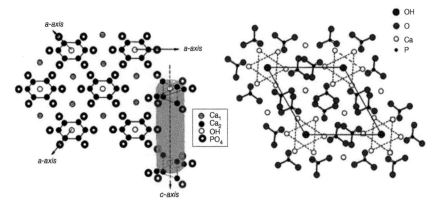

Figure 13.6 Crystal structure of hydroxyapatite powder. Source: Haider, A., Haider, S., Han, S.S., Kang I.-K. *RSC Advances* 7 : 7442–7458, 2017. https://pubs.rsc.org/image/article/2017/ra/c6ra26124h/c6ra26124h-f3_hi-res.gif (accessed 10 August 2020). CC BY 4.0.

carbon, zeolite, and titanium dioxide adsorbents. The utilization of eggshells as adsorbents to remove a diverse range of pollutants has been reported in many literature, for example, cyanide [38], dye, malachite green [39], methylene blue [40], carbon dioxide [41], hydrogen sulfide [42], acid mine drainage [43, 44], and heavy metals [45, 46]. In addition to purifying wastewater, another area of application is in gas purification. The powdered and calcined waste eggshells and animal bones could replace activated carbon as biofilter medium to remove odor and pollutants from exhaust gas from industrial processes [17]. By this, the emitted gas becomes cleaner, and atmospheric pollution is minimized.

13.3.4 Hydroxyapatite Material

HA is a major mineral matter of teeth and bones. The calcium apatite $[Ca_{10}(PO_4)_6(OH)_2]$ with molecular structure shown in Figure 13.6 has shown a good biocompatibility when used for bone repairs and regeneration in tissue engineering. The positively charged calcium (Ca^{2+}) and negatively charged phosphate (PO_4^{3-}) groups on the molecular structure provide the binding sites with tissues. The HA powder can be obtained either synthetically through wet chemical process, sol–gel technique, hydrothermal process, and microwave irradiation or naturally from animal bones biowaste, natural gypsum, calcite, etc. [1].

Natural bioceramic materials include eggshells and animal bones; the eggshell contains mainly $CaCO_3$. Animal bones containing apatite carbonate and tricalcium phosphate ($Ca_3(PO_4)_2$) can be used to produce HA powders. The two main routes used to prepare HA powder from eggshells are the wet method and the solid-state reactions [47]. However, the solid-state reaction is mostly used to prepare HA powders from animal bones. Figure 13.7 shows the flow diagrams for HA powder production from eggshells and animal bones. Ramesh et al. [48] reported HA powder production using bovine (cow), caprine (goat), and galline (chicken) bones via heat treatment under air atmosphere. After the pulverized bones were heat

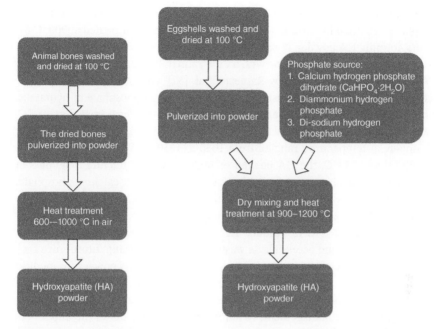

Figure 13.7 Production of hydroxyapatite powder from animal bones and eggshells. Source: Abarasi Hart.

treated, their XRD pattern revealed the presence of characteristic diffraction peaks of the HA phase. The HA powder obtained from bovine bone exhibited better thermal stability than caprine and galline bones, though both showed traces of tri-calcium phosphate (TCP) when the heat treatment was above 700 °C. The porous nature of the waste bones was also evident in their microstructures and supported by the observed low relative density. Unlike HA obtained from synthetic material and route, the HA from natural animal bones has organized crystal formation, and the carbonate ions seem to be excellent for bioresorbable bone substitutes [48, 49]. Therefore, HA powder obtained from natural animal bones are more suitable for medical applications than their synthetic counterparts [50].

On the other hand, HA powder has been synthesized successfully using waste eggshell biomaterial as a raw material, which is a renewable calcium source. The advantages of using natural biomaterials are that they are mostly from in vivo source, available in large quantities as waste, and they are value-added products, which reduces environmental burden. Eggshell-derived HA showed good changes in the cytocompatibility, antimicrobial activity, and drug release. Hence, they could serve as potential bone graft materials [51]. Additionally, the natural trace elements found in eggshell such as K, Na, Mg, Si, and Sr are incorporated into the crystalline structure of synthesized HA powder, thereby enhancing its compatibility to human bone [52]. Lala et al. [1] produced nano-sized HA powder using chicken eggshells and di-sodium hydrogen phosphate (Na_2HPO_4) as a source of phosphorus. In their preparation, the eggshells were first pulverized and calcined at 900 °C to convert the calcium carbonate ($CaCO_3$) to calcium oxide (CaO) powder. The CaO powder is then mixed with Na_2HPO_4 in a wet phase reaction. After 24 hours aging, the precipitate was dried for 48 hours at 70 °C, and thereafter heat treated at 900 °C for 2 hours, forming HA powder. It was found that the synthesized HA powder showed close similarity to the human bone with a Ca/P ratio of 1.74 and average particle size of 78 nm [1]. In another study by Ramesh et al. [52], nanostructured HA powder was synthesized with eggshell using di-calcium hydrogen phosphate di-hydrate as a phosphorus precursor. Both were mixed in solid state and heat treated at 800 °C, in which the $CaCO_3$ is converted to CaO, before being sintered for two hours in air atmosphere at temperature ranging from 1050 to 1350 °C. When the temperature is greater than 1250 °C, the HA powder exhibited excellent fracture toughness of 1.51 MPa m$^{1/2}$ and phase transformation into TCP and tetra-calcium phosphate. Furthermore, eggshell as a renewable calcium precursor and dicalcium phosphate dihydrate ($CaHPO_4 \cdot 2H_2O$) was milled together for one hour, followed by one hour heat treatment at 1200 °C to produce HA powder [47]. Hence, eggshells, which are readily available and accessible as waste, present a promising future since they can be used as a precursor biomaterial for the synthesis of HA powder for medical application in tissue engineering. Likewise, the use of animal bones and eggshell waste as a calcium source to make HA material suggests a viable material-recycling technology for waste management. In summary, Table 13.5 shows the differences between diverse natural bone sources of HA and eggshell based HA in terms of Ca/P ratio, particle size, morphology, phase composition, thermal stability, and trace elements. It is clear that the HA obtained from animal bones showed thermal stability, morphology, and trace elements composition similar to natural bones.

Table 13.5 Hydroxyapatite produced from several sources.

Parameters	Bovine bone	Fish bones	Eggshell	Natural bone
Ca/P ratio	>1.67	≥1.67	≥1.67	1.67
Particle size	Micro to nanometer range	Nanometer range	≥50 nm	Nanocomposite
Morphology	Needle, rod, plate, spherical, and so forth	Needle, rod, spherical, and so forth	Needle or rod	Needle or rod
Phase composition	Phase pure and tricalcium phosphate	Phase pure	Phase pure	Composite
Thermal stability	800–1000 °C	600–800 °C	<800 °C	>1200 °C
Trace elements	Na^+, Mg^{2+}, K^+, Ti^{2+}, etc.	Fe^{3+}, Cr^{3+}, Cu^{2+}, K^+, Mg^{2+}, etc.	CO_3^{2-} major Element Mg^{2+}, Na^+	Na^+, Mg^{2+}, K^+ Zn^{2+}, CO_3^{2-}

Source: Abdulrahman et al. [53]. https://www.hindawi.com/journals/jma/2014/802467/ (last accessed 24 February 2021) CC-BY 4.0.

13.4 Catalytic Applications of Eggshell and Animal Bones

Waste eggshell and animal bones can effectively be utilized as a catalyst or support/carrier for low-cost, recyclable, and renewable heterogeneous catalyst [17]. Their use in heterogeneous catalyst production is quite sustainable compared to exhaustible natural occurring zeolite, titania, and alumina. Calcium carbonate is the major component of eggshell and can be transformed into calcium oxide, which has shown high catalytic activity and good thermal stability particularly for biodiesel production through transesterification process. Likewise, the calcium phosphate of the animal bones can be converted to HA, which has demonstrated high catalytic activity and excellent thermal and chemical stability.

13.4.1 Catalytic Material Preparation from Eggshells and Animal Bones

Eggshells, animal, and fish bones are widely available waste products from meat production and poultry. The animal and fish bones mainly contain $Ca_3(PO_4)_2$ and $CaCO_3$, while the major component of eggshells is $CaCO_3$. Figure 13.8 represents the procedure followed to prepare heterogeneous catalyst from waste eggshell and animal bones biomaterials.

The waste eggshells and animal bones are washed with deionized water and acetic acid to remove dirt, membrane layer, fibrous matters, proteins, and other impurities and then dried at 100–200 °C for 5 hours [17]. The dried materials are pulverized into powder, which is sieved to obtain fine particles with size ranging from 0.1 to 80 μm depending on catalytic application. They can be acid treated to improve pore structure in order to enhance heat and mass transfer during reaction. Subsequently, the fine powder is calcined, the calcinations process involves heat treatment to decompose the major components of the powder into CaO (eggshell) and HA materials (animal bones) while also removing impurities under nitrogen or air. The calcination temperature depends on the application and can range from 500 to 1000 °C. After calcination, the derived catalytic material is placed in a desiccator to minimize the chances of coming in contact with humidity and carbon dioxide in the air. In some cases, these substances are considered poisonous to the catalyst's active site [16]. However, the calcination temperature plays a significant role on the textural

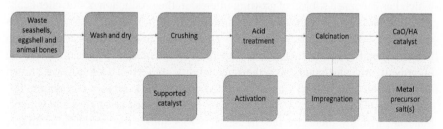

Figure 13.8 Catalytic material preparation procedure from eggshell and animal bones. Source: Abarasi Hart.

properties, pore network structure, and surface morphology of the derived catalyst. Hence, the catalytic activity of the derived catalyst is significantly dependent on the calcination temperature, which determines the textural and structural properties of the resulting materials, as well as their active sites intensity as shown in Figure 13.4 [18]. Temperature less than 600 °C is insufficient to convert the $CaCO_3$ component into CaO for eggshell [54], which is the catalyst suitable to catalyze transesterification for biodiesel production from vegetable oil [55]. The complete conversion of $CaCO_3$ to CaO is achievable at a calcination temperature greater than or equal to 900 °C [17, 54]. It is worthy to note that the formation of CaO improves the porosity or pore network structure of the catalyst due to the liberation of gaseous carbon dioxide. At the same time, it increases the total pore volume and pore size of the waste eggshell or animal bone-derived catalyst [18]. The calcination time at a given temperature also influences the evolution of CaO. When the calcination time is too short, the conversion of $CaCO_3$ into CaO may be incomplete, and the catalytic activity of the derived catalyst could be affected. However, at high temperature such as 900 °C, prolonged calcination time could cause sintering and result into shrinkage of the derived catalyst grains [17, 18]. As a result of the shrinkage, the surface area of the catalyst may be reduced [56]. It has been reported that the catalytic performance of calcined fish and animal bones was high due to the formation of β-TCP at high calcination temperature, due to the decomposition of HA [57]. After the calcination process, the proportion of calcium increases to about 98 wt% for the eggshell. However, other elements such as Mg, Fe, P, Na, K, Al, Cl, and Si are also present in noticeable amounts as impurities in the derived catalyst [11, 17]. These trace elements could improve the catalytic functionality depending on the chemical reactions.

Depending on the application, eggshell and animal bones can also be used as a support or carrier of active ingredients. In the case of supported catalyst synthesis, wet impregnation method is required using suitable active metal precursor mixed with the pulverized eggshell/animal bones or their calcined coarse particles in an aqueous or organic solvent. The obtained material after impregnation is then dried, calcined, and activated before it can be used as a catalyst. Interestingly, the catalytic activity of supported catalyst derived from eggshell/animal bone is higher than that of unsupported catalyst because of the improved active sites and basic strengths [18].

These natural bioceramic materials have prospectively proven useful in the development of a new generation of heterogeneous catalyst and solid support materials, which has become a subject of growing interest in the production of fine chemicals, petroleum refinery, energy, and environment pollution control. In heterogeneous catalysis, the support material also plays crucial roles in catalytic processes, which include metal-support interaction, impeding deactivation, dispersion of active ingredient, and mechanical and thermal stability. In Section 13.3.4, it was shown how eggshell and animal bones can be used for the synthesis of calcium HA $(Ca_{10}(PO_4)_6(OH)_2$, HA) material, which has been reported as catalyst and support material in many literature [58–70]. Figure 13.9 shows the synthesis technique from eggshell and a phosphate source and the scanning electron microscopy (SEM) images of synthesized and calcined HA. After synthesis, the images revealed

13.4 Catalytic Applications of Eggshell and Animal Bones

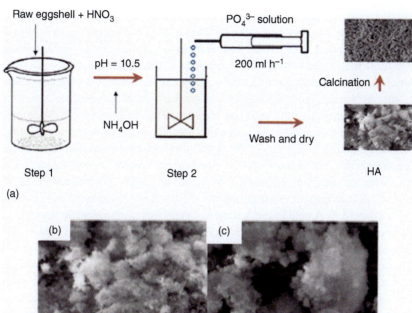

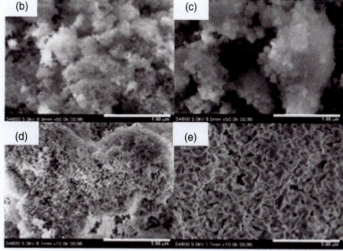

Figure 13.9 (a) Synthesis of HA from eggshell and SEM images, (b) phosphate source Na_2HPO_4 and dried, (c) phosphate source H_3PO_4 and dried, (d) with H_3PO_4 calcined at 700 °C and (e) with H_3PO_4 calcined at 950 °C. Source: Ibrahim et al. [70], pg 16: 7960–7975. / MDPI / CC BY 4.0.

spongy-like morphology (Figure 13.9b–c), and following calcination at 700 °C (Figure 13.9d), porous particles texture can be observed, while at 950 °C dense porous particles appeared (Figure 13.9e).

HA is a weak alkaline calcium phosphate having very low water solubility. As a catalyst or catalyst support material, it possesses strong adsorption capability. Examples of HA synthesized from animal bones include microporous biochars composed mainly of HA produced from waste animal bones through pre-pyrolysis, chemical treatment with NaOH, KOH, and K_2CO_3, and pyrolysis [71]. Also the valorization of waste pork bones has been reported for porous HA biochar production [72]. HA-derived material from waste animal bones is mostly produced through thermal treatment and has found application in catalysis, electrochemistry, and adsorption [73]. The textural characteristics of HA can be improved through chemical activation (using either H_3PO_4, H_2SO_4, NaOH, or K_2CO_3) and gasification reactions [71–73]. It has been reported in the literature that acid or alkali treatment can increase the porosity of HA derived material by about 30% [71]. Therefore, the morphological and textural features demonstrate its suitability as catalyst support material. HA possesses all the criteria of a catalyst support material such as possible high specific surface area depending on synthesis method [70], high thermal stability up to around 1000 °C, modular surface acidity and basicity as a function of the molar ratio of Ca to P, and very low water solubility [58–60]. Moreover, the high ion exchange ability of HA gives it the capability to form synergistic interactions with a variety of transition metals [61]. HA as a catalytic or support material can also be synthesized from eggshells with a suitable source of phosphate either using dry, wet, or sol–gel method (Figure 13.9) [17, 62] or directly from animal bones (Figure 13.7). A detailed review on HA synthesis methods, structure, and its application in heterogeneous catalysis has been reported by Fihri et al. [59].

The catalytic applications of HA synthesized from eggshell or animal bones include cross-coupling reactions, oxidation of alcohols and silanols, Knoevenagel condensation reaction, Claisen–Schmidt reaction, Friedel–Crafts reaction, Heck reaction, Diels–Alder, aldol reaction, Michael addition reaction, and transesterification reactions [58, 59]. Also it has been reported as a catalyst support in selective oxidation of alcohols [63, 64], dry reforming of methane [60, 64], partial oxidation of methane to syngas [65], glycerol steam reforming for hydrogen production [66], Fischer–Tropsch synthesis process [67], carbon monoxide oxidation [68], catalytic selective hydrogenation, and N_2O catalytic decomposition [69]. Aside from the use as a catalyst or a support material, other applications include environmental remediation, bone augmentation and replacement, and drug delivery [70]. Conversely, the liberation of hydrogen from the hydrolysis of ammonia-borane using HA-supported palladium (0) nanoclusters catalyst was investigated and reported by Rakap and Ozkar [74]. Furthermore, synthetic HA-supported bimetallic cobalt and nickel catalyst was evaluated for syngas production from dry reforming of methane, and it was found that CH_4 and CO_2 conversion in the range of 60–68% was observed at 700 °C and 73–79% at 750 °C, respectively, for long reaction times of 50–160 hours [60]. On the other hand, HA produced from pork bone via microwave irradiation showed good reusability and high yield of biodiesel when used as catalyst in transesterification reaction of jatropha oil with methanol [75]. In other words, eggshell and animal bones have demonstrated usefulness as sustainable and low-cost sources of catalyst, catalyst support material, and the synthesis of HA catalyst or support material with wide range of applications in fine chemicals synthesis, organic synthesis reactions, petroleum refining, energy, and environment pollution control.

13.4.2 Catalytic Applications of Catalyst Derived from Eggshell and Animal Bones

Every day, large number of eggshells and animal bones are treated as waste in landfill and thus becomes a source of environmental pollution. Metal oxides such as silicon oxide (SiO_2), aluminum oxide (Al_2O_3), cerium (IV) oxide (CeO_2), MnO_2, zirconium dioxide (ZrO_2), perovskite ($LaNiO_3$), spinel ($NiCo_2O_4$), titanium dioxide (TiO_2), etc. commonly used as catalysts support are known to be cost intensive, non-eco-friendly, and prepared via complex method.

On the other hand, eggshells and animal bones possess excellent intrinsic pores and hierarchical structures, which support their use as active catalyst, efficient support, or template. Unlike the aforementioned conventional materials used in heterogeneous catalysis, the preparation of catalyst from eggshells and animal bones is simple and offers low-cost benefit. Large-scale application of these materials in the production of heterogeneous catalysts will concurrently diminish their burden on the environment and enable high value-materials synthesis. Table 13.6 summarizes some of the catalytic applications of heterogeneous catalysts derived from waste eggshells and animal bones. The common areas of catalytic applications include transesterification reaction for biodiesel production, oxidative coupling of methane (OCM) to produce light hydrocarbons, gasification for hydrogen and syngas production, and WGS reaction for hydrogen production and other chemical reactions where their potential is yet to be fully exploited. In cases such as biodiesel production from waste cooking oil using methanol or ethanol and selective oxidation of methane, the catalysts derived from eggshell or animal bones are used as standalone. Additionally, the hierarchical and porous structure of eggshell and animal bones favors mass and heat transfer during chemical reaction [82]. They are inexpensive biomaterial, simple catalyst preparation process, and comparatively high catalytic activity and excellent stability.

13.4.2.1 Selective Catalytic Oxidation

The OCM reaction involves methane oxidation or pyrolysis with oxygen in the presence of a suitable heterogeneous catalyst at high temperature and moderate pressure as shown in Eq. (13.1) for ethylene production [81]. The reaction products include C_2—C_7, CO, CO_2, H_2, and H_2O. Production of C_2—C_7 hydrocarbons suggests this channel could be used to produce valuable chemicals and fuels from natural or shale gases.

$$2CH_4 + O_2 \rightarrow C_2H_4 + 2H_2O, \quad \Delta H°_{298} = -281 \text{ kJ mol}^{-1} \tag{13.1}$$

According to the literature, the presence of strongly bounded oxygen species on the surface of catalysts used in the OCM reaction ensures activation of methyl radicals generation and high basicity for rapid desorption of these radicals [116]. As a result, numerous perovskite-type oxides comprising alkaline, alkaline earth, and rare-earth metals have been reported to be active in OCM reaction [81, 116]. Furthermore, alkali metal oxides, predominantly calcium oxide (CaO), have shown a great catalytic property across a broad variety of reactions. In this regard, waste eggshell known for its rich $CaCO_3$ can be used as a cheap and sustainable source of CaO for OCM reaction. Eggshell-derived catalyst possesses a low specific surface area; this characteristic is good for selective oxidation reactions.

Lima and Perez-Lopez [81] investigated the production of light olefins from OCM reactions involving waste eggshell-derived catalyst (CaO) under the following

Table 13.6 Catalytic applications of materials derived from eggshells and animal bones.

Catalytic applications	Eggshell and animal bone-derived catalysts	Reference
Pollutant catalytic degradation:		
Dye, bacterial, and larvae	Yttria-decorated α-Ag_3VO_4/cuttlefish bone	[76]
Organic (2,4-dichlorophenol, 3-chlorophenol, etc.) pollutants	Swine bone biochar and ZnO/fluorapatite from animal bones	[77, 78]
Methylene blue (MB) dye pollutant	Mutton bone-derived HA/TiO_2	[79]
Selective catalytic oxidation		
Selective oxidation of methane/oxidative coupling of methane to light olefins	CaO derived from eggshell	[80, 81]
Benzene oxidation	Ag/CaO-derived from eggshell, Co_3O_4/eggshell	[19, 82, 83]
Oxidation of alkylarenes and alcohols	HA-derived from cow bone/MnO_2	[84]
Oxidative aromatization of 1,4-dihydropyridines	Animal bone-derived catalyst	[85]
Oxidation of glycerol	Au/biochar, Pd/biochar, Bi—Pt/biochar	[86]
Carbon dioxide capture	CaO derived from eggshell	[87, 88]
Energy production		
Hydrogen production via gasification	Catalyst derived from eggshell	[89]
Hydrogen production via water–gas shift	Transition metals supported on bone-derived HA	[90]
Biodiesel production by transesterification	HA, catalyst derived from eggshell, fish, and animal bones	[3, 12, 13, 91–103]
Coal gasification/gasification of lipid-extracted microalgae biomass	Catalyst derived from waste eggshell	[104, 105]
Production of other energy carriers:	Catalyst derived from eggshell	[106]
Production of dimethyl carbonate from propylene carbonate and methanol	Co—Ni/HA (HA derived from eggshell/animal bones)	[60, 64, 65]
Syngas production from dry reforming of methane		
Catalytic materials from eggshells and animal bones in organic synthesis		
Cross-coupling reactions	HA, calcined eggshell, and animal bones	[58, 59, 63, 107]
Oxidation of alcohols and silanols	Animal bone-derived catalyst	[108]
Knoevenagel condensation reaction	HA derived from animal bones	[109]
Claisen–Schmidt reaction	Eggshell-derived catalyst	[110]
Friedel–Crafts reaction	Eggshell-derived catalyst	[111]
Heck reaction		
Diels–Alder		
Aldol reaction		
Michael addition reaction		
Synthesis of benzimidazoles, benzoxazoles, and benzothiazoles		
Catalytic esterification reaction		
Oximes production from aldehydes and ketones		
Etherification reaction for glycerol oligomers production		

Table 13.6 (Continued)

Catalytic applications	Eggshell and animal bone-derived catalysts	Reference
Eggshell membranes (ESM) in fuel cell applications	Eggshell bioceramic material	[112, 113]
Lithium battery separator		
Electrode materials for supercapacitors		
Selective catalytic hydrogenation	Pd-HA, Cu-HA (HA derived from eggshell/animal bones)	[114, 115]
Furfural to tetrahydrofurfuryl alcohol		

conditions: flow rate 120 ml min^{-1} in the ratio 1 : 2 : 9 (CH$_4$: air : N$_2$), 0.1 g catalyst, and temperature in the range of 600–800 °C. The hydrocarbons produced are C$_2$—C$_3$ such as ethane, ethylene, propane, and propylene. At 800 °C temperature, methane conversion of 30% was achieved with selectivity toward ethylene and propylene of 53 and 3%, respectively. In comparison with other heterogeneous catalysts such as Na$_2$WO$_4$/TiO$_2$, Sr$_2$TiO$_4$_SP1, Na$_2$WO$_4$/SiO$_2$, and LaAlO$_3$-8, the eggshell-derived catalyst demonstrated in some cases superior performance, and in other cases it showed comparable performance. Consequently, Karoshi et al. [80] reported selective oxidation of methane into higher hydrocarbons (C$_2$—C$_7$) using a packed bed reactor of calcined eggshell in the temperature range of 650–750 °C. Unlike Li—MgO, Na—CaO, and CaO catalysts, which produced mostly ethane and ethylene, the use of calcined eggshell produced also hydrocarbons in C$_3$—C$_7$ range including aromatics with about 30% average fractional conversion of methane [80]. In a follow-up study, Karoshi et al. [117] reported Cu-impregnated chicken eggshell-derived catalysts (Cu loading; 2, 5, and 10% by weight) for selective oxidation of methane at 650 °C and atmospheric pressure. The eggshells-derived catalyst with 2 wt% Cu loading favored partial oxidation route to syngas production with 50% methane conversion achieved. However, further increase in Cu loading (5 and 10 wt%) favored the yield of C$_2$—C$_6$ hydrocarbons as a result of OCM reactions. Nevertheless, increasing copper loading on the calcined eggshell led to decreased conversion of methane, signifying that the catalytic properties could be manipulated by changing metal loading on the surface of the eggshell-derived material.

The results prove that waste eggshells can be used as catalyst support for metals to be used in oxidation reactions. In light of this result, Guo et al. [82] investigated waste eggshells as support for Ag/eggshell catalysts and its catalytic activity in benzene oxidation. The findings demonstrated that Ag/eggshell nanoparticles catalyst exhibited superior catalytic activity (~95% benzene conversion) over the use of pure Ag nanoparticles without eggshell carrier. The superior activity of Ag/eggshell nanoparticles can be attributed to the synergetic interaction between the Ag and eggshell. In addition, high dispersion of Ag nanoparticles on the surface of the eggshell-derived material further hones the catalytic properties [82]. Notably, after 200 hours of testing, the conversion of benzene remained at ~95%; therefore, it indicates excellent durability of the eggshell-derived catalyst. It was also discovered that the Ag/eggshell catalyst remarkably outperformed that of the Ag/commercial-$CaCO_3$ catalyst counterpart. The reason for this is that the ESM helps to control the size and dispersion of Ag nanoparticles during the preparation process due to the strong metal–protein bonding interactions, while agglomeration occurred with commercial $CaCO_3$ [82]. These findings reinforced waste eggshell-derived material as a promising support material for large-scale synthesis of noble metal catalyst with high catalytic performance and stability for volatile organic compound (VOC) oxidation. Considering the balance between the cost and catalytic performance, eggshell and animal bones derived materials could be a potential replacement of some conventional catalysts and support materials. Their abundant availability, low-cost biomaterial, and green and simple preparation approach make them eco-friendlier and more economical compared to conventional materials such as alumina, ceric oxide, titania, zirconia, and silica.

13.4.2.2 Gasification of Biomass for Hydrogen Production

The gasification of biomass produces synthetic gas or syngas, which mainly contains H_2, CH_4, CO, and CO_2, and two other fractions of tars (condensable hydrocarbons) and char (solid residues). Calcium oxide (CaO) has proven to be an effective catalyst in various gasification reactions for producing hydrogen. At a CaO/biomass ratio of 2, it has been discovered that carbon dioxide production decreased by about 93% compared with a steam gasification of biomass without CaO catalyst for hydrogen synthesis [118]. At the same time, the production of hydrogen increased via WGS reaction, while tar production decreased. The observed reduction in CO_2 production in the presence of CaO catalyst is due to absorption (Eq. (13.2)), causing disproportion in the reaction equilibrium. This forces the WGS reaction toward the production of CO_2, Eq. (13.3) [119].

$$CaO + CO_2 \rightarrow CaCO_3 \quad \Delta H°_{298} (-170.5 \text{ kJ mol}^{-1}) \quad (13.2)$$

$$CO + H_2O \rightarrow CO_2 + H_2 \quad \Delta H°_{298} (-41.09 \text{ kJ mol}^{-1}) \quad (13.3)$$

Waste eggshell and animal bones upon calcination can be used as a cheap source of CaO catalyst required to improve hydrogen production through catalytic gasification of biomass. The calcified material is rich in CaO with robust morphology and intrinsic pore network structure. Taufiq-Yap et al. [89] evaluated hydrogen synthesis from the gasification of wood using eggshell-derived catalyst. Hydrogen production increased from 57 to 73% as the eggshell-derived catalyst loading increased from 20 to 60%, in relation to the gasification reaction without catalyst. This is because of the enhanced WGS reaction that is promoted by the suppressed CO_2 production due to the reactive absorptive property of the eggshell-derived catalyst. Furthermore, it has been reported that the addition of eggshell catalyst favored tar and char gasification as well as the production of gaseous products [120]. In comparison to gasification without catalyst, the presence of the eggshell-derived catalyst lowered the CH_4 production temperature, in which the concentration increased with the catalyst loading.

Hydrogen production while reducing the carbon dioxide (CO_2) concentration in the gas stream during steam gasification of sawdust in a bubbling fluidized bed reactor (bed material: calcined eggshell-derived catalyst (CaO) of a mean size 600 µm and sand 250–500 µm) was reported by Salaudeen et al. [121]. Increasing the eggshell-derived catalyst provided more CaO to the process, which amplified the uptake of CO_2 through carbonation reaction, and as a consequence promoted hydrogen enrichment in the gas stream. The observed decrease in CO_2 production due to carbonation reaction with the eggshell-derived catalyst in biomass gasification is an indication of cleaner and low-cost improvement in the production of hydrogen. Similarly, Acharya et al. [118] observed an increase in hydrogen concentration from 22.29 to 55% and a significant reduction in CO_2 from 22.5 to 1.56% when the commercial CaO/biomass ratio in their experiments was increased from 0 to 2. Furthermore, the carbonation reaction promotes tar cracking, steam reforming, and char conversion and ultimately improves gas yield and quality. In terms of cost, availability, and sustainability, eggshell and animal bone-derived materials are better source for CaO catalyst than widely used limestone and dolomite.

13.4.2.3 Reactive Carbon Dioxide Capture (Calcium Looping)

The continuing burning of fossil fuels such as coal and crude oil leads to greenhouse gas (GHG) emissions particularly carbon dioxide (CO_2), which is a major global warming player. Presently, physical and chemical absorption using solvents such as selexol, rectisol, and mono-ethanol-amine (MEA) methods are the most commonly used technology to remove carbon dioxide from flue gas streams. But their application is restricted to low-temperature processes; they are cost intensive and energy intensive to regenerate [121].

Thermally treated eggshell and animal bone biocomposite materials have strong affinity for carbon dioxide, signifying their potential utilization in CO_2 separation from the effluent gas coming off power plants. These biomaterials offer an exclusive combination of sorbent strength and reactivity while maintaining a low cost. The reaction mechanism consists of a series of carbonation–calcination reactions (CCR): calcium oxide (CaO) reacts with CO_2, leading to calcium carbonate ($CaCO_3$), which upon calcination of the material results in the release of a pure CO_2 stream while the material is being regenerated [88]. Compared to carbon capture materials such as carbon fiber, zeolite, and activated carbon, eggshell-derived material is low cost, renewable, simple to prepare, and possesses excellent thermal stability. The pilot-scale demonstration of the process has been reported in the literature [122–124]. A typical calcium looping (i.e. lime, CaO) CO_2 adsorption process is shown in Figure 13.10.

In the pretreatment process with acetic acid, a mesoporous structure of the eggshell-derived material is developed, allowing the eggshell-derived sorbent to reach higher conversions over more CCR cycles while also removing the eggshell protein-rich membrane [87, 88]. The carbonation (Eq. (13.2)) of the eggshell-derived CaO through reaction with the carbon dioxide in flue gas leads to the formation of calcium carbonate ($CaCO_3$), while the calcination process (Eq. (13.4)) regenerates the CaO biocomposite material and releases pure stream of CO_2 for sequestration

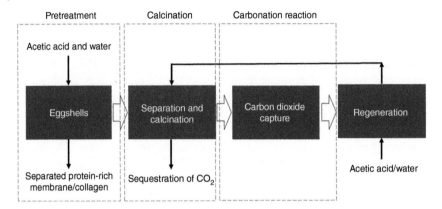

Figure 13.10 Schematic of the eggshell carbonation system for carbon dioxide capture. Source: Hart and Onyeaka [124]. https://www.intechopen.com/online-first/eggshell-and-seashells-biomaterials-sorbent-for-carbon-dioxide-capture (last accessed 24 February 2021) CC BY 4.0.

as shown in Figure 13.10.

$$CaCO_3 \rightarrow CaO + CO_2, \quad \Delta H° = 178 \text{ kJ mol}^{-1} \tag{13.4}$$

Sacia et al. [87] reported chicken eggshell-derived CaO sorbents for carbon dioxide capture from coal-fired power plant. In the work, the eggshell was pretreated with acetic acid to improve and expand the initial particle pore structure and surface area, favoring carbon dioxide diffusion into the biomaterial. Depending on the acetic acid pretreatment time, it was found that the conversion, which quantifies the extent of CaO utilization, lies between 70 and 80% in the first cycle and gradually drops to about 40% in the fifth cycle. It is also observed that regeneration restored the conversion in the range of 70–80% in which conversion increases on average after each subsequent regeneration, suggesting that periodic regeneration can effectively increase the reactivity of the spent eggshell-derived sorbent. In another study, it was discovered that after four regenerations the adsorption capacity of the eggshell-derived sorbent decreased from an average of 6824 mg CO_2 g^{-1} in the first cycle to 1608 mg CO_2 g^{-1} in the fourth cycle [125]. This suggests that the eggshell-derived material could accommodate about eight time its own weight of carbon dioxide from flue gas. This suggests that expanding the reuse of materials and catalysts obtained from waste eggshell and animal bones in the industries will reduce the risk of microbial problem associated with their decomposition while offering high economic and environmental benefits.

13.4.2.4 Water–Gas Shift (WGS) Reaction

Hydrogen production through WGS reaction (Eq. (13.3)) is usually equilibrium limited. To achieve high hydrogen productivity and uphold a high carbon monoxide conversion, the reaction could be tweaked to favor the product side by coupling carbon dioxide separation using membrane. As a result, hydrogen-selective membrane separation is commonly used to aid the forward reaction [126]. However, calcined eggshells and animal bones rich in calcium oxide with strong affinity for carbon dioxide could be used as a membrane material to enhance conversion and selectively absorb the produced carbon dioxide, resulting in a higher concentration of hydrogen in the exit stream. This has been proven by Chen et al. [127], using ceramic–carbonate dual-phase membrane in a single reactor for catalytic reverse water–gas shift (RWGS) reaction to produce CO with simultaneous CO_2 capture. The dual-phase membrane consists of a ceramic phase of oxygen ion conductor and molten carbonate phase. This carbonate material can be replaced with eggshell and animal bones, which are excellent bioceramic composite.

The yield of hydrogen can also be enhanced by incorporating in situ carbon dioxide removal to drive the temperature-dependent equilibrium limited WGS reaction forward. In light of this result, Iyer and Fan [128] investigated carbon dioxide removal from the reacting water–gas medium through its carbonation with CaO catalyst derived from eggshell. The eggshell-derived catalyst demonstrated capture capacity of about 66 wt% in the first cycle, dropping to 37 wt% after 15 cycles. The calcined eggshell sorbent membrane undergoes carbonation reaction (Eq. (13.2)). The reacted membrane is then isolated and calcined to release pure carbon dioxide for sequestration, while the CaO is recycled (Eq. (13.4)). The eggshell pellet demonstrated improved kinetics due enhanced mass transport characteristics when treated with acetic acid [128]. The coupling procedure avoids carbon dioxide emissions to the atmosphere. This catalytic application if scaled up could help the egg and livestock industries to dispose of their waste eggshells and bones, and at the same time, the wastes are converted into useful materials. Consequently, HA derived from calcined pork bone has been used as a catalytic support for transition metals such as Ni, Cu, Co, and Fe in WGS reaction for hydrogen production [90]. The bone-derived HA support exhibited good stability and provides suitable chemical properties for WGS reaction, promoting high selectivity toward hydrogen production and an inhibitory effect on the methanation. Therefore, eggshell and animal bone-derived catalyst or their synthesized support materials are an economically and environmentally interesting alternative to conventional heterogeneous catalysts.

13.4.2.5 Transesterification Reaction for Biodiesel Production

Biodiesel is an environmentally friendly fuel compatible with existing diesel engines. It is generally synthesized through transesterification of triglycerides (e.g. edible, nonedible, or waste oils) and alcohol such as methanol with a suitable catalyst mostly homogeneous. Unlike homogeneous catalyst, heterogeneous catalysts such as alkali metal oxides (e.g. CaO, ZnO, MgO, etc.) promote easy separation of products and catalyst after reaction [55]. Waste eggshells and animal bones are abundant source, low cost, and rich in CaO and $Ca_3(PO_4)_2$ when calcined. Based on this, derived catalyst from crushed goat bones calcined at 900 °C with crystallite size 41.5 nm and specific surface area 91 $m^2\ g^{-1}$ was tested for transesterification reaction using peanut oil and rapeseed oil and methanol at 60 °C temperature and four hours reaction time [14]. The bone-derived catalyst yields biodiesel of 94% (peanut oil) and 96% (rapeseed oil) under methanol-to-oil molar ratio of 20 : 1. The result of the bone-derived catalyst reusability test showed that the activity was maintained over six cycles of repeated use before gradual decrease in the catalytic activity. Likewise, biodiesel yield of 93% was observed, and complete deactivation was recorded after more than 15 cycles [14]. The decrease in catalytic activity could be attributed to changes in the surface structure and the formation of calcium hydroxide due to the water component in the reaction medium [17]. In addition to the low-cost and reusability benefits, the activity of the spent catalyst can be rejuvenated by thermal treatment in air at 600 °C [129]. Conversely, the biodiesel yield increased as the calcination temperature increased from uncalcined to 800 °C at step 200 °C and decreased with further increase to 1000 °C, with the yield for the derived catalyst at calcination temperature of 800 °C approximately identical with laboratory grade CaO [3]. This affirms that the activity of the eggshell- or bone-derived catalyst is dependent on the calcination temperature. It was also confirmed that the waste animal bone-derived catalyst performed equally well and comparable to laboratory-grade CaO [3, 98].

The summary of eggshells and animal bone-derived catalysts application in the transesterification of oil for biodiesel production is presented in Table 13.7. The production of biodiesel using waste chicken eggshells and waste cooking oil with methanol as an alcohol source was investigated and reported by Gupta and Rathod [93]. At optimized conditions of catalyst loading 1.50 wt%, temperature 60 °C, agitation 300 rpm, reaction time 50 minutes, and 10 : 1 methanol/oil, 96% yield of biodiesel was produced. Furthermore, it has been discovered that for biodiesel production through transesterification reaction, the catalytic activity of calcined eggshell is comparable with calcined pure CaO, suggesting that waste eggshell-derived material can substitute pure alkali metal oxides [16]. Consequently, as the calcination temperature increases from 600 to 900 °C, the catalytic activity also increases [120]. A detailed review of waste eggshells as catalyst has been reported elsewhere [120].

It has been reported that quail eggshell catalytic activity outweighed that of hen, because of its higher number of strong basic sites on the pores in the palisade layer [120]. Correia et al. [130] compared the catalytic activities of calcined crab shell and eggshell. It was discovered that the eggshell outperformed the crab shell due to the higher content of calcium on the surface of the calcined eggshell. Therefore, the catalytic activity depends on the type of eggshell or animal bone used in addition to the strength of their sites. Nonetheless, the derived catalyst surface area, particle size, and calcium oxide/phosphate content are some of the key parameters that could specifically affect the catalytic activity. These parameters are influenced by the calcination step. However, the calcium oxide and calcium phosphate catalysts derived from eggshell and animal bones have demonstrated analogous performance to homogeneous catalyst used in transesterification reaction for biodiesel production with added separation benefit [17, 55]. Consequently, the derived catalyst from these biomaterials is stable and reusable with catalytic activity sustained up to seven cycles. The formation of calcium hydroxide due to the produced water from transesterification reaction changes the surface structure, and glycerol coverage on the surface could retard activity over time [17]. But the catalytic activity can be regenerated

Table 13.7 Application of eggshells and animal bone-derived catalysts in biodiesel production.

Catalyst	Oil	Alcohol/oil ratio	Experimental condition	Biodiesel yield (%)	References
K/fish bones	Palm oil	8:1–16:1 methanol/oil	3–12 wt% catalyst and 2–6 h	51–96	[95]
Waste animal bones	Palm oil	18:1 methanol/oil	5–25 wt% catalyst, 200 rpm, 65°C, 4 h	96.8	[3]
Cu/fish bones	Oleic acid	0.8 ml min^{-1} ethanol	90°C, 1000 rpm, 1 h	91.9	[94]
Chicken bone and Fe/chicken bone	Used cooking oil	6:1 methanol/oil	3% (w/w) catalyst/oil, 3 h, 65°C	22	[13]
Waste chicken and fish bones	Used cooking oil	10:1 methanol/oil	1.98 w/v catalyst, 65°C, 1.54 h	89.5	[99]
Waste cow bone	Soybean oil	9:1–15:1 methanol/oil	10–20 wt% catalyst, 55–65°C, 500 rpm, 3 h	69–94.8	[98]
Chicken eggshell	Waste cooking oil	10:1 methanol/oil	1.50 wt% catalyst, 60°C, 300 rpm, 50 min	96.1	[93]
Duck and chicken eggshells	Palm oil	9:1 methanol/oil	20 wt% catalyst, 60°C, 4 h	92.9, 94.4	[130]
Eggshell	Sunflower oil	9:1 methanol/oil	3 wt% catalyst, 60°C, 3 h	97.8 conversion	[131]

through oxidative combustion at 600 °C. The different process factors that influence the performance of eggshells and animal bones as heterogeneous catalyst in biodiesel production include methanol/oil ratio, reaction time, temperature, and catalyst load.

13.4.2.6 Eggshell Membranes (ESM) in Fuel Cell Applications

In the development of batteries and supercapacitors for energy storage, electrode materials play a remarkable role in ensuring high performance. The ESM has unique properties as a biomaterial with microfibrous network within the microstructure. The eggshell allows interactions between the structure and other materials such as drugs, genes, and both proton- and anion exchange applications [132]. In light of this, ostrich ESM was tested in both anion exchange membrane (AEM) and proton-exchange membrane (PEM) fuel cells [132]. It proves practical for proton conductivity in different acidic solutions (0.48 mS cm^{-1}) and an anion conductivity of about 0.40 mS cm^{-1}. On the contrary, a supercapacitor electrode material was synthesized via carbonization of waste chicken ESMs biomaterial by Li et al. [113]. It was found that in spite of the moderately 221 m^2 g^{-1} specific surface area, the achieved capacitances were 297 and 284 F g^{-1} in basic and acidic electrolytes, respectively. Also at a current density of 4 A g^{-1}, the synthesized eggshell-derived electrode exhibited remarkable reusability stability as only 3% drop in capacitance was observed after 10 000 cycles. This is promising compared to complex microporous structures and disordered texture of activated carbon, which confines the efficiency of electron transfer, resulting in poor performance of the supercapacitor at high sweep rate [113]. Furthermore, the oxidation of the carbon generates oxygen on the surface, which causes instability and pseudocapacitance fading during cycling.

In 2018, Nguyen et al. [112] reported the use of eggshell-derived material as a membrane separator of anode and cathode in lithium-ion battery. The ESM functions as a lithium-ion transport facilitator. The result showed that the ESM is a comparable substitute to conventional separator with the capacities of 90 and 87 mAh g^{-1} at 1st and 150th cycles, respectively. This can be attributed to the porous morphology and the nonwoven textures of the eggshell-derived material. Waste ESM has also been investigated as a separator in promoting electricity recovery in air cathode microbial fuel cells [133]. It was found that depending on the current density, coulombic efficiency in the range of 67–95% can be achieved. Minakshi et al. [134] reported the use of waste eggshell-derived material as a cathode and its calcined form as an anode in an electrochemical cell. It was discovered that the uncalcined eggshell (i.e. CaCO$_3$) cathode showed a moderate discharge capacitance of 10 F g^{-1}, while the calcined (i.e. CaO) anode showed superb capacitance value of 47.5 F g^{-1}. However, at a current density of 0.15 A g^{-1}, the use of the calcined eggshell (i.e. CaO) electrode in both positive and negative regions exhibited 55 F g^{-1} with retention of approximately 100% after 1000 cycles. This reaffirms the thermal and chemical stability of eggshell-derived materials. Likewise, animal bone converted into hierarchical porous and textural carbon material had been reported as an electrode material. In 2011, Huang and coworkers utilized material derived from swine bone as precursor of electrode used in electrocatalytic electric double-layer capacitors, and they proposed that the hierarchical porous structure will accelerate electron transfer [135]. It was found that structure and texture of the animal bone-derived carbon material exhibited excellent capacitive performance with electrochemical test in 7 M KOH electrolyte.

13.4.2.7 Catalytic Materials from Eggshells and Animal Bones in Organic Synthesis

The design, synthesis, and development of catalytic materials from waste eggshells and animal bones offer cost-effective approach to recycling and producing value-added catalytic materials for organic synthesis processes. Their utilization will reduce solid waste environmental pollution while contributing to both organic and material syntheses [107]. Generally, eggshells and animal bones contain an oxide of alkaline earth metals such as calcium and other nonmetals such as phosphorus and carbonate. The considerable amount of calcium and phosphorous (bones) in both waste materials make them plausible candidates for the production of HA and β-TCP through thermal treatment processes. Calcium carbonate, calcium phosphate, calcium oxide (CaO), and HA are the most common catalytic materials derived from eggshells and animal bones. In light of this, CaO catalytic material derived from eggshell through calcination has been utilized in the synthesis of oximes from aldehydes and ketones using ethanol solvent [110]. The corresponding yields of aldoximes and ketoximes ranges from 70 to 96%, while the reusability and catalytic activities sustained up to seven times. Similarly, catalytic material derived from eggshell was used in the synthesis of glycerol oligomers via glycerol etherification reaction [111]. It was found that at 2 wt% catalyst loading, 220 °C temperature, and 24 hours reaction time, 85% conversion of glycerol and 43% yield of oligomers were obtained. Furthermore, eggshell derived-catalyst has been reported for the synthesis of Schiff base ligands (98% yield for 10–15 minutes) through the condensation reaction of various aromatic aldehydes with substituted aromatic amines [136], as a support material for Knoevenagel reaction with methanol at room temperature in which the yield ranges from 64 to 94% depending on reaction time [137], and as a catalyst in dimethyl carbonate synthesis (75% yield) from propylene carbonate and methanol at temperature 25 °C and pressure 1 atm [106]. In 2005, Montilla et al. [138] reported eggshell-derived catalyst (6 mg ml^{-1}) for the synthesis of lactulose through the isomerization of lactose obtained from milk ultrafiltrate; the reaction was stirred and refluxed at 98 °C in glycerol for 60 minutes.

On the other hand, HA-synthesized material using animal bone meal (ABM) has been reported as a catalyst and carrier doped with the following Lewis acids ($ZnBr_2$, $ZnCl_2$, and $CuBr_2$), as catalyst in the synthesis of benzimidazoles, benzoxazoles, and benzothiazoles via condensation of o-phenylenediamine, o-aminophenol, and o-aminothiophenol, respectively, with aromatic aldehydes [108]. A yield of 96% was observed using $ZnBr_2$/ABM catalyst and toluene as solvent, at reaction temperature 111 °C and time 15 minutes. Azzallou and coworkers [85] reported synthesized pyridine through aromatization of 1,4-dihydropyridines using ABM-derived catalyst. Monjezi and coworkers [139] reported the selective benzophenone production through liquid phase diphenylmethane oxidation using a well dispersed Co—Mn catalyst coated on calcined cow bone. In another study, 3-cyanopyridine was synthesized in a one-pot multicomponent reaction system of 1,3-dicarbonyl compounds, aromatic aldehydes, malononitrile, and alcohols, catalyzed by doped ABM-derived catalyst [140]. The advantages of the catalyst in the reaction include high yields in the range of 80–92% and short reaction times of 10–15 minutes. Animal bone-derived catalyst has been applied in crossed aldol condensation reaction and Suzuki–Miyaura cross-coupling reaction, which involves selective construction of carbon–carbon bonds, and is one of the most useful reactions for the preparation of several important compounds such as pharmaceuticals, polymers, and agrochemicals [141, 142]. For instance, isomeric position 4-monosubstituted pyrido[2,3-*d*] pyrimidines were synthesized via cross-coupling reactions using ABM-supported palladium (0) catalyst [141]. More also, ABM and KF or $NaNO_3$ doped ABMs had been investigated and reported as efficient and inexpensive catalysts for performing Claisen–Schmidt condensation and aza-Michael additions under mild conditions [143]. The same group also synthesized α,α′-bis(substituted benzylidene)cycloalkanones from crossed aldol condensation reaction of cycloalkanones and benzaldehydes in water catalyzed by ABM-derived catalytic material [142]. The ABM-derived catalyst showed no significant loss of activity, even when utilized up to five times. These results prove catalytic materials synthesized from eggshells and animal bones as green and renewable catalyst and as a consequent a cost-effective substitute for used as a catalyst in several organic synthesis reactions.

13.4.2.8 Other Catalytic Applications

Biochar has received great attention in the field of environmental catalysis especially pollutant degradation. In recent years, waste animal bone-derived biochar has gained increasing attention for its special surface character, pore volume, porous size, and surface charge distribution [77]. Also their HA and calcium phosphate component can act as a phosphorus doping precursor. Consequently, their hierarchical porous structure enhances mass transfer in their catalytic, adsorption, and performance processes. In this regard, swine bone was used as a precursor of biochar preparation and the bone-derived material applied into persulfate activation system, to achieve efficient 2,4-dichlorophenol (2,4-DCP) pollutant degradation [77]. It was observed that over 85% of 2,4-DCP can be degraded in 30 minutes and about 100% degradation of pollutant in two hours. Consequently, petroleum industry wastewater containing toxic, carcinogenic, and recalcitrant chlorophenols pollutant had been treated using synthesized material from waste buffalo animal bones as a photocatalytic degradation catalyst [78]. The photocatalytic degradation followed the Langmuir–Hinshelwood type kinetic model, and the animal bone-derived material proved to be a favorable photocatalytic material for degrading petroleum refineries wastewater pollutants. The photocatalytic activity toward chlorophenols can be summarized as follows: animal bone-derived catalyst > ZnO-animal bone material > ZnO.

It is well known that VOCs such as benzene, toluene, etc. are considered major pollutants from petroleum refining industries. Catalytic oxidation is considered as one of the approaches of degrading these pollutants. However, the economics of the process is adversely impacted by catalyst cost; hence, Co_3O_4 nanoparticles supported on eggshell-derived catalyst demonstrated excellent activity for benzene catalytic oxidation [19]. In 2013, Asgari et al. [144] tested and reported pyrolysis treated cow bone-derived material as a low-cost catalyst under normal conditions for the decomposition of ozone in a fluidized bed reactor. In another work, Wu and coworkers [145] reported SnO_2 nanoparticles supported on animal bone prepared through a facile hydrothermal approach and annealing process and compared with pure SnO_2 nanoparticles for photocatalytic degradation of rhodamine B dye. The result showed that better photocatalytic activity was achieved with SnO_2/bone catalyst in the degradation of rhodamine B dye than pure SnO_2 nanoparticles [145]. This is due to the larger specific surface area of SnO_2 nanoparticles supported on animal bone; the interface between SnO_2 and animal bone helps the separation of charges and is valuable for photocatalysis, improving light absorption capability of SnO_2/bone catalyst. Furthermore, the oxygen reduction reaction (ORR) by N-doped porous carbon derived from chicken and pig bones has been reported as an electrocatalyst, in which the results showed that the electrocatalytic performance in alkaline media was higher, achieved better stability, and excellent methanol tolerance than commercial Pt/C catalyst [146, 147]. In 2018, Matsagar et al. [148] demonstrated the combined catalysis of bone char and 1-methyl-3-(3-sulfopropyl)-imidazolium hydrogen sulfate acidic ionic liquid catalysts for selective conversion of glucose into 5-hydroxymethylfurfural (HMF), and the achieved selectivity was 54% in water. The bone char catalyst was stable under reaction conditions and exhibited the crystal structure of HA ($Ca_5(OH)(PO_4)_3$) material. In 2010, Duan and coworkers [149] examined carbon nanofiber with fishbone graphene alignment and commercial carbon nanotubes (CNTs) as supports for Ru catalyst for hydrogen production through ammonia decomposition. They found that the Ru nanoparticles supported on carbon nanofiber obtained from fish bone exhibited more activity than those on CNT support. A detailed review of catalysts derived from animal bone waste and their catalytic applications had been reported in the literature [150]. These applications would have positive impact on environment, economy, and waste management problem, and the derived materials from eggshell and animal bones have proven useful as catalysts/supports with cheap and environmentally friendly features.

13.5 Conclusions

The present linear economy mostly practiced globally, in which natural raw materials are extracted, processed into products, used, and then disposed of once their usefulness is outlived, resulting in environmental degradation. This economic model catalyzes fast depletion of natural resources as the world population and industrialization increases. To date, the beneficial use of eggshells and animal bones remains of utmost importance for waste management, pollution minimization, and environmental sustainability. In addition, this promote circular economy in which solid wastes are reused as raw materials to create new materials or energy instead of discarding them. However, an integrated approach to the management of these wastes focuses on raw material applications and other value-added products, which will have a significant socioeconomic, industrial, and environmental impact. This is potentially beneficial in reducing the cost associated with landfill disposal; likewise, the negative environmental impacts associated with these wastes will be diminished. The scale-up and expansion of their catalytic and adsorbent applications have proven to be renewable, greener, and cheaper alternatives to materials such as zeolite, alumina, titania, silica, and activated carbon. Huge tons of eggshells and animal bones are generated as waste per annum; these can be utilized as raw materials for HA material production, adsorbent/sorbent materials, calcium carbonate or phosphate source, paint production additive, and blended cement production and can serve as starting materials for other valuable products. In heterogeneous catalysis, they can serve as support materials, active catalyst, or promoter, and their synthesis methods are simple with significant catalytic performance. Also catalyst or its support material synthesized from eggshell and animal bones has found applications in fine chemicals synthesis, petroleum refining, organic synthesis, energy, and environment pollution control. The preparation of zeolite, silica, alumina, zirconia, cerium (IV) oxide, and titania is complex and involves high manufacturing cost. Therefore, eggshells and animal bones offer a low-cost alternative and sustainable source to these materials. The suitability of eggshells and animal bones is due to the negligible cost, readily and abundant accessibility, and the sufficient content of calcium carbonate and phosphate that allow them to be converted into calcium oxide catalyst, HA material/catalyst, support material for catalyst, or adsorbent. Considering the wide range of catalytic applications, high catalytic activity, and thermal stability, they could potentially reduce cost associated with heterogeneous catalyst manufactured by 40% while minimizing land pollution. Generally, this promotes the concept of waste to wealth and circular economy in our world of finite natural resources.

References

1 Das Lala, S., Deb, P., Barua, E. et al. (2019). Characterization of hydroxyapatite derived from eggshells for medical implants. *Materials Today: Proceedings* 15: 323–327.
2 Safitri, I.R., Supriyana, S., and Bahiyatun, B. (2017). Effect of egg shell flour on blood calcium levels in pregnant mice. *Belitung Nursing Journal* 3 (6): 791–795.
3 Obadiah, A., Swaroopa, G.A., Kumar, S.V. et al. (2012). Biodiesel production from Palm oil using calcined waste animal bone as catalyst. *Bioresource Technology* 116: 512–516.
4 Ononiwu, N.H. and Akinlabi, E.T. (2020). Effects of ball milling on particle size distribution and microstructure of eggshells for applications in metal

matrix composites. *Materials Today Proceedings* 2–6: 1049–1053. https://doi.org/10.1016/j.matpr.2020.02.209.

5 Quina, M.J., Soares, M.A.R., and Quinta-Ferreira, R. (2017). Applications of industrial eggshell as a valuable anthropogenic resource. *Resources Conservation and Recycling* 123: 176–186.

6 Guru, P.S. and Dash, S. (2014). Sorption on eggshell waste – a review on ultrastructure, biomineralization and other applications. *Advances in Colloid and Interface Science* 209: 49–67.

7 Yoo, S., Hsieh, J.S., Zou, P. et al. (2009). Utilization of calcium carbonate particles from eggshell waste as coating pigments for ink-jet printing paper. *Bioresource Technology* 100 (24): 6416–6421.

8 Shiferaw, N., Habte, L., Thenepalli, T. et al. (2019). Effect of eggshell powder on the hydration of cement paste. *Materials* 12 (15): 2483. doi:10.3390/ma12152483.

9 Barcelo, L., Kline, J., Walenta, G. et al. (2014). Cement and carbon emissions. *Materials and Structures* 47: 1055–1065.

10 Zaman, T., Mostari, M.S., Al Mahmood, M.A., and Rahman, M.S. (2018). Evolution and characterization of eggshell as a potential candidate of raw material. *Cerâmica* 64: 236–241.

11 Correia, L.M., Cecilia, J.A., Rodríguez-Castellón, E. et al. (2017). Relevance of the physicochemical properties of calcined quail eggshell (CaO) as a catalyst for biodiesel production. *Journal of Chemistry* 2017: 5679512. https://doi.org/10.1155/2017/5679512.

12 Tan, Y.H., Abdullah, M.O., Nolasco-Hipolito, C. et al. (2015). Waste ostrich- and chicken-eggshells as heterogeneous base catalyst for biodiesel production from used cooking oil: catalyst characterization and biodiesel yield performance. *Applied Energy* 160: 58–70.

13 Satraidi, H., Widayat, H., Prasetyaningrum, A. et al. (2019). Development of heterogeneous catalyst from chicken bone and catalytic testing for biodiesel with simultaneous processing. *IOP Conference Series: Materials Science and Engineering* 509 (1): 012125.

14 Jazie, A.A., Pramanik, H., and Sinha, A.S.K. (2013). Transesterification of peanut and rapeseed oils using waste of animal bone as cost effective catalyst. *Materials for Renewable and Sustainable Energy* 2 (11): 1–10. https://doi.org/10.1007/s40243-013-0011-4.

15 Laskar, I.B., Rajkumari, K., Gupta, R. et al. (2018). Waste snail shell derived heterogeneous catalyst for biodiesel production by the transesterification of soybean oil. *RSC Advances* 8 (36): 20131–20142.

16 Wong, Y.C. and Ang, R.X. (2018). Study of calcined eggshell as potential catalyst for biodiesel formation using used cooking oil. *Open Chemistry* 16 (1): 1166–1175.

17 Hart, A. (2020). Mini-review of waste shell-derived materials' applications. *Waste Management & Research* 38: 514–527.

18 Abdullah, S.H.Y.S., Hanapi, N.H.M., Azid, A. et al. (2017). A review of biomass-derived heterogeneous catalyst for a sustainable biodiesel production. *Renewable and Sustainable Energy Reviews* 70: 1040–1051.

19 Li, Z., Yang, D.P., Chen, Y. et al. (2020). Waste eggshells to valuable $Co_3O_4/CaCO_3$ materials as efficient catalysts for VOCs oxidation. *Molecular Catalysis* 483: 110766. https://doi.org/10.1016/j.mcat.2020.110766.

20 Kirboga, S. and Öner, M. (2013). Application of experimental design for the precipitationof calcium carbonate in the presence of biopolymer. *Powder Technology* 249: 95–104.

21 Zhang, X. and Vecchio, K.S. (2006). Creation of dense hydroxyapatite (synthetic bone) by hydrothermal conversion of seashells. *Materials Science and Engineering C* 26: 1445–1450.

22 Morris, J.P., Backeljau, T., and Chapelle, G. (2019). Shells from aquaculture: a valuable biomaterial, not a nuisance waste product. *Review in Aquaculture* 11 (1): 42–57.

23 Yan, N. and Chen, X. (2015). Don't waste seafood waste. *Nature* 524: 155–157.

24 Olivier, J.G.J., Janssens-Maenhout, G., Peters, J., and a H.W. (2012). *Trends in global CO_2 emissions: 2012 Report*, 1–40. The Hague: PBL Netherlands Environmental Assessment Agency.

25 Alkurdi, S.S.A., Al-Juboori, R.A., Bundschuh, J. et al. (2019). Bone char as a green sorbent for removing health threatening fluoride from drinking water. *Environmental International* 127: 704–719.

26 Yami, T.L., Du, J., Brunson, L.R. et al. (2015). Life cycle assessment of adsorbents for fluoride removal from drinking water in East Africa. *International Journal of Life Cycle Assessment* 20 (9): 1277–1286.

27 Alkurdi, S.S.A., Al-Juboori, R.A., Bundschuh, J. et al. (2020). Effect of pyrolysis conditions on bone char characterization and its ability for arsenic and fluoride removal. *Environmental Pollution* 262: 114221.

28 Wang, M., Liu, Y., Yao, Y. et al. (2020). Comparative evaluation of bone chars derived from bovine parts: Physicochemical properties and copper sorption behaviour. *Science of the Total Environment* 700: 134470.

29 Shahid, M.K., Kim, J.Y., and Choi, Y.G. (2019). Synthesis of bone char from cattle bones and its application for fluoride removal from the contaminated water. *Groundwater for Sustainable Development* 8: 324–331.

30 Mendes, K.F., de Sousa, R.N., Takeshita, V. et al. (2019). Cow bone char as a sorbent to increase sorption and decrease mobility of hexazinone, metribuzin, and quinclorac in soil. *Geoderma* 343: 40–49.

31 Bedin, K.C., de Azevedo, S.P., Leandro, P.K.T. et al. (2017). Bone char prepared by CO_2 atmosphere: preparation optimization and adsorption studies of Remazol Brilliant Blue R. *Journal of Cleaner Production* 161: 288–298.

32 Sellaoui, L., Mendoza-Castillo, D.I., Reynel-Ávila, H.E. et al. (2018). A new statistical physics model for the ternary adsorption of Cu^{2+}, Cd^{2+} and Zn^{2+} ions on bone char: experimental investigation and simulations. *Chemical Engineering Journal* 343: 544–553.

33 Rezaee, A., Rangkooy, H., Jonidi-Jafari, A. et al. (2013). Surface modification of bone char for removal of formaldehyde from air. *Applied Surface Science* 286: 235–239.

34 Cascarosa, E., Ortiz de Zarate, M.C., Sánchez, J.L. et al. (2012). Sulphur removal using char and ash from meat and bone meal pyrolysis. *Biomass and Bioenergy* 40: 190–193.

35 Yirong, C. and Vaurs, L.P. (2019). Wasted salted duck eggshells as an alternative adsorbent for phosphorus removal. *Journal of Environmental Chemical Engineering* 7 (6): 103443.

36 De Angelis, G., Medeghini, L., Conte, A.M. et al. (2017). Recycling of eggshell waste into low-cost adsorbent for Ni removal from wastewater. *Journal of Cleaner Production* 164: 1497–1506.

37 Pettinato, M., Chakraborty, S., Arafat, H.A. et al. (2015). Eggshell: a green adsorbent for heavy metal removal in an MBR system. *Ecotoxicology and Environmental Safety* 121: 57–62.

38 Eletta, O.A.A., Ajayi, O.A., Ogunleye, O.O. et al. (2016). Adsorption of cyanide from aqueous solution using calcinated eggshells: equilibrium and optimisation studies. *Journal of Environmental Chemical Engineering* 4 (1): 1367–1375.

39 Podstawczyk, D., Witek-Krowiak, A., Chojnacka, K. et al. (2014). Biosorption of malachite green by eggshells: mechanism identification and process optimization. *Bioresource Technology* 160: 161–165.

40 Tsai, W.T., Yang, J.M., Lai, C.W. et al. (2006). Characterization and adsorption properties of eggshells and eggshell membrane. *Bioresource Technology* 97 (3): 488–493.

41 Witoon, T. (2011). Characterization of calcium oxide derived from waste eggshell and its application as CO_2 sorbent. *Ceramics International* 37 (8): 3291–3298.

42 Habeeb, O.A., Yasin, F.M., and Danhassan, U.A. (2014). Characterization and application of chicken eggshell as green adsorbents for removal of H_2S from wastewaters. *IOSR Journal of Environmental Science, Toxicology and Food Technology* 8 (11): 7–12.

43 Kalyoncu Ergüler, G. (2015). Investigation the applicability of eggshell for the treatment of a contaminated mining site. *Minerals Engineering* 76: 10–19.

44 Muliwa, A.M., Leswifi, T.Y., and Onyango, M.S. (2018). Performance evaluation of eggshell waste material for remediation of acid mine drainage from coal dump leachate. *Minerals Engineering* 122: 241–250.

45 Arunlertaree, C., Kaewsomboon, W., Kumsopa, A. et al. (2007). Removal of lead from battery manufacturing wastewater by egg shell. *Songklanakarin Journal of Science and Technology* 29 (3): 857–868.

46 Park, H.J., Jeong, S.W., Yang, J.K. et al. (2007). Removal of heavy metals using waste eggshell. *Journal of Environmental Sciences* 19 (12): 1436–1441.

47 Wu, S.C., Hsu, H.C., Hsu, S.K. et al. (2015). Effects of heat treatment on the synthesis of hydroxyapatite from eggshell powders. *Ceramics International* 41 (9): 10718–10724.

48 Ramesh, S., Loo, Z.Z., Tan, C.Y. et al. (2018). Characterization of biogenic hydroxyapatite derived from animal bones for biomedical applications. *Ceramics International* 44 (9): 10525–10530.

49 Suchanek, W.L., Shuk, P., Byrappa, K. et al. (2002). Mechanochemical-hydrothermal synthesis of carbonated apatite powders at room temperature. *Biomaterials* 23 (3): 699–710.

50 Brzezińska-Miecznik, J., Haberko, K., Sitarz, M. et al. (2015). Hydroxyapatite from animal bones – Extraction and properties. *Ceramics International* 41 (3): 4841–4846.

51 Vidhya, G., Suresh Kumar, G., Kattimani, V.S. et al. (2019). Comparative study of hydroxyapatite prepared from eggshells and synthetic precursors by microwave irradiation method for medical applications. *Materials Today: Proceedings* 15: 344–352.

52 Ramesh, S., Natasha, A.N., Tan, C.Y. et al. (2016). Direct conversion of eggshell to hydroxyapatite ceramic by a sintering method. *Ceramics International* 42 (6): 7824–7829.

53 Abdulrahman, I., Tijani, H.I., Mohammed, B.A. et al. (2014). From garbage to biomaterials: an overview on egg shell based hydroxyapatite. *Journal of Materials* 2014: 802467. http://dx.doi.org/10.1155/2014/802467.

54 Smith, S.M., Oopathum, C., Weeramongkhonlert, V. et al. (2013). Transesterification of soybean oil using bovine bone waste as new catalyst. *Bioresource Technology* 143: 686–690.

55 Kayode, B. and Hart, A. (2019). An overview of transesterification methods for producing biodiesel from waste vegetable oils. *Biofuels* 10 (3): 419–437.

56 Roschat, W., Kacha, M., Yoosuk, B. et al. (2012). Biodiesel production based on heterogeneous process catalyzed by solid waste coral fragment. *Fuel* 98: 194–202.

57 Chakraborty, R., Bepari, S., and Banerjee, A. (2011). Application of calcined waste fish (*Labeorohita*) scale as low-cost heterogeneous catalyst for biodiesel synthesis. *Bioresource Technology* 102 (3): 3610–3618.

58 Zhang, D., Zhao, H., Zhao, X. et al. (2011). Application of hydroxyapatite as catalyst and catalyst carrier. *Progress in Chemistry* 23 (4): 687–694.

59 Fihri, A., Len, C., Varma, S.R. et al. (2017). Hydroxyapatite: a review of syntheses, structure and applications in heterogeneous catalysis. *Coordination Chemistry Reviews* 347: 48–76.

60 Phan, S.T., Sane, R.A., de Vasconcelos, R.B. et al. (2018). Hydroxyapatite supported bimetallic cobalt and nickel catalysts for syngas production from dry reforming of methane. *Applied Catalysis B: Environmental* 224: 310–321.

61 Boukha, Z., Gil-Calvo, M., de Rivas, B. et al. (2018). Behaviour of Rh supported on hydroxyapatite catalysts in partial oxidation and steam reforming of methane: on the role of the speciation of the Rh particles. *Applied Catalysis A: General* 556: 191–203.

62 Pham Minh, D., Rio, S., Sharrock, P. et al. (2014). Hydroxyapatite starting from calcium carbonate and orthophosphoric acid: synthesis, characterization, and applications. *Journal of Materials Science* 49: 4261–4269.

63 Mori, K., Hara, T., Mizugaki, T. et al. (2004). Hydroxyapatite-supported palladium nanoclusters: a highly active heterogeneous catalyst for selective oxidation of alcohols by use of molecular oxygen. *Journal of American Chemical Society* 126: 10657–10666.

64 Tran, Q.T., Pham Minh, D., Phan, S.T. et al. (2020). Dry reforming of methane over calcium-deficient hydroxyapatite supported cobalt and nickel catalysts. *Chemical Engineering Science* 228: 115975.

65 Jun, J.H., Jeong, K.S., Lee, T.J. et al. (2004). Nickel-calcium phosphate/hydroxyapatite catalysts for partial oxidation of methane to syngas: effect of composition. *Korean Journal of Chemical Engineering* 21: 140–146.

66 Zhao, J., Kunieda, M., Yang, G. et al. (2010). Nickel-hydroxyapatite as biomaterial catalysts for hydrogen production via glycerol steam reforming. *Key Engineering Materials* 447–448: 770–774.

67 Munirathinam, R., Pham Minh, D., and Nzihou, A. (2020). Hydroxyapatite as a new support material for cobalt-based catalysts in Fischer-Tropsch synthesis. *International Journal of Hydrogen Energy* 45 (36): 18440–18451.

68 Martínez-Hernández, H., Mendoza-Nieto, A., Pfeiffer, H. et al. (2020). Development of novel nano-hydroxyapatite doped with silver as effective catalysts for carbon monoxide oxidation. *Chemical Engineering Journal* 401: 125992.

69 Wei, X., Wang, Y., Xiao Li, X. et al. (2020). Co_3O_4 supported on bone-derived hydroxyapatite as potential catalysts for N_2O catalytic decomposition. *Molecular Catalysis* 491: 111005.

70 Ibrahim, A.-R., Li, X., Zhou, Y. et al. (2015). Synthesis of spongy-like mesoporous hydroxyapatite from raw waste eggshells for enhanced dissolution of ibuprofen loaded via supercritical CO_2. *International Journal of Molecular Sciences* 16: 7960–7975.

71 Iriarte-Velasco, U., Sierra, I., Zudaire, L. et al. (2015). Conversion of waste animal bones into porous hydroxyapatite by alkaline treatment: effect of the impregnation ratio and investigation of the activation mechanism. *Journal of Material Science* 50: 7568–7582.

72 Iriarte-Velasco, U., Sierra, I., Zudaire, L. et al. (2016). Preparation of a porous biochar from the acid activation of pork bones. *Food and Bioproducts Processing* 9 (8): 341–353.

73 Iriarte-Velasco, U., Ayastuy, L.J., Zudaire, L. et al. (2014). An insight into the reactions occurring during the chemical activation of bone char. *Chemical Engineering Journal* 251: 217–227.

74 Rakap, M. and Ozkar, S. (2011). Hydroxyapatite-supported palladium(0) nanoclusters as effective and reusable catalyst for hydrogen generation from the hydrolysis of ammonia-borane. *International Journal of Hydrogen Energy* 36: 7019–7027.

75 Buasri, A., Inkaew, T., Kodephun, L. et al. (2015). Natural hydroxyapatite (NHAp) derived from pork bone as a renewable catalyst for biodiesel production via microwave irradiation. *Key Engineering Materials* 659: 216–220.

76 Darwish, S.A., Sayed, A.M., and Sheb, A. (2020). Cuttlefish bone stabilized Ag_3VO_4 nanocomposite and its Y_2O_3-decorated form: waste-to-value development of efficiently eco-friendly visible-light-photoactive and biocidal agents for dyeing, bacterial and larvae depollution of Egypt's wastewater. *Journal of Photochemistry and Photobiology A: Chemistry* 401: 112749.

77 Zhou, X., Zeng, Z., Zeng, G. et al. (2020). Persulfate activation by swine bone char-derived hierarchical porous carbon: multiple mechanism system for organic pollutant degradation in aqueous media. *Chemical Engineering Journal* 383: 123091.

78 El-Gendy, S.N., El-Salamony, A.R., and Younis, A.S. (2016). Green synthesis of fluorapatite from waste animal bones and the photo-catalytic degradation activity of a new ZnO/green biocatalyst nano-composite for removal of chlorophenols. *Journal of Water Process Engineering* 12: 8–19.

79 Singh, N., Chakraborty, R., and Gupta, K.R. (2018). Mutton bone derived hydroxyapatite supported TiO_2 nanoparticles for sustainable photocatalytic applications. *Journal of Environmental Chemical Engineering* 6: 459–467.

80 Karoshi, G., Kolar, P., Shah, S.B. et al. (2015). Calcined eggshell as an inexpensive catalyst for partial oxidation of methane. *Journal of the Taiwan Institute Chemical Engineers* 57: 123–128.

81 Lima, D.S. and Perez-Lopez, O.W. (2020). Oxidative coupling of methane to light olefins using waste eggshell as catalyst. *Inorganic Chemistry Communications* 116: 107928. https://doi.org/10.1016/j.inoche.2020.107928.

82 Guo, Y., Yang, D.P., Liu, M. et al. (2019). Enhanced catalytic benzene oxidation over a novel waste-derived Ag/eggshell catalyst. *Journal of Materials Chemistry A* 7 (15): 8832–8844.

83 Bennett, A.J., Wilson, K., and Lee, F.A. (2016). Catalytic applications of waste derived materials. *Journal of Materials Chemistry A* 4: 3617–3637.

84 Shaabani, A., Afaridoun, H., and Shaabani, S. (2016). Natural hydroxyapatite-supported MnO_2: a green heterogeneous catalyst for selective aerobic oxidation of alkylarenes and alcohols. *Applied Organometallic Chemistry* 30: 772–776.

85 Azzallou, R., Ait Taleb, M., Mamouni, R. et al. (2018). Animal bone meal: a novel and efficient green catalyst for accelerated oxidation of hantzsch 1,4-dihydropyridines. *Journal of Materials and Environmental Sciences* 9 (5): 1598–1606.

86 Katryniok, B., Kimura, H., Skrzynska, E. et al. (2011). Selective catalytic oxidation of glycerol: perspectives for high value chemicals. *Green Chemistry* 13: 1960–1979.

87 Sacia, E.R., Ramkumar, S., Phalak, N. et al. (2013). Synthesis and regeneration of sustainable CaO sorbents from chicken eggshells for enhanced carbon dioxide capture. *ACS Sustainable Chemistry & Engineering* 1 (8): 903–909.

88 Vonder Haar, A.T. (2007). Engineering eggshells for carbon dioxide capture, hydrogen production, and as a collagen source. B.Eng Thesis. Ohio State University.

89 Taufiq-Yap, Y.H., Wong, P., Marliza, T.S. et al. (2013). Hydrogen production from wood gasification promoted by waste eggshell catalyst. *International Journal of Energy Research* 37: 1866–1871.

90 Iriarte-Velasco, U., Ayastuy, L.J., Boukha, Z. et al. (2018). Transition metals supported on bone-derived hydroxyapatite as potential catalysts for the Water-Gas Shift reaction. *Renewable Energy* 115: 641–648.

91 Tan, Y.H., Abdullah, M.O., and Nolasco-Hipolito, C. (2015). The potential of waste cooking oil-based biodiesel using heterogeneous catalyst derived from various calcined eggshells coupled with an emulsification technique: a review on the emission reduction and engine performance. *Renewable and Sustainable Energy Reviews* 47: 589–603.

92 Risso, R., Ferraz, P., Meireles, S. et al. (2018). Highly active Cao catalysts from waste shells of egg, oyster and clam for biodiesel production. *Applied Catalysis A: General* 567: 56–64.

93 Gupta, A.R. and Rathod, V.K. (2018). Waste cooking oil and waste chicken eggshells derived solid base catalyst for the biodiesel production: optimization and kinetics. *Waste Management* 79: 169–178.

94 Chakraborty, R. and RoyChowdhury, D. (2013). Fish bone derived natural hydroxyapatite-supported copper acid catalyst: Taguchi optimization of semi-batch oleic acid esterification. *Chemical Engineering Journal* 215–216: 491–499.

95 Shafinaz Abd Manaf, I., Jiun Yi, C., Rahim, M.H.A. et al. (2019). Utilization of waste fish bone as catalyst in transesterification of RBD palm oil. *Materials Today: Proceedings* 19: 1294–1302.

96 Mansir, N., Hwa Teo, S., Lokman Ibrahim, M. et al. (2017). Synthesis and application of waste egg shell derived CaO supported W-Mo mixed oxide catalysts for FAME production from waste cooking oil: effect of stoichiometry. *Energy Conversion and Management* 151: 216–226.

97 AlSharifi, M. and Znad, H. (2020). Transesterification of waste canola oil by lithium/zinc composite supported on waste chicken bone as an effective catalyst. *Renewable Energy* 151: 740–749.

98 Ayodeji, A.A., Blessing, I.E., and Sunday, F.O. (2018). Data on calcium oxide and cow bone catalysts used for soybean biodiesel production. *Data in Brief* 18: 512–517.

99 Tan, Y.H., Abdullah, M.O., Kansedo, J. et al. (2019). Biodiesel production from used cooking oil using green solid catalyst derived from calcined fusion waste chicken and fish bones. *Renewable Energy* 139: 696–706.

100 Marwaha, A., Rosha, P., Mohapatra, K.S. et al. (2018). Waste materials as potential catalysts for biodiesel production: current state and future scope. *Fuel Processing Technology* 181: 175–186.

101 Foroutan, R., Peighambardoust, J.S., Mohammadi, R. et al. (2020). One-pot transesterification of non-edible Moringa oleifera oil over a $MgO/K_2CO_3/HAp$ catalyst derived from poultry skeletal waste. *Environmental Technology & Innovation* https://doi.org/10.1016/j.eti.2020.101250.

102 Chakraborty, R., Mukhopadhyay, P., and Kumar, B. (2016). Optimal biodiesel-additive synthesis under infrared excitation using pork bone supported-Sb catalyst: engine performance and emission analyses. *Energy Conversion and Management* 126: 32–41.

103 Chingakham, C., Tiwary, C., and Sajith, V. (2019). Waste animal bone as a novel layered heterogeneous catalyst for the transesterification of biodiesel. *Catalysis Letters* 149: 1100–1110.

104 Fan, S., Xu, L.H., Kang, T.J. et al. (2017). Application of eggshell as catalyst for low rank coal gasification: experimental and kinetic studies. *Journal of the Energy Institute* 90 (5): 696–703.

105 Raheem, A., Liu, H., Ji, G. et al. (2019). Gasification of lipid-extracted microalgae biomass promoted by waste eggshell as CaO catalyst. *Algal Research* 42: 101601.

106 Gao, Y. and Xu, C. (2012). Synthesis of dimethyl carbonate over waste eggshell catalyst. *Catalysis Today* 190 (1): 107–111.

107 Konwar, M., Chetia, M., and Sarma, D. (2019). A low-cost, well-designed catalytic system derived from household waste "Egg Shell": applications in organic transformations. *Topics in Current Chemistry* 377 (6): https://doi.org/10.1007/s41061-018-0230-3.

108 Riadi, Y., Mamouni, R., Azzalou, R. et al. (2011). An efficient and reusable heterogeneous catalyst Animal Bone Meal for facile synthesis of benzimidazoles, benzoxazoles, and benzothiazoles. *Tetrahedron Letters* 52: 3492–3495.

109 Volli, V., Purkait, K.M., and Shu, C.-M. (2019). Preparation and characterization of animal bone powder impregnated fly ash catalyst for transesterification. *Science of the Total Environment* 669: 314–321.

110 Ait Taleb, M., Mamouni, R., Ait Benomar, M. et al. (2017). Chemically treated eggshell wastes as a heterogeneous and eco-friendly catalyst for oximes preparation. *Journal of Environmental Chemical Engineering* 5: 1341–1348.

111 Barros, F.J.S., Moreno-Tost, R., Cecilia, J.A. et al. (2017). Glycerol oligomers production by etherification using calcined eggshell as catalyst. *Molecular Catalysis* 433: 282–290.

112 Nguyen, V.H., Lee, D.H., Baek, S.Y. et al. (2018). Recycling different eggshell membranes for lithium-ion battery. *Materials Letters* 228: 504–508.

113 Li, Z., Zhang, L., Amirkhiz, B.S. et al. (2012). Carbonized chicken eggshell membranes with 3D architectures as high-performance electrode materials for supercapacitors. *Advanced Energy Materials* 2 (4): 431–437.

114 Li, C., Xu, G., Liu, X. et al. (2017). Hydrogenation of biomass-derived furfural to tetrahydrofurfuryl alcohol over hydroxyapatite-supported Pd catalyst under mild conditions. *Industrial & Engineering Chemistry Research* 56: 8843–8849.

115 Putrakumar, B., Seelam, P.K., Srinivasarao, G. et al. (2020). High performance and sustainable copper-modified hydroxyapatite catalysts for catalytic transfer hydrogenation of furfural. *Catalysts* 10: 1045.

116 Ivanova, Y.A., Sutormina, E.F., Rudina, N.A. et al. (2018). Effect of preparation route on Sr_2TiO_4 catalyst for the oxidative coupling of methane. *Catalysis Communications* 117: 43–48.

117 Karoshi, G., Kolar, P., Shah, S.B. et al. (2020). Valorization of eggshell waste into supported copper catalysts for partial oxidation of methane. *International Journal of Environmental Research* 14 (1): 61–70.

118 Acharya, B., Dutta, A., and Basu, P. (2010). An investigation into steam gasification of biomass for hydrogen enriched gas production in presence of CaO. *International Journal of Hydrogen Energy* 35 (4): 1582–1589.

119 Taufiq-Yap, Y.H., Sivasangar, S., and Salmiaton, A. (2012). Enhancement of hydrogen production by secondary metal oxide dopants on NiO/CaO material for catalytic gasification of empty palm fruit bunches. *Energy* 47 (1): 158–165.

120 Laca, A., Laca, A., and Díaz, M. (2017). Eggshell waste as catalyst: a review. *Journal of Environmental Management* 197: 351–359.

121 Salaudeen, S.A., Acharya, B., Heidari, M. et al. (2020). Hydrogen-rich gas stream from steam gasification of biomass: eggshell as a CO_2 sorbent. *Energy & Fuels* 34 (4): 4828–4836.

122 Wang, W., Ramkumar, S., Li, S. et al. (2010). Subpilot demonstration of the carbonation-calcination reaction (CCR) process: high-temperature CO_2 and sulfur capture from coal-fired power plants. *Industrial & Engineering Chemistry Research* 49 (11): 5094–5101.

123 Charitos, A., Rodríguez, N., Hawthorne, C. et al. (2011). Experimental validation of the calcium looping CO_2 capture process with two circulation fluidized bed carbonator reactors. *Industrial & Engineering Chemistry Research* 50 (16): 9685–9695.

124 Hart, A., and Onyeaka, H. (2021). Eggshell and seashells biomaterials sorbent for carbon dioxide capture. In: *Carbon Capture*, S.A.R. Khan Ed., IntechOpen, London, UK. DOI: https://doi.org/10.5772/intechopen.93870.

125 Banerjee, A., Panda, S., Sidhantha, M. et al. (2010). Utilisation of eggshell membrane as an adsorbent for carbon dioxide. *International Journal of Global Warming* 2 (3): 252–261.

126 Yin, H., Shang, J., Choi, J. et al. (2019). Generation and extraction of hydrogen from low temperature water-gas shift reaction by a ZIF-8-based membrane reactor. *Microporous and Mesoporous Materials* 280: 347–356.

127 Chen, T., Wang, Z., Liu, L. et al. (2020). Coupling CO_2 separation with catalytic reverse water-gas shift reaction via ceramic-carbonate dual-phase membrane reactor. *Chemical Engineering Journal* 379: 122182.

128 Iyer, M., and Fan, L. (2010). High temperature CO_2 capture using engineered eggshells: a route to carbon management. United States Patent, US 7,678,351 B2.

129 Hu, S., Wang, Y., and Han, H. (2011). Utilization of waste freshwater mussel shell as an economic catalyst for biodiesel production. *Biomass and Bioenergy* 35 (8): 3627–3635.

130 Correia, L.M., Saboya, R.M.A., de Sousa Campelo, N. et al. (2014). Characterization of calcium oxide catalysts from natural sources and their application in the transesterification of sunflower oil. *Bioresource Technology* 151: 207–213.

131 Buasri, A., Chaiyut, N., Loryuenyong, V. et al. (2013). Application of eggshell wastes as a heterogeneous catalyst for biodiesel production. *Sustainable Energy* 1 (2): 7–13.

132 Molavian, R.M., Abdolmaleki, A., Gharibi, H. et al. (2017). Safe and green modified ostrich eggshell membranes as dual functional fuel cell membranes. *Energy & Fuels* 31: 2017–2023.

133 Ma, M., You, S., Qu, J. et al. (2016). Natural eggshell membrane as separator for improved coulombic efficiency in air-cathode microbial fuel cells. *RSC Advances* 6: 66147–66151.

134 Minakshi, M., Higley, S., Baur, C. et al. (2019). Calcined chicken eggshell electrode for battery and supercapacitor applications. *RSC Advances* 9: 26981–26995.

135 Huang, W., Zhang, H., Huang, Y. et al. (2011). Hierarchical porous carbon obtained from animal bone and evaluation in electric double-layer capacitors. *Carbon* 49: 838–843.

136 Patil, S., Jadhav, S. D., and Shinde, S. K. (2012). CES as an efficient natural catalyst for synthesis of Schiff bases under solvent-free conditions: an innovative green approach. *Organic Chemistry International* 2012: 153159. doi:https://doi.org/10.1155/2012/153159.

137 Riadi, Y., Slimani, R., Haboub, A. et al. (2013). Calcined eggshell meal: new solid support for the Knoevenagel reaction in heterogeneous media. *Moroccan Journal of Chemistry* 1 (2): 24–28.

138 Montilla, A., del Castillo, M.D., Sanz, M.L. et al. (2005). Eggshell as catalyst of lactose isomerisation to lactulose. *Food Chemistry* 90: 883–890.

139 Monjezi, B.H., Yazdani, M.E., Mokfi, M. et al. (2014). Liquid phase oxidation of diphenylmethane to benzophenone with molecular oxygen over nano-sized Co-Mn catalyst supported on calcined cow bone. *Journal of Molecular Catalysis A: Chemistry* 383: 58–63.

140 Riadi, Y., Mamouni, R., Routier, S. et al. (2014). Ecofriendly synthesis of 3-cyanopyridine derivatives by multi-component reaction catalyzed by animal bone meal. *Environmental Chemistry Letters* 12: 523–527.

141 Riadi, Y., Geesi, M., Dehbi, O. et al. (2017). Novel animal-bone-meal-supported palladium as a green and efficient catalyst for Suzuki coupling reaction in water, under sunlight. *Green Chemistry Letters and Reviews* 10 (2): 101–106.

142 Riadi, Y., Mamouni, R., Azzalou, R. et al. (2010). Animal bone meal as an efficient catalyst for crossed-aldol condensation. *Tetrahedron Letters* 51: 6715–6717.

143 Riadi, Y., Abrouki, Y., Mamouni, R. et al. (2012). New eco-friendly animal bone meal catalysts for preparation of chalcones and aza-Michael adducts. *Chemistry Central Journal* 6 (60): https://doi.org/10.1186/1752-153X-6-60.

144 Asgari, G., Mohammadi, A.M.S., Mortazavi, S.B. et al. (2013). Investigation on the pyrolysis of cow bone as a catalyst for ozone aqueous decomposition: kinetic approach. *Journal of Analytical and Applied Pyrolysis* 99: 149–154.

145 Wu, Y., Wang, H., Cao, M. et al. (2015). Animal bone supported SnO_2 as recyclable photocatalyst for degradation of rhodamine B dye. *Journal of Nanoscience and Nanotechnology* 15: 6495–6502.

146 Song, H., Li, H., Wang, H. et al. (2014). Chicken bone-derived N-doped porous carbon materials as an oxygen reduction electrocatalyst. *Electrochimica Acta* 147: 520–526.

147 Wang, R., Wang, K., Wang, Z. et al. (2015). Pig bones derived N-doped carbon with multi-level pores as electrocatalyst for oxygen reduction. *Journal of Power Sources* 297: 295–301.

148 Matsagar, B.M., Van Nguyen, C., Hossain, M.S.A. et al. (2018). Glucose isomerization catalyzed by bone char and the selective production of 5-hydroxymethylfurfural in aqueous media. *Sustainable Energy and Fuels* 2: 2148–2153.

149 Duan, X., Zhou, J., Qian, G. et al. (2010). Carbon nanofiber-supported Ru catalysts for hydrogen evolution by ammonia decomposition. *Chinese Journal of Catalysis* 31 (8): 979–986.

150 Nasrollahzadeh, M., Bidgoli, S.S.N., Shafiei, N. et al. (2020). Low-cost and sustainable (nano)catalysts derived from bone waste: catalytic applications and biofuels production. *Biofuels Bioproducts and Biorefinery* 14: 1197–1227.

14

Natural Phosphates and Their Catalytic Applications

Karima Abdelouahdi[1], Abderrahim Bouaid[2], Abdellatif Barakat[3,4], and Abderrahim Solhy[3]

[1]*IMED-Lab, FST-Marrakech, University Cadi Ayyad, Av. A. Khattabi, BP 549, Marrakech, 40000, Morocco*
[2]*Chemical Engineering Department, Faculty of Chemistry, University of Complutense, Madrid, 28040, Spain*
[3]*IATE, Montpellier University, INRAE, Agro Institut, Montpellier, 34060, France*
[4]*Mohamed VI Polytechnic University, Lot 660 – Hay Moulay Rachid, Ben Guerir, 43150, Morocco*

14.1 Introduction

The mining sector contributes to the development of mining resources and the development of industries [1]. Phosphate ores occupy a prominent place in this sense, since the global natural phosphates (NP) market was estimated at 24.4 billion USD in 2018 and is expected to grow with a rate of 7.1% from 2019 to 2025 [2]. This growth is due to the ever-increasing demand for phosphate fertilizers by the agricultural sector to meet the food needs of the world population and also to the increasing market for food additives made from phosphates. The most important deposits of phosphate ores in the world are in Morocco, and its reserve is estimated at two thirds to three quarters of the world reserves [3], which gives this country a strategic role in the international trade of these ores and derivatives. Phosphate deposit origins and the minerals they contain are very varied [4]. Generally, there are two types of phosphate deposits: magmatic deposits and sedimentary deposits [5, 6]. The magmatic deposits are of volcanic origin. They are less abundant and, therefore, less important from an economic point of view. By contrast, the sedimentary deposits are of marine origin. They are formed over the time from the rest of skeletons of marine animals with the basic francolite structure (carbonate fluorapatite (FA)). It should be noted that c. 65 wt% of the composition of animal skeletons are made up of nanocrystalline compounds of apatitic structure (hard materials), and the rest is attributed to organic substances (soft materials) and water. Oolites, pisoliths, nodules, rosins, and corpoliths are the main forms of sedimentary phosphate grains. The mineralogical composition of NP is quite diverse; more than 200 species have been recorded [7–9]. These phosphates are generally in the form of francolite, modified by partial isomorphic substitutions [10]. Effort has been done in order to approximate the real structure of phosphate rocks [11–15]. The mineralogical and chemical study has allowed establishing the structural formula of francolite: $Ca_{10-a-b}Na_aMg_b(PO_4)_{6-x-y}(CO_3)_x(SO_4)_yF_{2+x+y-b}$ [16–18]. This ore has a pivotal role in several industrial activities such as its transformation into commodity products (phosphoric acid, sodium hexametaphosphate, monoammonium or diammonium phosphate, dipotassium phosphate, calcium phosphate, etc.) or into fertilizers. In addition, phosphoric acid and ammonium phosphate salts from the transformation of rock phosphates are the main building blocks for the synthesis of pure hydroxyapatite (HA). In parallel, applications of phosphate rocks in other fields have also been explored [19–21], including the use of phosphate rocks as efficient solid catalysts for many chemical reactions. The present chapter discusses the main achievement on the application of NP-based catalysts in organic chemistry.

14.2 Preparation and Characterization of Catalysts or Catalyst Supports from NP

The main phosphate rocks are associated with a wide variety of accessory minerals and impurities that cover almost all the elements of the periodic table [22–24]. Usually in a phosphate ore, one can distinguish between phosphate material and gangue [8]. The gangue consists of minerals (e.g. quartz, dolomite, calcite, sulfates, clays, and organic matter), which have low economic value, associated with the phosphate ores. From a crystallochemical point of view, these rocks mainly contain compounds of francolite structure (carbonate FA), but other minerals can be also present such as brushite ($CaHPO_4 \cdot 2H_2O$), monetite ($CaHPO_4$), whitlockite (β-$Ca_3(PO_4)_2$), crandallite ($CaAl_3(PO_4)_2(OH)_5 \cdot H_2O$), wavellite ($Al_3(OH)_3(PO_4)_2 \cdot 5H_2O$), taranakite ($K_2Al_6(PO_4)_6(OH)_2 \cdot 18H_2O$), millisite (Na, K)$CaAl_6(PO_4)_4(OH)_9 \cdot 3H_2O$), variscite ($AlPO_4 \cdot 2H_2O$), collinsite ($Ca_2Mg(PO_4)_2(H_2O)_2$), β-TCP ($Ca_3(PO_4)_2$), and strengite ($FePO_4 \cdot 2H_2O$) [25]. Also a given phosphate rock of the same deposit may contain apatites with different properties due to geological conditions and alterations after deposition.

Prior to its use as a catalyst or catalyst support, NP undergoes a series of treatments to increase its phosphate content ($P_2O_5\%$) and to eliminate impurities as much as possible. Sebti et al. developed a treatment process based on different steps of dry beneficiation, wet particle size separation, washing, drying, sieving, and calcination. The scheme of this process is presented in Figure 14.1, and a short commentary was also reported [26]. Wet particle size separation allows the recovery of the size fraction between 400 and 100 μm, which is rich in phosphate. This fraction is dried and then calcined at high temperature (700 °C) in order to reduce organic matter and carbonate levels. This sample is washed to remove the lime and magnesia formed during calcination. Then, the sample is crushed and sieved; only the fraction between 125 and 63 μm in size is recovered for a final heat treatment.

Most of the sedimentary phosphate minerals are part of the apatite family, which possesses a hexagonal crystal system (Figure 14.2) [27], and belongs to the P63/m space group [28]. The unit cell parameters reported in the literature are quite variable, but the typical values can be assumed as follows: $a = 9.3684$ Å and $c = 6.8841$ Å,

Figure 14.1 Process for the phosphate rock treatment developed by Solhy's group. Source: Abderrahim Solhy.

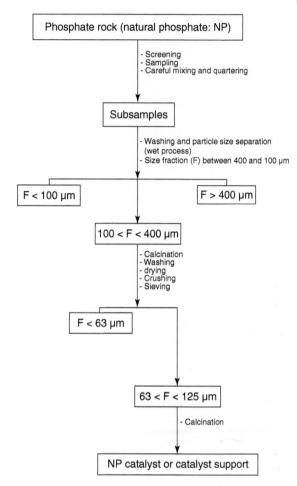

giving a unit cell volume of 523.25 Å^3 and a calculated density of 3.20 g cm^{-3}. The latter is in good agreement with the measured density of 3.15 g cm^{-3} [26].

Fourier transform infrared spectroscope (FTIR) analysis of the NP shows mainly the absorption bands, which can be assigned to the PO$_4^{3-}$ groups of the apatite crystalline structure [29]. The PO$_4^{3-}$ ion absorption bands are characterized by two absorption ranges between 1100 and 900 cm^{-1} and 600 and 500 cm^{-1}. The bands in the first domain correspond to the symmetrical and antisymmetrical vibrations of the P—O bond, while those in the second domain are assigned to bend vibrations of the O—P—O bond. In addition, the characteristic absorption bands at the vibration frequencies of OH$^-$ groups are located at 3500 and 1650 cm^{-1}. Low intensity bands at 1455, 1430, and 875 cm^{-1} are characteristic for the vibration of CO$_3^{2-}$ ions (and eventually silicate groups).

Raman spectroscopy of NP, reported in the literature, reveals the presence of 4 vibration modes relating to the P—O bonds of the PO$_4$ tetrahedra [30, 31]. These vibrations are observed at 400–450, 550–660, 950–1000, and 1020–1150 cm^{-1} on

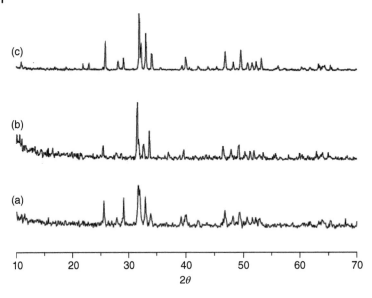

Figure 14.2 X-ray patterns of (a) raw phosphate (phosphate rock), (b) treated natural phosphate, and (c) synthetic fluorapatite. Source: [27]. Reproduced with permission of Elsevier.

Table 14.1 The main bands related to phosphates observed in Raman spectra of NP.

Bond	δ_s (P–O)	δ_{as} (P–O)	v_s (P–O)	v_{as} (P–O)	δ (C–O)
Bands (cm^{-1})	400–450	550–660	950–1000	1020–1150	1080

the most of the phosphate's Raman spectra (Table 14.1). These vibrations are relative to strong v_1 symmetric stretching mode of PO$_4^{3-}$ ions at 950–1000 cm^{-1}. The bands in the 1020–1150 cm^{-1} region result from antisymmetric v_3 vibrations. The band relative to the elongation vibrations of the carbonate group is also recorded around 1080 cm^{-1}. Other features shared by all spectra include weak bands in the 400–450 cm^{-1} region, which correspond to v_2 bending mode, and in the 550–660 cm^{-1} region, related to v_4 bending mode. Weak bands at less than 300 cm^{-1} are lattice modes.

NP have a multiple chemical composition according to their origins. Generally, chemical analyses or X-ray fluorescence (XRF) reveals that apart from phosphate expressed in bone phosphate of lime (BPL) or P$_2$O$_5$ (P$_2$O$_5$ (%) = 0.4576 × BPL), other oxides (CO$_2$, K$_2$O, CaO, MgO, SiO$_2$, SO$_3$, Na$_2$O, Fe$_2$O$_3$, Al$_2$O$_3$) and other trace elements (F, Cr, Cu, Sr, Zn, and V) can also be found [32–34]. The determination of the CaO/P$_2$O$_5$ molar ratio gives an idea on the quality of the phosphate rock, since calcium can come from apatite compounds but also from dolomite and calcite. The specific surface area of NP calculated by Brunauer–Emmett–Teller (BET) method is generally small, and does not exceed 1.5 m^2 g^{-1} with negligible pore volume [26].

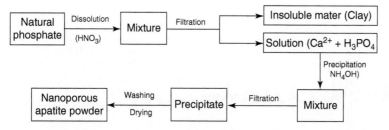

Figure 14.3 Preparation of nanostructured apatite by dissolution–precipitation process of NP. Source: [37] / With permission of Elsevier.

Scanning electron microscopy (SEM) analysis of an NP reveals that it consists of large, irregular, near-micron-sized particles that are not homogeneous in size and shape, and their surfaces have a crystalline envelope [10, 35]. The grains constituting the surface of the ore are in the form of large agglomerates, which penalizes the porosity of the material and subsequently confirms the low surface area of NP.

NP contain both acid and base sites on their surface, which can catalyze chemical reactions. The density of acid or basic sites can be determined by using, respectively, a basic or acidic molecule. Fraile et al. quantified the density of basic sites of an NP by means of phenol adsorption in cyclohexane [36]. Gas chromatography (GC) was used to follow the evolution of phenol adsorption on NP surface. The results obtained show that the NP have a relatively important basic character (0.54 mmol g^{-1}), despite its low specific surface area.

Finally, nanoporous HA could be synthesized from NP as starting materials, providing calcium and phosphate, via the dissolution–precipitation process [37–40]. El Asri et al. used an NP from Morocco to control HA synthesis according to the procedure shown in Figure 14.3 [37]. The authors studied the influence of temperature, maturation time, and the nature of the solvent on HA crystal growth. An NP fraction, with a size between 100 and 400 μm, is dissolved in nitric acid to obtain solvated Ca^{2+} and H_3PO_4. After filtration, the remaining solution is neutralized by NH_4OH (25%), leading to HA precipitation. The pH value of the precipitation must be maintained above 10 to avoid the formation of other types of calcium phosphates. The recovered and washed powder was dried at 100 °C. Two HA samples were prepared at room temperature under stirring for 24 hours using water (MNP1) and an ethanol–water mixture (MNP2) as solvent. A third sample (MNP3) was prepared in water at 75 °C under reflux for 3 hours.

X-ray diffraction (XRD) characterization showed a low crystallinity with broad diffraction peaks of the synthesized HA powders. All the diffraction peaks match well with the data reported for HA, with the most intense peak at 26° corresponding to the (002) plane and the series of three peaks at 31°, 32°, and 34° attributed to the (211), (112), and (300) plans (Figure 14.4). Elemental analysis showed that in all cases, the Ca/P molar ratio was higher than 1.67. The nitrogen adsorption/desorption isotherm analysis evidenced the type IV isotherm, according to the IUPAC classification, which proves the mesoporous structuration of these

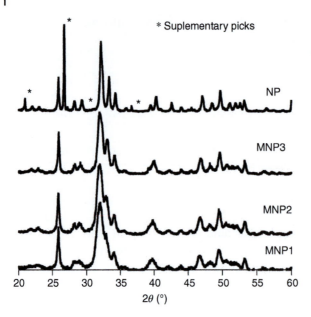

Figure 14.4 X-ray diffraction patterns of the as-prepared apatites from natural phosphate. "*" Indicates non-apatite peaks. Source: [37]. Reproduced with permission of Elsevier.

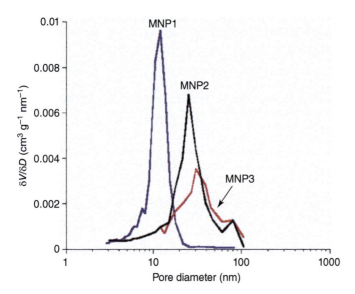

Figure 14.5 Pore size distribution ($\delta V/\delta D$) of HA synthesized from an NP. Source: [37]. Reproduced with permission of Elsevier.

materials. The sample prepared in water at room temperature (MNP1) exhibited the highest specific surface area (150 m² g⁻¹) with smallest average pore size (c. 10 nm). MNP2 synthesized in water/methanol mixture at room temperature had the specific surface area equal to 101 m² g⁻¹ with the average pore size around 20 nm, while MNP3 synthesized in water at 75 °C had the lowest specific surface area of 96 m² g⁻¹ and the largest average pore size (Figure 14.5).

14.3 Organic Synthesis Using NP and NP-Supported Catalysts

14.3.1 Condensation Reactions

14.3.1.1 Knoevenagel Reaction

Emil Knoevenagel reported in 1894 the condensation of diethyl malonate with formaldehyde, catalyzed by diethylamine to obtain tetraethyl propane-1,1,3,3-tetracarboxyalte [41]; since then, the interest in this condensation is growing steadily. It is one of the most important methods in the synthesis of functionalized alkenes [42]. It has been used to obtain a variety of compounds, such as synthetic natural products [43], drugs [44], fine chemicals [45], functional polymer materials [46], antihypertensive [47], etc. This condensation can be catalyzed by solid bases [48, 49]. Thus, Sebti et al. [50] have used NP as catalyst to synthetize a series of α,β-unsaturated carbonyl compounds such as malononitrile and methyl cyanoacetate, by the condensation of activated methylene with aromatic aldehydes (Scheme 14.1). In the absence of a solvent and with high catalyst contents, the yield reached 62 and 82% after 3 and 9 minutes of reaction, respectively. In the presence of methanol as solvent (see Scheme 14.1), the yield could reach up to 94%. The reaction between furfural and malononitrile exhibited the highest reaction rate and could be achieved after two minutes of reaction. On the other hand, the reaction between benzaldehyde and methyl cyanoacetate required more time (30 minutes) to achieve 80% yield.

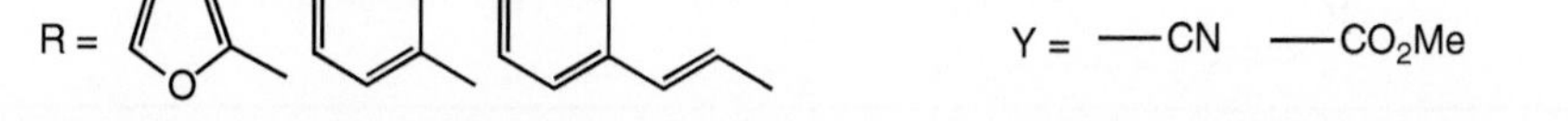

Scheme 14.1 Synthesis of alkenes via Knoevenagel condensation catalyzed by NP. Source: [50]. Reproduced with permission of Elsevier.

Under other conditions (Table 14.2), Sebti et al. [51] have realized this condensation by using (i) NP alone, (ii) NP impregnated by potassium fluoride, and (iii) NP modified by sodium nitrate. The reaction was carried out at room temperature in methanol. The results are shown in Table 14.2. The studied solid materials easily catalyze this condensation under the adopted reaction conditions, with high yield up to 98% in few minutes of reaction. A difference in the reaction kinetic is observed, which is mainly correlated to the nature's effect of the active methylene compounds (—CN > —CO$_2$Me > —CO$_2$Et). NP doped by potassium fluoride or sodium nitrate enhanced the yield and reduce the reaction time, in comparison with bare NP.

Table 14.2 Synthesis of alkenes via the Knoevenagel condensation by using NP, KF/NP, and $NaNO_3$/NP catalysts.

$$R-CHO + \underset{Y}{\overset{CN}{\diagdown\!\!\!\diagup}} \xrightarrow[\text{MeOH (1 ml) / r.t.}]{\text{Catal. (100 mg)}} R-CH=\underset{Y}{\overset{CN}{\diagdown\!\!\!\diagup}}$$

			Yields (%) (time (min))[a]		
Entry	R	Y	NP	KF/NP	$NaNO_3$/NP
1	Ph	—CN	98 (30)	90 (10)	96 (1)
2	2-furyl	—CN	88 (3)	91 (3)	98 (3)
3	Ph-CH=CH-	—CN	88 (1)	98 (1)	98 (1)
4	Ph	—CO_2Me	89 (60)	96 (60)	96 (30)
5	2-furyl	—CO_2Me	86 (5)	98 (1)	98 (1)
6	Ph-CH=CH-	—CO_2Me	84 (10)	80 (3)	85 (1)
7	Ph	—CO_2Et	21 (60)	68 (60)	94 (60)
8	2-furyl	—CO_2Et	84 (30)	93 (1)	96 (1)
9	Ph-CH=CH-	—CO_2Et	80 (30)	90 (10)	98 (10)

a) Yields in pure products isolated by distillation under vacuum, recrystallized with CCl_4, and identified by ^{1}H-NMR and IR spectroscopy.
Source: [51]. Reproduced with permission of Elsevier.

Another example consists of using of a micronized natural phosphate (NPm) for catalyzing the Knoevenagel condensation between aromatic aldehydes and active methylene compounds in methanol at room temperature [52]. The authors have used a Syrian NP as catalyst, which has a francolite structure. An air jet micronic pulverization process was used to obtain the fine NPm powder, which had higher specific surface area (28.7 m² g⁻¹) compared with the initial NP (15.2 m² g⁻¹). The comparison of the catalytic activity of NP and NPm in the Knoevenagel condensation of the aldehydes and active methylene compounds has clearly shown that NPm is more efficient than NP, explained by the effect of the specific surface area. A series

Table 14.3 Catalysis of Knoevenagel condensation by NP and micronized NP.

R—CHO + NC—CH₂—Y →[NP or NPm (100 mg), MeOH (1 ml) / r.t.] R—CH=C(Y)(CN)

Entry	R	Y	Yields (%) (time (min))[a] NP	NPm
1	Ph	—CN	69 (5)	93 (5)
2	Ph	—CO$_2$Me	68 (20)	89 (20)
3	Ph	—CO$_2$Et	57 (60)	96 (60)
4	Ph-CH=CH—	—CN	81 (1)	90 (1)
5	Ph-CH=CH—	—CO$_2$Me	59 (2)	80 (2)
6	Ph-CH=CH—	—CO$_2$Et	82 (10)	94 (10)
7	furyl	—CN	85 (1)	96 (1)
8	furyl	—CO$_2$Me	62 (2)	91 (2)
9	furyl	—CO$_2$Et	80 (10)	92 (10)

a) Yields in pure products isolated by distillation under vacuum and identified by ^{1}H-NMR and IR spectroscopy.
Source: [52] / Reproduced with permission of Springer Nature.

of unsaturated multifunctional compounds were obtained with excellent yields up to 96% (Table 14.3).

The synthesis of α,β-unsaturated arylsulfones was achieved by using an NP and NP-supported KF as catalysts [53]. NP catalyst was extracted from Khouribga region (Morocco) and underwent a series of pretreatment before use: attrition, sifting, calcinations (900 °C), washing, and recalcination. Its composition is as follows (wt%): P_2O_5 (34.24%), CaO (54.12%), F_2 (3.37%), SiO_2 (2.42%), SO_3 (2.21%), CO_2 (1.13%), Na_2O (0.92%), MgO (0.68%), Al_2O_3 (0.46%), Fe_2O_3 (0.36%), K_2O (0.04%), and trace elements at ppm level (Zn, Cu, Cd, V, U, and Cr). KF/NP was simply

Table 14.4 Synthesis of α,β-unsaturated arylsulfones via Knoevenagel condensation catalyzed by NP and KF/NP catalysts.

			Yields (%) (time (h))[a]	
Entry	R	Solvent (ml)	NP	KF/NP
1	H	MeOH (1)	56 (5)	64 (1)
2	H	EtOH (1)	58 (5)	60 (1)
3	H	MeOH/H_2O (0.9/0.1)	60 (5)	68 (1)
4	H	EtOH/H_2O (0.9/0.1)	68 (5)	77 (1)
5	H	EtOH (1)[b]	75 (5)	91 (1)
6	NO_2	EtOH (1)	72 (5)	82 (1)
7	NO_2	EtOH/H_2O (0.9/0.1)	87 (5)	95 (1)
8	NO_2	EtOH (1)[b]	94 (4)	97 (0.6)
9	Cl	EtOH (1)	66 (5)	75 (1)
10	Cl	EtOH/H_2O (0.9/0.1)	80 (5)	88 (1)
11	Cl	EtOH (1)[b]	93 (3)	97 (1)
12	Me	EtOH (1)	50 (5)	60 (1)
13	Me	EtOH/H_2O (0.9/0.1)	64 (5)	76 (1)
14	Me	EtOH (1)[b]	50 (5)	86 (1)
15	OMe	EtOH (1)	46 (5)	58 (1)
16	OMe	EtOH/H_2O (0.9/0.1)	60 (5)	70 (1)
17	OMe	EtOH (1)[b]	70 (5)	92 (1)

a) Yields in pure products isolated by distillation under vacuum and identified by ^{1}H-NMR and IR spectroscopy.
b) 50 mg of BTEAC under r.t.
Source: [53]. Reproduced with permission of Elsevier.

obtained by impregnation method. The condensation reaction was carried out in methanol, ethanol, or ethanol/water at room temperature. Table 14.4 evidences that NP was active in all the studied reaction and the yield could reach up to 94%. Moreover, recycling tests show that NP catalyst could be regenerated by simple filtration and calcination at 700 °C for 15 minutes. For each reaction, the impregnated catalyst (KF/NP) was found to be slightly more active than the bare NP. The addition of water to the reaction medium could accelerate the reaction kinetic at the ratio of ethanol/water = 90/10, which is supposed as the result of an interaction between water and the catalyst. However, above the ratio, a thin film can be formed around the catalyst surface, preventing the contact of the reagents with the catalyst, and so decreases the yield. Moreover, the impact of a quaternary ammonium salt (phase-transfer agent) such as benzyltriethylammonium chloride (BTEAC) in presence of each catalyst was also studied. BTEAC has no catalytic activity in the reaction, and only plays the role of phase-transfer agent. The addition of the phase-transfer agent had strong impact on the kinetic and the reaction yield with both NP and KF/NP catalysts. For example, by using benzaldehyde as reagent, the yield increased from 58 to 75% for 5 hours of reaction in the case of NP and from 60 to 91% for 1 hour of reaction in the case of KF/NP.

14.3.1.2 Claisen–Schmidt Condensation

Chalcones are widely described in the literature for its multiple pharmacological potentials, e.g. anti-infections [54], antiviral [55], antifungal [56], antioxidants [57], or antiinflammatory [58]. Natural chalcones such as retrochalcones (licochalcones), which can be isolated from *Glycyrrhiza inflata*, showed potent antibacterial activity especially to *Bacillus subtilis*, *Staphylococcus aureus*, and *Micrococcus luteus* [59]. The Claisen–Schmidt condensation is the simplest and the most commonly used method for chalcones synthesis [60]. It is an aldol reaction, between an acetophenone and a benzaldehyde, which can be catalyzed by a base or an acid, in a polar solvent. Due to the variety of acetophenones and benzaldehydes available commercially, a wide range of substituted chalcones can be obtained by this condensation. For instance, the Claisen–Schmidt condensation between aromatic aldehydes and acetophenone (equimolar mixture: 2.5 mmol) was studied in the absence of a solvent, using an NP as catalyst (2.5 g) [61]. Small amounts of water (0.25 ml) and also BTEAC as phase-transfer agent (0.06 g) were used in order to boost the catalytic performance of NP. Under the employed conditions, the reaction yield was more or less important even after a long reaction time of 24 hours (Table 14.5). In most of cases, water addition resulted in a positive effect, especially for NO_2 substituted aldehyde, which has a mesomeric effect (-M). Also the use of the phase-transfer agent had a high impact on the reaction yield. Finally, the combination of both water and BTEAC led to excellent yields for short reaction time (entries 8 and 12, Table 14.5).

In another work, the synthesis of chalcone [(*E*)-1,3-diphenylprop-2-en-1-one] via the condensation of benzaldehyde and acetophenone was studied by using NP as catalyst in various solvents such as methanol, ethanol, propan-1-ol, propan-2-ol, dioxane, dimethylformamide (DMF), acetonitrile, and tetrahydrofurane (THF) [61]. In most of cases, the yield was insignificant, except in the case of the use of methanol with 75% yield after 24 hours reaction time. However, when small amount of water was added to the reaction medium, the yield strongly increased for all the solvents used, in particular for alcohol solvents (96% after 24 hours reaction time with ethanol/water).

Sebti's group in collaboration with Mayoral's group have reported an efficient and convenient route for the synthesis of several chalcones by using $NaNO_3$-modified NP ($NaNO_3$/NP) as catalyst [62]. This material was prepared by adding NP to an aqueous solution of sodium nitrate under stirring at room temperature, followed

Table 14.5 Claisen–Schmidt condensation catalyzed by an NP catalyst under solvent-free conditions.

Entry	R^1	R^2	Water (ml)	BTEAC (g)	Yields (%) (time (h))[a]
1	H	H	0	0	10 (24)
2	H	H	0.25	0	12 (24)
3	H	H	0	0.06	60 (24)
4	H	H	0.25	0.06	77 (24)
5	H	Cl	0	0	0 (24)
6	H	Cl	0.25	0	21 (24)
7	H	Cl	0	0.06	91 (4)
8	H	Cl	0.025	0.06	91 (1)
9	NO_2	H	0	0	28 (24)
10	NO_2	H	0.25	0	75 (24)
11	NO_2	H	0	0.06	79 (24)
12	NO_2	H	0.25	0.06	92 (0.5)

a) Yields in pure chalcones isolated by distillation under vacuum and identified by ^{1}H-NMR and IR spectroscopy.

Source: [61]. Reproduced with permission of Elsevier.

by evaporation, drying, and calcination at 300, 500, or 900 °C. The condensation reaction was performed with an equimolar reaction mixture (2.5 mmol) of arylaldehydes and para-substituted acetophenone, in methanol as solvent (1–3 ml), with 0.1 g of NaNO$_3$/NP catalyst. Bare NP showed low yield (0–10%) after 24 hours reaction time. The impregnated catalyst allowed strongly increasing the reaction yield above 90% in most of cases (Table 14.6). The impact of the calcination temperature was also highlighted. The impregnation catalyst calcined at 900 °C led to the formation of new phases such as sodium phosphates and calcium oxide, which might be contributed to its high catalytic activity. On the other hand, the catalyst calcined at 300 and 500 °C, without these new phases, had lower catalytic activity.

The effect of the nature of catalyst additives, conditions of catalyst preparation, solvent nature, water addition, and phase-transfer agent was also studied in the synthesis of chalcones via the Claisen–Schmidt condensation NP-based catalysts [63]. The different X/NP catalysts (X = NaNO$_3$, NaCl, Na$_2$SO$_4$, or Na$_2$CO$_3$) were prepared by the impregnation of a NP with a solution of additive, followed by evaporation and calcination different temperatures for two hours. The condensation reaction

Table 14.6 Catalysis of Claisen–Schmidt condensation by sodium-modified NP in methanol.

Entry	Chalcones			Yield (%) (time (h)) [a]
	R^1	R^2	R^3	
1	H	H	H	98 (18)
2	Cl	H	H	94 (16)
3	H	NO_2	H	94 (16)
4	OCH_3	H	H	91 (36)
5	H	H	OCH_3	90 (24)
6	Cl	H	OCH_3	93 (48)
7	H	NO_2	OCH_3	81 (48)
8	OCH_3	H	OCH_3	70 (48)
9	H	H	NO_2	92 (16)
10	Cl	H	NO_2	94 (16)
11	H	NO_2	NO_2	86 (16)
12	OCH_3	H	NO_2	93 (16)

a) Yields in pure chalcones isolated by distillation under vacuum and identified by ^{1}H, ^{13}C NMR, IR, and MS.
Source: [62]. Reproduced with permission of Elsevier.

was performed with an equimolar mixture of aldehyde and acetophenone (2.5 mmol each) in methanol (1–3 ml) in the presence of 0.1 g of catalyst at room temperature under stirring. In the case of NaNO$_3$/NP, the calcination at 900 °C led to a modification of the crystallographic structure and the appearance of new crystallographic phases such as γ-Na$_3$PO$_4$, β-Na$_3$PO$_4$, and Na$_2$HPO$_4$. This resulted in a strongly basic catalyst, which has remarkable catalytic performance (up to 98% yield). On the other hand, the calcination at lower temperature (150, 300, or 500 °C) led to insignificant yield (2–4%) under the same conditions of the condensation reaction. On the other hand, the catalysts modified by other additives (NaCl, Na$_2$SO$_4$, or Na$_2$CO$_3$) had no catalytic activity under the same reaction conditions. The reaction rate was sensitive to the nature of reagents substituent groups (electron-attracting or electron-donating substituents), especially in the case of the use of *p*-substituted acetophenone. The solvent nature had also a pronounced effect in these reactions. The reaction did not take place in dichloromethane, THF, hexane, and butanol, while it took place in alcohols. For instance, the reaction yield reached 12% in isopropanol, 63% in ethanol, and 90% in methanol after 24 hours. The addition of water to methanol had a negative effect, while a positive effect was observed with ethanol. Finally, the addition of BTEAC as phase-transfer agent to the reaction mixture clearly improved the reaction yield, reaching more than 90% after 10 hours of reaction. By the same way, $LiNO_3$/NP (1/15 w/w) catalysts were also found to be strongly active in the Claisen–Schmidt condensation [64]. The addition of $LiNO_3$ onto the NP surface significantly enhanced the catalytic activity of the latter.

Macquarrie et al. reported the synthesis of 2′-hydroxychalcones and flavanones via Claisen–Schmidt condensation between substituted 2′-hydroxyacetophenone derivatives and arylaldehydes [65]. It should be pointed out the importance of flavanone derivatives from a pharmacological point of view since these compounds have various biological activities [66–69]. The condensation reaction was catalyzed by an NP-supported KF catalyst (KF/NP, 1/8 w/w ratio). For each condensation reaction, 10 mmol of acetophenone and 5 mmol of aldehyde were added to 0.6 g of n-tetradecane and heated to 180 °C (Scheme 14.2). At this temperature, 1 g of catalyst was added to start the reaction. After 55 minutes, 5 mmol of aldehyde was again added to the reaction mixture, and the reaction was kept for two hour of total reaction time. As exhibited in this paper, high acetophenone conversion and high chalcone and flavanone yields could be obtained, demonstrating the efficiency of this catalyst system.

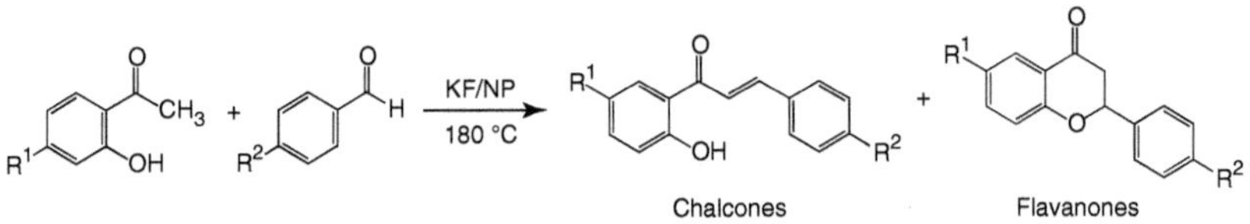

Scheme 14.2 Catalysis of Claisen–Schmidt condensation for synthesis of 2′-hydroxychalcones and flavanones. Source: Abderrahim Solhy.

14.3.1.3 Michael Addition

The Michael addition reaction has important place in organic synthesis [70, 71]. It has been applied in numerous C—X (X = C, O, N, S) bond-forming reactions to produce high added value compounds by using a basic catalyst [72]. The mechanism of the Michael addition for carbon–carbon bond-forming can be divided in three elemental steps: (i) the first step is the formation of the enolate by deprotonation of the pronucleophile by the base; (ii) the second step is the key carbon–carbon bond formation where the enolate reacts with the Michael acceptor via conjugate addition; and (iii) the last step consists in the reprotonation of the newly formed enolate with catalyst regeneration [73]. Sebti et al. achieved the Michael addition of nitro-alkanes on α,β-unsaturated carbonyl compounds in ethanol by using KF/NP as catalyst [74]. In most of cases, high yields (91–98%) could be reached at room temperature for a short reaction time of 2–4 hours.

NP coming from Khouribga region (Morocco) and KF/NP catalysts were also applied in the Michael addition of mercaptans on substituted chalcone derivatives [75]. For the reaction, 1.5 ml methanol, an equimolar mixture of the reagents (1 mmol each) and 100 mg of catalyst were added to reactor kept at room temperature. Both NP and KF/NF catalysts were highly efficient and selective (Table 14.7). The addition of KF mostly reduced the reaction time to reach the similar yield

Table 14.7 Michael addition for carbon–sulfur forming-bond catalyzed by using NP or KF/NP catalysts.

			Yield (%) (time (min)) [a]	
Entry	R^1	R^2	NP	KF/NP
1	H	Ph	95 (15)	96 (5)
2	H	2′-NH_2Ph	97 (10)	96 (2)
3	H	CH_2—CO_2Et	56 (45)	95 (5)
4	m-NO_2	Ph	92 (10)	93 (2)
5	m-NO_2	2′-NH_2Ph	94 (10)	92 (2)
6	m-NO_2	CH_2—CO_2Et	96 (20)	95 (5)
7	p-Cl	Ph	96 (15)	93 (5)
8	p-Cl	2′-NH_2Ph	96 (10)	97 (2)
9	p-Cl	CH_2—CO_2Et	86 (20)	97 (2)

a) Yields in pure products isolated by recrystallization with AcOEt/CH_2Cl_2 and identified by ^{1}H, ^{13}C NMR, and IR spectroscopy.
Source: [75]. Reproduced with permission of Elsevier.

in comparison with the bare NP catalyst. No undesirable products resulting from 1,2-addition, polymerization, or bis-addition were identified, and the catalysts were easily regenerated by a thermal treatment. These catalysts (NP, KF/NP) were also found to be active in the Michael addition for the synthesis of compounds containing C—N bond (Table 14.8) [76] and compounds containing C—C bond (1,4-addition of salicylaldehyde on malononitrile, leading to the formation of 4H-chromene) [77].

Lithium-doped NP (Li/NP) catalysts were also investigated in the Michael addition reaction [78]. The catalysts were synthesized by simple impregnation of NP with an aqueous solution of lithium nitrate, followed by evaporation, drying, and calcination at 900 °C. In the Michael-type 1,4-addition reaction of various nucleophiles (aromatic amines, thiols, and active methylene compounds) with substituted chalcone derivatives, Li/NP was more efficient than bare NP, because of the basicity due to Li addition (formation of Li_2O on the surface which increases the pH of the reaction medium) [78]. Various products resulting from the formation of carbon–carbon, carbon–nitrogen, and carbon–sulfur bonds have been obtained with high reaction yields, which depend on the reaction conditions and the substitution groups (electron-withdrawing or electron-donating groups) on chalcone

Table 14.8 Carbon–nitrogen bond formation via Michael addition catalyzed by NP or KF/NP catalysts.

			Yield (%) (time (h)) [a]	
Entry	R¹	R²	NP	KF/NP
1	H	Ph	98 (8)	95 (1)
2	Cl	Ph	71 (14)	95 (5)
3	OMe	Ph	93 (14)	95 (3.5)
4	Me	Ph	86 (14)	94 (4)
5	H	p-MeO—Ph	90 (10)	92 (1)
6	Cl	p-MeO—Ph	95 (19)	94 (5.5)
7	OMe	p-MeO—Ph	95 (19)	90 (5.5)
8	Me	p-MeO—Ph	87 (16)	93 (5)
9	H	Ph—CH$_2$	64 (24)	90 (24)
10	Cl	Ph—CH$_2$	75 (24)	92 (5)
11	OMe	Ph—CH$_2$	23 (24)	75 (24)
12	Me	Ph—CH$_2$	55 (24)	91 (24)

a) Yields in pure products isolated by chromatography and recrystallized in a mixture of n-hexane/ethyl acetate. The products were identified by ^{1}H, ^{13}C NMR, FTIR, and melting points.
Source: [76]. Reproduced with permission of Elsevier.

(Michael acceptor) and the nature of the nucleophile Michael donor [78]. No secondary reactions were observed except in a few cases where there were traces of polymerization products. The same research group studied the impact of NP milling by an air jet micronic pulverization process on its catalytic activity toward Michael 1,4-addition reaction [79]. This milling step allowed increasing the specific surface area of the NP catalyst and so improving its efficiency in the studied reaction.

14.3.2 Transesterification Reaction

The transesterification reaction is an organic reaction between an ester and an alcohol [80]. Two other types of transesterification, between an ester and a carboxylic acid or between an ester and another ester, are also possible [81]. The economic challenges involved transesterification focus more on the reaction between an ester and an alcohol [82]. This reaction is mainly used in two major industrial processes: the production of polyesters and the production of biodiesel [83–86]. Other applications exist particularly in the field of olfactory molecules' synthesis (linear esters and lactones often with a fruity smell). Transesterification reaction preserves the number and type of bonds between reagents and final products; it is a thermal transformation of which the reaction entropy, positive or negative, is not high. But it is a kinetically difficult reaction, since alcohols are poor nucleophiles and esters are poor electrophiles. The catalysis of this reaction can be acidic, basic, or enzymatic [87]. The acid catalysis of this reaction leads to the formation of undesirable carboxylic acid products, while enzyme catalysis is too expensive. The literature was mainly interested in the possibility of catalyzing this reaction with basic or acidic solid catalysts [88–90]. This field thus requires major research efforts in developing efficient solid catalysts. Bazi et al. investigated the transesterification reaction of methylbenzoate with various alcohols catalyzed by NP (Scheme 14.3) [91]. Different parameters of this reaction were examined, such as alcohol/ester molar ratio, solvent's effect, and catalyst loading, in order to optimize this transformation. The authors studied this reaction with toluene solvent or without solvent. The nature of alcohol is decisive for this catalytic process by NP [91]. Excellent yields were obtained, except where the authors used tert-butanol or 2-propanol in the presence or absence of toluene (no reaction). In the case of cyclohexanol, there was no reaction in the presence of toluene. In solvent-free conditions, the yield reached 94%. Three successive cycles of regeneration were performed. The NP catalyst could be easily regenerated by calcination at 700 °C. However, no mechanism study was conducted.

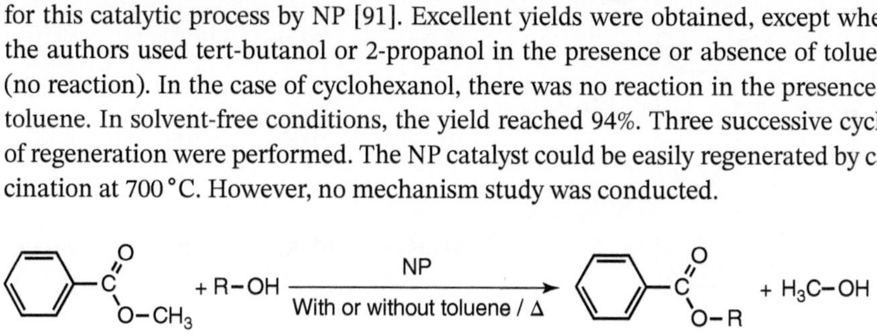

Scheme 14.3 Transesterification of methylbenzoate with various alcohols over natural phosphate. Source: Abderrahim Solhy.

In another study, Bazi et al. used several catalysts, namely, bare NP, KF, or $ZnCl_2$ supported on NP and NP modified by sodium doping [92]. The authors studied several parameters of the transesterification of methylbenzoate with butanol, chosen as a model reaction. In all cases, the yields obtained were high (above 90%) regardless of the catalyst. On the other hand, when ethanol was used, the yields obtained were 57% (at 48 hours of reaction), 72% (at 8 hours of reaction), 41% (at 8 hours of reaction), and 79% (at 12 hours of reaction), for NP, Na/NP, $ZnCl_2$/NP, and KF/NP, respectively (Table 14.9). The regeneration study of Na/NP and $ZnCl_2$/NP catalysts showed deactivation during catalyst cycling, without giving explanations or verification of the possibility of leaching especially for $ZnCl_2$ supported on NP. A heat treatment at 900 °C of Na/NP proved to be crucial for Na/NP to recover its catalytic performance.

14.3.3 Friedel–Crafts Alkylation

The Friedel–Crafts process remains the method of choice for the alkylation of aromatic compounds via the introduction of an alkyl group to arenes or heteroarenes [93]. The classic methodology for Friedel–Crafts alkylation was catalyzed by homogeneous catalysts (Lewis acids or strong mineral acids) [94–97]. But the

Table 14.9 Transesterification of methylbenzoate with various alcohols over NP-based catalysts.

		Ester conversion (%) (time (h)) [a]			
Entry	Products	NP	Na/NP	$ZnCl_2$/NP	KF/NP
1	O-C_2H_5 benzoate	57 (48)	72 (8)	41 (8)	97 (12)
2	O-$(CH_2)_2CH_3$ benzoate	74 (48)	76 (8)	75 (8)	80 (12)
3	O-C_4H_9 benzoate	100 (48)	97 (8)	99 (8)	100 (12)
4	O-C_6H_{13} benzoate	97 (48)	97 (8)	98 (8)	99 (12)
5	O-C_8H_{17} benzoate	98 (48)	99 (8)	99 (8)	99 (12)
6	benzyl benzoate	100 (48)	98 (8)	97 (8)	98 (12)

a) Yields in pure products isolated by filtration followed by evaporation under vacuum and purified by column chromatography in a mixture of n-hexane/ethylacetate. The products were identified by ^{1}H NMR.

Source: [92]. Reproduced with permission of Elsevier.

use of solid acid catalysts is the best way to overcome the problems related to homogeneous catalysis, e.g. catalyst separation [98–100]. Sebti's group investigated the Friedel–Crafts alkylation of benzene and toluene with benzyl chloride over NP-based catalysts (Table 14.10) [101]. NP was able to catalyze this reaction with high selectivity into monoalkyl compounds, due to the acidic sites of the catalyst. It is noteworthy that the reactivity of the aromatic nucleus increases with the number of electron donor groups. The impregnation of 1 g of NP with $ZnCl_2$ (1 mmol) or $CuCl_2$ (1 mmol) has significantly improved the catalytic performance (above 80% yield, 90% selectivity) in comparison with bare NP.

Tahir et al. explored the synthesis of arylhydantoins via Friedel–Crafts reaction by using both natural and synthetic phosphates-based catalysts [102]. Phenol, thiophene, and anisole reacted with 5-bromohydantoin previously prepared in

Table 14.10 Friedel–Crafts alkylation over NP-based catalysts.

Entry	Catalyst	Arene	Conversion (%) (time/min)a	Monoalkylation compound yield (%)		Dialkylation compound yield (%)^a
				Yield (%)^a	Ortho/para^b	
1	NP	Benzene	42	30	—	5
2		Toluene	100	91	42/58	5
3	Zn/NP	Benzene	100	92	—	7
4		Toluene	100	93	42/58	3
5	Cu/NP	Benzene	100	80	—	7
6		Toluene	100	95	40/60	4

a) Estimated by GC.
b) Ortho/para ratio determined by ^{1}H NMR.
Source: [101]. Reproduced with permission of CSJ Journals.

situ from bromination of hydantoin in dioxane (Scheme 14.4). Various catalysts were tested in this reaction such as NP, synthetic HA, FA, and these phosphates impregnated with ZnBr$_2$. It was reported in this study that the reaction between phenol and 5-bromohydantoin was able to proceed without catalyst (Table 14.11, entry 1), but the yield was mediocre, which did not exceed 31%. However, the reaction rate was significantly increased when using phosphates: HA, FA, or NP, and the obtained yields exceeded 80%, with heightened selectivity in para-alkylation product compared with ortho-alkylation. The reaction of 5-bromohydantoin with anisole and thiophene resulted in low yields especially when using FA and NP. HA was more active compared with FA and NP. The impregnation of these phosphates with ZnBr$_2$ clearly boosted the catalytic performance in the case of Zn/HA but slightly improved the other two catalytic systems (Zn/FA and Zn/NP). The authors reported that this difference between these three catalytic systems is related to the presence of P—OH group in HA's surface, which can probably act as a weak Brönsted acid sites. Unfortunately, this study did not elucidate the mechanism (identification of reaction intermediates) and the kinetic of the reaction.

14.3.4 Suzuki–Miyaura Coupling Reaction

The Suzuki–Miyaura cross-coupling reaction, catalyzed by palladium(0) or other transition metals, between an aromatic halide derivative (Cl, Br, I, OTf) and an arylboronic acid, in the presence of a base, is one of the most widely used reactions for the synthesis of biaryls [103, 104]. The reaction conditions are relatively

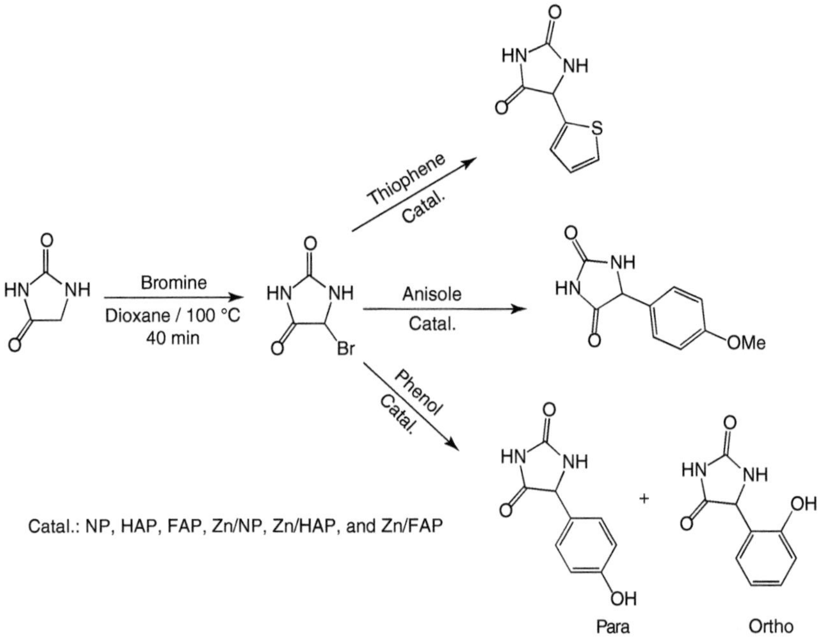

Scheme 14.4 Friedel–Craft reactions between 5-bromohydantoin and aromatic compounds. Source: [102]. Reproduced with permission of Elsevier.

mild, and the by-products formed are of low toxicity [105, 106]. Suzuki–Miyaura coupling is widely used today in the synthesis of natural products, drugs, polymers, etc. because it is a very versatile reaction and can tolerate many functional groups [107].

Pd/NP catalyst was prepared by impregnating NP with bis(benzonitrile)palladium (II) chloride ((PhCN)$_2$PdCl$_2$) as Pd precursor (Scheme 14.5). XPS has revealed the existence of Pd^{2+} coordinated with chlorine (Pd/Cl ratio was about 0.57 : 1) and also Pd^0 on the catalyst surface. carbon hydrogen and nitrogen (CHN) elemental analysis has confirmed the absence of carbon and nitrogen in Pd/NP catalyst. Finally, ICP-OES analysis of the filtrate after supporting Pd on the NP showed the absence of Ca, which demonstrated that the substitution of calcium by palladium in the apatite structure did not take place. In Suzuki–Miyaura coupling reaction of various aryl bromides efficiently with arylboronic acids in the presence of Pd/NP catalyst, reaction yield above 80% could be obtained (Table 14.12: entries 1–10). The reaction between 2-bromopyridine and phenylboronic acid, for the synthesis of aryl-substituted nitrogen heterocycles, was easily achieved with 70% yield. However, the cross-coupling reaction of 2-bromothiophene and phenylboronic acid did not take place, which is explained by a probable coordination of palladium with sulfur of the thiophene acting as scavenger molecule. Under the same reaction conditions, the reaction of 4-bromoacetophenone with phenylboronic acid in the presence of thiophene led to 27% yield. The aryl chlorides (1-nitro-4-chlorobenzene, 4-chlorobenzonitrile, and 4-chloroacetophenone) were also tested in order to

Table 14.11 Phosphate catalysts for the Friedel–Craft reactions in order to 5-arylhydantoins synthesis.

Entry	Arene	Catalyst [a]	Yield (%) (time (h)) [b]	Para/ortho [c]
1	C₆H₅–OH	—	31 (24)	86/14
2		HAP	92 (6)	72/28
3		FAP	83 (6)	83/17
4		NP	80 (6)	80/20
5	C₆H₅–OMe	HAP	66 (24)	—
6		FAP	38 (24)	—
7		NP	34 (24)	—
8		Zn/HAP	86 (24)	—
9		Zn/FAP	53 (24)	—
10		Zn/NP	48 (24)	—
11	thiophene	HAP	71 (24)	—
12		FAP	42 (24)	—
13		NP	46 (24)	—
14		Zn/HAP	94 (24)	—
15		Zn/FAP	58 (24)	—
16		Zn/NP	42 (24)	—

a) Reaction carried out in presence of 0.1 g of the catalyst.
b) Isolated yield estimated from the starting hydantoin. Products were identified by ^{1}H and ^{13}C NMR.
c) Determined by 1H NMR spectroscopy just for the case of the phenol use.
Source: [102]. Reproduced with permission of Elsevier.

investigate the capacity of this catalytic system (PdNP) since these compounds are difficult to react with aryl-boronic acid derivatives. These reactions were carried out in the presence of a phase transfer agent such as tetrabutylammonium bromide (TBAB) and a base such as NaOH. The yield obtained for 1-nitro-4-chlorobenzene, 4-chlorobenzonitrile, and 4-chloroacetophenone was 97, 87, and 81%, respectively. The regeneration study showed that the yield decreased slowly from 95% by fresh catalyst to 92 and 88% by the third and the fourth cycle, respectively. This same Pd/NP catalyst was also investigated in the microwave-assisted Suzuki–Miyaura cross-coupling [109]. This allowed reducing the reaction time in comparison with the process without microwave assistance [108]. The Pd/NP almost retained its catalytic activity during six successive recycling runs, since no Pd leaching was detected during these recycling tests.

14.3.5 Hydration of Nitriles

Amides are key intermediates for many organic transformations, particularly to synthesize engineering polymers [110] and biologically active molecules [111, 112]. It

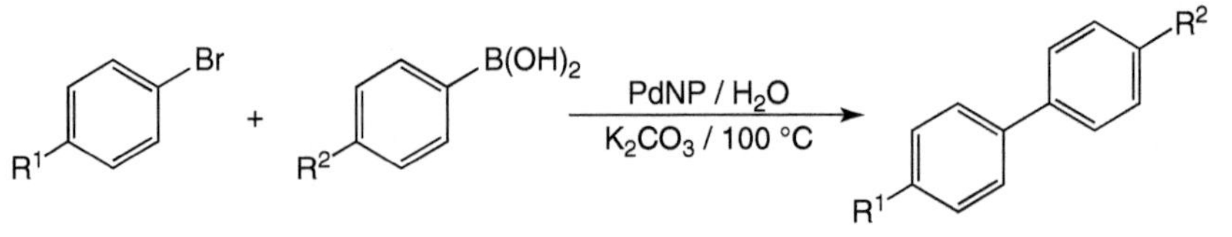

R^1 = H, CN, MeO, NO_2, R^2 = H, MeO
MeCO, OH, 2,6-Me

Scheme 14.5 Pd/NP-catalyzed Suzuki–Miyaura cross-coupling reaction of arylbromides with arylboronic acid. Source: [108]. Reproduced with permission of Elsevier.

Table 14.12 NP-supported Pd-catalyzed Suzuki–Miyaura cross-coupling reaction.

Entry	R^1	R^2	Yield (%)[a]
1	MeCO	H	95
2	NO_2	H	96
3	CN	H	90
4	MeO	H	90
5	OH	H	97
6	MeCO	OMe	85
7	NO_2	OMe	92
8	CN	OMe	93
9	OMe	OMe	80
10	OH	OMe	90

a) Determined by using GC.
Source: [108]. Reproduced with permission of Elsevier.

should be noted that amides are present in around 25% of top selling in pharmaceuticals industry [113]. Several approaches to amide synthesis have been developed [114–116], among others, the hydration of nitrile, which is one of the most widely transformations used for the synthesis of various amides [117]. Different groups have reported the hydration of nitriles into amides using different catalysts such as acids, bases, ionic liquids, and transition metals [118–120]. The development of effective and appropriate methods for the hydration of nitriles to amides under clean and soft conditions remains a current challenge [121, 122].

In this context, a broad range of amides, including nicotinamide (vitamin B3), was synthesized by hydration of the corresponding nitriles under eco-friendly conditions (water as solvent, reaction temperature of 100 °C) and by the catalytic action of NP and Na/NP materials [123]. Table 14.13 shows that important conversion of nitriles into amides from 79% (entry 6) to 98% (entry 10) could be achieved. In all cases, the modification of NP by sodium salt highly increased the efficiency of the synthesis.

Table 14.13 Selective hydration of various nitriles catalyzed by NP and Na/NP.

$$R-C\equiv N \xrightarrow{\text{Catal. (100 mg)}}_{\text{H}_2\text{O (5 ml) / 100 °C}} R-\overset{\text{O}}{\underset{}{C}}-NH_2$$

Entry	Product	Yield (%) (time (h))[a] NP	Na/NP
1	PhC(O)NH₂	13 (24)	93 (3)
2	PhCH₂C(O)NH₂	11 (24)	91 (16)
3	4-H₃C-C₆H₄-C(O)NH₂	15 (24)	91 (12)
4	2-CH₃-C₆H₄-C(O)NH₂	5 (24)	87 (48)
5	4-O₂N-C₆H₄-C(O)NH₂	48 (24)	92 (24)
6	4-Cl-C₆H₄-C(O)NH₂	5 (24)	79 (48)
7	2-Cl-C₆H₄-C(O)NH₂	7 (24)	60 (24)
8	4-pyridyl-C(O)NH₂	58 (24)	97 (2)
9	3-pyridyl-C(O)NH₂	65 (24)	96 (2)
10	2-pyridyl-C(O)NH₂	48 (24)	98 (2)

a) Yield in isolated products and identified ¹H NMR, ¹³C NMR, and FTIR spectroscopy.
Source: [123]. Reproduction with permission of Elsevier.

Pd/NP catalysts were also investigated in the selective hydration of nitriles in water at 160 °C, under microwave irradiation (Scheme 14.6) [124]. This catalyst was synthesized by impregnating NP with bis(benzonitrile) palladium(II) chloride in acetone, as previously reported [108]. These catalysts showed high catalytic performance in the hydration for a variety of benzonitrile derivatives [124]. Steric hindrance in the case of amino-benzonitriles was also a critical factor impacting the yield of this chemical transformation [124]. The conversion of heteroaromatic nitriles containing nitrogen, oxygen, and sulfur atoms into the corresponding amides was easily carried out with high yields. On the other hand, NP alone did not catalyze the hydration reaction. In addition, Pd/NP could be recycled at least four times in this reaction, with only a slight loss of activity. A reaction mechanism was proposed [124]. The catalytic cycle is first initiated by an exchange reaction between chloride ions and water molecules, which leads to the formation of a cationic Pd(II)-aqua species. The latter reacts with nitrile substrates producing a Pd(II)-bisnitrile species coordinated with the catalyst. Then a partial substitution of the nitrile by water takes place to give rise to Pd(II)-nitrile-water species as an intermediate. This intermediate is transformed into a Pd-iminolate species after its coordination with the water ligand, thus leading to the formation of its stable tautomeric form such as amide.

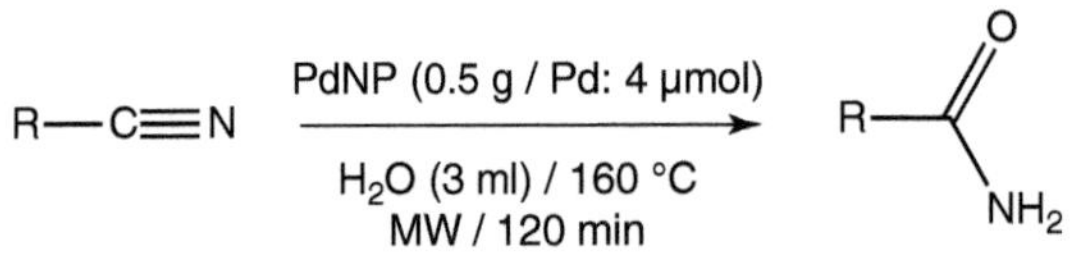

$$R-C\equiv N \xrightarrow[\text{H}_2\text{O (3 ml) / 160 °C}]{\text{PdNP (0.5 g / Pd: 4 µmol)}} R-\overset{O}{\underset{NH_2}{\|}}$$

Scheme 14.6 Selective hydration of various nitriles catalyzed by Pd/NP. Source: Abderrahim Solhy.

14.3.6 Synthesis of α-Hydroxyphosphonates

The α-hydroxyphosphonates are organophosphorus compounds, which are very significant in organic synthesis [125–129]. Their reactivity is mainly due to the hydroxyl group. These compounds are known for their interesting pharmacological activities [130, 131]. For example, trichlorfon and dichlorvos are widely used as insecticides [132]. These biological properties largely justify the development of scientific work on their synthesis and use [133]. Thus, Sebti et al. reported the synthesis of α-hydroxyphosphonates by reacting carbonyl compounds (aldehydes or ketones) with diethylphosphite in the presence of NP and KF/NP at room temperature and without solvent [134]. NP catalyzed this reaction, while NP impregnated with KF remarkably boosted the catalytic performance (Table 14.14). The same group also synthesized a series of α-hydroxyphosphonates using diethylphosphite and dimethylphosphite and aromatic aldehydes and ketones [27]. The reactions were carried out without solvents at room temperature over NP, KF/NP, or Na-modified NP. In all cases, high yield into desired products could be achieved. Na-modified NP was more active than NP or KF/NP and showed a good recyclability (Table 14.14).

Table 14.14 Synthesis of α-hydroxyphosphonates over NP or KF/NP.

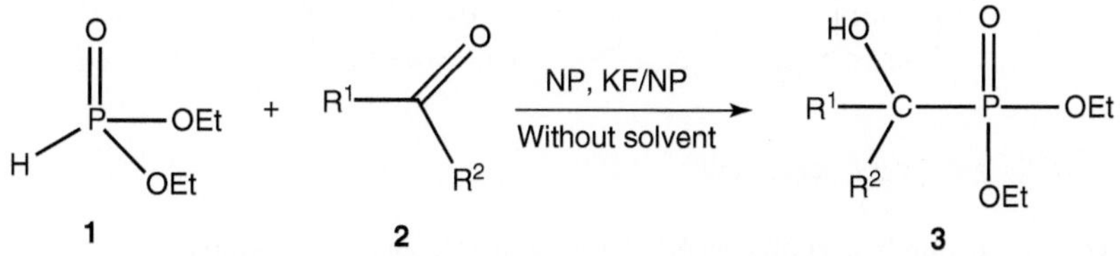

Reaction scheme: Diethyl phosphite (1) + R¹C(O)R² (2) → α-hydroxyphosphonate (3), NP, KF/NP, Without solvent.

3a: R¹ = Ph, R² = H
3b: R¹ = 4-Cl—C₆H₄, R² = H
3c: R¹ = PhCH=CH, R² = H
3d: R¹ = C₂H₅, R² = H
3e: R¹, R² = —C₅H₁₀—
3f: R¹ = Ph, R² = CH₃
3g: R¹ = CH₃, R² = CH₃

Entry	Product	Yield (%) (time (min))[a]	
		NP	KF/NP
1	3a	98 (30)	96 (05)
2	3b	98 (30)	97 (05)
3	3c	98 (15)	98 (05)
4	3d	89 (30)	88 (15)
5	3e	95 (60)	98 (05)
6	3f	58 (180)	82 (60)
7	3g	27 (240)	93 (30)

a) Yield in isolated products and identified ^{1}H NMR, ^{13}C NMR, and FTIR spectroscopy.
Source: [134] / With permission of Elsevier.

14.3.7 Multicomponent Reactions (MCRs)

14.3.7.1 Biginelli Reaction

MCRs are one-pot synthesis strategies implying three or more substrates in a highly regio- and stereoselective manner to provide structurally complex organic molecules for various applications [135]. In this regard, Biginelli reaction, which is the condensation of an aldehyde, a β-keto-ester, and urea, allows the synthesis of dihydropyrimidinones [136, 137]. Among them, the 3,4-dihydropyrimidin-2(1H)-ones compounds exhibit a wide range of pharmacological activities [138]. A variety of catalysts were used to achieve this condensation, including NP-based catalysts [139–142]. Thus, El Badaoui et al. studied the Biginelli reaction over NP-based catalyst, which were modified by impregnating with $ZnCl_2$, $CuCl_2$, $NiCl_2$, or $CoCl_2$ (Scheme 14.7) [142]. Unfortunately, these catalysts were not characterized in order to link the catalytic performance to their properties. Catalytic results clearly show the role of transition metal salts to boost the catalytic efficiency of NP [142]. A slight influence of the nature of the substituents on the aromatic aldehyde on the yields of this condensation was revealed [142].

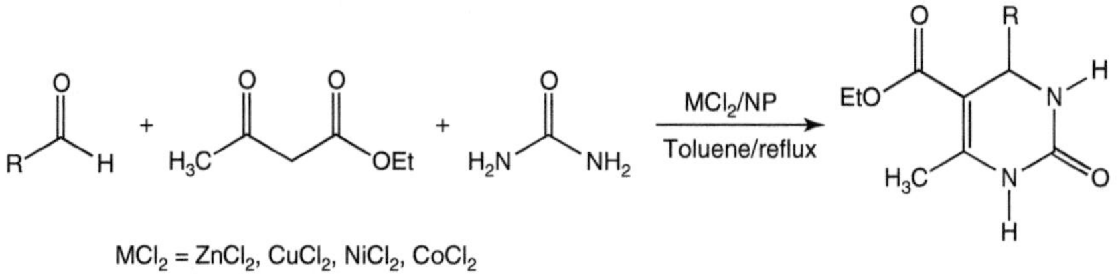

Scheme 14.7 Biginelli reaction over NP-based catalysts. Source: Abderrahim Solhy.

14.3.7.2 Synthesis of α-Aminophosphonates

The α-aminophosphonates are organophosphorus compounds that are of great interest because of their wide range of biological properties and structural analogies to α-amino acids [143]. The best known aminophosphonate is the famous glyphosate (N-(phosphonomethyl)glycine), which is the most widely used herbicide in the world [144]. The Kabachnik–Fields reaction stands for the most straightforward way to synthesize the α-aminophosphonates starting from aldehydes, amines, and dialkylphosphites [145, 146]. It has been postponed that this reaction could be promoted by several Lewis acid catalysts ($InCl_3$, $FeCl_3$, $Ce(OTf)_4$) [147–149] or solid acids (Amberlite-IR 120) [150]. Sebti's group studied this reaction using NP and KF/NP catalysts [151]. The reaction between carbonyl compound, amine, and dialkylphosphite was performed under solvent-free condition at room temperature in the presence of the catalysts. The results (Table 14.15) clearly showed the positive impact of NP impregnation with KF on the catalytic performance of NP, improving both the yield and the reaction rates. The functional groups (substituents) had no effect on this catalytic process under the adopted reaction conditions.

14.3.8 Oxidation Reactions

14.3.8.1 Oxidative Cleavage of Cycloalkanones

The oxidative cleavage consists in the replacement of C—C by C=O bonds [152]. The application of catalytic oxidative cleavage to cycloalkanones and α-ketols is widely studied [152, 153]. A concrete application of these oxidation reactions is the synthesis of adipic acid (1,6-hexanedioic acid) (>2 Mton year^{-1}), which is used for the manufacture of nylon and more generally for the synthesis of polyamides [154, 155]. The production of adipic acid on an industrial scale is achieved through the oxidation of cyclohexanone in the presence of nitric acid and vanadium(V)/copper(II) salts [156]. This process generates the gas NO_x (400 Kton year^{-1}) as by-products having greenhouse effect 200 times more dangerous than CO_2 [157]. To overcome these obstacles, new catalytic systems have recently been developed: e.g. vanadium-based heteropolyacids ($H_{3+n}[PMo_{12-n}V_nO_{40}]$: HPA-$n$ (n = 1–6)) or copper salts for the oxidation cleavage of cycloalkanones to obtain carboxylic acids [158]. The use of vanadyl ions supported on Nafion® to catalyze the oxidative cleavage of cycloalkanones and α-ketols in the presence of O_2 has also been reported [159]. In this sense, Dakkach et al. used NP-supported vanadium catalysts (V/NP) in order to catalyze the oxidative cleavage of C—C bonds of cycloalkanes and α-ketols

Table 14.15 One-pot synthesis of α-amino phosphonate via the three-component reaction catalyzed by NP-based materials.

4a: R_1 = Ph, R_2 = H, R_3 = Ph, R_4 = Et
4b: R_1 = 4-Me—C_6H_4, R_2 = H, R_3 = Ph, R_4 = Et
4c: R_1 = 4-MeO—C_6H_4, R_2 = H, R_3 = Ph, R_4 = Et
4d: R_1 = 4-Cl—C_6H_4, R_2 = H, R_3 = Ph, R_4 = Et
4e: R_1 = 4-NO_2—C_6H_4, R_2 = H, R_3 = Ph, R_4 = Et
4f: R_1 = 2-OH—C_6H_4, R_2 = H, R_3 = Ph, R_4 = Et
4g: R_1 = PhCH=CH, R_2 = H, R_3 = Ph, R_4 = Et
4h: R_1 = Ph, R_2 = H, R_3 = Ph, R_4 = Me
4i: R_1 = 4-MeO—C_6H_4, R_2 = H, R_3 = Ph, R_4 = Me
4j: R_1 = —$(CH_2)_5$—, R_3 = Ph, R_4 = Et
4k: R_1 = Ph, R_2 = H, R_3 = 2-Me—C_6H_4, R_4 = Et
4l: R_1 = 4—MeO—C_6H_4, R_2 = H, R_3 = 2-Me—C_6H_4, R_4 = Et
4m: R_1 = Ph, R_2 = H, R_3 = $PhCH_2CH_2$, R_4 = Et
4n: R_1 = —$(CH_2)_5$—, R_3 = $PhCH_2CH_2$, R_4 = Et

		Yield (%) (time (min))[a]	
Entry	Product	NP	KF/NP
1	4a	78 (30)	82 (14)
2	4b	80 (30)	84 (14)
3	4c	85 (30)	87 (14)
4	4d	72 (30)	77 (14)
5	4e	64 (30)	71 (14)
6	4f	73 (30)	67 (14)
7	4g	81 (30)	87 (14)
8	4h	80 (30)	83 (14)
9	4i	83 (30)	85 (14)
10	4j	77 (30)	76 (14)
11	4k	64 (30)	61 (14)
12	4l	62 (30)	68 (14)
13	4m	76 (30)	70 (14)
14	4n	59 (30)	64 (14)

a) Yield in isolated products.
Source: [151] / With permission of Elsevier.

in the presence of O_2 [160]. For the catalyst preparation, NP was impregnated with an aqueous solution of ammonium vanadate, followed by evaporation, drying, and calcination. Several catalysts were obtained by varying vanadium loading and the calcination temperature. The reaction of oxidative cleavage of several α-substituted cyclic ketones was performed in a batch reactor at 60 °C and 1 bar under oxygen atmosphere. The impact of vanadium loading, catalyst mass, and nature of solvent was investigated, which allowed determining the optimum operating conditions: 1 bar of O_2, 60 °C reaction temperature, $AcOH/H_2O$ (4.4/0.5: ml/ml) as solvent, and in the presence of 0.1 g of V/NP with 1 mmol of vanadium loaded on 10 g of NP. Under these conditions, high substrate conversion (e.g. 88–100%) and high yield

14 Natural Phosphates and Their Catalytic Applications

Table 14.16 Oxidative cleavage of α-substituted cyclic ketones over NP and V/NP catalysts.

Entry	Reagent	Time (h)	Conversion (%) [a]	Product (yield (%))
1	cyclopentanone-CO₂Me	8	97	MeO-C(O)-(CH₂)₃-COOH (96)
2	cyclohexanone-Ph	12	92	Ph-C(O)-(CH₂)₄-COOH (90)
3	cyclohexanone-Me	8	99	Me-C(O)-(CH₂)₄-COOH (98)
4	cyclohexanone-OH	24	88 [b]	OH-(CH₂)₄-C(O)-COOH (86)
5	cyclopentanone-OH	0.5	100	HO-C(O)-(CH₂)₃-COOH (99)

a) Conditions: reagent (0.8 mmol), AcOH/H$_2$O (4.5/0.5: ml/ml), 0.1 g of V/NP (1 mmol), $T = 60\,°C$, $p(O_2) = 0.1$ MPa.
b) 0.5 g of NP.
Source: [160]. Reproduced with permission of Elsevier.

into corresponding ketoacids (86–99%) could be achieved (Table 14.16). V/NP was found to be much more efficient than NP (entries 4 and 5), improving the conversion and the yield while reducing the reaction time. The O$_2$ consumption was close to the stoichiometry, which indicated that this catalytic system was regioselective.

These catalysts were also applied in the diastereoselective synthesis of pinonic acid, which is an important intermediate in the synthesis of chiral amino acids [161]. Thus, the oxidative cleavage of two commercial enantiomers of 2-hydroxy-3-pinanone ((1S,2S,5S)-(-)-2-hydroxy-3-pinanone and (1R,2R,5R)-(+)-2-hydroxy-3-pinanone) was performed over NP and V/NP (Scheme 14.8). The catalysts showed excellent diastereoselectivity into the desired product, with the formation of the single diastereoisomer of pinonic acid for each reagent at high conversions above 90%.

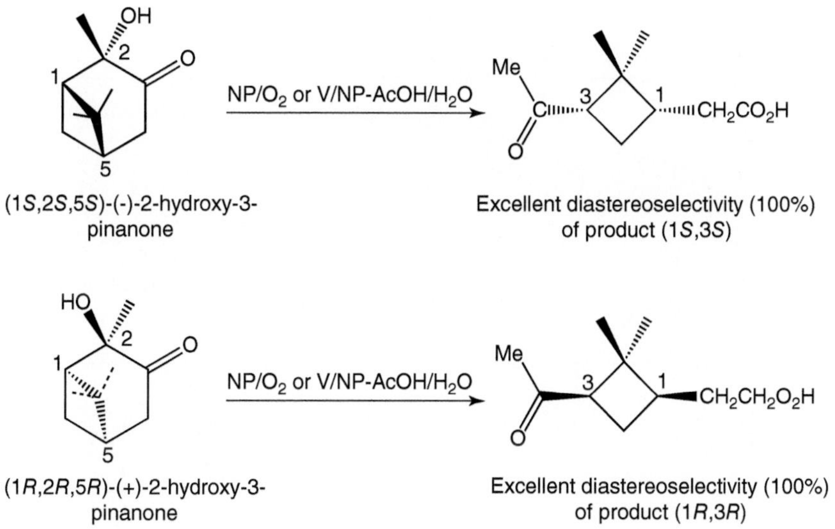

(1S,2S,5S)-(-)-2-hydroxy-3-pinanone → NP/O₂ or V/NP-AcOH/H₂O → Excellent diastereoselectivity (100%) of product (1S,3S)

(1R,2R,5R)-(+)-2-hydroxy-3-pinanone → NP/O₂ or V/NP-AcOH/H₂O → Excellent diastereoselectivity (100%) of product (1R,3R)

Scheme 14.8 Oxidative cleavage of 2-hydroxypinan-3-one (1S,2S,5S) and (1R,2R,5R) catalyzed by NP or V/NP. Source: [160]. Reproduced with permission of Elsevier.

14.3.8.2 Oxidation of Benzyl Alcohol

The oxidation of alcohols to aldehydes and ketones is a chemical transformation of primary industrial importance [162, 163]. Several methods have been developed to accomplish this particular reaction using catalysts from a variety of metals and supports, e.g. HA-supported chromium (Cr@HA) [164–169]. Laasri et al. investigated the oxidation of benzyl alcohol to benzaldehyde at moderate temperature (25–90 °C), 1 bar, and in presence of hydrogen peroxide as oxidant [170]. Various NP-based catalysts such as Ni/NP, Zr/NP, and Zr-Ni/NP and inorganic salts such as $ZrOCl_2 \cdot 8H_2O$, $Ni(NO_3)_2 \cdot 6H_2O$ were studied as catalysts in this reaction. NP-supported catalysts were prepared by impregnation, followed by evaporation, drying at 120 °C overnight, and calcination at 400 °C for 3 hours. For Ni—Zr/NP catalyst, XRD revealed the formation of NiO and $Zr_{1.5}Cl_2$ phases, without structural changes in NP support.

Different operating parameters were investigated including the effect of solvent (acetonitrile, water, toluene, and solvent-free condition), catalyst loading (0.05–0.2 g for 1 mmol of benzyl alcohol), reaction temperature (25–90 °C), type of catalyst, and reaction time (15–75 minutes). The optimum conditions were determined: 70 °C, 45 minutes, 0.08 g Zr—Ni/NP catalyst, and solvent-free condition, which allowed achieving 93% yield in benzaldehyde. A synergy effect between Ni and Zr seemed to be present in Zr—Ni/NP catalyst. However, recycling study showed that Zr—Ni/NP consecutively lost its activity, and the yield decreased from 93% with the fresh catalyst to 64% with the fifth recycling. A tentative mechanism was also proposed where H_2O_2 adsorbs on the metal Lewis acid sites before reacting with the benzyl alcohol to form intermediate. The latter undergoes two successive dehydration to form the aldehyde (Scheme 14.10).

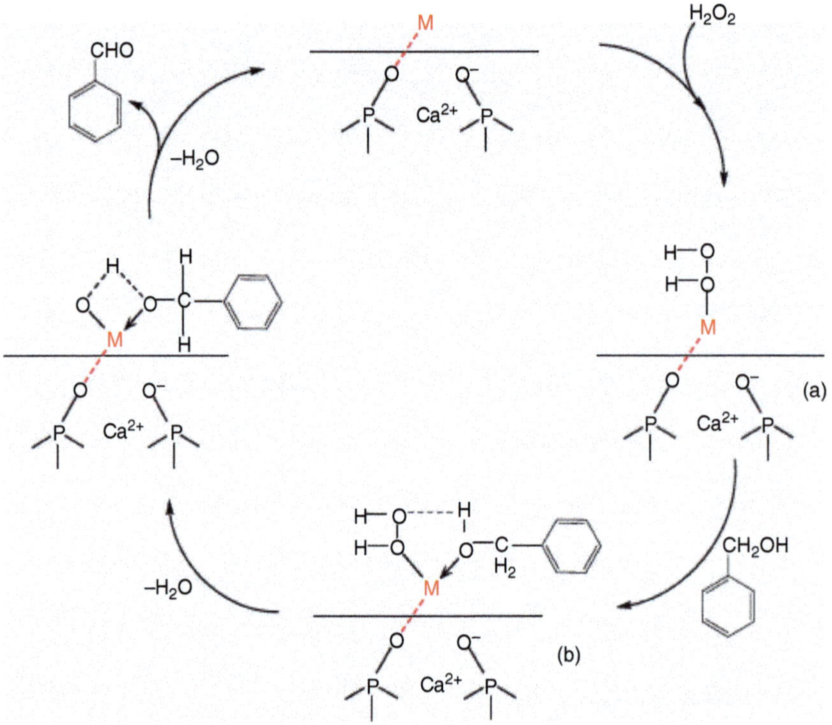

Scheme 14.9 Tentative mechanism in oxidation of benzylic alcohol to benzaldehyde by H_2O_2 over Zr–Ni/NP catalyst. Source: [170]. Reproduced with permission of Elsevier.

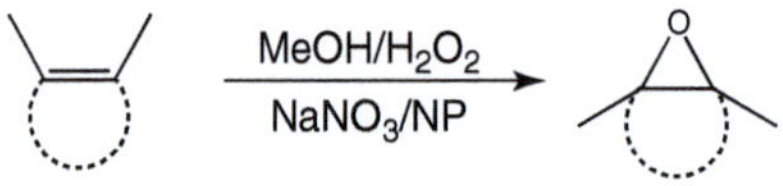

Scheme 14.10 Epoxidation of electron-deficient alkenes by H_2O_2 catalyzed by Na-modified NP. Source: Abderrahim Solhy.

14.3.8.3 Epoxidation of Electron-Deficient Alkenes

Epoxide rings (oxacyclopropane rings) are useful intermediates in organic synthesis that can be further cleaved to form anti vicinal diols [171, 172]. One way to synthesize oxacyclopropane rings is through the epoxidation of electron-deficient alkenes [173, 174]. The homogeneous catalysis of the epoxidation reaction of alkenes has been widely studied by using transition metal complexes [175–179]. The epoxidation of functionalized alkenes was easily carried out by solid catalysts such as zeolite [180], KF/Al_2O_3 [181], hydrotalcite in the presence of H_2O_2 [182], and Ti/SiO_2 [183]. NP-based catalysts have also been investigated in this reaction. For instant, Fraile et al. realized the epoxidation of a series of electron-deficient alkenes by H_2O_2 catalyzed by sodium-modified NP, in methanol as solvent (Scheme 14.10) [36]. BTEAC was also used as phase transfer agent, which had a positive effect on the epoxidation of mesityl oxide by increasing the selectivity into dioxolane [36]. This study shows that for most of alkenes investigated, the conversion was generally high (98–100%), with excellent selectivity into the corresponding dioxolanes. This demonstrated the high efficiency of the catalyst studied. However, low conversions (0–44%) were also observed for some alkenes, which is attributed to steric hindrance. The used catalyst could be easily regenerated by simple heat treatment at 500 °C, which allows recovering the initial catalytic activity of the fresh catalyst.

14.3.9 Hydrogenation Reactions

14.3.9.1 Selective Hydrogenation of Crotonaldehyde

Products resulting from the hydrogenation of α,β-unsaturated aldehydes into α,β-unsaturated alcohols are important in industry, because the products, e.g. terpene compounds, have application in sectors such as food, perfume, and pharmaceutical industries [184]. Noble metal catalysts (Pt, Ru, Pd, etc.) are active in the hydrogenation of C=C bonds, but not very active in the hydrogenation of C=O bonds, i.e. they easily lead to saturated aldehydes or alcohols whereas a good selectivity in unsaturated alcohol is too much sought after [185]. Selective catalysts are required because (i) of the thermodynamic aspect: it is more difficult to hydrogenate the C=O bond than the C=C bond, and (ii) the mechanism hydrogenation of the α,β-unsaturated compounds by heterogeneous catalysts is very complex and depends on the adsorption mode of the substrates on the catalyst surface [186, 187]. Effort has been devoted to the development of selective catalysts for the hydrogenation of α,β-unsaturated aldehydes into α,β-unsaturated [188]. The steric hindrance in the vicinity of the C=C double bond promotes a high selectivity toward unsaturated alcohol [189]. Thus, rhodium is relatively selective in the hydrogenation reactions of acetophenone [185] and cinnamaldehyde [190] but less selective in the hydrogenation of crotonaldehyde, citral, and cyclohexane [191, 192]. The hydrogenation of the α,β-unsaturated compounds is guided by the adsorption mode of the molecule, which is directly linked to the nature of the metal electronic structure [193]. The catalytic activity of different supported metal catalysts is determined by the ability to activate the C=C and C=O double bonds and also by the activity of the metal surface to activate hydrogen [194, 195]. Various supports have been investigated having the following characteristics: acidic (clays, zeolites, alumina, silica), basic (MgO), or redox (TiO_2, CeO_2) [196]. The choice of catalytic support is judicious to prepare a supported catalyst [197]. Metals supported on clays have been used in the hydrogenation of crotonaldehyde [198] and cinnamaldehyde [197]. In the hydrogenation of cinnamaldehyde, the selectivity to cinnamyl alcohol is very high by using Pt/K-10 (montmorillonite), where Pt particles were immobilized in the interfoliar space of the clay (K-10) [197]. In the case of crotonaldehyde, a maximum selectivity (40–45%) into crotyl alcohol was obtained by using Pt/H-B and Pt/K-10 as catalysts [198, 199]. Divakar et al. conducted a comparative study between two palladium-based catalysts (Pd/vermiculite and commercial Pd/C) for the hydrogenation of crotonaldehyde, cinnamaldehyde, and benzaldehyde, with high selectivity into unsaturated alcohols [200]. In this context,

Hidalgo-Carrillo et al. reported the use of NP-supported platinum catalyst in the selective hydrogenation of crotonaldehyde to unsaturated crotyl alcohol [201]. The catalysts (3–5 wt% Pt) were prepared by deposition–precipitation and impregnation methods, using, respectively, hexachloroplatinic acid and platinum acetylacetonate as platinum precursors. Prior to the catalytic test, the catalysts were reduced under H_2 at 400 °C. For a given hydrogenation experience, crotonaldehyde (0.5 M) in 20 ml of 1,4-dioxane/water (1 : 1 v/v) was set in contact with 100 mg catalyst under 0.414 MPa of H_2 at 30 °C during 8 h. Surprisingly, the catalysts pre-activated in hydrogen flow were less active (but more selective) than the non-reduced catalysts. The catalysts prepared by impregnation method were more active than those obtained with deposition–precipitation method, probably due to the higher Pt loading by the impregnation. The best result obtained with the catalysts elaborated by deposition–precipitation route was 18% conversion and 7% selectivity to crotyl alcohol, while the catalyst prepared by impregnation could achieve up to 80% conversion and 39% into crotyl alcohol. The addition of a second metal to form M-Pt/NP catalysts (M precursor was among $FeCl_2$, $FeCl_3$, $CoCl_2$, $ZnCl_2$, or $NiCl_2$) did not allow improving the catalytic efficiency. NP-based catalysts seem to be not competitive in comparison to the counterparts prepared on the conventional catalyst supports.

14.3.9.2 Reduction of Aromatic Nitro Compounds

Aromatic amines are widely considered as promising platforms for the production of high added value products such as phenacetin and paracetamol [202], azo compounds and hydroxyl amines [203], etc. The selective reduction of aromatic nitro compounds is among the most promising methods for the synthesis of aromatic amines [204, 205]. Several methods for the reduction of aromatic nitro compounds have been reported in the literature, including Bechamp reduction [206], catalytic hydrogenation [207, 208], reduction with metals [209], and electrolytic reduction [210, 211]. The major difficulty of this type of reaction is to selectively reduce the nitro group on substituted molecules in the presence of other organic functional groups. Platinum nanoparticles (average size of 3 nm) supported on polyaniline nanofibers were investigated in the catalytic reduction of nitrobenzene to aniline under H_2, achieving 94% yield [212]. This reduction can also be achieved via catalytic hydrogenation using hydrogen donors [213]. Sodium borohydride ($NaBH_4$) and lithium aluminum hydride ($LiAlH_4$) were also used as reducing agents in this reaction [214, 215]. However, the excess of boron hydride decreases the selectivity, leads to the formation of by-products, and increases wastes [216–218]. Sebti's group studied the reduction of nitro aromatics over an NP-supported $NiCl_2$ catalyst [219]. The latter was prepared by impregnation of NP with an aqueous solution of $NiCl_2$, followed by evaporation and drying at 120 °C before use. The reaction was carried out at room temperature and atmospheric pressure under stirring, using a mixture of nitro benzene (1 mmol), $NaBH_4$ (4 mmol), and catalyst (0.1 g) and water (3 ml). The test without catalyst or with bare NP showed no conversion, while the test with $NiCl_2$ salt showed only trace of the desired product. On the other hand, $NiCl_2$/NP achieved up to 75% conversion under the same reaction conditions, showing

Table 14.17 Reduction of various aromatic nitro compounds catalyzed by NP-supported Ni in water.

R–C₆H₄–NO₂ →(Ni/NP, NaBH₄/H₂O/r.t.) R–C₆H₄–NH₂

Entry	Nitro compounds	Product	Time (min)[a]	Conversion (%)[b]	Isolated yield (%)[c]
1	nitrobenzene	aniline	5	86	75
2	4-nitrophenol	4-aminophenol	10	86	82
3	2-methyl-6-nitroaniline	2-methylbenzene-1,6-diamine	5	94	74
4	4-chloro-2-nitroaniline	4-chlorobenzene-1,2-diamine	5	99	95
5	4-nitrobenzonitrile	4-aminobenzonitrile	10	86	86
6	3-nitrobenzaldehyde	(3-aminophenyl)methanol	5	98	90
7	4-nitrobenzaldehyde	(4-aminophenyl)methanol	5	98	95

a) Reaction conditions: nitro compounds (1 mmol), H₂O (3 ml), NaBH₄ (4 mmol), and Ni/NP (100 mg) at room temperature for the appropriate time.
b) Conversion determined by GC-MS.
c) Pure product obtained by column chromatography.
Source: [219] / Licensed under CC BY 3.0.

a possible synergy between $NiCl_2$ and the NP surface. The authors tested the reduction of nitrobenzene in methanol and ethanol. The replacement of water by organic solvents did not improve the conversion (65% in methanol, and no reaction in ethanol). The use of nitroaromatic compounds substituted by an aldehyde in meta position (entry 6) or para position (entry 7) showed that the reduction is not selective, since both functional groups of these molecule were hydrogenated (Table 14.17).

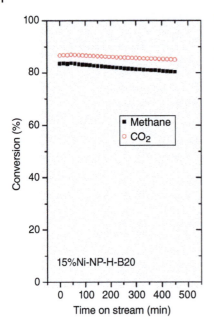

Figure 14.6 Conversion of CH_4 and CO_2 at 750 °C over Ni/NP catalyst. The NP support was pretreated with nitric acid and mechanical ball milling for 20 minutes. Source: [224]. Reproduced with permission of Elsevier.

14.3.10 Reforming of Methane

The reforming of methane to produce syngas using synthetic HA-based catalysts is presented in the chapter 8 of this book. Briefly, light hydrocarbon feedstocks (usually methane) can be converted with steam (steam reforming) or other oxidants (oxygen – partial oxidation of methane, CO_2 – dry reforming of methane) under adequate operating conditions into syngas [220–223]. Abba et al. studied the dry methane reforming reaction over NP-supported nickel catalysts [224]. For the catalyst preparation, NP support was washed with water, calcined at 900 °C for 2 hours, washed again, calcined again at 900 °C for 0.5 hours, and ground and sieved to obtain 63–125 µm particles. Then this support underwent the first treatment with 1 M nitric acid for 2 hours and drying at 120 °C, resulting to the modified NP-H support. This latter was then treated by mechanochemical ball milling for 20–60 minutes to increase the specific surface area and the surface reactivity of the support. Then these supports were impregnated by nickel acetate solution ($Ni(CH_3COO)_2$), drying overnight at 120 °C, and calcined at 400 °C for 2 hours. The Ni loading was fixed at 3 and 15 wt%. Prior to the catalytic test, the catalyst was reduced under H_2 flow at 500 °C for 1 hour. Dry reforming of methane was performed with a mixture of CH_4 : CO_2 : He (10 : 10 : 80 volume ratio) at 750 °C. The catalysts prepared with the support pretreated by nitric acid and by mechanical ball milling showed high CH_4 and CO_2 conversions (above 80%) and relatively good stability at 750 °C (Figure 14.6), which is related to a high dispersion of Ni on the modified NP support [224].

14.3.11 Photooxidation of VOC Model Compounds

Different types of volatile organic compounds (VOCs) can be distinguished depending on their molecular structures: linear hydrocarbons (saturated or unsaturated), monocyclic or polycyclic aromatic hydrocarbons, oxygenated VOCs (aldehydes, alcohols, ketones or esters), chlorinated VOCs, chlorofluorocarbons (CFC), and sulfur compounds [239, 240]. Many of these VOCs are well-known for their toxic effects; e.g. some VOCs are mutagenic or carcinogenic in indoor air (sick building syndrome) [241, 242]. Their elimination has become a public health priority [243, 244]. Air purification by VOCs elimination is a major environmental objective and can be achieved by several processes including photocatalytic oxidation [245]. In this process, the pollutant is oxidized over a catalyst in the presence of UV irradiation. Several semiconductor oxides or sulfides alone or impregnated with noble metal nanoparticles have been used as photocatalysts. These materials have a wide bandgap (Eg) sufficient to allow the photocatalytic process like TiO_2, ZnO, CdS, ZnS, WO_3, $SrTiO_3$, SnO_2, and Fe_2O_3 [225–227].

Hidalgo-Carrillo et al. [228] studied the photocatalytic oxidation of propan-2-ol as a model compound of VOCs over TiO_2/NP photocatalysts. Figure 14.7 shows the different steps for the photocatalyst preparation. Briefly, TiO_2 growth on NP surface was carried out by sol–gel method, using four different ageing approaches: by ultrasonic radiation, by reflux, by microwave radiation, and by magnetic stirring. The gels were calcined at 500 °C for 6 hours. Mesoporous photocatalysts were obtained with the specific surface area of 22–45 m² g⁻¹, containing 11.83–16.25 wt% TiO_2. In the

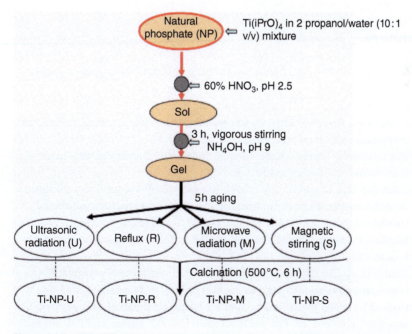

Figure 14.7 TiO_2/NP photocatalyst preparation. Source: [228]. Reproduced with permission of Elsevier.

photocatalytic oxidation of propan-2-ol under synthetic photo irradiation, NP had no photocatalytic activity. Among the TiO_2/NP photocatalysts, only the sample prepared with the ultrasonic radiation showed a higher photocatalytic activity than a mechanical mixture of NP support and a reference TiO_2.

14.4 Conclusions

The development of catalytic materials from NP becomes a reality, taking into account the related works during the last decades. NP meets different requirements of such a catalytic material to boost the activity, selectivity, and stability in different catalytic processes. NP can be used as a catalyst in various reactions, e.g. condensation, cross-coupling, alkylation reactions, hydration of aromatic nitriles, hydrogenation, reduction, and reforming of methane. NP can also be used as a catalyst support, since their apatitic structural flexibility to host the active phase allows developing various NP-based catalyst systems with controlled properties for targeted reactions. To date, the challenge is to work on the pretreatment of NP to minimize the impact of trace elements and thus harmonize the catalytic properties of NP-based catalytic materials. It is also recommended to enlarge the application of NP-based catalysts to other chemical processes, such as upgrading biomass into molecular building blocks and energy, since phosphate-based catalysts are still little explored.

References

1 Singh, R.K., Kumar, A., Garza-Reyes, J.A. et al. (2020). Managing operations for circular economy in the mining sector: an analysis of barriers intensity. *Resources Policy* 69: 101752. https://doi.org/10.1016/j.resourpol.2020.101752.

2 Grand View Research. https://www.grandviewresearch.com/industry-analysis/phosphate-rock-market (accessed 02 June 2021).

3 U.S. Geological Survey (2018). *Mineral Commodity Summaries*, 122–123. Reston, VA: U.S. Geological Survey.

4 Orris, G.J. and Chernoff, C.B. (2004). Review of world sedimentary phosphate deposits and occurrences. In: *Handbook of Exploration and Environmental Geochemistry*, vol. 8 (ed. J.R. Hein), 559–573. Amsterdam: Elsevier.

5 Nicolini, P. (1990). Gitologie et exploitation minière. *Technique et Document Lvoisier* 381.

6 Abouzeid, A.Z.M. (2008). Physical and thermal treatment of phosphate ores – an overview. *International Journal of Mineral Processing* 85: 59–84.

7 Cook, P.J. (1984). Spatial and temporal controls on the formation of phosphate deposits – a review. In: *Phosphate Minerals* (ed. J.O. Nriagu and P.B. Moore). Berlin, Heidelberg, Germany: Springer https://doi.org/10.1007/978-3-642-61736-2_7.

8 Slansky, M. (1980). Géologie des phosphates sédimentaires. Mémoire du BRGM, 114.

9 Ficher, A.G. and Jérome, D. (1973). Geochemistry of minerals containing phosphorus. In: *Environmental Phosphorus Hand Book* (ed. E.J. Griffith, A. Beeton, J.M. Spencer and D.T. Mitchell), 141. New York: John Wiley & Sons.

10 Bolourchifard, F. and Memar, A. (2015). The study of phosphate rock forming minerals (francolite) of Iran through the EDX-SEM to assessment of compositions in nano-scale. *Procedia Materials Science* 11: 108–113.

11 Gulbrandsen, R.A. (1966). Chemical composition of phosphorites of the hhosphoria formation. *Geochimica et Cosmochimica Acta* 30: 769–778.

12 McLellan, G.H. and Lehr, J.R. (1969). Crystal chemical investigations of natural apatites. *American Mineralogist* 54: 1374–1391.

13 McArthur, J.M. (1978). Systematic variations in the contents of Na, Sr, CO_3 and SO_4 in marine carbonate-fluorapatite and their relation to weathering. *Chemical Geology* 21: 89–112.

14 Smith, J.P. and Lehr, J.R. (1966). X-ray investigation of carbonate apatites. *Journal of Agricultural and Food Chemistry* 14 (4): 342–349.

15 Winand, L. (1968). Étude physico-chimique de divers carbonatapatites. *Bulletin de la Société chimique de France* 1718–1721.

16 Deans, T. (1938). Francolite from sedimentary ironstones of the coal measures. *Mineralogical Magazine* 25 (162): 135–139.

17 McConnell, D. (1937). The substitution of SiO_4- and SO_4-groups for PO_4-groups in the apatite structure; ellestadite, the end-member. *American Mineralogist* 22 (9): 977–989.

18 Sandell, E.B., Hey, M.H., and McConnell, D. (1939). The composition of francolite. *Mineralogical Magazine* 25 (166): 385–401.

19 Preston, J.S., Cole, P.M., Craig, W.M. et al. (1996). The recovery of rare earth oxides from a phosphoric acid by-product. Part 1: leaching of rare earth values and recovery of a mixed rare earth oxide by solvent extraction. *Hydrometallurgy* 41: 1–19.

20 Hammache, Z., Berbar, Y., Bensaadi, S. et al. (2020). Recovery of light rare earth elements by leaching and extraction from phosphate mining waste (fluorapatite and carbonate-fluorapatite). *Journal of African Earth Sciences* 171: 103937.

21 Satyavani, T.V.S.L., Srinivas Kumar, A., and Subba Rao, P.S.V. (2016). Methods of synthesis and performance improvement of lithium iron phosphate for high rate Li-ion batteries: a review. *Engineering Science and Technology, an International Journal* 19: 178–188.

22 Chen, M. and Graedel, T.E. (2015). The potential for mining trace elements from phosphate rock. *Journal of Cleaner Production* 91: 337–346.

23 Drouet, C. (2015). A comprehensive guide to experimental and predicted thermodynamic properties of phosphate apatite minerals in view of applicative purposes. *Journal of Chemical Thermodynamics* 81: 143–159.

24 Ihlen, P.M., Schiellerup, H., Gautneb, H. et al. (2014). Characterization of apatite resources in Norway and their REE potential – a review. *Ore Geology Reviews* 58: 126–147.

25 McKelvey, V.E. (1967). *Phosphate Deposits. Geological Survey Bulletin 1252-D. A Summary of Salient Features of the Geology of Phosphate Deposits, Their Origin, and Distribution*. Washington: United States Government Printing Office.

26 Sebti, S., Zahouily, M., Lazrek, H.B. et al. (2008). Phosphates: new generation of liquid-phase heterogeneous catalysts in organic chemistry. *Current Organic Chemistry* 12: 203–232.

27 Smahi, A., Solhy, A., Tahir, R. et al. (2008). Preparation of α-hydroxyphosphonates over phosphate catalyst. *Catalysis Communications* 9: 2503–2508.

28 Bertholus, M., Defranceschi, M. (2004). Les apatites des phosphates naturels. *Techniques de l'Ingénieur* AF6610.

29 Dakkach, M., Atlamsani, A., and Sebti, S. (2014). Natural phosphate as heterogeneous catalyst for oxidation of cyclic ketones to keto acids in environmentally friendly media. *Journal of Materials and Environmental Science* 5 (S1): 2122–2128.

30 Litasov, K.D. and Podgornykh, N.M. (2017). Raman spectroscopy of various phosphate minerals and occurrence of tuite in the Elga IIE iron meteorite. *Journal of Raman Spectroscopy* 48: 1518–1527.

31 Comodi, P., Liu, Y., and Frezzotti, M.L. (2001). Structural and vibrational behaviour of fluorapatite with pressure. Part II: in situ micro-Raman spectroscopic investigation. *Physics and Chemistry of Minerals* 28: 225–231.

32 Aklil, A., Mouflih, M., and Sebti, S. (2004). Removal of heavy metal ions from water by using calcined phosphate as a new adsorbent. *Journal of Hazardous Materials* A112: 183–190.

33 Mouflih, M., Aklil, A., Jahroud, N. et al. (2006). Removal of lead from aqueous solutions by natural phosphate. *Hydrometallurgy* 81: 219–225.

34 Hayumbu, P., Haselberger, N., Markowicz, A. et al. (1995). Analysis of rock phosphates by X-ray fluorescence spectrometry. *Applied Radiation and Isotopes* 46: 1003–1005.

35 Rentería-Villalobos, M., Vioque, I., Mantero, J. et al. (2010). Radiological, chemical and morphological characterizations of phosphate rock and phosphogypsum from phosphoric acid factories in SW Spain. *Journal of Hazardous Materials* 181: 193–203.

36 Fraile, J.M., Garcia, J.I., Mayoral, J.A. et al. (2001). Modified natural phosphates: easily accessible basic catalysts for the epoxidation of electron-deficient alkenes. *Green Chemistry* 3: 271–274.

37 El Asri, S., Laghzizil, A., Saoiabi, A. et al. (2009). A novel process for the fabrication of nanoporous apatites from Moroccan phosphate rock. *Colloids and Surfaces A: Physicochemical and Engineering Aspects* 350: 73–78.

38 El Asri, S., Laghzizil, A., Coradin, T. et al. (2010). Conversion of natural phosphate rock into mesoporous hydroxyapatite for heavy metals removal from

aqueous solution. *Colloids and Surfaces A: Physicochemical and Engineering Aspect* 362: 33–38.
39 Rhaiti, H., Laghzizil, A., Saoiabi, A. et al. (2012). Surface properties of porous hydroxyapatite derived from natural phosphate. *Materials Chemistry and Physics* 136: 1022–1026.
40 Saoiabi, S., El Asri, S., Laghzizil, A. et al. (2012). Lead and zinc removal from aqueous solutions by aminotriphosphonate-modified converted natural phosphates. *Chemical Engineering Science* 211–212: 233–239.
41 Knoevenagel, E. (1894). Ueber eine Darstellungsweise der Glutarsäure. *Berichte der Deutschen Chemischen Gesellschaft* 27: 2345–2346.
42 Weitkamp, J., Hunger, M., and Rymsa, U. (2001). Base catalysis on microporous and mesoporous materials: recent progress and perspectives. *Microporous and Mesoporous Materials* 48: 255–270.
43 Robertson, M.J., Deane, F.M., Robinson, P.J. et al. (2014). Synthesis of dynole 34-2, dynole 2-24 and dyngo 4a for investigating dynamin GTPase. *Nature Protocols* 9: 851–870.
44 Wei, L., Wang, J., Zhang, X. et al. (2017). Discovery of 2H-chromen-2-one derivatives as G protein-coupled receptor-35 agonists. *Journal of Medicinal Chemistry* 60: 362–372.
45 Raju, B.C. and Pathi, S. (2010). New and facile approach for the synthesis of (E)-α,β-unsaturated esters and ketones. *Chemistry: A European Journal* 16: 11840–11842.
46 Tabata, M., Boucard, V., Adès, D. et al. (2001). ESR and NMR studies of a novel conjugated donor-acceptor polymer containing magnetic spins: poly[bicarbazolylene-alt-phenylene-bis(cyanovinylene)]. *Macromolecules* 34: 8101–8106.
47 Marciniak, G., Delgado, A., Leclerc, G. et al. (1989). New 1,4-dihydropyridine derivatives combining calcium antagonism and alpha-adrenolytic properties. *Journal of Medicinal Chemistry* 32: 1402–1407.
48 Nagendrappa, G. (2011). Organic synthesis using clay and clay-supported catalysts. *Applied Clay Science* 53: 106–138.
49 Fihri, A., Len, C., Varma, R.S. et al. (2017). Hydroxyapatite: a review of syntheses, structure and applications in heterogeneous catalysis. *Coordination Chemistry Reviews* 347: 48–76.
50 Sebti, S., Saber, A., and Rhihil, A. (1994). Phosphate naturel et Phosphate trisodique: nouveaux catalyseurs solides de la condensation de Knoevenagel en milieu hétérogène. *Tetrahedron Letters* 35: 9399–9400.
51 Sebti, S., Smahi, A., and Solhy, A. (2002). Natural phosphate doped with potassium fluoride and modified with sodium nitrate: efficient catalysts for the Knoevenagel condensation. *Tetrahedron Letters* 43: 1813–1815.
52 Zahouily, M., Bahlaouan, B., Solhy, A. et al. (2003). Natural microphosphate as a catalyst for Knoevenagel condensation: specific surface effect. *Reaction Kinetics, Mechanisms and Catalysis* 78: 129–133.
53 Zahouily, M., Salah, M., Bahlaouan, B. et al. (2004). Solid catalysts for the production of fine chemicals: the use of natural phosphate alone and doped

base catalysts for the synthesis of unsaturated arylsulfones. *Tetrahedron* 60: 1631–1635.

54 Rizvi, S.U.F., Siddiqui, H.L., Johns, M. et al. (2012). Anti-HIV-1 and cytotoxicity studies of piperidyl-thienyl chalcones and their 2-pyrazoline derivatives. *Medicinal Chemistry Research* 21: 3741–3749.

55 Sharma, H., Patil, S., Sanchez, T.W. et al. (2011). Synthesis, biological evaluation and 3D-QSAR studies of 3-keto salicylic acid chalcones and related amides as novel HIV-1 integrase inhibitors. *Bioorganic & Medicinal Chemistry* 19: 2030–2045.

56 Lahtchev, K.V., Batovska, D.I., Parushev, S.P. et al. (2008). Antifungal activity of chalcones: a mechanistic study using various yeast strains. *European Journal of Medicinal Chemistry* 43: 2220–2228.

57 Aoki, N., Muko, M., Ohta, E. et al. (2008). C-geranylated chalcones from the stems of angelicakeiskei with superoxide-scavenging activity. *Journal of Natural Products* 71: 1308–1310.

58 Nowakowska, Z. (2007). A review of anti-infective and anti-inflammatory chalcones. *European Journal of Medicinal Chemistry* 42: 125–137.

59 Yadav, G.D. and Wagh, D.P. (2020). Claisen-Schmidt condensation using green catalytic processes: a critical review. *ChemistrySelect* 5: 9059–9085.

60 Marais, J.P.J., Ferreira, D., and Slade, D. (2005). Stereoselective synthesis of monomeric flavonoids. *Phytochemistry* 66: 2145–2176.

61 Sebti, S., Saber, A., Rhihil, A. et al. (2001). Claisen-Schmidt condensation catalysis by natural phosphate. *Applied Catalysis A: General* 206: 217–220.

62 Sebti, S., Solhy, A., Tahir, R. et al. (2001). Calcined sodium nitrate/natural phosphate: an extremely active catalyst for the easy synthesis of chalcones in heterogeneous media. *Tetrahedron Letters* 42: 7953–7955.

63 Sebti, S., Solhy, A., Tahir, R. et al. (2003). Application of natural phosphate modified with sodium nitrate in the synthesis of chalcones: a soft and clean method. *Journal of Catalysis* 213: 1–6.

64 Sebti, S., Solhy, A., Smahi, A. et al. (2002). Dramatic activity enhancement of natural phosphate catalyst by lithium nitrate. An efficient synthesis of chalcones. *Catalysis Communications* 3: 335–339.

65 Macquarrie, D., Nazih, R., and Sebti, S. (2002). KF/natural phosphate as an efficient catalyst for synthesis of 2′-hydroxychalcones and flavanones. *Green Chemistry* 4: 56–59.

66 Gupta, A., Birhman, K., Raheja, I. et al. (2016). Quercetin: a wonder bioflavonoid with therapeutic potential in disease management. *Asian Pacific Journal of Tropical Disease* 6: 248–252.

67 Santi, M.D., Arredondo, F., Carvalho, D. et al. (2020). Neuroprotective effects of prenylated flavanones isolated from *Dalea* species, in vitro and in silico studies. *European Journal of Medicinal Chemistry* 206: 112718.

68 Han, Q.-T., Xiao, K., Xiang, K.-L. et al. (2018). Scumoniliosides A and B, two new flavanone glucuronate esters from Scutellaria moniliorrhiza Komarov with anti-inflammatory activities. *Phytochemistry Letters* 25: 56–59.

69 Tournaire, C., Croux, S., Maurette, M.-T. et al. (1993). Antioxidant activity of flavonoids: efficiency of singlet oxygen ($^1\Delta_g$) quenching. *Journal of Photochemistry and Photobiology B: Biology* 19: 205–215.

70 Tokoroyama, T. (2010). Discovery of the Michael reaction. *European Journal of Organic Chemistry* 2010: 2009–2016.

71 Albanese, D.C.M. and Gaggero, N. (2014). N-heterocyclic carbene catalysis as a tool for gaining access to the 3,4-dihydropyran-2-one skeleton. *European Journal of Organic Chemistry* 26: 5631–5640.

72 Malkar, R.S., Jadhav, A.L., and Yadav, G.D. (2020). Innovative catalysis in Michael addition reactions for C—X bond formation. *Molecular Catalysis* 485: 110814.

73 Mather, B.D., Viswanathan, K., Miller, K.M. et al. (2016). Michael addition reactions in macromolecular design for emerging technologies. *Progress in Polymer Science* 31: 487–531.

74 Sebti, S., Boukhal, H., Hanafi, N. et al. (1999). Catalyse de la réaction de Michael par le fluorure de potassium supporté sur le phosphate naturel. *Tetrahedron Letters* 40: 6207–6209.

75 Abrouki, Y., Zahouily, M., Rayadh, A. et al. (2002). A natural phosphate and doped-catalyzed Michael addition of mercaptants to α,β-unsaturated carbonyl compounds. *Tetrahedron Letters* 43: 8951–8953.

76 Zahouily, M., Bahlaouan, B., Rayadh, A. et al. (2004). Natural phosphate and potassium fluoride doped natural phosphate: efficient catalysts for the construction of a carbon-nitrogen bond. *Tetrahedron Letters* 45: 4135–4138.

77 Zahouily, M., Bahlaouan, B., Aadil, M. et al. (2004). Natural phosphate doped with potassium fluoride: efficient catalyst for the construction of a carbon–carbon bond. *Organic Process Research & Development* 8: 275–278.

78 Zahouily, M., Mounir, B., Cherki, H. et al. (2007). Natural phosphate modified with lithium nitrate: a new efficient catalyst for the construction of carbon-carbon, carbon-sulfur, and carbon-nitrogen bonds. *Phosphorus, Sulfur, and Silicon and the Related Elements* 182: 1203–1217.

79 Zahouily, M., Mounir, B., Charki, H. et al. (2006). Investigation of the basic catalytic activity of natural phosphates in the Michael condensation. *Arkivoc* 2006 (xiii): 178–186.

80 Bandgar, B.P., Sadavarte, V.S., and Uppalla, L.S. (2001). Lithium perchlorate as an efficient catalyst for selective transesterification of β-Keto esters essentially under neutral conditions. *Synlett* 8: 1338–1340.

81 Wakasugi, K., Misaki, T., Yamada, K. et al. (2000). Diphenylammonium triflate (DPAT): efficient catalyst for esterification of carboxylic acids and for transesterification of carboxylic esters with nearly equimolar amounts of alcohols. *Tetrahedron Letters* 41: 5249–5252.

82 Iwasaki, T., Maegawa, Y., Hayashi, Y. et al. (2008). Transesterification of various methyl esters under mild conditions catalyzed by tetranuclear zinc cluster. *The Journal of Organic Chemistry* 73: 5147–5150.

83 Meher, L.C., Vidya Sagar, D., and Naik, S.N. (2006). Technical aspects of biodiesel production by transesterification – a review. *Renewable and Sustainable Energy Reviews* 10: 248–268.

84 Mahajan, S.S., Idage, B.B., Chavan, N.N. et al. (1996). Aromatic polyesters via transesterification of dimethylterephthalate/isophthalate with bisphenol-a. *Journal of Applied Polymer Science* 61: 2297–2304.

85 Vicente, G., Martínez, M., and Aracil, J. (2004). Integrated biodiesel production: a comparison of different homogeneous catalysts systems. *Bioresource Technology* 92: 297–305.

86 Ma, F. and Hanna, M.A. (1999). Biodiesel production: a review. *Bioresource Technology* 70: 1–15.

87 Ullmann, F. (2000). *Ullmann's Encyclopedia of Industrial Chemistry*. Weinheim, Germany: Wiley-VCH Verlag GmbH & Co. KGaA.

88 Chavan, S.P., Subbarao, Y.T., Dantale, S.W. et al. (2001). Transesterification of ketoesters using Amberlyst-15. *Synthetic Communications* 31: 289–294.

89 Choudary, B.M., Kantam, M.L., Reddy, C.V. et al. (2000). Mg—Al—O—*t*-Bu hydrotalcite: a new and efficient heterogeneous catalyst for transesterification. *Journal of Molecular Catalysis A: Chemical* 159: 411–416.

90 Suppes, G.J., Dasari, M.A., Doskocil, E.J. et al. (2004). Transesterification of soybean oil with zeolite and metal catalysts. *Applied Catalysis A: General* 257: 213–223.

91 Bazi, F., El Badaoui, H., Sokori, S. et al. (2006). Transesterification of methylbenzoate with alcohols catalyzed by natural phosphate. *Synthetic Communications* 36: 1585–1592.

92 Bazi, F., El Badaoui, H., Tamani, S. et al. (2006). Catalysis by phosphates: a simple and efficient procedure for transesterification reaction. *Journal of Molecular Catalysis A: Chemical* 256: 43–47.

93 You, S.-L., Cai, Q., and Zenga, M. (2009). Chiral Brønsted acid catalyzed Friedel–Crafts alkylation reactions. *Chemical Society Reviews* 38: 2190–2201.

94 Olah, G.A. (ed.) (1963). *Friedel-Crafts and Related Reactions*, vol. 1–4. New York: Wiley.

95 Tsuchimoto, T., Tobita, K., Hiyama, T. et al. (1996). Scandium(III) triflate catalyzed Friedel–Crafts alkylation with benzyl and allyl alcohols. *Synlett* 6: 557–559.

96 Isao, S., Meng, K.K., Miki, N. et al. (1997). Molybdenum-catalyzed aromatic substitution with olefins and alcohols. *Chemistry Letters* 26: 851–852.

97 Rueping, M. and Nachtsheim, B.J. (2010). A review of new developments in the Friedel–Crafts alkylation – From green chemistry to asymmetric catalysis. *Beilstein Journal of Organic Chemistry* 6 (6): 1–24.

98 Derouane, E.G., Crehan, G., Dillon, C.J. et al. (2000). Zeolite catalysts as solid solvents in fine chemicals synthesis: 2. Competitive adsorption of the reactants and products in the Friedel–Crafts acetylations of anisole and toluene. *Journal Catalysis* 194: 410–423.

99 Dörwald, F.Z. (2000). *Organic Synthesis on Solid Phase*. Weinheim, Germany: Wiley-VCH.

100 Guisnet, M. and Perot, G. (2001). The Fries rearrangement. In: *Fine Chemicals Through Heterogeneous Catalysis* (ed. R.A. Sheldon and H. van Bekkum), 211–216. Weinheim: Wiley-VCH.

101 Sebti, S., Rhihil, A., and Saber, A. (1996). Heterogeneous catalysis of the Friedel–Crafts alkylation by doped natural phosphate and tricalcium phosphate. *Chemistry Letters* 25: 721.

102 Tahir, R., Banert, K., and Sebti, S. (2006). Natural and synthetic phosphates: new and clean heterogeneous catalysts for the synthesis of 5-arylhydantoins. *Applied Catalysis A: General* 298: 261–264.

103 Miyaura, N. and Suzuki, A. (1995). Palladium-catalyzed cross-coupling reactions of organoboron compounds. *Chemical Reviews* 95: 2457–2483.

104 Han, F.-S. (2013). Transition-metal-catalyzed Suzuki-Miyaura cross-coupling reactions: a remarkable advance from palladium to nickel catalysts. *Chemical Society Reviews* 42: 5270–5298.

105 Marziale, A.N., Faul, S.H., Reiner, T. et al. (2010). Facile palladium catalyzed Suzuki-Miyaura coupling in air and water at ambient temperature. *Green Chemistry* 12: 35–38.

106 Lipshutz, B.H., Petersen, T.B., and Abela, A.R. (2008). Room-temperature Suzuki–Miyaura couplings in water facilitated by nonionic amphiphiles. *Organic Letters* 10: 1333–1336.

107 Koshvandi, A.T.K., Heravi, M.M., and Momeni, T. (2018). Current applications of Suzuki–Miyaura coupling reaction in the total synthesis of natural products: an update. *Applied Organometallic Chemistry* 32 (3): e4210.

108 Hassine, A., Sebti, S., Solhy, A. et al. (2013). Palladium supported on natural phosphate: catalyst for Suzuki coupling reactions in water. *Applied Catalysis A: General* 450: 13–18.

109 Hassine, A., Bouhrara, M., Sebti, S. et al. (2014). Natural phosphate-supported palladium: a highly efficient and recyclable catalyst for the Suzuki-Miyaura coupling under microwave irradiation. *Current Organic Chemistry* 18: 3141–3148.

110 Fonseca, A.C., Gil, M.H., and Simões, P.N. (2014). Biodegradable poly(ester amide)s – a remarkable opportunity for the biomedical area: review on the synthesis, characterization and applications. *Progress in Polymer Science* 39: 1291–1311.

111 Bhattacharya, A., Scott, B.P., Nasser, N. et al. (2007). Pharmacology and antitussive efficacy of 4-(3-trifluoromethyl-pyridin-2-yl)-piperazine-1-carboxylic acid (5-trifluoromethyl- pyridin-2-yl)-amide (JNJ17203212), a transient receptor potential vanilloid 1 antagonist in Guinea pigs. *Journal of Pharmacology and Experimental Therapeutics* 323: 665–674.

112 Rivara, S., Lodola, A., Mor, M. et al. (2007). *N*-(substituted-anilinoethyl)amides: design, synthesis, and pharmacological characterization of a new class of melatonin receptor ligands. *Journal of Medicinal Chemistry* 50: 6618–6626.

113 Brown, D.G. and Boström, J. (2016). Analysis of past and present synthetic methodologies on medicinal chemistry: where have all the new reactions gone? *Journal of Medicinal Chemistry* 59: 4443–4458.

114 Das, S., Wendt, B., Moller, K. et al. (2012). Two iron catalysts are better than one: a general and convenient reduction of aromatic and aliphatic primary amides. *Angewandte Chemie International Edition* 51: 1662–1666.

115 Marcia de Figueiredo, R., Suppo, J.S., and Campagne, J.M. (2016). Nonclassical routes for amide bond formation. *Chemical Reviews* 116: 12029–12122.

116 Lanigan, R.M. and Sheppard, T.D. (2013). Recent developments in amide synthesis: direct amidation of carboxylic acids and transamidation reactions. *European Journal of Organic Chemistry* 7453–7465.

117 Guo, B., de Vries, J.G., and Otten, E. (2019). Hydration of nitriles using a metal-ligand cooperative ruthenium pincer catalyst. *Chemical Science* 10: 10647–10652.

118 Ahmed, T.J., Knapp, S.M.M., and Tyler, D.R. (2011). Frontiers in catalytic nitrile hydration: nitrile and cyanohydrin hydration catalyzed by homogeneous organometallic complexes. *Coordination Chemistry Reviews* 255: 949–974.

119 Pattabiraman, V.R. and Bode, J.W. (2011). Rethinking amide bond synthesis. *Nature* 480: 471–479.

120 Lanigan, R.M., Starkov, P., and Sheppard, T.D. (2013). Direct synthesis of amides from carboxylic acids and amines using B(OCH$_2$CF$_3$)$_3$. *The Journal of Organic Chemistry* 78: 4512–4523.

121 Wang, H., Wang, Y., Xu, H. et al. (2019). Nanorod manganese oxide as an efficient heterogeneous catalyst for hydration of nitriles into amides. *Industrial & Engineering Chemistry Research* 58 (37): 17319–17324.

122 Liu, Y.-M., He, L., Wang, M.-M. et al. (2012). A general and efficient heterogeneous gold-catalyzed hydration of nitriles in neat water under mild atmospheric conditions. *ChemSusChem* 5: 1392–1396.

123 Bazi, F., El Badaoui, H., Tamani, S. et al. (2006). A facile synthesis of amides by selective hydration of nitriles using modified natural phosphate and hydroxyapatite as new catalysts. *Applied Catalysis A: General* 301: 211–214.

124 Hassine, A., Sebti, S., Solhy, A. et al. (2016). Natural phosphate-supported palladium for hydration of aromatic nitriles to amides in aqueous medium. *Current Organic Chemistry* 20: 1–7.

125 Olszewski, T.K. (2014). Environmentally benign syntheses of -substituted phosphonates: preparation of α-amino and α-hydroxyphosphonates in water, in ionic liquids, and under solvent-free conditions. *Synthesis* 46: 403–429.

126 Failla, S., Finocchiaro, P., and Consiglio, G.A. (2000). Syntheses, characterization, stereochemistry and complexing properties of acyclic and macrocyclic compounds possessing α-amino- or α-hydroxyphosphonate units: a review article. *Heteroatom Chemistry* 11: 493–504.

127 Gröger, H. and Hammer, B. (2000). Catalytic concepts for the enantioselective synthesis of α-amino and α-hydroxyphosphonates. *Chemistry – A European Journal* 6: 943–948.

128 Phillips, A.M.F. (2014). Organocatalytic asymmetric synthesis of chiral phosphonates. *Mini-Reviews in Organic Chemistry* 11: 164–185.

129 Sobhani, S. and Tashrifi, Z. (2010). Synthesis of α-functionalized phosphonates from α-hydroxyphosphonates. *Tetrahedron* 66 (7): 1429–1439.

130 Frechette, R.F., Ackerman, C., Beers, S. et al. (1997). Novel hydroxyphosphonate inhibitors of CD-45 tyrosine phosphatase. *Bioorganic & Medicinal Chemistry Letters* 7 (17): 2169–2172.

131 Eummer, J.T., Gibbs, B.S., Zahn, T.J. et al. (1999). Novel limonene phosphonate and farnesyl diphosphate analogues: design, synthesis, and evaluation as potential protein-farnesyl transferase inhibitors. *Bioorganic & Medicinal Chemistry* 7 (2): 241–250.

132 Braun, R., Schöneich, J., Weissflog, L. et al. (1982). Activity of organophosphorus insecticides in bacterial tests for mutagenicity and DNA repair – direct alkylation vs. metabolic activation and breakdown. I. Butonate, vinylbutonate, trichlorfon, dichlorvos, demethyl dichlorvos and demethyl vinylbutonate. *Chemico-Biological Interactions* 39 (3): 339–350.

133 Kiss, N.Z., Rádai, Z., and Keglevich, G. (2019). Green syntheses of potentially bioactive α-hydroxyphosphonates and related derivatives. *Phosphorus, Sulfur, and Silicon and the Related Elements* 194 (10): PBSi 2018.

134 Sebti, S., Rhihil, A., Saber, A. et al. (1996). Synthèse des α-hydroxyphosphonates sur des supports phosphatés en absence de solvant. *Teltrahedron Letters* 37 (23): 3999–4000.

135 Dömling, A., Wang, W., and Wang, K. (2012). Chemistry and biology of multicomponent reactions. *Chemical Reviews* 112 (6): 3083–3135.

136 Kappe, C.O. (1993). 100 years of the Biginelli dihydropyrimidine synthesis. *Tetrahedron* 49 (32): 6937–6963.

137 Silva, G.C.O., Correa, J.R., Rodrigues, M.O. et al. (2015). The Biginelli reaction under batch and continuous flow conditions: catalysis, mechanism and antitumoral activity. *RSC Advances* 5: 48506–48515.

138 Kappe, C.O. (2000). Biologically active dihydropyrimidones of the Biginelli-type – a literature survey. *European Journal of Medicinal Chemistry* 35 (12): 1043–1052.

139 Dondoni, A. and Massi, A. (2001). Parallel synthesis of dihydropyrimidinones using Yb(III)-resin and polymer-supported scavengers under solvent-free conditions. A green chemistry approach to the Biginelli reaction. *Tetrahedron Letters* 42 (45): 7975–7978.

140 Mitra, A.K. and Banerjee, K. (2003). Clay-catalysed synthesis of dihydropyrimidones under solvent-free conditions. *Synlett* 10: 1509–1511.

141 Sun, Q., Wang, Y.-Q., Ge, Z.-M. et al. (2004). A highly efficient solvent-free synthesis of dihydropyrimidones catalysed by zinc chloride. *Synthesis* 7: 1047–1051.

142 El Badaoui, H., Bazi, F., Tamani, S. et al. (2005). Lewis acid-doped natural phosphate: new catalysts for the one-pot synthesis of 3,4-dihydropyrimdin-2(1H)-one. *Synthetic Communications* 35: 2561–2568.

143 Palacios, F., Alonso, C., and de los Santos, J. M. (2005). Synthesis of β-aminophosphonates and -phosphinates. *Chemical Reviews* 105 (3): 899–932.

144 Richmond, M.E. (2018). Glyphosate: a review of its global use, environmental impact, and potential health effects on humans and other species. *Journal of Environmental Studies and Sciences* 8: 416–434.

145 Fiore, C., Sovic, I., Lukin, S. et al. (2020). Kabachnik-Fields reaction by mechanochemistry: new horizons from old methods. *ACS Sustainable Chemistry & Engineering* 8 (51): 18889–18902.

146 Cherkasov, R.A. and Galkin, V.I. (1998). The Kabachnik-Fields reaction: synthetic potential and the problem of the mechanism. *Russian Chemical Reviews* 67: 857–882.

147 Ranu, B.C., Hajra, A., and Jana, U. (1999). General procedure for the synthesis of α-amino phosphonates from aldehydes and ketones using indium(III) chloride as a catalyst. *Organic Letters* 1 (8): 1141–1143.

148 Rezaei, Z., Firouzabadi, H., Iranpoor, N. et al. (2009). Design and one-pot synthesis of α-aminophosphonates and bis(α-aminophosphonates) by iron(III) chloride and cytotoxic activity. *European Journal of Medicinal Chemistry* 44 (11): 4266–4275.

149 Sobhani, S. and Tashrifi, Z. (2009). One-pot synthesis of primary 1-aminophosphonates: coupling reaction of carbonyl compounds, hexamethyldisilazane, and diethyl phosphite catalyzed by Al(OTf)$_3$. *Heteroatom Chemistry* 20 (2): 109–115.

150 Bhattacharya, A.K. and Rana, K.C. (2008). Amberlite-IR 120 catalyzed three-component synthesis of α-amino phosphonates in one-pot. *Tetrahedron Letters* 49 (16): 2598–2601.

151 Zahouily, M., Elmakssoudi, A., Mezdar, A. et al. (2007). Natural phosphate and potassium fluoride doped natural phosphate catalyzed simple one-pot synthesis of a-amino phosphonates under solvent-free conditions at room temperature. *Catalysis Communications* 8: 225–230.

152 Suresh, A.K., Sharma, M.M., and Sridhar, T. (2000). Engineering aspects of industrial liquid-phase air oxidation of hydrocarbons. *Industrial & Engineering Chemistry Research* 39 (11): 3958–3997.

153 Wang, Y. (2006). Selective oxidation of hydrocarbons catalyzed by iron-containing heterogeneous catalysts. *Research on Chemical Intermediates* 32: 235–251.

154 Sato, K., Aoki, M., and Noyori, R. (1998). A green route to adipic acid: direct oxidation of cyclohexenes with 30 percent hydrogen peroxide. *Science* 281 (5383): 1646–1647.

155 Walter, I. (1895). Preparation of adipic acid and some of its derivatives. *Journal of the Chemical Society, Transactions* 67: 155–159.

156 Cornils, B. and Lappe, P. (2006). Dicarboxylic acids, aliphatic. In Elvers. Online ed. In: *Ullmann's Encyclopedia of Industrial Chemistry*. Weinheim, Germany: Wiley-VCH Verlag GmbH & Co.

157 Reimer, R.A., Slaten, C.S., Seapan, M. et al. (2000). Adipic Acid Industry – N_2O Abatement. Non-CO_2 *Greenhouse Gases: Scientific Understanding, Control and Implementation*, 347–358. Netherlands: Springer.

158 Brégeault, J.M., Saint-Antoine, B., and El Ali, B. (1988). Preparation of aliphatic carboxylic acids by oxidation of monocyclic ketones. *FR Brevet* 88: 11075.

159 El Aakel, L., Launay, F., Brégeault, J.M. et al. (2004). Nafion®-supported vanadium oxidation catalysts: redox versus acid-catalysed ring opening of 2-substituted cycloalkanones by dioxygen. *Journal of Molecular Catalysis A: Chemical* 212 (1–2): 171–182.

160 Dakkach, M., Atlamsani, A., and Sebti, S. (2012). Natural phosphate modified by vanadium: a new catalyst for oxidation of cycloalkanones and α-ketols with oxygen molecular. *Comptes Rendus Chimie* 15: 482–492.

161 Parkin, B.A. and Hedrick, G.W. (1958). The reaction of hydrazoic acid with pinonic acid and homoterpenyl methyl ketone. *Journal of the American Chemical Society* 80 (11): 2899–2902.

162 Hudlicky, M. (1990). *Oxidations in Organic Chemistry*. Washington, DC: ACS.

163 Larock, R.C. (1989). *Comprehensive Organic Transformation. A Guide to Functional Group Preparation*. New York: VCH.

164 Mallat, T. and Baiker, A. (2004). Oxidation of alcohols with molecular oxygen on solid catalysts. *Chemical Reviews* 104: 3037–3058.

165 Corma, A. and Garcia, H. (2002). Lewis acids as catalysts in oxidation reactions: from homogeneous to heterogeneous systems. *Chemical Reviews* 102: 3837–3892.

166 Adam, W., Gelalcha, F.G., Saha-Moller, C.R. et al. (2000). Chemoselective C—H oxidation of alcohols to carbonyl compounds with iodosobenzene catalyzed by (Salen)chromium complex. *The Journal of Organic Chemistry* 65: 1915–1918.

167 Kim, S.S. and Kim, D.W. (2003). Chemoselective oxidation of alcohols to aldehydes and ketones by iodosobenzene/(Salen) chromium complex. *Synlett* 1391–1394.

168 Zhang, S.H., Xu, L., and Trudell, M.L. (2005). Selective oxidation of benzylic alcohols and TBDMS ethers to carbonyl compounds with CrO_3—H_5IO_6. *Synthesis* 1757–1760.

169 Amer, W., Abdelouahdi, K., Ramananarivo, H.R. et al. (2013). Oxidation of benzyl alcohols into aldehydes under solvent-free microwave irradiation using new catalyst-support system. *Current Organic Chemistry* 17: 72–78.

170 Laasri, L. and El Makhfi, M. (2020). Ecofriendly oxidation of benzyl alcohol to benzaldehyde using Zr—Ni/Natural phosphate as an efficient and recyclable heterogeneous catalyst. *Materials Today: Proceedings* 31 (1): S156–S161.

171 Yudin, A. K. (2006). *Aziridines and Epoxides in Organic Synthesis*. Wiley-VCH Verlag GmbH & Co. KGaA: Weinheim, Germany. DOI:https://doi.org/10.1002/3527607862.

172 He, J., Ling, J., and Chiu, P. (2014). Vinyl epoxides in organic synthesis. *Chemical Reviews* 114 (16): 8037–8128.

173 Kamata, K., Sugahara, K., Yonehara, K. et al. (2011). Efficient epoxidation of electron-deficient alkenes with hydrogen peroxide catalyzed by $[\gamma\text{-}PW_{10}O_{38}V_2(\mu\text{-}OH)_2]^{3-}$. *Chemistry: A European Journal* 17 (27): 7549–7559.

174 Fraile, J.M., García, I., Marco, D. et al. (2000). Epoxidation of electron-deficient alkenes using heterogeneous basic catalysts. *Studies in Surface Science and Catalysis* 130: 1673–1678.

175 Estrada, J., Fernandez, I., Pedro, J.R. et al. (1997). Aerobic epoxidation of olefins catalysed by square-planar cobalt(III) complexes of bis-N,N'-disubstituted oxamides and related ligands. *Tetrahedron Letters* 38: 2377–2380.

176 Groves, J.T. and Stern, M.K. (1987). Olefin epoxidation by manganese (IV) porphyrins: evidence for two reaction pathways. *Journal of the American Chemical Society* 109: 3812–3814.

177 Yamada, T., Takai, T., Rhode, O. et al. (1991). Highly efficient method for epoxidation of olefms with molecular oxygen and aldehydes catalyzed by Nickel(II) complexes. *Chemistry Letters* 20 (1): 1–4.

178 Irie, R., Ito, Y., and Katsuki, T. (1991). Donor ligand effect in asymmetric epoxidation of unfunctionalized olefins with chiral salen complexes. *Synlett* 4: 265–266.

179 Yang, D., Wong, M.-K., and Yip, Y.-C. (1995). Epoxidation of olefins using methyl(trifluoromethyl)dioxirane generated in situ. *The Journal of Organic Chemistry* 60 (12): 3887–3889.

180 Antonioletti, R., Bonadies, F., Locati, L. et al. (1992). Zeolite-catalyzed epoxidation of allylic alcohols. *Tetrahedron Letters* 33 (22): 3205–3206.

181 Yadav, V.K. and Kapoor, K.K. (1994). Al_2O_3 supported KF: an efficient mediator in the epoxidation of electron deficient alkenes with *t*-BuOOH. *Tetrahedron Letters* 35 (50): 9481–9484.

182 Cativiela, C., Figueras, F., Fraile, J.M. et al. (1995). Hydrotalcite-promoted epoxidation of electron-deficient alkenes with hydrogen peroxide. *Tetrahedron Letters* 36 (23): 4125–4128.

183 Fraile, J.M., Garcia, J.I., Mayoral, J.A. et al. (1995). A new titanium-silica catalyst for the epoxidation of non-functionalized alkenes and allylic alcohols. *Journal of the Chemical Society, Chemical Communications* 5: 539–540.

184 Breitmaier, E. (2006) *Terpenes: Flavors, Fragrances, Pharmaca, Pheromones*, Weinheim, Germany. Wiley-VCH Verlag GmbH & Co. KGaA. DOI:https://doi.org/10.1002/9783527609949.

185 Bergault, I., Fouilloux, P., Joly-Vuillemin, C. et al. (1998). Kinetics and intraparticle diffusion modelling of a complex multistep reaction: hydrogenation of acetophenone over a rhodium catalyst. *Journal of Catalysis* 175 (2): 328–337.

186 Ronzon, E. and Del Angel, G. (1999). Effect of rhodium precursor and thermal treatment on the hydrogenation of 2-cyclohexenone on Rh/SiO_2 Catalysts. *Journal of Molecular Catalysis A: Chemical* 148 (1–2): 105–115.

187 Sordelli, L., Psaro, R., Vlaic, G. et al. (1999). EXAFS studies of supported Rh—Sn catalysts for citral hydrogenation. *Journal of Catalysis* 182 (1): 186–198.

188 Augustine, R.L. (1997). Selective heterogeneously catalyzed hydrogenations. *Catalysis Today* 37 (4): 419–440.

189 Singh, U.K. and Vannice, M.A. (2001). Liquid-phase citral hydrogenation over SiO_2-supported group VIII metals. *Journal Catalysis* 199 (1): 73–84.

190 Reyes, P., Rodrıgues, C., Pecchi, G. et al. (2000). Promoting effect of Mo on the selective hydrogenation of cinnamaldehyde on Rh/SiO_2 catalysts. *Catalysis Letters* 69: 27–32.

191 Aguirre, M.d.C., Reyes, P., Oportus, M. et al. (2002). Liquid phase hydrogenation of crotonaldehyde over bimetallic Rh-Sn/SiO$_2$ catalysts: effect of the Sn/Rh ratio. *Applied Catalysis A: General* 233 (1-2): 183–196.

192 Singh, U.K. and Vannice, M.A. (2001). Kinetics of liquid-phase hydrogenation reactions over supported metal catalysts – a review. *Applied Catalysis A: General* 213 (1): 1–24.

193 Masel, R.I. (1996). *Principles of Adsorption and Reaction on Solid Surfaces*. New York: Wiley.

194 Lashdaf, M., Krause, A.O.I., Lindblad, M. et al. (2003). Behaviour of palladium and ruthenium catalysts on alumina and silica prepared by gas and liquid phase deposition in cinnamaldehyde hydrogenation. *Applied Catalysis A: General* 241 (1–2): 65–75.

195 Toebes, M.L., Prinsloo, F.F., Bitter, J.H. et al. (2003). Influence of oxygen-containing surface groups on the activity and selectivity of carbon nanofiber-supported ruthenium catalysts in the hydrogenation of cinnamaldehyde. *Journal of Catalysis* 214 (1): 78–87.

196 Maki-Arvela, P., Hajek, J., Salmi, T. et al. (2005). Chemoselective hydrogenation of carbonyl compounds over heterogeneous catalysts. *Applied Catalysis A: General* 292 (292): 1–49.

197 Szollosi, G., Torok, B., Baranyi, L. et al. (1998). Chemoselective hydrogenation of cinnamaldehyde to cinnamyl alcohol over Pt/K-10 catalyst. *Journal of Catalysis* 179 (2): 619–623.

198 Kun, I., Szollosi, G., and Bartok, M. (2001). Crotonaldehyde hydrogenation over clay-supported platinum catalysts. *Journal of Molecular Catalysis A: Chemical* 169 (1-2): 235–246.

199 Koo-amornpattana, W. and Winterbottom, J.M. (2001). Pt and Pt-alloy catalysts and their properties for the liquid-phase hydrogenation of cinnamaldehyde. *Catalysis Today* 66 (2–4): 277–287.

200 Divakar, D., Manikandan, D., Rupa, V. et al. (2007). Palladium-nanoparticle intercalated vermiculite for selective hydrogenation of α,β-unsaturated aldehydes. *Journal of Chemical Technology & Biotechnology* 82 (3): 253–258.

201 Hidalgo-Carrillo, J., Sebti, J., Marinas, A. et al. (2012). XPS evidence for structure-performance relationship in selective hydrogenation of crotonaldehyde to crotyl alcohol on platinum systems supported on natural phosphates. *Journal of Colloid and Interface Science* 382: 67–73.

202 Clissold, S.P. (1986). Paracetamol and Phenacetin. *Drugs* 32: 46–59.

203 Ung, S., Falguières, A., Guy, A. et al. (2005). Ultrasonically activated reduction of substituted nitrobenzenes to corresponding N-arylhydroxylamines. *Tetrahedron Letters* 46: 5913–5917.

204 Orlandi, M., Brenna, D., Harms, R. et al. (2018). Recent developments in the reduction of aromatic and aliphatic nitro compounds to amines. *Organic Process Research & Development* 22 (4): 430–445.

205 Tafesh, A.M. and Weiguny, J. (1996). A review of the selective catalytic reduction of aromatic nitro compounds into aromatic amines, isocyanates, carbamates, and ureas using CO. *Chemical Reviews* 96 (6): 2035–2052.

206 Béchamp, A. (1854). De l'action des protosels de fer sur la nitronaphtaline et la nitrobenzine: nouvelle méthode de formation des bases organiques artificielles de Zinin. *Annales de Chimie et de Physique* 42: 186–196.

207 Downing, R.S., Kunkeler, P.J., and van Bekkum, H. (1997). Catalytic syntheses of aromatic amines. *Catalysis Today* 37 (2): 121–136.

208 Onopchenko, A., Sabourin, E.T., and Selwitz, C.M. (1979). Selective catalytic hydrogenation of aromatic nitro groups in the presence of acetylenes. Synthesis of (3-aminophenyl)acetylene via hydrogenation of (3-nitrophenyl)acetylene over cobalt polysulfide and ruthenium sulfide catalysts. *The Journal of Organic Chemistry* 44 (8): 1233–1236.

209 Du, Y., Chen, H., Chen, R. et al. (2004). Synthesis of p-aminophenol from p-nitrophenol over nano-sized nickel catalysts. *Applied Catalysis A: General* 277 (1–2): 259–264.

210 Cyr, A., Huot, P., Marcoux, J.-F. et al. (1989). The electrochemical reduction of nitrobenzene and azoxybenzene in neutral and basic aqueous methanolic solutions at polycrystalline copper and nickel electrodes. *Electrochimica Acta* 34 (3): 439–445.

211 Hashimoto, K., Kamiya, K., and Nakanishi, S. (2017). Cooperative electrocatalytic reduction of nitrobenzene to aniline in aqueous solution by copper modified covalent triazine framework. *Chemistry Letters* 47 (3): 304–307.

212 Chen, Y., Li, L., Zhang, L., and Han, J. (2018). In situ formation of ultrafine Pt nanoparticles on surfaces of polyaniline nanofibers as efficient heterogeneous catalysts for the hydrogenation reduction of nitrobenzene. *Colloid and Polymer Science* 296: 567–574.

213 Khan, F.A., Dash, J., Sudheer, C. et al. (2003). Chemoselective reduction of aromatic nitro and azo compounds in ionic liquids using zinc and ammonium salts. *Tetrahedron Letters* 44: 7783–7787.

214 Kurodaa, K., Ishidaa, T., and Haruta, M. (2009). Catalysis of gold nanoparticles within lysozyme single crystals. *Journal of Molecular Catalysis A: Chemical* 298: 7–11.

215 Anet, F.A.L. and Muchowski, J.M. (1960). Lithium aluminum hydride reduction of sterically hindered aromatic nitro compounds. *Canadian Journal of Chemistry* 38: 2526–2528.

216 Kurodaa, K., Ishidaa, T., and Haruta, M. (2009). Reduction of 4-nitrophenol to 4-aminophenol over Au nanoparticles deposited on PMMA. *Journal of Molecular Catalysis A: Chemical* 298 (1–2): 7–11.

217 Layek, K., Lakshmi-Kantam, M., Shirai, M. et al. (2012). Gold nanoparticles stabilized on nanocrystalline magnesium oxide as an active catalyst for reduction of nitroarenes in aqueous medium at room temperature. *Green Chemistry* 14: 3164–3174.

218 Zhang, Z., Shao, C., Sun, Y. et al. (2012). Tubular nanocomposite catalysts based on size-controlled and highly dispersed silver nanoparticles assembled on electrospun silicananotubes for catalytic reduction of 4-nitrophenol. *Journal of Materials Chemistry* 22: 1387–1395.

219 Baba, A., Ouahbi, H., Hassine, A. et al. (2018). Efficient reduction of aromatic nitro compounds catalyzed by nickel chloride supported on natural phosphate. *Mediterranean Journal of Chemistry* 7 (5): 317–327.

220 Acar, C. and Dincer, I. (2019). Review and evaluation of hydrogen production options for better environment. *Journal of Cleaner Production* 218: 835–849.

221 Zhai, X., Ding, S., Liu, Z. et al. (2011). Catalytic performance of Ni catalysts for steam reforming of methane at high space velocity. *International Journal of Hydrogen Energy* 36: 482–489.

222 Zhai, X., Cheng, Y., Zhang, Z. et al. (2011). Steam reforming of methane over Ni catalyst in micro-channel reactor. *International Journal of Hydrogen Energy* 36: 7105–7113.

223 Portugal, U.L. Jr., Marques, C.M.P., Araujo, E.C.C. et al. (2000). CO_2 reforming of methane over zeolite-Y supported ruthenium catalysts. *Applied Catalysis A: General* 193 (1–2): 173–183.

224 Abba, M.O., Gonzalez-DelaCruz, V.M., Colón, G. et al. (2014). In situ XAS study of an improved natural phosphate catalyst for hydrogen production by reforming of methane. *Applied Catalysis B: Environmental* 150-151: 459–465.

225 Mills, A. and Le Hunte, S. (1997). An overview of semiconductor photocatalysis. *Journal of Photochemistry and Photobiology A: Chemistry* 108: 1–35.

226 Serpone, N. (1997). Relative photonic efficiencies and quantum yields in heterogeneous photocatalysis. *Journal of Photochemistry and Photobiology A: Chemistry* 104: 1–12.

227 Zhao, J. and Yang, X. (2003). Photocatalytic oxidation for indoor air purification: a literature review. *Building and Environment* 38: 645–654.

228 Hidalgo-Carrillo, J., Sebti, J., Aramendía, M.A. et al. (2010). A study on the potential application of natural phosphate in photocatalytic processes. *Journal of Colloid and Interface Science* 344: 475–h481.

Index

a

ABM-derived catalytic material   466
acetaldehyde (AD)   108, 110, 147, 149, 178, 179, 180, 189, 323, 324, 327, 328, 352, 378
acetone to methyl isobutyl ketone   173–177
acid catalysts   2, 84, 498, 506
acid-base catalysts   2, 84, 232
acrylates   147
acrylic acid (AA)   86, 147, 149, 150, 311, 418
activated carbon   2, 6, 211, 243, 383, 446, 448, 464, 468
activation site and mechanism   227
adenosine 5'-triphosphate disodium salt hydrate   30
Aldol reaction   84, 352, 454, 491
adsorption
  of molecules   115, 376, 380
  pressure-swing adsorption   312
aerobic oxidation catalysis   211
Ag@mHA-Si-(S)) nanocatalyst   418
Ag/eggshell catalysts   457
agar-gelatin hybrid hydrogels   39
AgNP loaded 1,4-diazabicyclo[2.2.2] octane (DABCO) grafted mHA   419
AgNPs   418, 419
air treatment   374, 382

alcohol coupling reaction   327
alcohol oligomerization probability   181
aldol condensation   141, 148, 152, 178, 179, 252, 310, 323–325, 350–352, 466
alginic acid   23
alkali doped materials   186
alkali metal oxides   455, 462
alkali metals   165, 182
alkaline earth and alkali metal doped HA catalysts   300
alkaline earth metals   6, 165, 182, 214, 465
alkaline-earth apatites   96
alkane oxidative dehydrogenation reactions   201, 217
(E)-2-alkene-4-ynecarboxylic esters   311
alkenoic acids   209
alkyl lactates   147
alkyl phosphates   49, 209
allylamine hydrochloride   418
alpha-tricalcium phosphate (α-TCP)   28
Amberlyst CH28   176
amino-acids   103, 209, 506, 508
2-aminoethanol   29
2-amino-2-ethyl phosphate $(H_2PO_4(CH_2)_2NH_2)$ 49

Design and Applications of Hydroxyapatite-Based Catalysts, First Edition. Edited by Doan Pham Minh.
© 2022 WILEY-VCH GmbH. Published 2022 by WILEY-VCH GmbH.

o-aminophenol   465
α-aminophosphonates   506
3-aminopropyltriethoxylesilane (APTES)   50, 427, 428
2-aminopyridine   360
o-aminothiophenol   465
ammonia   1, 29, 30, 53, 184, 249, 261, 262, 269, 301, 308, 310, 312, 318–319, 467
ammonia temperature-programmed desorption (NH_3-TPD)   143, 147, 251, 310
ammonia-borane hydrolysis reaction   318
ammonium dihydrogen phosphate (($NH_4)_2HPO_4$)   4, 8, 20, 21, 23, 24, 31, 38, 53, 153, 328, 329
amorphous calcium phosphate (ACP)   75
amorphous HA   317
animal bone meal (ABM)   465
animal bones   440, 450
  and eggshell
  waste animal bone   307, 444, 453, 462, 466
anion-exchange membrane (AEM) fuel cells   464
apatite carbonate   442, 444, 445, 448
apatite carbonate type A ($Ca_{10}(PO_4)_6(CO_3)_2$)   444
apatite carbonate type B ($Ca_{10}(PO_4)_3(CO_3)_3(OH)_2$)   444
apatite mineral   443
apatite structure and model studies   76
apatite thermodynamics   95
apatite-based catalysts   75, 203
apatite/Pd(0) catalysts   211
apatite/Ru(III) catalysts   204
apatites   19, 345
  as catalyst supports   346
  as catalysts in C–C bond formation   347
apatitic calcium phosphates   74, 88, 89, 101, 102, 117
apatitic channels   78, 98, 105
apatitic crystal morphology   89
aragonite   442, 444
aromatic aldehydes   353, 465, 466, 487, 488, 491, 504
aromatic nitro compounds   512
atomic adsorption spectroscopy (AAS)   87
Au-Pd bimetallic system   309
Au/Al_2O_3 155
Au/HA catalysts   155, 309, 310
Au/HA-based catalysts   310
Au/hydrotalcite   155
Au/MgO   155
Au/SiO_2 155, 310
Au/TiO_2 155, 310
Au/ZnO   155, 310
β-azido alcohol   424

b

bacterial cellulose (BC)   38
ball-milling method   32–34
barbituric acid   353
benzaldehyde   154, 203, 208, 210, 212, 214, 230, 328, 353–355, 357, 358, 360, 423, 424, 466, 487, 491, 509–511
benzene to phenol   183–185
benzimidazoles   465
benzoic acid (BAcid)   209–210
benzophenone production   466
benzothiazoles   465
benzoxazoles   465
benzyl alcohol (BA)   29, 154–156, 203, 205, 207, 210–213, 230, 328, 509–510

g-benzyl-L-glutamate N-carboxyanhydride (BLG-NCA) 50
beta tri-calcium phosphate (β-TCP)
22, 23, 28, 34–36, 41, 44, 86, 98,
100, 452, 465, 482
bifunctional commercial ion exchange
resin 176
Biginelli condensation reaction 354
Biginelli reaction 505, 506
biguanide-functionalized
hydroxyapatite-
encapsulated-γ-Fe$_2$O$_3$
nanoparticles (HA-γ-Fe$_2$O$_3$)
303
bimetallic catalysts 284–286, 290, 309
bimetallic Ru-Zn/HA 6
bimetallic supported HA 308
biofuel additive 352
3-benzyl-2-phenylimidazo[1,2-*a*]
pyridine 363
1.3-butadiène 144, 145, 323
"biomimetic" apatites 76
bio-oil 141, 152, 186, 188
bioceramic composite 441, 461
biochar 6, 453, 466
biodegradable packaging 182
biodiesel 186
production
environmental impact 300
esterification reaction 307–308
other esterification reactions
308–312
overall transesterification reaction
300
trans-esterification reactions
301–307
bioethanol 145, 177, 180, 181, 189,
250, 329
biofilter (adsorbent) biomaterial 446
biofuels 180, 254, 325

biogas production technologies 269
biogasoline 145, 250
biomedical/biotechnology magnetic
fluids 414
biometric method 416
biomimetic apatites 76, 81–84, 88, 90,
92–95, 97, 98, 101, 102, 105, 113,
115
biphasic calcium phosphate (BCP) 23
bis-coumarin derivatives 426
α,α'-bis(substituted benzylidene)
cycloalkanones 466
1-(3,5-bis-(trifluoromethyl)phenyl-3-
propyl)thiourea 350
bone char 441, 446, 447, 467
bone-derived catalyst 452, 461, 462,
466
borohydride hydrolysis reaction 316
borohydrides 316
p-bromoacetophenone 348
Brønsted acid sites (HPO$_4^{2-}$) 3, 108,
144, 149, 151
Brunauer–Emmett–Teller (BET)
method 90, 210, 211, 246, 249,
391, 442, 484
bulk defective hydrogenphosphate 91,
94
bundle-like carbonated apatite 39
butadiene 144, 145, 250, 320, 323, 327
1,3-butadiene 144, 145, 250, 323
butan-2-ol model reaction 223
butanol 177
formation of 150
preparation 177
n-butanol 26, 27, 86, 108, 177, 178,
180, 249, 250, 299, 320–324, 352
1-butanol 8, 29, 144, 181, 250–252, 327
2-butanol 23, 114, 142–144
1-butene 143
2-buten-1-ol 149, 251, 324

butyraldehyde 149, 181
B-type hydroxyapatite 328

C

$CaCO_3$ nanoparticles 30
$Ca_{10-x}Co_x(PO_4)_6(OH)_2$ 285
Ca-deficient HA 47, 141, 143, 144, 146, 152, 154, 272, 281, 282, 317
   supported Co and Ni catalysts 154
$CaHPO_4$ 28, 29, 31, 33, 44, 482
$Ca(H_2PO_4)_2$ 31, 35
$Ca_{10-x-B}\square_{Ca,x+B\,(PO4)6-x-B}(HPO_4)_x(CO_3)_{A+B}(OH)_{2-x-2A-B}\square_{OH,x+A+B}$ 87
$Ca_{10-x}(HPO_4)_x(PO_4)_{6-x}(OH)_{2-x}$ 156
$Ca_{10-z}(HPO_4)_z(PO_4)_{6-z}(OH)_{2-z} \cdot nH_2O$ 47, 85
$Ca_{8.93}(HPO_4)_{1.56}(PO_4)_{4.44}(PO_4R)_{0.63}(OH)_{0.14}$ 49
calcination temperature 23, 36, 147, 148, 186, 273, 303, 305, 442, 444, 451, 452, 462, 492, 507
calcined eggshell sorbent membrane 461
calcined eggshell-derived catalyst (CaO) 459
calcite 442, 444, 448, 482, 484
calcium acetate (($CH_3COOH)_2Ca$) 4
calcium acetate ($Ca(CH_3COO)_2$) 22, 23
calcium carbonate ($CaCO_3$) 6–8, 20, 21, 23, 29–31, 33, 36, 46, 47, 154, 300, 304, 437, 438, 441–452, 459, 465, 468
calcium carbonate/oxide/phosphate 445
calcium hydroxyapatite ($Ca_{10}(PO_4)_6(OH)_2$, HA) 270, 299
calcium hydroxyapatite (HA, $Ca_{10}(PO_4)_6(OH)_2$) 270
calcium looping 459
calcium nitrate ($Ca(NO_3)_2$ 4, 22–25, 36, 39, 40, 49, 53, 171, 281, 304, 308, 393, 397, 419
calcium nitrate ($Ca(NO_3)_2 \cdot 4H_2O$) 4, 22–25, 36, 39, 49, 53, 171, 281, 304, 308, 393, 397, 419
calcium oxide (CaO) 34, 444, 445, 447, 449, 451, 455, 457, 458, 460, 461, 462, 465, 468, 492
calcium phosphate 20, 23, 25, 50, 73, 74, 78, 82, 90, 100–102, 104, 116, 163, 171, 215, 216, 413, 418, 437–438, 441, 442, 444–446, 448, 451, 462, 465, 466, 482
   hydrolysis 28
calcium supplement 445
calcium-deficient hydroxyapatite 7, 8, 46, 47, 212, 300, 303, 314, 329
$Ca_{10-y}Ni_y(OH)_2(PO_4)_6$ 277
$Ca_{9.5}Ni_{2.5}(PO_4)_6$ catalyst 173
$Ca_{10-y}Ni_y(PO_4)_6(OH)_2$ 285
$Ca_{3-x}Ni_x(PO_4)_2$ 277
$Ca_{9.5}Ni_{2.5}(PO_4)_6$ 171–173
$Ca(NO_3)_2$ 4, 8, 11, 20, 22–25, 27, 31, 35, 38, 39, 53, 153, 154, 305, 308, 327–329, 389
CaO/biomass ratio 458, 459
CaO—CeO_2/HA catalysts 305
$Ca(OH)_2$ 7, 20, 21, 31, 33, 35, 36, 46, 47, 49, 101, 143, 152, 300, 386
capillary microfluidic techniques 24
Ca/P ratio 29, 34, 38, 44, 46, 47, 83, 85, 86, 87, 142–148, 152–155, 157, 178, 189, 204, 212, 241, 242, 250, 251, 262, 281, 282, 310–312, 314, 323, 346, 352, 418, 450
   influence of Ca/P ratio 143–152
$Ca_{10}(PO_4)_{6-x}(SiO_4)_x(OH)_{2-x}$ 48
$Ca_{10}(PO_4)_6X_2$ 212
carbon dioxide
   temperature-programmed

desorption (CO_2-TPD) 147, 152, 153, 259, 310, 316
carbon monoxide oxidation 454
carbon nanofiber 2, 467, 468
carbon nanotubes (CNT) 2, 8, 284, 388, 467, 468
carbon-carbon bond formation 347, 494
carbon-carbon cross-coupling reaction 347
carbonate hydroxyapatite (CO_3HA) 182
carbonate substituted synthetic apatites 92
carbonated biomimetic apatites 93
carbonated hydroxyapatite 177
carbonation-calcination reactions 460
cassiterite (SnO_2) phases 311
catalysis 1, 414
  HAP analogs and their catalytic applications 425–426
  HAP catalysts and green chemistry 426–430
  magnetic HAP nanoparticles 420–425
  catalyst additives 2, 492
  catalytic oxidation 455–458, 467
catalytic selective hydrogenation 454
$Ca_{10}(VO_4)_6(OH)_2$ 289
$Ca_{10}(VO_4)_x(PO_4)_{6-x}(OH)_2$ solid solutions 86, 225, 227
CeO_2-doped HA 353
CeO_2-supported catalyst 286
cerium nitrate ($Ce(NO_3)_2 \cdot 6H_2O$) 53
cetyl trimethyl ammonium bromide (CTAB) 27, 222, 303, 308, 357, 386
chemical doping with nanoparticles 416
chemical precipitation 44, 154, 182, 241, 305, 315, 328, 417

chicken eggshells 437, 444, 449, 457, 460, 462
  derived CaO sorbents 460
chicken meat 438, 439
chitosan-doped HA 354
chlorapatite 76, 100, 104, 112
C_2-hydrocarbons 164
Claisen–Schmidt condensation 3, 466, 491–494
Claisen–Schmidt reaction 454
clean combustion 269, 319
$Co_{10}(PO_4)_6(OH)_2$ 222
Co(0)HA nanoclusters 317
co-precipitation method 20–22, 53, 110, 302–304, 327, 416, 425
Co/α-Al_2O_3 reference catalyst 290
Co/Al_2O_3 catalyst 12, 254, 280
Co/CeO_2 catalyst 281
Co/HA system 289
CO_2 emissions 180, 253
CO_2 methanation 253
CO_2–Fisher–Tropsch (FT) synthesis 253
cobalt phosphate 3, 401, 402
cobalt phthalocyanine/apatite 204, 210
cobalt pyrophosphate 3
cobalt-hydroxyapatite (Co/HA) catalyst 223, 328
$CoFe_2O_4$ magnetic oxide nanoparticles 427
collagen 437, 443
concentrated oil-in-water (O/W) emulsion technique 26
condensation reactions
  Claisen–Schmidt condensation 491–494
  Knoevenagel reactions 487–491
  Michael addition 494–496
conjugated carboxylic esters 311
conventional energy 311
CoO/$Sr_{10}(PO_4)_6(OH)_2$ 219

cooked waste fish bone (CWFB) 305
"core-shell" product 21
cross-coupling reaction 141, 273, 347–350, 361, 453, 454, 466, 499, 500, 502
crossed-aldol condensation reaction 466
crotonaldehyde 110, 149, 178, 179, 189, 320, 323, 511–512
Cu/CeO$_2$ 177
cyanoacetamide 354
β-cyanohydrin 424
3-cyanopyridine 466
β-cyclodextrin conjugated γ-Fe$_2$O$_3$/HA solid-liquid phase transfer catalyst 424
cyclooctane 29
cyclooctanol 29
cyclophosphates 3

d

data storage 415
decahydroquinoline (DHQ) 244
dehydrogenation reactions 52, 114, 201, 217–229, 241, 257–262, 328
Density Functional Theory (DFT) calculations 54, 83, 176, 378
deposition-precipitation (DP) method 309
Dess Martin periodinane (IBX) 203
2,6-diamino-pyran-3,5-dicarbonitriles 354, 363
di-ammonium hydrogen orthophosphate 308
1,4-diazabicyclo[2.2.2]octane 349, 419
di-calcium hydrogen phosphate di-hydrate 450
di-sodium hydrogen phosphate (Na$_2$HPO$_4$) 25, 39, 304, 388, 449, 450, 453, 493

diacetone alcohol (DAA) 175, 176
diadsorbed mesityl oxide 176
dialkyl phosphites 51
diammonium hydrogen phosphate ((NH$_4$)$_2$HPO$_4$) 4, 20–24, 31, 38, 39, 40, 53, 153, 328, 329, 393, 419
1,3-dicarbonyl compounds 466
2,4-dichlorophenol (2,4-DCP) pollutant degradation 466, 467
Diels–Alder reaction 108, 454
diethanolamine 29
diethyl 2,6-dimethyl-4-phenyl-1,4-dihydropyridine-3,5-dicarboxylate 357
diethyl ether (DEE) 327, 329
2,5-diformylfuran (DFF) 325, 429
1,4-dihydropyridines 357, 358, 364, 465
3,4-dihydropyrimidin-2(1H)-one 354, 363, 505
dimethylbarbituric acid 353
5,5-dimethylcyclohexane-1,3-dione 357
4,4-dimethyl-1,3-dioxane 144
dioctyl sulfosuccinate sodium salt (C$_{20}$H$_{37}$NaO$_7$S) 25
4-(1,3-diphenylprop-2-ynyl)morpholine 360, 364
dispersion of metals 214, 246, 261, 417
dissodium oleyphosphate (DSOP: C$_{18}$H$_{35}$Na$_2$O$_4$P) 49
dissolution-precipitation equilibrium 44, 45
dissolution-precipitation mechanism 112, 208, 376
divalent cobalt ions 219–224
1,3-diyne derivatives 350
1-dodecanol 211
dodecyl phosphate (C$_{12}$H$_{27}$O$_4$P) 25

doped photocatalytic HA 378–379
double decomposition method 21, 354
double emulsion droplets 24, 25
dry mechanosynthesis 34, 35, 36, 48
dry methods 20, 35–36, 417
dry reforming of methane (DRM) 4, 6, 7, 12, 115, 153, 154, 170, 231, 269, 454, 514
dual-phase membrane 461

e

eggshell and animal bones
  biodegradation 437
  biofilter (adsorbent) biomaterial 446–448
  calcium carbonate/oxide/phosphate 445
  calcium supplement 445–446
  catalytic applications of materials 456
  catalytic material preparation 451–454
  catalytic materials 465–466
  chemical composition and properties of 441–444
  eggshell membranes (ESM) in fuel cell applications 464–465
  gasification of biomass for hydrogen production 458–459
  hydroxyapatite material 448–450
  other catalytic applications 466–467
  reactive carbon dioxide capture 459–460
  selective catalytic oxidation 455–458
  transesterification reaction for biodiesel production 461–464
  water-gas-shift (WGS) reaction 460–461

eggshell membranes (ESM) in fuel cell applications 464–465
eggshell protein-rich membrane 459
electrocatalysis
  charge transport mechanism 384–385
  electrocatalytic sensors 386–396
  fuel cell application 396–401
  water oxidation 401–403
electron paramagnetic resonance (ESR) characterization 146
electrophotocatalysis 164
Eley–Rideal mechanism 166–168, 188
emulsion templating 24
endothermic steam reforming of methane (SRM) 214, 217, 269, 286–288, 291
energy additives 321
  from alcohols via Guerbet reaction 327–329
Energy Dispersive Spectrometry (EDS) 90
entropy 96, 97, 497
environment-friendly apatitic calcium phosphate 73
environmental footprint 438
environmental remediation 383, 415, 454
enynecarboxylic esters 311
enzymatic catalysts 2
enzymatic interesterification 427
esterification reaction 307–308, 312, 325, 326
ethane coupling 164
ethanol 29, 147
  coupling reaction 177–180
  to gasoline 180–182
  steam reforming reaction 289
ethanol/n-butanol Guerbet conversion reaction 86

4-ethylbenzaldehyde 357
2-ethyl-1-butanol 181, 327
2-ethyl-1-hexanol 181, 311
2-ethylhexyl acetate 311
2-ethylhexylphopshate (Phos) 209
ethyl lactate (EL) 147, 150, 182
ethyl tertiary butyl ether (ETBE) 320, 326
ethylene 8, 87, 144, 145, 150, 164–166, 178, 189, 217, 218, 222, 223, 250, 251, 257, 259, 311, 327, 397, 399, 455, 457
ethylene diamine tetracetic acid (EDTA) 87, 391, 394
ethylene glycol (EG) 311, 397, 399
exceptional ion-exchange capacity 6–7
exothermic methane partial oxidation (POM) 164, 169–173, 188, 201, 214–217, 230, 269, 277, 281, 284, 285, 287, 291

f

fatty acid methyl esters (FAME) 186, 300, 302, 303, 305, 306, 307
Fe-g-C_3N_4 based catalysts 185
Fe-g-C_3N_4/SBA-15 185
γ-Fe_2O_3@HA-Ag NPs 426
γ-Fe_2O_3@HA catalyst 244, 245
γ-Fe_2O_3@HA@Cu(0) 363, 364
γ-Fe_2O_3/HA/Cu(II) modified carbon paste electrode 430
γ-Fe_2O_3@HA-DABCO 349, 363
γ-Fe_2O_3@HA-DABCO-Pd 349
γ-Fe_2O_3@HA@melamine 357--359
γ-Fe_2O_3@HA-Ni^{2+} 353, 354, 363
γ-Fe_2O_3@HAp@Cu(0) 360
γ-Fe_3O_4@HA@Zn(II) 358, 360, 362, 363
Fischer–Tropsch synthesis process 214, 216, 242, 269, 454
fish bone 305, 308, 437, 438, 451, 467, 468

fluoride-containing carbonated apatites 94
formation enthalpy 96
fossil fuels 141, 186, 253, 269, 459
Fourier Transform Infrared Spectroscope (FTIR) analysis 48, 50, 54, 82, 87, 88, 90–94, 106, 107, 113, 153, 157, 242, 244, 258, 273, 274, 309, 377, 393, 483
free methyl radicals 165
freeze dried gelatin nanoparticles 25
Friedel–Crafts alkylation 497–499
Friedel–Crafts reactions 185, 454, 456, 498
Fritsch Pulverisette 6 planetary mill 34
fuel additives from furfural 324–326
fuel cell application 396–401, 464–465
fuel cell technologies 287
fural 324
furaldehyde 324
2-furanaldehyde 324
2-furancarboxaldehyde 324
2,5-furan dimethyl carboxylate (FDMC) 325–326
3-(furan-2-yl)-2-methylacryaldehyde 325, 326
furfural 3, 244, 245, 308, 309, 320, 324–326, 487
2-furfuraldehyde 324
furfural-n-propanol system 325
furfurol 324
furo[3,4-b]chromenes 357, 358, 361, 363

g

gas phase reactions
 alkane oxidative dehydrogenation reactions 217–227
 partial oxidation of methane 214–217

gasification of biomass for hydrogen production 458–459
gelatin nanoparticles 24, 25
Gibbs free energy 96, 97
Glaser–Hay reaction 350
glutaraldehyd (GLU) 427, 428
glycerol carbonate 320, 326, 327
glycerol to lactic acid 182–183, 261
(3-glycidyloxypropyl) trimethoxylesilane (GPTMS) 50
glycosaminoglycans (GAGs) 35
Glycyrrhiza inflata 491
gold supported HA (Au/HA) 155, 261, 308–310, 315, 386, 392
graphene 2, 386, 391, 467
green chemistry 73, 204, 353, 358, 364, 423, 426–430
greenhouse gas (GHG) emissions 141, 252, 257, 269, 300, 319, 459
Guerbet alcohols 144, 145, 251
Guerbet condensation 8
Guerbet coupling 110, 178, 180, 189, 246, 351
Guerbet coupling of ethanol 110
Guerbet cycle 150
Guerbet ethanol coupling 180
Guerbet reaction 84, 145, 148, 181, 249, 250, 320, 327–329, 350–352
Guerbet-mechanism 178
Gum Acacia assisted hydroxyapatite (GA-HA) 246

h

Haber-Bosch process 1
HA-core-shel-γ-Fe_2O_3 magnetic nanoparticles 353
HaβTCP ceramic (α-TCP/β-TCP/HA) 28
HA/γ-Fe_2O_3 nanoparticles 303, 428, 430
HA-γ-Fe_2O_3 materials 210
HAm-T catalysts 147
Hammett indicator method 186, 303
Hantzsch condensation reaction 357
HAP analogs and their catalytic applications 425–426
HAP catalysts and green chemistry 426–430
hazardous environmental pollutants 141
Heck reaction 154, 454
heliophotocatalysis 375
heterocyclic alcohols 209, 212
heterogeneous catalysis 1–13, 20, 51, 55, 73, 84, 114, 117, 141, 157, 204, 230, 345, 364, 380, 382, 452, 454, 455, 468
heterogeneous catalysts 2, 73, 141, 143, 174, 186, 201, 203, 204, 207, 211, 229, 255, 280, 301, 303, 304, 308, 311, 316, 320, 347, 353, 358, 360, 413, 418–420, 425, 429, 438, 440, 443, 444, 451, 453, 455, 457, 462, 464, 468
hexagonal calcium hydroxyapatite (HA) 76
hexamethyldisilazane [$(CH_3)_3Si]_2NH$ (HMDS) 49
hexanoic acid (HAcid) 209
1-hexanol 29, 181, 311
n-hexanol 27
high-density polyethylene (HDPE) 441
high-resolution transmission electron microscopy (HRTEM) 86, 89, 90, 226, 274
high-temperature hydroxyapatite compounds 115
higher energy-density fuel additive 351
highly crystalline carbonate-fluorapatite phase 302
hollow $CaCO_3$ microspheres 30

homogeneous catalysts 2, 174, 413, 425, 461, 462, 497
homogeneous-heterogeneous reaction network 165
H_3PO_4 20, 23, 25, 30, 31, 49, 95, 101, 143, 154, 301, 304, 386, 392, 453, 454, 485
hybrid HA/gelatin nanoparticles 24
hydrated (biomimetic) apatites 90
hydrocarbon fuel 328
hydrogen production 1, 287
  ammonia-borane hydrolysis reaction 318–319
  borohydride hydrolysis reaction 316–318
  water-gas-shift reactions 312–316
hydrogen-selective membrane separation 460
hydrogenation reactions
  aromatic nitro compounds reduction 512–514
  biomass-derived compounds to fuels and fine chemicals 242–245
  crotonaldehyde 511–512
  of benzene, phenol and diols 247–249
  of carbon dioxide
    alcohol synthesis 255
    CO_2 methanation 253
    CO_2–Fisher–Tropsch (FT) synthesis 253–255
    fuel-type compounds 252
    hydrogen-transfer processes 257
    water-gas shift and reverse water-gas shift 255–256
  of olefins and nitro compounds 245–247
hydrolysis methods 28–29
hydrophile-lipophile balance (HLB) value 27
hydrotalcite 152, 155, 204, 212, 302, 510
hydrotalcite-hydroxyapatite (HT-HA) 302, 303
hydrothermal method 20, 30–31, 182, 241, 251, 289, 308, 324, 379, 380, 390, 401
hydrothermal processing 416, 417
hydrothermal-impregnation method 311
hydroxyapatite ($Ca_{10}(PO_4)_6(OH)_2$) 19, 73, 163, 241, 270, 413, 437, 441, 443
hydroxyapatite (HA) 19, 76, 202, 204, 313, 350, 437
  ball-milling method 32–34
  co-precipitation method 20–22
  dehydrogenation reactions 257–262
  dry methods 35–36
  emulsion templating 24
  exceptional ion-exchange capacity 6–7
  features 241
  formulation of 8–11
  high affinity with organic compounds 8
  high thermal stability 4–6
  in hydrogenation reactions
    biomass-derived compounds to fuels and fine chemicals 242–245
    of benzene, phenol and diols 247–249
    of carbon dioxide 252–256
    of olefins and nitro compounds 245–247
    selective catalytic reduction of nitric oxide 249
  hydrolysis methods 28–29
  hydrothermal method 30–31

influence of Ca/P ratio 143–152
as macro ligands for organometallic moieties 347
microwave-assisted synthesis 31–32
nature of interactions with transition metal catalysts 271–274
nonstoichiometric calcium-deficient or calcium-rich hydroxyapatites 46–47
opportunities and challenges 11–13
possible high porous volume and high specific surface area 4
sol-gel method 22–23
solubility 44–45
sonochemistry 34–35
stoichiometric and non-stoichiometric 242
structure 42–43
substitutions in the structure of 47–48
supported bimetallic Co-Ni catalysts 3–4
supported bimetallic Ni-Co catalysts 285
supported bimetallic Pd-Co catalyst 314
supported Cu catalysts 184
supported palladium (0) nanoclusters catalyst 454
surface functional groups 45–46
surface modification and functionalization 49
thermal stability 44
tunable acid-base properties 7–8
hydroxyapatite based catalysts 51
 acetone to methyl isobutyl ketone 173–177
 apatites as catalysts in C–C bond formation 347–361
 apatites as catalyst supports 346–347
 benzene to phenol 183–185
 bimetallic catalysts 284–286
 composition in reforming reactions 281–284
 ethanol coupling reaction 177–180
 ethanol to gasoline 180–182
 glycerol to lactic acid 182–183
 HA as macro-ligands for catalytic moieties 347
 influence of Ca/P ratio 152–156
 noble metal catalysts 286–287
 oxidative coupling of methane 164–169
 partial oxidation of methane (POM) 169–173
 reforming of other hydrocarbons 287–290
 stoichiometric and non-stoichiometric apatites 346
 suitability of 274–281
 transesterification 186–188
hydroxyapatite biomaterial 441
hydroxyapatite material 448–450
hydroxyapatite supported palladium (0) nanoclusters (Pd(0)/HA) 318, 319
hydroxyapatite supported Pd(II) catalyst (Pd/HA) 261, 311
hydroxyapatite-based materials
 apatite structure and model studies 76–81
 apatite thermodynamics 95–98
 electrocatalysis
  charge transport mechanism 384–385
  electrocatalytic sensors 386–396
  fuel cell application 396–401
  water oxidation 401–403
 influence of substitution on surface reactivity 109–110
 interfacial tension 103–106

low temperature ion immobilization and adsorption properties 110–117
nature of acid and base sites 106–109
photocatalysis
  dope 378–379
  multiphasic 379–382
  principles 373–376
  single phase 377–378
  structure and properties 376–377
relevance of apatites in catalysis 84–87
solubility and evolution in solution 100–102
specificities of non-stoichiometric and/or biomimetic apatites 81–84
structural and compositional characterization 87–95
surface charge 102–103
hydroxyapatite-bound lanthanum complex (LaHA) 52
hydroxyethylmethacrylate (HEMA) 49
4-hydroxyfuran-2(5H)-one 357
hydroxyisobutyric acid 144
α-hydroxyisobutyric acid 144
5-hydroxymethylfurfural (HMF) 3, 429, 467
5-hydroxymethyl-2-furfural (HMF) 3, 325, 326, 429, 467
α-hydroxyphosphonates 51, 504–505
α-hydroxyphosphonates synthesis 504–505

i

imidazo[1,2-*a*]pyridine derivatives 360
imidazo[1,2-a]pyridine 360, 363, 364
imidazole-based ionic liquid 426
incipient wetness impregnation method 183, 303, 306
inductively coupled plasma (ICP) emission spectroscopy 87
inorganic-organic hybrid nanocatalyst 429
interfacial tension 103–106
interrupted coupling products 181
inverse mini-emulsion technique 25
in situ and *operando* characterization techniques 228
ion exchange procedure 111, 249, 253
ion scattering spectroscopy (ISS) 86
ion-exchange resins 6–7
ionic chromatography 87
ionic-tagged mHA 419
ionometry using specific electrodes 87
Ir/TiO_2 170
iron modified HA 324
isobutene 224, 320
isopropyl alcohol (IPA) 174
2-(isopropoxymethyl) furan (2-IPMF) 326

k

β-ketonitriles 351
KH_2PO_4 23
KIO_3/HA catalyst 304
Knoevenagel condensation reaction 3, 425, 426, 454, 487–491
Knoevenagel reaction 3, 353, 465, 487–491
Kyoto Protocol 180

l

$LaAlO_3$-8 457
labile carbonates 92, 94
lactic acid (LA) 23, 86, 147, 149, 182–183, 261

Langmuir–Hinshelwood mechanism 166–168, 188
Langmuir–Hinshelwood type kinetic model 466, 467
Langmuir-type isotherm 115, 117
lauryl dimethylaminoacetic acid 32
lead-doped hydroxyapatite 112
Lewis acidity 85, 108
Lewis acids 85–87, 106, 108, 144, 149, 151, 308, 314, 395, 465, 466, 497, 506
Lewis to Brønsted acid sites (L/B) 3
light olefins 455, 457
linear alkylbenzenesulfonate 32
lipase catalysts 427
lipase immobilized HA/γ-Fe$_2$O$_3$ biocatalyst 428
liquid phase reactions
  apatite-based catalysts 203–204
  apatite/Pd(0) catalysts 211–214
  apatite/Ru(III) catalysts 204–211
lithium-doped NP (Li/NP) catalysts 495

m

macro-porous HA ceramics 10
maghemite (γ-Fe$_2$O$_3$) 420
magmatic deposits 481
magnesium carbonate 442
magnetic γ-Fe$_2$O$_3$ 423
magnetic γ-Fe$_2$O$_3$ nanocrystallites 210, 424
magnetic HA 360, 414–427, 430
  catalysis 420–430
magnetic HAP nanoparticles 420–425
magnetic hydroxyapatite (mHA) 50, 281, 350–352, 413, 415, 417–419, 421, 426, 427, 429, 430
magnetic Mo-HA@γ-Fe$_2$O$_3$ nano catalyst 429

magnetic nanocatalyst 349, 357, 423, 424
magnetic natural HA (HA/Fe$_3$O$_4$ NP's) 358, 426
magnetic resonance imaging 415
magnetic separable solid catalyst 303
magnetic Zn/HA/MgFe$_2$O$_4$ 421
magnetite (Fe$_3$O$_4$) 382, 420, 423, 424, 426, 427
malononitrile 353, 354, 466, 487, 495
Mars and Van Krevelen mechanism 209, 218, 231
"maturation" phenomenon 83
mesityl oxide (MO) 110, 174, 511
mesoporous ionic-modified γ-Fe$_2$O$_3$@HA-DABCO-Pd containing palladium nanoparticles 349
mesoporous nano-hydroxyapatite (mn-HA) 32, 33
metal cations 3, 6, 75, 77, 96, 111, 142, 143, 176, 219–225, 272, 376, 383, 404
metal ion modifications 219–225
metal loaded spongy carbonate-fluorapatite 301
metal modified hydroxyapatites 114, 218–221, 227, 229, 230, 232
metal nanoparticles 3, 12, 143, 157, 246, 261, 386, 403, 515
metallic copper-based catalysts 182
"Metastable Equilibrium Solubility" (MES) 102
methacrylic acid (MA) 311
methane adsorption equilibrium constant 167
2-methoxyaldehyde 354
4-methoxybenzaldehyde 353
methyl acetate (MAA) 310

2-methyl-3-butyn-2-ol (MBOH)   84, 108, 145, 146
methyl ethyl ketone   143, 323, 324
methyl glycolate (MG)   329
methyl isobutyl ketone (MIBK or 4-methyl-2-pentanone)   173–177
methyl lactate (ML)   147, 150, 310
methyl lactate (MLA)   147, 150, 310
N-methyl morpholine-N-oxide (NMO)   203
methyl pyruvate (MPA)   155, 156, 309, 310
methyl-2-furoate   309, 325, 326
1-methyl-3-(3-sulfopropyl)-imidazolium hydrogen sulfate acidic ionic liquid catalysts   467
methyl-tert-butyl ether (MTBE)   326
methylbenzyl alcohols (MB-OH)   328
methylisobutylketone   352
methylmercaptane   377
methylmethacrylate (MMA)   49, 310
methylphenylcarbinol   144
Michael addition reaction   454, 494, 495
micro arc oxidation   416
microporous biochars   453
microwave assisted co-precipitation   416
microwave irradiation   241, 416, 417, 425, 448, 454, 504
microwave-assisted synthesis   31–32
mineral hydroxyapatite   437
miniemulsions   25
mixed metal oxides   177
mixed metal-based HA catalysts   300
Mizoroki-Heck cross-coupling reaction   348, 349
MnO_2/NAp   204
MnOx/HA catalysts   154
mole fraction of oligomer   181
N-monoalkylated hydroxylamines   418
mono-ethanol-amine (MEA)   459
monoammonium dihydrogen phosphate ($NH_4H_2PO_4$)   4, 8, 11, 23, 35, 154, 389
monometallic supported porous HA nanorods   308
4-monosubstituted pyrido[2,3-d] pyrimidines   466
montmorillonite   204, 511
multi-layered biomaterial composite   437
multicatalysis   12
multicomponent reactions (MCRs)   352–361
  α-amino-phosphonates synthesis   506
  Biginelli reaction   505–506
multiphasic HA-containing photocatalyst
  composites   381
  electronic transfers   380
  TiO_2 biphasic composites   380–381
  TiO_2 multiphasic composites   381
multiwall carbon nanotubes (MWCNTs)   388, 398, 399, 402, 403
$M_{10}(XO_4)_6(Y)_2$ 19
$M_5(XO_4)_3Y$ 75

n

$NaLaCa_3(PO_4)_3OH$ 425
$NaLaSr_3(PO_4)$ 425
nanocomposite $CoFe_2O_4$/HA   428, 429
nanocrystalline apatites   76, 83, 98, 113
nanocrystalline HA   23, 311
nanoemulsion technique   24
nanoparticle infiltration   416
nanoporous hydroxyapatite   485
nano-sized HA powder   449
nano-SnO_2 grafted natural-hydroxyapatite photocatalyst   311

nano-tin oxide grafted natural-hydroxyapatite (SnO$_2$/HA)   311
natural and doped HA-catalysts   186
natural apatite (NAp)   203, 204
natural bio-ceramic materials   448, 452, 453
natural gas (NG)   153, 164, 177, 201, 253, 269, 312
natural hydroxyapatite (nHA)   316
  -supported copper solid acid catalyst   308
natural phosphates (NP)
  catalysts or catalyst supports   482–486
  mineralogical composition of   481
  organic synthesis
    condensation reactions   487–491, 496
    Friedel–Crafts alkylation   497–499
    hydration of nitriles   501–504
    hydrogenation reactions   511–514
    α-hydroxyphosphonates synthesis   504–505
    MCRs   505–506
    oxidation reactions   506–511
    reforming of methane   514
    Suzuki–Miyaura cross-coupling reaction   499–501
    transesterification reaction   496–497
    VOCs   515–516
natural quarry calcium carbonate   438
Na$_2$WO$_4$–Mn/SiO$_2$ 166
Na$_2$WO$_4$/SiO$_2$ 457
Na$_2$WO$_4$/TiO$_2$ 457
neem (Azadirachta indica) oil   305
(NH$_4$)$_2$HPO$_4$ 4, 20, 21, 23, 24, 31, 53, 153, 328, 329
Ni hydroxyapatite system   114, 290
Ni strontium hydroxyapatite system   214

Ni-added strontium phosphate catalysts   284
Ni-Ca-HA catalysts   277
Ni-Ca-PO$_4$/HA catalysts   281
Ni/CeO$_2$ catalyst   280
Ni$_{SA}$/HA catalysts   7
Ni/ZrO$_2$ DRM catalyst   276
nickel calcium hydroxyapatite system   215
nickel/HA/cobalt ferrite (Ni/HA/CoFe$_2$O$_4$) novel nanocomposite catalyst   420
nitric acid production   1
β-nitro-alcohol   424
noble metal catalysts   182, 270, 286–287, 290, 458, 511
N$_2$O catalytic decomposition   454
N$_2$O decomposition   153
non-active μ-oxo dimers   204
non-alkoxide sol-gel process   23
non-apatitic calcium phosphates   31
non-noble metal based heterogeneous catalyst   429
non-stoichiometric aluminum phosphate–hydroxyapatite   329
non-stoichiometric apatites   13, 28, 76–77, 79, 81–82, 84–85, 90, 94, 98, 100–101, 105, 113–114, 346
non-stoichiometric non-carbonated apatites   87
non-stoichiometric calcium-deficient or calcium-rich hydroxyapatites   46–47
normal hydrogen electrode (NHE)   401
NP12 (poly(oxyethylene)12 nonylphenol ether   25
NP5 (poly(oxyethylene)5 nonylphenol ether   25
n-type doping   378
nucleophilic carbon-carbon bond forming reactions   347, 350–352

o

octacalcium phosphate
  ($Ca_8H_2(PO_4)_6 \cdot 5H_2O$ (OCP)) 47, 75, 101, 444
n-octane 29, 52, 202, 219, 226, 258
octane number 145, 326
octanol-1 oxidation 209
1-octanol 29, 207
olefins 181, 201, 217–218, 245–247, 250, 257, 269, 328, 455, 457
oleic acid esterification 308
oligomer 3, 181, 465
one-pot three-component A^3-coupling reaction 353–354, 356–358, 360, 362–363
operando impedance spectroscopy 228
organic carbonates agents 326–327
organic pollutant degradation 164
organic synthesis, NP and NP-supported catalysts
  condensation reactions
    Claisen–Schmidt condensation 491–494
    Knoevenagel reaction 487–491
    Michael addition 494–496
  Friedel–Crafts alkylation 497–499
  hydration of nitriles 501–504
  hydrogenation reactions 511–514
  α-hydroxyphosphonates synthesis 504–505
  MCRs 505–506
  oxidation reactions 506–511
  reforming of methane 514
  Suzuki–Miyaura cross-coupling reaction 499–501
  transesterification reaction 496–497
  volatile organic compounds (VOCs) 515–516
orthophosphates (PO_4^{3-}) 3, 21, 87, 308, 390
orthophosphoric acid 6, 21, 25, 419
Ostwald–Brauer process 1
"over-stoichiometric" compounds 85
oxidation reactions
  benzyl alcohol 509–511
  epoxidation of electron-deficient alkenes 510–511
  oxidative cleavage of cycloalkanones 506–509
oxidative coupling of methane 164–169, 270, 455
oxidative dehydrogenation (ODH) 52, 79, 108, 201, 217–227, 231, 257, 259, 261, 299
oxidative dehydrogenation of ethane (ODHE) 202, 219, 222, 231, 259
oxidative dehydrogenation of propane (ODHP) 79, 108, 163, 202, 219, 224–225, 228–229, 257, 299
oxidative dehydrogenation reactions (ODH) 52, 201, 217–218
oxygen adsorption equilibrium constant 167
oxygen evolution reaction (OER) 401–402
oxygen reduction reaction (ORR) 467
oxygenated additives 320, 326–327
oxyhydroxyapatite 4–5, 109
oxyhydroxyapatite formation 109

p

palladium 185, 211, 247, 248, 313–314, 318, 323, 326, 347–350, 381, 383, 396–397, 404, 466, 499–500, 504, 511
palladium (II) ion exchanged hydroxyapatite (Pd^{2+}-HAP) 318
palm kernel oil 303
partial oxidation of methane (POM) 112, 164, 169–173, 188, 201, 214–217, 230, 269, 277, 281, 284–285, 287, 290–291, 454, 514
partially substituted hydroxyapatites 299

$Pb_{10}(PO_4)_6(OH)_2$ 112
$Pb_{10}(PO_4)_6Cl_2$ 112
Pd nanoclusters 213, 247
Pd_0NPs@nano-HA 348
Pd_{II}NPs@nano-HA containing palladium nanoparticles 348
$Pd_{2060}(NO_3)_{360}(OAc)_{360}O_{80}$ 213
$Pd_{561}phen_{60}$-$(OAc)_{180}$ 213
$Pd(OAc)_2$ 214, 347
Pd-Co(1)/Al_2O_3 315
Pd/Co(1)/HA catalyst 315
$PdCl_2(PhCN)_2$ 6, 154, 347
Pechmann reaction 426
1,4-pentanediol 242–243
2,3-pentanedione (2,3-PD) 147
permanently polarized HA (p-HA) 255
phenol (carbolic acid) 183
phenylacetylene 350, 358, 360, 423–424
phenylboronic acid 348, 500
o-phenylenediamine 465
(R,S)-1-phenylethanol 428–429
phenylglyoxylic acid 361
phenylpropiolic acid 361
phosphate deposits 481
phosphate ores 481–482
phosphate-bearing apatites 95
phospho-vanado-molybdic acid complex 87
phosphorous pentoxide (P_2O_5) 13, 23, 482, 484, 489
photocatalysis 164
 dope 378–379
 principles 373–376
 single phase 377–378
 structure and properties 376–377
photocatalytic water splitting 374
$PO(Me_3O)_3$ 23
$PO(OEt)_3$ 22–23
point of zero charge (PZC/pzc) 102, 418

poly(acrylic acid) 418
poly(N-isopropylacrylamide) (PNIPAM) 50
poly[(butylene-co-ethylene)-b-(ethyleneoxide)] [P(B/E-b-EO)] 25
polyfunctional catalysts 2
polyhydroquinoline 357, 359, 364
polyphosphates 3
polyvinylpyrrolidone (PVP) 38
"poorly-crystalline" apatites 82
porous graphitic carbon nitride (g-C3N4) 185
porous HA 26, 308, 381, 425, 453
potassium hydroxide catalyst 305
potato starch hydrolysis 1
potential-determining 102
poultry meat 438–439
 production 438
precipitated amorphous calcium phosphates 89
precipitation combined with mechanical grinding 416
pressure-swing adsorption (PSA) 312
proline 209–210
prolinol 209–210
1,2-propanediol 327
propionic acid (PA) 147, 390
protein 90, 379–380, 386, 413, 437, 441, 444, 451, 460
protide compounds 316–317
proton-exchange membrane (PEM) fuel cells 464
pseudo sol-gel method 303
pseudo-sol-gel microwave-assisted protocol 32
Pt-nanoparticles (Pt NPs) 249
p-type doping 378
p-xylene 25, 203
pyrano[2,3-d]pyrimidine derivatives 353

pyrano[2,3-d]pyrimidinones 353
  derivative 353
[1,2-a]pyridine derivatives 363
pyrido[2,3-d]pyrimidines 355
  derivatives 353
pyrimidinobenzimidazoles 427
pyrophosphates ($P_2O_7^{4-}$) 3, 87, 100–101
pyruvaldehyde (PA) 155
pyruvic acid (PA) 147, 150

r

racemic acyclic propargylamine 358
rare earth metals 165, 347, 455
reactive carbon dioxide capture 459–460
rectisol 459
recyclable, in situ prepared, catalytic, natural HA supported MnO_2 420
recyclable, non-functionalized Fe_3O_4-HA magnetic nanorods 423
redox catalysts 2
reduced grapheme oxide (rGO) 391–393, 403
reusable magnetic nanocatalyst 424
reverse water-gas shift (RWGS) reaction 253, 255–256, 277, 461
Rh/Al_2O_3 170, 287
Rh/MgO-Al_2O_3 catalyst 286
Rh/SiO_2 170, 216
Rh/TiO_2 170
rhodium supported hydroxyapatite 230
rhombohedra calcite ($CaCO_3$) 444
Ru-based oxidation catalysts 210
Ru-CoHA catalyst 209
Ru/HA-γ-Fe_2O_3 211
Ru/HA catalyst 242–243, 247, 261–262, 315, 318
$RuCl_2(PPh_3)_3$ 207
RuHA-γ-Fe_2O_3 catalyst 325
RuNPs@nano-HA 319
ruthenium complexes 177
ruthenium supported HA (Ru/HA) 315
ruthenium-hydroxyapatite-encapsulated superparamagnetic γ-Fe_2O_3 nanocrystallites 424

s

Schizochytrium algae oil 306
sealed Teflon-lined autoclave 30
sedimentary deposits 481
selected area electron diffraction (SAED) 33, 90
selective catalytic oxidation 455–458
selective hydrogenation 164, 241–262, 329, 454, 511–512
selective oxidation catalysis 201
selexol 459
semiconductor photocatalytic functions 185
semiconductors (SC) 374, 378, 380
simple salt approximation (SSA) 96
simulated biofluid method 416
single crystal HA 317
SiO_2-based materials 2
slightly-milky nanoemulsions 24
SnO_2/HA catalyst 311
sodium lauryl ether sulfate 32
sodium-doped nanohydroxyapatite (nHA) 306, 316
sodium-modified hydroxyapatite (NaHA) 303
sol-gel method 22–23, 303, 454, 515
solar-type energy 311
solar-type/ultrasound batch reactor 311
solid base catalysts 186, 303–304, 306
solid base metal oxides 177
solid waste powders 441
sonochemistry 34

Sonogashira cross-coupling reaction 349–351
soybean oil 302–304
soybean oil biodiesel 303–304
specificities of non-stoichiometric and/or biomimetic apatites 81–84
spongy carbonate-fluorapatite 301
$Sr_{10}(PO_4)_6(OH)_2$ 222, 289
$Sr_{10}(VO_4)_6(OH)_2$ 289
$Sr_2TiO_4\_SP1$ 457
standard formation enthalpy 96
steam reforming of methane (SRM) 169, 214, 217, 269, 286–288, 291
stoichiometric (non-carbonated) apatite 87
stoichiometric apatites 75, 79, 86, 94, 98, 101, 105, 113, 212, 346
stoichiometric calcium phosphate hydroxyapatite 3
strong electrostatic adsorption method 7, 279, 284
strong metal-support interactions (SMSI) 114–115, 244, 273–274, 290
strontium phosphate hydroxyapatite 324
substituted vanadate anions 110
sulfuric acid production 1
sulphonated-chitosan encapsulated HA-Fe_3O_4 427
surface charge 102–103, 106, 111, 376–377, 418, 466
surface electronic transfers 374
surface hydroxyl groups 165, 273, 277, 427
surface ion immobilization 111
surface/interfacial (free) energy or "tension" 103
Suzuki–Miyaura cross-coupling reaction 347–348, 466, 499–501
Swern reagent 203

syngas production 4, 201, 455, 457
synthetic HA supported bimetallic cobalt and nickel catalyst 454

t

temperature programmed desorption (TPD) experiments 54, 85, 143, 310, 393, 394
tert-butyl hydroperoxide (TBHP) 309, 325
tetracalcium phosphate (TTCP) 4–5, 44, 98
5,6,7,9-tetrahydro-1*H*-furo[3,4-*b*] chromene 360, 361
tetraethoxysilane (TEOS) 50
tetrahydrofurfuryl alcohol (THFAL) 326
Therm'AP model 96
thermal behavior 5, 98–100
thermally stimulated depolarization current (TSDC) measurements 385
thermodynamic dissolution-precipitation equilibrium 44–45
thermogravimetric analyser (TGA) 50, 444
thiourea-functionalized mHA 350–352, 418, 429–430
three-stage homogeneous catalysis process 174
TiO_2@HA 51, 54
toluene 29, 207–208, 210–211, 355, 465, 467, 497–498, 509
transesterification reaction 186, 300–307, 326–327, 428–429, 454–455, 461–464, 496–497
  for biodiesel production 462
transition metal ions (TMI) 19, 201–202, 218, 347, 402, 404
transition metal-substituted HA 165

transition metals   6, 20, 165, 203, 211, 270–271, 286, 290, 313, 316, 347, 374, 454, 461, 499, 502
trianisylcarbinol   143
tricalcium phosphate ($Ca_3(PO_4)_2$ (TCP))   4–5, 28–29, 35, 44, 98, 216–217, 227, 300, 274, 282, 287, 381, 442, 444–445, 448–451, 462
β-tricalcium phosphate (β-TCP)   22, 86, 101
triethanolamine   29
triethoxyphosphine ($PO(OEt)_3$)   22–23
trimethyl phosphite (($CH_3O)_3P$)   23
trimethylcarbinol   144
trimethylsilyl cyanide (TMSCN)   350, 429
triphenylcarbinol   143
tritolycarbinol   143
trivalent iron and chromium cations   224–225
tunable acid-base properties   7–8, 218, 364
turnover frequency (TOF)   110, 203, 207–211, 213, 245, 262, 319, 348, 401
Tween 80   26–27
TX-100   26–27

u

ultrasonic irradiation   35, 416
ultrasound-assisted precipitation method   35, 381
ultrasound-wave energy   311
uncalcined eggshell   464
α,β-unsaturated aromatic enones   350
UV-Vis-NIR techniques   272–273

v

γ-valerolactone hydrogenation   243
vanadate anions   110, 225–227

vanadium impregnated hydroxyapatites   226
vanadium oxy-hydroxy-apatite $Ca_{10}(PO_4)_{6-x}(VO_4)_x(OH)_{2-2y}O_y$ solid solution   228
vanadium pentoxide   1, 226
vanadium substituted $Ca_{10}(VO_4)_x(PO_4)_{6-x}(OH)_2$ stoichiometric apatite solid solution   86
vanadium substituted hydroxyapatite   226–227, 231
vegetable oil deodorization distillate (VODD)   302–303
versatile mHAs   430
vicinal-diiodoalkenes   311
volatile organic compounds (VOCs)   8, 141, 146–147, 299, 458, 467, 515–516

w

waste animal bone   307, 444, 453, 462, 466
waste chicken eggshells   462
waste cooking oil (WCO)   301–302, 304, 455, 462
waste eggshells   438–440, 442–444, 447–449, 451, 455, 457–459, 461–462, 465
water oxidation   401–403
water treatment   376, 382, 383, 446
water-gas-shift (WGS) reaction   153, 252, 255, 312–316, 458, 460–461
water-in-oil-in-water (W/O/W) double emulsion drops   24
water-soluble polar organic molecules   209
Wavelength Dispersive Spectrometry (WDS)   90
wet impregnation   111, 153, 219, 226, 231, 243, 306, 308, 315, 327, 452

World poultry meat and chicken meat production 439

x
xonotlite nanowires 30

z
zeolites 2, 6, 8, 78, 141, 152, 177–178, 204, 318, 320, 446, 448, 451, 459, 468, 510–511

zeta potential 102, 111, 419
Zn(II) anchored magnetic HA nanocatalyst 423
Zn-bound hydroxyapatite 112
$Zn^{II}/HA/Fe_3O_4$ 423–424
$Zn(II)/HAP/Fe_3O_4$ 422
zwitterionic poly(3-carboxy-N,N-dimethyl-N-(3'-acrylamido-propyl) propanaminium inner salt) (PCBAA) 38